Thermochemistry of Organic and Organometallic Compounds

J. D. COX

National Physical Laboratory
Teddington, Middlesex, England

and

G. PILCHER

University of Manchester
Manchester, England

1970

ACADEMIC PRESS · LONDON and NEW YORK

ACADEMIC PRESS INC. (LONDON) LTD
Berkeley Square House
Berkeley Square,
London, W1X 6BA

U.S. Edition published by

ACADEMIC PRESS INC.
111 Fifth Avenue,
New York, New York 10003

Library of Congress Catalog Card Number: 75–92407
SBN: 12—194350—X

Printed in Great Britain by
HARRISON AND SONS LIMITED
BY APPOINTMENT TO HER MAJESTY THE QUEEN,
PRINTERS, LONDON, HAYES (MIDDX.) AND HIGH WYCOMBE

Preface

The application of thermodynamics to practical problems arising in research and development requires the ready availability of reliable numerical data. So far as organic thermochemistry is concerned, data are disseminated through the world's scientific literature, and are not always easy to find or to assess for their reliability: such compilations as exist tend to be critical in character but limited in scope, or uncritical but comprehensive. With the encouragement of many fellow researchers in thermochemistry we set out, in equal partnership, to remedy the situation by compiling standard heats of formation and vaporization at 25°C for organic and organometallic compounds. The results of our labours are presented in tabular form in Chapter 5 of this book. The values there given are taken from work published in the modern period of organic thermochemistry only (approximately 1930 onwards); values of low precision or those obtained by suspect methods are excluded; error limits are included; selected values are offered in the case of two or more independent determinations. The resulting compilation is therefore both critical and comprehensive.

Most current textbooks of thermodynamics give scant attention to modern organic thermochemistry, so we deemed it right to preface the data tables with a detailed discussion of the principles of the accurate determination of heats of formation and heats of vaporization (or sublimation) of organic and organometallic compounds. Proofs of thermodynamic relations are omitted, as we assume readers are already familiar with the basic equations of chemical thermodynamics. The data tables are followed by a Chapter whose purpose is to demonstrate the application of organic thermochemical data to practical problems. In the final Chapter we discuss the relevance of thermochemical data to problems of bonding and structure in organic compounds; a compilation of recommended bond energy terms is included. In sum, the book offers critically evaluated data that will be of use to scientists of many disciplines and technologists in many industries.

The presence of some errors in a compilation of this size is almost inevitable: we would appreciate hearing from readers who discover significant errors of omission or commission.

In the preparation of this book we have had the benefit of discussion or correspondence with colleagues from many countries of the world and these we thank, particularly those who communicated pre-publication data. We would like to put on record the encouragement we received throughout the

v

preparation of the book from Dr. H. A. Skinner of the University of Manchester, and Dr. E. F. G. Herington of the National Physical Laboratory. Our grateful thanks are also due to Mrs. S. M. Larcombe and Miss M. Crossley for typing the manuscript, and to Mr. D. P. Biddiscombe and Mr. R. A. Fletcher for help with the checking.

July 1969

J. D. Cox
Teddington
G. Pilcher
Manchester

Acknowledgments

The authors are grateful to the following for permission to reproduce Figures from previously published material: The American Chemical Society, Figs. 6 and 19; The Faraday Society for Figs. 7, 8, 9, 12, 15. and 16; the International Union of Pure and Applied Chemistry and Butterworths Scientific Publications for Fig. 10; Einar Munksgaard for Figs. 11 and 18; John Wiley & Sons for Fig. 13; and the U.S. Government Printing Office for Fig. 17.

Contents

Chapter 1

The Science of Thermochemistry

1.1. Historical Introduction

Thermochemistry is concerned with the energy changes of chemical reactions and of associated physical processes involving substances of defined composition, and may be regarded as a branch of the larger science of thermodynamics. Thermochemistry originated in the period 1780–1840, when Lavoisier, de Laplace, and Hess made the first measurements of the heat evolved in chemical reactions. In the second half of the nineteenth century thermochemistry was dominated by the work of two schools, those of Thomsen in Copenhagen and Berthelot in Paris. At that time it was thought that the affinity, or driving force, of a chemical reaction was determined solely by the heat of the reaction, and in an attempt to amass values of chemical affinities the heats of a very great number of reactions were measured by Thomsen and by Berthelot. With the realization, near the turn of the century, that the entropy change of a reaction was an additional factor in determining the driving force of the reaction, the study of thermochemistry seems to have lost impetus. Notwithstanding the important work of Richards, and others, impetus was not regained until the late 1920's when it was realized that the further development of chemical technology depended on the existence of a reliable body of physicochemical data. It was apparent that the accuracy of the thermochemical data extant at that time, as summarized in the International Critical Tables (26/1) and by Kharasch (29/1), was barely sufficient for the calculation of heat balances of chemical reactions of interest to technology, and was totally inadequate for the meaningful calculation of equilibrium constants (see Section 2.4). It was realized that the accuracy of existing thermochemical data would have to be increased by about one order of magnitude before the data could be coupled with entropy values, then becoming available for the first time, to yield reliable values of equilibrium constants. The "classical" period of thermochemistry may be considered to have ended with this realization, and the "modern" period to have been inaugurated with the publication of a series of papers from the National Bureau of Standards, Washington D.C., which set entirely new standards of accuracy. In 1931, Rossini published his accurate measurement of the heat of

1

formation of water, which when corrected to the standard state (39/1) and to the 1961 atomic weight scale is the value accepted today. Other notable contributions from the Washington laboratory were those of Washburn (33/1), who gave the first detailed thermodynamic analysis of the processes occurring in a bomb-calorimeter combustion experiment; of Jessup and Green (34/1), who gave an accurate value for the energy of combustion of the primary-standard substance, benzoic acid; and of Dewey and Harper (38/1), Jessup (38/2), and Prosen and Rossini (44/1, 44/2), who determined the heat of formation of carbon dioxide. Between them, these American workers laid the foundations of modern organic thermochemistry.

The standards of accuracy attained by Rossini and his colleagues were very high, but the accuracy attained in the mass of thermochemical work since published by authors throughout the world has not always been comparable. Our prime purpose in preparing the present book was to collate and digest this mass of published data, paying particular attention to the accuracy of the data. We hope we have thereby revealed those organic compounds on which further thermochemical measurements are needed to meet the requirements of modern chemical technology. We have concerned ourselves almost exclusively with measurements published after 1930, as in our view most of the measurements published before 1930 should be regarded as of historical interest only.

1.2. The aims of thermochemistry

Fundamental to thermochemistry is the measurement of the heats of chemical reactions. Such measurements may in turn be used to predict the heats of other chemical reactions as indicated by Rossini (36/2) in his succinct definition of the principal aim of thermochemistry, namely "to provide the experimental data for compiling a table of values from which may be calculated the heat of every possible chemical reaction". A table of the type envisaged by Rossini can be rapidly constructed once relations between heats of reaction and molecular structural features of the reactants and products have been established, and it will be the purpose of Chapters 6 and 7 to examine how far we have progressed in being able to calculate the heat of any organic reaction from available thermochemical data. The interests of the thermochemist, however, are not restricted to the determination or prediction of heats of reaction. He is concerned to derive the heats of formation of compounds from their constituent elements, and to relate the results to the chemical binding energies in the molecules. For this purpose, thermochemical data for the ideal-gas state are essential, since the effect of intermolecular forces must be removed from consideration. The thermochemist is therefore concerned with the measurement of heats of vaporization, as will be explained more fully in Chapter 4. Furthermore, he will often wish to use his data for determining the feasibility of proposed new reactions; for this purpose, heat data are combined with entropy data to yield Gibbs energies and equilibrium constants. The application of thermochemical data is not confined within the bounds of thermodynamics—the physical chemist studying reaction kinetics, for example, often has need of heats of formation, which must be determined by the thermochemist. The application of thermochemistry to kinetics is outside the scope of this book, however, and is in any case well documented. (Trotman–Dickenson, 55/7; Cottrell, 58/6; Kerr, 66/7; Benson, 68/11).

1.3. Organic and inorganic thermochemistry

This book is concerned with the thermochemistry of organic and organo-metallic compounds, and it is important to consider how the thermochemistry of these substances differs from that of inorganic substances. Our decision to concentrate on organic and organometallic compounds was dictated by the fact that their thermochemistry was inadequately documented, whereas many modern compilations of inorganic thermochemical data are available. The differences between organic and inorganic thermochemistry are how-ever more deep-seated. In the first place, organic thermochemistry is con-cerned with all states of matter, but particularly the gas state, at temperatures up to a maximum of 1000°K, few organic compounds being stable beyond that temperature. On the other hand, inorganic thermochemistry is concerned largely with materials in the solid state, and the maximum temperature of interest extends far beyond 1000°K (Kubaschewski, 58/1). In the second place, the experimental methods of thermochemistry differ somewhat as between organic and inorganic compounds. Combustion calorimetry is the commonest method of determining heats of formation of organic compounds, whereas the methods of reaction calorimetry, especially those employing solutions, are commonly used for inorganic compounds. The differences between the two sides of thermochemistry should not however be over-stressed, as there is much common ground. For example, the organic thermo-chemist will readily admit his dependence on the data of inorganic thermo-chemistry, for he frequently employs inorganic reagents, whose heats of formation must be known; the inorganic thermochemist for his part is making increasing use of the "organic" technique of combustion calorimetry, as a review by Holley shows (64/5).

In preparing our compilation of thermochemical data we critically examined original papers on heats of formation of organic and organometallic com-pounds, published between 1930 and 1967. Where necessary, we corrected the results to the 1961 atomic-weight scale (62/1) and to modern energy units, introduced the latest values for subsidiary thermochemical quantities, and incorporated such other corrections as appeared justified. The resulting values were admitted to the tables in Chapter 5 only if we had reasonable confidence in their validity; uncertainties were calculated by a uniform procedure. Values of heats of vaporization (or sublimation) at 25°C were selected from the original literature and are listed in Chapter 5 against the corresponding values of heats of formation.

Although we prepared the compilation by studying original papers, we have from time to time made use of other compilations of thermodynamic data. These are now listed, with summaries of their scope, and some comments on their criticality.

(a) Kharasch (29/1): an uncritical summary of heats of combustion of organic compounds published up to 1929.

(b) Bichowsky and Rossini (36/1): a comprehensive, critical, survey of the thermochemistry of inorganic compounds and C_1 and C_2 organic compounds published up to 1933.

(c) American Petroleum Institute (47/1; 53/1): thermochemical and thermo-dynamic data on alkanes, alkenes, alkynes, and the simpler alicyclic and aromatic hydrocarbons. The data, though based on experimental values, have been smoothed, and some listed values were obtained by inter-polation or extrapolation. The data should thus be used with caution for testing theoretical predictions, because some listed values are them-selves predicted ones. Addenda are issued from time to time.

(d) Manufacturing Chemists Association (61/2): a compilation similar in character to that of the American Petroleum Institute but with a wider scope; it embraces organic compounds containing oxygen, nitrogen, the halogens, etc. Addenda are issued from time to time as part of the activities of the Thermodynamics Research Center Data Project, College Station, Texas.

(e) Landbolt–Börnstein (61/1): a compilation of thermodynamic data published up to 1959 for inorganic and organic compounds. Though comprehensive, it is uncritical, and values of widely differing accuracy are listed. Values are given in joules.

(f) National Bureau of Standards, Circular 500 (50/1) and Technical Notes 270 (65/1; 68/4): Circular 500 lists internally consistent, selected, values of the thermodynamic properties of inorganic and C_1 and C_2 organic compounds; experimental uncertainties are not given. The Technical Notes 270, of which three had been published at the time of writing, will progressively replace Circular 500.

(g) JANAF Thermochemical Tables (61/3): the thermodynamic functions of a selection of inorganic and simple organic compounds in the condensed and ideal-gas states over very wide temperature ranges; species stable only at high temperatures are included. Some of the heats of formation are estimated values. Addenda are issued from time to time.

(h) Termicheskie Konstanty Veshchestv (65/54): a comprehensive, critical, review of the thermodynamic properties, including those related to phase transitions and the critical state, of inorganic and C_1 and C_2 organic compounds; uncertainties are quoted.

(i) Skinner (64/1): a review of the heats of formation and heats of vaporization of compounds containing carbon–metal bonds. Reasons for selecting values from the often discrepant data are given.

(j) Skinner (64/5) and Holley (64/5): selected values for the heats of formation of "key" compounds, based on measurements published up to 1963.

1.4. The units of thermochemistry

1.4.1. THE SYSTEME INTERNATIONAL D'UNITES

It is necessary to define the units to be used in the calculations given in subsequent Chapters. With certain exceptions explained below, the units are based on those of the 1960 Système International d'Unités (60/6) for which the accepted abbreviation is SI. According to the SI there are six primary units:

the unit of length—the metre (m)

the unit of mass—the kilogramme (kg)

the unit of time—the second (s)

the unit of electric current—the ampere (A)

the unit of temperature—the degree Kelvin (°K)

the unit of luminous intensity—the candela (cd).

Various supplementary and derived units are recognized by the SI, of which the following are relevant for the purposes of this book:

the unit of force—the newton ($N = kg\ m\ s^{-2}$)

the unit of energy—the joule ($J = N\ m = kg\ m^2\ s^{-2}$)

the unit of power—the watt ($W = J\ s^{-1} = kg\ m^2\ s^{-3}$)

the unit of electromotive force—the volt ($V = W/A = kg\ m^2\ s^{-3}\ A^{-1}$)

the unit of electrical resistance—the ohm ($\Omega = V/A = kg\ m^2\ s^{-3}\ A^{-2}$).

A list of the internationally agreed names of multiples and submultiples of the units is given by Anderton and Bigg (65/7).

1.4.2. UNITS OF ENERGY

As stated above, the SI unit of energy is the joule, and the joule has been in use, principally by physicists, for many years. Chemists, however, have been, and still are, reluctant to abandon the calorie as their favoured energy unit, notwithstanding several attempts to convert them to the use of the joule. Some compilations of thermochemical data, e.g. the Landolt–Börnstein Tabellen (61/1), employ the joule but the majority, including the latest National Bureau of Standards tables (65/1; 68/4), employ the calorie. We, too, have adopted the calorie, because it is the unit most familiar to our readers, whom we expect to be chemists in the main. Probably the joule will be increasingly used in future original publications in thermochemistry, and the calorie may well be obsolescent within a few years. In revising published data for incorporation into the tables of Chapter 5, we took care to express values in terms of one particular form of the calorie only, namely the *thermochemical calorie*, which is related to the joule by definition: 1 thermochemical calorie (cal) = 4·184 J exactly. Hence, the conversion of all thermochemical data in this book from calories to joules (or from kcal to kJ) merely involves multiplication by 4·184.

We do not propose to discuss earlier definitions of the calorie, since all are now obsolete, but for the benefit of readers scanning original papers we list the following equivalents:

1 International Tables calorie (cal$_{IT}$) = 4·186 8 J

1 15° calorie (cal$_{15}$) = 4·185 5 J

1 calorie as used in U.S.A., 1930–1948 = 4·183 3 International J

$$= 4·184 \text{ J}$$

$$= 1 \text{ thermochemical calorie.}$$

Other units of energy and their joule equivalents are:

1 60°F British thermal unit (Btu) = 1 054·54 J

1 International Steam Table Btu = 1 055·06 J

1 cm^3 atmosphere (cm^3 atm) = 0·101 328 J

1 electron volt (eV) = 1·602 10 × 10^{-19} J

1 kilowatt hour (kW h) = 3·6 × 10^6 J.

1.4.3. UNITS OF PRESSURE

In the SI the unit of pressure is the newton per square metre (N m^{-2}). However, the standard atmosphere is so deeply entrenched as the unit of pressure in chemical thermodynamics that we have adopted it in this book. 1 Standard atmosphere is that pressure which supports a 760-mm column of liquid of uniform density $13.5951 \text{ g cm}^{-3}$ under acceleration of 980.665 cm s^{-2}:

$$1 \text{ standard atmosphere (atm)} = 101\,325 \text{ N m}^{-2}.$$

When low pressures are under discussion, it is often convenient to use the torr as the pressure unit:

$$1 \text{ torr} = 1/760 \text{ atm} = 133.32 \text{ N m}^{-2}.$$

1.4.4. TEMPERATURE SCALE

The size of the degree Kelvin ($^\circ$K) is defined by assigning the value 273.16°K to the temperature of the triple point of water. The degree Celsius ($^\circ$C) is the same size as the degree Kelvin, and the scales are related by

$$^\circ C = {}^\circ K - 273.15.$$

The fundamental scale of temperature is Kelvin's thermodynamic scale, but the scale to which the experimental work described in this book refers is the International Practical Scale of Temperature of 1948 (IPTS 1948), defined by certain fixed points (66/6). From 1st January 1969, the scale recommended for experimental use is the IPTS 1968 (69/2), which differs slightly from IPTS 1948, but is identical with the thermodynamic scale within the accuracy of current measurements.

1.4.5. ATOMIC WEIGHTS AND THE MOLE

The 1961 Table of Relative Atomic Weights, agreed by both the International Union of Pure and Applied Chemistry and the International Union of Pure and Applied Physics, will be used throughout the book. The basis of the Table is that the relative nuclidic mass of ^{12}C is taken as 12 exactly (62/1).

The *mole*, a concept long used in chemistry to represent the gramme formula weight of a compound, was redefined in 1965 by the International Union of Pure and Applied Chemistry as follows: "A mole is an amount of a substance, of specified chemical formula, containing the same number of formula units (atoms, molecules, ions, electrons, quanta, or other entities) as there are atoms in 0.12 kilogrammes (exactly) of the pure nuclide ^{12}C". This liberal definition is adopted in this book, and in particular we consider "other entities" to include chemical reactions in which 1 mole of a nominated substance is formed or transformed (*cf.* Rossini, 50/2). Thus with respect to 1 mole of a substance X, the following may be spoken of as "1 mole of reaction":

$$2A + 3B \rightarrow X + 2.5Y.$$

1.4.6. THE GAS CONSTANT, R

The value used throughout the book is

$$R = 8 \cdot 314\,3 \pm 0 \cdot 001\,2 \text{ J deg}^{-1} \text{ mol}^{-1}$$
$$= 1 \cdot 987\,17 \pm 0 \cdot 002\,9 \text{ cal deg}^{-1} \text{ mol}^{-1}$$
$$= 82 \cdot 056 \pm 0 \cdot 012 \text{ cm}^3 \text{ atm deg}^{-1} \text{ mol}^{-1}.$$

1.4.7. SYMBOLS, NOMENCLATURE AND SIGN CONVENTION

The recommendations of the International Union of Pure and Applied Chemistry have been followed with regard to symbols and nomenclature for thermodynamic quantities. Thus *internal energy* is given the symbol U; *enthalpy* or *heat content* (contracted to *heat* when applied to a process) is given the symbol H; *Gibbs energy*, the quantity formerly called *Gibbs free energy*, is given the symbol G; *entropy* is given the symbol S; *heat capacity* is given the symbol C.

The sign of energy changes etc. is always determined from the standpoint of the *system*, e.g. if energy is evolved from a system the energy change, ΔU, for the process is negative.

Chapter 2

Heats of Reaction and Formation

2.1. Heats of reaction

The measurement of *heats of reaction* $\Delta H_{r,T}$ is fundamental to thermochemistry. For a chemical reaction at constant pressure, which may be written in the general form

Reactants $(v_A A + v_B B + v_C C + ...) \rightarrow$ Products $(v_X X + v_Y Y + v_Z Z ....),$

$\Delta H_{r,T}$ corresponds to the difference in heat content between products and reactants for 1 mole of a nominated species at temperature T. It is necessary to complete the specification of the process by defining (a) the physical states of A, B, C, ..., X, Y, Z, ... to which the measurement refers, and (b) the temperature to which $\Delta H_{r,T}$ refers. When heat is evolved from the reacting system $\Delta H_{r,T}$ is negative, and the reaction is said to be *exothermic*; when heat is absorbed by the reacting system $\Delta H_{r,T}$ is positive, and the reaction is said to be *endothermic*.

For reactions carried out at constant volume the measured quantity is the energy of reaction, $\Delta U_{r,T}$. Both heat content, H, and energy, U, are extensive properties, and in this book the use of capitals for extensive quantities such as H and U implies that we are dealing with *molar* quantities. The relationship between $\Delta H_{r,T}$ and $\Delta U_{r,T}$ is given by

$$\Delta H_{r,T} = \Delta U_{r,T} + p \Delta V_{r,T}, \tag{1}$$

where $\Delta V_{r,T}$ is the difference in molar volume between products and reactants, at constant pressure, p.

Since it is often important to be able to calculate $\Delta H_{r,T}$ from $\Delta U_{r,T}$, it is instructive to consider the magnitude of the term $p \Delta V_{r,T}$ in equation (1). For a reaction involving condensed phases only, consider the typical values $p = 1$ atm and $\Delta V_{r,T} = 4$ cm^3; hence $p \Delta V_{r,T} = 4$ cm^3 atm $\approx 10^{-1}$ cal. If, as is commonly found, the precision with which $\Delta U_{r,T}$ may be determined experimentally is no better than ± 10 cal (and for many reactions it is in the range $\pm 10^2$ to $\pm 10^3$ cal), the term $p \Delta V_{r,T}$ is negligible compared with uncertainties in the measurement of $\Delta U_{r,T}$, when the reaction involves condensed phases only; in this circumstance $\Delta H_{r,T} = \Delta U_{r,T}$ to a close

9

approximation. However, for reactions in which one or more of the participating species is gaseous, $\Delta V_{r,\,T}$ may well be several orders of magnitude greater than $\Delta V_{r,\,T}$ for a condensed-phase reaction, so that the term $p\Delta V_{r,\,T}$ will often be significant and will need to be evaluated. In principle, the evaluation requires a knowledge of the equation of state for the gaseous species taking part in the reaction, but in practice the error resulting from the assumption of ideal-gas behaviour of each species will generally be much less than the error in the measurement of $\Delta U_{r,\,T}$, at least for reactions involving non-associated gases at moderate pressures. Generally, therefore, it will be reasonable to write equation (1) in the form

$$\Delta H_{r,\,T} = \Delta U_{r,\,T} + \Delta n_g RT, \tag{2}$$

where Δn_g is the increase in the number of moles of *gaseous* species resulting from the reaction. For example, for the gas-state reaction

$$C_2H_2(g) + 2H_2(g) \rightarrow C_2H_6(g),$$

$\Delta n_g = -2$, and $\Delta H_{r,\,T} = \Delta U_{r,\,T} - 1 \cdot 19 \text{ kcal}$, at $T = 298 \cdot 15°\text{K}$.

2.2. Standard states and reference states

2.2.1. DEFINITIONS OF THE STANDARD STATE AND THE REFERENCE STATE

The thermochemist is concerned with matter in the gas, liquid, and solid states. Since for any substance the heat contents of these states differ considerably, it is important to specify the state as precisely as possible in any description of a thermochemical measurement. Better still, the measured value should be corrected to $\Delta H_{r,T}$ for a reaction in which all participants are in their *standard states*, defined as follows:

(a) For a gas, the standard state is that of the hypothetical ideal gas at one atmosphere pressure; in this state the heat content is the same as that of the real gas at the same temperature and at zero pressure.

(b) For a liquid, the standard state is that of the pure substance under a pressure of 1 atm.

(c) For a solid, the standard state is that of the pure crystalline substance, under a pressure of 1 atm.

It will be noted that temperature is not part of the definition of the standard state, and should therefore be quoted separately. In this book we follow the convention of explicitly specifying temperature only when it is other than 25°C (298·15°K), i.e. a temperature-dependent thermodynamic quantity written without a temperature descriptor relates to 25°C.

Symbols for thermodynamic quantities relating to substances in their standard states, or to reactions with all participants in their standard states, are distinguished by the superscript ° e.g. $S°$, $\Delta H_r°$, $\Delta U_{r, 400°K}°$.

The *reference state* for a substance at a specified temperature is its stable physical state at that temperature and at a pressure of 1 atm. It is possible for a substance at temperature T_1 to be in its reference state though not in a standard state, or to be in a standard state though not in its reference state, or to be in both a reference and a standard state (i.e. a *standard reference state*). The last circumstance obtains when

(a) a substance that exists as a gas at $T_1°K$ and 1 atm pressure (i.e. a substance that has a critical temperature $< T_1°K$, or has a saturation vapour pressure > 1 atm at $T_1°K$) is at an extremely low pressure, hereafter referred to as "zero" pressure;

(b) a substance that is liquid at 1 atm pressure (i.e. T_1 is above the melting point, and the saturation vapour pressure < 1 atm at $T_1°K$) is maintained at 1 atm pressure;

(c) a substance that is a crystalline solid at 1 atm pressure (i.e. T_1 is below the melting point, and the sublimation pressure < 1 atm at $T_1°K$) is maintained at 1 atm pressure. Strictly the reference state of a polymorphic solid is that of its most thermodynamically stable crystalline form, but in instances where this form is not readily accessible (e.g. black phosphorus) another form may be designated as the reference state.

2.2.2. EXAMPLES OF SUBSTANCES IN THEIR STANDARD STATES AND REFERENCE STATES

There follow some examples of the application of the definitions of standard states and reference states given in Section 2.2.1.

(a) At $298 \cdot 15°K$ and 1 atm pressure, isobutene (normal boiling point $226 \cdot 25°K$) is a gas. The standard reference state is therefore that of the ideal gas, equivalent in terms of heat content to the real gas at zero pressure. But calorimetric experiments with isobutene will be carried out at some finite pressure p_1, not at zero pressure, so the calculation of the difference in heat content of isobutene between the standard state and the real state at pressure p_1 is important. The basis of the calculation is provided by the thermodynamic equation

$$\left(\frac{\partial H}{\partial p}\right)_T = -T\left(\frac{\partial V}{\partial T}\right)_p + V. \tag{3}$$

By integration of equation (3) for a particular temperature T,

$$H_{\text{real}} - H_{\text{ideal}} = H_{p=p_1} - H_{p=0} = \int_0^{p_1} \left[-T \left(\frac{\partial V}{\partial T} \right)_p + V \right] \, dp. \qquad (4)$$

The integral in equation (4) can be evaluated from the equation of state of isobutene gas (66/6), which near 298°K and 1 atm pressure may be written as

$$V = \frac{RT}{p} - \frac{1 \cdot 78 \times 10^9}{T^{2.6}} \text{ cm}^3.$$

Hence, when $p_1 = 1$ atm and $T = 298 \cdot 15°K$,

$$H_{p=1} - H_{p=0} = \int_0^1 \left(\frac{-3 \cdot 6 \times 1 \cdot 78 \times 10^9}{T^{2.6}} \right) \, dp$$

$$= -2362 \text{ cm}^3 \text{ atm}$$

$$= -57 \text{ cal.}$$

It is evident that $(H_{\text{real}} - H_{\text{ideal}})$ though small by comparison with most heats of reaction is certainly not negligible in terms of the precision of modern reaction calorimetry. In accurate calorimetric work, therefore, it would be necessary to allow for the difference in heat content between the real and the standard state of isobutene, and a similar statement applies generally to heat-of-reaction measurements with any moderately imperfect gas.

(b) At 298·15°K and 1 atm pressure the standard reference state of benzene is that of the liquid, since the melting point of the substance is 278·7°K and the normal boiling point is 353·3°K. Calorimetric measurements on benzene at 298·15°K are most likely to be made on the liquid, but it is also feasible to make measurements on the real gas at pressures below 94 torr (the saturation vapour pressure). It would then be necessary to evaluate the difference in heat content between the real gas and the hypothetical ideal gas, according to equation (4). The results would apply to benzene in its standard state as a gas at 298·15°K.

(c) At 298·15°K, 4-chlorophenol is below its melting point (316°K). Hence the standard reference state is that of the crystalline solid under 1 atm pressure. 4-Chlorophenol can however be readily maintained at room temperature as an undercooled liquid, and thermochemical measurements can be made with it in that state (53/4); assuming the prevailing pressure is 1 atm, the measurements will refer to the substance in the standard state as a liquid at 298·15°K.

For most organic liquids and solids the term in square brackets in equation (4) is of the order of 10^2 cm^3. Hence it is easily shown that departures from the standard pressure of 1 atm will affect the heat contents of condensed phases to a calorimetrically significant extent only when a heat of reaction under a high pressure is measured. In a bomb-calorimetric experiment, gas pressures as high as 40 atm are employed (Section 3.2); such an experiment is actually conducted at constant volume, and the thermodynamic relation for converting the energies of the phases into those of the substances in their standard states is then

$$\left(\frac{\partial U}{\partial p}\right)_T = -T \left(\frac{\partial V}{\partial T}\right)_p - p \left(\frac{\partial V}{\partial p}\right)_T. \tag{5}$$

A further comment on the effect of pressure on the energies of condensed phases is made in Section 3.2.5, but in general pressure is a less significant variable in thermochemistry than is temperature. The dependence of heat and energy quantities on temperature is discussed in Section 2.3.

2.2.3. STANDARD HEATS OF REACTION

If all the participants in a chemical reaction are in their standard states, the heat of reaction is referred to as the *standard heat of reaction*, $\Delta H^\circ_{r,\,T}$. For example, ΔH°_r for the addition of chlorine to tetrachloroethylene refers to the following reaction, carried out at 298·15°K:

C$_2$Cl$_4$ (liquid under 1 atm pressure) + Cl$_2$ (real gas at zero pressure)

→ C$_2$Cl$_6$ (crystalline solid under 1 atm pressure).

The shorthand way of writing this equation is

$$C_2Cl_4(l) + Cl_2(g) \rightarrow C_2Cl_6(c).$$

2.2.4. STANDARD HEATS OF FORMATION

Special significance attaches to the standard heat of the reaction in which a compound in its standard state is formed from its elements in their standard states. This heat quantity is termed the *standard heat of formation* of the compound, ΔH°_f, and is almost always quoted for a temperature of 298·15°K. (The heat of formation of an element in its standard state must by this definition be zero.) For example, ΔH°_f of liquid ethyl bromide is the heat of the reaction

$$2\,C\,(c, graphite) + 2\cdot5\,H_2(g) + 0\cdot5\,Br_2(l) \rightarrow C_2H_5Br(l).$$

The first law of thermodynamics indicates that for a reaction

$$v_A A + v_B B + \ldots \rightarrow v_X X + v_Y Y + \ldots,$$

$$\Delta H_r^\circ = \sum \Delta H_f^\circ \text{ (products)} - \sum \Delta H_f^\circ \text{ (reactants)}, \tag{6}$$

where $\quad \sum \Delta H_f^\circ \text{ (reactants)} = v_A (\Delta H_f^\circ, A) + v_B (\Delta H_f^\circ, B) + \ldots \dagger$

and $\quad \sum \Delta H_f^\circ \text{ (products)} = v_X (\Delta H_f^\circ, X) + v_Y (\Delta H_f^\circ, Y) + \ldots \dagger$

The important equation (6) will be used in later Sections,

(a) for the calculation of ΔH_f° of a compound, when ΔH_r° of a reaction involving that compound is known (the compound may be amongst either the reactants or the products) and the heats of formation of *all other* participants in the reaction are known,

(b) for the calculation of ΔH_r° for reactions in which the heats of formation of *all* the participants are known.

2.3. The temperature dependence of heats of reaction

2.3.1. KIRCHHOFF'S EQUATION

The molar heat capacity at constant pressure, C_p, of any substance is defined by the relation

$$\left(\frac{\partial H}{\partial T} \right)_p = C_p. \tag{7}$$

Hence, the change in heat content of a substance from temperature T_1 to temperature T_2 is given by

$$H_{T_2} - H_{T_1} = \int_{T_1}^{T_2} C_p \, dT. \tag{8}$$

From the first law of thermodynamics it can readily be shown that the change in a heat of reaction consequent upon a change in temperature from T_1 to T_2 is given by

$$\Delta H_{r, T_2} - \Delta H_{r, T_1} = \int_{T_1}^{T_2} \Delta C_p \, dT, \tag{9}$$

where $\quad \Delta C_p = \sum C_p \text{ (products)} - \sum C_p \text{ (reactants)}. \tag{10}$

---

† Throughout the book it is to be understood that summation of thermodynamic quantities for products or reactants is carried out with use of the appropriate stoicheiometric co-efficients.

Equation (9), known as Kirchhoff's equation, permits the calculation of the heat of reaction at any temperature T_2 when the heat of reaction at another temperature T_1 is known. Some examples of the use of equation (9) are given in Section 2.3.2, but first it is necessary to consider how $\Delta H_{r,\,T_1}$ for a single temperature T_1 is itself derived.

Let us suppose that certain reactants are contained in a thermally isolated calorimeter, all at temperature T_{initial}, and that one mole of a chemical reaction at constant pressure is caused to take place. The temperature of the calorimeter and its contents (now the products) changes to T_{final}. If the restriction of thermal isolation of the calorimeter were removed, the same overall result could be achieved in three consecutive steps, as follows:

(a) The initial system is brought from temperature T_{initial} to some reference temperature T_1, involving a change in heat content of the system equal to† $C_{p,\,\text{initial}}\,(T_1 - T_{\text{initial}})$.

(b) The chemical reaction is carried out in such a way that the temperature remains constant at T_1 (an *isothermal* reaction); the change in heat content of the system is $\Delta H^{\circ}_{r,\,T_1}$.

(c) The final system is brought from temperature T_1 to T_{final}, involving a change in heat content of the system equal to $C_{p,\,\text{final}}\,(T_{\text{final}} - T_1)$.

The net change in heat content ΔH for the three steps would therefore be given by

$$\Delta H = C_{p,\,\text{initial}}\,(T_1 - T_{\text{initial}}) + \Delta H^{\circ}_{r,\,T_1} + C_{p,\,\text{final}}\,(T_{\text{final}} - T_1). \tag{11}$$

Now, the net change in heat content in the original experiment was zero, because the calorimeter was isolated. Hence ΔH in equation (11) is zero also, because from the first law of thermodynamics the change in heat content for a constant-pressure process is independent of the path followed.

As the heat capacity of the calorimeter itself is a common component of the quantities $C_{p,\,\text{final}}$ and $C_{p,\text{initial}}$, we may write

$$C_{p,\,\text{final}} - C_{p,\text{initial}} = \sum C_p\,(\text{products}) - \sum C_p\,(\text{reactants}) = \Delta C_p. \tag{12}$$

By combination of equations (11) and (12) and remembering that ΔH in equation (11) is zero, it follows that

$$\Delta H_{r,\,T_1} = -C_{p,\,\text{initial}}\,(T_{\text{final}} - T_{\text{initial}}) + \Delta C_p\,(T_1 - T_{\text{final}}), \tag{13}$$

or, if equation (13) is recast,

$$\Delta H_{r,\,T_1} = -C_{p,\,\text{final}}\,(T_{\text{final}} - T_{\text{initial}}) + \Delta C_p\,(T_1 - T_{\text{initial}}). \tag{14}$$

---

† It is assumed that T_1 is sufficiently close to both T_{initial} and T_{final} that changes in heat contents can be evaluated as the product of a heat capacity and a temperature change. Reference to equation (8) will show that this requires the C_p terms to be independent of temperature, which is almost always a close approximation to the truth over temperature ranges of a few degrees.

The measurement of the quantity $-C_{p,\,initial}\,(T_{final}-T_{initial})$ in equation (13) is discussed in Section 2.5.1. The derivation of the other term in equation (13), namely $\Delta C_p\,(T_1-T_{final})$, remains to be discussed. In principle, ΔC_p can be determined from very accurate measurements of the energy equivalent of the calorimetric system before and after the reaction has taken place, using electrical heating. In practice, however, ΔC_p is almost always calculated from the heat capacities of reactants and products, which are often available from compilations such as Landolt–Börnstein (61/1). By arranging that $T_{final} = T_1$, equation (13) reduces to

$$\Delta H_{r,\,T_1} = -C_{p,\,initial}\,(T_{final}-T_{initial}) \tag{15}$$

and knowledge of ΔC_p is then unnecessary; alternatively by arranging that $T_{initial} = T_1$, then from equation (14)

$$\Delta H_{r,\,T_1} = -C_{p,\,final}\,(T_{final}-T_{initial}). \tag{16}$$

2.3.2.   THE APPLICATION OF KIRCHHOFF'S EQUATION

By use of equation (13) or (14), the heat of a reaction at a reference temperature T_1 may be derived. We now show how equation (9), Kirchhoff's equation, may be employed for calculating the heat of reaction at temperature T_2 from the value at T_1. Such calculations are made with varying degrees of rigour.

(a) The simplest assumption is that ΔC_p is a constant, independent of T; equation (9) then leads to

$$\Delta H_{r,\,T_2}-\Delta H_{r,\,T_1} = \Delta C_p\,(T_2-T_1). \tag{17}$$

The assumption that ΔC_p is a constant will hold quite well if T_2 is within 10 degC of T_1, and will often hold with reasonable accuracy when T_2 and T_1 are 100 degC or more apart; the assumption will be least in error if the value of ΔC_p for the mean temperature $\frac{1}{2}\,(T_1+T_2)$ is used in the calculations.

(b) A more rigorous way to evaluate the integral of equation (9) is to calculate ΔC_p for various temperatures in a range that includes T_1 and T_2, and to express ΔC_p as an analytical function of T. For example, C_p for many gases and vapours at low pressure can be accurately expressed in terms of T by an equation of the type

$$C_p = a+bT+cT^2, \tag{18}$$

so that for a reaction in which all participants are gaseous

$$\Delta C_p = A+BT+CT^2, \tag{19}$$

and from equation (9)

$$\Delta H_{r, T_2} - \Delta H_{r, T_1} = A\,(T_2 - T_1) + \frac{B}{2}\,(T_2{}^2 - T_1{}^2) + \frac{C}{3}\,(T_2{}^3 - T_1{}^3). \quad (20)$$

For example, for the reaction (67/1) expressed by

$$C_2H_5CHO\,(g) + H_2(g) \rightarrow C_3H_7OH(g),$$

$$\Delta C_p = -8{\cdot}92 + 1{\cdot}71 \times 10^{-2}T - 1{\cdot}02 \times 10^{-5}T^2\ \text{cal deg}^{-1}.$$

Hence $\Delta H_{r,\ 500°K} - \Delta H_{r,\ 298°K} = -759\ \text{cal.}$

If the value of ΔC_p at the mean temperature (399°K) is used in equation (17) we obtain

$$\Delta H_{r,\ 500°K} - \Delta H_{r,\ 298°K} = -751\ \text{cal.}$$

In this instance, the value of $(\Delta H_{r, T_2} - \Delta H_{r, T_1})$ calculated by method (a) agrees closely with the value calculated by the more exact method (b). However, the data needed for application of method (b) are often unavailable and calculations are most commonly made by method (a) alone.

Instead of using equation (9), one may calculate the change in ΔH_r between temperatures T_1 and T_2 from relative heat contents $[H^\circ - H_0^\circ]$ or $[H^\circ - H_{298}^\circ]$, when tabulated values of these functions are available for all participants in a reaction. The necessary equation, deduced from the first law of thermodynamics, is

$$\Delta H_{r,\ T_2}^\circ - \Delta H_{r,\ T_1}^\circ = \sum [H_{T_2}^\circ - H_0^\circ]\ (\text{products}) - \sum [H_{T_1}^\circ - H_0^\circ]\ (\text{products})$$

$$- \sum [H_{T_2}^\circ - H_0^\circ]\ (\text{reactants}) + \sum [H_{T_1}^\circ - H_0^\circ]\ (\text{reactants}).$$
$$(21)$$

An alternative equation can be written with H_{298}° in place of H_0°. Values of relative heat contents are to be found in compilations such as the JANAF tables (61/3), but the number of organic compounds for which tabulated heat contents are available is not large, so that equation (21) finds less application than Kirchhoff's equation.

2.4. Gibbs energies of reaction, and equilibrium constants

From the equation which defines Gibbs energy, namely

$$G = H - TS, \quad (22)$$

the following equation for the change in standard Gibbs energy for a chemical reaction may be deduced

$$\Delta G^{\circ}_{r,\,T} = \Delta H^{\circ}_{r,\,T} - T\Delta S^{\circ}_{r,\,T}, \tag{23}$$

where $\Delta S^{\circ}_{r,\,T}$ is the standard entropy change resulting from the reaction at temperature T.

It is shown in standard texts that the equilibrium constant K_p of a reaction involving ideal gases† is given by

$$\Delta G^{\circ}_{r,\,T} = -RT \ln K_p, \tag{24}$$

where

$$K_p = \frac{\prod\limits_{i} p^{\nu_i} \text{ (products)}}{\prod p^{\nu_i} \text{ (reactants)}}. \tag{25}$$

Combination of equations (23) and (24) leads to the important equation

$$K_p = \exp\left(\frac{-\Delta H^{\circ}_{r,\,T}}{RT}\right) \exp\left(\frac{\Delta S^{\circ}_{r,\,T}}{R}\right). \tag{26}$$

As stated in Section 1.2, an important use of thermochemical data is for the calculation of equilibrium constants of chemical reactions; the calculation is effected by combining heats of reaction with entropies of reaction by means of equation (26). Suppose that the feasibility of a certain reaction for the synthesis of a given compound is under consideration. A thermodynamic calculation will indicate the yield of the compound that can be obtained from the equilibrium reaction, and for this purpose K_p must be evaluated. If K_p can be derived with an uncertainty of $\pm 50\%$ from extant values of $\Delta H^{\circ}_{r,\,T}$ and $\Delta S^{\circ}_{r,\,T}$, these data will probably be acceptable, at least for preliminary calculations of K_p. It is instructive to consider to what extent standard errors in $\Delta H^{\circ}_{r,\,T}$ and $\Delta S^{\circ}_{r,\,T}$, here denoted as $\bar{s}_{\Delta H}$ and $\bar{s}_{\Delta S}$ respectively, affect the overall error in K_p, here denoted as $\bar{s}_K$. It follows from the theory of the variance of a function (66/6) that

$$\bar{s}_K^2 = \left(\frac{\partial K_p}{\partial \Delta H^{\circ}_{r,\,T}}\right)^2 \bar{s}_{\Delta H}^2 + \left(\frac{\partial K_p}{\partial \Delta S^{\circ}_{r,\,T}}\right)^2 \bar{s}_{\Delta S}^2, \tag{27}$$

so long as temperature is measured with sufficient accuracy that it need not be considered as an independent variable. Consider for the moment that errors in K_p result from errors in $\Delta H^{\circ}_{r,\,T}$ only, i.e. that the error in $\Delta S^{\circ}_{r,\,T}$ is

---

† For the present discussion, it suffices to consider only ideal-gas reactions. Equilibria involving real gases and condensed phases are referred to in Section 2.5.2.

insignificant. Then equation (27) simplifies to

$$\bar{s}_K = \pm \left(\frac{dK_p}{d\Delta H^\circ_{r,\,T}} \right) \bar{s}_{\Delta H}. \tag{28}$$

Hence from equations (26) and (28)

$$\bar{s}_K = \pm \frac{\bar{s}_{\Delta H}}{RT} \left[\exp \left(\frac{-\Delta H^\circ_{r,\,T}}{RT} \right) \exp \left(\frac{\Delta S^\circ_{r,\,T}}{R} \right) \right], \tag{29}$$

so that

$$\frac{\bar{s}_K}{K_p} = \pm \frac{\bar{s}_{\Delta H}}{RT}. \tag{30}$$

Table 1 lists values of $(\pm \bar{s}_K / K_p)$ for values of $\bar{s}_{\Delta H}$ at three temperatures. It is seen that to determine K_p to $\pm 50\%$ (i.e. $|\bar{s}_K / K_p| \not> 0.5$), $\bar{s}_{\Delta H}$ should not exceed ± 0.3 kcal at $298°K$, ± 0.5 kcal at $500°K$ and ± 1 kcal at $1000°K$. These important results should be kept in mind during any discussion of the accuracy needed in experimental thermochemistry, as for example in Section 3.1.

TABLE 1. Errors in K_p resulting from errors in $\Delta H^\circ_{r,\,T}$.

$\pm s_{\Delta H}$ kcal	$\pm s_K / K_p$		
	$T = 298°K$	$T = 500°K$	$T = 1000°K$
0·1	0·17	0·1	0·05
0·3	0·5	0·3	0·15
0·5		0·5	0·25
1·0			0·5

If we now consider the situation that arises when errors in K_p result from errors in $\Delta S^\circ_{r,\,T}$ only, i.e. when the error in $\Delta H^\circ_{r,\,T}$ is insignificant, then equation (27) simplifies to

$$\bar{s}_K = \pm \left(\frac{dK_p}{d\Delta S^\circ_{r,\,T}} \right) \bar{s}_{\Delta S} \tag{31}$$

$$= \pm \frac{\bar{s}_{\Delta S}}{R} \left[\exp \left(\frac{-\Delta H^\circ_{r,\,T}}{RT} \right) \exp \left(\frac{\Delta S^\circ_{r,\,T}}{R} \right) \right], \tag{32}$$

so that

$$\frac{\bar{s}_K}{K_p} = \pm \frac{\bar{s}_{\Delta S}}{R}. \tag{33}$$

It is seen from equation (33) that for K_p to be known with a standard error not exceeding $\pm 50\%$, $\Delta S^\circ_{r, T}$ must be known with a standard error not exceeding ± 1 cal deg^{-1}; temperature does not appear explicitly in equation (33). Molar entropies of many substances at 298°K can be determined by low-temperature calorimetry with standard errors of less than 0·5 cal deg^{-1}, and molar ideal-gas entropies of substances of simple molecular structure can in favourable cases be determined by statistical mechanics with an error of less than 0·1 cal deg^{-1}. In general, therefore, ΔS°_r at 298°K will be known with sufficient precision to permit meaningful calculation of K_p. By contrast, few extant values of ΔH°_f of organic compounds have sufficient precision to permit calculation of ΔH°_r with an error of less than 0·3 kcal, which as we have seen is the criterion for the meaningful calculation of K_p at 298°K. To remedy this situation is a challenge to the thermochemist: the strategy he must adopt is discussed in the next Section.

2.5. The principles of the accurate determination of ΔH°_r.

2.5.1. CALORIMETRY—THE "FIRST-LAW" METHOD

For any process occurring in any system, the first law of thermodynamics states that

$$\Delta u = q + w + w', \tag{34}$$

where q is the heat absorbed from the surroundings, w is the mechanical work and w' is any other type of work that is done on the system by the surroundings†. In the particular case of a process occurring at constant volume in a calorimeter, $w = 0$ and if $w' = 0$ also, $\Delta u = q$; in the case of a process occurring at constant pressure in a calorimeter $w = -p\Delta v$, and if $w' = 0$, $\Delta u = q - p\Delta v$, i.e. $\Delta h = q$. Some textbooks give the impression that a calorimeter is an instrument for measuring q, and indeed a few calorimeters in which q is measured have been described [Section 3.3.4, paragraphs (b) and (c)]. In most types of calorimeter, however, the amount of heat exchanged with the surroundings is kept minimal, and in an *adiabatic* calorimeter q is in fact reduced to zero. Adiabatic calorimeters are used extensively for measurement of heats of phase change and heat capacities, but have not as yet found great favour for the accurate measurement of the heats of organic reactions. The most commonly used instrument for this purpose is the *isothermal-environment*‡ calorimeter, in which the surroundings are held at a constant temperature and the calorimeter is insulated to some extent from these surroundings.

---

† Positive values of q, w, and w' indicate that energy has been supplied to the system from the surroundings. Negative values indicate that energy has been lost from the system to the surroundings.
‡ Sometimes termed *isoperibol*.

The rate of heat exchange between calorimeter and surroundings is then small and calculable by the known laws of heat flow.

In a process occurring in an adiabatic calorimeter $q = 0$. Hence, when such a calorimeter is used to study a chemical reaction conducted at constant volume, $\Delta u = 0$, i.e. occurrence of a chemical reaction within the calorimeter brings about no change in the internal energy of the system as a whole. What happens is that the energy released by the chemical reaction raises the temperature of the system, which now has a different chemical composition from the original one. This temperature rise is measured in a calorimetric experiment of the type described, and to derive Δu_r from it the calorimeter must be calibrated, and the number of moles of reaction, n, must be determined.

When an adiabatic calorimeter is used to study a chemical reaction conducted at constant pressure, again $q = 0$ and since in this case $q = \Delta h$ (see above), it is evident that there is no change in the heat content of the system as a whole. Let us now examine a constant-pressure calorimetric experiment more closely, this time admitting the possibility that the calorimeter may not be truly adiabatic. We suppose that the calorimeter is equipped with an electric heater, and contains reactants, the temperature of the whole system being $T'_{initial}$. First a potential, E volts, is applied to the heater, causing a current, i ampères, to flow for t seconds; the electrical energy thus supplied to the calorimeter causes the temperature of the latter to rise by an amount $\Delta\theta'$. However, some heat exchange with the surroundings will have occurred during the input of electrical energy, and we must correct for this by calculating the temperature rise $\Delta\theta'_{corr}$ which would have occurred if the calorimeter had been adiabatic (see Section 3.2.1). When $\Delta\theta'_{corr}$ has been calculated we have in effect the situation where $q = 0$ and w is effectively zero, but now $w' = Eit$ joules. In this experiment, therefore, $w' = \Delta h$ and

$$\Delta h = Eit. \tag{35}$$

We now define a quantity $\mathscr{E}_{initial}$, the *energy equivalent* of the calorimeter in its initial condition (i.e. containing reactants), by the equation

$$\mathscr{E}_{initial} = \frac{Eit}{\Delta\theta'_{corr}}. \tag{36}$$

With the determination of $\mathscr{E}_{initial}$ the calorimeter has been calibrated. Let us now cool the calorimeter back to a temperature $T_{initial}$ near to $T'_{initial}$, and initiate a chemical reaction, which causes the temperature to rise by $\Delta\theta$, approximately equal to $\Delta\theta'$. Again we must allow for heat exchanged with the surroundings and correct $\Delta\theta$ to $\Delta\theta_{corr}$, which is the temperature rise that

would have occurred if the calorimeter had been adiabatic. As already explained, $\Delta h = 0$ for the system as a whole, but the heat content of the final system is less than the heat content of the initial system *at the same temperature* by an amount equal to $-n\Delta H_{r,\ T_{initial} \rightarrow T_{final}}$.
Hence

$$-n\Delta H_{r,\ T_{initial} \rightarrow T_{final}} = \mathscr{E}_{initial}\Delta\theta_{corr}. \tag{37}$$

By combination of equations (36) and (37) we obtain

$$\Delta H_{r,\ T_{initial} \rightarrow T_{final}} = \frac{-Eit}{n}\left(\frac{\Delta\theta_{corr}}{\Delta\theta'_{corr}}\right). \tag{38}$$

The only quantity on the right-hand side of equation (38) that is not immediately known is n; this may be calculated either from the number of moles of reactant(s) lost or from the number of moles of product(s) formed. The value of $\Delta H_{r,\ T_{initial} \rightarrow T_{final}}$ then found is identical with the term $-C_{p,\ initial}(T_{final} - T_{initial})$ in equation (13); hence $\Delta H_{r,\ T_1}$ can be evaluated from equation (13), as described in Section 2.3.1, or if $T_{final} = T_1$ from equation (15). Alternatively, $\Delta H_{r,\ T_1}$ can be derived from equations (14) or (16), provided that the energy equivalent is measured for the final system; for this purpose it would be necessary to determine the energy equivalent by an electrical calibration of the calorimeter + products after the reaction had taken place.

In the derivation of equation (38) it was tacitly assumed that the reaction under consideration was exothermic, i.e. that $\Delta\theta$ was positive. If an endothermic reaction is studied (an uncommon event in organic thermochemistry), equation (38) is still valid, but $\Delta\theta_{corr}$ will of course be negative. Other assumptions made in the derivation of equation (38) were that $T_{initial} \approx T'_{initial}$, and that $\Delta\theta \approx \Delta\theta'$. These assumptions were made to ensure that $\mathscr{E}_{initial}$, which is in reality very slightly temperature-dependent, was transferable from the calibration experiment to the chemical experiment; the closer the above approximations are to equalities the nearer will $\mathscr{E}_{initial}$ be to a constant, dependent on the nature of the calorimetric system only. To keep the error due to these assumptions less than 0.01% it is recommended that

$$\left|\frac{T_{initial} - T'_{initial}}{\Delta\theta}\right| \not> 0.25$$

and that

$$\left|\frac{\Delta\theta - \Delta\theta'}{\Delta\theta}\right| \not> 0.25.$$

We have now seen how a calorimeter may be calibrated electrically, and then used to measure the energy or heat of a chemical reaction. There is an important alternative to the electrical method of calibration, namely the use of a chemical reaction for which $\Delta U^\circ_{r,\,T_1}$ or $\Delta H^\circ_{r,\,T_1}$ has itself been accurately determined by means of an electrically calibrated calorimeter, usually in a national standardizing laboratory. Thus in bomb calorimetry the combustion of benzoic acid in oxygen (see Section 3.2.1) and in flame calorimetry the burning of hydrogen in oxygen (see Section 3.2.6) are internationally recognized methods of calibration; benzoic acid and hydrogen are termed *primary-standard substances* in this context. For calorimeters other than the oxygen-bomb and oxygen-flame types no primary standard substances are yet recognized, and electrical calibration should therefore be employed.

The calorimeters that have been considered so far in this Chapter were somewhat idealized, and it is now necessary to discuss those features of real calorimeters upon which the accuracy of a measurement of $\Delta H^\circ_{r,\,T}$, depends. The essential requirements are as follows.

(a) A calorimeter must have a thermometer capable of sensing the moving temperature of the calorimeter without permitting significant, unmeasured transfer of heat to the surroundings. The types of thermometer that broadly satisfy these requirements and are used in calorimetry are:

platinum (or other metal) resistance, coupled to a resistance bridge or potentiometer — discrimination† *ca.* 10^{-4} degC;

mercury-in-glass, especially Beckmann type — discrimination† *ca.* 10^{-3} degC;

thermojunction(s), coupled to a potentiometer — discrimination† *ca.* $(3 \times 10^{-2}/x)$ degC, where x is the number of junctions;

thermistor, coupled to a resistance bridge — discrimination† *ca.* 5×10^{-5} degC;

quartz-crystal oscillator — discrimination† *ca.* 10^{-4} degC. (As this type of thermometer has only recently been developed, few examples of its use in calorimetry exist, but its potential is considerable.)

The merits and demerits of the various types of thermometer are discussed in standard texts (53/9, 53/10, 41/2, 55/6, 62/6, 64/47); Boucher (67/40) has summarized the use of thermistors in calorimetry.

(b) A calorimeter is usually designed so that all parts come quickly to the same temperature. Use of a block of highly conducting metal (e.g. silver, copper) is one method of meeting this requirement. An alternative is to use a vessel filled with fluid (often water) with provision for stirring the fluid

---

† The values quoted for discrimination are those that can be attained by the use of good commercially available equipment. Better discrimination can be attained by use of specialized equipment.

and/or rotating or shaking the vessel by mechanisms that do not permit significant heat exchange with the surroundings. Use of a stirring, rotating, or shaking mechanism leads to the development of frictional heat in the calorimeter and this must be allowed for. The most satisfactory procedure is to agitate the calorimeter or its contents at the same constant speed and for the same time in the calibration experiment as in the chemical experiment. Furthermore, if the rates of temperature rise in the two experiments are made as nearly equal as is practicable, the error from failure to achieve complete thermal equilibrium throughout the calorimeter is minimized.

(c) Calorimeters that are to be calibrated electrically must have means for leading in current to the electric heater within the calorimeter. The experimental arrangement should be such that significant heat transfer does not take place through the leads, and that significant amounts of heat are not dissipated in the current-carrying leads. The heater should be constructed from wire having a very low temperature-coefficient of resistance, so that the resistance change due to self-heating will be negligible. The potential applied to the heater should be from a stable source: an electronically stabilized power pack or a high-capacity accumulator is suitable. The potential drop across the heater will normally be measured by means of a potentiometer and the current through the heater will normally be calculated from the measured potential drop across a standard resistance in series with the calorimeter heater; there is some merit in continuously recording the two measured potentials, since both are liable to short-term fluctuations. For the measurement of the heating time an accurate chronometer or a scaler referred to a standard-frequency source is suitable. Detailed discussion of the electrical calibration of a calorimeter has been given by Challoner, Gundry, and Meetham (55/2).

(d) Means for initiating the chemical reaction must be incorporated. For oxygen-combustion reactions this is generally done by supplying an electric spark or "hot spot" to ignite either the main reactant or some auxiliary substance (e.g. cotton thread or plastic strip), which will in turn ignite the main reactant; materials that are difficult to ignite may require the use of an arc or a miniature electric furnace within the calorimeter. Many devices for mixing liquid or gaseous reactants in a calorimeter have been described, such as diaphragms that may be punctured, ampoules that may be broken, buckets that may be tipped. For gases, remotely operated valves may be used. Whatever initiating system is employed, the energy put into the calorimeter must be determined and deducted from the measured, gross energy/heat of reaction.

Puncturing, breaking, or tipping devices require the expenditure of very little energy, often less than 1 J. Energies of electrical-ignition systems vary greatly, depending on their design. Systems involving a thin, fusible platinum wire may consume less than 1 J (62/7), but systems involving thicker wire may consume up to 100 J, and the measurement of the ignition energy by an integrator circuit (55/3) is desirable. The electric arc used by Kolesov, Zenkov and Skuratov (62/8) to cause chlorotrifluoromethane to react with sodium consumed *ca.* 1000 J, and this is probably the maximum amount of ignition energy that could be accepted in a calorimetric experiment.

(e) Heat exchange between the calorimeter and its surroundings must be under close control. In an adiabatic apparatus, the calorimeter will probably be suspended in a vacuum and surrounded by a shield held at a temperature within ±0·001 degC of that of the calorimeter surface. Control of the shield temperature is achieved by use of differential thermocouples, or of thermistor elements in two arms of a bridge circuit, with one sensor on the outer surface of the calorimeter and the other on the inner surface of the shield; any out-of-balance signal is amplified and used to control the input of electrical energy to heaters around the shield. Isothermal-environment calorimeters are usually separated from their surrounding jackets by an air gap of *ca.* 1 cm or by an evacuated interspace. The temperature of the jacket is held constant to within ±0·005 degC or better throughout the experiment. Methods for computing the heat exchange between a calorimeter and its isothermal surroundings are referred to in Section 3.2.1.

(f) Means for calculating the number of moles of reaction must be devised. For reactions which can be shown to go to completion without formation of by-products, careful weighing of the reactant(s) may suffice. Calculation of the amount of reaction from the mass of substance put into the calorimeter does however presuppose that the substance is pure—this is a matter of such importance in calorimetry that a separate Section, 2.6, is devoted to discussion of it. The alternative method of determining the amount of reaction is to analyse quantitatively the products of reaction, after the physical part of the calorimetric experiment has been completed. Frequently the chemical part of a calorimetric experiment is the most challenging part for an experimenter, because the overall accuracy of the experiment may well be limited by the accuracy of his chemical analysis. Probably the best procedure, where it can be applied, is for a determination to be made of both the number of moles of reactant(s) and the number of moles of product(s).

(g) Even when the number of moles of reaction has been determined, there remains one important consideration before $\Delta H^{\circ}_{r, T_1}$ can be calculated: almost certainly, the chemical reaction which took place in the calorimeter was not one in which all reactants and products were in their standard states. Therefore small, but significant, corrections must be made to the measured value of $\Delta H_{r, T_1}$ to allow for the departures of the participants in the reaction from their defined standard states. We shall return to this topic in Section 3.2.5.

2.5.2. VARIATION OF EQUILIBRIUM CONSTANTS WITH TEMPERATURE—THE "SECOND-LAW" METHOD

The determination of the equilibrium constants of chemical reactions from heat and entropy data was discussed in Section 2.4. In the present Section and in Section 2.5.3 we shall discuss the reverse procedure. From the Gibbs–Helmholtz equation

$$\left(\frac{\partial G}{\partial T}\right)_p = \frac{G-H}{T}, \tag{39}$$

it follows that

$$\frac{\partial(\Delta G^{\circ}_{r, T}/T)}{\partial T} = \frac{-\Delta H^{\circ}_{r, T}}{T^2}. \tag{40}$$

Hence, by combination of equations (24) and (40),

$$\frac{\mathrm{d}\ln K_p}{\mathrm{d}T} = \frac{\Delta H^{\circ}_{r, T}}{RT^2}. \tag{41}$$

Equation (40) implies that $\Delta H^{\circ}_{r, T}$ may be derived from measurements of the temperature dependence of the quantity $\Delta G^{\circ}_{r, T}/T$, which may in turn be derived from measurements of the electromotive force E°_{T} of a reversible cell at various temperatures, since

$$-\Delta G^{\circ}_{r, T} = z E^{\circ}_{T} F.\dagger \tag{42}$$

Unfortunately, there are few examples of the determination of the heats of organic reactions by this approach. Of much greater importance is the determination of the temperature dependence of equilibrium constants, followed by application of equation (41)—this method of determining $\Delta H^{\circ}_{r, T}$ is often called the *second-law method*.

Suppose that values of K_p for a gas reaction have been measured at several temperatures. Various procedures for the derivation of ΔH°_r may be

---

† F is Faraday's constant $= 96\ 487 \cdot 0$ A s mol^{-1}; z is the number of unit charges per ion.

employed.

(a) The procedure most commonly found in the literature, though the least rigorous, is to assume that ΔH_r° is independent of temperature (i.e. $\Delta C_p = 0$). If values of K_p for only two temperatures are available, it then follows from equation (41) that

$$\Delta H_r^\circ = \frac{R \ln(K_{p,1}/K_{p,2})}{(1/T_2 - 1/T_1)} = \frac{2 \cdot 303\, R\, T_1 T_2 \log_{10}(K_{p,1}/K_{p,2})}{T_1 - T_2}. \qquad (43)$$

If values of K_p are available for more than two temperatures, it is preferable to plot $\log_{10} K_p$ against $1/T$. According to the integrated form of equation (41), viz.

$$\log_{10} K_p = \frac{-\Delta H_r^\circ}{2 \cdot 303\, RT} + \text{constant}, \qquad (44)$$

such a plot should be linear, and ΔH_r° may be calculated from the slope. Clearly, the error arising from the assumption that ΔH_r° is independent of temperature will be less the shorter the temperature range of the experiments; but the shorter the temperature range the greater will be the error in determining the slope of the $\log_{10} K_p$ against $1/T$ plot. A reasonable compromise between these opposing considerations is to use a temperature range of ca. 50 degC, and to assign the calculated value of $\Delta H_{r,\,T}^\circ$ to the mid-temperature of the range. It is often found that if much wider temperature ranges are used in a second-law plot, the points lie on a curve. In some instances the curvature may be a manifestation of the non-constancy of ΔH_r°, but in other instances found in the literature the curvature is almost certainly due to failure to measure true equilibrium constants at one, or both, end(s) of the temperature range. The most common cause of error at low temperatures is a failure to attain equilibrium, arising from the slowness of reaction, whilst the most common cause of error at high temperatures is the failure to allow for by-product formation.

(b) If values of K_p can be determined with high accuracy over a wide temperature range, then an analytical expression for $\ln K_p$ as a function of temperature can in principle be developed, and $\Delta H_{r,\,T}^\circ$ at any temperature can be determined with the aid of equation (41) by differentiation of the $\ln K_p(T)$ expression. In practice, it is better to start with an analytical expression for ΔC_p° as a function of temperature and to develop by standard thermodynamic arguments the corresponding expression for $\log_{10} K_p$ as a function of temperature. For example, equation (19) may be used as the starting point of such procedure. It can then be shown that

$$\Delta H^\circ_{r,\,T} = \Delta H^* + AT + (B/2)T^2 + (C/3)T^3 , \qquad (45)$$

$$\Delta G^\circ_{r,\,T} = \Delta H^* - 2\cdot303\,AT\,\log_{10}T + IT - (B/2)T^2 - (C/6)T^3 , \quad (46)$$

$$\text{and } \Delta S^\circ_{r,\,T} = A - I + 2\cdot303\,A\,\log_{10}T + BT + (C/2)T^2 . \qquad (47)$$

It follows from combination of equations (24) and (46) that a plot of the function $[-2\cdot303\,R\,\log_{10}K_p + 2\cdot303\,A\,\log_{10}T + (B/2)T + (C/6)T^2]$ against $1/T$ should be a straight line, from which the constants ΔH^* and I may be evaluated. With the availability of these constants, $\Delta H^\circ_{r,\,T}$, $\Delta S^\circ_{r,\,T}$, and $\Delta G^\circ_{r,\,T}$ may be calculated for any value of T in the temperature range for which the constants of the heat-capacity equation are valid†. Although this procedure for evaluating $\Delta H^\circ_{r,\,T}$ is preferable to that described in (a), it depends upon the availability of an equation for ΔC°_p as a function of T and is therefore less generally applicable than procedure (a).

Equations (45), (46), and (47) were used by Buckley and Cox (67/1) to determine the thermodynamic functions for the reversible reaction

$$C_3H_7OH\ (g) \rightleftharpoons C_2H_5CHO\ (g) + H_2(g),$$

for which values of ΔC°_p were quoted in Section 2.3.2. The value of $\Delta H^\circ_{r,\,528^\circ K}$ derived by the rigorous procedure was 16·56 kcal, and the value derived by assuming $\Delta H^\circ_{r,\,T}$ to be constant was 16·55 kcal; the close concordance arises from the small magnitude of ΔC°_p for this reaction.

So far in this discussion of chemical equilibria it has been assumed that we are concerned with gases which behave ideally—for permanent gases at pressures not exceeding 1 atm, and for most vapours at pressures not exceeding 10% of the saturation vapour pressure, the error arising from this assumption will be negligible. But for equilibria involving gases or vapours at partial pressures outside these limits, the appropriate form of equation (25) is

$$K_p = \frac{\prod_i f^{v_i}\ (\text{products})}{\prod_i f^{v_i}\ (\text{reactants})} , \qquad (48)$$

where f is the *fugacity* of the real gas, defined by the equation

$$G_{\text{real gas}} - G_{\text{ideal gas},\ p=1} = RT\,\ln f . \qquad (49)$$

---

† Although equation (45) indicates that $\Delta H^\circ_{r,\,0^\circ K} = \Delta H^*$, the extrapolation to 0°K from the temperature range for which the constants were derived is so long that the equality is invalid. ΔH^* and I are both arbitrary constants.

The fugacity f_1 of a real gas at any pressure p_1 and a temperature T_1 is determined from the thermodynamic equation

$$\ln f_1 = \ln p_1 - \int_0^{p_1} \left(\frac{1}{p} - \frac{V_g}{RT_1} \right) dp, \tag{50}$$

where V_g is the molar volume of the real gas, which must be known as a function of p. The application of equations (48) and (50) is rare in thermodynamic studies of organic reactions, since almost all quantitative equilibrium studies have been made with sufficiently low partial pressures of gases or vapours for equation (25) to be applicable. However, the calculation of high partial pressures from equilibrium constants is a matter of importance to chemical engineers. We therefore consider briefly the evaluation of the integral in equation (50), selecting two of the many equations of state that have been proposed, namely the Berthelot equation†

$$V_g = \frac{RT}{p} + \left(\frac{9}{128} \right) \left(\frac{RT_c}{p_c} \right) \left(1 - \frac{6T_c^2}{T^2} \right), \tag{51}$$

and the pressure-implicit virial equation

$$V_g = \frac{RT}{p} + B + Cp + Dp^2 + \dots . \tag{52}$$

Combination of equations (50) and (51) leads to

$$\ln f_1 = \ln p_1 + \left(\frac{9p_1}{128T_1} \right) \left(\frac{T_c}{p_c} \right) \left(1 - \frac{6T_c^2}{T_1^2} \right) \tag{53}$$

and combination of equations (50) and (52) leads to

$$\ln f_1 = \ln p_1 + \frac{1}{RT_1} \left(Bp_1 + \frac{Cp_1^2}{2} + \frac{Dp_1^3}{3} + \dots \right). \tag{54}$$

Use of equation (53) requires knowledge of the critical temperature T_c and critical pressure p_c of the gas, whilst use of equation (54) requires knowledge of the virial coefficients B, C, etc. in the equation of state of the gas.

---

† T_c is the critical temperature and p_c is the critical pressure.

Equations (53) and (54) are applicable to the calculation of the fugacity of a pure vapour. Equations for the calculation of the fugacities of mixed vapours are to be found in ref. 64/11.

When equilibria involving one or more condensed phases are under consideration, it is usual to write a, *activity*, in place of f, fugacity, in equation (48), since the concept of activity is more readily applicable to all states of matter—for a liquid mixture obeying Raoult's law activity is equivalent to mole fraction, and for a pure solid phase activity is taken as unity. To pursue the question of calculating activities of solutes in liquid mixtures that do not follow Raoult's law would be pedantic, since there are few examples of the determination of $\Delta H^\circ_{r,\,T}$ of organic compounds from equilibrium measurements in solution and most of these relate to liquid-phase isomerizations, in which systems departures from Raoult's law would be expected to be slight (59/4).

2.5.3. COMBINATION OF EQUILIBRIUM CONSTANTS WITH OTHER THERMODYNAMIC QUANTITIES—THE "THIRD-LAW" METHOD

A measured value of an equilibrium constant at a single temperature T_1 affords a value of $\Delta G^\circ_{r,\,T_1}$ through equation (24). It is then possible to calculate the corresponding value of $\Delta H^\circ_{r,\,T_1}$ by equation (23), provided that $\Delta S^\circ_{r,\,T_1}$ is known. Now

$$\Delta S^\circ_{r,\,T_1} = \sum \Delta S^\circ_{f,\,T_1} \text{ (products)} - \sum \Delta S^\circ_{f,\,T_1} \text{ (reactants)} \tag{55}$$

$$= \sum S^\circ_{T_1} \text{ (products)} - \sum S^\circ_{T_1} \text{ (reactants)}. \tag{56}$$

Hence, the derivation of $\Delta H^\circ_{r,\,T_1}$ from a single value of an equilibrium constant depends upon the availability of values for the standard entropies of all reactants and products. For ideal gases, the method of statistical mechanics (50/2) may be used to calculate standard entropies, when information on the fundamental vibrational frequencies, the molecular geometry, and the barriers to internal rotation of groups is available. If a substance in its standard reference state at temperature T_1 is liquid or solid, the standard entropy may be calculated from calorimetric measurements made between T_1 and temperatures near $0°K$, by application of the third law of thermodynamics (50/2). The derivation of $\Delta H^\circ_{r,\,T_1}$ by this route is sometimes called the *third-law method*.

An alternative way of applying the third-law method is to use values of the *Gibbs energy function* $-(G^\circ - H^\circ_{T\text{ref}})/T$ of all reactants and products, instead of values of standard entropies. Thus we may write

$$-\frac{\Delta G^{\circ}_{r,T}}{T} = \Delta\left[-\left(\frac{G^{\circ} - H^{\circ}_{T\,\text{ref}}}{T}\right)\right] - \frac{\Delta H^{\circ}_{r,\,T\text{ref}}}{T};\qquad(57)$$

whence
$$\Delta H^{\circ}_{r,\,T\text{ref}} = -RT_1 \ln K_p + T_1 \sum -\left(\frac{G^{\circ} - H^{\circ}_{T\text{ref}}}{T_1}\right)\ (\text{reactants}) -$$

$$T_1 \sum -\left(\frac{G^{\circ} - H^{\circ}_{T\text{ref}}}{T_1}\right)\ (\text{products}).\qquad(58)$$

The Gibbs energy function for ideal gases is calculable by statistical mechanics, given the availability of the basic information detailed in the preceding paragraph. Some authors calculate $-(G^{\circ} - H^{\circ}_0)/T$ (i.e. $T_{\text{ref}} = 0°\text{K}$), and some calculate $-(G^{\circ} - H^{\circ}_{298})/T$ (i.e. $T_{\text{ref}} = 298°\text{K}$). The latter functions are more useful in the present context, since when $T_{\text{ref}} = 298°\text{K}$ equation (58) provides us with a valuable way of calculating $\Delta H^{\circ}_{r,\,298°\text{K}}$ directly. An example of the application of equation (58) follows. It is taken from the work of Goy and Pritchard (65/5), who studied the equilibrium

$$CH_4(g) + I_2(g) \rightleftharpoons CH_3I(g) + HI(g).$$

Their data for the temperature range 585–748°K are shown in Table 2.

TABLE 2. Data for the equilibrium $CH_4(g) + I_2(g) \rightleftharpoons CH_3I(g) + HI(g)$.

$T, °\text{K}$	$-\ln K_p$	$\Delta\left[\dfrac{-(G^{\circ}_T - H^{\circ}_{298})}{T}\right]$ cal deg^{-1}	ΔH°_r at 298°K kcal
585	9·1386	3·364	12·592
607	8·7644	3·378	12·623
616	8·5449	3·385	12·545
620	8·5887	3·387	12·682
688	7·5053	3·436	12·626
708	7·2735	3·451	12·677
717	7·1084	3·457	12·607
732	6·9693	3·468	12·677
748	6·7823	3·480	12·685

The second column of Table 2 shows experimental values of $-\ln K_p$, and the third column shows values of the change in the Gibbs energy function for the reaction, calculated by application of statistical mechanics, in its simple-harmonic-oscillator, rigid-rotator approximation (58/9), in respect of each species participating in the reaction. The final column of Table 2 shows values of $\Delta H^{\circ}_{r, 298^{\circ}K}$ calculated for each temperature; impressive concordance is seen. After application of corrections for anharmonic vibration in methane and methyl iodide, Goy and Pritchard gave the mean value, $\Delta H^{\circ}_r = +12 \cdot 67 \pm 0 \cdot 05$ kcal. It is of interest that a simple second-law treatment of the data, according to the method of Section 2.5.2 (a), followed by conversion of the resulting value of $\Delta H_{r, 666^{\circ}K}$ into the value for $298^{\circ}K$, gave $\Delta H^{\circ}_r = +12 \cdot 37 \pm 0 \cdot 20$ kcal. The third-law result is obviously to be preferred.

2.6. The purity of samples used in calorimetric measurements

2.6.1. THE EFFECTS OF IMPURITIES ON THE MEASUREMENT OF $\Delta \dot{H}^{\circ}_r$.

It was indicated in Section 2.5.1 that the purity of the sample employed in a measurement of ΔH°_r can significantly affect the overall accuracy of the measurement. There are at least three reasons for this. Firstly, the use of an impure specimen implies that the substance is not in its standard state, and if it is a real-gas mixture or a non-ideal liquid or solid solution, a correction for the heat of demixing the impurities from the host must in principle be applied. Secondly, when an impure specimen is used the effective molecular weight is not that calculated from atomic-weight tables (unless the impurities are isomeric with the host), and the determination of the number of moles of reaction will be in error. Thirdly, any impurities present in the reactant(s) may themselves undergo reaction and contribute to the measured heat change. The accuracy of measured equilibrium constants may also be affected by the use of impure reactants, because the impurities may affect the kinetics or the course of the reaction to an extent which precludes attainment of the desired equilibrium.

We now consider in more detail the effects of impurities on calorimetric measurements of ΔH°_r (or ΔU°_r).

(a) *Heat of demixing.* Heats of mixing of one organic compound with another of similar structure are normally very small in relation to heats of reaction, so that no significant error is likely to arise from neglect of a correction for the heat of demixing, say, $0 \cdot 01$ moles of impurity from $0 \cdot 99$ moles of structurally similar host. If, however, the impurity forms a loose compound with the host (e.g. a solvate), or if the impurity is water and the host is an organic compound with which water associates (e.g. an amine or an alcohol), neglect of a correction for heat of demixing may not be justifiable.

(b) *Erroneous molecular weight.* Clearly the effect of impurities on the calculation of the number of moles of reaction from the mass of reactant taken, and hence on the calculated value of ΔH_r°, will be small when the heat of reaction per unit mass of the impurity is similar to that of the host, and relatively large when the heat of reaction per unit mass of the impurity is nil. The latter situation arises fairly often in combustion calorimetry, because water is a common impurity in organic compounds and contributes no heat on combustion of the sample. Thus 0·1 weight per cent of water would cause 0·1 per cent error in a ΔU_c° value, based on the mass of sample taken and its assumed molecular weight. Such an error would have a serious effect on the derived value of ΔH_f°, but may be avoided in one of two ways: either the proportion of water in the sample can be determined by analysis and an allowance for it made, or ΔU_c° can be calculated from the number of moles of a product, e.g. by determination of the number of moles of carbon dioxide formed.

(c) *Heat of reaction of the impurities.* As indicated in (b), no great error will arise in ΔH_r° from use of an impure sample if the heat of reaction of the impurity is similar to that of the host. This is the usual situation in combustion and reaction calorimetry, provided that the sample has been subjected to some purification before the calorimetric experiments are attempted; with the exception of water, residual impurities are likely to be structurally similar to the host, and therefore to have similar heats of reaction. If the impurities can be identified and their proportion determined, then their contribution to the total heat of reaction can be calculated and an appropriate correction applied in the calculation of ΔH_r°. For example, Smith and co-workers (64/6) measured the energy of combustion of a sample of 2, 2′-dichlorobiphenyl containing 0·5% of 2-chlorobiphenyl as impurity, and made a correction for this impurity in the calculation of ΔH_c°.

2.6.2. PURIFICATION OF ORGANIC COMPOUNDS

Although the effects of impurities on the measurement of ΔH_r° may be mitigated as described in Section 2.6.1 it is better to avoid these effects altogether by use of specimens of the reactant(s) that have been purified to the point where residual impurities have negligible effect on the heat measurement. The purification work may prove to be more difficult and time-consuming than the calorimetric work, but unless the specimen is shown to be adequately pure, there can be no guarantee that a calorimetric measurement made with it will be meaningful. This consideration was prominent among those which led us to reject most measurements of ΔH_f° made before 1930, because few satisfactory methods of purification were available before that time and virtually no methods of proving purity had been developed. Nowadays, any or all of the following methods are available for the purification of organic compounds.

(a) Chemical purification—the classic techniques of the preparative organic chemist in which a reagent is used to segregate a wanted compound from its impurities by formation of a derivative or complex (e.g. a picrate) from which the compound can subsequently be regenerated.

(b) Fractional distillation—a technique applicable to those liquids that can be boiled without decomposition in the pressure range $10^3 - 10^{-3}$ torr; the possibility of azeotrope formation between impurity and host should not be overlooked. (Coulson and Herington, 58/7; Weissberger, 65/8).

(c) Fractional sublimation—a technique applicable to volatile solids at pressures below their triple-point pressures; it is a useful preliminary method for segregating solids from non-volatile impurities e.g. inorganic salts (Weissberger, 65/8).

(d) Fractional crystallization from solvents—a familiar technique that is often more effective if the solvent is changed from stage to stage of the purification; care must be taken to remove all solvent from the final product, and in particular the possible formation of a solvate between sample and solvent must not be overlooked.

(e) Fractional crystallization from the melt—a technique applicable to solids that are stable in the molten state; the method is usually most effective when the specimen has already been brought to a state of reasonable purity (99%) by other methods. (Herington, 63/8; Zief and Wilcox, 67/4).

(f) Zone refining—a convenient method of purification by repeated crystallization from the melt; the method is applicable on the 0·1–1000-g scale to solids that do not undergo a large volume change on melting and are not too viscous in the molten state. (Herington, 63/8).

(g) Gas–liquid chromatography—an alternative to fractional distillation or sublimation as a method for purifying volatile liquids and solids; the use of the method for purifying large samples is facilitated by the availability of automatic equipment for injecting the sample on to the chromatographic columns and isolating the wanted fractions (Ambrose, 61/9).

(h) Column chromatography, thin-layer chromatography, paper chromatography, electrophoresis, solvent extraction—these elegant purification techniques have not yet found great use in organic thermochemistry, but are likely to be increasingly employed as thermal measurements on progressively more complex organic compounds are attempted.

The choice of methods of purification to be used in a given instance depends in part on the physical properties of the sample, and in part on the chemical nature of the impurities. It is advantageous to identify the impurities at the outset (e.g. by spectroscopy or gas–liquid chromatography) and to tailor the choice of purification methods accordingly. For example, a sample of an alkane contaminated with an alkene may be effectively purified simply by treatment with sulphuric acid, whereas a sample of an alkane contaminated with a homologue or isomer will need purification by a physical technique such as fractional distillation or gas–liquid chromatography. Whatever purification method is adopted, the progress of purification should be monitored by an appropriate analytical technique. As soon as it is considered that the residual impurity content of the sample will not materially affect the accuracy of subsequent calorimetric work, a definitive determination of the purity of the specimen should be attempted. In some instances it may be necessary to stop the purification process prematurely, because of diminution in the size of sample: a definitive determination of purity would then be particularly important.

Calorimetric measurements have sometimes been made on two or more samples of one compound that have been purified by different methods; concordance between the measurements has then been taken as proof of the insignificance of impurities. Although such claims might well be wrong, the procedure is to be preferred to the use of a single, uncharacterised specimen, a practice which is virtually inexcusable now that so many powerful analytical techniques are available. The most commonly used techniques for proof of purity form the subject of the next Section.

2.6.3. CRITERIA OF PURITY

(a) *Chemical analysis.* Results of ultimate analyses are often quoted as evidence of the purity of specimens. However, for specimens to be used in combustion calorimetry, ultimate analysis is a poor criterion of purity, because the precision of the analysis is likely to be less than that of the calorimetry. Analysis of reactive groups, e.g. hydrolysable chlorine, is likely to be a better criterion of purity.

(b) *Physical constants.* Constants such as melting point, vapour pressure (especially the normal boiling point), refractive index, and density are often quoted and compared with literature values: concordance is claimed as proof of purity of the specimen. Where it can be convincingly shown that the measured physical property is sensitive to the presence of impurities, the method is valuable. Otherwise, the method is of little value and may even be misleading, due to the possibility that the specimen used to obtain the literature value was itself impure, or that systematic errors were present in the measurements of the respective investigators.

(c) *Spectroscopy.* Infrared spectroscopy, nuclear magnetic resonance spectroscopy, and to a smaller extent ultraviolet spectroscopy, are valuable techniques for identifying and quantitatively analysing impurities in material to be subjected to purification (63/7). Spectroscopic techniques are also useful for checking the progress of purification, but with a few exceptions they cease to be useful when the proportion of impurity falls below 0·5%, owing to insufficient sensitivity.

Mass spectrometry is a powerful technique for detecting impurities, especially those impurities that produce ions of higher molecular weight than the host substance does; in favourable cases 0·05% of impurity may be determined. By use of a modern high-resolution mass spectrometer, identification of the impurity ions may be possible from the accurate values of molecular weight which the instrument affords.

(d) *Gas–liquid chromatography.* Modern gas-chromatographic equipment employing sensitive detectors is capable of detecting down to 0·1% of impurities in a specimen that is sufficiently volatile and stable (61/9). The method is therefore useful for characterising highly purified specimens for use in combustion calorimetry, since the aim in purifying such specimens is generally to reduce the total impurity content to 0·2% or less. Gas–liquid chromatography is also well suited to monitoring the progress of purification because it is rapid, demands only a very small sample and is, in a series of comparative measurements, precise.

The shortcoming of gas–liquid chromatography as an analytical method is its dependence on the assumption that all impurities are separated from the main component by the chromatographic column. Thus although a specimen can be said to be *impure* if its chromatogram shows two or more peaks, it cannot with certainty be said to be *pure* if its chromatogram shows one peak, because the choice of stationary phase, column temperature etc. may have been inadequate for resolving impurities. A mixture of organic compounds with different functional groups, or a mixture of homologues, can in general be separated without difficulty, but a mixture of positional isomers may be difficult to separate, requiring trial of a range of stationary phases or column-temperature programmes.

When conditions have been found for producing a chromatogram with Gaussian-shaped, well separated peaks, several methods are available for making a quantitative analysis of the specimen. The simplest method is to assume that the detector responds linearly to the concentration of organic compound in the carrier gas, and that the response factor is the same for all organic components in the specimen; then the areas under the chromatogram peaks will be proportional to the amount of each component. But, since the assumptions involved may be unjustified, and since the method requires the main peak to be "on-scale", which may limit the discrimination of the minor components, the method should be used with caution. It is much more satisfactory to calibrate the equipment by deliberate addition of known amounts of an impurity to the specimen under examination. The added impurity may be chemically identical with the component being determined, or may be a chemically similar compound (a *marker*) which produces an adjacent peak in the chromatogram; both possibilities are illustrated in the chromatogram shown in Fig. 1. Here C is the main peak, allowed to run "off-scale",

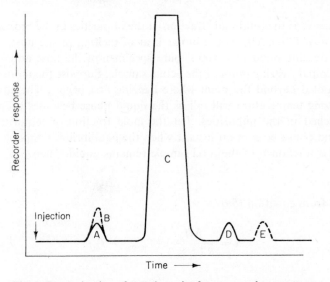

Fig. 1. Determination of sample-purity from a gas chromatogram.

of a sample which shows two impurity peaks A and D. Suppose that the impurity giving peak A is identifiable by its retention time. The proportion of this impurity can be deduced from the area of the augmented peak A + B produced by adding a known proportion of this impurity to the sample. For the determination of the impurity giving rise to peak

D, a known proportion of a marker giving rise to peak E has been added to the sample: the quantitative analysis is based on the relative areas under peaks D and E. The total impurity content is then estimated, but owing to the possibility that other impurity peaks were unresolved from peak C, the value is to be regarded as a minimum estimate of impurity. Greater certainty of the correctness of the analysis would be obtained if experiments with longer columns and other stationary phases gave no indication of the presence of more than two impurities.

(e) *Depression of the freezing point.* The least equivocal method for determining the total impurity content of a sample of an organic compound that can be melted without decomposition is based on measurements of the freezing characteristics of the sample. If the impurities are soluble in the melt but insoluble in the solid host, van't Hoff's theory of freezing-point depressions indicates that

$$x^* = \frac{\Delta H_m}{R T_{m,\,\text{pure}}^2} \, (T_{m,\,\text{pure}} - T_{m,\,\text{initial}}), \tag{59}$$

where x^* is the total mole fraction of the impurities in the molten sample ($x^* \not> ca. \, 0{\cdot}01$), ΔH_m is the molar heat of melting of the host, $T_{m,\,\text{pure}}$ is the freezing point of a 100% pure specimen of the host and $T_{m,\,\text{initial}}$ is the initial freezing point of the actual sample. Suppose the molten sample is cooled beyond the point where freezing first begins. The equilibrium freezing temperature will fall as the liquid phase becomes progressively enriched in the impurities. Let the mole fraction of impurities in the liquid phase be x at an instant when the equilibrium temperature is T_m and a fraction F of the total sample remains liquid. Then

$$x = x^*/F \tag{60}$$

and from equation (59)

$$x = \frac{\Delta H_m}{R T_{m,\,\text{pure}}^2} \, (T_{m,\,\text{pure}} - T_m). \tag{61}$$

Combination of equations (60) and (61) leads to

$$T_m = T_{m,\,\text{pure}} - \frac{x^* R T_{m,\,\text{pure}}^2}{F \, \Delta H_m}. \tag{62}$$

Hence a plot of T_m versus $1/F$ should be linear. $T_{m,\,\text{pure}}$ can be found by extrapolation to $1/F = 0$, and x^* can then be deduced from the slope of the plot if ΔH_m is known. The various experimental methods that have been described for the determination of x^* differ according to how equilibrium is maintained between solid and liquid phases during freezing, how T_m and F are measured, and how ΔH_m (or the related quantity $RT^2_{m, \text{pure}}/\Delta H_m$, the *molar cryoscopic constant*) is treated. Before discussing a selection of these experimental methods, we must emphasize that equation (62) depends upon the following assumptions.

(i)   The system obeys the laws of ideal dilute solutions which underlie equation (59); experience shows that this assumption is reasonable for organic samples in which the mole fraction of impurities is less than 0·01.

(ii) The impurities are completely soluble in the liquid phase; this will generally be the case for a highly purified sample, but it should be remembered that water, a common impurity, may be fairly soluble in an organic liquid at room temperature yet virtually insoluble at, say, $-40°C$. The precipitation of ice from such a liquid as it is cooled to its freezing point may pass unnoticed, and an over-optimistic estimate of the sample purity may be obtained.

(iii) Solid solutions are not formed between impurities and host. Formation of solid solutions tends to occur when the molecular dimensions of the impurities are similar to those of the host substance, or when the latter has a small heat of melting. The presence of solid solutions can be detected by thermal methods, and corroborated by analysis of the solid phase. The derivation of x^* when solid solutions are formed is explained by Mastrangelo and Dornte (55/11); solid solutions and the non-uniform distribution of impurities are discussed by McCullough and Waddington (57/1).

(iv) Equilibrium between solid and liquid phases is maintained during the transit through the freezing or melting region. Maintenance of equilibrium is highly dependent on the experimental technique used to study the transition, but no technique will be successful if decomposition takes place on melting.

The experimental determination of freezing-point depression can be effected under conditions of either freezing or melting. The most precise method employs a calorimeter, usually of the adiabatic type, to study the melting characteristics of the sample. In well defined stages, measured amounts of electrical energy are supplied to the sample, initially just below the temperature of incipient melting, and the equilibrium temperature at the end of each stage is measured. The amount of electrical energy needed to melt the whole sample (which yields ΔH_m) is determined by summation, and the value of F at each stage of melting is deduced by the method of proportional parts. A typical plot of T_m versus $1/F$, taken from the work of Counsell, Hales, and Martin (65/9) on butyl alcohol, is shown in Fig. 2. Values of T_m in this work were measured

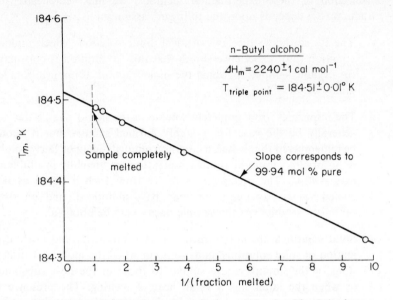

Fig. 2. Determination of sample purity by a study of melting behaviour.

to 0·001 degC by platinum-resistance thermometry. Fig. 2 shows a linear plot extrapolated to $T_{m, \text{pure}} = 184\cdot51°K$, and giving $x^* = 0\cdot00060$, by application of equation (62) to the measured slope. The purity of the sample was therefore 99·94 moles per cent.

The adiabatic-calorimeter method generally permits attainment of equilibrium between the phases, because melting can be carried out very slowly—over a period of several days, if desired. A drawback of the method is the need to employ expensive and complicated equipment for best results. Somewhat simplified calorimeters have been described by Aston and co-workers (47/2) and by Pilcher (57/2), and highly simplified versions have been described by Tunnicliff and Stone (55/1), Brooks and Pilcher (59/1), and Clarke, Johnston, and de Sorbo (53/2).

Numerous non-calorimetric techniques for determination of freezing-point depressions have been described; most are *dynamic* methods, since they involve continuous freezing or melting. If heat is continuously and uniformly abstracted from a sample during freezing, or supplied to a sample during melting, then the measurement of the actual amount of heat flowing out or in can be replaced by measurement of the time–temperature relationship, above, through, and below the freezing point. Success in the use of a dynamic method depends largely on whether or not the phases can be kept in quasi-equilibrium during the transition; for this purpose agitation of the sample may be necessary.

Rossini and his colleagues (41/1, 44/3, 53/3) used a dynamic freezing method in which the sample (ca. 90 cm^3) was agitated by means of a reciprocating stirrer; amounts of impurity were estimated to $\pm 10\%$ with a sensitivity of 0·01 mol %. Handley and Herington (50/3, 57/9) have described the use of a pulsating gas pressure applied to one surface of a freezing liquid sample (ca. 15 cm^3) contained in a U-shaped vessel; this method of agitation is superior to mechanical stirring because it promotes attainment of thermal equilibrium up to the point when 75% of the sample is solid, whereas stirring becomes ineffective when about 60% of the sample is solid. Handley and Herington recommend use of the *comparative procedure* for deriving x^*, as originally described by Schwab and Wichers (41/2). In this procedure, the freezing-point depression of the sample ΔT, equal to $T_{m,\ F=1} - T_{m,\ F=\frac{1}{2}}$, is first determined, then small amounts of impurity x_1^* and x_2^* are deliberately added to the sample, and the corresponding freezing-point depressions ΔT_1 and ΔT_2 are measured; x^* can be determined from either of the following equations:

$$x^* = \frac{x_1^* \, \Delta T}{\Delta T_1 - \Delta T}, \tag{63a}$$

$$x^* = \frac{x_2^* \, \Delta T_1 - x_1^* \, (\Delta T_2 - \Delta T_1)}{\Delta T_2 - \Delta T_1}. \tag{63b}$$

As may be seen from equations (63a) and (63b), knowledge of ΔH_m, or of the cryoscopic constant, is unnecessary with the Schwab and Wichers procedure.

Handley (57/9) has used a dynamic melting method for establishing the purity of specimens which tend to crystallize sluggishly; the sample (ca. 15 cm^3) is contained in a metal vessel containing many conducting metal vanes which are intended to promote thermal equilibrium. By use of a small, thin-glass cell with a re-entrant well for a thermocouple, Gunn (62/11) has measured melting curves on samples as small as 0·3 cm^3. Smit (56/2, 57/3) has described the use of a thin film of solid (ca. 0·5 cm^3) around the thermometer in a dynamic melting apparatus incorporating automatic control of the temperature difference between the sample and its surroundings. This and many other methods of purity control by thermal methods were reviewed at a symposium in Amsterdam in 1957 (57/4), at which it was shown that not all cryoscopic procedures are reliable for determining purity. To some extent, the applicability of a given method depends on the properties of the substance under examination, but for benzene at least Glasgow and co-workers (57/10) have proved that the purity of a specimen as determined by the calorimetric melting method does not differ from the purity as determined by a dynamic freezing method, so long as the measurements are made with sufficient care.

Amongst thermal methods for purity determination described since the 1957 symposium, two† have novel features. The dielectric-constant method of Ross and Frolen (63/9) represents an attempt to utilize a measurable property other than heat content or temperature during the equilibrium melting of a solid specimen. The measured quantity is the electrical capacitance of a capacitor in which the sample is the dielectric; F is calculated from the change in capacitance on melting, and a plot of T_m against $1/F$ is used for the calculation of x^*. The method of differential scanning calorimetry, for which commercial equipment is now available, is a dynamic one, though ΔH_m is determined calorimetrically. One type of apparatus is said to give a "meaningful measure of purity in about the time normally required for a simple melting-point determination" (66/8). Aside from its rapidity, the method has the merit of requiring only a few milligrammes of sample.

2.6.4. THE FINAL STATE AFTER A CALORIMETRIC EXPERIMENT

We have discussed the importance of purifying reactants and of proving their purity, in order that the initial state in a calorimetric reaction may be rigorously defined. After completion of a calorimetric experiment, the final state must be defined with equal rigour, and for this purpose quantitative analysis of both the main product(s) and the by-product(s) is necessary. Volumetric analysis, gravimetric analysis, and instrumental analysis all play their part, and when solid products are formed it may be necessary to identify the actual phase(s) by means of X-ray diffraction measurements. Typical analytical procedures will be described in Chapter 3.

With the establishment of the chemical and physical nature of the reaction products, thermal corrections for the formation of by-products and for the non-standard states of all the products may be made. Experimental work on the measurement of ΔH_r° will then be complete.

---

†*Note added in proof.* Another novel method, based on nuclear magnetic resonance, has been described by Herington and Lawrenson (68/19).

2.7. Recommended method of expressing uncertainties in thermochemistry

It is a truism that estimation of the uncertainty attaching to the measured value of a physicochemical constant is almost as important as the measurement itself. In thermochemistry, as in all measurement science, an attempt is made to devise apparatus and techniques that minimize the effects of *systematic* (*persistent*) *errors*. Bias in a measured value of a heat quantity arising from systematic errors is often discernible only when the result is compared with results obtained by radically different methods. We have already discussed the variety of routes to ΔH_f° that may be followed, and in the literature of organic thermochemistry there are many examples of good concordance between values obtained by different routes. We may say, therefore, that the well tried experimental methods to be described in Chapter 3 do not involve systematic errors that are significantly greater than the errors arising from chance variability. Errors of the latter kind, *random errors*, may be estimated by established statistical methods (56/1; 66/6); the recommended method of expressing the result is the subject of the remainder of this Section.

It was proposed by Rossini (36/2), and further argued by Rossini and Deming (39/2) that *the uncertainty interval* is the most appropriate measure of the random error of an experimentally determined quantity in thermochemistry. By definition, an uncertainty interval is twice the corresponding *standard deviation of the mean* (s.d.m.) $\bar{s}$, where $\bar{s}$ is given by

$$\bar{s} = \left[\frac{\sum (x - \bar{x})^2}{n(n-1)} \right]^{\frac{1}{2}}. \tag{64}$$

Here n is the number of replicate measurements of a quantity, of which the mean value is $\bar{x}$ and any individual value is x. Rossini recommends that in experimental thermochemistry n should lie between 4 and 12.

If in a series of measurements of ΔH_r° in an electrically calibrated calorimeter the mean value of the energy equivalent, $\mathscr{E}$, is associated with an s.d.m. $\bar{s}_{\mathscr{E}}$, and the mean value of the temperature change per unit amount of reaction, α, is associated with an s.d.m. $\bar{s}_\alpha$, then the overall value of $\bar{s}_{\Delta H}$ for the series of measurements of ΔH_r° is given by

$$\frac{\bar{s}_{\Delta H}}{\Delta H_r^\circ} = \left[\left(\frac{\bar{s}_{\mathscr{E}}}{\mathscr{E}} \right)^2 + \left(\frac{\bar{s}_\alpha}{\alpha} \right)^2 \right]^{\frac{1}{2}}. \tag{65}$$

If the calorimeter is calibrated with a primary-standard substance for which the heat of reaction per unit mass, β, itself has an s.d.m. $\bar{s}_\beta$, the overall value of $\bar{s}_{\Delta H}$ for a series of measurements of ΔH_r° is given by

$$\frac{\bar{s}_{\Delta H}}{\Delta H_r^\circ} = \left[\left(\frac{\bar{s}_\beta}{\beta} \right)^2 + \left(\frac{\bar{s}_{\mathscr{E}}}{\mathscr{E}} \right)^2 + \left(\frac{\bar{s}_\alpha}{\alpha} \right)^2 \right]^{\frac{1}{2}}, \tag{66}$$

where $\mathscr{E}$ and $\bar{s}_{\mathscr{E}}$ are now determined from a series of chemical calibration experiments with the primary-standard substance. The general method for computing overall standard deviations for experiments in which auxiliary substances undergo reaction in a calorimeter has been discussed by Bjellerup (61/8). Methods of computing *propagated* standard deviations (e.g. the error in ΔG_r° arising from error in the measurement of K_p and use of equation (24)) are dealt with in standard texts on the statistical treatment of experimental observations (50/5); the specific case of deriving the standard deviations of thermodynamic quantities calculated from $K_p(T)$ data has been discussed by Clarke and Glew (66/9).

Clearly it is desirable that thermochemists should express the precision of their experimental observations in a uniform way. In the past this has not been done, as mean errors, probable errors, standard deviations of a single observation, standard deviations of the mean, and uncertainty intervals are all to be found in the thermochemical literature. In recalculating published experimental data for incorporation into Chapter 5, we have tried to express overall errors in terms of uncertainty intervals, wherever sufficient data were given to permit their calculation. For the future, we would urge thermochemists to use uncertainty intervals as the normal method for expressing precision. This mode of expressing precision implies that if the respective mean values of two investigators' measurements of the same quantity differ by more than the sum of the uncertainty intervals it is probable, though not certain, that one or both sets of measurements contains a systematic error. In that event, the experimental results should be scrutinized, and the possible presence of errors in calibration constants and auxiliary data should be considered.

Chapter 3

Experimental Aspects of the Measurement of Heats of Reaction

3.1. Considerations regarding choice of experimental method

The principles of the derivation of heats of formation of organic compounds from measured heats of reaction or equilibrium constants were discussed in Chapter 2. Now we consider experimental aspects of the measurements, under the headings of combustion calorimetry, reaction calorimetry, and chemical equilibria. The term "combustion calorimetry" refers to the measurement of the energy (at constant volume) or heat (at constant pressure) of that reaction in which the carbon skeleton of a compound is completely broken down when the compound burns in gaseous oxygen†. The term "reaction calorimetry" refers to the measurement of the energy or heat of any reaction other than combustion, usually one in which the carbon skeleton remains intact.

The choice of experimental method for the derivation of ΔH_f° for a given organic compound is governed to a large extent by the reactivity of the compound. For compounds of low reactivity, such as the alkanes, the use of reaction calorimetry is virtually precluded (unless a clean reaction can be found which gives the compound as a *product*), and combustion calorimetry is the natural method to use. Reaction calorimetry finds application to compounds that readily undergo clean reactions at moderate temperatures, without formation of by-products; chemical-equilibrium measurements find application to compounds that undergo reversible reactions, often at temperatures higher than could conveniently be used in reaction calorimetry (i.e. above 200°C). Since combustion calorimetry is the one method applicable to virtually all organic compounds, why is it not invariably employed? The question is best considered in terms of a numerical example—let us examine the derivation of ΔH_f° of methyl iodide by each of the three main approaches:

---

† A few measurements of the heat of reaction when a compound "burns" in gaseous fluorine have been reported (see Section 3.2.7). Since such reactions involve complete destruction of the compound's carbon skeleton, they too may be regarded as combustions.

45

(a) *Combustion calorimetry*

$$CH_3I \text{ (l)} + 1.75 \, O_2(g) \rightarrow CO_2(g) + 1.5 \, H_2O \text{ (l)} + 0.5 \, I_2(c)$$

$$\Delta H_c^\circ = -193.1 \, \text{kcal}$$

$$\Delta H_f^\circ [CO_2(g) + 1.5 \, H_2O \text{ (l)}] = -196.5 \, \text{kcal}$$

$$\therefore \; \Delta H_f^\circ [CH_3I \text{ (l)}] = -196.5 - (-193.1) = -3.4 \, \text{kcal};$$

(b) *Reaction calorimetry*

$$CH_3I \text{ (l)} + 0.5 \, H_2(g) \rightarrow CH_4(g) + 0.5 \, I_2(c)$$

$$\Delta H_r^\circ = -14.5 \, \text{kcal}$$

$$\Delta H_f^\circ [CH_4(g)] = -17.9 \, \text{kcal}$$

$$\therefore \; \Delta H_f^\circ [CH_3I \text{ (l)}] = -17.9 - (-14.5) = -3.4 \, \text{kcal};$$

(c) *Chemical equilibrium*

$$CH_3I \text{ (g)} + HI \text{ (g)} \rightleftharpoons CH_4(g) + I_2(g)$$

$$\Delta H_r^\circ = -12.7 \, \text{kcal}$$

$$\Delta H_f^\circ [CH_4(g) - HI \text{ (g)}] + \Delta H_s[I_2(c)] - \Delta H_v[CH_3I \text{ (l)}] = -16.1 \, \text{kcal}$$

$$\therefore \; \Delta H_f^\circ [CH_3I \text{ (l)}] = -16.1 - (-12.7) = -3.4 \, \text{kcal}.$$

Suppose we require to measure the standard heat of formation of methyl iodide with an uncertainty interval not exceeding $0.3 \, \text{kcal mol}^{-1}$. The uncertainties in the ΔH_f° (products) terms in methods (a) and (b) are small enough to be neglected, so that the requirement is for ΔH_c° or ΔH_r° to be measured with an uncertainty interval of no more than $0.3 \, \text{kcal mol}^{-1}$. Hence the allowable uncertainty interval is only 0.15% of the measured heat if combustion calorimetry is used, but is 2% of the measured heat if reaction calorimetry is used. If the third-law chemical-equilibrium method is used, then provided that auxiliary data are known to high enough accuracy, the allowable uncertainty interval amounts to 50% of the measured value of K_p (see Section 2.4). Thus the heat-of-combustion route to ΔH_f° is the most demanding of the three possible routes in terms of precision, and indeed it remains competitive with the other two routes only because uncertainty intervals much lower than 0.15% can be attained without difficulty. In general, there is no clear-cut "best" method for determining the heats of formation of organic compounds: the selection of the method to be applied to a given compound will be governed by the chemical reactivity of that compound, the accuracy to which the subsidiary data involved are known, and the personal preferences of the investigator, based on his past experience and the availability of equipment.

Once the general approach to the derivation of ΔH_f° of a particular compound has been decided, a design of apparatus needs to be selected: the criteria governing the selection are now the physical state of the compound and the accuracy to be aimed at. Considerable guidance in the selection of apparatus for combustion and reaction calorimetry, though not for equilibrium studies, is provided by the two volumes of "Experimental Thermochemistry" (56/1, 62/2), prepared under the auspices of the International Union of Pure and Applied Chemistry. We shall assume that our readers have access to these volumes, and will confine our discussion to matters not fully covered in "Experimental Thermochemistry".

3.2. Combustion calorimetry

For the combustion calorimetry of liquids and solids, a pressure vessel (*bomb*) is employed—the quantity derived from an experiment at constant volume in such a vessel is the standard molar energy of combustion ΔU_c°. For the combustion calorimetry of gases a bomb may also be employed, though for work of high precision a constant-pressure flame calorimeter is preferred— the quantity derived from an experiment with a flame calorimeter is the standard molar heat of combustion ΔH_c°. The derivation of ΔH_c° from ΔU_c°, or *vice versa*, may be readily affected by means of equation (2).

3.2.1. THE STATIC-BOMB CALORIMETER

In the most frequently used design of bomb calorimeter the bomb remains static inside a water-filled calorimeter can. A typical bomb is illustrated in Fig. 3. It is constructed of a special stainless alloy and may be lined with platinum; the internal volume is *ca.* 300 cm^3, and the wall thickness, 8 mm, is sufficient to withstand 50 atm pressure.

The two gas valves in the bomb-head, only one of which, G, is illustrated, are of the needle pattern with replaceable seatings of gold; valves of the Schräder pattern are used in other types of combustion bomb. Of the two electrical connectors shown protruding from the bomb-head, one connects to electrode F, and the other, which is electrically insulated from the bomb-head, connects to electrode E. A looped fuse-wire K (Section 2.5.1) is shown with its fusible section positioned *ca.* 1 cm above the bottom of platinum crucible L (drawn cut-away), in which the sample for combustion is placed. The bomb head is sealed to the bomb-body by hand tightening of the large nut D, which compresses O-ring C located in a groove; other types of bomb employ flat

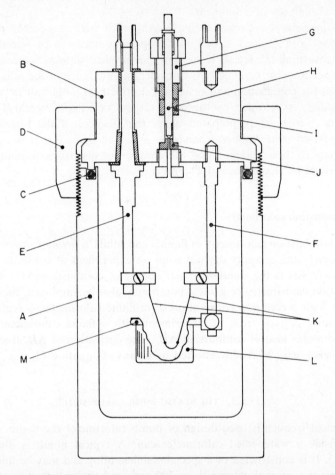

Fig. 3. A combustion bomb. A, bomb body; B, bomb head; C, O-ring; D, Sealing nut; E, Firing-electrode, insulated from bomb head; F, Firing-electrode, connected to bomb head; G, Oxygen-outlet valve; H, valve needle; I, packing gland; J, replaceable valve seat; K, platinum firing leads; L, platinum crucible; M, crucible support ring.

gaskets of soft metal or plastic instead of an O-ring. The disposition of the bomb in a calorimeter can of modern design is shown in Fig. 4. The can A, capacity *ca.* 3000 cm³, is equipped with a close-fitting lid E, a paddle stirrer J driven by a constant-speed motor, and a platinum-resistance thermometer I.

Hermetic sealing of the can by the lid is accomplished by an O-ring D round the rim of the can and an oil seal H round the rotating stirrer shaft. The

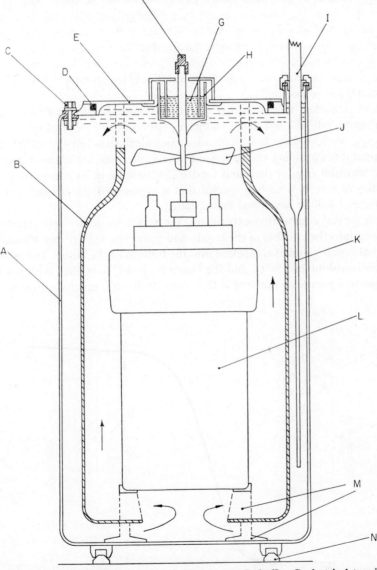

Fig. 4. A static-bomb calorimeter. A, calorimeter can; B, baffle; C, electrical terminals; D, O-ring gasket; E, Calorimeter lid; F, insulating connector between stirrer shaft and motor; G, oil; H, oil-cup vapour seal; I, platinum resistance thermometer, entering calorimeter through a vapour-tight connection; J, paddle; K, thin metal sheath, containing platinum resistance; L, bomb; M, bomb supports; N, insulating supports for calorimeter.

calorimeter is completely surrounded by a water-filled jacket, the temperature of which is kept constant to ± 0.001 degC. The outer surface of the can, highly polished to minimize heat losses by radiation, is separated from the inner surface of the thermostatically controlled jacket, also polished, by an air-filled gap 1 cm wide. Heat transfer between the can and the jacket takes place largely by conduction through the air, since convection in a 1-cm air gap is slight and radiation between two reflecting surfaces differing in temperature by *ca.* 2 degC is small; heat transfer thus takes place effectively according to Newton's law of cooling.

Most published measurements of ΔU_c° by static-bomb calorimetry have been made with the aid of an isothermal-environment calorimeter. A few, however, have been made with an adiabatic calorimeter (39/4), and this type of apparatus is finding increasing use in routine measurements of energies of combustion (e.g. of fuels and foodstuffs), because of its convenience and rapidity of use. It is to be expected that a high-precision, adiabatic, bomb calorimeter will be developed before long.

In a typical combustion experiment, the substance under investigation is weighed into the crucible of the bomb, and water (*ca.* 0·3 cm³ per 100 cm³ of internal bomb volume) is pipetted into the bottom of the bomb. The ignition system is assembled (56/1), and the bomb is closed, then filled with purified oxygen to a measured pressure in the range 10–50 atm, generally 30 atm. The

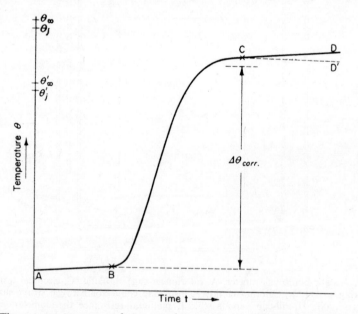

Fig. 5. Time−temperature curve for an experiment with an isothermal-environment calorimeter.

bomb is placed in the calorimeter, which is next charged with a measured amount of water, closed, and placed in its isothermal environment. The three distinct stages of the ensuing calorimetric experiment are illustrated in Fig. 5. Throughout the experiment, measurements of temperature at fixed time intervals†, say every 30 seconds, are made. The nearly straight portion AB of the temperature–time curve represents the *initial (rating) period* or *fore-period*, during which heat is being received by the calorimeter from its jacket, since the temperature of the latter (θ_j) is higher. At point B, conveniently 20 minutes after the start, the substance in the bomb is ignited. Within *ca.* 10 seconds the curve begins to deviate from the initial rating curve and follows the course BC for *ca.* 10 minutes; BC represents the *main period*, or *reaction period*. The end of the main period is identified by the beginning of a second nearly straight portion, CD, representing the *final (rating) period*, the length of which is commonly *ca.* 15 minutes.

It is found experimentally that AB and CD are exponential curves governed by the same differential equation:

$$d\theta/dt = k\,(\theta_\infty - \theta), \tag{67}$$

where k is a constant (the *heat-transfer coefficient*) and θ_∞ is the *convergence temperature*, which would be reached by either curve AB or curve CD after infinite time. That θ_∞ slightly exceeds θ_j is due to the steady input of heat to the calorimeter by mechanical stirring. Two general methods for correcting for the heat of stirring and for the heat exchanged between the calorimeter and its environment during the main period are in use; they are known as the Dickinson and Regnault–Pfaundler methods respectively, and are described in textbooks such as "Experimental Thermochemistry" (56/1) and "Tochnaya Kalorimetriya" (64/47). Each method provides a value of $\Delta\theta_{corr}$ (Fig. 5), which is the rise in temperature of the calorimeter that would have been observed if there had been no heat of stirring and no heat exchange with the surroundings.

In Fig. 5, θ_j is shown as lying above the temperature at point C. This arrangement is preferable if an unsealed calorimeter can is used, since any water vaporized will not condense on surfaces in the interspace, if these surfaces are warmer than the water in the can. But if a sealed calorimeter can is used, as drawn in Fig. 4, θ_j may be any convenient temperature. A convenient jacket temperature between B and C is shown as θ'_j in Fig. 5; the final period CD' is now characterized by a *fall* in temperature towards θ'_∞.

When $\Delta\theta_{corr}$ is to be derived by a graphical method, it is common practice to plot the quantity actually measured by the thermometric system (e.g. electrical resistance if a resistance thermometer and bridge are used), rather

---

† Many workers prefer to measure the times corresponding to fixed temperature intervals.

than temperature itself; conversion to true temperatures, by the use of appropriate calibration constants, is effected in the final stages of the calculations. The calculation of $\Delta\theta_{corr}$ is greatly facilitated by use of a digital computer, to which the experimental data may be directly supplied without intermediate logging. The principles of the subsequent calculation of ΔU_c° have been described in Section 2.5.1.

The two feasible methods of calibrating a calorimeter were discussed in Section 2.5.1. Relatively few thermochemists choose to calibrate their bomb calorimeters electrically: most prefer to calibrate by combustion of a highly purified sample of benzoic acid, the internationally agreed primary-standard substance (see Section 3.2.4). When benzoic acid is used for calibration, it is not difficult to arrange that the temperature–time curve for the calibration experiment matches the corresponding curve for the substance under investigation quite closely; systematic errors are minimized thereby. When electrical calibration is used and the calibrating heater is immersed in the water of the calorimeter, the temperature–time curve has a nearly straight portion BC with small-radius curves at either end. The possibility then arises that the temporal relation between the temperature of the calorimeter's outer surface, which determines the rate of heat exchange with the surroundings, and the temperature measured by the thermometer will differ as between a calibration and a combustion experiment. The avoidance of systematic errors from this and other sources is clearly a matter of importance, and has been discussed by Challoner, Gundry and Meetham (55/2). Detailed descriptions of bomb-calorimetric procedure when benzoic acid is employed for calibration have been given by Jessup (60/7).

Several types of static-bomb calorimeter differing radically in design from the conventional type have been described. Most are *aneroid* instruments. In these the calorimeter is not water-filled but consists of a block of metal of high thermal conductivity. Magnus and Becker (51/1) used a copper block in which was machined a cavity serving as a combustion chamber; the cavity had a pressure-tight lid and the internal surfaces of the chamber were gold-plated. Temperature rises were measured using a multi-junction thermopile, arranged with one set of junctions on the metal block and another set on an aluminium block suspended below it in an evacuated space. Pilcher and Sutton's aneroid calorimeter (55/3) consisted of a conventional steel combustion bomb fitted into an aluminium block. By measurement of the temperature of the block with a platinum-resistance thermometer, a precision of 0·01% was attained in combustion experiments; the shortcoming of the instrument was the excessive time, *ca.* 60 minutes, needed to reach thermal equilibrium in the main period. A highly precise aneroid bomb calorimeter constructed from a block of silver, has been used by Meetham and Nicholls (60/1) and by Hawtin and co-workers (66/1). The energy equivalent is less

3. THE MEASUREMENT OF HEATS OF REACTION

than that of a conventional bomb calorimeter, so that only *ca.* 100 mg of organic sample is needed to produce a 2 degC temperature rise; the bomb is equipped for electrical calibration. Aneroid static-bomb calorimeters that can be used for yet smaller samples have been described by Ponomarev and Alekseeva (61/10), Mackle and O'Hare (63/11), and Calvet and Tachoire (60/8, 62/9). In Ponomarev and Alekseeva's design the bomb is a nickel cylinder which fits into a spherical copper block. The temperature of the block is measured by means of a platinum-resistance thermometer embedded in it; when sufficiently sensitive electrical instruments are used for measuring resistance changes, only *ca.* 20 mg of organic substance need be burned in the bomb. Mackle and O'Hare's bomb is of copper, the wall thickness being sufficient to withstand the internal high pressure of oxygen only if the interspace between the bomb and its jacket is itself pressurized. Because the bomb, which is also the calorimeter, has a very small energy equivalent only a few milligrammes of substance need be burned, which is obviously advantageous for work with compounds that are difficult to isolate or synthesize in quantity. A disadvantage of the method is the relatively large correction for heat transfer, occasioned by the use of pressurized air in the interspace. Calvet and Tachoire's bomb is a small thick-walled steel cylinder which fits into a micro-calorimeter of the type to be described in Section 3.3.4; only a few milligrammes of sample need be burned.

Although by use of aneroid calorimeters errors arising from weighing water into the calorimeter, lack of constancy in the rate of stirring, and evaporation of water from the calorimeter are all avoided, aneroid calorimeters have not yielded results of significantly greater precision than those from conventional bomb calorimeters. It seems probable that the factor limiting precision in bomb calorimetry is the determination of the chemistry of the combustion reaction.

3.2.2. CHEMICAL ASPECTS OF BOMB COMBUSTIONS

Numerous investigators have shown that the static-bomb calorimeter is a satisfactory instrument for application to compounds containing the elements C and H, or C, H and O, since in a well designed experiment nearly quantitative conversion of the compound into carbon dioxide and water takes place according to the equation:

$$C_aH_bO_c + \left(\frac{4a+b-2c}{4}\right) O_2\,(g) \rightarrow a\,CO_2\,(g) + \frac{b}{2}\,H_2O\,(l). \qquad (68)$$

For compounds containing C, H and N, or C, H, O and N, static-bomb calorimetry is also a satisfactory technique, though the stoicheiometry of the combustion process is more complex:

$$C_aH_bO_cN_d + \left(\frac{4a+b-2c+5y}{4}\right) O_2 \text{ (g)}$$

$$\rightarrow a\, CO_2 \text{ (g)} + y\left[HNO_3\left\{\left(\frac{b-y}{2y}\right) H_2O\right\}\right] \text{(l)} + \frac{d-y}{2}N_2 \text{ (g)}. \quad (69)$$

In practice, the value of y is generally such that $y/d \approx 0.15$, but the exact value of y for each experiment must be determined by analysis of the nitric acid present in solution after combustion†. Once y is known, it is normal practice to calculate the energy of the reaction

$$\frac{y}{2}N_2 \text{ (g)} + \frac{5y}{4}O_2 \text{ (g)} + \frac{b}{2}H_2O \text{ (l)} \rightarrow y\left[HNO_3\left\{\left(\frac{b-y}{2y}\right) H_2O\right\}\right] \text{(l)}, \quad (70)$$

and to subtract this energy from the measured energy of the bomb reaction. The energy of the following reaction is thereby obtained:

$$C_aH_bO_cN_d + \left(\frac{4a+b-2c}{4}\right) O_2 \text{ (g)} \rightarrow a\, CO_2 + \frac{b}{2}H_2O \text{ (l)} + \frac{d}{2}N_2 \text{ (g)}. \quad (71)$$

The derivation of meaningful values of ΔU_c° for reactions of the types expressed by equations (68) and (71) requires that the physical states of the compounds burned are accurately definable. Thus, if the compound is a liquid at $T_{initial}$ all the compound initially present in the bomb should be in the liquid state and none in the vapour state, and if the compound is a solid at $T_{initial}$ all the compound should be in the solid state and none in the vapour or undercooled liquid states. In instances when the liquid or solid is so involatile that the amount of substance vaporizing into the bomb would be negligible, it is often permissible to place the substance unenclosed in the crucible of the bomb; exceptions arise when the substance is liable to undergo significant reaction with oxygen or water vapour before ignition takes place. In such instances, and in instances when the substance would volatilize to a significant extent, enclosure of the substance in a suitable container is necessary. For liquids, enclosure in thin-walled glass ampoules has long been employed; such ampoules are sealed with a flame. With the exercise of some

---

† Nitric acid must also be determined after the combustion of a compound containing no nitrogen, since a small amount of the acid is formed from gaseous nitrogen adventitiously present in the compressed oxygen.

dexterity (60/7), an ampoule may be virtually completely filled with liquid and will then withstand subsequent compression in the bomb, whereas an ampoule containing a visible air or vapour bubble may collapse under pressure; a filled, sealed ampoule must be kept at a temperature close to the intended initial temperature of a combustion experiment, otherwise premature cracking of the ampoule may occur. Some skill and luck are needed to make a glass ampoule that will crack, but not shatter violently, when subjected to the thermal shock of the ignition system. Cracking will permit the liquid charge to burn smoothly within the crucible, whereas shattering will cause burning drops of liquid to be thrown on to the walls of the bomb, with premature extinction of the flame. A disadvantage of glass ampoules is their tendency to fuse in the crucible, trapping and charring some of the liquid charge in the melt.

In recent years it has become common practice to enclose liquid samples in plastic containers, instead of glass ampoules. Bags made by welding plastic sheet (62/2, p. 17) are the easiest containers to construct but some workers prefer to fabricate plastic ampoules by moulding or machining techniques; polyethylene, polypropylene, and polyethylene glycol terephthalate ("polyester") are the most frequently used plastic materials. Plastic containers are usually filled by pipetting or injecting (e.g. with a hypodermic syringe) the liquid into the container, followed by heat sealing of the aperture. Plastic bags may also be used for enclosing volatile or reactive *solids*. A technique for encapsulation of a liquid in a thick-walled polyethylene ampoule has been devised by Head and co-workers (69/1); the ampoule is sealed by means of a jet of hot nitrogen.

When a plastic container filled with a liquid or solid sample is ignited, the container burns away completely and in so doing aids the combustion of the sample by raising the temperature of the crucible. The energy of combustion of the plastic must be subtracted from the measured gross energy of combustion, and separate experiments to determine the energy of combustion of the plastic alone must be made. Likewise, the energy contributed by combustion of any material forming part of the fuse system must be subtracted from the gross energy of combustion. A great many fuse systems incorporating combustible material have been described. In most of them a thread (e.g. cotton or plastic) is in contact with metal fuse-wire at one end and the sample or ampoule at the other end; fusion of the wire by passage of a high current causes the organic fuse to ignite, and this in turn ignites the sample, or cracks the ampoule and ignites the contents.

Just as the initial state of a combustion experiment must be carefully defined, so must the final state. In particular the amounts of the major and minor products of combustion should be determined. Some calorimetrists perform no analyses of the products, except for nitric acid, and assume that

ΔU_c° for the combustion of a compound containing C, H, O, and N can be calculated from the mass of sample taken. But such an assumption is fraught with the risks that (a) the true composition of the sample may be different from that assumed (see Section 2.6.1), (b) the sample may not burn completely, and (c) there may be products other than those shown in equations (68) and (69). For work of high accuracy, chemical analysis of the products of some, and preferably all, experiments is essential. The most important determination is that of carbon dioxide, which is effected gravimetrically by passage of the gaseous bomb products through tubes containing sodium hydroxide on asbestos. The analytical technique of Rossini (56/1, pp. 65–66) is capable of a precision of 0·01 % if care is exercised. The quantitative analysis of water formed in bomb combustion reactions is never attempted, but nitric acid formed is determined as a matter of routine. Usually a solution prepared by rinsing the interior surfaces of the bomb with distilled water is titrated against alkali, but analytical reagents specific for NO_3^-, or the method of reduction of NO_3^- to NH_3, using Devarda's alloy, may be used. In principle, elemental nitrogen in the bomb gases may be determined by conventional gas analysis or by mass spectrometry, but the relatively low accuracy of these methods makes them of little real value.

The minor products of the combustion of a C, H, O, N compound may include the following: carbon, carbon monoxide, partial oxidation products of the compound (e.g. aldehydes), nitrogen oxides, nitrous acid solution†. The amount of carbon on the bomb walls or suspended in the liquid products is determined by collection on a filter, followed by combustion in a microanalytical apparatus; the amount of carbon in the platinum crucible can conveniently be determined from the loss in mass of the dried crucible on ignition. Carbon monoxide in the bomb gases may be determined by one of the standard specific methods (e.g. oxidation with iodine pentoxide or palladium chloride), or from the amount of carbon dioxide formed when the exit gases from the sodium hydroxide–asbestos absorber are passed over hot copper oxide. The presence of partial oxidation products is sometimes indicated by an acrid odour found in the bomb after the compressed gases have been released; it is prudent to discount the result of the experiment when this occurs. The possible presence of nitrogen oxides in the gaseous products from combustion of a nitrogen-containing compound may be checked by releasing the bomb gases through a solution of an oxidizer (e.g. acid potassium permanganate), and then analysing for NO_3^- by Devarda's method. Nitrous acid in the liquid combustion products may be determined colorimetrically

---

† It is assumed that the interior surfaces and fittings of the bomb are of materials that are unattacked by the flame or the combustion products; if this is not so, metal oxides etc. may be among the minor products.

by means of the Griess–Ilosvay reagent. Once the amounts of the minor products of combustion have been determined, the appropriate calorimetric corrections to the measured energy of combustion should be made, using values of ΔU_f° of the minor products, taken from tables (50/1; 65/1).

We now consider the chemistry of the bomb-combustion reactions of compounds containing elements additional to C, H, O and N. Rarely do such reactions proceed in such a way that the additional elements L, M, ... present in the compounds are found in the products in a single oxidation state. More commonly L, M, ... are found partly in the elemental state, and partly in a range of oxidation states; if water is present in the bomb, as it normally is, the element or other products may be partly in solution, or may undergo secondary reactions with the water, e.g. combustion of an organo-arsenic compound produces As, As_2O_3 (partly in solution), As_2O_4, As_2O_5 (largely in solution as H_3AsO_4), CO_2, H_2O, small amounts of HNO_3 (in solution), and C. When it is remembered that the products in solution are likely to be at different concentrations in different parts of the bomb, the task of accurately defining the chemical and physical states of the products is seen to be formidable. This consideration severely hindered the application of combustion calorimetry to compounds of the "difficult" elements, until the development of the moving-bomb calorimeter, described in the next Section.

3.2.3. THE MOVING-BOMB CALORIMETER

In this form of combustion calorimeter the bomb is equipped with a mechanism which causes it to rotate or oscillate; with a water-filled calorimeter it is usual for the bomb to rotate inside the calorimeter can, but with an aneroid instrument it is usual for the bomb and calorimeter to move together. During assembly of the bomb, water or a liquid reagent (10–50 cm^3) is placed in the bottom of the bomb. After closure, the bomb is fixed in position and remains fixed during the initial rating period and the ignition. Then, soon after the flame has become extinguished (ca. 30 seconds after ignition), the rotation or oscillation mechanism is started, causing the liquid in the bomb to wash all internal surfaces and to dissolve all soluble products of combustion; the energy changes resulting from dissolution are part of the overall energy change and are usually complete within the normal reaction period of 10 minutes. In a well designed experiment a homogeneous solution is produced in the bomb, and by subsequent chemical analysis an accurate definition of the final state can be given. Examples of the use of liquid reagents in the bomb will be given later in this Section.

The pioneers in the development of a moving-bomb calorimeter were Popoff and Schirokich (33/2), who described an instrument in which the bomb oscillated through an angle, with a period of 12 seconds. Further experiments with moving-bomb calorimeters were made by Sunner (44/4; 46/1), leading eventually to the development at the Bureau of Mines Laboratory, Bartlesville, of an elaborate design in which the bomb can rotate about two axes at right angles, simultaneously (54/1; 56/1; 56/3); the tumbling motion thereby imparted to the liquid within the bomb promotes rapid dissolution of combustion products. A typical rotating bomb with water-filled calorimeter is illustrated in Fig. 6. As may be seen, a drive-shaft enters the calorimeter

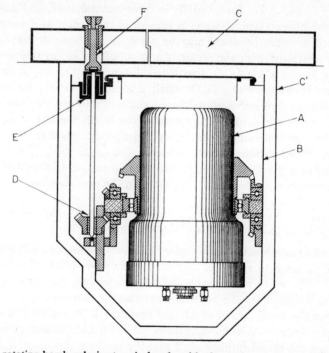

Fig. 6. A rotating-bomb calorimeter. A, bomb, with clamped-on ball-race and ring gear, shown in inverted position; B, calorimeter can; C, C', water-filled constant-temperature jacket; D, drive shaft for rotating mechanism; E, liquid-filled rotatable seal; F, connector to drive shaft, made of poor thermal conductor.

through an oil seal and imparts a twofold motion (axial and end-over-end) to the bomb by means of a system of gears immersed in the water of the calorimeter. Naturally, some frictional heat is generated by the rotation and must be allowed for—either a definite number of rotations in the reaction period is permitted, and a correction for the heat of rotation, determined in a separate experiment, is applied, or rotation is commenced at a particular time in the reaction period and continued through the final period, when an explicit correction need not be made (56/3).

Aneroid moving-bomb calorimeters have been described by Keith and Mackle (58/2), and Geiseler and Schaffernicht (67/2). Keith and Mackle's instrument is a copper bomb (lined internally with platinum), which is separated by an air-gap from a thermostatically controlled metal block. The bomb can be oscillated mechanically through 360 degrees around its cylindrical axis, and the whole assembly can be simultaneously rocked through an angle of 90 degrees. Temperatures are measured by means of a copper resistance wire wound in grooves on the outside of the bomb. Geiseler and Schaffernicht's bomb calorimeter is a massive copper sphere containing a platinum-lined cylindrical combustion chamber; the temperature of the sphere is measured by means of an embedded thermistor. The sphere, contained within a vacuum chamber, can be made to oscillate about a horizontal axis through an angle of 360 degrees.

Experimental details on the construction and operation of rotating-bomb calorimeters are given in "Experimental Thermochemistry" and in the original papers already cited. We shall therefore largely confine our discussion to the chemical aspects of rotating-bomb calorimetry, giving separate consideration to (a) organosulphur compounds, (b) organohalogen compounds, and (c) organometallic compounds.

(a) *Organosulphur compounds.* It has long been known that when an organosulphur compound is burned in a bomb most of the combined sulphur is converted into sulphuric acid, but some may be converted into sulphur dioxide. Almost complete conversion of combined sulphur into sulphuric acid is assured by leaving more than the usual amount of gaseous nitrogen in the bomb, so that the nitrogen oxides formed in the flame may catalyse the further oxidation of sulphur dioxide, as in the lead-chamber process for manufacturing sulphuric acid. The admixture of a nitrogen-containing compound (e.g. urea) with the substance to be burned is also an effective means for securing oxidation to the S^{VI} state (63/10).

Much of the early work on the combustion calorimetry of sulphur compounds was performed with static bombs, but Becker and Roth (34/3) showed that the sulphuric acid solution produced is at varying concentrations at different parts of the bomb; since the heat of dilution of sulphuric acid is relatively large, the calorimetric uncertainty arising from this multiplicity of acid concentrations was shown to impose a limitation on the overall accuracy. Huffman and Ellis (35/1) attempted to reduce the uncertainty by not adding water to the bomb before combustion, in the hope that sulphuric acid would be formed as a fine mist and be deposited in uniform concentration on the bomb walls. However, the problem of bringing sulphuric acid to a single, definite concentration is more satisfactorily solved by use of a moving bomb. If water, say

10 cm^3, is placed in the bomb initially, rotation of the bomb after combustion will produce a solution of sulphuric acid of definite concentration, for which the appropriate value of ΔH_f° can be selected from reference tables. Formation of significant amounts of sulphur dioxide may usually be avoided by ensuring that the compressed oxygen in the bomb contains at least $2 \cdot 5\%$ of nitrogen and that the H:S atomic ratio in the material for combustion exceeds 2; in the event that the H:S atomic ratio in the compound under investigation is less than 2, the amount of combined hydrogen in the combustible system should be increased by the introduction of known amounts of a hydrogen-rich compound such as a paraffin—an additive of this sort is known as an *auxiliary substance*. The main combustion reaction can then be represented by the equation:

$$C_aH_bO_cS_e + \left(\frac{4a+b-2c+6e}{4}\right) O_2(g) + \left[\frac{2e(n+1)-b}{2}\right] H_2O\ (l)$$

$$\to aCO_2(g) + e\left[H_2SO_4(nH_2O)\right](l), \tag{72}$$

in which the empirical formula $C_aH_bO_cS_e$ is calculated to include any auxiliary substance used. Analysis of the combustion products will normally include gravimetric determination of carbon dioxide, titrimetric determination of total acid in the bomb solution, and gravimetric determination of the SO_4^{2-} content of an aliquot of the neutralized solution, as barium sulphate. Nitric acid may be estimated from the number of equivalents of total acid minus the number of equivalents of SO_4^{2-}, or may be determined on an aliquot of the neutralized bomb solution by Devarda's method. Any sulphur dioxide formed as a minor product will be partly in the gas phase of the bomb and partly in the sulphuric acid solution; that part which is dissolved will be expelled from solution by flushing the bomb with oxygen after the compressed gases have been released. Hence the determination can be confined to the gas phase—the reducing effect of the gases can be determined by a suitable reagent (e.g. acid potassium dichromate solution), or a colorimetric test for sulphur dioxide can be employed. Tests for nitrogen oxides in the bomb gases should also be applied, but experience shows that despite the catalytic importance of these gases, their amount is usually insignificant calorimetrically. The bomb solution should be analysed for nitrous acid by the Griess–Ilosvay reagent.

The problem of indeterminate dilution states is solved by the moving-bomb method at the expense of introducing a source of error that is normally insignificant in the static-bomb method. This error arises from the use of so much solution in the bomb that the energies of solution of the gases present, particularly carbon dioxide, constitute a significant part of the gross energy effect and may not always be calculable from available data with the accuracy demanded. One way of minimizing the error is to perform a calibration experiment in which the final state is similar to that in the combustion experiment with the sulphur compound. This procedure, known as the method of *comparison experiments*, involves the combustion in the calibration experiment of sufficient primary-standard benzoic acid, mixed where necessary with a reference substance (Section 3.2.4), to produce the same partial pressure of carbon dioxide as in a combustion experiment with the sulphur compound; it also involves placing initially in the bomb the same amount of sulphuric acid solution, of the same concentration, as is formed in the combustion experiment. The method of comparison experiments is not of course confined to the rotating-bomb calorimetry of sulphur compounds, but is applicable to the rotating-bomb calorimetry of many other types of compound.

(b) *Organohalogen compounds.* Compounds of carbon, hydrogen, and a halogen Hal, when burned in a bomb give carbon dioxide, water, and the halogen-containing compounds shown in Table 3. It is seen that a single halogen-containing product is given by iodo-compounds alone; Roth (44/5) made tests for oxidized iodine species in combustion products but found none. Hence the combustion of an organo-iodine compound produces elemental iodine in what is virtually its standard state (crystalline), since the amount dissolved in the water of the bomb is scarcely significant. As such a combustion reaction is so straightforward, it can be conducted satisfactorily in a static bomb, the only special precaution needed being the lining of the internal surfaces of the bomb with platinum, to prevent attack on the metal of the bomb by iodine vapour. A detailed discussion is to be found in "Experimental Thermochemistry" (56/1, p. 221).

TABLE 3. Combustion of organohalogen compounds.

	Percentage of combined Hal found as:		
	HHal	Hal$_2$	CHal$_4$
Fluorine compounds	20—100	—	80—0
Chlorine compounds	80—85	20—15	—
Bromine compounds	3—10	97—90	—
Iodine compounds	—	100	—

As shown in Table 3, organohalogen compounds other than iodo-compounds may afford more than one halogen-containing combustion product. It is convenient to consider chloro- and bromo-compounds together, since in both instances some elemental halogen and some hydrogen halide is formed. The combustion calorimetry of fluoro-compounds will be considered last.

It was shown many years ago by Smith and co-workers at the University of Lund that the energies of combustion of organochlorine compounds could be measured with reasonable precision if arsenious oxide solution were placed initially in the bomb, so that elemental chlorine produced by combustion would be reduced by the subsequent reaction:

$$Cl_2(g) + 0.5 [As_2O_3(nH_2O)](l) \rightarrow [0.5 As_2O_5 + 2HCl] \left\{ \left(\frac{n}{2} - 1 \right) H_2O \right\}(l).$$

(73)

Thus, aqueous hydrochloric acid was the sole chlorine-containing product of the overall reaction, which could be written:

$$C_aH_bO_cCl_f + \left(\frac{4a+b-f-2c}{4} \right) O_2(g) + \left(\frac{2fn-b+f}{2} \right) H_2O(l)$$

$$\rightarrow aCO_2(g) + f[HCl(nH_2O)](l).$$

(74)

Smith's work was done without the aid of a moving-bomb calorimeter, but rapid reduction of the chlorine produced by combustion was achieved by supporting the arsenious oxide solution on a spirally wound rope of quartz fibres, pressed against the platinum-lined inner wall of the bomb; the volume of solution used was 20 cm^3, and its concentration was 0.35 M, ensuring that an excess of As^{III} was available. At the end of an experiment, chemical analysis of the solution within the bomb was undertaken: nitric acid, unreacted As^{III}, and dissolved gold and platinum were separately determined, and appropriate thermal corrections made, of which the biggest was for oxidation of As^{III} to As^V. A revision of the corrections, with special emphasis on those for dissolved carbon dioxide, was later made in a paper summarizing 25 years' work at Lund (53/4). A somewhat different experimental technique was employed by Hubbard, Knowlton, and Huffman (54/2), who made use of hydrazine dihydrochloride solution as chlorine reductant, and of glass-fibre cloth as support for the liquid reagent. A discussion of the relative advantages and disadvantages of the Smith and the Hubbard techniques is given in

"Experimental Thermochemistry" (56/1, p. 181). Since then, Smith, Scott, and McCullough (64/7) have directly compared the effectiveness of arsenious oxide and hydrazine dihydrochloride solutions as reductants, using a moving-bomb calorimeter to promote rapid chemical reaction as soon as combustion was finished. Arsenious oxide solution was preferred, partly because the stoicheiometry of its oxidation is more straight-forward (hydrazine dihydrochloride undergoes oxidation to two products, nitrogen and ammonium chloride), partly because it is more stable (hydrazine dihydrochloride solution is liable to decompose when in contact with metallic platinum in the bomb) and partly because the thermal correction is less; the one advantage of the hydrazine dihydrochloride reagent is its greater solubility in water. The importance of measuring the pH of the final bomb solution when arsenious oxide is used as reductant was pointed out by Sellers, Sunner, and Wadsö (64/9), since the proportions of dissolved As_2O_3 and As_2O_5 present as H_3AsO_3 and H_3AsO_4 respectively are dependent on total acidity, and this fact has repercussions on the thermal corrections to be made.

Table 4 shows values of $-\Delta U_c^\circ/M$ and their uncertainty intervals for various organochlorine compounds, as determined by different techniques; the values refer to a reaction of the type expressed by equation (74) in which $n = 600$. The results obtained by the static-bomb methods are seen to be in substantial agreement with those obtained by the moving-bomb methods. Likewise, the results obtained by use of As_2O_3 are mostly in accord with those obtained by use of $N_2H_4 . 2HCl$, though a discrepancy greater than the sum of the uncertainty intervals is apparent in the very precise measurements on 2, 3, 5, 6-tetrachlorodimethyl-benzene.

TABLE 4. $-\Delta U_c^\circ/M$ (cal/g) values for some organochlorine compounds at 25°C.

Compound	Static-bomb methods		Moving-bomb methods	
	Quartz spiral As_2O_3	Glass cloth $N_2H_4 . 2HCl$	As_2O_3	$N_2H_4 . 2HCl$
1-Chloropentane	7498 $\pm[18]^a$		7500 $\pm 8^c$	
Chlorobenzene	6604 $\pm[17]^a$	$6598 \cdot 0 \pm 1 \cdot 7^b$	$6600 \cdot 5 \pm 2 \cdot 1^f$	
1, 2-Dichlorobenzene	$4814 \cdot 6 \pm 2 \cdot 0^e$	$4813 \cdot 8 \pm 1 \cdot 2^b$	$4816 \cdot 8 \pm 4 \cdot 0^c$	
4-Chlorobenzoic acid	4682 $\pm[12]^a$		$4679 \cdot 7 \pm 2 \cdot 5^c$	$4678 \cdot 4 \pm 4 \cdot 8^g$
2, 3, 5, 6-Tetrachloro-dimethylbenzene			$3846 \cdot 3 \pm 0 \cdot 6^d$	$3842 \cdot 4 \pm 0 \cdot 9^d$

a = Ref. 53/4; b = Ref. 54/2; c = Ref. 54/3; d = Ref. 64/7; e = Ref. 58/32; f = Ref. 67/29; g = Ref. 68/2.

It is evident from Table 3 that the bomb combustion of an organo-bromine compound results in the formation of substantial amounts of elemental bromine. But as with chloro-compounds, the use of an arsenious oxide solution in the bomb leads to the quantitative reduction of the halogen, so that the overall reaction is:

$$C_aH_bO_cBr_g + \left(\frac{4a+b-g-2c}{4}\right)O_2(g) + \left(\frac{2gn-b+g}{2}\right)H_2O(l)$$

$$\rightarrow aCO_2(g) + g[HBr(n\ H_2O)](l). \tag{75}$$

It should be noted, however, that Bjellerup, a leading worker in the field, prefers to quote values of ΔU_c^o that relate to the reaction:

$$C_aH_bO_cBr_g + \left(\frac{4a+b-2c}{2}\right)O_2(g) \rightarrow aCO_2(g) + \frac{g}{2}Br_2(l) + \frac{b}{2}H_2O(l). \tag{76}$$

After combustion, analysis of the solution for nitric acid, unreacted As^{III}, and dissolved platinum and gold is carried out. Experimental details of calorimetric and analytical procedures are given in both volumes of "Experimental Thermochemistry" (56/1, p. 205; 62/2, p. 41); all measurements on organobromine compounds by Bjellerup were obtained with the aid of a rotating bomb and are quite precise.

It is shown in Table 3 that the combustion of an organofluorine compound may afford two fluorine-containing products†, namely hydrogen fluoride and tetrafluoromethane. The hydrogen fluoride will be largely in solution, and since a concentrated solution of hydrofluoric acid is very corrosive it is normal practice to place sufficient water in the bomb initially to ensure that the final concentration of acid does not exceed 5 M; the use of a rotating bomb is advantageous for ensuring that the solution is uniform in concentration. The interior surface of the bomb should be of platinum and any bomb fitting that may come in contact with the flame or solution should be of platinum (or noble-metal alloy) also. Good, Scott, and Waddington (56/3) suggested that if the bomb is held in an upside-down position while the charge is burning, the water added to the bomb initially will protect the inner

---

† The combustion of a fluoro-compound containing no hydrogen, in the absence of water, is known to yield carbonyl fluoride among the products, but there are no examples of precise calorimetric measurements with systems of this type.

surfaces of the bomb head from the effects of the flame; adoption of this technique merely requires the use of a gimbal mounting for the platinum crucible. Use of glass ampoules for enclosing fluoro-compounds for combustion is precluded, because the hydrofluoric acid produced would attack the remnants of the ampoule in a slow reaction lasting through the final period. Instead, polyester bags may often be employed for enclosing organofluorine compounds, since the permeation of these compounds through polyester film (say 0·025 mm thick) is generally very slow; polyester has the additional advantage of providing a hydrogen source. Such a source is desirable in the combustion calorimetry of a compound $C_aH_bO_cF_h$ in which h/b exceeds unity, since combustion in the absence of an auxiliary hydrogen source tends to produce tetrafluoromethane in relatively large amounts, according to the equation

$$C_aH_bO_cF_h + \left(\frac{4a+b-2c-h}{4}\right)O_2(g) + \left(\frac{(h-4z)(1+2n)-b}{2}\right)H_2O(l)$$

$$\rightarrow (a-z)\,CO_2(g) + z\,CF_4(g) + (h-4z)\,[HF(nH_2O)](l)\,. \tag{77}$$

By physical admixture of the fluorine compound with a hydrogen-containing auxiliary substance of known energy of combustion, or by use of a hydrogen-containing plastic container for the fluorine compound, the formation of tetrafluoromethane may usually be suppressed, at least to an extent that ensures that the error in the correction for formation of tetrafluoromethane is acceptably small. The correction in question is for the energy of the reaction

$$z CF_4(g) + z(4n+2)\,H_2O(l) \rightarrow z CO_2(g) + 4z[HF(nH_2O)](l)\,. \tag{78}$$

By addition of equations (77) and (78), equation (79) is obtained—the reaction expressed by this equation takes place when an organofluorine compound having $(h/b) < 1$ is burned in the presence of water.

$$C_aH_bO_cF_h + \left(\frac{4a+b-2c-h}{4}\right)O_2(g) + \left(\frac{h(1+2n)-b}{2}\right)H_2O(l)$$

$$\rightarrow a CO_2(g) + h[HF(nH_2O)](l)\,. \tag{79}$$

The determination of the amount of tetrafluoromethane formed in the combustion of an organofluorine compound of high fluorine content is a matter of some importance. The chemical inertness of tetrafluoromethane implies that any direct method of analysis must be by some physical technique. Mass-spectrometric analysis of the gaseous combustion products has been tried (56/3; 64/8) but the precision was insufficient. Indirect methods of analysis have therefore been used. Good, Scott and Waddington (56/3) derived the amount of tetrafluoromethane from the difference between the number of moles of combined fluorine taken and the number of moles of F^- found in the bomb solution; the estimation of F^- was effected either gravimetrically by precipitation of lead chlorofluoride, or volumetrically by thorium nitrate titration. Cox, Gundry, and Head (64/8) preferred to derive the amount of tetrafluoromethane from the difference between the number of moles of combined carbon taken and the number of moles of carbon dioxide formed by combustion. Obviously, both these indirect methods of analysis depend on the assumed composition of the organofluorine compound burned, so that the use of very pure samples is essential. A precaution necessary in the gravimetric determination of carbon dioxide produced by combustion of an organofluorine compound is the removal of hydrogen fluoride gas, or hydrofluoric acid spray, from the released bomb gases; this is effected by passing the gases through a scrubber containing sodium fluoride solution, which retains hydrogen fluoride as involatile HF_2^-. It is also necessary to determine the amount of dissolved platinum in an aliquot of the bomb solution, another aliquot being used for determination of NO_3^- by Devarda's method.

In the combustion calorimetry of organofluorine compounds the comparison method may be used; the energy equivalent of the calorimeter is determined by combustion of benzoic acid with relatively large amounts of hydrofluoric acid in the bomb. Alternatively, the thermal corrections for solution of carbon dioxide in hydrofluoric acid may be derived from the data of Cox and Head (62/10).

(c) *Organometallic compounds (including organic compounds of B, Si, etc.)* Most published measurements of ΔU_c° of these compounds were made by static-bomb calorimetry, and it is clear that the investigators were faced with severe experimental problems, often resulting in low accuracy. One problem is the avoidance of incomplete combustion†; thus, Long and Norrish (49/1) carried out 50 combustion experiments on diethylzinc, with 20 variations in technique, without achieving a fully satisfactory combustion. Another problem is the difficulty of defining the final state of a reaction in which the products are ill-defined chemically and physical-

---

† By "incomplete combustion" is meant either the circumstance in which some of the compound is left unreacted by premature extinction of the flame, or that in which carbon carbon monoxide, or a carbide is formed.

ly. In fact, in only a few instances have static-bomb combustion experiments on organometallic compounds been sufficiently straightforward for the results to be acceptable by the standards used for judging measurements on C,H,O,N compounds—the work of Skinner and co-workers on organotin compounds (63/1; 64/2; 66/3) falls in this category, since conditions were found for essentially complete combustion, with tin dioxide and a small amount of metallic tin as combustion products. Generally, however, the use of a moving bomb is to be regarded as *de rigueur* for work of high accuracy. Even with a moving bomb available, the experimentalist must exercise considerable artifice, exploring such variables as sample size, container material, auxiliary substance, oxygen pressure, amount and nature of liquid reagent, and crucible material, shape and mass, in order to find conditions ensuring a thermodynamically definable state after combustion. Technical details are to be found in "Experimental Thermochemistry" (62/2, p.57). Discussion here is limited to two examples, which illustrate methods for attainment of a well defined final state by use of (i) a liquid reagent, and (ii) an auxiliary substance.

(i) Good and co-workers (56/4; 59/2) studied the combustion of tetra-alkyl-lead compounds, with a solution of nitric acid and arsenious oxide placed in the rotating bomb before combustion. The purpose of the nitric acid was to dissolve metallic lead and lead oxides formed on combustion, whilst the purpose of the arsenious oxide was to reduce any Pb^{4+} to Pb^{2+}. Comparison experiments were carried out in which a sample of hydrocarbon oil (of known energy of combustion) was burned in the bomb, with a sample of crystalline lead nitrate placed initially in a separate crucible and a solution of nitric acid and arsenious oxide placed initially in the bomb. The final state, after combustion of the oil and dissolution of the lead nitrate, was virtually identical with that obtained in the experiments on the organolead compounds. With the aid of the comparison experiments, ΔH_f° of the organolead compounds could be calculated with respect to ΔH_f° of crystalline lead nitrate, obviating the need to determine ΔH_f° of lead nitrate in aqueous solution containing other dissolved species.

(ii) The combustion of an organosilicon compound tends to be frustrated by formation of a skin of silica on the surface of the burning substance, which extinguishes the flame prematurely. Thompson (53/5) tried to overcome this difficulty by choosing conditions such that the combustion reaction was explosive. A more subtle approach was adopted by Good and his colleagues (64/3), who showed that in the combustion of hexamethyldisiloxane the use of α,α,α-trifluorotoluene as an auxiliary substance resulted in the formation of gaseous silicon tetrafluoride in place of silica, and a complete, clean combustion was obtained. Water was placed initially in the bomb, so that the sole silicon-containing product was a homogeneous solution of hexafluorosilicic acid in excess hydrofluoric acid:

$$5\,C_6H_5CF_3(l)+C_6H_{18}OSi_2(l)+49{\cdot}5\,O_2(g)+406\,H_2O(l)$$

$$\rightarrow 41\,CO_2(g)+2\,\{[H_2SiF_6+1{\cdot}5\,HF]\,(212\,H_2O)\}(l). \qquad (80)$$

The heat of formation of the hexafluorosilicic acid solution was determined in separate experiments in which elemental silicon mixed with a sample of polyvinylidene fluoride of known energy of combustion was burned, with a solution of hydrofluoric acid placed initially in the bomb. It was thus possible to calculate ΔH_f° of hexamethyldisiloxane with respect to the particular sample of elemental silicon employed in the measurements.

Good's determination (64/3) of ΔH_f° of hexafluorosilicic acid solution, when combined with the heat of solution of amorphous silica in hydrofluoric acid solution, leads to a value for ΔH_f° of silica that is considerably more negative than the value obtained by combustion of silicon in a static bomb (52/1). Because of this we mistrust values of ΔH_f° of silicon *compounds* determined by static-bomb combustion calorimetry, and have excluded such values from the Tables of Chapter 5.

3.2.4. THE PRIMARY-STANDARD SUBSTANCE, AND REFERENCE SUBSTANCES FOR COMBUSTION CALORIMETRY

The calibration of a bomb calorimeter by combustion of a specially purified sample of benzoic acid was referred to in Sections 2.5.1, and 3.2.1. Benzoic acid has been internationally accepted since 1934 (34/2) as the substance to be used for this purpose; it is called the *primary-standard substance*, and was selected for this role because it is readily obtainable in a state of high purity, stable in air, not appreciably volatile at room temperature, non-hygroscopic, easily compressible into a pellet, and readily ignited. Possibly the one disadvantage of benzoic acid is its tendency under some experimental conditions to leave a slight residue in the crucible after combustion; this failing may often be overcome by use of a thinner-gauge platinum crucible or by reduction of the thermal conductivity of the crucible support system, so that the crucible attains a higher temperature during the combustion.

It may be argued that in principle a single, homogeneous batch of benzoic acid should be used as the primary standard by all bomb calorimetrists throughout the world. In practice, the size of the batch needed for this purpose would be prohibitively large, since a portion of the material is used to destruction every time a calibration experiment is performed. The most practicable approach to the ideal is the preparation of a batch of benzoic acid that is likely to last for several years, followed by the determination of the energy of combustion of the material by a national standardizing laboratory; the latter will determine the energy of combustion in terms of the SI unit of energy, the joule, making use of electrical standards which the laboratory maintains. The National Bureau of Standards, Washington DC, has offered for sale batches of benzoic acid certified in this way, for many years past. In Britain, commercial chemical suppliers offer for sale benzoic acid that has been purified by them, but certified with respect to its energy of combustion by the National Physical Laboratory, Teddington. In Russia, two grades of benzoic acid, called respectively K–1 (99·997 moles per cent pure) and K–2 (99·99 moles per cent pure) are issued and certified by the Mendeleev All-Union Metrology Research Institute, VNIIM (64/10). In China, the issue and certification of benzoic acid are undertaken by the Institute of Chemistry, Academia Sinica (66/11).

The certificates issued by the standardizing laboratories normally quote $\Delta U_{cert}/M$, the energy evolved per gramme† when (a) the standard sample of benzoic acid is burned in a bomb at constant volume in pure oxygen at an initial pressure of 30 atm at 25°C, (b) the internal volume of the bomb in litres is one third of the number of grammes of sample burned, (c) the internal volume of the bomb in litres is one third of the number of grammes of water initially placed in the bomb, and (d) the combustion reaction is referred to 25°C. If the user of the benzoic acid sample wishes to depart from any of the conditions (a) to (d) he may correct the value of the energy of combustion from certificate conditions to his own conditions by multiplying by a factor f, given by

$$f = 1 + 10^{-6}[20(p_i - 30) + 42(m_s/v_b - 3) + 30(m_w/v_b - 3) - 45(\theta - 25)] . \quad (81)$$

Here p_i is the initial pressure (atm) of oxygen, m_s is the mass (g) of sample burned, v_b is the internal volume (l) of the bomb, m_w is the mass (g) of water placed initially in the bomb, and θ is the temperature (°C) to which the combustion reaction is referred. The error in f will not exceed 15×10^{-6} f $20 < p_i < 40$, $2 < m_s/v_b < 4$, $2 < m_w/v_b < 4$, $20 < \theta < 30$.

It is becoming increasingly common for investigators to burn thermo-chemical-standard benzoic acid under conditions that differ very markedly

---

† The certificate will specify whether mass *in vacuo* or in air of a certain density is intended.

from the certificate conditions. This situation may arise when benzoic acid is used as an auxiliary substance or in comparison experiments. The effective energy of combustion per gramme of the benzoic acid sample is then best calculated from the standard energy of combustion per gramme, $\Delta U_c^\circ/M$, which in turn is calculated from $\Delta U_{cert}/M$: the relation between ΔU_{cert} and ΔU_c° is given by

$$|\Delta U_{cert}/M| - |\Delta U_c^\circ/M| = 20.4 \, \text{J g}^{-1} . \qquad (82)$$

The calculation of the energy of combustion of the benzoic acid under non-certificate conditions requires application of the reduction to standard states (see Section 3.2.5) in reverse.

Values of $-\Delta U_{cert}/M$ published since 1942 have lain between 26 431·7 and 26 438·1 J g^{-1}. In a statistical analysis of the values Hawtin (66/10) proposed an overall weighted mean of 26 434·4 $\pm$ 1·2 J g^{-1}, but does not appear to have included Meetham and Nicholls's value 26 431·7 $\pm$ 2·2 J g^{-1} (60/1), which is noteworthy as having been determined with an aneroid isothermal-environment bomb calorimeter. Published since Hawtin's review are the values 26 433·3 $\pm$ 5·0 J g^{-1} (Hu, Yen and Geng, 66/11) and 26 433·3 $\pm$ 3·0 J g^{-1} (Peters and Tappe, 67/42); the latter value was obtained by means of a bomb operated isothermally within a Bunsen calorimeter containing diphenyl ether. [see Section 3.3.4, paragraph (a)]. The value recommended in 1961 by the Council of VNIIM for use throughout the U.S.S.R. (64/10) was 26 434 J g^{-1}, and this does indeed appear to be the most probable value based on published measurements†.

The adoption of succinic acid as a secondary-standard substance for calibrating bomb calorimeters was recommended in 1936 by the International Commission on Thermochemistry, but so far as is known no certified sample has been issued by a standardizing laboratory. Succinic acid has, however, found use as a *reference-standard substance* for testing calorimetric (and analytical) procedures, and as an auxiliary substance [Section 3.2.3 paragraph (a)]. Reference standards containing elements other than C, H, and O have also been proposed, and although no internationally agreed list of reference standards has been drawn up, it is possible to assemble a tentative list of suitable substances, based on published information. This has been done in Table 5. The properties of a substance to be used as a reference standard should ideally conform to the criteria given above for the properties of the primary standard; an additional criterion is that the substance should if possible be amenable to purification by zone refining. Although hippuric acid is included in Table 5, there is some doubt as to its suitability as a reference standard (61/11); possibly urea, which has been used

---

† See note on page 98.

as an auxiliary substance (63/10), would be a better nitrogen-containing reference standard.

Auxiliary substances are used in combustion calorimetry for a variety of purposes, e.g. in comparison experiments to control the amount of carbon dioxide produced in relation to the energy evolved, in combustion of sulphur- and fluorine-containing compounds to control the stoicheiometry of the reactions, and in the combustion of difficult-to-burn materials as aids to ignition. Fractions of hydrocarbon oils of low volatility are often used for these purposes, and since their energies of combustion must then be determined with very high precision they may also be regarded as secondary standards, even though they are not necessarily highly pure substances.

TABLE 5. Some reference standards for combustion calorimetry.

Reference standards		$-\Delta U_c^o/M$ J g^{-1}	Literature Reference
Name	Type		
Succinic acid	C, H, O	12 633 ±4	38/4
		12 638 ±2	55/3
		12 635 ±2	59/5
		12 634 ±4	64/12
Hippuric acid	C, H, N	23 542 ±9	51/5
		23 529 ±7	38/4
		23 543 ±9	61/11
Thianthrene	C, H, S	33 467 ±5	63/14
		33 464 ±10	54/1
		33 469 ±10	66/11
4-Fluorobenzoic acid	C, H, F	21 862 ±6	64/8
		21 850 ±9	56/3
Pentafluorobenzoic acid	C, H, F	12 061 ±5	64/8
4-Chlorobenzoic acid	C, H, Cl	19 580 ±11	54/3
4-Bromobenzoic acid	C, H, Br	15 372 ±10	59/7

3.2.5. REDUCTION OF BOMB-CALORIMETRIC DATA TO STANDARD STATES

Standard states were defined in Section 2.2.1, and their significance in thermochemistry has been stressed throughout the book. For thermochemical reactions occurring at a constant pressure near 1 atm, the thermal corrections for the departure of actual states from standard states are readily calculated [Section 2.2.2, paragraph (a)], but the corresponding corrections for a combustion reaction occurring in a bomb are more difficult to calculate. A thorough discussion of the corrections for the bomb combustion of C,H and C,H,O compounds was given by Washburn (33/1), whose approach has been extended by other authors to compounds containing the elements N, S and/or the halogens, in addition to C, H, and O. As details of the methods for correction (often called *reduction*) to standard states are given in the two volumes of "Experimental Thermochemistry", it is unnecessary to give sample calculations here. Instead, we emphasize the physical effects for which corrections must be made, and draw attention to the auxiliary data needed for making the corrections.

(a) *Compression of condensed phases.* Since the substance(s) for combustion, the water initially present in the bomb, and the aqueous phase finally present are all under a pressure much greater than 1 atm, a thermal correction for the compression energy stored in these species is required. From the exact equation (5) the approximate correction for one mole of a species j is given by

$$\int_{1}^{p_B} \left(\frac{\partial U_j}{\partial p} \right)_T dp \approx [-TV_j \alpha_j p]_{p=1}^{p=p_B}, \tag{83}$$

where α_j is the coefficient of cubical expansion of j and p_B is the pressure in the bomb.

(b) *Compression of the gaseous phase.* Since neither the initial nor the final gaseous state in the bomb is at very low pressure, a thermal correction for the non-ideality of the gases is required. The correction is obtained by integration of equation (5), the limits being p_B and zero. The evaluation of the integral requires knowledge of the equation of state of gaseous oxygen and of the gas mixture constituting the final state. Equations of state in the Beattie–Bridgman form for pure oxygen and oxygen+carbon dioxide mixtures were determined by Rossini and Frandsen (32/4); for gas mixtures containing nitrogen, it is assumed that the volumetric behaviour of the nitrogen content is identical with that of an equal number of moles of oxygen, and for gas mixtures containing carbon tetrafluoride the volumetric behaviour is assumed to be calculable from the equations of state of pure carbon tetrafluoride (62/2, p.27) and oxygen+carbon dioxide mixture.

(c) *Solution of the gases in the aqueous phase.* In the initial state some oxygen will be dissolved in the water placed in the bomb, and in the final state some of the mixed gas phase will be dissolved in the aqueous phase; thermal corrections for dissolution of the gases are required. For this purpose, the Henry's law constant and ΔU_{soln}° for each gas dissolved in the water or aqueous phase must be known; the volumes of water and solution, and the partial pressure of each gas must be calculated. For water, and solutions of nitric acid, sulphuric acid (56/1 pp.86, 106), hydrofluoric acid (62/10), hydrochloric acid and As_2O_3 (68/7) the Henry's law constants and values of ΔU°_{soln} are known. For other aqueous solutions the necessary data are fragmentary or completely lacking, and cannot always be estimated with any confidence. In that circumstance the method of comparison experiments is valuable, because the need to correct for dissolution of the gases is obviated.

(d) *Vaporization of water.* When water is placed in the bomb initially, the gas-space becomes saturated with water vapour and remains so throughout the experiment. But the concentration of water vapour may change between the initial and final conditions because (i) the partial pressure of water vapour over the final aqueous phase may differ significantly from that over pure water, (ii) the partial pressure of water vapour in the presence of a compressed gas is dependent on the nature and pressure of the gas, and (iii) the rise in temperature of the bomb contents will cause an increase in the partial pressure of water vapour (see paragraph (f) below.) The necessary thermal corrections require knowledge of the heat of vaporization of water at 25°C and the number of moles of water vapour present in the initial and final states, the calculation of which in turn requires knowledge of the bomb's internal volume, the vapour pressure of the solution, and the magnitude of the vapour-pressure enhancement effect. The heat of vaporization is well established and vapour-pressure data for most aqueous solutions likely to be of interest are known; data for the vapour-pressure enhancement effect are less well known, that for oxygen as the compressing gas being estimated from data for nitrogen and air (56/1, pp. 83, 89).

(e) *Dilution of the aqueous phase.* The energy of formation of the final aqueous phase will depend on its concentration. In a series of replicate experiments it is almost impossible to obtain the same final concentration in each experiment, so it is customary to make a thermal correction for dilution of the aqueous phase to an arbitrarily selected reference concentration, e.g. HCl (600 H_2O) in the combustion of organochlorine compounds. The necessary values for the energy of dilution from the calculated concentration of the final solution to the reference concentration may usually be obtained from standard tables (65/1).

(f) *Non-isothermal reaction.* The need to assign a measured heat or energy of reaction to a single reference temperature was stressed in Section 2.3.1. For the specific case of bomb-combustion calorimetry it is usual to correct $\Delta U^{\circ}_{c, T_1}$ to 25°C (298·15°K), and by an argument similar to that used to derive equation (13) we may write†

$$\Delta U^{\circ}_c = -C_{p,\text{ initial}}(T_{\text{final}} - T_{\text{initial}}) + \Delta C_v(298\cdot15 - T_{\text{final}}). \qquad (84)$$

The derivation of the term ΔC_v requires knowledge of the constant-volume heat capacities of the substance(s) to be burned, the water placed in the bomb, the oxygen, the gas mixture produced, and the final aqueous solution. The small term for the extra vaporization of water between T_{initial} and T_{final} (see paragraph (d) above) does not appear explicitly in equation (84), because it is generally considered to be a part of the first term on the right-hand side; this is legitimate so long as calibration and combustion experiments are made over approximately the same temperature range.

It is seen that much auxiliary information is needed to complete a reduction to standard states. Before 1933 reductions to standard states were not attempted. Since that time, most authors have included a reduction to standard states in their calculations of ΔU°_c, but some have not and a few have applied the reduction incorrectly; readers of original papers should be mindful of these facts. The net effect of a standard-states reduction is almost always to make $|\Delta U^{\circ}_c/M| < |\Delta U_B/M|$, where ΔU_B is the molar energy change of the bomb reaction after application of all thermal corrections (e.g. for side reactions) *except* those arising from the standard-states reduction. Of the six physical effects listed above as contributing to the reduction, effects (b) and (c) are generally the most significant, and in rotating-bomb calorimetry involving use of large volumes of liquid reagents effect (c) predominates. The labour of applying a full standard-states reduction is undeniably tedious; hence, the use of a digital computer for the purpose is advantageous. Alternatively. for compounds of empirical formula $C_aH_bO_c$, an abridged correction to standard states may be made with the aid of Washburn's approximate formula (33/1)

$$\% \text{ correction} = \frac{-0\cdot3\,ap_{\text{initial}}}{-\Delta U_B}\left[1 - \frac{1\cdot1(b-2c)}{4a} + \frac{2}{p_{\text{initial}}}\right], \qquad (85)$$

where p_{initial} is expressed in atm. For most C,H,O compounds equation (85) gives a value for the standard-states correction that is within 15% of the rigorously calculated value.

---

† In equation (84) the "heat capacity" (more properly the "energy equivalent") of the calorimetric system is written with subscript p because it is in a constant-pressure environment. The change in heat capacity of the bomb contents is however written with subscript v, because for the gases the interior of the bomb is a constant volume and for the condensed-phase bomb contents the distinction between C_p and C_v is unimportant.

Once ΔU_c° has been calculated, it is customary to derive ΔH_c° from it by application of equation (2), and finally to calculate ΔH_f° of the compound burned by application of equation (6) in the form

$$\Delta H_f^\circ \text{ (compound burned)} = \sum \Delta H_f^\circ \text{ (products of combustion)} - \Delta H_c^\circ -$$

$$\sum \Delta H_f^\circ \text{ (reactants other than the compound burned).} \qquad (86)$$

The last term on the right-hand side of equation (86) is zero when oxygen is the only reactant other than the compound burned, but will be non-zero if, for example, some water is used up in the combustion reaction (*cf.* equation (79) for the case when $b < h$).

3.2.6. THE FLAME CALORIMETER

The flame calorimeter is used to measure the heats of combustion of gases and volatile liquids at constant pressure; the discussion in this Section is confined to combustion in oxygen. The method was used by Rossini (31/1; 31/2) to determine the heat of combustion of hydrogen, and also to measure the heats of combustion of methanol, ethanol, and some lower members of the aliphatic hydrocarbon series. More recently, a flame calorimeter has been developed by Pilcher *et al.* (63/2) and used for measurements on ethers (64/4; 65/3). In Pilcher's design the water-filled calorimeter vessel is similar to that used to contain a static bomb, but in place of a bomb the reaction vessel shown in Fig. 7 is employed. Oxygen is admitted to the reaction vessel at A, and the gaseous substance for combustion at D. The substance can be premixed with oxygen just below the flame jet by admission of oxygen at C. Such premixing increases the flame speed, but if necessary the flame speed can be decreased by premixing of the substance with an inert gas, e.g. argon. If the substance is a liquid at room temperature it is carried into the burner as vapour by means of a stream of inert gas. Gases leaving the burner chamber pass through a heat-exchange spiral before leaving the calorimeter. In a typical experiment, the fore-period is *ca.* 20 minutes, the main-period is *ca.* 40 minutes (20 minutes for reaction and 20 minutes for thermal equilibration), and the final period is *ca.* 50 minutes.

In Rossini's work the calorimeter was calibrated electrically, but in Pilcher's work the calorimeter was calibrated by combustion of hydrogen in oxygen.

It is not a practical proposition in flame calorimetry to determine the amount of reaction from the mass of substance burned. The result is normally based on the mass of carbon dioxide produced, but as a check on the completeness of the combustion reaction the mass of water formed may also be measured. In Pilcher's calorimeter the gases are ignited by a high-voltage spark across a 4-mm gap between platinum points. The energy of ignition is determined from an experiment in which the compound is allowed to burn

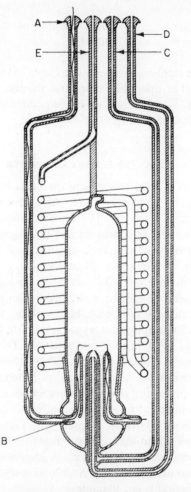

Fig. 7. A flame calorimeter. A, tube for supply of secondary oxygen, B, orifice for entry of secondary oxygen; C, tube for supply of primary oxygen; D, tube for supply of organic vapour; E, gas-exit tube.

for only about 60 seconds, and the amount of reaction is determined by CO_2 analysis. The amount of heat contributed by combustion of the compound is calculated from normal experiments of 20 minutes' reaction-time, and the energy of ignition is calculated by difference. The quantity derived from measurements by flame calorimetry is ΔH_c, since the reaction is conducted at virtually constant pressure. The calculation of ΔH_c° presents no difficulty [cf. Section 2.2.2, paragraph (a)], but a significant correction is required to allow for the vaporization of some of the water formed.

Apart from measurements on cyanogen (51/2; 51/3) and chloroethane (51/4), measurements by flame calorimetry have so far been limited to C,H and C,H,O compounds. Before the technique can be more widely applied, many analytical problems must be solved.

3.2.7. COMBUSTION CALORIMETRY INVOLVING FLUORINE

Combustion calorimetry has been discussed in previous Sections with reference to oxygen as the oxidant. But combustion (i.e. rapid and complete degradation of the carbon skeleton of an organic compound) can also be effected by use of gaseous fluorine as the oxidant, e.g.

$$C_aH_bF_h + \left(\frac{4a+b-h}{2}\right) F_2(g) \rightarrow aCF_4(g) + bHF(g). \qquad (87)$$

Because of the exceptional reactivity of fluorine its use as a reagent in calorimetry involves the solution of formidable problems: (a) the gas is difficult to purify, (b) there is a limitation on the materials of construction with which fluorine may safely be allowed to come in contact, (c) there are difficulties in analysing the products of reaction, partly due to the inertness of products like CF_4, and (d) auxiliary data on the thermodynamic properties of fluorine and its compounds are often poorly established. Notwithstanding these difficulties, Jessup and Armstrong (62/2, p.129) have developed techniques for fluorine flame calorimetry, whilst Hubbard and his colleagues (62/2, p.94) have developed techniques for fluorine bomb calorimetry. The emphasis of work in fluorine calorimetry has been on inorganic substances, so that the application of fluorine calorimetry to organic compounds is so far limited to very few examples. Jessup et al. (55/8) showed that methane diluted with helium burned smoothly in an atmosphere of fluorine in a copper and nickel-alloy flame calorimeter, and Domalski and Armstrong (67/3) showed that polytetrafluoroethylene burned cleanly in fluorine at 20 atm pressure in a nickel bomb. In each case the stoicheiometry of the combustion reaction was substantially as given by equation (87). The frustration of earlier workers' attempts to burn graphite in fluorine to form tetrafluoromethane is now known to have been due to partial formation of other carbon fluorides, but recent experiments by Settle, Greenberg and Hubbard (67/41; 68/1), using a two-compartment nickel bomb, have met with success.

With the solving of at least some of the practical problems listed above, and with the better establishment of the thermodynamic properties of key fluorine compounds, fluorine-combustion calorimetry may come to play a bigger role in thermochemical studies of organic compounds. In particular, the technique should provide a route to the determination of ΔH_f° for C,H,F compounds and possibly for C,H,N,F compounds also.

3.3. Reaction calorimetry

By custom, the term "reaction calorimetry" is considered to apply to the measurement of heats of reactions other than combustion reactions. As illustrated by an example in Section 3.1, the magnitude of ΔH_r° is likely to be at least one order lower than the magnitude of ΔH_c°, so that the accuracy of measurement of ΔH_r° can generally be lower than the accuracy of measurement of ΔH_c° to produce values of ΔH_f° of comparable quality. Calibration is therefore a less exacting process for a reaction calorimeter than it is for a bomb or flame calorimeter, and the electrical calibration method can be applied by experimenters having relatively modest electrical-measurement equipment. Indeed, the electrical method is the only calibration method for reaction calorimeters that is at present internationally recognised, though there are indications that the neutralization of tris(hydroxymethyl)amino-methane ("THAM") with hydrochloric acid (64/46; 65/10) may come to be recognised as a suitable reaction for calibrating some reaction calorimeters.

The most common types of reaction studied calorimetrically may be classified as "hydrogenation", "halogenation", "hydrolysis", and "polymerization". When a reaction-calorimetric experiment of any of these types is being designed, variables such as temperature, concentration, and catalyst should be so chosen that the reaction will proceed cleanly to completion, and the products will be amenable to precise analysis. Rapid reactions are the easiest to study, but special techniques have been devised for slow reactions and will be described in Section 3.3.4. The overall accuracy of a reaction-calorimetric experiment is often limited by the accuracy with which the amount of reaction can be determined, and the need for proper analysis to demonstrate the absence of side reactions cannot be overstressed.

Even when the experimenter has selected an appropriate design of calorimeter and optimum operational conditions, he has still to decide how he will combat the problem of heat-exchange between the calorimeter and its surroundings. He may decide to use an isothermal environment, and allow for heat-exchange by the methods referred to in Section 3.2.1. Alternatively, he may opt to employ an adiabatic or a truly isothermal apparatus, and so obviate the need to correct for heat exchange. Examples of the use of the various techniques will be found in the ensuing Sections. It is feasible to discuss reaction calorimetry under the headings of the reaction types given in a previous paragraph. However, since the choice of calorimeter design is dictated more by the physical states of the reactants than by the reaction type, the discussion that follows is sub-divided according to physical states.

3.3.1. GAS–GAS REACTIONS

Kistiakowsky *et al.* (35/2; 38/13) measured heats of hydrogenation and halogenation of olefins in the gaseous phase at 80°C, with an accuracy of about 0·1%. Their calorimeter is no longer in use but a brief description of it is warranted because of the importance of the results obtained with it. The calorimeter contained 1500 cm³ of diethylene glycol and was equipped

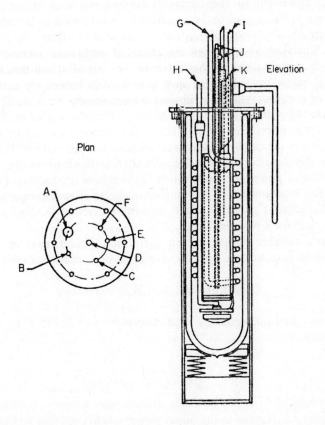

Fig. 8. An isothermal calorimeter for gaseous reactions. A and K, condenser; B and G product exit; C and H, nitrogen coolant; D and J reactant entry; E and I, heater leads; F, thermocouple well.

with a stirrer, a calibrating heater, and multijunction thermocouples for temperature measurement. It was operated adiabatically.

The reaction vessel, submerged in the calorimeter, contained a glass annular catalyst-chamber into which the reacting gases were passed at a constant rate. When a steady state had been achieved, i.e. when the calorimeter temperature was rising linearly with time, the effluent gases were diverted for a measured time through a combustion furnace; the amount of carbon dioxide formed was determined and used to calculate the number of moles of reaction per unit time. The calorimeter was calibrated by measurement of the electrical power needed to produce the same rate of temperature rise. With the aid of this calorimeter Kistiakowsky and co-workers measured many values of $\Delta H^{\circ}_{r,\,353^{\circ}K}$; correction of the values to 298·15°K can be made without difficulty by means of equation (9).

For measurement of the heats of gaseous reactions at temperatures higher than 353°K, Lacher and co-workers have developed an ingenious form of isothermal calorimeter (49/2, 56/5, 57/5); it is illustrated in Fig. 8. As with the Kistiakowsky calorimeter, the heart of the Lacher calorimeter was a catalyst-chamber into which the reactant gases flowed at constant rate. The catalyst-chamber was immersed in a liquid, dibutyl ether for reactions at 376°K and tetra-2-ethylhexyl orthosilicate for reactions at 521°K. The liquids were stirred by a stream of nitrogen bubbles, which also cooled the liquids by causing evaporation. The cooling effect was exactly balanced by electrical heating. When the exothermically reacting gases were passed into the catalyst chamber, less electrical power was required to maintain isothermal conditions, the decrease of power being equivalent to the rate of production of chemical energy. For measurement of the number of moles of reaction per unit time, either the reactant or the product gases were diverted through a suitable analysis train for a measured interval.

Lacher's calorimeter has proved useful for measurement of heats of chlorination, bromination, hydrobromination, and reductions such as

$$CH_3Cl(g) + H_2(g) \rightarrow CH_4(g) + HCl(g).$$

The values obtained for $\Delta H^{\circ}_{r,\,T}$ require correction to 298·15°K by means of equation (9).

3.3.2. GAS–LIQUID REACTIONS

For reactions such as the addition of hydrogen, halogen, or diborane to unsaturated compounds in the liquid phase, either undiluted or in solution, it is experimentally convenient to place the liquid reagent in a vessel submerged in a liquid-filled calorimeter and to lead in the gaseous reactant.

For example, Skinner and co-workers (57/7) used the apparatus shown in Fig. 9 to measure the heats of catalytic hydrogenation of unsaturated compounds in solution. The calorimeter was a water-filled Dewar vessel equipped with a calibrating heater; temperatures were measured with a thermistor. A

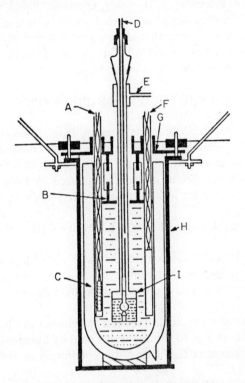

Fig. 9. A Dewar-vessel calorimeter for study of hydrogenation reactions. A, heater; B, ceiling; C, Dewar vessel; D, hydrogen inlet; E, vibro-shaker arm; F, thermistor; G, calorimeter lid; H, copper can; I, glass reaction vessel.

stirrer was unnecessary because the water in the calorimeter was sufficiently agitated by the movement of the reaction vessel, which was connected to a vibratory shaker. Initially, the reaction vessel contained 15 cm^3 of solvent (ethanol or acetic acid), 0·1 g of platinum oxide, or palladium oxide, and the unsaturated compound, solid or liquid, sealed in a thin glass ampoule. Air was pumped out from the vessel, and on admission of hydrogen the metal oxide was reduced to metal. The drift of temperature with time was observed in a fore-period, then the reaction was initiated by breaking the ampoule; rapid hydrogenation was effected by vibratory agitation of the heterogeneous system. The volume of hydrogen absorbed was measured, and the number of moles of reaction deduced from it. The calorimeter was calibrated electrically after completion of the reaction experiment.

Without vigorous agitation of the reaction vessel and resultant dispersion of the dense catalyst, hydrogenation would have proceeded too slowly for study in a simple calorimeter. However, many gas–liquid reactions proceed fast enough without the aid of solid catalysts to be amenable for study in a simple calorimeter similar to that shown in Fig. 9, but with a paddle stirrer for agitating the liquid phase instead of a vibrator. Thus Pedley, Skinner, and Chernick (57/6) studied reactions of the type:

$$SnR_4(l) + Br_2(g) \rightarrow SnR_3Br(l) + RBr(g).$$

Dry nitrogen transported bromine vapour, from a weighable evaporator containing liquid bromine, into the reaction vessel containing the tin alkyl. The amount of reaction was calculated from the loss in weight of the evaporator. A similar experimental approach was used by Fowell and Mortimer (61/4) to measure the heat of hydrolysis of butyl-lithium:

$$C_4H_9Li(l) + H_2O(g) \rightarrow LiOH(c) + C_4H_{10}(g).$$

The reaction of butyl-lithium with *liquid* water would have been too violent for calorimetric study, but when nitrogen saturated with water vapour was bubbled into the liquid phase the rate of hydrolysis was satisfactorily controlled.

As an example of a gas–liquid reaction calorimeter not based on a Dewar vessel we illustrate in Fig. 10 the apparatus used by Skinner et al. (61/5, 61/6, 63/3) for measurement of the heats of hydroboration of olefins:

$$6R.CH = CH_2(l) + B_2H_6(g) \rightarrow 2(R.CH_2.CH_2)_3B(\text{solution}).$$

The silver-plated copper calorimeter was equipped with a stirrer, a calibrating heater, and a thermistor. Gaseous diborane could be admitted through a sintered–glass disc covered with mercury, which acted as a valve. Rapid reaction of diborane and olefin took place in diethyleneglycol dimethyl ether solution. The major contributing factor to the overall uncertainty was the uncertainty in determining the amount of reaction.

3.3.3. LIQUID—LIQUID AND SOLID—LIQUID REACTIONS

Most hydrolysis, oxidation, reduction, and complexing reactions fall under this heading. A glass Dewar-vessel calorimeter, basically similar to that described in Section 3.3.2, is the form of calorimeter most frequently used in this work. It is favoured because it is cheap and simple to construct, and being made of glass involves no problem of corrosion by the more common types of chemical reagent. Since the overall accuracy in liquid-phase reaction

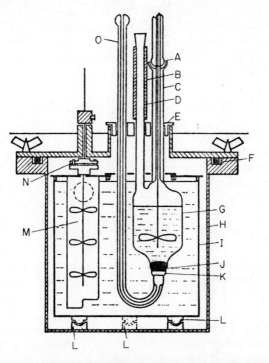

Fig. 10. A calorimeter for study of hydroboration reactions. A and C, capillary-tube ball and socket joint; B, electric heater; D, gas-exit tube; E, polyethylene; F, O-ring; G, glass reaction vessel; H, outer jacket for calorimeter; I, silver-plated copper calorimeter; J, mercury; K, glass frit (sinter); L, insulating pegs; M, paddle stirrer; N, stirrer connection; O, entry tube for diborane.

calorimetry is frequently limited by the determination of the chemistry of the reaction, the inherent defects of the Dewar vessel as a calorimeter are often unimportant. The defects in question are (i) that the effective boundary of the calorimeter is not defined, and could be different for the electrical calibration and the reaction experiment, and (ii) that the calorimeter takes a long time to equilibrate after a change of temperature. For precise thermo-chemical studies of reactions in solution, Sunner and Wadsö (59/3) tested several vacuum-jacketed calorimeters, designed to avoid the defects of glass Dewar vessels. One of their most successful and versatile designs is shown in Fig. 11. The calorimeter vessel consists of a thin-walled brass can, chromium-plated, with the interior parts coated with polytetrafluoroethylene. An ampoule-breaking spike is set in the base, and a calibrating heater and a thermistor for temperature measurement are located in separate re-entrant wells. The liquid reagent (100 cm³) in the calorimeter is stirred by rotation of an ampoule holder. Into the latter is fitted a small glass cylindrical ampoule, which contains the second reagent, liquid or solid; alternatively, a liquid

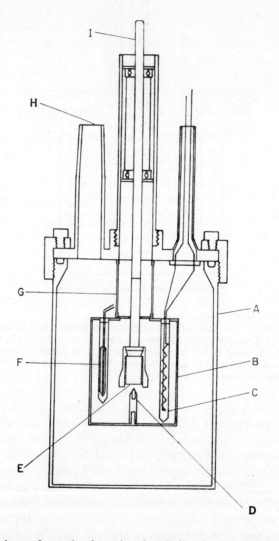

Fig. 11. A calorimeter for study of reactions in solution. A, outer jacket; B, calorimeter can; C, electric heater for calibration; D, spike; E, glass ampoule in holder; F, thermistor. G, glass tube for supporting calorimeter; H, connection to vacuum pump; I, stirrer shaft.

reagent can be introduced into the calorimeter *via* a special pipette (63/13). The calorimeter vessel is suspended by a thin-walled glass tube from an outer can immersed in a thermostatically controlled bath; the space between the calorimeter and the outer can is evacuated to 10^{-4} torr to reduce heat transfer.

The calorimeter was used by Wadsö (58/4, 62/4) to measure the heats of hydrolysis of a number of O-, S-, and N-acetyl compounds. As an example, we consider the measurement of the heat of hydrolysis of ethyl acetate in ethanol + water solution containing excess sodium hydroxide. The reaction actually studied is defined by equation (88), but to determine the *standard* heat of hydrolysis of ethyl acetate (equation (92)), it was necessary to measure in addition the heats of solution represented by equations (89), (90), and (91).

$$CH_3.CO_2C_2H_5(l) + H_2O \text{ (NaOH solution)}$$
$$\rightarrow [C_2H_5OH + CH_3CO_2H] \text{ (NaOH solution)} \quad (88)$$
$$\Delta H_r = -10.58 \pm 0.02 \text{ kcal.}$$

$$C_2H_5OH(l) + NaOH \text{ solution} \rightarrow C_2H_5OH \text{ (NaOH solution)} \quad (89)$$
$$\Delta H_r = -0.05 \pm 0.00 \text{ kcal.}$$

$$CH_3CO_2H(l) + NaOH \text{ solution} \rightarrow CH_3CO_2H \text{ (NaOH solution)} \quad (90)$$
$$\Delta H_r = -11.61 \pm 0.05 \text{ kcal.}$$

$$H_2O(l) + NaOH \text{ solution} \rightarrow H_2O \text{ (NaOH solution)} \quad (91)$$
$$\Delta H_r = -0.19 \pm 0.01 \text{ kcal.}$$

$$CH_3.CO_2C_2H_5(l) + H_2O(l) \rightarrow C_2H_5OH(l) + CH_3CO_2H(l) \quad (92)$$

Thus
$$\Delta H_r^\circ = -10.58 - (-0.05) - (-11.61) - 0.19$$
$$= +0.89 \pm 0.06 \text{ kcal.}$$

In this example, a small-magnitude standard heat of reaction† was derived from the measured heats of four processes, and it is clear that the precision of each separate measurement had to be high to ensure good overall precision. The importance of using a versatile, high-precision calorimeter for this type of work is obvious.

---

† An unusual feature of the given example is that the calculated overall heat is positive, whereas the measured heats of the four stages were all negative. A calorimeter of the type described can, however, be used for *direct* measurement of positive (endothermic) heats.

3.3.4. CALORIMETERS FOR SLOW REACTIONS, AND MICROCALORIMETERS

The liquid-phase reaction calorimeters described in Sections 3.3.2 and 3.3.3 are suitable for the study of reactions that proceed to completion within several (say $\not> 30$) minutes, inasmuch as the corrections for heat exchange are acceptably small during reaction periods of that duration. But many organic reactions of interest to thermochemists require hours for completion. One approach to the problem of slow reactions is to adapt a calorimeter that is normally operated with an isothermal environment, to operate adiabatically. Another approach is to develop a special type of calorimeter, designed from the outset to permit operation over long periods of time. The Bunsen calorimeter and the microcalorimeters of Calvet and of Benzinger are of this type; these calorimeters are now described.

(a) *The Bunsen calorimeter.* The operation of this type of calorimeter is dependent on the transfer of reaction heat to a partially melted solid and the measurement of the volume change resulting from the extra melting (exothermic reaction) or freezing (endothermic reaction). Since the measurement takes place at the melting point of a pure substance, the calorimeter is a truly isothermal one, not to be confused with the iso-thermal-environment type of calorimeter, which is often loosely referred to as "isothermal". A Bunsen calorimeter can be constructed to have quite high sensitivity (0·002 cal) and can be used for experiments of up to *ca.* 6 hours' duration. This form of calorimeter has been applied to the study of relatively few organic reactions, but is a favoured instru-ment for the measurement of heats of polymerization (57/8). The original Bunsen calorimeter operated at 0°C with ice as the working substance. It is now more usual to employ highly pure diphenyl ether (m.p. 26·9°C) as the working substance: the resulting value of $\Delta H^\circ_{r, \, 300°K}$ requires a correction to 298·15°K that is generally small enough to be ignored. Benzene and naphthalene have also been used as working substances.

The Bunsen calorimeter shown in Fig. 12 consists of three concentric tubes, of which the inner one constitutes the reaction vessel; the space between the outer pair of tubes is evacuated. The charge of diphenyl ether surrounding the inner tube must be of high purity; fortunately this substance can be brought to a purity of 99·99 moles per cent simply by repeated crystallization from the melt, with rejection at each stage of a small unfrozen portion. The calorimeter is immersed in a bath held at a temperature near 26·9°C; the actual temperature of the bath can be adjusted to reduce the rate of melting virtually to zero. Before a calori-metric experiment is started, a mantle of frozen diphenyl ether is formed from a melt of the ether by internal cooling of the reaction compartment. Care is taken to ensure that the mantle is uniform in thickness and free from cracks. If it is supported on protrusions or vanes, heat quantities of the order of 300 cal may then be absorbed without disintegration of the mantle; an unsupported mantle can absorb much less heat before disintegrating.

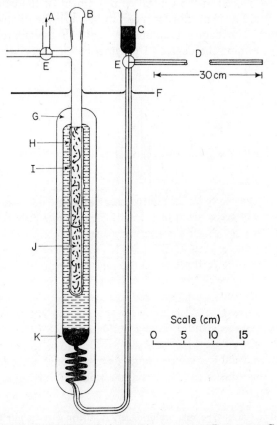

Fig. 12. A Bunsen-type isothermal calorimeter. A, to vacuum; B, stopper; C, mercury reservoir; D, precision-bore capillary; E, 3-way tap; F, thermostat bath; G, vacuum jacket; H, liquid diphenyl ether; I, solid diphenyl ether; J, reaction compartment; K, mercury.

The reactants are placed in the inner compartment and the calorimeter is allowed to equilibrate. After initiation of the reaction, the volume change of the melting ether results in the movement of mercury in a capillary and the heat evolved is deduced from the mass of mercury displaced. The factor relating displacement of mercury to the amount of heat generated was determined by Jessup (55/4) as $79 \cdot 10 \pm 0 \cdot 01$ J $(g \, Hg)^{-1}$ and by Peters and Tappe (67/43) as $79 \cdot 32 \pm 0 \cdot 01$ J $(g \, Hg)^{-1}$. Displaced mercury can be directly weighed, or by use of a precision-bore capillary tube the weight can be deduced from the linear movement of the meniscus, observed with a travelling microscope.

When a Bunsen calorimeter is used for measuring heats of polymerization (57/8), the reaction may advantageously be carried out inside a dilatometer placed in the central compartment of the calorimeter, with an inert liquid as heat-transfer medium between the dilatometer and the calorimeter. The contraction of the liquid in the dilatometer can be used to determine the degree of polymerization associated with a given output of heat.

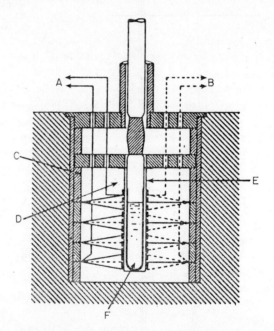

Fig. 13. Cell of a Calvet-type microcalorimeter. A, thermoelectric pile for detection; B, Peltier effect; C, external chamber; D, internal chamber; E, silver socket; F, coppered glass cell.

(b) *The Calvet microcalorimeter.* The outstanding feature of this type of calorimeter is its amenability to measurement of the heat generated in very slow reactions (rates as low as $0 \cdot 001 \, \mathrm{cal \, h^{-1}}$), and of heat effects that are very small *in toto*—hence the description "microcalorimeter". The principle of operation of the Calvet microcalorimeter is unlike that of any other calorimeter described so far in this book: no attempt is made to confine heat within the calorimeter, and it is in fact the heat flux between the calorimeter and its surroundings that is measured.

A microcalorimeter cell is shown in Fig. 13. The cell is completely surrounded by a thermostatically controlled block (a *heat sink*), and heat generated in the cell is conducted to the block through thermocouple wires. Provided these thermocouples are uniformly distributed, the total electromotive force (e.m.f.) produced will be directly proportional to the integrated temperature difference across the thermopile. The rate of heat transfer will also be directly proportional to the integrated difference. Hence the total e.m.f. will be directly proportional to the rate of heat transfer, and the area under a curve of e.m.f. against time will be proportional to the total amount of heat flowing from the cell (exothermic reaction) or to it (endothermic reaction).

The practical utility of such a system must obviously be limited by the temperature stability of the surrounding heat sink, and in order to remove this limitation it is normal practice to employ twin cells. The two cells are identical in construction, but one contains the reacting system under investigation and the other an inert filling. Let us suppose that the two cells, 1 and 2, are at temperatures θ_1 and θ_2, which differ from the temperature θ_0 of their common environment. Let e_1 and e_2 be the e.m.f. outputs from each cell. Hence

$$e_1 = k(\theta_1 - \theta_0) \tag{93}$$

and
$$e_2 = k(\theta_2 - \theta_0), \tag{94}$$

because the identical construction of the two cells ensures that the proportionality constant k is the same for each cell. If the thermocouples from the two cells are connected in opposition to one another, the net e.m.f. recorded, $(e_1 - e_2)$, is given by

$$e_1 - e_2 = k(\theta_1 - \theta_2); \tag{95}$$

i.e. $(e_1 - e_2)$ is not dependent on θ_0, so that fluctuations in the tempera ture of the heat sink are unimportant, provided that the two cells sense the fluctuations equally. This proviso is met by symmetrical disposition of the twin cells within the surrounding thermostatted block. An arrangement of this type forms the basis for several types of Calvet microcalorimeter, some of which are available commercially—one model is suitable for operation in the temperature range 0–200°C, and another in the temperature range 0–800°C. These calorimeters may be calibrated electrically. Evans, McCourtney, and Carney (68/5) have described a form of Calvet microcalorimeter that is particularly well suited to the measurement of heats of reaction in solution. Two liquid reactants (*ca.* 1 cm^3 of each) are contained initially in separate compartments of a cylindrical tantalum cell; the liquids are caused to mix by reciprocal rotation of the entire experimental assembly. The calorimeter may be calibrated either chemically (e.g. by an acid + base reaction in dilute solution) or electrically.

To date, the Calvet microcalorimeter has found most application in measurement of physical or biological heat effects, rather than chemical heat effects (56/1, 62/2, 63/12). As one of the few examples of its use in organic thermochemistry, we cite the measurement of the standard heat of esterification of acetic acid with ethanol, studied by Bérenger-Calvet (27/1); the work was done with an early, single-cell model of microcalorimeter, and the precision attained was much lower than that attainable with a modern, twin-cell form of Calvet microcalorimeter. The value found, $-1 \cdot 07$ kcal mol^{-1}, should be identical in magnitude with the standard heat of hydrolysis of ethyl acetate but of opposite sign (cf. Wadsö's value given in Section 3.3.3).

Now that Calvet microcalorimeters can be obtained commercially, increasing application of them to the measurement of the heats of very slow organic reactions is to be expected. Already it has been shown (66/22) that a commercial instrument is suitable for measuring the heat of a polymerization reaction lasting 24–36 hours; in this study, the microcalorimeter was calibrated by means of the known heat of the reaction between tris(hydroxymethyl)aminomethane and $0 \cdot 1$ M hydrochloric acid.

(c) *The Benzinger microcalorimeter*—The principle of operation of this equipment is similar to that of the Calvet microcalorimeter, but the construction is radically different. Benzinger's apparatus, though usable for measuring heat evolved over a period of a few hours, is best suited to measurement of small amounts of heat generated by moderately rapid reactions—hence its alternative designation as a "heat-burst" microcalorimeter. Two views of the heart of the Benzinger microcalorimeter are shown in Fig. 14. The annular reaction vessel shown permits the mixing of a small volume ($0 \cdot 1 - 0 \cdot 3$ cm^3) of one liquid reactant with a larger volume (15 cm^3) of another. Mixing is effected by rotation of the entire unit, either axially or end-over-end. The apparatus is often calibrated by mixing dilute solutions of sodium hydroxide and hydrochloric acid, using the accurately known value of ΔH_r° for this reaction.

The Benzinger microcalorimeter, which is now commercially available, has been chiefly used hitherto for measuring the heats of biochemical reactions in dilute solution. Some of the reactions in question were reversible, with equilibrium constants that did not differ very much from unity. For such reversible reactions, Benzinger (56/6) has pointed out that by measurement of the heats of both the forward and the backward (opposite sign) reactions to the point of equilibrium, it is possible to derive not only the value of ΔH_r° but also the values of ΔG_r° and ΔS_r°. Consider the reversible reaction

$$aA + bB + cC \ldots \rightleftharpoons xX + yY + zZ \ldots .$$

Suppose that the measured heat of the reaction between A, B and C ... (in molar proportions $a:b:c\ldots$) up to the point of equilibrium is Q_1 cal

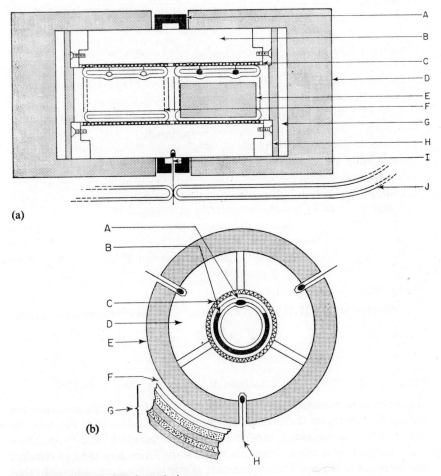

(a)

(b)

Fig. 14. Benzinger-type microcalorimeter.
(a) A, ring suspending the block; B, heat sink; C, thermopile; D, circular caps; E, reaction vessel; F, control vessel; G, air space; H, end discs; I, wire-support for block; J, Dewar vessels.
(b) A, reactant; B, second reactant; C, thermopile; D, heat sink; E, circular caps; F, air space, Dewar vessels; G, insulation; H, wire-support for block.

per mole of A and that the measured heat of the reaction between X, Y and $Z \dots$ (in molar proportions $x:y:z \dots$) up to the same equilibrium point is $-Q_2$ cal per (x/a) mole of X. It can then be shown that with respect to one mole of component A

$$\Delta H_r^\circ = Q_1 - Q_2 \tag{96}$$

and $$\Delta G_r^\circ = -RT \ln \left\{ \frac{[Q_1/(Q_1-Q_2)]^{x+y+z \cdots} [X_0]^x [Y_0]^y [Z_0]^z \cdots}{[-Q_2/(Q_1-Q_2)]^{a+b+c \cdots} [A_0]^a [B_0]^b [C_0]^c \cdots} \right\} \tag{97}$$

where $[A_0]$, $[B_0]$, $[C_0] \dots$, $[X_0]$, $[Y_0]$, $[Z_0] \dots$ are the initial activities of components A, B, $C \dots$, X, Y, $Z \dots$ in the respective experiments. Thus the heat change, the Gibbs energy change, and the entropy change of a reversible reaction may be determined solely from measurements of heats of reaction and *initial* activities of reactants at a single temperature, without measurement of *equilibrium* activities. The application of equations (96) and (97) to actual reacting systems is dependent on the availability of a calorimeter that can yield meaningful results with (i) reactants at low concentration (so that concentration may be equated with activity), (ii) both endothermic and exothermic reactions, (iii) values of Q_1 and Q_2 that may differ by several orders of magnitude. For reversible reactions that proceed to equilibrium in solution within a few hours and have values of K between 10^{-3} and 10^3 at 25°C, the Benzinger microcalorimeter fulfils these requirements. The hydrolysis of the amide group in glutamine (59/5; 59/6) provides an example. For the reaction

Glutamine (c) $+ H_2O(l) \rightleftharpoons [\text{Glutamate}^{+ - -} + NH^+]$ (aqueous solution),

the values $\Delta H_r^\circ = -5 \cdot 16$ kcal and $\Delta G_r^\circ = -3 \cdot 42$ kcal were found.

3.4. Heats of reaction from experimental studies of chemical equilibria

The theoretical principles for deriving heats of reaction from equilibria were discussed in Sections 2.5.2 and 2.5.3. We recapitulate that the Second-Law method for obtaining ΔH_r° requires measurement of the temperature dependence of equilibrium constants, whereas the Third-Law method requires measurement of the equilibrium constant at one temperature only, given that data on the change in the entropy, or the change in the Gibbs energy function $[(G^\circ - H_{\text{ref}}^\circ)/T]$, are available for the reaction.

In experiments aimed at the accurate measurement of equilibrium constants it is important that true thermodynamic equilibrium should be established. This is often difficult to achieve in organic systems, because reaction rates are sometimes so slow that inordinate times are needed to reach an equilibrium state, and because side reactions may occur. Frequently success in achieving true equilibrium in an organic system is dependent on finding a catalyst of high specificity, which will speed up the rate of the desired reaction and minimize the occurrence of side reactions. The search for such a catalyst may involve an investigator in considerable labour, but fortunately some guidance on the most promising catalysts for study is available from chemical (especially petrochemical) technology. That true equilibrium has been established in a given system can be proved experimentally by showing that values of K_p are independent of the initial concentrations of the reacting species, and independent of time beyond that apparently needed for achieving equilibrium. In particular, it is important to prove that the value of K_p obtained in a "forward" reaction (i.e. with initial concentrations of reactants greater than their equilibrium concentrations) is identical with the value of K_p obtained in a "backward" reaction (i.e. with initial concentrations of reactants less than their equilibrium concentrations).

We now give some examples, which illustrate how heats of reaction are derived from equilibrium studies.

3.4.1. GAS-PHASE EQUILIBRIA

(a) *Static measurements*—The reversible reaction

$$CH_4(g) + I_2(g) \rightleftharpoons CH_3I(g) + HI(g)$$

was studied over a wide temperature range by Goy and Pritchard (65/5). Known quantities of methane and iodine were sealed in a 300-cm^3 borosilicate-glass flask fitted with a break-seal. The flask was heated in a furnace, the temperature of which was kept constant to ± 1 degC. The experiment was terminated by breaking the seal and distilling the reaction mixture into a cooled trap containing cyanogen iodide, which converted hydrogen iodide to hydrogen cyanide. The gaseous mixture of methyl iodide + hydrogen cyanide obtained by warming the trap was analysed by gas–liquid chromatography. In a preliminary series of experiments at 630°K, with concentrations of iodine and of methane each covering a 20-fold range, it was shown that K_p was independent of reactant concentration. Values of K_p were then measured over the temperature range 585–748°K. Both the Second-Law and the Third-Law methods were used to calculate ΔH_r°, as explained in Section 2.5.3. There is excellent agreement between the results of Goy and Pritchard's work and the results of a similar equilibrium study by Golden, Walsh, and Benson (65/11).

By determination of each of the products and from knowledge of the initial amounts of the reactants, Goy and Pritchard were able to check the stoicheiometry of the methane + iodine reaction. Some investigators, however, have relied on measurements of total pressure only in their studies of gas-phase equilibria in reactions in which $\Delta n_g \neq 0$, e.g. $AB(g) = A(g) + B(g)$. Such reliance involves the assumption that the reaction follows a certain stoicheiometric course, without by-product formation; a systematic error in K_p will result if the assumption is false.

(b) *Flow measurements*—Reversible reactions of the type

$$R_1.CO.R_2(g) + H_2(g) \rightleftharpoons R_1.CH(OH).R_2(g)$$

were studied by Buckley and Herington (65/6) for the systems in which $R_1 = R_2 = CH_3$; $R_1 = CH_3, R_2 = C_2H_5$; $R_1 = R_2 = C_2H_5$. Equilibrium was achieved by causing a gaseous mixture of a ketone + the corresponding secondary alcohol + excess hydrogen to flow through a catalyst-bed containing aluminium oxide, zinc, and copper, supported on copper turnings. Experiments were so designed that the feed mixture to the catalyst contained sometimes more, sometimes less, ketone than the equilibrium mixture; derived values of K_p were shown to be independent of the direction from which equilibrium was approached. The apparatus used is illustrated in Fig. 15.

By means of the electrolytic doser E the liquid mixture of ketone + alcohol stored in C was fed at a known rate into a flash-vaporizer, whence the vapour was carried by a stream of hydrogen from L into a glass catalyst-chamber M, surrounded by a thermostatically controlled (± 0.1 degC) metal block J. Samples of the gas mixture leaving the catalyst-chamber were isolated in sampling valve K and subsequently transferred to an adjacent gas–liquid chromatography column for analysis. It was shown that the reactions were effectively "frozen" in the absence of catalyst, so that fine control of the temperature of the sampling valve and of the exit lines from the catalyst-chamber was unnecessary; it was only necessary to keep the lines and valve hot enough to prevent condensation of the organic vapours.

A rigorous Second-Law method was used for treating the data. Values of $\Delta C_p(T)$ for the reactions were calculated from experimental heat-capacity data and expressed in the form of equation (19). Values of the constants ΔH^* and I in equations (45), (46), and (47) were derived by the method indicated in Section 2.5.2. It is thus possible to calculate reliable values of $\Delta H^\circ_{r,T}$, $\Delta G^\circ_{r,T}$, and $\Delta S^\circ_{r,T}$ for the gas-phase hydrogenation reactions studied, for values of T between, say, 270 and 600°K.

As an illustration of the precision of this experimental approach to the determination of thermodynamic quantities, we briefly discuss Buckley and Herington's results for the hydrogenation of acetone. From their data, $\Delta H^\circ_{r,\ 353^\circ K}$ is calculated to be $-13\cdot37\pm0\cdot10$ kcal, in close agreement with the value measured calorimetrically by Dolliver *et al.* (38/3), namely $-13\cdot41\pm0\cdot06$ kcal. Also from Buckley and Herington's data the value $\Delta S^\circ_r = +27\cdot22\pm0\cdot25$ cal deg^{-1} may be derived for $T = 298\cdot15^\circ K$. Coupled with values of $S^\circ(g)$ for hydrogen (50/1) and isopropyl alcohol (63/4), this value of ΔS°_r leads to $S^\circ(g)$ for acetone $= 70\cdot04\pm0\cdot35$ cal deg^{-1}, which is more precise than the value, $70\cdot6\pm1$ cal deg^{-1} determined by Kelley (29/2) from measurements of heat capacity.

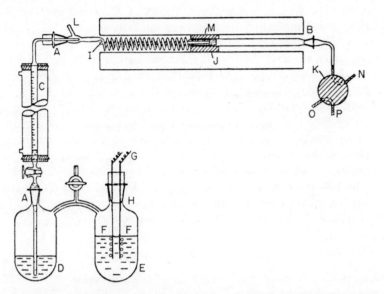

Fig. 15. Apparatus for study of catalysed gas-phase equilibria by a flow technique. A conical glass joints; B, conical glass joint with polytetrafluoroethylene sleeve; C, burette; D, vessel containing mercury; E, vessel containing potassium hydroxide solution; F, nickel electrodes; G, tungsten connectors; H, cemented joint; I, thermocouple pocket; J, thermostatically controlled metal block; K, gas-sampling valve; L, hydrogen inlet; M, catalyst; N, outlet to gas collection system; O, inlet for carrier gas; P, outlet to gas-liquid chromatography column.

3.4.2. GAS—CONDENSED-PHASE EQUILIBRIA

The Phase Rule indicates that when only one component of a reversibly reacting system is gaseous, the other components being condensed phases of negligible vapour pressure, the pressure at a given temperature is constant, the *dissociation pressure*, p_d. Hence, experimental studies of equilibria in systems of this type are similar to experimental studies of the saturation vapour pressures of single components, a subject that is discussed in Section 4.3. As an example of the measurement of dissociation pressures we consider Ivin's investigation (55/5) of the equilibrium

methyl methacrylate(g) $\rightleftharpoons$ poly(methyl methacrylate) (c).

The polymerization reaction was studied in the temperature range 96–142°C, and the approach to equilibrium was made from both sides, i.e. with initial gas pressures sometimes more, sometimes less, than the equilibrium dissociation pressures. Establishment of equilibrium was accelerated by ultraviolet irradiation of the reactants. Pressures in the range 44–760 torr were measured by means of a glass Bourdon gauge. Values of K_p were calculated on the assumption that methyl methacrylate vapour behaves as an ideal gas—i.e. equation (25), rather than the more rigorous equation (48), was employed. For the calculation of $\Delta H^\circ_{r, T_1}$, the Second-Law method was used in its zero-th approximation [$\Delta C_p = 0$; see Section 2.5.2, paragraph (a)]. For this purpose, a plot of $\log_{10} p_d$ against $1/T$ was made, and $\Delta H^\circ_{r, 392°K}$ was evaluated from equation (44) as $-2.303\ R$ times the slope; the value found was -21.4 ± 0.5 kcal. By introduction of the known heat of vaporization of methyl methacrylate, the heat of polymerization of the *liquid* monomer was calculated to be -13.4 ± 0.5 kcal. This latter heat quantity was measured calorimetrically by Ekegren *et al.* (50/4), and by Dainton *et al.* (60/2), who obtained the values -13.9 ± 0.3 kcal (at 350°K) and -13.8 ± 0.2 kcal (at 300°K), respectively. The values obtained by calorimetry are clearly compatible with that obtained by the equilibrium study.

3.4.3.   LIQUID-PHASE EQUILIBRIA

The calculation of equilibrium constants of reactions taking place in homogeneous liquid media requires knowledge of the equilibrium activities of the reacting species (equation (48), with "activity" written in place of "fugacity"). In practice, concentrations (mole fractions, x) can be used in lieu of activities when the reacting species are structurally similar organic liquids, since such liquids form nearly ideal solutions. Most reported liquid-phase equilibrium studies have related to isomerization or disproportionation reactions, so that the assumption of equivalence between activities and mole fractions is unlikely to have led to much error. The following example, taken from the work of Haraldson *et al.* (60/3), illustrates the application of the equilibrium method to elucidate the thermodynamics of a disproportionation reaction. The equilibrium

$R_1.SS.R_1$(solution) + $R_2.SS.R_2$(solution) $\rightleftharpoons$ $2R_1.SS.R_2$(solution)

was studied in hexane solution at 25 and 60°C. The establishment of equilibrium was catalysed both by addition of sodium hydroxide and by irradiation with a mercury lamp; results by the two techniques were identical. The solutions were analysed by gas–liquid chromatography, and it was established that no further disproportionation of the disulphides took place in the chromatographic column. The compounds studied were $CH_3.SS.CH_3$, $C_2H_5.SS.C_2H_5$, i-$C_3H_7.SS.i$-C_3H_7, t-$C_4H_9.SS.t$-C_4H_9. For the equilibrium

$$CH_3.SS.CH_3(\text{solution}) + C_2H_5.SS.C_2H_5(\text{solution})$$
$$\rightleftharpoons 2CH_3.SS.C_2H_5 \text{ (solution)}$$

K_x was found to be approximately 4, corresponding to a random distribution of alkyl groups, but for the equilibrium

$$CH_3.SS.CH_3(\text{solution}) + t\text{-}C_4H_9.SS.t\text{-}C_4H_9(\text{solution})$$
$$\rightleftharpoons 2CH_3.SS.t\text{-}C_4H_9 \text{ (solution)}$$

K_x was found to be approximately 24; this was interpreted in terms of steric hindrance in t-butyl disulphide. Haraldson and co-workers found that the equilibrium constants of all the systems they studied were independent of temperature in the range 25–60°C. This finding implies that ΔH_r° is zero for each system, so that disproportionation equilibria in the disulphide series are determined entirely by entropy considerations.

Note added in proof

The following values of $-\Delta U_{cert}/M$ for a batch of thermochemical-standard benzoic acid (National Bureau of Standards batch 39i) have recently been obtained: $26\,434 \cdot 4 \pm 3 \cdot 3\,\mathrm{J\,g^{-1}}$ (Churney and Armstrong, 68/20), $26\,432 \cdot 7 \pm 1 \cdot 6\,\mathrm{J\,g^{-1}}$ (Mosselman and Dekker, 69/3), and $26\,434 \cdot 4 \pm 1 \cdot 8\,\mathrm{J\,g^{-1}}$ (Head *et al.*, 69/4). The concordant results of these three sets of workers give added confidence in the selection of $26\,434\,\mathrm{J\,g^{-1}}$ as the most probable value of $-\Delta U_{cert}/M$.

Chapter 4

Methods of Determining Heats of Vaporization and Heats of Sublimation

4.1. The significance of heats of vaporization and sublimation in organic thermochemistry

The standard heat of formation of an organic compound in the solid or liquid state depends on both the chemical binding forces within the molecule and the forces between molecules. For discussion of the chemical binding forces alone, as for example in Chapter 7, it is necessary to remove the inter-molecular forces from consideration. This can be achieved by conversion of the value of ΔH_f° for a condensed state into the value for the hypothetical ideal-gas state, wherein the effect of intermolecular forces is zero; ΔH_f° for a substance whose reference state is the gas already relates to the ideal-gas state. The derivation of the heat of formation of an ideal gas from the value for a real gas was discussed in Section 2.2.2, paragraph (a); for a liquid the derivation of ΔH_f° (ideal gas) requires knowledge of the standard heat of vaporization of the liquid at 25°C, ΔH_v°, and for a solid it requires knowledge of the standard heat of sublimation of the solid at 25°C, ΔH_s°. Thus, from the first law of thermodynamics it is evident that

$$\Delta H_f^\circ(\text{ideal gas}) = \Delta H_f^\circ(\text{l}) + \Delta H_v^\circ \tag{98}$$

and

$$\Delta H_f^\circ(\text{ideal gas}) = \Delta H_f^\circ(\text{c}) + \Delta H_s^\circ. \tag{99}$$

Heats of vaporization and sublimation are also important in organic thermochemistry for practical calculations of the heats of gas-phase reactions at high temperatures, based on heats of condensed-phase reactions measured at 25°C. The present Chapter is devoted to a consideration of methods for measuring heats of vaporization and sublimation, or for estimating these quantities when experimental data are lacking.

4.2. Theoretical aspects of the determination of heats of vaporization and sublimation

4.2.1. DEFINITIONS OF STANDARD HEATS OF VAPORIZATION AND SUBLIMATION

The standard heat of vaporization at temperature T_1, $\Delta H^\circ_{v,\,T_1}$, is the heat change for the isothermal process:

$$\text{liquid (standard state)} \rightarrow \text{gas (standard state)}.$$

The standard heat of sublimation at temperature T_1, $\Delta H^\circ_{s,\,T_1}$, is the heat change for the isothermal process:

$$\text{crystalline solid (standard state)} \rightarrow \text{gas (standard state)}.$$

As elsewhere in this book the absence of an explicitly stated temperature as subscript to the symbols ΔH°_v or ΔH°_s implies that the assigned temperature is 298·15°K; other temperatures would be specified thus: $\Delta H^\circ_{v,\,400°K}$.

The standard state of a liquid at a specified temperature T_1 is that of the pure liquid under a pressure of 1 atm (Section 2.2.1), and the standard state of a gas is that of the ideal gas at temperature T_1 at 1 atm pressure, in which state the heat content is the same as that of the real gas at zero pressure (Section 2.2.1). The standard heat of vaporization at temperature T_1 can therefore be related to the heat of vaporization, measured at the saturation vapour pressure of the liquid, by the cycle:

liquid under 1 atm pressure $\xrightarrow{\quad \Delta H^\circ_{v,\,T_1} \quad}$ real gas at zero pressure
at $T_1°$K $\qquad\qquad\qquad\qquad\qquad\qquad$ at $T_1°$K

$\Big\downarrow \Delta H_{1,\,T_1}$ $\qquad\qquad\qquad\qquad\qquad\qquad$ $\Big\uparrow \Delta H_{3,\,T_1}$

liquid under its saturation $\xrightarrow{\quad \Delta H_{2,\,T_1} \quad}$ real gas at its saturation
vapour pressure $\mathfrak{p}$ at $T_1°$K $\qquad\qquad$ vapour pressure $\mathfrak{p}$ at $T_1°$K.

From the first law:

$$\Delta H^\circ_{v,\,T_1} = \Delta H_{1,\,T_1} + \Delta H_{2,\,T_1} + \Delta H_{3,\,T_1}. \tag{100}$$

Let us now consider the heat quantities, $\Delta H_{1,\,T_1}$, $\Delta H_{2,\,T_1}$, and $\Delta H_{3,\,T_1}$. $\Delta H_{2,\,T_1}$ is the heat of vaporization, $\Delta H_{v,\,T_1}$, as measured experimentally at $T_1°$K; it applies to the reversible, isothermal process of vaporization under the saturation vapour pressure. The quantities $\Delta H_{1,\,T_1}$ and $\Delta H_{3,\,T_1}$ may both be evaluated by integration of the appropriate form of equation (3):

$$\Delta H_{1,\,T_1} = \int_1^{\mathfrak{p}} \left[-T\left(\frac{\partial V}{\partial T}\right)_p + V \right] dp, \tag{101}$$

and

$$\Delta H_{3,\,T_1} = \int\limits_{\mathfrak{p}}^{0} \left[-T \left(\frac{\partial V}{\partial T} \right)_p + V \right] \mathrm{d}p, \tag{102}$$

for $T = T_1$.

In the remainder of this Section we restrict the discussion to the particular case of $T = 298 \cdot 15°\text{K}$, since this is the temperature to which heats of formation are customarily referred. For 1 mole of an organic liquid the value of the integrand in equation (101) is likely to be less than $100\,\text{cm}^3$ and may be considered independent of pressure. Clearly ΔH_1 will be largest for $\mathfrak{p} = 0$ (i.e. for an involatile liquid), and will then approach $100\,\text{cm}^3\,\text{atm} \approx 2\,\text{cal}$. Since a typical value of ΔH_2 is $10\,000\,\text{cal}$ (with an uncertainty in the range 5–200 cal), it is evident that ΔH_1 can safely be neglected in the evaluation of ΔH_v° from equation (100). For 1 mole of many organic vapours the value of the integrand in equation (102) will be in the range 10^3–$10^4\,\text{cm}^3$, approximately independent of pressure. The significance of ΔH_3 in comparison with the experimental error in the measurement of ΔH_2 will depend largely on the magnitude of $\mathfrak{p}$. As a rough rule, for substances having $\mathfrak{p} < 0 \cdot 1\,\text{atm}$ the values of ΔH_3 will be negligible, but for substances having $1\,\text{atm} > \mathfrak{p} > 0 \cdot 1$ atm a significant error may well be incurred if ΔH_3 in equation (100) is neglected and ΔH_v° is taken as equal to ΔH_2. Thus for n-pentane ($\mathfrak{p} = 0 \cdot 678\,\text{atm}$ at 25°C) $\Delta H_3 \approx 70\,\text{cal}$, and $\Delta H_2 = 6\,316 \pm 10\,\text{cal}$. Since, however, most organic liquids have $\mathfrak{p} < 0 \cdot 1\,\text{atm}$ at 25°C it is usually permissible to write $\Delta H_v^\circ = \Delta H_2 = \Delta H_v$.

Considerations similar to the above apply to sublimation processes, but for most organic solids at 25°C the sublimation pressure $\mathfrak{p}_s \ll 0 \cdot 1\,\text{atm}$, and the uncertainties in the measurement of the heat of sublimation are likely to exceed $100\,\text{cal mol}^{-1}$. The terms corresponding to ΔH_1 and ΔH_3 in the form of equation (100) appropriate for sublimation may therefore generally be neglected, and we may almost always write $\Delta H_s^\circ = \Delta H_2 = \Delta H_s$.

4.2.2. TEMPERATURE DEPENDENCE OF STANDARD HEATS OF VAPORIZATION AND SUBLIMATION

It often happens that an experimental value of $\Delta H_{v,\,T_1}^\circ$ is available for a temperature T_1, other than $298 \cdot 15°\text{K}$. For calculation of ΔH_v° we may write

$$\Delta H_v^\circ = \Delta H_{v,\,T_1}^\circ + \int\limits_{T_1}^{298.15} [C_p^\circ(\mathrm{g}) - C_p^\circ(\mathrm{l})]\,\mathrm{d}T . \tag{103}$$

In practice, the term $[C_p^\circ(\mathrm{g}) - C_p^\circ(\mathrm{l})]$ is sufficiently independent of temperature over, say, a 100-degC range to permit expression of equation (103) in the form

$$\Delta H_v^\circ = \Delta H_{v,\,T_1}^\circ + (298 \cdot 15 - T_1)\,[C_p^\circ(\text{g}) - C_p^\circ(\text{l})]\,. \tag{104}$$

Experimental values of both $C_p^\circ(\text{g})$ and $C_p^\circ(\text{l})$ are available for many of the commoner organic compounds, and for these the application of equation (104) presents no difficulties. For many of the less common compounds C_p° data are unavailable but it is often possible to estimate the heat capacities by group-contribution methods, as described by Benson and Buss (58/8) and in texts such as those of Janz (58/9) and Reid and Sherwood (66/12). Where the critical temperature T_c of a compound is known and $\Delta H_{v,\,T_{bp}}$ is known for the normal boiling point, Watson's (31/4) empirical equation (105) provides a useful means of calculating $\Delta H_{v,\,T_1}$:

$$\Delta H_{v,\,T_1}^\circ = \Delta H_{v,\,T_{bp}}^\circ \left[\frac{1 - (T_1/T_c)}{1 - (T_{bp}/T_c)} \right]^{0.38}. \tag{105}$$

Kharbanda (55/10) has published a nomogram for the application of equation (105). In circumstances where neither the group-contribution methods nor the Watson method can be applied, Sidgwick's rule may be employed, viz. $C_p^\circ(\text{g}) - C_p^\circ(\text{l}) \approx -13 \text{ cal deg}^{-1}$.

Equation (103) expresses the temperature dependence of heats of vaporization in a rigorous way. However, empirical ways of expressing this dependence [e.g. by equations such as (106)] are often adopted by investigators who measure heats of vaporization at several temperatures:

$$\Delta H_{v,\,T}^\circ = \alpha + \beta T + \gamma T^2\,. \tag{106}$$

Where such equations are available for a given liquid, ΔH_v° for $T = 298 \cdot 15^\circ \text{K}$ can of course be calculated without knowledge of heat capacities.

For the temperature dependence of heats of sublimation an equation analogous to equation (103) may be written:

$$\Delta H_s^\circ = \Delta H_{s,\,T_1}^\circ + \int_{T_1}^{298.15} [C_p^\circ(\text{g}) - C_p^\circ(\text{c})]\,\mathrm{d}T\,. \tag{107}$$

Unfortunately, values of $C_p^\circ(\text{g})$ for vapours of organic crystalline solids are rarely available. However, the integrand in equation (107) is generally smaller than that in equation (103), and since the experimental error in measurement of $\Delta H_{s,\,T_1}^\circ$ is often quite large, most investigators ignore the integral in equation (107) and write $\Delta H_s^\circ = \Delta H_{s,\,T_1}^\circ$.

4.2.3. PRINCIPLES OF METHODS FOR DETERMINING HEATS OF VAPORIZATION AND SUBLIMATION

The two methods most frequently used are the direct calorimetric method, and the indirect vapour-pressure method. The principles of these methods will now be explained; practical considerations will be discussed in Section 4.3.

(a) *Calorimetric measurement.*—The principle of the calorimetric method is straightforward: the amount of electrical energy needed to vaporize isothermally a measured quantity of liquid or solid is determined. Generally, it is possible to choose experimental conditions such that the vaporization takes place reversibly against a pressure equal to the saturation vapour pressure, and in this case the measured quantity is $\Delta H_{v,\,T_1}$, or $\Delta H_{s,\,T_1}$. Vaporization into a vacuum (*free evaporation*), which would yield $\Delta U_{v,\,T_1}$, or $\Delta U_{s,\,T_1}$, is hardly ever studied experimentally with organic substances. However, data have been reported by Morawetz and Sunner (63/15; 67/7) for conditions intermediate between those of reversible and free evaporation; these measurements are discussed further in Section 4.3.3.

(b) *Vapour-pressure measurement.*—Heats of vaporization or sublimation may be calculated from the Clapeyron equation in the forms appropriate for the respective states:

for liquids

$$\Delta H_{v,\,T_1} = T_1 \left(\frac{dp}{dT} \right) [V_g - V_l], \tag{108}$$

and for solids

$$\Delta H_{s,\,T_1} = T_1 \left(\frac{dp_s}{dT} \right) [V_g - V_{cr}]. \tag{109}$$

Hence knowledge of the temperature dependence of saturation vapour (or sublimation) pressures and of the molar volumes† of the coexisting phases is needed for the application of equations (108) and (109); the values of (dp/dT), V_g, V_l and V_{cr} to be used are those for the temperature T_1. V_l and V_{cr} are mostly small compared with V_g and can usually be estimated with sufficient accuracy from an experimental value for the density of the condensed phase at a temperature within, say, 50 degC of T_1. In the complete absence of liquid-density data, V_l can be estimated

---

† Subscript g here relates to the gas phase at the saturation vapour (or sublimation) pressure and subscripts l and cr respectively relate to the liquid and crystalline solid phases under this pressure.

by methods given by Reid and Sherwood (66/12). For the rigorous calculation of V_g, the equation of state of the vapour in the region of the saturation line must be known. For example, if values of the virial coefficients $B, C, D, \ldots$ are available for temperature T_1, then V_g can be calculated from equation (52). Values of the second virial coefficients B for organic vapours are steadily being accumulated (66/6); information on the higher virial coefficients $C, D, \ldots$ is at present sparse, but these coefficients contribute so little to V_g at pressures below 1 atm (except for a few highly associated vapours) that lack of values for them is unimportant in the present context. If no equation-of-state data are available for a given vapour, procedures for estimating V_g may be employed (66/12); for example, the Berthelot equation, (51), often finds favour for estimating V_g. Many authors, however, make no allowance for gas-law deviations, and assume that V_g can be calculated from the ideal-gas equation applied to the saturated vapour:

$$V_g = RT_1/p. \tag{110}$$

A further common assumption is that V_l and V_{cs} can be neglected in comparison with V_g, so that equations (108) and (109) can respectively be written as

$$\Delta H_{v, T_1} = RT_1^2 \left(\frac{d \ln p}{dT} \right)$$

$$= -R \left(\frac{d \ln p}{d (1/T)} \right), \tag{111}$$

and

$$\Delta H_{s, T_1} = RT_1^2 \left[\frac{d \ln p_s}{dT} \right)$$

$$= -R \left(\frac{d \ln p_s}{d (1/T)} \right). \tag{112}$$

These equations are forms of the Clausius–Clapeyron equation.

Equations (108), (109), (111), and (112) all contain differential coefficients whose evaluation requires knowledge of the dependence of p on temperature. Hence it is necessary to measure vapour (or sublimation) pressures over a temperature range and to fit the results to a vapour-pressure equation. The func-

TABLE 6. Some vapour-pressure equations, and derived differential coefficients

Equation No. (and common name)	Vapour-pressure equation	(dp/dT)	$-[d(\ln p)/d(1/T)]$
(113)	$\log_{10}p = A - (B/T)$	$\dfrac{2\cdot303\,Bp}{T^2}$	$2\cdot303\,B$
(114) ("Antoine")	$\log_{10}p$ $= A - B/(T+C-273\cdot15)$	$\dfrac{2\cdot303\,Bp}{(T+C-273\cdot15)^2}$	$\dfrac{2\cdot303\,BT^2}{(T+C-273\cdot15)^2}$
(115) ("Rankine")	$\log_{10}p = A - (B/T) - C\log_{10}T$	$p\left(\dfrac{2\cdot303\,B}{T^2} - \dfrac{C}{T}\right)$	$2\cdot303\,B - CT$
(116) ("Cragoe")	$\log_{10}p = A - (B/T) + DT + ET^2$	$2\cdot303p[(B/T^2) + D + 2ET]$	$2\cdot303(B + DT^2 + 2ET^3)$
(117) ("Martin")	$\log_{10}p = A - (B/T) + DT + ET^2$ $+ FT^3 - C\log_{10}(G-T)$	$2\cdot303p[(B/T^2) + D + 2ET + 3FT^2]$ $+ Cp/(G-T)$	$2\cdot303(B + DT^2 + 2ET^3 + 3FT^4) + CT^2/(G-T)$
(118) ("Cox")	$\log_{10}p = A[1-(T_{bp}/T)]$, where $\log_{10}A = a + bT + cT^2$, and T_{bp} is the temperature of the normal boiling point in °K.	$2\cdot303p\left[\dfrac{T_{bp}}{T^2} + \left(1-\dfrac{T_{bp}}{T}\right)(2\cdot303b + 4\cdot606cT)\right]$ $\times\exp[2\cdot303(a+bT+cT^2)]$	$2\cdot303\left[T_{bp} + \left(1-\dfrac{T_{bp}}{T}\right)(2\cdot303bT^2 + 4\cdot606cT^3)\right]$ $\times\exp[2\cdot303(a+bT+cT^2)]$

tional forms that have been proposed for vapour-pressure equations are legion: the more important ones are shown in Table 6, together with the required expressions for (dp/dT) and $[d(\ln p)/d(1/T)]$.

4.3. Experimental aspects of the determination of heats of vaporization and sublimation

Of the many published experimental methods for the measurement of heats of vaporization we describe some that are known to be reliable and reasonably precise. Techniques involving boiling, such as those described in Section 4.3.1, paragraphs (a) and (b), are applicable to liquids only, but the techniques described in Section 4.3.1, paragraph (c), and Section 4.3.3 have been applied to both liquids and solids.

4.3.1. ADIABATIC CALORIMETRY INVOLVING EVAPORATION AT A CONTROLLED PRESSURE

(a) Mathews and Fehlandt (31/3) described a simple calorimeter for measuring the heat of vaporization at the normal boiling point of a liquid, with a precision of *ca.* 0·3%. From one arm of a balance a glass vaporizer containing a 30-cm^3 sample of the liquid was suspended by a platinum wire in the vapour of the same liquid, boiling continuously under reflux. When the liquid sample in the vaporizer had reached its boiling point, a measured amount of electrical energy was supplied to a heater immersed in the sample; the quantity of liquid vaporized was determined from the loss in weight of the vaporizer + contents. Surrounding the vaporizer by vapour at the normal boiling point ensured that evaporation took place with little transfer of heat to the surroundings; any departure from adiabaticity was determined from the measured slow rate of weight change before and after the input of electrical energy. Results obtained by this method require correction to 25°C, by application of equation (104).

(b) Pitzer (41/4) described an experimental technique that afforded values of both $\Delta H_{v, T}$ and $C_p(g)$; vapour generated by boiling the liquid in a calorimeter was passed through a second calorimeter for measurement of $C_p(g)$, then condensed and returned at the boiling temperature to the first calorimeter. The method therefore involved continuous cycling of vapour. More accurate calorimeters working on this principle have been described by Waddington, Todd, and Huffman (47/6) and Hales, Cox, and Lees (63/5). Fig. 16 shows the apparatus of the last-mentioned workers. A flow of vapour was generated by dissipation of electrical energy in a heating coil E immersed in the liquid under examination;

the liquid was contained in the vacuum-jacketed glass vessel A, sub-merged in an oil-bath, which was thermostatically controlled at the temperature of the boiling liquid. The steadiness of boiling was monitored by thermocouples in re-entrant wells B and C, located to sense the

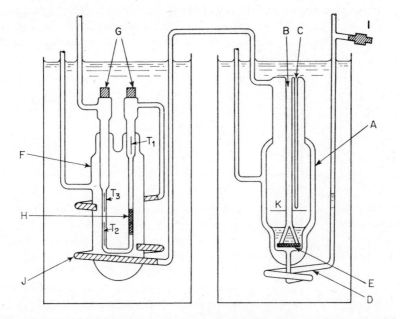

Fig. 16. Apparatus for measurement of heats of vaporization of liquids, and heat capacities of vapours. A, vacuum-jacketed boiler for measurement of heats of vaporization; B, well for heater leads and thermocouple; C, well for thermocouple; E, and H, electrical heaters; D, heat-exchanger for liquid returning to boiler; F, vacuum-jacketed calorimeter for measurement of vapour heat capacities; G, through-seals for heater and thermometer leads; I, screw-plug for emptying boiler; K, anti-splash baffle; J, heat-exchanger for vapour entering the calorimeter; T_1, T_2 and T_3, platinum resistance thermometers.

temperatures of the boiling liquid and the vapour respectively; any necessary adjustments to the boiling rate were made by alteration of the potential applied to the immersion heater E. After passing through a heat-capacity calorimeter, the vapour was condensed and normally returned to A *via* the heat-exchange coil D. However, for measurement of the vapour flow-rate, vapour could be diverted through a 3-way valve for a measured time, condensed, and collected in a weighed trap. During this diversion, liquid from a reservoir was run into the heat-exchange coil. The level of liquid in A was thereby maintained constant, and no correction was required for vapour needed to fill the liquid space. (*Cf.* paragraph (c) below).

Experiments were carried out at different boiling temperatures by control of the pressure of inert gas in the apparatus. The precision of measurement of $\Delta H_{v,\,T}$ by the recycling technique is *ca.* 0·1%.

(c) The heats of vaporization at 25°C of 59 hydrocarbons were determined with a precision of *ca.* 0·1% by Osborne and Ginnings (47/5) by means of the electrically heated calorimeter shown schematically in Fig. 17. The calorimeter C was of gold-plated brass; it had interior vanes to

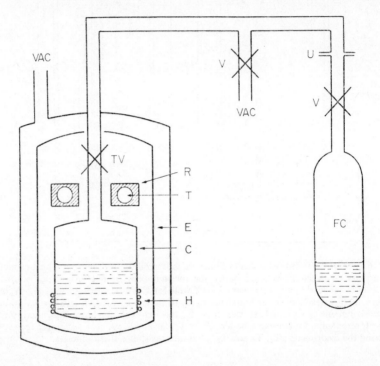

Fig. 17. Adiabatic calorimeter for measurement of heats of vaporization or sublimation by controlled evaporation. C, gold-plated brass calorimeter; E, adiabatic shield; FC, fluid container; H, electric heater; R, reference metal block; T, resistance thermometer; TV, throttle valve; U, union; V, valve.

promote thermal equilibrium, and silver gauze across the exit-tube from the calorimeter to prevent entrainment of liquid drops in the vapour leaving the calorimeter. The vessel was suspended in an evacuated chamber. Evaporation took place at reduced pressure, under conditions that were virtually adiabatic: thermocouples detected any difference in temperature between the calorimeter and an adiabatic shield E, and any out-of-balance was corrected by adjustment of either the throttle valve TV or the current in the calorimeter heater H. The number of moles n of vapour removed from the calorimeter by dissipation of a measured quantity of electrical energy ξ was determined by condensation of the vapour in the weighed trap FC. Although n moles of vapour, volume nV_g, were removed from the calorimeter, the vapour had to occupy the volume $n(V_g + V_l)$, not nV_g. Hence ξ exceeded $n\Delta H_v$ by an amount equal to the energy needed to vaporize sufficient liquid to give a vapour volume of nV_l. It then follows, by application of the Clapeyron equation, that

$$\Delta H_v = \frac{\xi}{n} - 298 \cdot 15 V_l \left(\frac{\mathrm{d}p}{\mathrm{d}T} \right). \tag{119}$$

4.3.2. ADIABATIC CALORIMETRY INVOLVING EVAPORATION INTO A GAS STREAM

Wadsö (66/13) has described a calorimeter for measuring heats of vaporization at 25°C of liquids having vapour pressures between 0·5 and 200 torr. The calorimeter is shown in Fig. 18. Evaporation of the liquid from compartment b of the calorimeter was caused by a flow of nitrogen at reduced pressure entering through g and leaving through f. The liquid (50–150 mg) was introduced into the calorimeter through tube i, and with tubes f, g, and i closed by plastic caps the calorimeter + contents were weighed. As vaporization proceeded, the resultant cooling was compensated by input of sufficient measured electrical energy into heater k to maintain constancy of tempera-

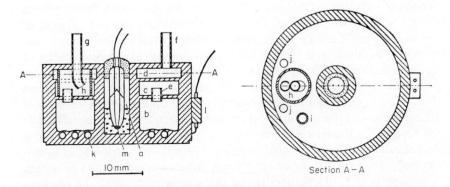

Fig. 18. Calorimeter for measurement of heats of vaporization into a gas stream. a, bore of annular silver calorimeter; b, c, d, compartments of calorimeter; e, silver connecting tube; f, gas-outlet tube; g, gas-inlet tube; h, back-suction trap; i, sample filling tube; j, holes; k, electric heater; l, heater connection-block; m, thermistor.

ture, determined with the aid of a thermistor m. The amount of liquid vaporized was determined from the loss of weight of the whole calorimeter. The calorimeter was supported in a can surrounded by a water bath kept at 25°C, with a temperature variation of less than 10^{-3} degC. Vaporization took place isothermally at 25°C and corrections for heat exchange with the surroundings were negligible. An earlier version of the Wadsö calorimeter is described in ref. 60/4, and a more recent modification in ref. 68/6.

Values of ΔH_v determined by Wadsö are in excellent agreement with values determined by Osborne and Ginnings by the method of Section 4.3.1 paragraph (c), as indicated in Table 7.

TABLE 7. Comparison of some values of ΔH_v (25°C).

Compound	ΔH_v, kcal	
	Wadsö (60/4; 66/13)	Osborne and Ginnings (47/5)
Water	10·52	10·51
Cyclohexane	7·89	7·90
Methylcyclohexane	8·42	8·45
n-Octane	9·94	9·92
1, 3, 5-Trimethylbenzene	11·36	11·35

4.3.3. ADIABATIC CALORIMETRY INVOLVING EVAPORATION AT VERY LOW PRESSURE

The vaporization calorimeters described so far all involve measurement of *heats* of vaporization under conditions whereby the vapour must perform volume work against an external pressure. In principle, the measurement of *energies* of vaporization is possible, involving evaporation against zero pressure. Morawetz and Sunner (63/15) have made calorimetric studies of the vaporization of liquids and the sublimation of solids against very low pressures, using a special form of Knudsen cell [see Section 4.3.6, paragraph (b)]. The cell had an internal volume of 1 cm³ and required *ca.* 0·3 g of sample. It was of silver, and contained a small electrical heater embedded in the base and a thermistor to measure the temperature; orifices of various diameters and thicknesses could be fitted. The cell was suspended in a continuously evacuated chamber (10^{-4} torr), which was submerged in a bath controlled at 25°C. As evaporation of the sample took place, the resultant cooling of the cell was compensated by dissipation of measured amounts of electrical energy in the heater. The number of moles vaporized, n, was calculated from the mass loss of the cell.

Morawetz (68/15) found that (i) when the mean free path λ of the vaporizing species was very small compared with the distance d between the surface of the sample and the orifice, the measured quantity approached $n\Delta H_v$; (ii) when $\lambda > d$, the measured quantity approached $n\Delta U_v = n(\Delta H_v - 0.59)$ kcal; (iii) when λ was somewhat less than d, the measured quantity approximated to $n(\Delta H_v - 0.23)$ kcal, corresponding to the condition of maximum attainable Mach number of vapour flow, before the onset of non-continuum flow [condition (ii)]. By selection of orifice sizes, it was possible to determine which of the three vaporization modes applied to a given experiment.

Morawetz considers that ΔH_v of substances having vapour pressures at 25°C in the range 200–0.01 torr can be determined by his technique to ± 0.04 kcal, and ΔH_v of substances having vapour pressures at 25°C in the range 10^{-2}–10^{-3} torr can be determined to ± 0.08 kcal.

4.3.4. MEASUREMENT OF VAPOUR-PRESSURE–TEMPERATURE RELATIONS IN THE PRESSURE RANGE ABOVE 10 TORR (~ 1 kN m^{-2})

Application of equation (108) or (111) to the determination of heats of vaporization requires the vapour-pressure–temperature relation for a liquid to be determined experimentally over a temperature range of at least 20 degC. Numerous experimental methods are available for this purpose, and readers are referred to reviews by Thomson (46/2; 59/8) and Milazzo (56/7). The experimental method most favoured by workers engaged in precise measurement of vapour pressure of organic liquids is the differential ebulliometric method, now to be described.

Reliable types of ebulliometer were developed by Swietoslawski (45/2); the variant used by Osborn and Douslin (66/5) is shown in Fig. 19. In the differential method, two ebulliometers of the type shown in Fig. 19 are employed: the liquid under examination is boiled under reflux in one ebulliometer and a liquid of accurately known vapour pressure (usually water, though benzene has also been used) is boiled under reflux in the other. The reflux condensers of the two ebulliometers are connected through cold-traps to a common, stable pressure of inert gas. The temperature of the boiling liquid in each ebulliometer is measured with a platinum resistance thermometer inserted in the re-entrant well E; the boiling action causes the liquid to be pumped over the outer surface of the well. The temperature of the condensing vapour in each ebulliometer is also measured, using a platinum resistance thermometer inserted in the re-entrant well D. With highly purified compounds that boil without significant decomposition, the difference in temperature between the boiling liquid and the condensing vapour is generally less than 0.005 degC. The rate of boiling is observed by means of the drop counter B, and it is normal practice to prove experimentally that the measured temperatures are substantially independent of

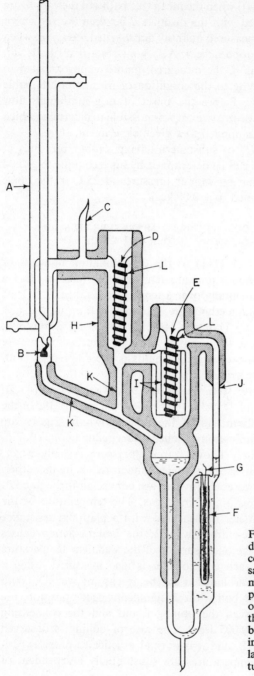

Fig. 19. Ebulliometer for precise determination of vapour pressure. A, condenser; B, drop counter; C, sample seal-off point; D, well for measurement of condensation temperature; E, well for measurement of boiling temperature; F, glass thread for promotion of steady boiling; G, electric heater; H, insulation; I, glass baffles; J, percolator tube; K, condensate return tubes; L, glass spirals.

boiling rate. The vapour pressure of the liquid at a given measured temperature is calculated from the measured boiling-temperature of the reference liquid, with the aid of very precise data for the vapour pressure of the latter, e.g. when water is the reference liquid the data of Osborne, Stimson, and Ginnings (39/3) are employed. The practical lower limit for use of water as a reference liquid has often been considered to be *ca.* 60 torr, because water tends to "bump" when boiled under lower pressures. However, by careful design of the boiling chamber of the ebulliometer, Ambrose (68/8) has shown that bump-free boiling of water may be achieved at pressures well below 60 torr, with *ca.* 15 torr as the lower limit.

Vapour pressures can be measured almost as precisely by static methods as by the ebulliometric method described above. Static measurements of vapour pressures in the range 10–2000 torr can be made by means of mercury manometers; manometric liquids other than mercury can seldom be used for accurate vapour-pressure measurements on organic compounds, because the vapour is likely to dissolve in the manometric liquid, unless the latter is identical with the substance under investigation, as in the *isoteniscope* method (63/55). The inclined-piston gauge of Douslin and Osborn (65/12) is a particularly precise instrument for measurement of vapour pressure below 30 torr. Other techniques employing pressure-sensitive mechanical devices (diaphragms, bellows, spirals, spoons), though they may be sensitive, are generally less reliable; these devices are best used as null-detectors, the balancing pressure of gas being measured by a static method (36/4).

Precise vapour-pressure data are generally fitted to an Antoine equation (equation (114), Table 6) if the data cover a temperature range of, say, 20–80 degC. If the data cover a temperature range much wider than 80 degC a better fit is likely to be given by equation (115), (116), (117), or (118). If the data cover a temperature range narrower than 20 degC, or if the data are approximate, equation (113) is the most convenient one to use. Equation (113) is obtained by integration of the Clausius–Clapeyron equation (111), with the assumption that $\Delta H_{v,T}$ is independent of temperature over the temperature range of the measurements. Hence use of equation (113) for representing vapour-pressure data implies not only that $\Delta H_{v,T}$ is a constant over a certain temperature range but also that the vapour behaves as an ideal gas and that the molar volume of the liquid is negligible. To some extent errors arising from failure of these assumptions cancel one another, so that equation (113) often provides a reasonably good fit of vapour-pressure data, even for substances such as the carboxylic acids, which are known to have very imperfect vapours. When equation (113) is used for the calculation of $\Delta H_{v,T}$, the value obtained should be assigned to the mid-point of the temperature range of the experimental data; correction of the value to that for 25°C may be made as described in Section 4.2.2.

4.3.5. MEASUREMENT OF VAPOUR-PRESSURE–TEMPERATURE RELATIONS FOR PRESSURES IN THE RANGE 10–10^{-2} TORR ($\sim 10^3$–1 N m^{-2})

Two of the assumptions underlying the use of the two-constant equation (113), namely that the saturated vapour behaves as an ideal gas and that the molar volume of the condensed phase is negligible compared with the vapour volume, become less erroneous the lower the saturation vapour (or sublimation) pressure. It follows that the error arising from application of equation (113) to vapour-pressure data in the range 10–10^{-2} torr will be smaller than the corresponding error for vapour pressures above 10 torr.

Some of the experimental methods referred to in Section 4.3.4, when suitably refined, are applicable also in the pressure range 10–10^{-2} torr. For example, Biddiscombe and Martin (58/10) used a static method to measure the saturation vapour pressures of phenols down to 10^{-2} torr: a thin corrugated silver diaphragm was used as a null-indicator, in conjunction with a sensitive displacement detector, and the balancing pressure of nitrogen was measured with a McLeod gauge. The following three experimental methods are specially applicable to the 10–10^{-2} torr pressure range.

(a) *The gas-saturation method.*—A tube containing the organic liquid or solid dispersed on glass fibres is submerged in a constant-temperature bath. Nitrogen is passed at a slow, constant rate through the tube and on emerging is analysed for its organic content; the total gas pressure in the tube and the flow rate are determined. The usefulness of the method depends on the availability of a sensitive analytical procedure for determining the amount of organic substance transported by the carrier gas. Thus for measurements on phenols, Biddiscombe and Martin (58/10) and Andon *et al.* (60/5) passed the saturated gas stream through bubblers containing sodium hydroxide solution, and later analysed the solutions for their phenolate contents by ultraviolet-absorption spectroscopy. The saturation vapour pressures were calculated with the assumption that Dalton's law of partial pressures applied to the saturated nitrogen stream. The fact that calculated values of p were independent of flow rate showed that saturation had been achieved.

(b) *The Rodebush method* (27/2).—A cell containing the substance under examination is closed by a quartz lid suspended from a quartz cantilever. Also suspended from the cantilever is a soft-iron armature, surrounded coaxially by a solenoid. The weight of the quartz lid is counterbalanced by the force of the vapour plus the magnetic force of the solenoid, the current in which is measured. Calibration of a Rodebush gauge may be effected either by means of nitrogen pressure, measured independently with a McLeod gauge, or by means of weights added to the lid, the effective area of which must then be measured. Unfortunately, it is difficult to avoid errors from sticking of the lid and magnetic hysteresis in the armature.

The method has been applied to organic compounds by Sears and Hopke (48/1), and by Balson (47/7) whose measurements on phenol are in agreement with those made by the silver-diaphragm and gas-saturation methods (58/10), referred to above.

(c) *The gas–liquid chromatographic method.*—Since minute amounts of organic vapours can be analysed quantitatively by gas–liquid chromatography, the method can in principle be used to determine the number of moles of saturated vapour in a known volume, and by application of equation (110) to calculate the vapour pressure. Mackle *et al.* (60/48; 64/45) and Geiseler and Jannasch (66/47) have shown that for determination of $\Delta H_{v,\,T}$ it is more convenient, and probably more accurate, to measure as a function of temperature a quantity that is *proportional* to vapour pressure, knowledge of the proportionality constant being unnecessary. Thus an amount of vapour, initially in equilibrium with a reservoir of liquid held at constant temperature, is isolated in a loop of constant volume, then swept by means of a carrier gas into a gas–liquid chromatograph. The area under the peak of the chromatogram displayed on a recorder chart is measured, and the experiment is repeated with the liquid reservoir at different measured temperatures. Clearly the peak-areas $a(T)$ will be directly proportional to the vapour pressures at the respective temperatures, provided that the detector+amplifier +recorder respond linearly to the concentration of substance in the carrier gas. By suitable choice of equipment (60/48) the necessary linear response can be obtained. If $\mathfrak{p}(T) = \kappa\,a(T)$, equation (113) may be written in the form

$$\log_{10} a(T) = A - \log_{10}\kappa - B/T$$

$$= A' - B/T. \tag{120}$$

Assuming that equation (113) suitably represents the vapour-pressure–temperature variation of the substance, a plot of $\log_{10} a(T)$ against $1/T$ will therefore be a straight line of slope $-B$. From equation (111) and the fourth column of Table 6, it follows that

$$\Delta H_{v,\,T} = 2\cdot303\,RB. \tag{121}$$

Mackle and McClean (64/45) and Geiseler and Jannasch (66/47) studied organic compounds of various types by the above technique and showed that, with one exception, the values of $\Delta H_{v,\,T}$ obtained agreed with the known values to within 0·3 kcal.

4.3.6. MEASUREMENT OF VAPOUR-PRESSURE—TEMPERATURE RELATIONS IN THE PRESSURE RANGE BELOW 10^{-2} TORR (~ 1 N m^{-2})

In this low-pressure range, equation (113) is entirely adequate for expressing the temperature dependence of vapour (or sublimation) pressure. Therefore $\Delta H_{v,T}$ (or $\Delta H_{s,T}$) can be deduced from equation (121), in which B is the negative slope of a linear plot of either $\log_{10} p$, or the logarithm of a quantity proportional to p [see Section 4.3.5, paragraph (c)], against $1/T$. Two of the most important experimental methods for the low-pressure range are now described.

(a) *The torsion–effusion method.*—The principle of this method, first described by Neumann and Völker (32/2), is illustrated in Fig. 20. The

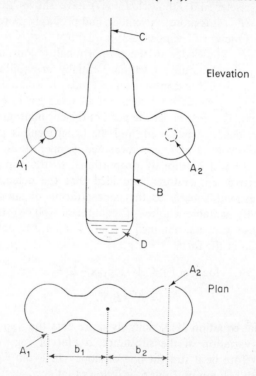

Fig. 20. Cell for measurement of low vapour pressures by the torsion-effusion method A$_1$ and A$_2$, effusion holes; B, effusion cell; C, torsion fibre; D, substance under investigation.

substance under study is contained in a small cell having two holes (areas a_1 and a_2) located at distances b_1 and b_2 from a vertical fibre, by which the cell is suspended in an evacuated chamber. The recoil force arising from vapour effusing through the holes causes the fibre to twist through an angle ϕ, and to be deflected from the vertical if the two recoil forces are unequal. From the kinetic theory of gases it can be deduced that

$$p = \frac{\tau \phi}{a_1 b_1} \text{ or } \frac{\tau \phi}{a_2 b_2}, \text{ whichever is the less;} \qquad (122)$$

here τ is the torsional constant of the fibre and can be determined from the period of oscillation of a couple consisting of metal tubes, in place of the effusion vessel. The magnitudes of a_1, a_2, b_1, and b_2 can be determined with the aid of a travelling microscope, and with τ known, the derivation of p at a particular temperature requires the measurement of the angle of twist ϕ at that temperature. Vapour pressures down to 5×10^{-6} torr can be measured by this technique. If the object of the measurements is to derive $\Delta H_{v, T}$ (or $\Delta H_{s, T}$), rather than p, then τ, a_1, a_2, b_1, and b_2 need not be known, so long as they remain constant for a series of measurements of ϕ at different temperatures; $\log_{10} \phi(T)$ is plotted against $1/T$, and $\Delta H_{v, T}$ (or $\Delta H_{s, T}$) is deduced from the slope of the linear plot.

For work of high accuracy it is necessary to introduce into equation (122) a correction for the finite depths of the effusion holes (55/9).

(b) *The Knudsen effusion method.*—This method is the one most extensively used for measuring vapour pressures below 10^{-2} torr. The substance is placed in a small cylindrical cell, contained in an evacuated chamber. In the lid of the cell there is a hole, area a_3, through which the vapour effuses. The rate of mass loss $-\dot{m}$ is often determined by continuous weighing of the cell, using a silica spring balance of typical sensitivity 10 cm g^{-1} (53/7). A more sensitive technique for determining $-\dot{m}$ depends on use of a radioactively labelled effusing species. Thus Carson, Stranks, and Wilmshurst (58/5) studied ^{203}Hg-labelled diphenylmercury, and Carson, Cooper, and Stranks (62/5) studied tritium-labelled tetraphenyllead and tetraphenyltin; the effusate was collected on a cooled target, the count-rate of which was subsequently measured.

Knudsen showed that if the depth of the effusion hole is negligible compared with its diameter, and if the latter is much less than the mean free path of the effusing molecules,

$$p = - \frac{\dot{m}}{a_3} (2\pi RT/M)^{\frac{1}{2}}. \qquad (123)$$

It is seen that derivation of p from equation (123) requires knowledge of the molecular weight M of the effusing vapour. In this respect the

Knudsen-effusion method differs from the torsion-effusion method, since equation (122) does not contain M. Bradley and Cleasby (53/8) derived the effective values of M for some hydrogen-bonded compounds in the vapour state, by an experimental technique combining the torsion and Knudsen methods. For the compounds studied M was found to correspond to the monomeric molecular weight, and it seems reasonable to assume that the predominant vapour species of most organic compounds at low pressures is the monomer.

A correction for the finite depth of the effusion hole, known as the Clausing correction (32/3), should be made to equation (123) in work of high accuracy.

When $-\dot{m}$ is determined by direct weighing, p is calculable in absolute units from equation (123). When the radioactive method is employed, and $\Delta H_{v,\,T}$ (or $\Delta H_{s,\,T}$) is the quantity sought, it suffices to permit effusion to proceed for a standard time at several temperatures, then to measure the count-rates of the target, $\dot{G}(T)$ and to derive $\Delta H_{v,\,T}$ (or $\Delta H_{s,\,T}$) from the slope of a plot of $\log_{10}(\dot{G}\sqrt{T})$ against $1/T$.

4.4. Estimation of the heats of vaporization of organic liquids

Since the calculation of $\Delta H_f^\circ(g)$ from a value of $\Delta H_f^\circ(l)$ requires knowledge of ΔH_v°, it is desirable that values of ΔH_v° should be available for every liquid for which ΔH_f° has been measured. In fact, experimental values of ΔH_v° are lacking for more than half of these liquids, so it is clearly important that means should be available for estimating ΔH_v° values. The more important methods of estimation are now described; they are intercompared in Section 4.4.4.

4.4.1. ESTIMATION OF $\Delta H_{v,\,T}$ FROM MEAGRE VAPOUR-PRESSURE DATA

Where the vapour pressure of a liquid has been measured over a reasonable temperature range, $\Delta H_{v,\,T}$ can be derived by the methods described in Section 4.2.3. paragraph (b), but where the vapour pressure has been measured over a narrow range of temperature, or at two very widely spaced temperatures, or at one temperature only (e.g. the normal boiling point), one of the following methods may be used to derive $\Delta H_{v,\,T}$.

(a) *Trouton's and related methods.*—From Trouton's observation that $\Delta H_{v,\,T_{bp}}/T_{bp}$ is approximately $0\cdot021$ kcal deg^{-1} for many liquids, the following simple rule is available for estimating $\Delta H_{v,\,T_{bp}}$ (kcal), solely from a measurement of T_{bp}:

$$\Delta H_{v,\,T_{bp}} = 0\cdot021\,T_{bp}. \tag{124}$$

The corresponding value of ΔH_v at 25°C can then be found by the methods given in Section 4.2.2.

It has long been known that the "constant" in equation (124) is merely an average for certain relatively low-boiling liquids, and that $\Delta H_{v,\,T_{bp}}/T_{bp}$ is greater than 0·021 kcal deg^{-1} for higher-boiling liquids and for associated liquids. Modified Trouton-type rules have been proposed by Pitzer, Guggenheim, Barclay and Butler, Everett (66/12, p.151), and Hildebrand (64/21), but these rules are more relevant to the discussion of liquid structure than to the estimation of $\Delta H_{v,\,T}$ from meagre data. For the latter purpose, the equations of Fishtine (63/16) and Wadsö (66/14) are more useful. For application of Fishtine's equation,

$$\Delta H_{v,\,T_{bp}} = 10^{-3}K'T_{bp}(8·75 + R\ln T_{bp})\text{ kcal},\qquad(125)$$

an experimental value of T_{bp} and the appropriate value of the parameter K' (66/12, p.152) are required. Wadsö's equations, which are improved versions of those proposed by Klages (49/4), give ΔH_v at 25°C directly in terms of T_{bp}: for slightly associated or non-associated liquids,

$$\Delta H_v = 5·0 + 0·041\,(T_{bp} - 273)\text{ kcal},\qquad(126)$$

and for alcohols,

$$\Delta H_v = 6·0 + 0·055\,(T_{bp} - 273)\text{ kcal}.\qquad(127)$$

(b) *Methods involving knowledge of critical constants.*—If experimental values of T_c and p_c are available in addition to T_{bp}, $\Delta H_{v,\,T_{bp}}$ may be estimated from one of the empirical equations (128), (129) and (130)†, known respectively as the Giacalone, Klein–Fishtine, and Riedel–Plank–Miller equations:

$$\Delta H_{v,\,T_{bp}} = \frac{2·303\,RT_{bp}T_c\log_{10}p_c}{T_c - T_{bp}}\ ;\qquad(128)$$

$$\Delta H_{v,\,T_{bp}} = \frac{2·303\,RT_{bp}T_cK''(\log_{10}p_c)}{T_c - T_{bp}}\,[1 - (T_c^3/p_cT_{bp}^3)]^{\frac{1}{2}},\qquad(129)$$

where $K'' = 1·045$ for liquids with $T_{bp} > 300°K$;

$$\Delta H_{v,\,T_{bp}} = 2·303\,RT_cG[1 - (0·97\,T_c/p_cT_{bp})]\,[1 + (T_{bp}/T_c)^2 + K'''(1 + 2T_{bp}/T_c)]\qquad(130)$$

---

† $\Delta H_{v,T_{bp}}$ from equations (128), (129), and (130) will be in calories if R is taken as 1·98717 cal deg^{-1} mol^{-1}; p_c in equations (128), (129), (130), (131), and (132) is in atmospheres.

where $G = 0.2471 + 0.1965\,a$

$$a = (T_{bp} \ln p_c)/(T_c - T_{bp})$$

$$K''' = (a/2.303\,G) - (1 + T_{bp}/T_c).$$

According to Reid and Sherwood (66/12), equation (128) gives an estimate of $\Delta H_{v,\,T_{bp}}$ with an average error of 2·8 % for 94 liquids for which $\Delta H_{v,\,T_{bp}}$ is already known; the corresponding average errors from use of the more complex equations (129) and (130) are 1·9% and 1·7% respectively, and an average error of 1·7% also results from use of the convenient equation due to Chen (65/13), which is derived from Pitzer's acentric-factor correlation (66/12). The Chen equation gives $\Delta H_{v,\,T_1}$ at any temperature at which the saturation vapour pressure, p_1 atm, is known:

$$\Delta H_{v,\,T_1} = \frac{T_1[7.9(T_1/T_c) - 7.82 - 7.11 \log_{10}(p_1/p_c)]}{1000(1.07 - T_1/T_c)} \quad \text{kcal.} \qquad (131)$$

Similar to, but slightly less accurate than, Chen's equation, and restricted to the estimation of $\Delta H_{v,\,T_{bp}}$, is Riedel's (54/4) equation:

$$\Delta H_{v,\,T_{bp}} = \frac{T_{bp}(5 \log_{10} p_c - 2.17)}{1000(0.93 - T_{bp}/T_c)} \quad \text{kcal.} \qquad (132)$$

4.4.2. ESTIMATION OF $\Delta H_{v,T}$ FROM PHYSICAL PROPERTIES OTHER THAN VAPOUR PRESSURE

Many attempts have been made to correlate heats of vaporization with other physical properties of fluids. The only attractive correlation for estimating $\Delta H_{v,\,T}$ is that involving the densities of coexisting liquids and vapours. The method was proposed by Bowden and Jones (48/2) and further developed by Wright (60/9). Wright defines a quantity [L], the *normal lyoparachor*, by the equation

$$[L] = M \Delta h_{v,T_{bp}}^{0.8}/(\rho_l - \rho_g), \qquad (133)$$

where $\Delta h_{v,\,T_{bp}}$ is the heat of vaporization (cal g^{-1}) of a liquid at its normal boiling point, M is its molecular weight, and ρ_l and ρ_g are respectively the densities (g cm^{-3}) of the liquid and vapour at the normal boiling point. [L] is said to be a function of molecular structure, and values for various structural contributions to [L] have been quoted by Wright (60/9). Application of equation (133) to the estimation of heats of vaporization therefore depends on the availability of the required structural contributions to $[L]$ and of a value for ρ_l; the value of ρ_g can always be estimated with sufficient accuracy from the ideal-gas equation.

4.4.3. ESTIMATION OF ΔH_v FROM MOLECULAR STRUCTURAL PARAMETERS

The value of ΔU_v (and hence of ΔH_v) for a given liquid depends on the magnitude of the intermolecular forces in the liquid. Since this magnitude depends directly on the shape, size, and polarity of the molecules, and indirectly on the number, type, and steric arrangements of atoms in the molecules, some correlation between ΔH_v and molecular structure may be expected. The following discussion is concerned with structural correlations that can be used for estimating values of ΔH_v at 298°K.

(a) *Homologous series.*—Examination of ΔH_v data for several homologous series shows that ΔH_v increases linearly with the number of carbon atoms, n, per molecule. This fact is demonstrated particularly well by data for hydrocarbons (47/5), alcohols, bromides (66/14), and thiols (60/10). For any one of these homologous series, the increment in ΔH_v per methylene group becomes effectively constant for $n \geqslant 2$. Moreover, the methylene increments for the four series are fairly similar, indicating that the methylene increment may be roughly constant, $1\cdot1_2$ kcal, for all homologous series. This finding offers a method for estimating ΔH_v for a substituted straight-chain alkane when an experimental value of ΔH_v for at least one member ($n \geqslant 2$) of the same homologous series is available.

(b) *Chain branching.*—Examination of values of ΔH_v for positional isomers shows that chain branching has a significant effect on ΔH_v. Methods of correlating the numbers and relative positions of chain branches with values of various physicochemical properties were developed by Wiener (47/9; 48/3) and Platt (47/10; 52/2). The Wiener–Platt correlation was later extended by Greenshields and Rossini (58/11) to a range of hydrocarbon properties including ΔH_v°. Their equation relates ΔH_v° for a normal alkane to ΔH_v° for an isomeric, branched-chain compound:

$$\Delta H_v^\circ \text{ (branched)} - \Delta H_v^\circ \text{ (normal)} = 0\cdot118C_3 - 0\cdot307C_4 + 0\cdot164\Delta P_3$$

$$+ 3\cdot081\Delta W/(n^2 - n) \text{ kcal.} \qquad (134)$$

The parameters have the following significance: C_3 and C_4 are respectively the numbers of tertiary and quaternary carbon atoms in the branched compound; $\Delta P_3 = P_3(\text{branched}) - P_3(\text{normal})$, where P_3 is the total number of pairs of carbon atoms three bonds apart; $\Delta W = W$ (branched) $- W$ (normal), where W, the Wiener number, is the total number of bonds between all pairs of carbon atoms.

As an example of the application of equation (134) we consider the three isomeric pentanes†:

(i) for $CH_3-CH_2-CH_2-CH_2-CH_3$, $C_3 = 0$, $C_4 = 0$, $P_3 = 2$, $W = 20$;

(ii) for $\begin{array}{c} CH_3 \\ \diagdown \\ CH_3 \diagup \end{array} CH-CH_2-CH_3$, $C_3 = 1$, $C_4 = 0$, $P_3 = 2$, $W = 18$;

(iii) for $\begin{array}{c} CH_3 \diagdown \quad \diagup CH_3 \\ C \\ CH_3 \diagup \quad \diagdown CH_3 \end{array}$, $C_3 = 0$, $C_4 = 1$, $P_3 = 0$, $W = 16$.

Hence

$$\Delta H_v^\circ(\text{ii}) - \Delta H_v^\circ(\text{i}) = (0 \cdot 118 \times 1) + (3 \cdot 081 \times -2/20) = -0 \cdot 19 \text{ kcal}$$

and

$$\Delta H_v^\circ(\text{iii}) - \Delta H_v^\circ(\text{i}) = (-0 \cdot 307 \times 1) + (0 \cdot 164 \times -2) + (3 \cdot 081 \times -4/20)$$
$$= -1 \cdot 25 \text{ kcal}.$$

The experimental values (53/1) are $-0 \cdot 42$ and $-1 \cdot 05$ kcal respectively (cf. paragraph (c) below).

It is not possible to test the Greenshields–Rossini method adequately with respect to the heats of vaporization of branched-chain *substituted* alkanes $C_nH_{2n+1}X$, because of insufficiency of experimental data. It seems likely, though, that equation (134) can be used for estimating ΔH_v° for a branched-chain compound in which the substituent X is bonded to a carbon atom two or more bonds away from the site of the chain branching.

Wadsö (66/14) has assembled some experimental data on ΔH_v of branched-chain substituted alkanes, as shown in Table 8. To a first approximation, the values $-0 \cdot 4_8$ and $-1 \cdot 3$ kcal can be used to calculate the change in ΔH_v for the respective isomerization processes specified in Table 8, whatever the nature of substituent X.

---

† A quick way to compute the Wiener number for any alkane is to multiply the number of carbon atoms on one side of a bond by the number on the other and to sum the products for all the carbon–carbon bonds.

TABLE 8.

Values for the change in ΔH_v at 25°C (kcal) arising from isomerization of an alkyl group

Isomerization process	$\Delta (\Delta H_v)$ for X =					
	CH$_3$	OH	SH	Br	CO$_2$CH$_3$	SCOCH$_3$
CH$_3$(CH$_2$)$_2$X→(CH$_3$)$_2$CHX	−0·39	−0·48	−0·57	−0·41	−0·45	−0·44
CH$_3$(CH$_2$)$_3$X→(CH$_3$)$_3$CX	−1·05	−1·36	−1·35	—	−1·33	−1·23

(c) *Structural contributions to* ΔH_v.—Laidler and Lovering (56/8; 60/10; 62/12) have proposed a scheme which relates ΔH_v to molecular structure for many types of aliphatic, alicyclic, and benzenoid compounds. Contributions to ΔH_v are allotted to the various types of bond, distinction being drawn between primary, secondary, tertiary and benzenoid C—H bonds; these bonds will be designated here by the symbols $(C—H)_p$, $(C—H)_s$, $(C—H)_t$ and $(C_b—H)$ respectively. The contributions to ΔH_v arising from a C—C bond in which both carbon atoms are sp^3 hybridized and from an external C—C bond in benzyl derivatives are both regarded as zero. However, the contributions allotted to benzene-ring bonds, here designated by the symbol $(C_b—C_b)$ are not zero. Laidler and Lovering's values for these and certain other structural contributions to ΔH_v are listed in Table 9; their values for bonds occurring in C$_5$ and C$_6$ cyclanes and in sulphur compounds may be found in refs. 62/12 and 60/10.

TABLE 9. Bond and group contributions to ΔH_v at 25°C.

Bond type	Contribution to ΔH_v, kcal	Group type	Contribution to ΔH_v, kcal
$(C—H)_p$	0·494	C—OH	6·76
$(C—H)_s$	0·579	C—CHO	5·21
$(C—H)_t$	0·518	C—CO—C	6·76
C$_b$—H	0·350	C—O—C	1·25
$(C_b—C_b)$	0·968	C—CO$_2$H	18·8
		C—ONO$_2$	6·44
		C—NO$_2$	7·04

It is of interest to compare the relative values of ΔH_v estimated by the Laidler–Lovering scheme for the three isomeric pentanes with those estimated by the Greenshields–Rossini scheme. Laidler and Lovering's scheme gives $\Delta H_v(ii) - \Delta H_v(i) = -0·32$ kcal and $\Delta H_v(iii) - \Delta H_v(i) = -0·51$ kcal. (*cf.* paragraph (b) above). Evidently neither the Laidler–Lovering scheme nor the Greenshields–Rossini scheme is particularly successful in evaluating the effect of chain-branching on ΔH_v of the pentanes.

4.4.4. COMPARISON OF METHODS FOR ESTIMATING ΔH_v

A dozen or so methods for estimating heats of vaporization have been presented in the three preceding Sections, and many more methods are to be found in the literature. The reader may well wish to know which of them is the most convenient and reliable, and with this in mind we have tested the methods on six organic compounds of differing structural types; in each case the experimental value of ΔH_v at 25°C was known to better than ± 0.1 kcal. The results of the test are shown in Table 10. In instances where the method gave a value of $\Delta H_{v,\,T_{bp}}$, the corresponding value of ΔH_v was calculated by application of equation (105).

In considering the relative accuracies of the estimation methods, as indicated by the figures in Table 10, it should be remembered that like is not necessarily being compared with like, since the respective methods require varying amounts of input data and afford values of heats of vaporization that apply to differing temperatures; where correction has been made

TABLE 10.

Comparison of estimated values of ΔH_v (at 25°C) with experimental values

Source† of ΔH_v	ΔH_v, kcal					
	n-heptane	2, 2-dimethyl-pentane	toluene	chloro-benzene	pentan-2-one	n-butanol
Experimental value	8·74	7·75	9·08	9·63	9·14	12·50
Trouton's eqn. (124)	8·9$_7$	8·2$_9$	9·3$_5$	9·9$_5$	9·0$_7$	9·7$_7$
Fishtine's eqn. (125)	8·7$_6$	7·9$_7$	9·1$_5$	10·1$_0$	9·4$_0$	12·5$_5$
Wadsö's eqns. (126), (127)	9·0$_3$	8·2$_5$	9·5$_3$	10·4$_0$	9·1$_9$	12·4$_7$
Giacalone's eqn. (128)	8·9$_7$	8·0$_4$	9·3$_1$	9·9$_5$	9·3$_3$	11·4$_1$
Klein–Fishtine's eqn. (129)	8·8$_2$	7·9$_0$	9·2$_8$	9·9$_5$	9·2$_9$	11·5$_1$
Riedel–Plank–Miller's eqn. (130)	8·7$_6$	7·8$_0$	9·2$_4$	9·8$_9$	9·2$_7$	11·6$_8$
Chen's eqn. (131)	8·8$_6$	7·8$_2$	9·3$_4$	10·2$_3$	9·3$_9$	12·6$_4$
Riedel's eqn. (132)	8·8$_1$	7·8$_3$	9·2$_9$	9·9$_5$	9·3$_4$	11·8$_9$
Wright's eqn. (133)	8·6$_5$	7·8$_3$	9·3$_6$	10·1$_8$	9·6$_7$	—
CH_2-incremental method‡	8·8$_0$	7·7$_9$	8·9$_8$	—	8·8$_9$	12·4$_9$
Laidler–Lovering method	8·7$_5$	8·2$_4$	9·0$_4$	—	12·0$_4$	11·7$_2$

† Necessary data (boiling points, critical parameters, vapour pressures, densities) were taken from reputable compilations.
‡ ΔH_v was calculated from the value for the next *higher* homologue.

to 25°C an extra uncertainty enters the calculations. Certain conclusions can nevertheless be drawn:

(i) Trouton's method is rather inaccurate.

(ii) The Fishtine method, and to a lesser extent the Wadsö method, are fairly accurate; both give good estimations of ΔH_v of n-butanol.

(iii) The methods dependent on knowledge of critical parameters (equations (128)–(132)) are fairly accurate, though all except Chen's method give low values for ΔH_v of n-butanol.

(iv) Wright's method gives good estimations for the two isomeric heptanes but poor estimations for three other compounds; it is not applicable to n-butanol, as a value for the – OH contribution to the normal lyoparachor is lacking.

(v) The methylene-incremental method is fairly accurate, but the low value for pentan-2-one should be noted; other data suggest that the methylene increment for ketones in general is appreciably lower than 1·1 kcal, the value used in the calculations.

(vi) The Laidler–Lovering method gives good estimations for the hydrocarbons, but poor estimations for the oxygen-containing compounds; revision of some of the parameters, in the light of more recent experimental data, might improve the method's usefulness.

Since Fishtine's method (equation (125)) is fairly reliable for non-polar, polar, and hydrogen-bonded compounds alike, and requires little input data, it is recommended as the preferred method for estimating $\Delta H_{v, T_{bp}}$, and thence ΔH_v. Somewhat less reliable, but equally undemanding of input data, is Wadsö's method (equations (126) and (127)), which gives values of ΔH_v directly.

4.5. Estimation of the heats of sublimation of organic solids

4.5.1. DIRECT ESTIMATION OF $\Delta H_{s,T}$

The intermolecular forces binding the molecules together in molecular crystals (i.e. non-ionic crystalline solids) are no different in kind from the forces binding the molecules together in organic liquids, but the molecules in a molecular crystal are much more ordered than those in a liquid. One would expect, therefore, that correlations between ΔH_s and the molecular structures of solid organic compounds would be rather imperfect unless factors related to the geometry of the crystal lattices were taken into account. As yet, no ΔH_s-structure correlation embracing both intramolecular and intermolecular structural factors has been proposed for organic crystalline compounds. The nearest approach is the correlation proposed by Bondi (63/17), who gave structural contributions to the heat of sublimation at the temperature, T_{tr}, of the lowest first-order (isothermal) solid-state transition†. Bondi justified his choice of T_{tr} as reference temperature by pointing out that the molecules in some organic crystalline solids have appreciable mobility (orientational and sometimes translational) between T_{tr} and the melting point T_m. Such solids have liquid-like characteristics in this temperature range and cannot fairly be compared with solids having more rigid crystalline structures up to T_m.

For n-alkanes, Bondi found a linear relationship between $\Delta H_{s, T_{tr}}$ and carbon number, the methylene-increment being 2·0 kcal. He showed that values of $\Delta H_{s, T_{tr}}$ for branched-chain alkanes are related to the values for the straight-chain isomers by a simple rule, and that values for alkyl derivatives RX are given by

$$\Delta H_{s, T_{tr}} (RX) = \Delta H_{s, T_{tr}} (R) + \Delta H_{s, T_{tr}} (X), \tag{135}$$

where the terms for the radicals R and X are given in extensive tabulations (63/17). Equation (107) may be used to convert values of $\Delta H_{s, T_{tr}}$ into values of ΔH_s at 25°C.

4.5.2. INDIRECT ESTIMATION OF $\Delta H_{s,T}$

Equation (136) provides an indirect route for deriving the heat of sublimation at the temperature of the melting point:

$$\Delta H_{s, T_m} = \Delta H_{v, T_m} + \Delta H_m. \tag{136}$$

If an experimental value of ΔH_m is available, but an experimental value of $\Delta H_{v, T_m}$ is not, an estimated value of the latter can be obtained by the methods described in Section 4.4 and combined with ΔH_m to yield $\Delta H_{s, T_m}$. If an experimental value of $\Delta H_{v, T_m}$ is available, but an experimental value of ΔH_m is not, an estimated value of the latter will again yield $\Delta H_{s, T_m}$. Unfortunately, methods of estimating ΔH_m are not well established, but since ΔH_m is commonly only *ca.* 20–30% of $\Delta H_{s, T_m}$ the accuracy required of a method for estimating ΔH_m is not high. Possibly the best estimation method is that of Bondi (63/17) who proposed the relation:

$$\Delta H_m = T_m (S'_{tr} + S'_m). \tag{137}$$

---

† For molecular crystals having *no* solid-state transitions, $T_{tr} = T_m$; for those having non-isothermal solid-state transitions, Bondi gives no guidance as to whether his method applies at the transition temperature or at the melting point.

Here T_m is the melting point and ΔH_m the heat of melting of a compound RX, and S'_{tr} and S'_m are the entropies of solid-state transition and melting respectively of a compound R.CH$_3$, known as the *homomorph* of compound RX. Values of ΔH_m compiled by Bondi (67/44) largely substantiate the validity of equation (137) for estimating values of ΔH_m from data for hydrocarbon homomorphs.

There is evidence that in some homologous series the methylene-increment in ΔH_m is approximately constant at 1·0 kcal. Combination of this value with the methylene-increment in ΔH_v, given in Section 4.4.3, paragraph (a), indicates that the methylene increment in ΔH_s should be *ca.* 2·1 kcal. (*Cf.* the value 2·0 kcal for hydrocarbons, quoted in Section 4.5.1.) This finding permits the establishment of linear relationships between carbon number and ΔH_s in various homologous series, as proposed by Davies and co-workers (59/37, 60/47, 61/55, 65/4) and Swain and co-workers (68/3).

Chapter 5

Thermochemical Data on Organic and Organometallic Compounds

The tabulated thermochemical data in this Chapter are based on a systematic search of the literature for the period 1930–1966, and a partial search of the literature for 1967–1968. Some data published in the period under review have, however, been deliberately excluded from the compilation, for example where the measurements were made by techniques now known to be very unreliable (e.g. static-bomb combustion measurements on organosilicon compounds), or where the experimental precision was low. Generally, results were excluded if the molar uncertainty intervals were more than 2 kcal for organic compounds or more than 5 kcal for organometallic compounds, although latitude was exercised in respect of a result of lower precision when the measurement was the only one available. Where authors have repeated their own earlier measurements, we have usually reported their later measurements only.

The tables in this Chapter contain not only the results of thermochemical measurements but also some indications of how the measurements were made. Of necessity, this information is conveyed in highly condensed form, and the abbreviations and conventions used are now explained. Readers are particularly advised to refer to Section 5.1, wherein the order of arrangement of compounds is explained.

5.1. Column 1 of the data tables: "Formula"

In this column is given the compound's molecular formula, which determines its position in the tables. The organic compounds (those containing C and one or more of the elements H, O, N, S, F, Cl, Br, I) are listed before the organometallic compounds, which in this context include organic derivatives of B, Si, P, Se and the metalloids, in addition to organic derivatives of the metals proper.

128

The organic compounds are listed in groups determined by the sequence C... H... O... N... S... Hal†, which causes the groups to be placed in the following order: CH, CHO, CHN, CHON, CHS, CHOS, CHNS, CHONS, CHHal, CHOHal, CHNHal, CHSHal, CHONHal, CHOSHal, CHONSHal. Except for the CH compounds, each group is divided into sub-groups, according to the number of atoms other than C and H that are present in the molecular formula. Thus we have the sub-groups C_aH_bO, $C_aH_bO_2$, $C_aH_bO_3$ etc, and these are listed sequentially. Within a given sub-group, compounds are listed in order of increasing number of carbon atoms; with a fixed number of carbon atoms the listing is in order of increasing number of hydrogen atoms, starting from zero, e.g. CO, CH_2O, CH_4O. The halogens are treated as a single entity, Hal, so far as group and sub-group sequences are concerned, but within a given table compounds of the individual halogens are listed in the order F... Cl... Br... I. For example, in the table of $C_aH_bO_cHal_j$ compounds, the order of arrangement for $a = 1$, $b = 0$, $c = 1$, $j = 2$ would be COF_2, COFCl, COFBr, COFI, $COCl_2$, COClBr, COClI, $COBr_2$, COBrI, COI_2.

We have made no elaborate rules for the order of listing of isomers. For hydrocarbons we have adopted the ordering scheme used by the American Petroleum Institute, and for other groups of isomers we have devised a simple scheme which brings together isomers of similar functional type, e.g. the carboxylic acids, the esters, the diols etc.

The organometallic compounds are listed according to the position of the metal in the Periodic Table: Group I... II... III... IV... V... VI... VII... VIII. Within a given Periodic-Table Group the order is that of increasing atomic weight. When more than one metal element is present, the compound is listed under the element which comes later in the Periodic Table. For a particular element the compounds are listed according to the rules given above for organic compounds.

5.2. Column 2 of the data tables: "g.f.w."

This column gives the gramme formula weights of the compounds, calculated from the 1961 Table of Relative Atomic Weights (62/1), based on $^{12}C = 12$ exactly. For all compounds but polymers the g.f.w. is the same as the molecular weight.

5.3. Column 3 of the data tables: "Name"

In naming compounds we have been guided by the IUPAC rules, though

---

† Abbreviation for the halogens.

for compounds of complex structure we have tended to use trivial names, when these are well known.

It is perhaps surprising that several precise thermochemical measurements have been reported for substances that were ambiguously named in the publication. We have attempted to make a positive identification of these substances from the quoted physical properties, but where this has not proved possible we have given the name used in the original publication, leaving the reader to draw his own conclusions.

5.4. Column 4 of the data tables: "Determination of ΔH_r°"

5.4.1. THE SUB-COLUMN HEADED "State"

This defines the physical state of the sample to which the quoted determination of ΔH_r° applies. The code used is as follows: c = crystalline solid; l = liquid; g = gas (see Sections 2.2.1 and 2.2.2). Occasionally other physical states are mentioned in the tables: am = amorphous solid; aq = aqueous solution at a very low concentration ("infinite dilution").

5.4.2. THE SUB-COLUMN HEADED "Purity, Mol %"

Where an investigator has made a convincing quantitative assessment of purity the result is given here: a figure with no symbols attached refers to a purity assessment by measurement of freezing-point depression (Section 2.6.3 paragraph (e)), whilst a figure + a lettered abbreviation refer to a purity assessment by some other technique. These same abbreviations are used to indicate where tests for purity were made, without leading to a precise numerical result: glc = gas–liquid chromatography; gsc = gas–solid chromatography; ir = infrared spectroscopy; uv = ultraviolet spectroscopy; ms = mass spectrometry; an = chemical analysis.

Occasionally a heat of reaction has been measured using samples from more than one source, or purified in more than one way; this is indicated by ... s, where the number preceding the s is the number of different samples examined.

5.4.3. THE SUB-COLUMN HEADED "Type"

The type of measurement used to determine ΔH_r° is described by the following code: SB = energy of combustion using a static-bomb calorimeter; FC = heat of combustion using a flame calorimeter; RB = energy of combustion using a rotating-bomb calorimeter; H = heat of hydrogenation; R = heat of reaction, other than combustion or hydrogenation; E = chemical equilibrium study.

Where appropriate, further information on the precise nature of the reaction affording ΔH_r° is given in the sub-column headed "Remarks".

5.4.4. THE SUB-COLUMN HEADED "No. of expts."

Here, the number of replicate experiments made to determine a given value of ΔH_r° is recorded.

5.4.5. THE SUB-COLUMN HEADED "Det'n. of react'n."

This sub-column indicates how the determination of the amount of chemical reaction in a measurement of ΔH_r° was carried out. In addition to the abbreviations defined in Section 5.4.2, the following abbreviations are used: m = determination from the mass(es) of reactant(s) taken; CO_2 = determination from the mass of carbon dioxide formed by combustion of the reactant; P/CO_2 = determination from the mass of carbon dioxide formed by combustion of the product; H_2O = determination from the mass of water formed by combustion of the reactant; H_2 = determination from the volume of hydrogen absorbed by the reactant; TP = determination [of equilibrium constants] from the measured total pressure.

5.4.6. THE SUB-COLUMN HEADED "$-\Delta H_r^\circ$, kcal/g.f.w."

This sub-column gives the standard heat change when a given substance in the physical state specified under "State" undergoes a reaction specified under "Type" and in some instances further specified under "Remarks". The numerical values are in kcal/g.f.w. (synonomous with kcal/mol for all compounds but polymers), and apply to reactions at 25°C with all reactants and products in their standard states. The minus sign in the title of the sub-column should be carefully noted—it implies that a positive numerical entry in the sub-column relates to an exothermic reaction, and a negative entry to an endothermic reaction. Often a numerical entry differs somewhat from that given in the original publication. In instances where the differences are not merely due to changes in units or atomic weights, reasons for the differences are given under "Remarks".

The $\pm$ terms attaching to values of $-\Delta H_r^\circ$ are uncertainty intervals, calculated wherever possible by the method described in Section 2.7. Sometimes, however, too little detail was found in an original publication to permit identification of the statistical significance of the $\pm$ values cited by the authors; in such cases we generally assumed that the authors' $\pm$ values *were* uncertainty intervals. Where we had grounds for suspecting that a published $\pm$ term was a significant underestimate of the *overall* error (e.g. where uncertainties relating to the determination of the amounts of reaction or to the heats of formation of some of the participants had apparently been ignored), we have indicated our suspicion by means of a star. A $\pm$term in square brackets [] represents our own guess of the likely uncertainty, made wherever a publication gave no numerical indication of precision.

5.4.7. THE SUB-COLUMN HEADED "Ref."

This sub-column gives the literature reference to the measurement of ΔH_r°. A reference number such as 33/12 refers to the twelfth entry for the year 1933 in the bibliography towards the end of the book.

5.4.8. THE SUB-COLUMN HEADED "Remarks"

The three principal functions of these remarks are (a) to give further information, where necessary, on the chemical reaction to which a given value of ΔH_r° refers, (b) to show why significant changes have been made to published values of ΔH_r° (and where these changes are due to use of different ancillary ΔH_f° data, the values used are given), (c) to give subsidiary information to help the reader make a value judgement of the data listed.

Certain ancillary ΔH_f° data that were extensively used in our calculations are not repeated under "Remarks". These values are as follows: $CO_2(g)$, $\Delta H_f^\circ = -94\cdot051$ kcal; $H_2O(l)$, $\Delta H_f^\circ = -68\cdot315$ kcal; $HNO_3(100\ H_2O)$ (l), $\Delta H_f^\circ = -49\cdot440$ kcal; $H_2SO_4(115\ H_2O)$ (l), $\Delta H_f^\circ = -212\cdot192$ kcal $\pm 0\cdot09$†; $HF(g)$, $\Delta H_f^\circ = -64\cdot8$ kcal; $HF(20\ H_2O)$ (l), $\Delta H_f^\circ = -76\cdot95$ kcal; $HCl(g)$, $\Delta H_f^\circ = -22\cdot062$ kcal; $HCl(600\ H_2O)$ (l), $\Delta H_f^\circ = -39\cdot823$ kcal; $Br_2(g)$, $\Delta H_f^\circ = +7\cdot387$ kcal; $HBr(g)$, $\Delta H_f^\circ = -8\cdot70$ kcal; $I_2(g)$, $\Delta H_f^\circ = +14\cdot923$ kcal; $HI(g)$, $\Delta H_f^\circ = +6\cdot33$ kcal; $H_3PO_4(c)$, $\Delta H_f^\circ = -305\cdot7$ kcal; $B_2O_3(c)$, $\Delta H_f^\circ = -304\cdot1 \pm 0\cdot4$ kcal; $B_2O_3(am)$, $\Delta H_f^\circ = -299\cdot7 \pm 0\cdot4$ kcal; $H_3BO_3(c)$, $\Delta H_f^\circ = -261\cdot47 \pm 0\cdot20$ kcal. With the exception of the values for HF $(20\ H_2O)$ (l), $B_2O_3(c)$, $B_2O_3(am)$, and $H_3BO_3(c)$, the above values were taken from the National Bureau of Standards Technical Notes 270–1 and 270–2 (65/1). The above value for $HF(20\ H_2O)$ (l) seems to us more plausible than that in ref. 65/1; the above values for the three boron compounds, which differ little from those in ref. 65/1, were taken from the work of Good and Månsson (66/4). Heats of formation of substances in solution at concentrations different from those specified above were calculated with the aid of heats of dilution from ref. 65/1.

Other values of ΔH_f° needed in the calculations were mostly drawn from

---

† The quoted uncertainty interval is our own selected value, based on the work described in refs. 60/36 and 63/10; our selected value of ΔH_f° is identical with that in ref. 65/1.

the National Bureau of Standards Circular 500 (50/1) or Technical Notes 270–1 and 270–2 (65/1). But where the values from these sources appeared to have been outdated by subsequent measurements, we have used values recommended by Skinner (64/5) and Holley (64/5). Where values of ΔH_f° are given in " Remarks", the values in kcal are placed in braces, { }, immediately after the relevant molecular formula in an equation† representing the chemical reaction used to determine ΔH_r°. An equation is written in full only when a non-standard reaction was used; the absence of a chemical equation under "Remarks" implies that the reaction studied was one of the standard reactions now to be described.

(a) *Combustion of* $C_aH_bO_cN_d$ *compounds.*—The values of ΔH_r° for combustion of compounds of this type, including those with c and/or $d = 0$, relate to the standard reaction:

$$C_aH_bO_cN_d + \left(\frac{4a+b-2c}{4}\right)O_2 \rightarrow a\,CO_2 + \frac{b}{2}H_2O(l) + \frac{d}{2}N_2. \qquad (138)$$

Values of the quantity C/An given under "Remarks" refer to the ratio [mass of CO_2 observed]/[mass of CO_2 expected] for the bomb combustion of a $C_aH_bO_cN_d$ compound, and values of the quantity C/H An refer to the ratio

$$\frac{[\text{mass of } CO_2 \text{ observed}]}{[\text{mass of } CO_2 \text{ expected}]} \Bigg/ \frac{[\text{mass of } H_2O \text{ observed}]}{[\text{mass of } H_2O \text{ expected}]}$$

for the flame combustion of a $C_aH_bO_cN_d$ compound. The closeness of a given value of C/An or C/H An to unity is an indication of the completeness of combustion.

(b) *Combustion of* $C_aH_bO_cN_dS_e$ *compounds.*—The standard reaction for combustion of compounds of this type, including those with c and/or $d = 0$, is:

$$C_aH_bO_cN_dS_e + \left(\frac{4a+b-2c+6e}{4}\right)O_2 + \left(116e - \frac{b}{2}\right)H_2O(l)$$

$$\rightarrow a\,CO_2 + \frac{d}{2}N_2 + e\,[H_2SO_4(115\,H_2O)](l). \qquad (139)$$

---

† It should be noted that the suffix (g) has been omitted from the symbols for hydrogen, oxygen, nitrogen and carbon dioxide in these equations, because these substances are always gaseous in thermochemistry.

(c) *Combustion of* $C_aH_bO_cN_dHal_f$ *compounds.*—The standard reaction for combustion of organochlorine compounds is:

$$C_aH_bO_cN_dCl_f + \left(\frac{4a+b-2c-f}{4}\right)O_2 + \left(\frac{1201f-b}{2}\right)H_2O(l)$$

$$\rightarrow a\,CO_2 + \frac{d}{2}N_2 + f[HCl(600\,H_2O)](l). \tag{140}$$

The dilution state $HCl\,(600\,H_2O)$ specified in equation (140) was adopted by Smith and co-workers in their summarizing paper (53/4); where a different dilution state has been used by other authors, the appropriate equation is written in full. A special comment is necessary concerning those measurements made by Smith *et al.* using the covered-crucible technique—it is now apparent that results obtained by this technique for compounds having normal boiling points less than *ca.* 180°C were unreliable, so these results were not included in the data tables.

As explained in Section 3.2.3, paragraph (b), the reaction corresponding to equation (140) takes place only if a solution containing a reductant is present in the bomb. The reductant (abbreviation, "red.") employed in particular combustion experiments on chlorine - or bromine-compounds is specified under "Remarks". The standard reaction for combustion of organobromine compounds is:

$$C_aH_bO_cN_dBr_g + \left(\frac{4a+b-2c}{4}\right)O_2 \rightarrow a\,CO_2 + \frac{b}{2}H_2O(l)$$

$$+ \frac{d}{2}N_2 + \frac{g}{2}Br_2(l). \tag{141}$$

This form of equation was used by Bjellerup, and has been adopted in this Chapter. It should be noted, however, that the equation does not correspond to the final state of a bomb-combustion experiment in which a reductant is placed initially in the bomb, since the bromine-containing product is in reality $HBr(nH_2O)$. Further comments concerning thermochemical corrections to combustion data for organochlorine and organobromine compounds are given below under the definition of the code-letter **J**.

The standard reaction for combustion of organo-iodine compounds is:

$$C_aH_bO_cN_dI_i + \left(\frac{4a+b-2c}{4}\right) O_2 \rightarrow a\, CO_2 + \frac{b}{2}\, H_2O(l)$$

$$+ \frac{d}{2}\, N_2 + \frac{i}{2}\, I_2(c). \qquad (142)$$

It is impracticable to write a standard equation for the combustion of organofluorine compounds, because the dilution state of the hydrofluoric acid formed and the amount of tetrafluoromethane produced both vary greatly from case to case. In correlating the thermal data for organofluorine compounds we adopted rounded values for the stoicheiometric coefficients of CF_4 in the equations representing the combustion reactions. Corrections to this rounded value of v were made using $\Delta U_r^\circ = \Delta H_r^\circ = -41\cdot9$ kcal for the reaction:

$$CF_4(g) + 82\, H_2O(l) \rightarrow CO_2 + 4[HF(20\, H_2O)](l). \qquad (143)$$

(d) *Gas-phase hydrogenation of unsaturated compounds.*—The standard reaction is:

Unsaturated reactant $(g) + n\, H_2 \rightarrow$ Saturated product(g).

The value of n is that needed to saturate *all* olefinic or acetylenic linkages in the reactant. Equations for hydrogenation reactions involving liquid phases, or for hydrogenation reactions of bonds other than olefinic or acetylenic, are written in full.

It has been mentioned already that many values of ΔH_r° in the data tables differ from the values originally published. Where the differences are very small ones (often arising from changes in the atomic-weight scale), no special comment is made. Code letters A–J are used to describe differences arising from other sources. The meanings of these code letters are as follows.

A indicates that we have corrected the calibration factor of a calorimeter, by use of a more accurate value for the heat of reaction of the calibrating substance.

B indicates that we have corrected weights in air, as given in the original publication, to weights in vacuum, with an assumed air density of $0\cdot0012$ g cm^{-3}.

C indicates that we have corrected an observed energy of combustion to the value for the process with all participants in their standard states. Some publications give insufficient information to permit rigorous reduction to standard states, and in those circumstances we have used the approximate equation (85) for $C_aH_bO_c$ compounds and a modified form of equation (85) for $C_aH_bO_cN_d$ compounds; the modification in question consisted of the addition of the term $1 \cdot 1d/2a$ to the terms within the square brackets of equation (85).

D indicates that we have corrected an error in the derivation of ΔH_r° from ΔU_r°. Instances have been found where the $\Delta n_g RT$ term (see equation (2)) had been applied with the wrong sign, or where Δn_g must have been wrongly calculated.

E indicates that we have corrected a value of ΔH_r° measured at a temperature other than 25°C to the value for 25°C. When the necessary heat-capacity data were known the correction was made by means of equation (9). Otherwise, the heat capacities of solid or liquid reactants were estimated by Kopp's Law, using the following values (cal $\deg^{-1}$) for atomic contributions to the molar heat capacity: C, $1 \cdot 8$; H, $2 \cdot 6$; O, $4 \cdot 0$; N, $6 \cdot 3$.

F indicates that we have applied a correction to literature values of ΔU_c° that were published in summary only, using a factor derived from fully published work on other compounds by the same author(s).

G indicates that we have recalculated the value of ΔH_c° for an organo-sulphur compound to the value corresponding to formation of sulphuric acid in the dilution state H_2SO_4 ($115\,H_2O$) (l). Necessary heats of dilution were taken from ref. 65/1.

H indicates that we have changed a published value of ΔH_r° as a result of correspondence with the original author(s).

I indicates that we have corrected an arithmetical error in a published value, in instances where the published experimental data made it patently clear that such a correction was necessary.

J indicates that we have used a new value for ΔH_f° of a reaction participant that is not explicitly stated to have been present. Thus, when only small amounts of CF_4 were formed in the combustion of an organofluorine compound, the equation given under "Remarks" will show *no* CF_4 among the products, yet a revised correction for formation of CF_4 was made. [See the text relating to equation (143)]. Also, the corrections for oxidation of As_2O_3 to As_2O_5 are not explicitly given for measurements of the energies of combustion of organochlorine and organobromine compounds in the presence of As_2O_3. The value $\Delta U_r^\circ = -76 \cdot 4\,\text{kcal}$ for the reaction $As_2O_3(c) + O_2 \rightarrow As_2O_5(c)$ was the basis of the corrections; it was derived from the work of Bjellerup, Sunner and Wadsö (57/47) and Sellers, Sunner and Wadsö (64/9), coupled with the value of ΔH_f° of HBr (nH_2O) from ref. 65/1. Again, values of

ΔH_c° for organobromine compounds contain a hidden correction for the oxidation of hydrobromic acid to elemental bromine; the value $\Delta U_r^\circ = -5\cdot11$ kcal (65/1) for the reaction

$$HBr(300\ H_2O)\ (l) + 0\cdot25\ O_2 \rightarrow 300\cdot5\ H_2O(l) + 0\cdot5\ Br_2(l)$$

was employed.

5.5. Column 5 of the data tables: "ΔH_f° (l or c), kcal/g.f.w."

This column gives values of ΔH_f° at 25°C, derived by use of the appropriate form of equation (6), for compounds whose reference state at 25°C is either the liquid or the crystalline solid. In instances when ΔH_r° was *measured* for a liquid or crystalline state, as indicated by the symbol l or c in the sub-column headed "State", the $\Delta H_f^\circ(l$ or c) value is derived directly from ΔH_r°, but in instances when ΔH_r° was measured for the gas state a value of $\Delta H_f^\circ(g)$ is placed in column 7 and any entry in column 5 will have been derived from that in column 7 with the aid of a value of ΔH_v° or ΔH_s°.

The sources of ancillary ΔH_f° data needed for the calculation of $\Delta H_f^\circ(l$ or c) have already been indicated in Section 5.4.8. Often, the uncertainty intervals of these ancillary data are unknown, but where they are known they have been combined with the uncertainty intervals of ΔH_r° to afford the overall uncertainty intervals of ΔH_f°, given as $\pm$ terms in column 5; where the $\pm$ terms of ΔH_r° and ΔH_f° are shown as identical, the error contribution from the ancillary data is either negligible or unknown.

As stated in Section 2.2.4, a value of ΔH_f° relates to the formation of a compound from its constituent elements, both compound and elements being in standard states at 25°C. Though most authors adopt standard *reference* states of the elements at 25°C as the basis for their calculations, a few do not. To avoid doubt on this score, there follows a list of the reference states of the elements at 25°C that were used in the calculations in the data tables:

C, c (graphite); H_2, g; O_2, g; N_2, g; S, c (rhombic); F_2, g; Cl_2, g; Br_2, l; I_2, c; Li, c; Mg, c; Zn, c; Cd, c (α); Hg, l; B, c (β); Al, c; Ga, c; Tl, c; Si. c; Ge, c; Sn, c (I, white); Pb, c; V, c; P, c (α, white); As, c (α, grey metallic); Sb, c (III); Bi, c; Cr, c; Mo, c; W, c; Se, c (hexagonal, black); Mn, c (IV); Fe, c; Co, c; Ni, c.

Where more than one value of $\Delta H_f^\circ(l$ or c) is given in the data tables, we have offered a *selected value*. The selection was made on experimental evidence only, without regard to any speculation as to what a given value "ought to be". Sometimes we have selected a single value as being the most reliable of a number of values; at other times we have taken a mean, using weighting factors derived from the respective uncertainty intervals.

5.6. Column 6 of the data tables: "Determination of ΔH_v"

5.6.1. THE SUB-COLUMN HEADED "Type"

The symbols in this sub-column define the type of measurement or estimation procedure that was used to determine ΔH_v° at 25°C, as explained below.

(a) *Determination of ΔH_v by calorimetry.*—C1 indicates that calorimetric measurements were made at 25°C, or at another temperature with conversion of the results to 25°C by means of experimental data using equation (103) or (106). C2 indicates that calorimetric measurements were made at a temperature other than 25°C, with conversion of the results to 25°C by means of estimated heat-capacity data [equation (103)] or Watson's procedure [equation (105)].

(b) *Determination of ΔH_v from vapour-pressure–temperature measurements.*— V1 indicates that accurately measured vapour-pressure–temperature data were fitted to the Antoine equation, (114), or to equations such as (115)–(118); ΔH_v was then calculated from the Clapeyron equation, (108), with due allowance for vapour imperfection and the molar volume of the liquid.

V2 indicates that the same procedure was adopted as in V1 except that equation (111) was used for the calculation of ΔH_v, i.e. vapour imperfection and the molar volume of the liquid were ignored.

V3 indicates that measured vapour-pressure–temperature data, with 25°C within the range of the measurements, were fitted to equation (113), and ΔH_v was calculated from equation (111).

V4 indicates that the same procedure as in V3 was applied to experimental measurements for a temperature range that did not include 25°C, and the resulting value of ΔH_v was converted to 25°C.

V5 indicates that the same procedure as in V3 was applied to a published vapour pressure equation, based on measurements for an unstated temperature range.

A similar coding, but with prefix S instead of V, is used to indicate values of ΔH_s that were derived from sublimation-pressure–temperature data for solids. Values of ΔH_s that were obtained by adding ΔH_v of the liquid to ΔH_m of the solid are distinguished by the subscript m where the value of ΔH_m was experimental, and by the subscript me where the value of ΔH_m was estimated.

(c) *Determination of ΔH or ΔH_s by an estimation procedure.*— E1 indicates that ΔH_v was estimated from the values for compounds of similar molecular structure, e.g. by use of homologous-series increments or Wadsö's chain-branching parameters, as described in Section 4.4.3 paragraphs (a) and (b).

E2 indicates that ΔH_v of a branched-chain compound was estimated by the Greenshields–Rossini method, using a reliable value of ΔH_v for a straight-chain isomer. [Section 4.4.3 paragraph (b)].

E3 indicates that the Greenshields–Rossini method was applied outside the stated range of its applicability.

E4 indicates that an estimated value of ΔH_v or ΔH_s was given in the reference cited, with no statement of the method of estimation.

E5 indicates that ΔH_v was estimated by Fishtine's method (equation (125)).

E6 indicates that ΔH_v was estimated by the Laidler–Lovering method. (Section 4.4.3 paragraph (c)).

E7 indicates that ΔH_v was estimated by Wadsö's method. (equations (126) and (127)).

E8 indicates that ΔH_v was estimated by Trouton's rule (Section 4.4.1 paragraph (a)), using ΔH_v of a compound of similar molecular structure to establish the "constant" in equation (124).

5.6.2. THE SUB-COLUMN HEADED "ΔH_v°, kcal/g.f.w."

This sub-column lists standard heats of vaporization at 25°C, with their uncertainty intervals; a value marked (sub.) is that for the standard heat of sublimation at 25°C.

No attempt has been made to list all available experimental values—only the most convincing value for a given compound is listed, or occasionally two or more values of approximately equal quality.

5.6.3. THE SUB-COLUMN HEADED "Ref."

This sub-column gives the reference number in the bibliography at the end of the book for the experimental data from which ΔH_v° or ΔH_s° was derived.

5.7. Column 7 of the data tables: "$\Delta H_f^\circ(g)$ kcal/g.f.w."

As explained in Section 5.5, an entry in column 7 for a compound studied thermochemically in the gas state was derived directly from the value of ΔH_c°. For compounds studied thermochemically in the liquid or the crystalline state the value of $\Delta H_f^\circ(g)$ in column 7 was derived by application of either equation (98) or equation (99), whichever was appropriate.

THERMOCHEMISTRY OF ORGANIC AND ORGANOMETALLIC COMPOUNDS

C_aH_b 1 kcal = 4·184 kJ

| | | | | | | No. | Detn. | | |
1 Formula	2 g.f.w.	3 Name	State	Purity mol %	Type	of expts.	of reactn.	$-\Delta H_c^\circ$ kcal/g.f.w.	Re
C	12·01115	Graphite	c	5s	SB	20	CO_2	94·050±0·024	38,
C		Graphite	c	4s	SB	24	CO_2	94·037±0·013	38,
C		Graphite	c	2s	SB	17	CO_2	94·065±0·013	44,
C		Graphite	c	3s	SB	52	m	94·040±0·012	66,
C		Graphite	c					**94·051±0·011**	44, 65,
C	12·01115	Diamond	c		SB	15	CO_2	94·505±0·023	38,
C		Diamond	c		SB	5	m	94·477±0·030	66,
C		Diamond	c						
CH_4	16·0430	Methane	g		FC	5	H_2O	212·79±0·06	31/
CH_4		Methane	g		SB		m	213·14±0·26*	32,
CH_4		Methane							
C_2H_2	26·0382	Acetylene	g		H	3	P/CO_2	74·58±0·15	39,
C_2H_4	28·0542	Ethylene	g		FC	8	H_2O	337·28±0·07	37,
C_2H_4		Ethylene	g		H	16	P/CO_2	32·58±0·05	35,
C_2H_4		Ethylene	g		E		an	32·6 ±1·2	51,
C_2H_4		Ethylene							
C_2H_6	30·0701	Ethane	g	>99·8	FC	7	H_2O	372·81±0·11	34,
C_3H_4	40·0653	Propyne	g		H	3	P/CO_2	69·22±0·15	39,
C_3H_4	40·0653	Allene	g		H	5	P/CO_2	70·54±0·25	36,
C_3H_4		Allene	g		E		ir	0·90±0·50	59/
C_3H_4		Allene							
C_3H_4	40·0653	Cyclopropene	g		FC			485·0 ±0·6	62/
C_3H_6	42·0813	Propene	g		FC	7	H_2O	491·81±0·15	37
C_3H_6		Propene	g		H	7	P/CO_2	29·87±0·10	35,
C_3H_6		Propene	g		E		an	29·5 ±1·2	51,
C_3H_6		Propene	g		FC	3	CO_2	491·83±0·27	68,
C_3H_6		Propene							
C_3H_6	42·0813	Cyclopropane	g	99·96±0·12	FC	14	H_2O	499·83±0·13	49,
C_3H_8	44·0972	Propane	g	>99·8	FC	8	H_2O	530·58±0·13	34

1 kcal$=4\cdot184$ kJ

ΔH_r° / Remarks	5 ΔH_f° (l or c) kcal/g.f.w.	6 Determination of ΔH_v			7 ΔH_f° (g) kcal/g.f.w.
		Type	ΔH_v° kcal/g.f.w.	Ref.	
Selected value					
	$+0\cdot454\pm0\cdot027$				
	$+0\cdot426\pm0\cdot034$				
Selected value	$+0\cdot444\pm0\cdot017$				
					$-17\cdot89\pm0\cdot07$
ss of CH_4 from PVT measurements		E			$-17\cdot54\pm0\cdot26*$
Selected value					$-17\cdot89\pm0\cdot07$
H_6(g) $\{-20\cdot24\pm0\cdot12\}$		E			$+54\cdot34\pm0\cdot19$
					$+12\cdot55\pm0\cdot08$
H_6(g) $\{-20\cdot24\pm0\cdot12\}$		E			$+12\cdot34\pm0\cdot13$
H_4(g)$+H_2 \rightleftharpoons C_2H_6$(g) $\{-20\cdot24\pm0\cdot12\}$ 2nd Law					$+12\cdot4$ $\pm1\cdot3$
Selected value					$+12\cdot45\pm0\cdot10$
H An $0\cdot99968\pm0\cdot00018$					$-20\cdot24\pm0\cdot12$
H_8(g) $\{-24\cdot83\pm0\cdot14\}$		E			$+44\cdot39\pm0\cdot21$
H_8(g) $\{-24\cdot83\pm0\cdot14\}$		E			$+45\cdot71\pm0\cdot29$
$I_3C\equiv CH$(g) $\{+44\cdot39\pm0\cdot21\}$ $\rightleftharpoons CH_2=C=CH_2$(g). 2nd Law		E			$+45\cdot29\pm0\cdot60$
Selected value					$+45\cdot63\pm0\cdot27$
					$+66\cdot2$ $\pm0\cdot6$
H An $0\cdot99997\pm0\cdot00018$					$+4\cdot71\pm0\cdot16$
H_8(g) $\{-24\cdot83\pm0\cdot14\}$		E			$+5\cdot04\pm0\cdot16$
H_6(g)$+H_2 \rightleftharpoons C_3H_8$(g) $\{-24\cdot83\pm0\cdot14\}$ 2nd Law					$+4\cdot7$ $\pm1\cdot3$
					$+4\cdot73\pm0\cdot28$
Selected value					$+4\cdot88\pm0\cdot16$
H An $0\cdot99989\pm0\cdot00006$					$+12\cdot73\pm0\cdot14$
H An $0\cdot99992\pm0\cdot00019$					$-24\cdot83\pm0\cdot14$

THERMOCHEMISTRY OF ORGANIC AND ORGANOMETALLIC COMPOUNDS

C_aH_b

1 kcal $= 4 \cdot 184$ kJ

Determinati

1 Formula	2 g.f.w.	3 Name	State	Purity mol %	Type	No. of expts.	Detn. of reactn.	$-\Delta H_r^\circ$ kcal/g.f.w.	Re
C_4H_6	$54 \cdot 0924$	1-Butyne	g	$99 \cdot 88 \pm 0 \cdot 07$	FC	6	CO_2	$620 \cdot 64 \pm 0 \cdot 20$	51/
C_4H_6	$54 \cdot 0924$	2-Butyne	g	$99 \cdot 96 \pm 0 \cdot 04$	FC	5	CO_2	$615 \cdot 84 \pm 0 \cdot 23$	51/
C_4H_6		2-Butyne	g		H	3	P/CO_2	$65 \cdot 10 \pm 0 \cdot 30$	39/
C_4H_6		2-Butyne							
C_4H_6	$54 \cdot 0924$	1, 2-Butadiene	g	$99 \cdot 94 \pm 0 \cdot 05$	FC	6	CO_2	$619 \cdot 93 \pm 0 \cdot 13$	49/ 51/
C_4H_6	$54 \cdot 0924$	1, 3-Butadiene	g	$99 \cdot 83 \pm 0 \cdot 06$	FC	8	CO_2	$607 \cdot 16 \pm 0 \cdot 18$	49/ 51/
C_4H_6		1, 3-Butadiene	g		H	5	P/CO_2	$56 \cdot 57 \pm 0 \cdot 10$	36/
C_4H_6		1, 3-Butadiene							
C_4H_6	$54 \cdot 0924$	Cyclobutene	g		FC	3	CO_2	$618 \cdot 60 \pm 0 \cdot 36$	68/
C_4H_6	$54 \cdot 0924$	Methylenecyclopropane	g		FC	3	CO_2	$629 \cdot 07 \pm 0 \cdot 43$	68/
C_4H_6	$54 \cdot 0924$	1-Methylcycloprop-1-ene	g		FC	3	CO_2	$639 \cdot 36 \pm 0 \cdot 27$	68/
C_4H_6	$54 \cdot 0924$	Bicyclobutane	g		FC	2	CO_2	$633 \cdot 05 \pm 0 \cdot 19$	68/
C_4H_8	$56 \cdot 1084$	1-Butene	g	$99 \cdot 88$	FC	6	CO_2	$649 \cdot 33 \pm 0 \cdot 18$	51/
C_4H_8		1-Butene	g		H	9	P/CO_2	$30 \cdot 10 \pm 0 \cdot 10$	35/
C_4H_8		1-Butene							
C_4H_8	$56 \cdot 1084$	cis-2-Butene	g	$99 \cdot 74 \pm 0 \cdot 10$	FC	7	CO_2	$647 \cdot 65 \pm 0 \cdot 29$	51/
C_4H_8		cis-2-Butene	g		H	5	P/CO_2	$28 \cdot 33 \pm 0 \cdot 10$	35/
C_4H_8		cis-2-Butene	g		E		glc	$1 \cdot 20 \pm 0 \cdot 15$	63/1 64/1
C_4H_8		cis-2-Butene							
C_4H_8	$56 \cdot 1084$	trans-2-Butene	g	$99 \cdot 32 \pm 0 \cdot 08$	FC	5	CO_2	$646 \cdot 90 \pm 0 \cdot 23$	51/
C_4H_8		trans-2-Butene	g	$99 \cdot 9$	H	8	P/CO_2	$27 \cdot 38 \pm 0 \cdot 10$	35/
C_4H_8		trans-2-Butene	g		E		glc	$2 \cdot 80 \pm 0 \cdot 20$	63/1 64/1
C_4H_8		trans-2-Butene							
C_4H_8	$56 \cdot 1084$	2-Methylpropene	g	$99 \cdot 75 \pm 0 \cdot 10$	FC	5	CO_2	$645 \cdot 19 \pm 0 \cdot 25$	51/
C_4H_8		2-Methylpropene	g		H	6	P/CO_2	$28 \cdot 15 \pm 0 \cdot 10$	35/
C_4H_8		2-Methylpropene							
C_4H_8	$56 \cdot 1084$	Cyclobutane	l	$> 99 \cdot 8$	SB	6	m	$650 \cdot 35 \pm 0 \cdot 12$	50/6 52/
C_4H_{10}	$58 \cdot 1243$	n-Butane	g	$99 \cdot 78 \pm 0 \cdot 08$	FC	6	CO_2	$687 \cdot 42 \pm 0 \cdot 15$	51/

1 kcal = 4·184 kJ

ΔH_r Remarks	5 ΔH_f° (l or c) kcal/g.f.w.	6 Determination of ΔH_v			7 ΔH_f° (g) kcal/g.f.w.
		Type	ΔH_v° kcal/g.f.w.	Ref.	
H An 0·99952±0·00019	+33·91±0·21	Cl	5·58±0·02	50/7	+39·49±0·21
H An 0·99980±0·00017	+28·33±0·24	Cl	6·36±0·01	41/5	+34·69±0·24
C₄H₁₀(g) {−30·36±0·16} E	+28·38±0·34				+34·74±0·34
Selected value	+28·35±0·20 (l)				+34·71±0·20
H An 0·99994±0·00019	+33·22±0·14	Cl	5·56±0·01	47/11	+38·78±0·14
H An 1·00024±0·00022	+20·74±0·19	Cl	5·27±0·01	45/4	+26·01±0·19
C₄H₁₀(g) {−30·36±0·16} E	+20·94±0·19				+26·21±0·19
Selected value	+20·84±0·15 (l)				+26·11±0·15
					+37·45±0·37
					+47·92±0·44
					+58·21±0·28
					+51·90±0·20
H An 0·99962±0·00009	−5·05±0·19	Cl	4·92±0·02	46/3	−0·13±0·19
C₄H₁₀(g) {−30·36±0·16} E	−5·18±0·18				−0·26±0·18
Selected value	−5·12±0·13 (l)				−0·20±0·13
H An 1·00007±0·00014	−7·21±0·30	Cl	5·40±0·02	44/8	−1·81±0·30
C₄H₁₀(g) {−30·36±0·16} E	−7·43±0·18				−2·03±0·18
s-2-butene (g) ⇌ trans-2-butene (g) {−2·95±0·18} 2nd Law					−1·75±0·23
−7·15±0·23 Selected value	−7·26±0·20 (l)				−1·86±0·20
H An 0·99984±0·00024	−7·72±0·24	Cl	5·16±0·02	45/5	−2·56±0·24
C₄H₁₀(g) {−30·36±0·16} E	−8·14±0·18				−2·98±0·18
Butene (g) {−0·20±0·13} ⇌ trans-2-Butene (g). 2nd Law	−8·16±0·25				−3·00±0·25
Selected value	−8·15±0·18				−2·99±0·18
H An 0·99956±0·00021	−9·19±0·27	V1	4·92±0·03	53/1	−4·27±0·26
-C₄H₁₀(g) {−32·41±0·13} E	−9·18±0·18				−4·26±0·17
Selected value	−9·18±0·16				−4·26±0·15
	+0·89±0·13	Cl	5·89±0·01	53/11	+6·78±0·14
H An 0·99992±0·00009	−35·38±0·16	Cl	5·02±0·02	40/1	−30·36±0·16

C_aH_b 1 kcal = 4·184 kJ

								Determinati	
1 Formula	2 g.f.w.	3 Name	State	Purity mol %	Type	No. of expts.	Detn. of reactn.	$-\Delta H_r^\circ$ kcal/g.f.w.	Ref
C_4H_{10}	58·1243	2-Methylpropane	g	99·88±0·06	FC	6	CO_2	685·37±0·11	51/
C_4H_{10}		2-Methylpropane	g		E			2·32±0·20	45/
C_4H_{10}		2-Methylpropane							
C_5H_6	66·1036	cis-Pent-3-ene-1-yne	l		H	3	H_2	95·6 ±1·1	59/
C_5H_6	66·1036	trans-Pent-3-ene-1-yne	l		H	3	H_2	96·0 ±0·4	59/
C_5H_6	66·1036	Cyclopentadiene	g		H	4	P/CO_2	50·38±0·20	36/
C_5H_8	69·1195	1, 2-Pentadiene	g	99·66±0·15	FC	5	CO_2	777·14±0·15	55/
C_5H_8	68·1195	1, cis-3-Pentadiene	g	99·92±0·04	FC	5	CO_2	768·94±0·30	55/
C_5H_8		1, cis-3-Pentadiene	g		E		glc	1·01±0·18	65/
C_5H_8		1, cis-3-Pentadiene							
C_5H_8	68·1195	1, trans-3-Pentadiene	g	99·92±0·03	FC	3	CO_2	761·64±0·15	55/
C_5H_8	68·1195	1, 4-Pentadiene	g	99·93±0·05	FC	5	CO_2	768·94±0·30	55/
C_5H_8		1, 4-Pentadiene	g	99·90	H	4	P/CO_2	60·22±0·15	36/
C_5H_8		1, 4-Pentadiene							
C_5H_8	68·1195	2, 3-Pentadiene	g	99·85±0·07	FC	5	CO_2	775·32±0·16	55/
C_5H_8	68·1195	2-Methyl-1, 3-butadiene (Isoprene)	g	99·96±0·03	FC	6	CO_2	761·62±0·23	55/
C_5H_8		2-Methyl-1, 3-butadiene (Isoprene)	l		SB	8	CO_2	754·82±0·38	38/
C_5H_8		2-Methyl-1, 3-butadiene (Isoprene)							
C_5H_8	68·1195	Cyclopentene	l	99·98±0·02	SB	5	CO_2	744·55±0·14	61/
C_5H_8		Cyclopentene	l		SB			744·46±0·17	49/
C_5H_8		Cyclopentene	g		H	4	P/CO_2	26·67±0·06	37/
C_5H_8		Cyclopentene							
C_5H_8	68·1195	Vinylcyclopropane	l	99·97, glc	SB			772·8 ±	62/
C_5H_8	68·1195	Spiropentane	g	99·87	FC	4	CO_2	787·77±0·17	55/
C_5H_8		Spiropentane	l		SB	3	m	778·78±0·31	50/
C_5H_8		Spiropentane							

1 kcal = 4.184 kJ

ΔH_r°

Remarks		ΔH_f° (l or c) kcal/g.f.w.	Type	ΔH_v° kcal/g.f.w.	Ref.	ΔH_f° (g) kcal/g.f.w.
		5	— **6** Determination of ΔH_v			**7**
H An 0.99991 ± 0.00015		-37.02 ± 0.13	C1	4.61 ± 0.02	40/2	-32.41 ± 0.13
$_4H_{10}(g)\ \{-30.36\pm0.16\} \rightleftharpoons$ iso-C_4H_{10} (g)		-37.29 ± 0.26				-32.68 ± 0.26
Selected value		-37.02 ± 0.13				-32.41 ± 0.13
$H_6(l)+3H_2(g)$ $=$n-C_5H_{12}(l) $\{-41.46\pm0.19\}$		$+54.1\ \pm1.2$				
$H_6(l)+3H_2(g)$ $=$n-C_5H_{12}(l) $\{-41.46\pm0.19\}$		$+54.5\ \pm0.5$				
clopentane(g) $\{-18.44\pm0.20\}$	E	$+25.16\pm0.30$	V3	6.78 ± 0.06	65/14	$+31.94\pm0.28$
H An 0.99988 ± 0.00021						$+33.62\pm0.16$
H An 1.00003 ± 0.00013 $-1,3$-C_5H_8 (g) $\rightleftharpoons$ trans-$1,3$-C_5H_8 (g) $\{+18.12\pm0.16\}$ 2nd Law						$+19.78\pm0.22$ $+19.13\pm0.24$
Selected value						$+19.13\pm0.24$
						$+18.12\pm0.16$
H An 0.99952 ± 0.00027 C_5H_{12}(g) $\{-35.14\pm0.15\}$						$+25.42\pm0.31$ $+25.08\pm0.22$
Selected value						$+25.25\pm0.20$
H An 1.00007 ± 0.00019						$+31.80\pm0.17$
H An 0.99963 ± 0.00008		$+11.70\pm0.25$	V2	6.40 ± 0.06	36/4	$+18.10\pm0.24$
	E	$+11.32\pm0.39$				$+17.72\pm0.40$
Selected value		$+11.66\pm0.21$				$+18.06\pm0.20$
andard-state corr. appears to be 0.30 kcal too large, but original result given.		$+1.03\pm0.15$	V1	6.71 ± 0.07	50/8	$+7.74\pm0.17$
		$+0.94\pm0.18$				$+7.65\pm0.20$
yclopentane(g) $\{-18.44\pm0.20\}$		$+1.52\pm0.24$				$+8.23\pm0.22$
Selected value		$+1.52\pm0.24$				$+8.23\pm0.22$
.B.S. measurement		$+29.3\ \pm$				
H An 0.99957 ± 0.00007		$+37.67\pm0.18$	C1	6.58 ± 0.01	50/10	$+44.25\pm0.18$
		$+35.26\pm0.32$				$+41.84\pm0.33$
Selected value		$+37.67\pm0.18$				$+44.25\pm0.18$

C_aH_b

1 kcal = 4·184 kJ

1 Formula	2 g.f.w.	3 Name	State	Purity mol %	Type	No. of expts.	Detn. of reactn.	$-\Delta H_r^\circ$ kcal/g.f.w.	Re
C_5H_{10}	70·1355	1-Pentene	g		E		glc	2·60±0·20	66/
C_5H_{10}	70·1355	cis-2-Pentene	g		E,H	2	P/CO_2	28·14±0·20	66/ 36/
C_5H_{10} C_5H_{10}		cis-2-Pentene cis-2-Pentene	l		SB	2	m	797·48±0·07	46/
C_5H_{10}	70·1355	trans-2-Pentene	g		E,H	2	P/CO_2	27·21±0·20	66/ 36/
C_5H_{10} C_5H_{10}		trans-2-Pentene trans-2-Pentene	l		SB	2	m	797·30±0·10	46/
C_5H_{10}	70·1355	2-Methyl-1-butene	g		H	2	P/CO_2	28·25±0·10	36/
C_5H_{10}	70·1355	3-Methyl-1-butene	g		H	4	P/CO_2	30·19±0·06	37/
C_5H_{10}	70·1355	2-Methyl-2-butene	g		H	5	P/CO_2	26·68±0·06	36/
C_5H_{10}	70·1355	Ethylcyclopropane	l	99·98, glc	SB			808·8 ±	62/
C_5H_{10}	70·1355	Methylcyclobutane	l		SB	5	m	801·20±0·32	50/
C_5H_{10} C_5H_{10} C_5H_{10} C_5H_{10}	70·1355	Cyclopentane Cyclopentane Cyclopentane Cyclopentane	l l l l	99·98±0·01 >99·8	SB SB SB SB	5 6 4	CO_2 m m	786·54±0·17 786·62±0·30 786·87±0·14	46/ 47/1 52/
C_5H_{12} C_5H_{12} C_5H_{12} C_5H_{12}	72·1514	n-Pentane n-Pentane n-Pentane n-Pentane	g l g	>99·8 99·86±0·07 >99·87	FC SB FC	8 14 8	H_2O CO_2 CO_2	845·27±0·22 838·71±0·18 844·99±0·23	34/ 44/ 67/
C_5H_{12} C_5H_{12} C_5H_{12} C_5H_{12}	72·1514	2-Methylbutane 2-Methylbutane 2-Methylbutane 2-Methylbutane	g g g g	>99·9 99·99	FC SB FC E	11 3 4	CO_2 m CO_2	843·38±0·15 843·11±0·84* 843·31±0·22 1·86±[0·20]	39/ 36/ 67/ 45/
C_5H_{12}		2-Methylbutane							
C_5H_{12} C_5H_{12} C_5H_{12}	72·1514	2,2-Dimethylpropane 2,2-Dimethylpropane 2,2-Dimethylpropane	g g	99·7 99·98±0·01	FC FC	10 7	H_2O CO_2	840·59±0·22 839·88±0·23	39/ 67/

1 kcal = 4·184 kJ

ΔH_r° Remarks	5 ΔH_f° (l or c) kcal/g.f.w.	6 Determination of ΔH_v Type	ΔH_v° kcal/g.f.w.	Ref.	7 ΔH_f° (g) kcal/g.f.w.
entene (g) $\rightleftharpoons$ trans-2-Pentene (g) {$-7\cdot93\pm0\cdot26$} 2nd Law	$-11\cdot42\pm0\cdot32$	V1	$6\cdot09\pm0\cdot03$	53/1	$-5\cdot33\pm0\cdot31$
C_5H_{12}(g) {$-35\cdot14\pm0\cdot15$}. Eqn. (66/15) used to determine composition of (cis+trans) hydrogenated: cis→trans $-0\cdot9\pm0\cdot15$	$-13\cdot41\pm0\cdot27$	V1	$6\cdot41\pm0\cdot03$	53/1	$-7\cdot00\pm0\cdot26$
F	$-14\cdot35\pm0\cdot11$				$-7\cdot94\pm0\cdot12$
Selected value	$-13\cdot41\pm0\cdot27$				$-7\cdot00\pm0\cdot26$
C_5H_{12}(g) {$-35\cdot14\pm0\cdot15$}. Eqn. (66/15) used to determine composition of (cis+trans) hydrogenated: cis→trans $-0\cdot9\pm0\cdot15$	$-14\cdot31\pm0\cdot27$	V1	$6\cdot38\pm0\cdot03$	53/1	$-7\cdot93\pm0\cdot26$
F	$-14\cdot53\pm0\cdot12$				$-8\cdot15\pm0\cdot13$
Selected value	$-14\cdot31\pm0\cdot27$				$-7\cdot93\pm0\cdot26$
C_5H_{12}(g) {$-36\cdot80\pm0\cdot15$} E	$-14\cdot73\pm0\cdot18$	C1	$6\cdot18\pm0\cdot01$	49/7	$-8\cdot55\pm0\cdot18$
C_5H_{12}(g) {$-36\cdot80\pm0\cdot15$} E	$-12\cdot31\pm0\cdot19$	V1	$5\cdot70\pm0\cdot03$	53/1	$-6\cdot61\pm0\cdot18$
C_5H_{12}(g) {$-36\cdot80\pm0\cdot15$} E	$-16\cdot59\pm0\cdot18$	C1	$6\cdot47\pm0\cdot01$	49/7	$-10\cdot12\pm0\cdot18$
B.S. measurement	$-3\cdot0\ \pm$				
	$-10\cdot63\pm0\cdot33$				
	$-25\cdot29\pm0\cdot19$	C1	$6\cdot85\pm0\cdot01$	59/12	$-18\cdot44\pm0\cdot20$
An 0·99964	$-25\cdot21\pm0\cdot31$				$-18\cdot36\pm0\cdot32$
	$-24\cdot96\pm0\cdot17$				$-18\cdot11\pm0\cdot18$
Selected value	$-25\cdot29\pm0\cdot19$				$-18\cdot44\pm0\cdot20$
H An 0·99970±0·00022	$-41\cdot27\pm0\cdot24$	C1	$6\cdot39\pm0\cdot05$	47/5	$-34\cdot88\pm0\cdot23$
	$-41\cdot44\pm0\cdot19$				$-35\cdot05\pm0\cdot21$
H An 1·00015±0·00018	$-41\cdot55\pm0\cdot25$				$-35\cdot16\pm0\cdot24$
Selected value	$-41\cdot49\pm0\cdot17$				$-35\cdot10\pm0\cdot15$
H An 1·00001±0·00016	$-42\cdot80\pm0\cdot18$	C1	$6\cdot03\pm0\cdot05$	51/7	$-36\cdot77\pm0\cdot17$
	$-42\cdot98\pm0\cdot85$*				$-36\cdot95\pm0\cdot85$*
H An 0·99998±0·00024	$-42\cdot87\pm0\cdot24$				$-36\cdot84\pm0\cdot23$
C_5H_{12}(g) {$-35\cdot14\pm0\cdot15$} $\rightleftharpoons$ iso-C_5H_{12} (g). 2nd Law	$-43\cdot03\pm[0\cdot30]$				$-37\cdot00\pm[0\cdot30]$
Selected value	$-42\cdot88\pm0\cdot17$				$-36\cdot85\pm0\cdot15$
H An 0·99995±0·00014					$-39\cdot56\pm0\cdot24$
H An 1·00036±0·00022					$-40\cdot27\pm0\cdot25$
Selected value					$-40\cdot27\pm0\cdot25$

C_aH_b

1 kcal=4·184 kJ

Determinati

1 Formula	2 g.f.w.	3 Name	State	Purity mol %	Type	No. of expts.	Detn. of reactn.	$-\Delta H_r^\circ$ kcal/g.f.w.	Re
C_6H_6	78·1147	1, 5-Hexadiyne	l		H	4	H_2	139·3 ±1·0	59/
C_6H_6	78·1147	Benzene	l	99·96	SB	11	CO_2	780·98±0·10	45/
C_6H_6		Benzene	l		SB	6	m	780·82±0·09	46/
									47/
C_6H_6		Benzene	l	99·98±0·02	SB	7	CO_2	780·97±0·12	M
C_6H_6		Benzene	g		H	5	P/CO_2	49·06±0·15	36/
C_6H_6		Benzene							
C_6H_8	80·1307	1, 3-Cyclohexadiene	g		H	4	P/CO_2	54·88±0·10	36/
C_6H_{10}	82·1466	1, 5-Hexadiene	g		H	4	P/CO_2	60·03±0·10	36/
C_6H_{10}		1, 5-Hexadiene	l		SB	5	m	918·81±0·07	46/
C_6H_{10}	82·1466	2, 3-Dimethyl-1, 3- butadiene	g		H	4	P/CO_2	53·39±0·15	37/
C_6H_{10}	82·1466	1-Methylcyclopentene	l	99·89±0·08	SB	5	CO_2	896·86±0·14	61/
C_6H_{10}		1-Methylcyclopentene	l	99·89±0·08	SB	8	CO_2	897·22±0·14	M1
C_6H_{10}		1-Methylcyclopentene							
C_6H_{10}	82·1466	3-Methylcyclopentene	l	[99·5±0·4]	SB	5	CO_2	900·22±0·13	61/
C_6H_{10}	82·1466	4-Methylcyclopentene	l	99·8±0·1	SB	5	CO_2	901·68±0·24	61/
C_6H_{10}	82·1466	Cyclohexene	l	99·98±0·02	SB	4	CO_2	896·62±0·12	61/
C_6H_{10}		Cyclohexene	l	99·98±0·02	SB	7	CO_2	896·80±0·13	M1
C_6H_{10}		Cyclohexene	g	99·9	H	4	P/CO_2	28·35±0·10	36/
C_6H_{10}		Cyclohexene	g		H	5	m	29·06±0·44	49/
C_6H_{10}		Cyclohexene							
C_6H_{10}	82·1466	Bicyclopropyl	l	99·5	SB	5	CO_2	928·78±0·80	66/
C_6H_{10}	82·1466	Methylenecyclopentane	l	97·5±0·5	SB	5	CO_2	901·08±0·19	61/1
C_6H_{10}	82·1466	Bicyclo [3, 1, 0] hexane	l		SB			907·12±0·46	67/1
C_6H_{12}	84·1625	1-Hexene	l		H	4	H_2	30·10±0·40	57/
C_6H_{12}		1-Hexene	l	99·2	H	3	H_2	30·20±0·20	59/1
C_6H_{12}		1-Hexene							
C_6H_{12}	84·1625	cis-2-Hexene	l	99·94±0·05	SB	5	CO_2	954·15±0·31	60/1

1 kcal=4·184 kJ

ΔH°_r / Remarks	5 ΔH°_f (l or c) kcal/g.f.w.	6 Determination of ΔH_v			7 ΔH°_f (g) kcal/g.f.w.
		Type	ΔH°_v kcal/g.f.w.	Ref.	
H_6(l)+4H_2(g) =n-C_6H_14(l) {−47·46±0·18}	+91·8 ±1·1				
	+11·73±0·12	C1	8·09±0·01	47/5	+19·82±0·12
F	+11·57±0·11				+19·66±0·11
'An 0·99997±0·00010	+11·72±0·13				+19·81±0·13
clohexane(g) {−29·50±0·15} E	+11·47±0·21				+19·56±0·21
Selected value	+11·72±0·13				+19·81±0·13
clohexane(g) {−29·50±0·15} E					+25·38±0·19
C_6H_14(g) {−39·92±0.19} E					+20·11±0·27
F	+12·93±0·11				
3-Dimethylbutane(g) {−42·61±0·24} E					+10·78±0·30
/An 0·99994±0·00008	−9·02±0·17	E7	8·1±0·4		−0·9 ±0·5
	−8·66±0·17				−0·6 ±0·5
Selected value	−8·66±0·17				−0·6 ±0·5
	−5·66±0·16	E7	7·7±0·4		+2·0 ±0·5
	−4·20±0·25	E7	7·7±0·4		+3·5 ±0·5
/An 0·99995±0·00005	−9·26±0·14	V1	8·00±0·08	50/8	−1·26±0·17
	−9·08±0·15				−1·08±0·18
yclohexane(g) {−29·50±0·15} E	−9·15±0·19				−1·15±0·18
yclohexane(g) {−29·50±0·15} E	−8·46±0·51				−0·46±0·50
Selected value	−9·08±0·15				−1·08±0·18
/An 0·9995±0·0005	+22·90±0·81	E4	8·0±0·3	66/16	+30·9 ±0·9
	−4·80±0·21				
	+1·24±0·46		7·85±0·10	67/10	+9·09±0·49
_6H_12(l)+H_2=n-C_6H_14(l) {−47·46±0·18}	−17·36±0·47	V1	7·34±0·05	56/9	−10·02±0·48
_6H_12(l)+H_2=n-C_6H_14(l) {−47·46±0·18}	−17·26±0·32				−9·92±0·35
Selected value	−17·29±0·29				−9·95±0·30
	−20·05±0·32	V1	7·54±0·05	56/9	−12·51±0·33

C_aH_b

$$1 \text{ kcal} = 4 \cdot 184 \text{ kJ}$$

1 Formula	2 g.f.w.	3 Name	State	Purity mol %	Detn. Type	No. of expts.	Detn. of reactn.	$-\Delta H_r^\circ$ kcal/g.f.w.	Re
C_6H_{12}	84·1625	*trans*-2-Hexene	l	99·94±0·11	SB	8	CO_2	953·76±0·37	60/1
C_6H_{12}	84·1625	*cis*-3-Hexene	l	99·90±0·08	SB	4	CO_2	955·33±0·31	60/1
C_6H_{12}	84·1625	*trans*-3-Hexene	l	99·95±0·03	SB	4	CO_2	953·63±0·31	60/1
C_6H_{12}	84·1625	2-Methyl-1-pentene	l	99·92±0·07	SB	4	CO_2	952·70±0·30	60/1
C_6H_{12}	84·1625	3-Methyl-1-pentene	l	99·75±0·20	SB	4	CO_2	955·52±0·35	60/1
C_6H_{12}	84·1625	4-Methyl-1-pentene	l	99·85±0·12	SB	3	CO_2	955·07±0·42	60/1
C_6H_{12}	84·1625	2-Methyl-2-pentene	l	99·96±0·03	SB	4	CO_2	950·65±0·34	60/1
C_6H_{12}	84·1625	3-Methyl-*cis*-2-pentene	l	99·89±0·09	SB	4	CO_2	951·62±0·34	60/1
C_6H_{12}	84·1625	3-Methyl-*trans*-2-pentene	l	99·89±0·08	SB	4	CO_2	951·60±0·30	60/1
C_6H_{12}	84·1625	4-Methyl-*cis*-2-pentene	l	99·93±0·07	SB	5	CO_2	953·40±0·27	60/1
C_6H_{12}	84·1625	4-Methyl-*trans*-2-pentene	l	99·92±0·07	SB	4	CO_2	952·32±0·33	60/1
C_6H_{12}	84·1625	2-Ethyl-1-butene	l	99·95±0·04	SB	4	CO_2	953·38±0·34	60/1
C_6H_{12}	84·1625	2, 3-Dimethyl-1-butene	l	99·86±0·13	SB	4	CO_2	951·35±0·39	60/1
C_6H_{12}		2, 3-Dimethyl-1-butene	g		H	4	P/CO_2	27·75±0·10	36/5
C_6H_{12}		2, 3-Dimethyl-1-butene							
C_6H_{12}	84·1625	3, 3-Dimethyl-1-butene	l	99·91±0·06	SB	4	CO_2	953·10±0·36	60/1
C_6H_{12}		3, 3-Dimethyl-1-butene	g		H	3	P/CO_2	30·10±0·15	37/2
C_6H_{12}		3, 3-Dimethyl-1-butene							
C_6H_{12}	84·1625	2, 3-Dimethyl-2-butene	l	99·94±0·05	SB	4	CO_2	949·72±0·34	60/1
C_6H_{12}		2, 3-Dimethyl-2-butene	g		H	4	P/CO_2	26·39±0·10	36/5
C_6H_{12}		2, 3-Dimethyl-2-butene							
C_6H_{12}	84·1625	Methylcyclopentane	l	99·84±0·05	SB	6	CO_2	941·14±0·18	46/5
C_6H_{12}		Methylcyclopentane	l		SB	6	m	940·78±0·40	39/7, 40/3
C_6H_{12}		Methylcyclopentane	l	99·99±0·01	SB	8	CO_2	941·35±0·16	M10
C_6H_{12}		Methylcyclopentane	l		E			3·51±[0·40]	39/8
C_6H_{12}		Methylcyclopentane							

1 kcal = 4·184 kJ

ΔH_r°		6 Determina tion of Δ_v			7
Remarks	5 ΔH_f° (l or c) kcal/g.f.w.	Type	ΔH_v° kcal/g.f.w.	Ref.	ΔH_f° (g) kcal/g.f.w.
	$-20\cdot44\pm0\cdot38$	V1	$7\cdot56\pm0\cdot05$	56/9	$-12\cdot88\pm0\cdot39$
	$-18\cdot87\pm0\cdot32$	V1	$7\cdot49\pm0\cdot05$	56/9	$-11\cdot38\pm0\cdot33$
	$-20\cdot57\pm0\cdot32$	V1	$7\cdot56\pm0\cdot05$	56/9	$-13\cdot01\pm0\cdot33$
	$-21\cdot50\pm0\cdot31$	V1	$7\cdot31\pm0\cdot05$	56/9	$-14\cdot19\pm0\cdot32$
	$-18\cdot68\pm0\cdot36$	V1	$6\cdot85\pm0\cdot05$	56/9	$-11\cdot83\pm0\cdot37$
	$-19\cdot13\pm0\cdot43$	V1	$6\cdot88\pm0\cdot05$	56/9	$-12\cdot25\pm0\cdot44$
	$-23\cdot55\pm0\cdot35$	V1	$7\cdot57\pm0\cdot05$	56/9	$-15\cdot98\pm0\cdot36$
	$-22\cdot58\pm0\cdot35$	V1	$7\cdot69\pm0\cdot05$	56/9	$-14\cdot89\pm0\cdot36$
	$-22\cdot60\pm0\cdot31$	V1	$7\cdot51\pm0\cdot05$	56/9	$-15\cdot09\pm0\cdot32$
	$-20\cdot80\pm0\cdot28$	V1	$7\cdot06\pm0\cdot05$	56/9	$-13\cdot74\pm0\cdot29$
	$-21\cdot88\pm0\cdot34$	V1	$7\cdot18\pm0\cdot05$	56/9	$-14\cdot70\pm0\cdot35$
	$-20\cdot82\pm0\cdot35$	V1	$7\cdot43\pm0\cdot05$	56/9	$-13\cdot39\pm0\cdot36$
3-Dimethylbutane(g) $\{-42\cdot61\pm0\cdot24\}$ E	$-22\cdot85\pm0\cdot40$ $-21\cdot85\pm0\cdot28$	V1	$6\cdot99\pm0\cdot05$	56/9	$-15\cdot86\pm0\cdot41$ $-14\cdot86\pm0\cdot27$
Selected value	$-22\cdot18\pm0\cdot50$				$-15\cdot19\pm0\cdot50$
2-Dimethylbutane(g) $\{-44\cdot48\pm0\cdot24\}$ E	$-21\cdot10\pm0\cdot37$ $-20\cdot76\pm0\cdot30$	V1	$6\cdot38\pm0\cdot04$	56/9	$-14\cdot72\pm0\cdot38$ $-14\cdot38\pm0\cdot29$
Selected value	$-20\cdot89\pm0\cdot26$				$-14\cdot51\pm0\cdot25$
3-Dimethylbutane(g) $\{-42\cdot61\pm0\cdot24\}$ E	$-24\cdot48\pm0\cdot35$ $-24\cdot02\pm0\cdot27$	V1 C1	$7\cdot80\pm0\cdot01$	56/9, 55/13	$-16\cdot68\pm0\cdot35$ $-16\cdot22\pm0\cdot27$
Selected value	$-24\cdot22\pm0\cdot25$				$-16\cdot42\pm0\cdot25$
A	$-33\cdot06\pm0\cdot20$ $-33\cdot42\pm0\cdot41$	C1	$7\cdot58\pm0\cdot01$	47/5, 59/12 60/11	$-25\cdot48\pm0\cdot20$ $-25\cdot84\pm0\cdot41$
An $0\cdot99992\pm0\cdot00012$ yclohexane(l) $\{-37\cdot40\pm0\cdot15\}$ ⇌Me cyclopentane(l). 2nd Law	$-32\cdot85\pm0\cdot18$ $-33\cdot89\pm[0\cdot50]$				$-25\cdot27\pm0\cdot18$ $-26\cdot31\pm[0\cdot50]$
Selected value	$-32\cdot85\pm0\cdot18$				$-25\cdot27\pm0\cdot18$

THERMOCHEMISTRY OF ORGANIC AND ORGANOMETALLIC COMPOUNDS

C_aH_b

1 kcal = 4·184 kJ

1 Formula	2 g.f.w.	3 Name	State	Purity mol %	Type	No. of expts.	Detn. of reactn.	$-\Delta H_r^\circ$ kcal/g.f.w.	Ref
C_6H_{12}	84·1625	Cyclohexane	l	99·997	SB	6	CO_2	936·88±0·17	46/
C_6H_{12}		Cyclohexane	l	>99·8	SB	5	m	936·80±0·17	52/
C_6H_{12}		Cyclohexane	l		SB	6	CO_2	936·53±0·24	47/1
C_6H_{12}		Cyclohexane	l		SB	5	m	936·81±0·31	39/
$C_6H_{1_-}$		Cyclohexane	l	99·999±0·006	SB	8	CO_2	936·88±0·17	M1
C_6H_{12}		Cyclohexane							
C_6H_{14}	86·1785	n-Hexane	l	99·90±0·06	SB	12	CO_2	994·99±0·20	44/
C_6H_{14}		n-Hexane	l	>99	SB	6	CO_2	995·16±0·33	37/
C_6H_{14}		n-Hexane	l	99·98±0·01	SB	8	CO_2	995·05±0·16	M1
C_6H_{14}		n-Hexane							
C_6H_{14}	86·1785	2-Methylpentane	l	99·99	SB	8	CO_2	993·60±0·22	41/
C_6H_{14}	86·1785	3-Methylpentane	l		SB	11	CO_2	994·14±0·22	41/
C_6H_{14}	86·1785	2, 3-Dimethylbutane	l	2s	SB	11	CO_2	992·94±0·22	41/
C_6H_{14}	86·1785	2, 2-Dimethylbutane	l	2s	SB	8	CO_2	991·41±0·22	41/
C_7H_8	92·1418	1, 3, 5-Cycloheptatriene	g		H	2	P/CO_2	72·11±0·30	39/
C_7H_8	92·1418	Toluene	l	99·96	SB	11	CO_2	934·50±0·12	45/
C_7H_8		Toluene	l		SB	7	m	934·72±0·11	46/
C_7H_8		Toluene	l	99·999	SB	11	CO_2	934·55±0·12	M1
C_7H_8		Toluene							
C_7H_{10}	94·1578	1-Methylene-2-cyclohexene	l		SB			996·9 ±0·4	63/1
C_7H_{10}	94·1578	1-Methyl-2-cyclohexadiene	l		SB			985·9 ±0·9	63/1
C_7H_{10}	94·1578	1, 3-Cycloheptadiene	g		H	3	P/CO_2	50·77±0·15	39/
C_7H_{12}	96·1737	1-Ethylcyclopentene	l	99·5±0·2	SB	5	CO_2	1054·32±0·19	61/1
C_7H_{12}	96·1737	Vinylcyclopentane	l	99·92±0·08	SB	5	CO_2	1059·93±0·25	61/1
C_7H_{12}	96·1737	Ethylidenecyclopentane	l	99·94±0·05	SB	5	CO_2	1054·69±0·19	61/1
C_7H_{12}	96·1737	1-Methylcyclohexene	l	99·86±0·08	SB	5	CO_2	1048·85±0·16	61/1
C_7H_{12}	96·1737	1-Methylcyclohexene	l		SB			1050·0 ±0·9	63/1
C_7H_{12}		1-Methylcyclohexene							

1 kcal = 4·184 kJ

ΔH_f° Remarks	5 ΔH_f° (l or c) kcal/g.f.w.	6 Determination of ΔH_v			7 ΔH_f° (g) kcal/g.f.w.
		Type	ΔH_v° kcal/g.f.w.	Ref.	
	$-37\cdot32\pm0\cdot19$	Cl	$7\cdot90\pm0\cdot01$	47/5	$-29\cdot42\pm0\cdot19$
	$-37\cdot40\pm0\cdot19$				$-29\cdot50\pm0\cdot19$
An 0·99975	$-37\cdot67\pm0\cdot25$				$-29\cdot77\pm0\cdot25$
A	$-37\cdot39\pm0\cdot32$				$-29\cdot49\pm0\cdot32$
An 1·00000±0·00011	$-37\cdot32\pm0\cdot19$				$-29\cdot42\pm0\cdot19$
Selected value	$-37\cdot40\pm0\cdot15$				$-29\cdot50\pm0\cdot15$
	$-47\cdot52\pm0\cdot22$	Cl	$7\cdot54\pm0\cdot01$	47/5	$-39\cdot98\pm0\cdot22$
	$-47\cdot35\pm0\cdot34$				$-39\cdot81\pm0\cdot34$
An 0·99998±0·00008	$-47\cdot46\pm0\cdot18$				$-39\cdot92\pm0\cdot18$
Selected value	$-47\cdot46\pm0\cdot18$				$-39\cdot92\pm0\cdot18$
An > 0·9998	$-48\cdot91\pm0\cdot24$	Cl	$7\cdot14\pm0\cdot01$	47/5, 49/8	$-41\cdot77\pm0\cdot24$
An > 0·9998	$-48\cdot37\pm0\cdot24$	Cl	$7\cdot24\pm0\cdot01$	47/5, 49/8	$-41\cdot13\pm0\cdot24$
An > 0·9998	$-49\cdot57\pm0\cdot24$	Cl	$6\cdot96\pm0\cdot01$	47/5, 49/8	$-42\cdot61\pm0\cdot24$
An > 0·9998	$-51\cdot10\pm0\cdot24$	Cl	$6\cdot62\pm0\cdot01$	47/5, 46/6	$-44\cdot48\pm0\cdot24$
cloheptane(g) {$-28\cdot21\pm0\cdot18$} E	$+34\cdot65\pm0\cdot37$	V1	$9\cdot25\pm0\cdot05$	56/10	$+43\cdot90\pm0\cdot36$
F	$+2\cdot88\pm0\cdot14$	Cl	$9\cdot08\pm0\cdot01$	47/5, 62/15	$+11\cdot96\pm0\cdot14$
	$+3\cdot10\pm0\cdot13$				$+12\cdot18\pm0\cdot13$
An 0·99993±0·00009	$+2\cdot93\pm0\cdot14$				$+12\cdot01\pm0\cdot14$
Selected value	$+2\cdot91\pm0\cdot10$				$+11\cdot99\pm0\cdot10$
	$-3\cdot0\ \pm0\cdot5$				
	$-14\cdot0\ \pm0\cdot9$				
cloheptane(g) {$-28\cdot21\pm0\cdot18$} E					$+22\cdot56\pm0\cdot24$
	$-13\cdot93\pm0\cdot22$				
andard state correction revised	$-8\cdot32\pm0\cdot27$				
	$-13\cdot56\pm0\cdot22$				
	$-19\cdot40\pm0\cdot19$	V1	$9\cdot06\pm0\cdot05$	60/12	$-10\cdot34\pm0\cdot20$
	$-18\cdot3\ \pm0\cdot9$				$-9\cdot2\ \pm0\cdot9$
Selected value	$-19\cdot40\pm0\cdot19$				$-10\cdot34\pm0\cdot20$

Determinat

1 Formula	2 g.f.w.	3 Name	State	Purity mol %	Type	No. of expts.	Detn. of reactn.	$-\Delta$., kcal/gf.w.	R
C_7H_{12}	96·1737	Methylenecyclohexane	l		SB			1053·6±0·9	63/
C_7H_{12}	96·1737	Cycloheptene	g		H	2	P/CO_2	26·02±0·15	39
C_7H_{12}	96·1737	Bicyclo [4, 1, 0] heptane	l		SB			1059·45±0·63	67/
C_7H_{12}	96·1737	Bicyclo [2, 2, 1] heptane	c		SB	5	CO_2	1046·24±0·52	63/
C_7H_{14}	98·1896	1-Heptene	l	99·84±0·01	SB	4	CO_2	1113·37±0·28	61/
C_7H_{14}		1-Heptene	l		SB	5	m	1111·61±0·19	46/
C_7H_{14}		1-Heptene	l	99·85	H	3	m	29·71±0·13	42
C_7H_{14}		1-Heptene	l	99·85	H	2	m	29·5 ±0·3	36
C_7H_{14}		1-Heptene	g	99·85	H	5	P/CO_2	29·89±0·06	36
C_7H_{14}		1-Heptene							
C_7H_{14}	98·1896	5-Methyl-1-hexene	l		SB	5	m	1112·66±0·12	46
C_7H_{14}	98·1896	3-Methyl-cis-3-hexene	l	99·85±0·10	SB	4	CO_2	1108·85±0·26	61/
C_7H_{14}	98·1896	3-Methyl-trans-3-hexene	l	99·85±0·10	SB	3	CO_2	1109·62±0·26	61/
C_7H_{14}	98·1896	2, 4-Dimethyl-1-pentene	l	99·88±0·09	SB	4	CO_2	1108·60±0·32	61/
C_7H_{14}	98·1896	4, 4-Dimethyl-1-pentene	l	99·85±0·08	SB	4	CO_2	1110·13±0·42	61/
C_7H_{14}		4, 4-Dimethyl-1-pentene	g		H	3	P/CO_2	29·29±0·10	37/
C_7H_{14}		4, 4-Dimethyl-1-pentene							
C_7H_{14}	98·1896	2, 4-Dimethyl-2-pentene		99·95±0·04	SB	4	CO_2	1107·14±0·26	61/
C_7H_{14}	98·1896	4, 4-Dimethyl-cis-2-pentene		99·85±0·11	SB	4	CO_2	1111·39±0·33	61/
C_7H_{14}	98·1896	4, 4-Dimethyl-trans-2-pentene		99·81±0·03	SB	4	CO_2	1107·47±0·26	61/
C_7H_{14}	98·1896	3-Methyl-2-ethyl-1-butene	l	99·85±0·10	SB	3	CO_2	1109·30±0·33	61/
C_7H_{14}	98·1896	2, 3, 3-Trimethyl-1-butene	l	99·95±0·04	SB	4	CO_2	1108·43±0·32	61/
C_7H_{14}	98·1896	1, 1-Dimethyl-2-ethylcyclopropane	l		SB	8	CO_2	1115·0 ±0·2	60/

1 kcal$=4\cdot184$ kJ

ΔH_r°		5 ΔH_f° (l or c) kcal/g.f.w.	6 Determination of ΔH_v			7 ΔH_f° (g) kcal/g.f.w.
	Remarks		Type	ΔH_v° kcal/g.f.w.	Ref.	
		$-14\cdot7\ \pm0\cdot9$				
cloheptane(g) $\{-28\cdot21\pm0\cdot18\}$	E					$-2\cdot19\pm0\cdot24$
		$-8\cdot80\pm0\cdot64$		$9\cdot14\pm0\cdot10$	67/10	$+0\cdot34\pm0\cdot66$
An $0\cdot99982$		$-22\cdot01\pm0\cdot54$				
		$-23\cdot19\pm0\cdot30$	V1	$8\cdot52\pm0\cdot04$	50/8	$-14\cdot67\pm0\cdot31$
	F	$-24\cdot95\pm0\cdot22$				$-16\cdot43\pm0\cdot23$
H_{14}(l)$+H_2=$n-C_7H_{16}(l) $\{-53\cdot59\pm0\cdot22\}$		$-23\cdot88\pm0\cdot28$				$-15\cdot36\pm0\cdot29$
H_{14}(l)$+H_2=$n-C_7H_{16}(l) $\{-53\cdot59\pm0\cdot22\}$		$-24\cdot1\ \pm0\cdot5$				$-15\cdot6\ \pm0\cdot5$
C_7H_{16}(g) $\{-44\cdot85\pm0\cdot22\}$		$-23\cdot48\pm0\cdot26$				$-14\cdot96\pm0\cdot25$
	Selected value	$-23\cdot33\pm0\cdot25$				$-14\cdot81\pm0\cdot25$
	F	$-23\cdot90\pm0\cdot16$	E1	$8\cdot2\ \pm0\cdot2$	53/1	$-15\cdot70\pm0\cdot27$
		$-27\cdot71\pm0\cdot28$	V1	$8\cdot73\pm0\cdot05$	60/12	$-18\cdot98\pm0\cdot30$
		$-26\cdot94\pm0\cdot28$	V1	$8\cdot58\pm0\cdot05$	60/12	$-18\cdot36\pm0\cdot30$
		$-27\cdot96\pm0\cdot34$	V1	$7\cdot93\pm0\cdot05$	60/12	$-20\cdot03\pm0\cdot35$
		$-26\cdot43\pm0\cdot44$	V1	$7\cdot47\pm0\cdot05$	60/12	$-18\cdot96\pm0\cdot45$
2-Dimethylpentane(g) $\{-49\cdot20\pm0\cdot37\}$	E	$-27\cdot38\pm0\cdot41$				$-19\cdot91\pm0\cdot40$
	Selected value	$-26\cdot91\pm0\cdot50$				$-19\cdot44\pm0\cdot50$
		$-29\cdot42\pm0\cdot28$	V1	$8\cdot22\pm0\cdot05$	60/12	$-21\cdot20\pm0\cdot30$
		$-25\cdot17\pm0\cdot35$	V1	$7\cdot81\pm0\cdot05$	60/12	$-17\cdot36\pm0\cdot36$
		$-29\cdot09\pm0\cdot28$	V1	$7\cdot87\pm0\cdot05$	60/12	$-21\cdot22\pm0\cdot30$
		$-27\cdot26\pm0\cdot35$	V1	$8\cdot25\pm0\cdot05$	60/12	$-19\cdot01\pm0\cdot36$
		$-28\cdot13\pm0\cdot34$	V1	$7\cdot70\pm0\cdot05$	60/12	$-20\cdot43\pm0\cdot35$
		$-21\cdot6\ \pm0\cdot2$				

THERMOCHEMISTRY OF ORGANIC AND ORGANOMETALLIC COMPOUNDS

C_aH_b

1 kcal $= 4 \cdot 184$ kJ

1 Formula	2 g.f.w.	3 Name	State	Purity mol %	Type	No. of expts.	Detn. of reactn.	$-\Delta H_r^?$ kcal/g.f.w.	R
C_7H_{14}	$98 \cdot 1896$	1, 1-Dimethyl-cyclopentane	1	$99 \cdot 97 \pm 0 \cdot 02$	SB	5	CO_2	$1095 \cdot 44 \pm 0 \cdot 25$	49
C_7H_{14}	$98 \cdot 1896$	cis-1, 2- Dimethyl-cyclopentane	1	$99 \cdot 98 \pm 0 \cdot 02$	SB	7	CO_2	$1097 \cdot 06 \pm 0 \cdot 30$	49
C_7H_{14}	$98 \cdot 1896$	trans-1, 2-Dimethyl-cyclopentane	1	$99 \cdot 87 \pm 0 \cdot 10$	SB	7	CO_2	$1095 \cdot 64 \pm 0 \cdot 27$	49
C_7H_{14}	$98 \cdot 1896$	cis-1, 3-Dimethyl-cyclopentane	1	$99 \cdot 65 \pm 0 \cdot 09$	SB	5	CO_2	$1095 \cdot 90 \pm 0 \cdot 26$	49/ 55/
C_7H_{14}	$98 \cdot 1896$	trans-1, 3-Dimethyl-cyclopentane	1	$99 \cdot 59 \pm 0 \cdot 23$	SB	5	CO_2	$1096 \cdot 39 \pm 0 \cdot 33$	49/ 55/
C_7H_{14}	$98 \cdot 1896$	Ethylcyclopentane	1	$99 \cdot 92 \pm 0 \cdot 03$	SB	6	CO_2	$1097 \cdot 50 \pm 0 \cdot 22$	46,
C_7H_{14}		Ethylcyclopentane	1		SB	5	m	$1096 \cdot 94 \pm 0 \cdot 37$	40,
C_7H_{14}		Ethylcyclopentane							
C_7H_{14}	$98 \cdot 1896$	Methylcyclohexane	1	$99 \cdot 90 \pm 0 \cdot 08$	SB	6	CO_2	$1091 \cdot 13 \pm 0 \cdot 23$	46,
C_7H_{14}		Methylcyclohexane	1		SB	5	m	$1090 \cdot 89 \pm 0 \cdot 45$	40,
C_7H_{14}		Methylcyclohexane							
C_7H_{14}	$98 \cdot 1896$	Cycloheptane	1	$> 99 \cdot 8$	SB	6	m	$1099 \cdot 13 \pm 0 \cdot 14$	52/
C_7H_{14}		Cycloheptane	1		SB	10	m	$1099 \cdot 17 \pm 0 \cdot 40$	47/
C_7H_{14}		Cycloheptane	1		SB	> 2	CO_2	$1098 \cdot 7 \pm 0 \cdot 2$	61/
C_7H_{14}		Cycloheptane							
C_7H_{16}	$100 \cdot 2056$	n-Heptane	1	$99 \cdot 88$	SB	20	CO_2	$1151 \cdot 33 \pm 0 \cdot 21$	44/
C_7H_{16}		n-Heptane	1	> 99	SB	6	CO_2	$1151 \cdot 15 \pm 0 \cdot 45$	37/ 45/
C_7H_{16}		n-Heptane	1	$99 \cdot 88$	SB		m	$1151 \cdot 28 \pm 0 \cdot 35$	41/
C_7H_{16}		n-Heptane							51/
C_7H_{16}	$100 \cdot 2056$	2-Methylhexane	1		SB		m	$1150 \cdot 03 \pm 0 \cdot 23$	41/ 51/
C_7H_{16}	$100 \cdot 2056$	3-Methylhexane	1		SB		m	$1150 \cdot 76 \pm 0 \cdot 45$	41/ 51/
C_7H_{16}	$100 \cdot 2056$	3-Ethylpentane	1	$[99 \cdot 4]$	SB	5	CO_2	$1151 \cdot 21 \pm 0 \cdot 31$	41/
C_7H_{16}		3-Ethylpentane	1		SB		m	$1151 \cdot 09 \pm 0 \cdot 23$	41/
C_7H_{16}		3-Ethylpentane							51/

Determina

$$1 \text{ kcal} = 4\cdot184 \text{ kJ}$$

ΔH_r°			6 Determination of ΔH_v			
Remarks		5 ΔH_f° (l or c) kcal/g.f.w.	Type	ΔH_v° kcal/g.f.w.	Ref.	7 ΔH_f° (g) kcal/g.f.w.
		$-41\cdot12\pm0\cdot27$	V1	$8\cdot08\pm0\cdot05$	53/1	$-33\cdot04\pm0\cdot28$
		$-39\cdot50\pm0\cdot32$	V1	$8\cdot55\pm0\cdot05$	53/1	$-30\cdot95\pm0\cdot33$
		$-40\cdot92\pm0\cdot29$	V1	$8\cdot26\pm0\cdot05$	53/1	$-32\cdot66\pm0\cdot30$
		$-40\cdot66\pm0\cdot28$	C1	$8\cdot18\pm0\cdot05$	59/12, 60/11	$-32\cdot48\pm0\cdot30$
		$-40\cdot17\pm0\cdot35$	V1	$8\cdot25\pm0\cdot05$	53/1	$-31\cdot92\pm0\cdot36$
	A Selected value	$-39\cdot06\pm0\cdot24$ $-39\cdot62\pm0\cdot38$ $-39\cdot06\pm0\cdot24$	V1	$8\cdot72\pm0\cdot05$	53/1	$-30\cdot34\pm0\cdot25$ $-30\cdot90\pm0\cdot39$ $-30\cdot34\pm0\cdot25$
	A Selected value	$-45\cdot43\pm0\cdot25$ $-45\cdot67\pm0\cdot46$ $-45\cdot43\pm0\cdot25$	C1	$8\cdot45\pm0\cdot01$	47/5	$-36\cdot98\pm0\cdot25$ $-37\cdot22\pm0\cdot46$ $-36\cdot98\pm0\cdot25$
	Selected value	$-37\cdot43\pm0\cdot18$ $-37\cdot39\pm0\cdot41$ $-37\cdot9\ \pm0\cdot2$ $-37\cdot42\pm0\cdot16$	V1	$9\cdot21\pm0\cdot05$	56/10	$-28\cdot22\pm0\cdot20$ $-28\cdot18\pm0\cdot42$ $-28\cdot7\ \pm0\cdot2$ $-28\cdot21\pm0\cdot18$
	A Selected value	$-53\cdot55\pm0\cdot24$ $-53\cdot73\pm0\cdot46$ $-53\cdot60\pm0\cdot36$ $-53\cdot59\pm0\cdot22$	C1	$8\cdot74\pm0\cdot01$	47/5, 47/6	$-44\cdot81\pm0\cdot24$ $-44\cdot99\pm0\cdot46$ $-44\cdot86\pm0\cdot36$ $-44\cdot85\pm0\cdot22$
	A	$-54\cdot85\pm0\cdot27$	V1	$8\cdot33\pm0\cdot05$	53/1	$-46\cdot52\pm0\cdot28$
	A	$-54\cdot12\pm0\cdot46$	V1	$8\cdot39\pm0\cdot05$	53/1	$-45\cdot73\pm0\cdot47$
	A Selected value	$-53\cdot67\pm0\cdot33$ $-53\cdot79\pm0\cdot26$ $-53\cdot67\pm0\cdot33$	C1	$8\cdot42\pm0\cdot01$	47/5	$-45\cdot25\pm0\cdot33$ $-45\cdot37\pm0\cdot26$ $-45\cdot25\pm0\cdot33$

C_aH_b THERMOCHEMISTRY OF ORGANIC AND ORGANOMETALLIC COMPOUNDS

$$1 \text{ kcal} = 4 \cdot 184 \text{ kJ}$$

Determina

1 Formula	2 g.f.w.	3 Name	State	Purity mol %	Type	No. of expts.	Detn. of reactn.	$-\Delta.$, kcal/gf.w.	R
C_7H_{16}	100·2056	2, 2-Dimethylpentane	l	99·85±0·05	SB	5	CO_2	1147·93±0·35	41
C_7H_{16}		2, 2-Dimethylpentane	l		SB		m	1147·37±0·23	41
C_7H_{16}		2, 2-Dimethylpentane							51
C_7H_{16}	100·2056	2, 3-Dimethylpentane	l		SB	6	CO_2	1149·17±0·34	41
C_7H_{16}		2, 3-Dimethylpentane	l		SB		m	1149·57±0·23	41
C_7H_{16}		2, 3-Dimethylpentane							51
C_7H_{16}	100·2056	2, 4-Dimethylpentane	l	99·4±0·4	SB	5	CO_2	1148·81±0·26	41
C_7H_{16}		2, 4-Dimethylpentane	l		SB		m	1148·29±0·23	41
C_7H_{16}		2, 4-Dimethylpentane							51
C_7H_{16}	100·2056	3, 3-Dimethylpentane	l	99·97±0·02	SB	5	CO_2	1148·91±0·26	41
C_7H_{16}		3, 3-Dimethylpentane	l		SB		m	1148·25±0·23	41
C_7H_{16}		3, 3-Dimethylpentane							51
C_7H_{16}	100·2056	2, 2, 3-Trimethylbutane	l	99·1±0·2	SB	6	CO_2	1148·35±0·30	41
C_7H_{16}		2, 2, 3-Trimethylbutane	l		SB		m	1148·48±0·35	41
C_7H_{16}		2, 2, 3-Trimethylbutane	l		SB	4	m	1148·18±0·14	51 46
C_7H_{16}		2, 2, 3-Trimethylbutane							
C_8H_6	102·1370	Phenylacetylene	l		H	4	H_2	70·7 ±1·0	58/
C_8H_8	104·1530	Cyclo-octatetraene	l	99·92±0·05	SB	5	CO_2	1086·50±0·30	50/
C_8H_8		Cyclo-octatetraene	l	96·7	SB	3	m	1085·0 ±1·4	54
C_8H_8		Cyclo-octatetraene							
C_8H_8	104·1530	Cubane	c		SB	5	m		66/
C_8H_8	104·1530	Styrene	l	99·98	SB	5	CO_2	1050·40±0·20	45
C_8H_8		Styrene	l	99·95	SB	7	CO_2	1050·58±0·28	47
C_8H_8		Styrene	l		SB	3	m	1051·5 ±2·0	29
C_8H_8		Styrene	l		SB	3	m	1046·8 ±1·5	38
C_8H_8		Styrene	g		H	4	P/CO_2	76·50±0·25	37
C_8H_8		Styrene							
C_8H_{10}	106·1689	Octa-1, 7-diyne	l		H	5	H_2	139·7 ±1·2	57

1 kcal = 4·184 kJ

ΔH_r° Remarks	5 ΔH_f°(l or c) kcal/g.f.w.	6 Determination of ΔH_v Type	Δ_v kcal/g.f.w.	Ref.	7 ΔH_f° (g) kcal/g.f.w.
	$-56{\cdot}95\pm0{\cdot}37$	C1	$7{\cdot}75\pm0{\cdot}01$	47/5	$-49{\cdot}20\pm0{\cdot}37$
A	$-57{\cdot}51\pm0{\cdot}25$				$-49{\cdot}76\pm0{\cdot}25$
Selected value	$-56{\cdot}95\pm0{\cdot}37$				$-49{\cdot}20\pm0{\cdot}37$
	$-55{\cdot}71\pm0{\cdot}36$	C1	$8{\cdot}18\pm0{\cdot}01$	47/5	$-47{\cdot}53\pm0{\cdot}36$
A	$-55{\cdot}31\pm0{\cdot}26$				$-47{\cdot}13\pm0{\cdot}26$
Selected value	$-55{\cdot}51\pm0{\cdot}20$				$-47{\cdot}33\pm0{\cdot}20$
	$-56{\cdot}07\pm0{\cdot}29$	C1	$7{\cdot}86\pm0{\cdot}01$	47/5	$-48{\cdot}21\pm0{\cdot}29$
A	$-56{\cdot}59\pm0{\cdot}26$				$-48{\cdot}73\pm0{\cdot}26$
Selected value	$-56{\cdot}07\pm0{\cdot}29$				$-48{\cdot}21\pm0{\cdot}29$
	$-55{\cdot}97\pm0{\cdot}29$	C1	$7{\cdot}89\pm0{\cdot}01$	47/5	$-48{\cdot}08\pm0{\cdot}29$
A	$-56{\cdot}63\pm0{\cdot}26$				$-48{\cdot}74\pm0{\cdot}26$
Selected value	$-55{\cdot}97\pm0{\cdot}29$				$-48{\cdot}08\pm0{\cdot}29$
	$-56{\cdot}53\pm0{\cdot}33$	C1	$7{\cdot}66\pm0{\cdot}01$	47/5, 47/6	$-48{\cdot}87\pm0{\cdot}33$
A	$-56{\cdot}40\pm0{\cdot}37$				$-48{\cdot}74\pm0{\cdot}37$
F	$-56{\cdot}70\pm0{\cdot}18$				$-49{\cdot}04\pm0{\cdot}18$
Selected value	$-56{\cdot}53\pm0{\cdot}33$				$-48{\cdot}87\pm0{\cdot}33$
H_6(l)+$2H_2$ =Ethylbenzene(l) {$-2{\cdot}95\pm0{\cdot}19$}	$+67{\cdot}7\ \pm1{\cdot}1$				
	$+60{\cdot}83\pm0{\cdot}32$	V1	$10{\cdot}30\pm0{\cdot}08$	49/10	$+71{\cdot}13\pm0{\cdot}33$
	$+59{\cdot}3\ \pm1{\cdot}4$				$+69{\cdot}6\ \pm1{\cdot}4$
Selected value	$+60{\cdot}83\pm0{\cdot}32$				$+71{\cdot}13\pm0{\cdot}33$
rrections made for incomplete combustion	$+129{\cdot}5\pm0{\cdot}8$	S4	$19{\cdot}2\pm0{\cdot}4$ (sub.)	66/17	$+148{\cdot}7\pm1{\cdot}0$
	$+24{\cdot}73\pm0{\cdot}22$	V1	$10{\cdot}50\pm0{\cdot}10$	46/15	$+35{\cdot}23\pm0{\cdot}26$
rrection made for ethylbenzene impurity	$+24{\cdot}91\pm0{\cdot}30$				$+35{\cdot}41\pm0{\cdot}33$
ABCE	$+25{\cdot}8\ \pm2{\cdot}1$				$+36{\cdot}3\ \pm2{\cdot}1$
C	$+21{\cdot}1\ \pm1{\cdot}6$				$+31{\cdot}6\ \pm1{\cdot}6$
hylcyclohexane(g) {$-41{\cdot}03\pm0{\cdot}37$} E	$+24{\cdot}97\pm0{\cdot}47$				$+35{\cdot}47\pm0{\cdot}45$
Selected value	$+24{\cdot}80\pm0{\cdot}23$				$+35{\cdot}30\pm0{\cdot}25$
H_{10}(l)+$4H_2$=n-C_8H_{18}(l) {$-59{\cdot}78\pm0{\cdot}25$}	$+79{\cdot}9\ \pm1{\cdot}3$				

C_aH_b

1 kcal = 4·184 kJ

								Determina	
1 Formula	2 g.f.w.	3 Name	State	Purity mol %	Type	No. of expts.	Detn. of reactn.	$-\Delta H_r^\circ$ kcal/g.f.w.	R
C_8H_{10}	106·1689	Ethylbenzene	l	99·86	SB	10	CO_2	1091·03±0·17	4
C_8H_{10}		Ethylbenzene	l		SB	5	m	1090·43±0·24	53
C_8H_{10}		Ethylbenzene	g		H	4	P/CO_2	48·18±0·10	37
C_8H_{10}		Ethylbenzene							
C_8H_{10}	106·1689	o-Xylene	l	99·89	SB	6	CO_2	1088·16±0·24	4
C_8H_{10}		o-Xylene	l		SB	5	m	1087·83±0·12	46
C_8H_{10}		o-Xylene	g		H	2	P/CO_2	46·51±0·20	37
C_8H_{10}		o-Xylene							
C_8H_{10}	106·1689	m-Xylene	l	99·77	SB	8	CO_2	1087·92±0·15	45
C_8H_{10}		m-Xylene	l		SB	6	m	1087·37±0·13	46
C_8H_{10}		m-Xylene							
C_8H_{10}	106·1689	p-Xylene	l	99·92	SB	4	CO_2	1088·16±0·22	45
C_8H_{10}		p-Xylene	l		SB	4	m	1087·82±0·12	46
C_8H_{10}		p-Xylene							
C_8H_{10}	106·1689	Dimethylfulvene	l		SB	5	m	1115·5 ±1·2	57/
C_8H_{12}	108·1848	Oct-3-yne-1-ene	l		H	4	H_2	93·4 ±1·5	58/
C_8H_{12}	108·1848	Bicyclo [2, 2, 2] octene-2	c		SB			1156·72±0·08	67/
C_8H_{14}	110·2008	Allylcyclopentane	l	99·90±0·10	SB	5	CO_2	1214·87±0·23	61/
C_8H_{14}	110·2008	1-Ethylcyclohexene	l	99·90±0·09	SB	5	CO_2	1205·11±0·23	61/
C_8H_{14}	110·2008	Ethylidenecyclohexane	l	99·87±0·06	SB	5	CO_2	1205·88±0·14	61/
C_8H_{14}	110·2008	Vinylcyclohexane	l	99·95±0·04	SB	5	CO_2	1209·42±0·18	61/
C_8H_{14}	110·2008	Cyclo-octene	g		H	3	P/CO_2	23·28±0·15	39/
C_8H_{14}	110·2008	Bicyclo [4, 2, 0] octane	l		SB			1214·37±0·59	67/1
C_8H_{14}	110·2008	Bicyclo [5, 1, 0] octane	l		SB			1216·34±0·54	67/1
C_8H_{14}	110·2008	cis-Bicyclo [3, 3, 0] octane	l		SB	7	m	1198·86±0·66	36/
C_8H_{14}	110·2008	trans-Bicyclo [3, 3, 0]-octane	l		SB	4	m	1205·0 ±1·2	36/
C_8H_{14}	110·2008	Bicyclo [2, 2, 2] octane	c		SB			1195·49±0·10	67/1

$$1 \text{ kcal} = 4.184 \text{ kJ}$$

ΔH_r° Remarks		5 ΔH_f° (l or c) kcal/g.f.w.	6 Determination of ΔH_v			7 ΔH_f° (g) kcal/g.f.w.
			Type	ΔH_v° kcal/g.f.w.	Ref.	
		$-2\cdot95\pm0\cdot19$	C1	$10\cdot10\pm0\cdot01$	47/5,	$+7\cdot15\pm0\cdot19$
		$-3\cdot55\pm0\cdot25$			45/7	$+6\cdot55\pm0\cdot25$
ylcyclohexane(g) $\{-41\cdot03\pm0\cdot37\}$	E	$-2\cdot95\pm0\cdot40$				$+7\cdot15\pm0\cdot40$
Selected value		$-2\cdot95\pm0\cdot19$				$+7\cdot15\pm0\cdot19$
		$-5\cdot82\pm0\cdot26$	C1	$10\cdot38\pm0\cdot01$	47/5	$+4\cdot56\pm0\cdot26$
	F	$-6\cdot15\pm0\cdot15$				$+4\cdot23\pm0\cdot15$
-1, 2-dimethylcyclohexane (g)	E	$-5\cdot00\pm0\cdot49$				$+5\cdot38\pm0\cdot49$
$\{-41\cdot13\pm0\cdot44\}$						
Selected value		$-5\cdot82\pm0\cdot26$				$+4\cdot56\pm0\cdot26$
		$-6\cdot06\pm0\cdot18$	C1	$10\cdot20\pm0\cdot01$	47/5	$+4\cdot14\pm0\cdot18$
	F	$-6\cdot61\pm0\cdot17$				$+3\cdot59\pm0\cdot17$
Selected value		$-6\cdot06\pm0\cdot18$				$+4\cdot14\pm0\cdot18$
		$-5\cdot82\pm0\cdot24$	C1	$10\cdot13\pm0\cdot01$	47/5	$+4\cdot31\pm0\cdot24$
	F	$-6\cdot16\pm0\cdot15$				$+3\cdot97\pm0\cdot15$
Selected value		$-5\cdot82\pm0\cdot24$				$+4\cdot31\pm0\cdot24$
	C	$+21\cdot5\pm1\cdot2$	V5	$10\cdot6\pm0\cdot5$	57/11	$+32\cdot1\pm1\cdot3$
$H_{12}(l)+3H_2=n\text{-}C_8H_{18}(l)\ \{-59\cdot78\pm0\cdot25\}$		$+33\cdot6\pm1\cdot6$				
		$-5\cdot58\pm0\cdot13$				
		$-15\cdot74\pm0\cdot25$				
		$-25\cdot50\pm0\cdot25$	V1	$10\cdot34\pm0\cdot05$	60/12	$-15\cdot16\pm0\cdot26$
		$-24\cdot73\pm0\cdot18$				
		$-21\cdot19\pm0\cdot21$				
clo-octane(g) $\{-29\cdot73\pm0\cdot26\}$	E					$-6\cdot45\pm0\cdot30$
		$-16\cdot24\pm0\cdot60$		$9\cdot85\pm0\cdot20$	67/10	$-6\cdot39\pm0\cdot65$
		$-14\cdot27\pm0\cdot55$		$10\cdot42\pm0\cdot10$	67/10	$-3\cdot85\pm0\cdot58$
	BCE	$-31\cdot75\pm0\cdot70$				
	BCE	$-25\cdot6\pm1\cdot2$				
		$-35\cdot12\pm0\cdot15$				

THERMOCHEMISTRY OF ORGANIC AND ORGANOMETALLIC COMPOUNDS

C_aH_b

1 kcal $= 4\cdot184$ kJ

Determina▶

1 Formula	2 g.f.w	3 Name	State	Purity mol %	Type	No. of expts.	Detn. of reactn.	$-\Delta_r$ kcal/g.f.w.	R
C_8H_{16}	112·2167	1-Octene	l	99·77±0·13	SB	4	CO_2	1269·82±0·27	61/
C_8H_{16}	112·2167	2, 2-Dimethyl-*cis*-3-hexene	l	99·86±0·12	SB	4	CO_2	1268·71±0·66	61/
C_8H_{10}	112·2167	2, 2-Dimethyl-*trans*-3-hexene	l	99·85±0·10	SB	4	CO_2	1264·29±0·38	61/
C_8H_{10}	112·2167	2-Methyl-3-ethyl-1-pentene	l	99·81±0·08	SB	4	CO_2	1265·98±0·30	61/
C_8H_{16}	112·2167	2, 4, 4-Trimethyl-1-pentene	l	99·95±0·03	SB	4	CO_2	1264·13±0·30	61/
C_8H_{16}		2, 4, 4-Trimethyl-1-pentene	l		H		m	28·6 ±0·8	36/
C_8H_{16}		2, 4, 4-Trimethyl-1-pentene	g	99·5	H	3	P/CO_2	26·99±0·06	37/
C_8H_{16}		2, 4, 4-Trimethyl-1-pentene							
C_8H_{16}	112·2167	2, 4, 4-Trimethyl-2-pentene	l	99·93±0·05	SB	5	CO_2	1264·90±0·49	61/▶
C_8H_{16}	112·2167	1, 1-Dimethyl-2-propylcyclopropane	l		SB	8	m	1271·2 ±0·4	60/▶
C_8H_{16}	112·2167	n-Propylcyclopentane	l	99·81±0·10	SB	7	CO_2	1253·74±0·28	46/
C_8H_{16}	112·2167	Ethylcyclohexane	l	99·89±0·08	SB	5	CO_2	1248·23±0·35	46/.
C_8H_{16}	112·2167	1, 1-Dimethyl-cyclohexane	l	99·93±0·03	SB	5	CO_2	1246·65±0·45	47/1
C_8H_{16}	112·2167	*cis*-1, 2-Dimethyl-cyclohexane	l	99·98±0·02	SB	5	CO_2	1248·31±0·43	47/1
C_8H_{16}	112·2167	*trans*-1, 2-Dimethyl-cyclohexane	l	99·92±0·07	SB	6	CO_2	1246·77±0·44	47/1
C_8H_{16}	112·2167	*cis*-1, 3-Dimethyl-cyclohexane	l	99·94±0·05	SB	6	CO_2	1245·66±0·41	47/1
C_8H_{16}	112·2167	*trans*-1, 3-Dimethyl-cyclohexane	l	99·88±0·07	SB	4	CO_2	1247·38±0·40	47/1.

1 kcal $= 4\cdot184$ kJ

ΔH_r° Remarks	5 ΔH_f° (l or c) kcal/g.f.w.	Type	6 Determination of ΔH_v ΔH_v° kcal/g.f.w.	Ref.	7 ΔH_f° (g) kcal/g.f.w.
	$-29\cdot11\pm0\cdot30$	V1	$9\cdot70\pm0\cdot05$	50/8	$-19\cdot41\pm0\cdot31$
	$-30\cdot22\pm0\cdot67$	V1	$8\cdot88\pm0\cdot05$	60/12	$-21\cdot34\pm0\cdot68$
	$-34\cdot64\pm0\cdot41$	V1	$8\cdot91\pm0\cdot05$	60/12	$-25\cdot73\pm0\cdot42$
	$-32\cdot95\pm0\cdot33$	V1	$8\cdot98\pm0\cdot05$	60/12	$-23\cdot97\pm0\cdot34$
	$-34\cdot80\pm0\cdot33$	V1	$8\cdot55\pm0\cdot04$	60/12	$-26\cdot25\pm0\cdot34$
H_{16}(l)$+H_2$ $=$2, 2, 4-Trimethylpentane(l) $\{-61\cdot94\pm0\cdot37\}$	$-33\cdot3\pm0\cdot9$				$-24\cdot7\pm0\cdot9$
2, 4-Trimethylpentane(g) $\{-53\cdot54\pm0\cdot37\}$	$-35\cdot10\pm0\cdot39$				$-26\cdot55\pm0\cdot39$
Selected value	$-34\cdot92\pm0\cdot30$				$-26\cdot37\pm0\cdot30$
	$-34\cdot03\pm0\cdot51$	V1	$8\cdot96\pm0\cdot04$	60/12	$-25\cdot07\pm0\cdot52$
	$-27\cdot7\pm0\cdot4$				
	$-45\cdot19\pm0\cdot30$	C1	$9\cdot82\pm0\cdot01$	47/5	$-35\cdot37\pm0\cdot30$
	$-50\cdot70\pm0\cdot37$	C1	$9\cdot67\pm0\cdot01$	47/5	$-41\cdot03\pm0\cdot37$
	$-52\cdot28\pm0\cdot46$	V1	$9\cdot05\pm0\cdot05$	53/1	$-43\cdot23\pm0\cdot47$
	$-50\cdot62\pm0\cdot44$	C1	$9\cdot49\pm0\cdot01$	47/5	$-41\cdot13\pm0\cdot44$
	$-52\cdot16\pm0\cdot45$	C1	$9\cdot17\pm0\cdot01$	47/5	$-42\cdot99\pm0\cdot45$
	$-53\cdot27\pm0\cdot42$	C1	$9\cdot14\pm0\cdot01$	47/5	$-44\cdot13\pm0\cdot42$
	$-51\cdot55\pm0\cdot41$	C1	$9\cdot37\pm0\cdot01$	47/5	$-42\cdot18\pm0\cdot41$

C_aH_b

1 kcal $= 4 \cdot 184$ kJ

1 Formula	2 g.f.w.	3 Name	State	Purity mol %	Type	No. of expts.	Detn. of reactn.	$-\Delta H_r^\circ$ kcal/g.f.w.	R
C_8H_{16}	$112 \cdot 2167$	cis-1, 4-Dimethyl- cyclohexane	l	$99 \cdot 94 \pm 0 \cdot 04$	SB	5	CO_2	$1247 \cdot 40 \pm 0 \cdot 41$	47/
C_8H_{16}	$112 \cdot 2167$	trans-1, 4-Dimethyl- cyclohexane	l	$99 \cdot 89 \pm 0 \cdot 08$	SB	5	CO_2	$1245 \cdot 78 \pm 0 \cdot 41$	47/
C_8H_{16}	$112 \cdot 2167$	Cyclo-octane	l	$> 99 \cdot 8$	SB	5	m	$1258 \cdot 84 \pm 0 \cdot 22$	52/
C_8H_{16}		Cyclo-octane	l		SB	5	m	$1258 \cdot 44 \pm 0 \cdot 38$	47/
C_8H_{16}		Cyclo-octane	l		SB	3	m	$1255 \cdot 3 \pm 2 \cdot 4$	33/
C_8H_{16}		Cyclo-octane							
C_8H_{18}	$114 \cdot 2327$	n-Octane	l	$99 \cdot 65 \pm 0 \cdot 05$	SB	10	CO_2	$1307 \cdot 54 \pm 0 \cdot 25$	44/
C_8H_{18}		n-Octane	l	$99 \cdot 95$	SB	6	CO_2	$1307 \cdot 38 \pm 0 \cdot 34$	37/
C_8H_{18}		n-Octane	l	$99 \cdot 95$	SB	3	m	$1307 \cdot 8 \pm 1 \cdot 3$	45/ 33/
C_8H_{18}		n-Octane							
C_8H_{18}	$114 \cdot 2327$	2-Methylheptane	l	> 97	SB	7	CO_2	$1306 \cdot 29 \pm 0 \cdot 34$	45/
C_8H_{18}	$114 \cdot 2327$	3-Methylheptane	l	> 97	SB	6	CO_2	$1306 \cdot 93 \pm 0 \cdot 31$	45/
C_8H_{18}	$114 \cdot 2327$	4-Methylheptane	l	$> 98 \cdot 5$	SB	5	CO_2	$1307 \cdot 10 \pm 0 \cdot 31$	45/
C_8H_{18}	$114 \cdot 2327$	3-Ethylhexane	l	> 98	SB	5	CO_2	$1307 \cdot 40 \pm 0 \cdot 30$	45/
C_8H_{18}	$114 \cdot 2327$	2, 2-Dimethylhexane	l	$> 98 \cdot 5$	SB	4	CO_2	$1304 \cdot 65 \pm 0 \cdot 30$	45/
C_8H_{18}	$114 \cdot 2327$	2, 3-Dimethylhexane	l	> 95	SB	7	CO_2	$1306 \cdot 87 \pm 0 \cdot 38$	45/
C_8H_{18}	$114 \cdot 2327$	2, 4-Dimethylhexane	l	> 98	SB	5	CO_2	$1305 \cdot 81 \pm 0 \cdot 31$	45/
C_8H_{18}	$114 \cdot 2327$	2, 5-Dimethylhexane	l	> 99	SB	5	CO_2	$1305 \cdot 01 \pm 0 \cdot 38$	45/
C_8H_{18}	$114 \cdot 2327$	3, 3-Dimethylhexane	l	$> 99 \cdot 5$	SB	5	CO_2	$1305 \cdot 69 \pm 0 \cdot 30$	45/
C_8H_{18}	$114 \cdot 2327$	3, 4-Dimethylhexane	l	> 98	SB	5	CO_2	$1307 \cdot 05 \pm 0 \cdot 38$	45/
C_8H_{18}	$114 \cdot 2327$	2-Methyl- 3-ethylpentane	l	> 99	SB	5	CO_2	$1307 \cdot 59 \pm 0 \cdot 33$	45/
C_8H_{18}	$114 \cdot 2327$	3-Methyl- 3-ethylpentane	l	$> 99 \cdot 4$	SB	5	CO_2	$1306 \cdot 81 \pm 0 \cdot 33$	45/
C_8H_{18}	$114 \cdot 2327$	2, 2, 3-Trimethylpentane	l	$> 99 \cdot 5$	SB	5	CO_2	$1305 \cdot 84 \pm 0 \cdot 38$	45/
C_8H_{18}	$114 \cdot 2327$	2, 2, 4-Trimethylpentane	l	$> 99 \cdot 8$	SB	6	CO_2	$1305 \cdot 30 \pm 0 \cdot 35$	45/

Determinat

1 kcal = 4·184 kJ

$H°$	Remarks	5 ΔH_f° (l or c) kcal/g.f.w.	6 Determination of ΔH_v			7 ΔH_f° (g) kcal/g.f.w.
			Type	ΔH_v° kcal/g.f.w.	Ref.	
		$-51\cdot53\pm0\cdot42$	Cl	$9\cdot33\pm0\cdot01$	47/5	$-42\cdot20\pm0\cdot42$
		$-53\cdot15\pm0\cdot42$	Cl	$9\cdot05\pm0\cdot01$	47/5	$-44\cdot10\pm0\cdot42$
		$-40\cdot09\pm0\cdot25$	V1	$10\cdot36\pm0\cdot05$	56/10	$-29\cdot73\pm0\cdot28$
		$-40\cdot49\pm0\cdot39$				$-30\cdot13\pm0\cdot40$
	BCE	$-43\cdot6\ \pm2\cdot4$				$-33\cdot2\ \pm2\cdot4$
	Selected value	$-40\cdot09\pm0\cdot25$				$-29\cdot73\pm0\cdot28$
		$-59\cdot70\pm0\cdot28$	Cl	$9\cdot92\pm0\cdot01$	47/5	$-49\cdot78\pm0\cdot28$
		$-59\cdot86\pm0\cdot36$				$-49\cdot94\pm0\cdot36$
	CE	$-59\cdot4\ \pm1\cdot3$				$-49\cdot5\ \pm1\cdot3$
	Selected value	$-59\cdot78\pm0\cdot25$				$-49\cdot86\pm0\cdot25$
		$-60\cdot95\pm0\cdot36$	Cl	$9\cdot48\pm0\cdot01$	47/5	$-51\cdot47\pm0\cdot36$
		$-60\cdot31\pm0\cdot33$	Cl	$9\cdot52\pm0\cdot01$	47/5	$-50\cdot79\pm0\cdot33$
		$-60\cdot14\pm0\cdot33$	Cl	$9\cdot48\pm0\cdot01$	47/5	$-50\cdot66\pm0\cdot33$
		$-59\cdot84\pm0\cdot32$	Cl	$9\cdot48\pm0\cdot01$	47/5	$-50\cdot36\pm0\cdot32$
		$-62\cdot59\pm0\cdot32$	Cl	$8\cdot91\pm0\cdot01$	47/5	$-53\cdot68\pm0\cdot32$
		$-60\cdot37\pm0\cdot40$	Cl	$9\cdot27\pm0\cdot01$	47/5	$-51\cdot10\pm0\cdot40$
		$-61\cdot43\pm0\cdot33$	Cl	$9\cdot03\pm0\cdot01$	47/5	$-52\cdot40\pm0\cdot33$
		$-62\cdot23\pm0\cdot40$	Cl	$9\cdot05\pm0\cdot01$	47/5	$-53\cdot18\pm0\cdot40$
		$-61\cdot55\pm0\cdot32$	Cl	$8\cdot97\pm0\cdot01$	47/5	$-52\cdot58\pm0\cdot32$
		$-60\cdot19\pm0\cdot40$	Cl	$9\cdot32\pm0\cdot01$	47/5	$-50\cdot87\pm0\cdot40$
		$-59\cdot65\pm0\cdot35$	Cl	$9\cdot21\pm0\cdot01$	47/5	$-50\cdot44\pm0\cdot35$
		$-60\cdot43\pm0\cdot35$	Cl	$9\cdot08\pm0\cdot01$	47/5	$-51\cdot35\pm0\cdot35$
		$-61\cdot40\pm0\cdot40$	Cl	$8\cdot82\pm0\cdot01$	47/5	$-52\cdot58\pm0\cdot40$
		$-61\cdot94\pm0\cdot37$	Cl	$8\cdot40\pm0\cdot01$	47/5	$-53\cdot54\pm0\cdot37$

THERMOCHEMISTRY OF ORGANIC AND ORGANOMETALLIC COMPOUNDS

C_aH_b

$$1 \text{ kcal} = 4 \cdot 184 \text{ kJ}$$

Determina

1 Formula	2 g.f.w.	3 Name	State	Purity mol %	Type	No. of expts.	Detn. of reactn.	$-\Delta H_f^{\circ}$ kcal/g.f.w.	R
C_8H_{18}	114·2327	2, 3, 3-Trimethylpentane	l	>99	SB	5	CO_2	1306·65±0·36	4·
C_8H_{18}	114·2327	2, 3, 4-Trimethylpentane	l	>99·5	SB	5	CO_2	1306·29±0·41	4
C_8H_{18}	114·2327	2, 2, 3, 3-Tetramethyl-butane	c	>98	SB	4	CO_2	1303·04±0·46	4!
C_9H_8	116·1641	Indene	l	99·90	SB	5	m	1146·16±0·30	61
C_9H_8		Indene	g		H	6	P/CO_2	69·93±0·50	3‍
C_9H_8		Indene							
C_9H_{10}	118·1801	Indane	l	99·89	SB	5	m	1190·64±0·47	61
C_9H_{10}		Indane	g		H	4	P/CO_2	45·06±0·25	37
C_9H_{10}		Indane	g		E		an	21·63±0·14	50
C_9H_{10}		Indane							
C_9H_{10}	118·1801	α-Methylstyrene	l	99·6	SB	6	CO_2	1204·87±0·26	51
C_9H_{10}	118·1801	Phenylcyclopropane	l		SB			1212·0 ±0·2	61
C_9H_{10}		Phenylcyclopropane	l	99·95, glc	SB			1213·3 ±	62
C_9H_{10}		Phenylcyclopropane							
C_9H_{12}	120·1960	n-Propylbenzene	l	99·65	SB	11	CO_2	1247·19±0·16	45
C_9H_{12}	120·1960	Isopropylbenzene	l	99·95±0·02	SB	7	CO_2	1246·52±0·23	45
C_9H_{12}	120·1960	1-Methyl-2-ethylbenzene	l	99·76±0·07	SB	5	CO_2	1245·26±0·23	45
C_9H_{12}	120·1960	1-Methyl-3-ethylbenzene	l	99·78±0·15	SB	7	CO_2	1244·71±0·25	45
C_9H_{12}	120·1960	1-Methyl-4-ethylbenzene	l	99·95±0·02	SB	7	CO_2	1244·45±0·31	45
C_9H_{12}	120·1960	1, 2, 3-Trimethylbenzene	l	99·92±0·04	SB	6	CO_2	1242·36±0·27	45
C_9H_{12}	120·1960	1, 2, 4-Trimethylbenzene	l	99·64±0·20	SB	5	CO_2	1241·58±0·23	45
C_9H_{12}	120·1960	1 3, 5-Trimethylbenzene	l	99·93±0·03	SB	6	CO_2	1241·19+0·31	45

$$1 \text{ kcal} = 4\cdot184 \text{ kJ}$$

ΔH_r° Remarks	ΔH_f° (l or c) kcal/g.f.w.	6 Determination of ΔH_v Type	ΔH_v° kcal/g.f.w.	Ref.	7 ΔH_f° (g) kcal/g.f.w.
	$-60\cdot59\pm0\cdot38$	C1	$8\cdot90\pm0\cdot01$	47/5	$-51\cdot69\pm0\cdot38$
	$-60\cdot95\pm0\cdot43$	C1	$9\cdot01\pm0\cdot01$	47/5	$-51\cdot94\pm0\cdot43$
	$-64\cdot20\pm0\cdot48$	S1 (sub.)	$10\cdot37\pm0\cdot05$	52/4	$-53\cdot83\pm0\cdot49$
$+$ *trans*-hexahydroindane(g) E	$+26\cdot44\pm0\cdot31$ $+26\cdot4\ \pm1\cdot2$	V2	$12\cdot64\pm0\cdot20$	61/15	$+39\cdot08\pm0\cdot37$ $+39\cdot0\ \pm1\cdot2$
$\{-30\cdot9\pm1\cdot0\}$ Selected value	$+26\cdot44\pm0\cdot31$				$+39\cdot08\pm0\cdot37$
$+$ *trans*-hexahydroindane(g) E	$+2\cdot61\pm0\cdot46$ $+2\cdot4\ \pm1\cdot1$	V2	$11\cdot81\pm0\cdot20$	61/15	$+14\cdot42\pm0\cdot53$ $+14\cdot2\ \pm1\cdot1$
$\{-30\cdot9\pm1\cdot0\}$ .dene(g) $\{+39\cdot08\pm0\cdot37\}+H_2$ $\rightleftharpoons C_9H_{10}$(g). 2nd Law	$+5\cdot6\ \pm0\cdot5$*				$+17\cdot45\pm0\cdot42$*
Selected value	$+2\cdot61\pm0\cdot46$				$+14\cdot42\pm0\cdot53$
An 0·99948: correction made for 0·4% styrene	$+16\cdot84\pm0\cdot28$				
B.S. measurement	$+24\cdot0\ \pm0\cdot3$ $+25\cdot3\ \pm$				
Selected value	$+24\cdot7\ \pm0\cdot7$				
	$-9\cdot16\pm0\cdot19$	C1, V1	$11\cdot05\pm0\cdot01$	47/5, 65/16	$+1\cdot89\pm0\cdot19$
	$-9\cdot83\pm0\cdot26$	C1	$10\cdot79\pm0\cdot01$	47/5	$+0\cdot96\pm0\cdot26$
	$-11\cdot09\pm0\cdot26$	V1	$11\cdot40\pm0\cdot05$	53/1	$+0\cdot39\pm0\cdot27$
	$-11\cdot64\pm0\cdot28$	V1	$11\cdot21\pm0\cdot05$	53/1	$-0\cdot43\pm0\cdot29$
	$-11\cdot90\pm0\cdot33$	V1	$11\cdot14\pm0\cdot05$	53/1	$-0\cdot76\pm0\cdot34$
	$-13\cdot99\pm0\cdot29$	C1	$11\cdot73\pm0\cdot01$	47/5	$-2\cdot26\pm0\cdot29$
	$-14\cdot77\pm0\cdot26$	C1	$11\cdot46\pm0\cdot01$	47/5	$-3\cdot31\pm0\cdot26$
	$-15\cdot16\pm0\cdot33$	C1	$11\cdot35\pm0\cdot01$	47/5, 60/4	$-3\cdot81\pm0\cdot33$

THERMEOCHEMISTRY OF ORGANIC AND ORGANOMETALLIC COMPOUNDS

C_aH_b 1 kcal$=4\cdot184$ kJ

Determinat

1 Formula	2 g.f.w.	3 Name	State	Purity mol %	Type	No. of expts.	Detn. of reactn.	$-\Delta H_c^\circ$ kcal/g.f.w.	R
C_9H_{16}	124·2279	cis-Hexahydroindane	l	99·95±0·02	SB	5	CO_2	1351·60±0·36	60/
C_9H_{16}		cis-Hexahydroindane	l		SB	3	m	1349·6 ±1·4	37,
C_9H_{16}		cis-Hexahydroindane							
C_9H_{16}	124·2279	trans-Hexahydroindane	l	99·71±0·11	SB	5	CO_2	1350·86±0·41	60/
C_9H_{16}		trans-Hexahydroindane	l		SB	3	m	1348·2 ±1·4	37,
C_9H_{16}		trans-Hexahydroindane							
C_9H_{16}	124·2279	Cyclohexylcyclopropane	l	99·95, glc	SB			1317·8 ±	62/
C_9H_{16}	124·2279	Bicyclo [6, 1, 0] nonane	l		SB			1373·55±0·61	67/
C_9H_{16}	124·2279	Spiro [4, 4] nonane	l		SB	>5	m	1358·6 ±0·3	64/
C_9H_{18}	126·2438	n-Propylcyclohexane	l	>99·7	SB	6	CO_2	1404·34±0·27	46,
C_9H_{18}	126·2438	cis-1, 3, 5-Trimethyl-cyclohexane	g		H	4	P/CO_2	46·88±0·20	37,
C_9H_{18}	126·2438	Cyclononane	l	>99·8	SB	5	m	1417·99±0·22	52,
C_9H_{20}	128·2598	n-Nonane	l	99·3±0·2	SB	4	CO_2	1463·89±0·27	44,
C_9H_{20}		n-Nonane	l		SB	6	CO_2	1463·67±0·45	37/
C_9H_{20}		n-Nonane							45,
C_9H_{20}	128·2598	3, 3-Diethylpentane	l	99·99±0·01	SB	6	CO_2	1463·79±0·38	47/
C_9H_{20}	128·2598	2, 2, 3, 3-Tetramethyl-pentane	l	99·94±0·02	SB	6	CO_2	1463·10±0·36	47/
C_9H_{20}	128·2598	2, 2, 3, 4-Tetramethyl-pentane	l	99·99±0·01	SB	5	CO_2	1463·24±0·28	47/
C_9H_{20}	128·2598	2, 2, 4, 4-Tetramethyl-pentane	l	99·89±0·08	SB	7	CO_2	1462·69±0·31	47/
C_9H_{20}		2, 2, 4, 4-Tetramethyl-pentane	l		SB		m	1462·2 ±0·6	59/
C_9H_{20}		2, 2, 4, 4-Tetramethyl-pentane							
C_9H_{20}	128·2598	2, 3, 3, 4-Tetramethyl-pentane	l	99·96±0·04	SB	5	CO_2	1463·18±0·39	47/

$$1 \text{ kcal} = 4 \cdot 184 \text{ kJ}$$

ΔH_r°		5 ΔH_f° (l or c) kcal/g.f.w.	6 Determination of ΔH_v			7 ΔH_f° (g) kcal/g.f.w.
Remarks			Type	ΔH_v° kcal/g.f.w.	Ref.	
		$-41 \cdot 38 \pm 0 \cdot 38$	V2	$11 \cdot 00 \pm 0 \cdot 30$	60/15	$-30 \cdot 38 \pm 0 \cdot 48$
	CE	$-43 \cdot 4 \ \pm 1 \cdot 4$				$-32 \cdot 4 \ \pm 1 \cdot 5$
	Selected value	$-41 \cdot 38 \pm 0 \cdot 38$				$-30 \cdot 38 \pm 0 \cdot 48$
		$-42 \cdot 12 \pm 0 \cdot 43$	V2	$10 \cdot 70 \pm 0 \cdot 30$	60/15	$-31 \cdot 42 \pm 0 \cdot 52$
	CE	$-44 \cdot 8 \ \pm 1 \cdot 4$				$-34 \cdot 1 \ \pm 1 \cdot 5$
	Selected value	$-42 \cdot 12 \pm 0 \cdot 43$				$-31 \cdot 42 \pm 0 \cdot 52$
B.S. measurement		$-75 \cdot 2 \ \pm$				
		$-19 \cdot 43 \pm 0 \cdot 62$		$12 \cdot 04 \pm 0 \cdot 10$	67/10	$-7 \cdot 39 \pm 0 \cdot 64$
An $0 \cdot 9993 \pm 0 \cdot 0002$		$-34 \cdot 4 \ \pm 0 \cdot 3$				
		$-56 \cdot 95 \pm 0 \cdot 30$	C1, V1	$10 \cdot 78 \pm 0 \cdot 01$	47/5, 65/17	$-46 \cdot 17 \pm 0 \cdot 30$
3, 5-Trimethylbenzene(g) $\{-3 \cdot 81 \pm 0 \cdot 33\}$						$-50 \cdot 69 \pm 0 \cdot 39$
		$-43 \cdot 30 \pm 0 \cdot 26$	E5	$11 \cdot 57 \pm 0 \cdot 30$	57/45	$-31 \cdot 73 \pm 0 \cdot 40$
		$-65 \cdot 72 \pm 0 \cdot 30$	C1	$11 \cdot 10 \pm 0 \cdot 01$	47/5	$-54 \cdot 62 \pm 0 \cdot 30$
		$-65 \cdot 94 \pm 0 \cdot 47$				$-54 \cdot 84 \pm 0 \cdot 47$
	Selected value	$-65 \cdot 76 \pm 0 \cdot 25$				$-54 \cdot 66 \pm 0 \cdot 25$
		$-65 \cdot 82 \pm 0 \cdot 41$	E2	$10 \cdot 41 \pm 0 \cdot 10$	61/17	$-55 \cdot 41 \pm 0 \cdot 44$
		$-66 \cdot 51 \pm 0 \cdot 39$	E2	$9 \cdot 84 \pm 0 \cdot 10$	61/17	$-56 \cdot 67 \pm 0 \cdot 42$
		$-66 \cdot 37 \pm 0 \cdot 31$	E2	$9 \cdot 76 \pm 0 \cdot 10$	61/17	$-56 \cdot 61 \pm 0 \cdot 36$
		$-66 \cdot 92 \pm 0 \cdot 34$	E2	$9 \cdot 12 \pm 0 \cdot 10$	61/17	$-57 \cdot 80 \pm 0 \cdot 39$
		$-67 \cdot 4 \ \pm 0 \cdot 6$				$-58 \cdot 3 \ \pm 0 \cdot 6$
	Selected value	$-66 \cdot 92 \pm 0 \cdot 34$				$-57 \cdot 80 \pm 0 \cdot 39$
		$-66 \cdot 43 \pm 0 \cdot 42$	E2	$10 \cdot 00 \pm 0 \cdot 10$	61/17	$-56 \cdot 43 \pm 0 \cdot 45$

THERMOCHEMISTRY OF ORGANIC AND ORGANOMETALLIC COMPOUNDS

C_aH_b

1 kcal = 4·184 kJ

Determinati

1 Formula	2 g.f.w.	3 Name	State	Purity mol %	Type	No. of expts.	Detn. of reactn.	$-\Delta H_r^{\circ}$ kcal/g.f.w.	Re
$C_{10}H_8$	128·1753	Azulene	c		SB	10	m	1264·5 ±0·8	55/ 57/
$C_{10}H_8$	128·1753	Naphthalene	c	99·97±0·03	SB	6	CO_2	1232·54±0·38	60/
$C_{10}H_8$		Naphthalene	c		SB	7	m	1230·20±0·74	63/
$C_{10}H_8$		Naphthalene	c		SB	3	m	1229·3 ±0·6	52
$C_{10}H_8$		Naphthalene	c	99·98±0·01	SB	4	CO_2	1232·35±0·22	66/
$C_{10}H_8$		Naphthalene	c		SB		m	1233·4 ±0·5	39
$C_{10}H_8$		Naphthalene	c		SB	6	m	1230·26±0·36	35
$C_{10}H_8$		Naphthalene	c		SB	5	m	1231·5 ±1·2	32
$C_{10}H_8$		Naphthalene	c		SB	6	m	1231·10±0·36	31
$C_{10}H_8$		Naphthalene	c		SB	14	m	1230·9 ±1·2	31
$C_{10}H_8$		Naphthalene							
$C_{10}H_{10}$	130·1912	1, 2-Dihydronaphthalene	l		H	3	m	24·08±0·20	42
$C_{10}H_{10}$	130·1912	1, 4-Dihydronaphthalene	l		H	3	m	27·11±0·10	42
$C_{10}H_{12}$	132·2071	1, 2, 3, 4-Tetrahydro-naphthalene	l		SB	3	m	1334·1 ±2·6	51
$C_{10}H_{12}$		1, 2, 3, 4-Tetrahydro-naphthalene	g		E		glc	28·8 ±1·2	58/
$C_{10}H_{12}$		1, 2, 3, 4-Tetrahydro-naphthalene							
$C_{10}H_{12}$	132·2071	α-Dicyclopentadiene	c		SB	6	m	1378·3 ±1·4	34
$C_{10}H_{12}$	132·2071	Dicyclopentadiene	g		E			−17·3 ±1·0	41
$C_{10}H_{14}$	134·2231	n-Butylbenzene	l		SB		CO_2	1403·46±0·27	46
$C_{10}H_{14}$	134·2231	Isobutylbenzene	l		SB		CO_2	1402·04±0·29	46
$C_{10}H_{14}$	134·2231	s-Butylbenzene	l		SB		CO_2	1402·85±0·27	46
$C_{10}H_{14}$	134·2231	t-Butylbenzene	l		SB		CO_2	1401·82±0·27	46
$C_{10}H_{14}$	134·2231	1, 2, 3, 4-Tetramethyl-benzene	l		SB	3	m	1392·8 ±1·4	33
$C_{10}H_{14}$	134·2231	1, 2, 3, 5-Tetramethyl-benzene	l		SB	3	m	1389·9 ±1·4	33.

1 kcal = 4·184 kJ

ΔH_r°		5 ΔH_f° (l or c) kcal/g.f.w.	6 Determination of ΔH_v			7 ΔH_f° (g) kcal/g.f.w.
Remarks			Type	ΔH_v° kcal/g.f.w.	Ref.	
		+50·7 ±0·8	S3	22·8 ±0·1* (sub.)	62/16	+73·5 ±0·9
		+18·77±0·40	S3	17·42±0·07 (sub.)	63/6	+36·19±0·41
icrobomb: sample mass *ca.*12mg		+16·43±0·75				+33·85±0·76
	C	+15·5 ±0·6				+32·9 ±0·6
ʼAn 0·99955±0·00013		+18·58±0·25				+36·00±0·28
⊃nverted to modern units in 60/16		+19·6 ±0·5				+37·0 ±0·5
	C	+16·49±0·38				+33·91±0·39
	BCE	+17·7 ±1·2				+35·1 ±1·2
		+17·33±0·38				+34·75±0·39
	BCE	+17·1 ±1·2				+34·5 ±1·2
Selected value		+18·63±0·22				+36·05±0·25
$_0H_{10}$(l)+H_2=Tetrahydronaphthalene(l) {−6·1±1·4}		+18·0 ±1·5				
$_0H_{10}$(l)+H_2=Tetrahydronaphthalene(l) {−6·1±1·4}		+21·0 ±1·5				
		−16·3 ±2·6	V3	13·4±[0·4]	55/16	−2·9 ±2·7
ıphthalene(g) {+36·05±0·25}+2H_2 ⇌$C_{10}H_{12}$(g). 2nd Law	E	−6·1 ±1·4				+7·3 ±1·3
Selected value		−6·1 ±1·4				+7·3 ±1·3
	CE	+27·9 ±1·4				
cyclopentadiene(g) ⇌2 Cyclopentadiene (g) {+31·94±0·28}. 2nd Law						+46·6 ±1·2
		−15·26±0·30	V1	11·98±0·01	46/7	−3·28±0·30
		−16·68±0·32	V1	11·54±0·03	46/7	−5·14±0·33
		−15·87±0·30	V1	11·72±0·05	46/7	−4·15±0·31
		−16·90±0·30	V1	11·50±0·05	46/7	−5·40±0·31
	CE	−25·9 ±1·4				
	CE	−28·8 ±1·4				

1 Formula	2 g.f.w.	3 Name	State	Purity mol %	Type	No. of expts.	Detn. of reactn.	$-\Delta H_r^\circ$ kcal/g.f.w.	4 Determinatio
$C_{10}H_{14}$	134·2231	1, 2, 4, 5-Tetramethyl-benzene	c		SB	3	m	1388·4 ±1·4	33/
$C_{10}H_{14}$		1, 2, 4, 5-Tetramethyl-benzene	c		SB	5	CO_2	1389·23±0·25	64/1
$C_{10}H_{14}$		1, 2, 4, 5-Tetramethyl-benzene							
$C_{10}H_{14}$	134·2231	Dihydro-α-dicyclo-pentadiene	c		SB	5	m	1413·0 ±1·4	34/
$C_{10}H_{16}$	136·2390	cis-Dec-3-ene-1-yne	l		H	4	H_2	95·62±0·50	59/1
$C_{10}H_{16}$	136·2390	trans-Dec-3-ene-1-yne	l		H	3	H_2	95·92±0·20	59/1
$C_{10}H_{16}$	136·2390	α-Pinene	l	99·6	SB	5	CO_2	1483·1 ±0·5	54/
$C_{10}H_{16}$	136·2390	β-Pinene	l	98·7	SB	6	CO_2	1485·2 ±0·7	54/
$C_{10}H_{16}$	136·2390	+-Limonene	l	98·0	SB	3	CO_2	1474·0 ±0·5	54/
$C_{10}H_{16}$	136·2390	Dipentene	l	99·3	SB	4	CO_2	1474·9 ±0·5	54/
$C_{10}H_{16}$	136·2390	cis-Allo-ocimene	l	93·7	SB	5	CO_2	1481·3 ±1·2	54/
$C_{10}H_{16}$	136·2390	Myrcene	l		SB	5	CO_2	1490·5 ±0·5	54/
$C_{10}H_{16}$	136·2390	α-Phellandrene	g		H	2	P/CO_2	52·92±0·30	37/2
$C_{10}H_{16}$	136·2390	α-Terpinene	g		H	2	P/CO_2	50·22±0·30	37/2
$C_{10}H_{16}$	136·2390	Tetrahydro-α-dicyclopentadiene	c		SB	6	m	1454·2 ±1·5	34/
$C_{10}H_{18}$	138·2550	Dicyclopentyl	l	99·63	SB	>2	CO_2	1513·5 ±0·7	61/1
$C_{10}H_{18}$	138·2550	cis-Decahydro-naphthalene	l	99·93±0·05	SB	6	CO_2	1502·92±0·22	60/1
$C_{10}H_{18}$		cis-Decahydro-naphthalene	l	99·68	SB	4	m	1502·8 ±0·3	41/9
$C_{10}H_{18}$		cis-Decahydro-naphthalene	l		SB	4	m	1496·6 ±2·3	37/4
$C_{10}H_{18}$		cis-Decahydro-naphthalene							

ΔH_r°		6 Determination of ΔH_v				
Remarks		5 ΔH_f° (l or c) kcal/g.f.w.	Type	ΔH_v° kcal/g.f.w.	Ref.	7 ΔH_f° (g) kcal/g.f.w.
	CE	$-30\cdot3 \pm 1\cdot4$				
		$-29\cdot49 \pm 0\cdot28$				
Selected value		$-29\cdot49 \pm 0\cdot28$				
	CE	$-5\cdot7 \pm 1\cdot4$				
$_0H_{16}(l) + 3H_2$ $= n-C_{10}H_{22}(l)$ $\{-71\cdot92 \pm 0\cdot25\}$		$+23\cdot70 \pm 0\cdot60$				
$_0H_{16}(l) + 3H_2$ $= n-C_{10}H_{22}(l)$ $\{-71\cdot92 \pm 0\cdot25\}$		$+24\cdot00 \pm 0\cdot30$				
	A	$-3\cdot9 \pm 0\cdot5$	V2	$10\cdot70 \pm 0\cdot30$	54/7	$+6\cdot8 \pm 0\cdot6$
	A	$-1\cdot8 \pm 0\cdot7$	V2	$11\cdot08 \pm 0\cdot30$	54/7	$+9\cdot3 \pm 0\cdot8$
	A	$-13\cdot0 \pm 0\cdot5$	V4	$11\cdot5 \pm 0\cdot5$	47/15	$-1\cdot5 \pm 0\cdot7$
	A	$-12\cdot1 \pm 0\cdot5$	V4	$11\cdot5 \pm 0\cdot5$	47/15	$-0\cdot6 \pm 0\cdot7$
purity believed to be *trans* isomer	A	$-5\cdot7 \pm 1\cdot2$				
	A	$+3\cdot5 \pm 0\cdot5$				
Methyl-4-isopropylcyclohexane(g) $\{-55\cdot12 \pm 0\cdot80\}$	E	$-14\cdot3 \pm 1\cdot1$	V3	$12\cdot1 \pm 0\cdot5$	47/15	$-2\cdot20 \pm 0\cdot90$
Methyl-4-isopropylcyclohexane(g) $\{-55\cdot12 \pm 0\cdot80\}$	E					$-4\cdot90 \pm 0\cdot90$
	CE	$-32\cdot8 \pm 1\cdot5$				
An $1\cdot0002 \pm 0\cdot0002$		$-41\cdot8 \pm 0\cdot7$				
		$-52\cdot43 \pm 0\cdot26$	V1	$12\cdot00 \pm 0\cdot50$	56/9	$-40\cdot43 \pm 0\cdot56$
	A	$-52\cdot5 \pm 0\cdot3$				$-40\cdot5 \pm 0\cdot6$
	CE	$-58\cdot7 \pm 2\cdot3$				$-46\cdot7 \pm 2\cdot4$
Selected value		$-52\cdot43 \pm 0\cdot26$				$-40\cdot43 \pm 0\cdot56$

THERMOCHEMISTRY OF ORGANIC AND ORGANOMETALLIC COMPOUNDS

C_aH_b 1 kcal = 4·184 kJ

Determinati

1 Formula	2 g.f.w.	3 Name	State	Purity mol %	Type	No. of expts.	Detn. of reactn.	$-\Delta H_r^\circ$ kcal/g.f.w.	Re
$C_{10}H_{18}$	138·2550	trans-Decahydro-naphthalene	l	99·97±0·03	SB	5	CO_2	1500·23±0·22	60/
$C_{10}H_{18}$		trans-Decahydro-naphthalene	l	97·88	SB	4	m	1500·7 ±0·3	41/
$C_{10}H_{18}$		trans-Decahydro-naphthalene							
$C_{10}H_{18}$	138·2550	Bicyclo [5, 3, 0] decane	l		SB			1511·49±0·89	67/
$C_{10}H_{18}$	138·2550	Spiro [4, 5] decane	l		SB	>5	m	1507·5 ±0·3	64/
$C_{10}H_{20}$	140·2709	1-Decene	l	99·91±0·07	SB	7	CO_2	1582·12±0·44	61/
$C_{10}H_{20}$	140·2709	2, 2, 5, 5-Tetramethyl-cis-3-hexene	l		SB	7	CO_2	1584·57±0·44	61/
$C_{10}H_{20}$	140·2709	2, 2, 5, 5-Tetramethyl-trans-3-hexene	l		SB	4	CO_2	1574·06±0·62	61/
$C_{10}H_{20}$	140·2709	n-Butylcyclohexane	l	>99·7	SB	6	CO_2	1560·78±0·29	46/
$C_{10}H_{20}$	140·2709	1-Methyl-4-isopropyl-cyclohexane	g		H	3	P/CO_2	53·62±0·30	37/
$C_{10}H_{20}$	140·2709	Cyclodecane	l		SB	4	m	1574·26±0·21	60/
$C_{10}H_{20}$		Cyclodecane	l		SB	4	m	1577·5 ±1·5	33/
$C_{10}H_{20}$		Cyclodecane							
$C_{10}H_{22}$	142·2868	n-Decane	l	99·94±0·04	SB	5	CO_2	1620·13±0·36	44/
$C_{10}H_{22}$		n-Decane	l		SB	6	CO_2	1620·03±0·27	37/ 45/
$C_{10}H_{22}$		n-Decane							
$C_{10}H_{22}$	142·2868	2-Methylnonane	l		SB	5	m	1617·94±0·56	40/
$C_{10}H_{22}$	142·2868	5-Methylnonane	l		SB	5	m	1618·40±0·38	40/
$C_{11}H_{10}$	142·2024	1-Methylnaphthalene	l	99·97±0·03	SB	6	CO_2	1389·59±0·40	60/
$C_{11}H_{10}$	142·2024	2-Methylnaphthalene	c	99·92±0·06	SB	5	CO_2	1386·88±0·35	60/
$C_{11}H_{10}$		2-Methylnaphthalene	c	2s	SB	8	m	1384·50±0·52	39/
$C_{11}H_{10}$		2-Methylnaphthalene							
$C_{11}H_{14}$	146·2342	1-Methyl-1, 2, 3, 4-tetra-hydronaphthalene	l		SB	5	m	1496·6 ±6·0	51/

1 kcal $=4\cdot184$ kJ

ΔH_r°	Remarks	5 ΔH_f° (l or c) kcal/g.f.w.	6 Determination of ΔH_v ΔH_v° Type kcal/g.f.w.	Ref.	7 ΔH_f° (g) kcal/g.f.w.
		$-55\cdot12\pm0\cdot26$	V1 $11\cdot60\pm0\cdot50$	56/9	$-43\cdot52\pm0\cdot56$
	A	$-54\cdot7\ \pm0\cdot3$			$-43\cdot1\ \pm0\cdot6$
	Selected value	$-55\cdot12\pm0\cdot26$			$-43\cdot52\pm0\cdot56$
		$-43\cdot87\pm0\cdot91$	$12\cdot50\pm0\cdot20$	67/10	$-31\cdot37\pm0\cdot94$
An $1\cdot0001\pm0\cdot0005$		$-47\cdot9\ \pm0\cdot3$			
		$-41\cdot54\pm0\cdot47$	V1 $12\cdot06\pm0\cdot05$	50/8	$-29\cdot48\pm0\cdot48$
		$-39\cdot09\pm0\cdot47$			
		$-49\cdot60\pm0\cdot64$			
		$-62\cdot88\pm0\cdot32$	V1 $11\cdot96\pm0\cdot05$	65/17	$-50\cdot92\pm0\cdot33$
-limonene(g) $\{-1\cdot5\pm0\cdot7\}$	E				$-55\cdot12\pm0\cdot80$
		$-49\cdot40\pm0\cdot26$	E5 $12\cdot52\pm0\cdot30$	57/45	$-36\cdot88\pm0\cdot40$
	BCE	$-46\cdot2\ \pm1\cdot5$			$-33\cdot7\ \pm1\cdot6$
	Selected value	$-49\cdot40\pm0\cdot26$			$-36\cdot88\pm0\cdot40$
		$-71\cdot85\pm0\cdot39$	C1 $12\cdot28\pm0\cdot01$	47/5	$-59\cdot57\pm0\cdot39$
		$-71\cdot95\pm0\cdot31$			$-59\cdot67\pm0\cdot31$
	Selected value	$-71\cdot92\pm0\cdot25$			$-59\cdot64\pm0\cdot25$
	A	$-74\cdot04\pm0\cdot58$	E2 $11\cdot92\pm0\cdot10$	61/17	$-62\cdot12\pm0\cdot60$
	A	$-73\cdot58\pm0\cdot41$	E2 $11\cdot78\pm0\cdot10$	61/17	$-61\cdot80\pm0\cdot45$
		$+13\cdot45\pm0\cdot42$			
$H_m\ 2\cdot83\pm0\cdot01$ (57/13)		$+10\cdot74\pm0\cdot37$			
	A	$+8\cdot36\pm0\cdot53$			
	Selected value	$+10\cdot74\pm0\cdot37$			
		$-16\cdot2\ \pm6\cdot0$			

THERMOCHEMISTRY OF ORGANIC AND ORGANOMETALLIC COMPOUNDS

C$_a$H$_b$

$1 \text{ kcal} = 4 \cdot 184 \text{ kJ}$

1 Formula	2 g.f.w.	3 Name	State	Purity mol %	Type	No. of expts.	Detn. of reactn.	$-\Delta H_r^\circ$ kcal/g.f.w.	Ref
C$_{11}$H$_{16}$	148·2502	Pentamethylbenzene	c		SB	3	m	1547·7 ±1·5	33/
C$_{11}$H$_{16}$		Pentamethylbenzene	c		SB	4	m	1548·79±0·61	46/
C$_{11}$H$_{16}$		Pentamethylbenzene	c		SB	5	CO$_2$	1549·14±0·25	64/
C$_{11}$H$_{16}$		Pentamethylbenzene							
C$_{11}$H$_{20}$	152·2821	Dicyclopentylmethane	l	99·60	SB	>2	CO$_2$	1668·7 ±0·5	61/
C$_{11}$H$_{20}$	152·2821	Cyclopentylcyclohexane	l		SB	2	CO$_2$	1662·7 ±[1·0]	61/
C$_{11}$H$_{20}$	152·2821	2-Methyl-trans-decahydronaphthalene	l		SB	>2	CO$_2$	1654·4 ±0·5	61/
C$_{11}$H$_{20}$	152·2821	9-Methyl-cis-decahydronaphthalene	l	>98, glc	SB	5	CO$_2$	1659·43±0·45	60/1
C$_{11}$H$_{20}$	152·2821	9-Methyl-trans-decahydronaphthalene	l	>98, glc	SB	5	CO$_2$	1658·04±0·41	60/1
C$_{11}$H$_{20}$	152·2821	Spiro [5, 5] undecane	l		SB	>5	CO$_2$	1658·4 ±0·5	64/1
C$_{11}$H$_{22}$	154·2980	1, 1-Dimethyl-2-hexylcyclopropane	l		SB	8	m	1739·9 ±0·4	60/1
C$_{11}$H$_{22}$	154·2980	Cyclo-undecane	l		SB	5	m	1729·75±0·26	60/1
C$_{11}$H$_{24}$	156·3139	n-Undecane	l		SB	6	CO$_2$	1776·13±0·61	37/3 45/
C$_{11}$H$_{24}$	156·3139	2, 2, 5, 5-Tetramethyl-heptane	l	99·4±0·1	SB		m	1770·4 ±0·7	59/1
C$_{11}$H$_{24}$	156·3139	3, 3, 5, 5-Tetramethyl-heptane	l		SB		m	1776·5 ±0·7	59/1
C$_{11}$H$_{24}$	156·3139	2, 2, 4, 4, 5-Penta-methylhexane	l		SB		m	1775·7 ±0·7	59/1
C$_{12}$H$_8$	152·1976	Biphenylene	c		SB	5	CO$_2$	1486·3 ±1·4	62/1
C$_{12}$H$_8$	152·1976	Acenaphthylene	c		SB	6	m	1446·5 ±1·1	65/1
C$_{12}$H$_8$		Acenaphthylene	c		SB			1447·9 ±1·0	66/1
C$_{12}$H$_8$		Acenaphthylene							

1 kcal = 4·184 kJ

ΔH_r°		5 ΔH_f° (l or c) kcal/g.f.w.	6 Determination of ΔH_v			7 ΔH_f° (g) kcal/g.f.w.
Remarks			Type	ΔH_v° kcal/g.f.w.	Ref.	
CE		−33·4 ±1·5				
		−32·29±0·63				
		−31·94±0·29				
Selected value		−31·94±0·29				
/An 1·0003±0·0002		−49·0 ±0·5				
/An 1·0002±0·0001		−55·0 ±[1·0]				
/An 0·9998±0·0005		−63·3 ±0·5				
impurity believed to be *trans* isomer		−58·28±0·47				
impurity believed to be *cis* isomer		−59·67±0·44				
/An 0·9996±0·0005		−59·3 ±0·5				
		−46·1 ±0·4				
		−56·28±0·30	E5	13·41±0·30	57/45	−42·87±0·42
		−78·21±0·63	E1	13·46±0·05		−64·75±0·64
		−83·9 ±0·7	E3	11·66±0·15		−72·2 ±0·8
		−77·8 ±0·7	E3	11·71±0·15		−66·1 ±0·8
		−78·6 ±0·7	E3	11·48±0·15		−67·1 ±0·8
/An 0·9958		+84·4 ±1·4	S4	30·8±[0·5] (sub.)	55/20	+115·2 ±1·5
		+44·6 ±1·1	S3	17·0±0·3 (sub.)	65/18	+61·6 ±1·2
		+46·0 ±1·0				+63·0 ±1·1
Selected value		+44·6 ±1·1				+61·6 ±1·2

THERMOCHEMISTRY OF ORGANIC AND ORGANOMETALLIC COMPOUNDS

C_aH_b

1 kcal = 4·184 kJ

Determinati

1 Formula	2 g.f.w.	3 Name	State	Purity mol %	No. of Type	Detn. of expts.	reactn.	$-\Delta H_r^\circ$ kcal/g.f.w.	Re
$C_{12}H_{10}$	154·2135	Biphenyl	c	3s	SB	10	m	1493·43±0·37	51/
$C_{12}H_{10}$		Biphenyl	c		SB	5	m	1493·35±0·96	63/
$C_{12}H_{10}$		Biphenyl	c		SB	3	m	1493·5 ±1·5	35/
$C_{12}H_{10}$		Biphenyl	c	99·99±0·01	SB	5	CO_2	1494·22±0·33	66/
$C_{12}H_{10}$		Biphenyl							
$C_{12}H_{10}$	154·2135	Acenaphthene	c		SB	5	m	1487·0 ±0·6	65/
$C_{12}H_{14}$	158·2454	Hexacyclo [7, 2, 1, $0^{2.5}, 0^{3.10}, 0^{4.8}, 0^{6.12}$]- dodecane	c	99·9	SB	10	m	1619·03±0 30	58/
$C_{12}H_{14}$	158·2454	1-Phenylcyclohexene	l		SB	3	m	1602·8 ±1·6	35/
$C_{12}H_{16}$	160·2613	Dicyclohexadiene	l		SB	6	m	1681·4 ±1·7	34/
$C_{12}H_{18}$	162·2773	Dodeca-3, 9-diyne	l		H	6	H_2	131·15±0·50	57/
$C_{12}H_{18}$	162·2773	Dodeca-5, 7-diyne	l		H	5	H_2	127·25±0·70	57/
$C_{12}H_{18}$	162·2773	Hexamethylbenzene	c		SB	6	m	1704·42±0·72	46/
$C_{12}H_{18}$		Hexamethylbenzene	c		SB	3	m	1705·2 ±1·7	33/
$C_{12}H_{18}$		Hexamethylbenzene	c		SB	5	CO_2	1704·84±0·34	64/1
$C_{12}H_{18}$		Hexamethylbenzene							
$C_{12}H_{20}$	164·2932	Tetrahydrodicyclo- hexadiene	l		SB	5	m	1760·9 ±1·8	34/
$C_{12}H_{22}$	166·3091	Cyclohexylcyclohexane	l		SB	3	m	1801·4 ±1·8	35/4
$C_{12}H_{22}$	166·3091	Cyclopentylcycloheptane	l	99·23	SB	>2	CO_2	1826·0 ±0·9	61/1
$C_{12}H_{22}$	166·3091	Spiro [5, 6] dodecane	l		SB	>5	CO_2	1819·6 ±0·2	64/1
$C_{12}H_{24}$	168·3251	Cyclododecane	c		SB	5	m	1875·10±0·31	60/1
$C_{12}H_{26}$	170·3410	n-Dodecane	l	>99·94	SB	6	CO_2	1932·73±0·39	44/6
$C_{12}H_{26}$		n-Dodecane	l	99·87	SB	4	m	1933·7 ±1·9	33/4
$C_{12}H_{26}$		n-Dodecane	l	99·87	SB	6	CO_2	1933·22±0·75	37/3 45/1
$C_{12}H_{26}$		n-Dodecane							

1 kcal = 4·184 kJ

ΔH_r° Remarks	5 ΔH_f° (l or c) kcal/g.f.w.	6 Determination of ΔH_v			7 ΔH_f° (g) kcal/g.f.w.
		Type	ΔH_v° kcal/g.f.w.	Ref.	
crobomb: sample mass *ca.* 12mg	+23·24±0·39	S3	19·50±0·50	53/8	+42·74±0·63
	+23·16±0·97		(sub.)		+42·66±1·10
BCE	+23·3 ±1·5				+42·8 ±1·6
An 0·99981±0·00015	+24·03±0·36				+43·53±0·60
Selected value	+24·03±0·36				+43·53±0·60
	+16·8 ±0·6	S3	20·6±0·2	65/18	+37·4 ±0·7
			(sub.)		
An 0·99996	+12·21±0·33				
BCE	−4·0 ±1·6				
CE	+6·3 ±1·7				
$_2$H$_{18}$(l)+4H$_2$ =n-C$_{12}$H$_{26}$(l) {−83·88±0·35}	+47·27±0·62				
$_2$H$_{18}$(l)+4H$_2$ =n-C$_{12}$H$_{26}$(l) {−83·88±0·35}	+43·37±0·80				
	−39·03±0·74	S3	17·86±0·50	65/52	−21·17±0·89
CE	−38·3 ±1·7		(sub.)		−20·4 ±1·9
	−38·61±0·37				−20·75±0·62
Selected value	−38·61±0·37				−20·75±0·62
CE	−50·9 ±1·9				
BCE	−78·7 ±1·9				
An 1·0003±0·0002	−54·1 ±0·9				
An 1·0004±0·0001	−60·5 ±0·3				
	−73·29±0·36	E5$_m$	18·26±0·40	57/45	−55·03±0·54
			(sub.)		
	−83·98±0·43	E1	14·64±0·05		−69·34±0·44
CE	−83·0 ±1·9				−68·4 ±1·9
	−83·49±0·77				−68·85±0·78
Selected value	−83·88±0·35				−69·24±0·37

C_aH_b

1 kcal = 4·184 kJ

Determina

1 Formula	2 g.f.w.	3 Name	State	Purity mol %	Type	No. of expts.	Detn. of reactn.	$-\Delta H_r^\circ$ kcal/g.f.w.	R
$C_{12}H_{26}$	170·3410	3, 3, 6, 6-Tetramethyl-octane	l	98·7±0·2	SB		m	1927·6 ±0·8	59/
$C_{13}H_{12}$	168·2406	2-Methylbiphenyl	l		SB	3	m	1658·4 ±1·7	35
$C_{13}H_{12}$	168·2406	3-Methylbiphenyl	l		SB	3	m	1653·0 ±1·7	35
$C_{13}H_{12}$	168·2406	4-Methylbiphenyl	c		SB	3	m	1645·8 ±1·7	35
$C_{13}H_{12}$	168·2406	Diphenylmethane	l		SB	5	m	1653·37±0·66	46
$C_{13}H_{12}$		Diphenylmethane	l		SB	5	m	1653·98±0·32	50/
$C_{13}H_{12}$		Diphenylmethane	c		SB	5	m	1656·2 ±0·2	46
$C_{13}H_{12}$		Diphenylmethane							
$C_{13}H_{26}$	182·3522	n-Heptylcyclohexane	l	97·1	SB	5	m	2026·40±0·55	40
$C_{13}H_{26}$	182·3522	Cyclotridecane	l		SB	4	m	2036·75±0·36	60/
$C_{13}H_{28}$	184·3681	4, 4, 6, 6-Tetramethyl-nonane	l		SB		m	2090·4 ±0·8	59/
$C_{13}H_{28}$	184·3681	3, 5-Dimethyl-3, 5-diethylheptane	l		SB		m	2092·9 ±0·8	59/
$C_{14}H_{10}$	178·2358	Diphenylethyne (Tolane)	c		SB	5	m	1732·95±0·24	53/
$C_{14}H_{10}$		Diphenylethyne (Tolane)	c		H	5	H_2	64·2 ±1·1	58/
$C_{14}H_{10}$		Diphenylethyne (Tolane)							
$C_{14}H_{10}$	178·2358	Anthracene	c		SB	5	m	1683·99±0·67	46/
$C_{14}H_{10}$		Anthracene	c		SB	9	m	1687·1 ±0·7	52/
$C_{14}H_{10}$		Anthracene	c		SB	5	m	1689·0 ±0·5	52/
$C_{14}H_{10}$		Anthracene	c		SB	6	m	1686·13±1·03	63/
$C_{14}H_{10}$		Anthracene	c		SB	6	m	1685·61±0·58	39/
$C_{14}H_{10}$		Anthracene	c		SB	10	m	1689·85±0·68	35/
$C_{14}H_{10}$		Anthracene	c		SB	5	m	1686·0 ±1·7	32/
$C_{14}H_{10}$		Anthracene	c		SB	4	m	1686·73± 0·30	31/
$C_{14}H_{10}$		Anthracene	c		SB	4	m	1689·66±0·38	51/
$C_{14}H_{10}$		Anthracene	c	99·95±0·01	SB	4	CO_2	1689·17±0·41	66/
$C_{14}H_{10}$		Anthracene							

1 kcal $=4\cdot184$ kJ

ΔH_r° Remarks	5 ΔH_f° (l or c) kcal/g.f.w.	6 Determination of ΔH_v			7 ΔH_f° (g) kcal/g.f.w.
		Type	ΔH_v° kcal/g.f.w.	Ref.	
	$-89\cdot1\ \pm0\cdot8$	E3	$13\cdot06\pm0\cdot15$		$-76\cdot0\ \pm0\cdot9$
BCE	$+25\cdot8\ \pm1\cdot7$				
BCE	$+20\cdot4\ \pm1\cdot7$				
BCE	$+13\cdot2\ \pm1\cdot7$				
$H_m\ 4\cdot36\pm0\cdot25$ (50/12)	$+20\cdot82\pm0\cdot68$	S3	$19\cdot7\pm0\cdot2^*$ (sub.)	59/36	$+36\cdot2\ \pm0\cdot8$
	$+21\cdot43\pm0\cdot35$				$+36\cdot8\ \pm0\cdot5$
F	$+23\cdot6\ \pm0\cdot3$				$+43\cdot3\ \pm0\cdot4$
Selected value	$+21\cdot43\pm0\cdot35$ (l)				$+36\cdot8\ \pm0\cdot5$
A	$-84\cdot36\pm0\cdot58$	V1	$15\cdot50\pm0\cdot05$	53/1	$-68\cdot86\pm0\cdot59$
	$-74\cdot01\pm0\cdot41$	E5	$15\cdot13\pm0\cdot30$	57/45	$-58\cdot88\pm0\cdot50$
	$-88\cdot7\ \pm0\cdot8$	E3	$13\cdot82\pm0\cdot15$		$-74\cdot9\ \pm0\cdot9$
	$-86\cdot2\ \pm0\cdot8$	E3	$14\cdot16\pm0\cdot15$		$-72\cdot0\ \pm0\cdot9$
	$+74\cdot66\pm0\cdot29$				
H_{10}(c)$+2H_2$	$+76\cdot5\ \pm1\cdot2$				
$=$Bibenzyl(c) $\{+12\cdot32\pm0\cdot31\}$ Selected value	$+74\cdot66\pm0\cdot29$				
	$+25\cdot70\pm0\cdot69$	S4	$24\cdot4\pm0\cdot5$ (sub.)	53/8	$+50\cdot0\ \pm1\cdot2$
	$+28\cdot8\ \pm0\cdot7$	S3	$24\cdot7\pm[0\cdot7]$ (sub.)	58/39	$+53\cdot1\ \pm1\cdot2$
	$+30\cdot7\ \pm0\cdot5$	S4	$24\cdot1\pm[1\cdot0]$ (sub.)	58/40	$+55\cdot0\ \pm1\cdot1$
Microbomb: sample mass $ca.$ 12mg	$+27\cdot84\pm1\cdot04$	S4	$25\cdot0\pm[1\cdot0]$ (sub.)	49/4	$+52\cdot1\ \pm1\cdot4$
A	$+27\cdot32\pm0\cdot60$	S4	$22\cdot3\pm0\cdot2^*$ (sub.)	38/14	$+51\cdot6\ \pm1\cdot2$
CE	$+31\cdot56\pm0\cdot70$	S4	$23\cdot5\pm0\cdot5$ (sub).	64/44	$+55\cdot9\ \pm1\cdot2$
CE	$+27\cdot7\ \pm1\cdot7$				$+52\cdot0\ \pm1\cdot9$
	$+28\cdot43\pm0\cdot36$				$+52\cdot7\ \pm1\cdot1$
E	$+31\cdot37\pm0\cdot41$				$+55\cdot7\ \pm1\cdot1$
/An $0\cdot99987\pm0\cdot00011$	$+30\cdot88\pm0\cdot44$				$+55\cdot2\ \pm1\cdot1$
Selected value	$+30\cdot88\pm0\cdot44$		$24\cdot3\pm1\cdot0$		$+55\cdot2\ \pm1\cdot1$

C_aH_b

1 Formula	2 g.f.w.	3 Name	State	Purity mol %	No. of Type	Detn. of expts.	reactn.	$-\Delta H_r^\circ$ kcal/g.f.w.	Re
$C_{14}H_{10}$	178·2358	Phenanthrene	c		SB	3	m	1685·3 ±0·7	52/
$C_{14}H_{10}$		Phenanthrene	c		SB	4	m	1675·70±0·61	39/
$C_{14}H_{10}$		Phenanthrene	c		SB	6	m	1685·20±0·64	35/
$C_{14}H_{10}$		Phenanthrene	c		SB	5	m	1675·1 ±1·7	32/
$C_{14}H_{10}$		Phenanthrene	c		SB	5	m	1684·81±0·34	51/
$C_{14}H_{10}$		Phenanthrene	c		SB	5	CO_2	1686·06±0·30	66/
$C_{14}H_{10}$		Phenanthrene							
$C_{14}H_{12}$	180·2517	9, 10-Dihydroanthracene	c		SB	4	m	1742·47±0·28	51/
$C_{14}H_{12}$	180·2517	1, 1-Diphenylethene	l	99·89	SB	6	m	1767·81±0·28	50/
$C_{14}H_{12}$	180·2517	cis-1, 2-Diphenylethene (cis-Stilbene)	l	99·60	SB	8	m	1770·46±0·35	50/
$C_{14}H_{12}$		cis-1, 2-Diphenylethene (cis-Stilbene)	l		H	3	m	31·44±0·20	42/
$C_{14}H_{12}$		cis-1, 2-Diphenylethene (cis-Stilbene)							
$C_{14}H_{12}$	180·2517	trans-1, 2-Diphenylethene (trans-Stilbene)	c	99·90	SB	5	m	1760·18±0·25	50/
$C_{14}H_{12}$		trans-1, 2-Diphenylethene (trans-Stilbene)	c		SB	6	m	1759·53±0·67	39/
$C_{14}H_{12}$		trans-1, 2-Diphenylethene (trans-Stilbene)	c		H	2	m	18·59±0·20	42/
$C_{14}H_{12}$		trans-1, 2-Diphenylethene (trans-Stilbene)	c		SB	4	m	1759·28±0·93	68/
$C_{14}H_{12}$		trans-1, 2-Diphenylethene (trans-Stilbene)							
$C_{14}H_{14}$	182·2677	3, 3′-Dimethylbiphenyl	l		SB	3	m	1799·7 ±1·8	35/
$C_{14}H_{14}$	182·2677	4, 4′-Dimethylbiphenyl	c		SB	3	m	1798·3 ±1·8	35/
$C_{14}H_{14}$	182·2677	1, 1-Diphenylethane	l		SB	5	m	1806·55±0·48	53/
$C_{14}H_{14}$	182·2677	1, 2-Diphenylethane (Bibenzyl)	c		SB	5	m	1805·64±0·72	46/
$C_{14}H_{14}$		1, 2-Diphenylethane	c		SB	3	m	1807·5 ±1·8	33/
$C_{14}H_{14}$		1, 2-Diphenylethane	c		SB	5	m	1806·79±0·24	53/
$C_{14}H_{14}$		1, 2-Diphenylethane	c	99·92±0·02	SB	4	CO_2	1807·24±0·26	66/
$C_{14}H_{14}$		1, 2-Diphenylethane							

$$1 \text{ kcal} = 4\cdot184 \text{ kJ}$$

Remarks	5 ΔH_f° (l or c) kcal/g.f.w.	6 Determination of ΔH_v			7 ΔH_f° (g) kcal/g.f.w.
		Type	ΔH_v° kcal/g.f.w.	Ref.	
	$+27\cdot0\ \pm0\cdot7$	S4	$20\cdot7\pm0\cdot5$ (sub.)	53/8	$+48\cdot7\ \pm1\cdot2$
A	$+17\cdot41\pm0\cdot63$	S4	$22\cdot2\pm[1\cdot0]$ (sub.)	49/4	$+39\cdot1\ \pm1\cdot2$
CE	$+26\cdot91\pm0\cdot68$	S4	$22\cdot9\pm[0\cdot7]$ (sub.)	58/39	$+48\cdot6\ \pm1\cdot2$
CE	$+16\cdot8\ \pm1\cdot7$				$+38\cdot5\ \pm1\cdot9$
E	$+26\cdot52\pm0\cdot38$				$+48\cdot2\ \pm1\cdot1$
An $0\cdot99931\pm0\cdot00016$	$+27\cdot77\pm0\cdot34$				$+49\cdot5\ \pm1\cdot1$
Selected value	$+27\cdot77\pm0\cdot34$		$21\cdot7\pm1\cdot0$		$+49\cdot5\ \pm1\cdot1$
E	$+15\cdot87\pm0\cdot32$	S3	$22\cdot3\pm[1\cdot0]$ (sub.)	58/39	$+38\cdot2\ \pm[1\cdot1]$
C	$+41\cdot21\pm0\cdot32$	V5	$17\cdot5\pm[1\cdot0]$	56/44	$+58\cdot7\ \pm[1\cdot1]$
C	$+43\cdot86\pm0\cdot39$	V3	$16\cdot5\pm0\cdot3$	52/6	$+60\cdot36\pm0\cdot50$
$_4H_{12}(l)+H_2$ $=$Bibenzyl(c) {$+12\cdot32\pm0\cdot31$}	$+43\cdot76\pm0\cdot37$				$+60\cdot26\pm0\cdot48$
Selected value	$+43\cdot81\pm0\cdot30$				$+60\cdot31\pm0\cdot42$
C	$+33\cdot58\pm0\cdot30$				
A	$+32\cdot93\pm0\cdot69$				
$_4H_{12}(c)+H_2$ $=$Bibenzyl(c) {$+12\cdot32\pm0\cdot31$}	$+30\cdot91\pm0\cdot38$				
	$+32\cdot68\pm0\cdot95$				
Selected value	$+32\cdot72\pm0\cdot24$				
BCE	$+4\cdot8\ \pm1\cdot8$				
BCE	$+3\cdot4\ \pm1\cdot8$				
	$+11\cdot63\pm0\cdot51$				
	$+10\cdot72\pm0\cdot74$	S3	$20\cdot1\pm0\cdot1^*$ (sub.)	59/36	$+30\cdot8\ \pm0\cdot8$
CE	$+12\cdot6\ \pm1\cdot8$				$+32\cdot7\ \pm1\cdot8$
	$+11\cdot87\pm0\cdot29$				$+32\cdot0\ \pm0\cdot3$
/An $0\cdot99992\pm0\cdot00014$	$+12\cdot32\pm0\cdot31$				$+32\cdot4\ \pm0\cdot3$
Selected value	$+12\cdot32\pm0\cdot31$				$+32\cdot4\ \pm0\cdot3$

C_xH_y

1 kcal = 4·184 kJ

Determina‖

1 Formula	2 g.f.w.	3 Name	State	Purity mol %	No. of Type expts.	Detn. of reactn.	$-\Delta H_r^\circ$ kcal/g.f.w.	R
$C_{14}H_{20}$	188·3155	1, 8-Cyclotetradecadiyne	c		SB 5	m	2035·18±0·40	64
$C_{14}H_{24}$	192·3474	*trans-anti-trans-*perhydroanthracene	c		SB 8	m	2066·38±0·54	63
$C_{14}H_{24}$	192·3474	*trans-syn-trans-*perhydroanthracene	c		SB 9	m	2057·48±0·70	63
$C_{14}H_{26}$	194·3633	1, 2-Dicyclohexylethane	l		SB 3	m	2113·5 ±2·1	35
$C_{14}H_{26}$	194·3633	Dicycloheptyl	l		SB >2	CO_2	2136·7 ±0·6	61
$C_{14}H_{28}$ $C_{14}H_{28}$ $C_{14}H_{28}$	196·3793	Cyclotetradecane Cyclotetradecane Cyclotetradecane	c c		SB 4 SB 5	m m	2184·19±0·88 2183·67±0·31	64 60
$C_{15}H_{14}$	194·2788	1, 1-Diphenylcyclopropane	l		SB	m	1933·3 ±0·8	61
$C_{15}H_{14}$	194·2788	*cis*-1, 2-Diphenylcyclopropane	l		SB	m	1931·7 ±0·2	61
$C_{15}H_{14}$	194·2788	*trans*-1, 2-Diphenylcyclopropane	l		SB	m	1928·7 ±0·6	61
$C_{15}H_{18}$	198·3107	1, 4-Dimethyl-7-isopropylazulene	c		SB 6	m	2040·2 ±0·9	55
$C_{15}H_{18}$	198·3107	α-Tricyclopentadiene	c		SB 6	m	2055·8 ±2·0	34
$C_{15}H_{20}$	200·3267	Dihydro-α-tricyclopentadiene	c		SB 6	m	2091·6 ±2·0	34
$C_{15}H_{20}$	200·3267	Dihydro-β-tricyclopentadiene	c		SB 5	m	2089·4 ±2·0	34
$C_{15}H_{22}$	202·3426	Tetrahydro-α-tricyclopentadiene	c		SB 6	m	2135·6 ±2·1	34
$C_{15}H_{22}$	202·3426	Tetrahydro-β-tricyclopentadiene	c		SB 5	m	2132·0 ±2·1	34

$$1 \text{ kcal} = 4 \cdot 184 \text{ kJ}$$

ΔH_r°					
		5	6 Determination of ΔH_v		7
Remarks		ΔH_f° (l or c) kcal/g.f.w.	Type $\quad$ Δ_v kcal/g.f.w. $\quad$ Ref.		ΔH_f° (g) kcal/g.f.w.
		$+35 \cdot 32 \pm 0 \cdot 56$	S4 $39 \cdot 68 \pm 0 \cdot 76$ 64/16 (sub.)		$+75 \cdot 00 \pm 0 \cdot 95$
		$-70 \cdot 11 \pm 0 \cdot 58$	S3 $17 \cdot 38 \pm 0 \cdot 79$ 63/21 (sub.)		$-52 \cdot 73 \pm 0 \cdot 98$
		$-79 \cdot 01 \pm 0 \cdot 73$	S3 $20 \cdot 89 \pm 0 \cdot 57$ 63/21 (sub.)		$-58 \cdot 12 \pm 0 \cdot 93$
	BCE	$-91 \cdot 3 \ \pm 2 \cdot 1$			
	E	$-68 \cdot 1 \ \pm 0 \cdot 7$			
		$-88 \cdot 93 \pm 0 \cdot 98$ $-89 \cdot 45 \pm 0 \cdot 37$	S3 $32 \cdot 21 \pm 0 \cdot 35$ 64/16 (sub.)		$-56 \cdot 72 \pm 1 \cdot 04$ $-57 \cdot 23 \pm 0 \cdot 50$
	Selected value	$-89 \cdot 34 \pm 0 \cdot 32$			$-57 \cdot 13 \pm 0 \cdot 48$
An $1 \cdot 0004 \pm 0 \cdot 0001$		$+44 \cdot 3 \ \pm 0 \cdot 9$			
An $1 \cdot 0002 \pm 0 \cdot 0001$		$+42 \cdot 7 \ \pm 0 \cdot 3$			
An $1 \cdot 0000 \pm 0 \cdot 0001$		$+39 \cdot 7 \ \pm 0 \cdot 7$			
		$+14 \cdot 6 \ \pm 0 \cdot 9$			
	CE	$+30 \cdot 2 \ \pm 2 \cdot 0$			
	CE	$-2 \cdot 3 \ \pm 2 \cdot 1$			
	CE	$-4 \cdot 5 \ \pm 2 \cdot 1$			
	CE	$-26 \cdot 6 \ \pm 2 \cdot 2$			
	CE	$-30 \cdot 2 \ \pm 2 \cdot 2$			

C_aH_b

1 kcal $= 4 \cdot 184$ kJ

1 Formula	2 g.f.w.	3 Name	State	Purity mol %	No. of Type expts.	Detn. of reactn.	$-\Delta H_r^\circ$ kcal/g.f.w.	R
$C_{15}H_{30}$	$210 \cdot 4064$	n-Decylcyclopentane	l	$99 \cdot 80 \pm 0 \cdot 18$	SB 6	CO_2	$2348 \cdot 01 \pm 0 \cdot 66$	60/
$C_{15}H_{30}$		n-Decylcyclopentane	l	$99 \cdot 80 \pm 0 \cdot 18$	SB 6	CO_2	$2347 \cdot 54 \pm 0 \cdot 46$	55/
$C_{15}H_{30}$		n-Decylcyclopentane						
$C_{15}H_{30}$	$210 \cdot 4064$	Cyclopentadecane	c	2s	SB 10	m	$2345 \cdot 61 \pm 0 \cdot 38$	60/
$C_{15}H_{30}$		Cyclopentadecane	c		SB 6	m	$2347 \cdot 7 \pm 2 \cdot 3$	33
$C_{15}H_{30}$		Cyclopentadecane						
$C_{15}H_{32}$	$212 \cdot 4223$	5, 5, 7, 7-Tetramethyl-undecane	l		SB	m	$2400 \cdot 2 \pm 1 \cdot 0$	59/
$C_{15}H_{32}$	$212 \cdot 4223$	4, 6-Dimethyl-4, 6-diethylnonane	l		SB	m	$2404 \cdot 7 \pm 1 \cdot 0$	59/
$C_{16}H_{10}$	$202 \cdot 2581$	Diphenylbutadiyne	c		SB 5	m	$1970 \cdot 30 \pm 0 \cdot 24$	53/
$C_{16}H_{10}$	$202 \cdot 2581$	Fluoranthene	c		SB 6	m	$1892 \cdot 4 \pm 1 \cdot 3$	65/
$C_{16}H_{10}$		Fluoranthene	c		SB		$1891 \cdot 77 \pm 0 \cdot 09$	67/
$C_{16}H_{10}$		Fluoranthene						
$C_{16}H_{10}$	$202 \cdot 2581$	Pyrene	c	3s	SB 11	m	$1873 \cdot 81 \pm 0 \cdot 84$	39/
$C_{16}H_{10}$		Pyrene	c		SB		$1873 \cdot 83 \pm 0 \cdot 09$	67/
$C_{16}H_{10}$		Pyrene						
$C_{16}H_{14}$	$206 \cdot 2900$	Di-o-tolylethyne	c		SB 5	m	$2039 \cdot 14 \pm 0 \cdot 48$	53/
$C_{16}H_{14}$	$206 \cdot 2900$	Di-p-tolylethyne	c		SB 5	m	$2036 \cdot 99 \pm 0 \cdot 48$	53/
$C_{16}H_{14}$	$206 \cdot 2900$	cis, cis-1, 4-Diphenyl-butadiene	c		SB 5	m	$2030 \cdot 53 \pm 0 \cdot 48$	53/
$C_{16}H_{14}$	$206 \cdot 2900$	trans, trans-1, 4-Diphenylbutadiene	c		SB 5	m	$2025 \cdot 75 \pm 0 \cdot 24$	53/
$C_{16}H_{14}$		trans, trans-1, 4-Diphenylbutadiene	c		H 2	m	$44 \cdot 04 \pm 0 \cdot 14$	42/
$C_{16}H_{14}$		trans, trans-1, 4-Diphenylbutadiene	c		SB 5	m	$2022 \cdot 7 \pm 1 \cdot 6$	48/
$C_{16}H_{14}$		trans, trans-1, 4-Diphenylbutadiene						
$C_{16}H_{14}$	$206 \cdot 2900$	2, 7-Dimethyl-phenanthrene	c		SB 4	m	$1991 \cdot 72 \pm 0 \cdot 40$	63/2
$C_{16}H_{14}$	$206 \cdot 2900$	4, 5-Dimethyl-phenanthrene	c		SB 6	m	$2004 \cdot 28 \pm 1 \cdot 40$	63 2

1 kcal = $4 \cdot 184$ kJ

H_r° Remarks		5 ΔH_f° (l or c) kcal/g.f.w.	6 Determination of ΔH_v			7 ΔH_f° (g) kcal/g.f.w.
			Type	ΔH_v° kcal/g.f.w.	Ref.	
n $1 \cdot 00006 \pm 0 \cdot 00009$		$-87 \cdot 48 \pm 0 \cdot 69$				
		$-87 \cdot 95 \pm 0 \cdot 51$				
	Selected value	$-87 \cdot 81 \pm 0 \cdot 45$				
		$-89 \cdot 88 \pm 0 \cdot 44$	S3	$17 \cdot 84 \pm 0 \cdot 10$ (sub.)	57/45	$-72 \cdot 04 \pm 0 \cdot 47$
	BCE	$-87 \cdot 8 \ \pm 2 \cdot 4$				$-70 \cdot 0 \ \pm 2 \cdot 4$
	Selected value	$-89 \cdot 88 \pm 0 \cdot 44$				$-72 \cdot 04 \pm 0 \cdot 47$
		$-103 \cdot 6 \ \pm 1 \cdot 1$	E3	$15 \cdot 98 \pm 0 \cdot 15$		$-87 \cdot 6 \ \pm 1 \cdot 2$
		$-99 \cdot 1 \ \pm 1 \cdot 1$	E3	$16 \cdot 15 \pm 0 \cdot 15$		$-82 \cdot 9 \ \pm 1 \cdot 2$
		$+123 \cdot 91 \pm 0 \cdot 30$				
		$+46 \cdot 0 \ \pm 1 \cdot 3$	S4	$24 \cdot 4 \pm 0 \cdot 5$ (sub.)	65/18	$+70 \cdot 4 \ \pm 1 \cdot 4$
		$+45 \cdot 38 \pm 0 \cdot 19$				$+69 \cdot 78 \pm 0 \cdot 56$
	Selected value	$+45 \cdot 38 \pm 0 \cdot 19$				$+69 \cdot 78 \pm 0 \cdot 56$
	A	$+27 \cdot 42 \pm 0 \cdot 86$	S4	$22 \cdot 50 \pm 0 \cdot 60$ (sub.)	53/8	$+49 \cdot 92 \pm 1 \cdot 06$
		$+27 \cdot 44 \pm 0 \cdot 19$				$+49 \cdot 94 \pm 0 \cdot 64$
	Selected value	$+27 \cdot 44 \pm 0 \cdot 19$				$+49 \cdot 94 \pm 0 \cdot 64$
		$+56 \cdot 12 \pm 0 \cdot 52$				
		$+53 \cdot 97 \pm 0 \cdot 52$				
		$+47 \cdot 51 \pm 0 \cdot 52$				
		$+42 \cdot 73 \pm 0 \cdot 30$				
$_6 H_{14}$(c)$+2H_2=1, 4$-Diphenylbutane(c) $\{-2 \cdot 36 \pm 0 \cdot 52\}$		$+41 \cdot 68 \pm 0 \cdot 60$				
	E	$+39 \cdot 7 \ \pm 1 \cdot 7$				
	Selected value	$+42 \cdot 73 \pm 0 \cdot 30$				
		$+8 \cdot 70 \pm 0 \cdot 44$	S4	$25 \cdot 5 \pm 0 \cdot 2$ (sub.)	65/19	$+34 \cdot 20 \pm 0 \cdot 50$
		$+21 \cdot 26 \pm 1 \cdot 42$	S4	$25 \cdot 0 \pm 0 \cdot 3$ (sub.)	65/19	$+46 \cdot 26 \pm 1 \cdot 46$

THERMOCHEMISTRY OF ORGANIC AND ORGANOMETALLIC COMPOUNDS

C_aH_b 1 kcal$=4\cdot184$ kJ

								Determina
1 Formula	2 g.f.w.	3 Name	State	Purity mol %	Type	No. of expts.	Detn. of reactn.	$-\Delta H_r^{\circ}$ kcal/g.f.w.
$C_{16}H_{14}$	$206\cdot2900$	9, 10-Dimethyl-phenanthrene	c		SB	5	m	$1994\cdot4\ \pm2\cdot0$ 66
$C_{16}H_{16}$	$208\cdot3059$	1, 1-Di-o-tolylethene	c		SB	5	m	$2072\cdot60\pm0\cdot48$ 53
$C_{16}H_{16}$	$208\cdot3059$	cis-1, 2-Di-o-tolylethene	c		SB	5	m	$2071\cdot88\pm0\cdot48$ 53
$C_{16}H_{16}$	$208\cdot3059$	trans-1, 2-Di-o-tolylethene	c		SB	5	m	$2069\cdot25\pm0\cdot48$ 53
$C_{16}H_{16}$	$208\cdot3059$	1, 1-Di-p-tolylethene	c		SB	5	m	$2071\cdot88\pm0\cdot48$ 53
$C_{16}H_{16}$	$208\cdot3059$	cis-1, 2-Di-p-tolylethene	c		SB	5	m	$2074\cdot75\pm0\cdot48$ 53
$C_{16}H_{16}$	$208\cdot3059$	trans-1, 2-Di-p-tolylethene	c		SB	5	m	$2065\cdot19\pm0\cdot48$ 53
$C_{16}H_{16}$	$208\cdot3059$	2, 2-Metacyclophane	c		SB			$2070\cdot1\ \pm1\cdot6$ 67
$C_{16}H_{16}$	$208\cdot3059$	2, 2-Paracyclophane	c		SB	7	m	$2088\cdot2\ \pm0\cdot9$ 66
$C_{16}H_{18}$	$210\cdot3219$	1, 1-di-o-tolylethane	l		SB	5	m	$2113\cdot41\pm0\cdot48$ 53
$C_{16}H_{18}$	$210\cdot3219$	1, 2-di-o-tolylethane	c		SB	5	m	$2110\cdot09\pm0\cdot48$ 53
$C_{16}H_{18}$	$210\cdot3219$	1, 1-di-p-tolylethane	l		SB	5	m	$2111\cdot72\pm0\cdot48$ 53
$C_{16}H_{18}$	$210\cdot3219$	1, 2-di-p-tolylethane	c		SB	5	m	$2109\cdot58\pm0\cdot48$ 53
$C_{16}H_{18}$	$210\cdot3219$	1, 4-Diphenylbutane	c		SB	5	m	$2117\cdot29\pm0\cdot48$ 53
$C_{16}H_{26}$	$218\cdot3856$	n-Decylbenzene	l	$99\cdot88\pm0\cdot10$	SB	6	CO_2	$2341\cdot37\pm0\cdot88$ 60
$C_{16}H_{26}$		n-Decylbenzene	l	$99\cdot88\pm0\cdot10$	SB	6	CO_2	$2340\cdot58\pm0\cdot42$ 55
$C_{16}H_{26}$		n-Decylbenzene						
$C_{16}H_{28}$	$220\cdot4016$	2, 2-Perhydropara-cyclophane	c		SB			$2402\cdot9\ \pm2\cdot7$ 67
$C_{16}H_{32}$	$224\cdot4334$	1-Hexadecene	l	$99\cdot93\pm0\cdot06$	SB	5	CO_2	$2520\cdot56\pm0\cdot73$ 60
$C_{16}H_{32}$		1-Hexadecene	l	$99\cdot84\pm0\cdot18$	SB	6	CO_2	$2519\cdot17\pm0\cdot44$ 55
$C_{16}H_{32}$		1-Hexadecene						

1 kcal = 4·184 kJ

H_r°		5 ΔH_f° (l or c) kcal/g.f.w.	6 Determination of ΔH_v			7 ΔH_f° (g) kcal/g.f.w.
Remarks			Type	ΔH_v° kcal/g.f.w.	Ref.	
		+11·4 ±2·0	S4	28·55±0·29 (sub.)	66/20	+40·0 ±2·1
		+21·26±0·52				
		+20·54±0·52				
		+17·91±0·52				
		+20·54±0·52				
		+23·41±0·52				
		+13·85±0·52				
		+18·8 ±1·6	S4	22·0±0·5 (sub.)	67/10	+40·8 ±1·7
		+36·9 ±0·9	S4	23·0±1·0 (sub.)	66/21	+59·9 ±1·4
	C	−6·24±0·52				
	C	−9·56±0·52				
	C	−7·93±0·52				
	C	−10·07±0·52				
		−2·36±0·52				
An 0·99990±0·00008		−51·54±0·91	E1	19·06±0·10	53/1	−32·48±0·92
		−52·33±0·57				−33·27±0·59
	Selected value	−52·07±0·50				−33·01±0·51
		−58·3 ±2·7		21·0±0·3 (sub.)	67/10	−37·3 ±2·8
An 0·99982±0·00011		−77·30±0·78	E1	19·14±0·05		−58·16±0·79
		−78·69±0·50				−59·55±0·51
	Selected value	−78·23±0·60				−59·09±0·61

THERMOCHEMISTRY OF ORGANIC AND ORGANOMETALLIC COMPOUNDS

C_aH_b

1 kcal = 4·184 kJ

Determina

1 Formula	2 g.f.w.	3 Name	State	Purity mol %	Type	No. of expts.	Detn. of reactn.	$-\Delta H_r^\circ$ kcal/g.f.w.	F
$C_{16}H_{32}$	224·4334	n-Decylcyclohexane	l	99·88±0·11	SB	6	CO_2	2497·87±0·80	60
$C_{16}H_{32}$		n-Decylcyclohexane	l	99·86±0·11	SB	5	CO_2	2497·90±0·43	55
$C_{16}H_{32}$		n-Decylcyclohexane							
$C_{16}H_{32}$	224·4334	Cyclohexadecane	c	2s	SB	9	m	2501·44±0·45	60
$C_{16}H_{34}$	226·4494	n-Hexadecane	l		SB	9	m	2557·15±0·97	39
$C_{16}H_{34}$		n-Hexadecane	l	96·0 ±0·4	SB	7	CO_2	2557·58±0·68	4
$C_{16}H_{34}$		n-Hexadecane	l	99·96±0·04	SB	6	CO_2	2557·15±0·42	55
$C_{16}H_{34}$		n-Hexadecane							
$C_{17}H_{34}$	238·4605	Cycloheptadecane	c		SB	4	m	2657·35±0·48	60
$C_{18}H_{12}$	228·2963	Triphenylene	c		SB	4	m	2138·01±0·52	51
$C_{18}H_{12}$		Triphenylene	c		SB			2136·53±0·11	67
$C_{18}H_{12}$		Triphenylene							
$C_{18}H_{12}$	228·2963	Chrysene	c		SB	6	m	2137·53±0·49	51/
$C_{18}H_{12}$	228·2963	3, 4-Benzphenanthrene	c		SB	5	m	2147·00±0·45	51/
$C_{18}H_{12}$	228·2963	1, 2-Benzphenanthrene	c		SB	4	m	2143·64±0·56	51/
$C_{18}H_{12}$	228·2963	Tetracene	c		SB	4	m	2140·76±0·32	51/
$C_{18}H_{14}$	230·3123	5, 12-Dihydrotetracene	c		SB	5	m	2196·56±0·33	51/
$C_{18}H_{14}$	230·3123	Diphenylfulvene	c	99·8	SB	4	m	2242·2 ±3·0	57/
$C_{18}H_{18}$	234·3442	[18]-Annulene	c		SB	3	CO_2	2346·8 ±4·0	65/
$C_{18}H_{18}$	234·3442	3, 4, 5, 6-Tetramethyl-phenanthrene	c	>99, glc	SB	5	m	2313·35±1·17	65/
$C_{18}H_{18}$	234·3442	2, 4, 5, 7-Tetramethyl-phenanthrene	c	>99, glc	SB	4	m	2310·67±1·47	65/

$1 \text{ kcal} = 4 \cdot 184 \text{ kJ}$

H_r° Remarks	5 ΔH_f° (l or c) kcal/g.f.w.	6 Determination of ΔH_v Type	ΔH_v° kcal/g.f.w.	Ref.	7 ΔH_f° (g) kcal/g.f.w.
$\ln 1 \cdot 00019 \pm 0 \cdot 00012$	$-99 \cdot 99 \pm 0 \cdot 84$	V1	$18 \cdot 83 \pm 0 \cdot 10$	65/17	$-81 \cdot 16 \pm 0 \cdot 86$
	$-99 \cdot 96 \pm 0 \cdot 49$				$-81 \cdot 13 \pm 0 \cdot 51$
Selected value	$-99 \cdot 97 \pm 0 \cdot 42$				$-81 \cdot 14 \pm 0 \cdot 44$
	$-96 \cdot 42 \pm 0 \cdot 51$	S3	$19 \cdot 54 \pm 0 \cdot 10$ (sub.)	57/45	$-76 \cdot 88 \pm 0 \cdot 52$
A	$-109 \cdot 02 \pm 1 \cdot 00$	V3	$19 \cdot 22 \pm 0 \cdot 30$	49/11, 49/12	$-89 \cdot 80 \pm 1 \cdot 05$
	$-108 \cdot 59 \pm 0 \cdot 72$				$-89 \cdot 37 \pm 0 \cdot 79$
$\ln 1 \cdot 00003 \pm 0 \cdot 00005$	$-109 \cdot 02 \pm 0 \cdot 48$				$-89 \cdot 80 \pm 0 \cdot 59$
Selected value	$-109 \cdot 02 \pm 0 \cdot 48$				$-89 \cdot 80 \pm 0 \cdot 59$
	$-102 \cdot 87 \pm 0 \cdot 54$	S3	$15 \cdot 80 \pm 0 \cdot 15$ (sub.)	57/45	$-87 \cdot 07 \pm 0 \cdot 58$
E	$+35 \cdot 20 \pm 0 \cdot 55$	S4	$25 \cdot 6 \pm [1 \cdot 0]$ (sub.)	67/39	$+63 \cdot 4 \pm 1 \cdot 2$
	$+33 \cdot 72 \pm 0 \cdot 22$	S4	$28 \cdot 2 \pm [1 \cdot 0]$ (sub.)	58/39	$+61 \cdot 9 \pm 1 \cdot 1$
Selected value	$+33 \cdot 72 \pm 0 \cdot 22$		$28 \cdot 2 \pm [1 \cdot 0]$ (sub.)		$+61 \cdot 9 \pm 1 \cdot 1$
E	$+34 \cdot 72 \pm 0 \cdot 53$	S4	$28 \cdot 1 \pm [1 \cdot 0]$ (sub.)	67/39	$+62 \cdot 8 \pm 1 \cdot 2$
E	$+44 \cdot 19 \pm 0 \cdot 50$	S4	$25 \cdot 4 \pm [1 \cdot 0]$ (sub.)	67/39	$+69 \cdot 6 \pm 1 \cdot 2$
E	$+40 \cdot 83 \pm 0 \cdot 60$				
E	$+37 \cdot 95 \pm 0 \cdot 37$	S4	$29 \cdot 8 \pm [1 \cdot 0]$ (sub.)	67/39	$+69 \cdot 8 \pm 1 \cdot 1$
E	$+25 \cdot 44 \pm 0 \cdot 39$	S4	$27 \cdot 7 \pm [1 \cdot 0]$ (sub.)	58/39	$+53 \cdot 1 \pm 1 \cdot 1$
C	$+71 \cdot 1 \pm 3 \cdot 0$	E1	$25 \cdot 0 \pm 2 \cdot 0$ (sub.)	57/11	$+96 \cdot 1 \pm 4 \cdot 0$
	$+39 \cdot 0 \pm 4 \cdot 0$				
D	$+5 \cdot 6 \pm 1 \cdot 2$	S4	$31 \cdot 9 \pm 0 \cdot 9$ (sub.)	65/19	$+37 \cdot 5 \pm 1 \cdot 5$
D	$+2 \cdot 9 \pm 1 \cdot 5$	S4	$27 \cdot 3 \pm 0 \cdot 4$ (sub.)	65/19	$+30 \cdot 2 \pm 1 \cdot 6$

THERMOCHEMISTRY OF ORGANIC AND ORGANOMETALLIC COMPOUNDS

C_aH_b

1 kcal = 4·184 kJ

1 Formula	2 g.f.w.	3 Name	State	Purity mol %	Type	No. of expts.	Detn. of reactn.	$-\Delta H_r^{\circ}$ kcal/g.f.w.	F
$C_{18}H_{30}$	246·4398	1, 2, 4-Tri-t-butyl-benzene	c		SB			2657·5 ±0·8	61
$C_{18}H_{30}$	246·4398	1, 3, 5-Tri-t-butyl-benzene	c		SB			2640·7 ±0·9	61
$C_{18}H_{36}$	252·4876	n-Dodecylcyclohexane	l	98·8	SB	5	m	2810·84±0·84	4(
$C_{18}H_{38}$	254·5036	n-Octadecane	c	95	SB	8	m	2855·28±1·14	4(
$C_{19}H_{16}$	244·3394	Triphenylmethane	c		SB	9	m	2372·43±0·95	4(
$C_{19}H_{16}$		Triphenylmethane	c		SB	6	m	2374·4 ±0·3	46
$C_{19}H_{16}$		Triphenylmethane							4;
$C_{20}H_{12}$	252·3186	Perylene	c		SB			2334·60±0·11	67
$C_{20}H_{16}$	256·3505	Triphenylethene	c	99·88	SB	5	m	2483·32±0·40	50
$C_{20}H_{16}$	256·3505	5, 8-Dimethylbenzo(c)-phenanthrene	c		SB	4	m	2453·41±1·28	63
$C_{20}H_{16}$	256·3505	1, 12-Dimethylbenzo(c)-phenanthrene	c		SB	5	m	2464·46±0·80	63
$C_{20}H_{16}$	256·3505	3', 6-Dimethyl-1, 2-benzanthracene	c		SB	3	m	2445·75±0·48	63
$C_{20}H_{16}$	256·3505	1', 9-Dimethyl-1, 2-benzanthracene	c		SB	3	m	2460·75±0·48	63
$C_{20}H_{16}$	256·3505	9, 10-Dimethyl-1, 2-benzanthracene	c		SB	4	m	2461·9 ±1·0	66
$C_{20}H_{16}$	256·3505	5, 6-Dimethylchrysene	c		SB	6	m	2459·2 ±1·5	66
$C_{20}H_{18}$	258·3665	1, 1, 1-Triphenylethane	c		SB	5	m	2533·42±0·48	53
$C_{20}H_{18}$	258·3665	1, 1, 2-Triphenylethane	c		SB	6	m	2526·97±0·48	53
$C_{20}H_{24}$	264·4143	Tetracyclopentadiene	c		SB	5	m	2730·7 ±2·7	34

$$1 \text{ kcal} = 4 \cdot 184 \text{ kJ}$$

H_r°			6 Determination of ΔH_v			
Remarks		5 ΔH_f° (l or c) kcal/g.f.w.	Type	ΔH_v° kcal/g.f.w.	Ref.	7 ΔH_f° (g) kcal/g.f.w.
		$-60 \cdot 1 \ \pm 0 \cdot 9$				
		$-76 \cdot 9 \ \pm 1 \cdot 0$				
	A	$-111 \cdot 75 \pm 0 \cdot 88$	V3	$22 \cdot 33 \pm 0 \cdot 30$	49/11	$-89 \cdot 42 \pm 0 \cdot 93$
rection made for 5% liquid due to premelting		$-135 \cdot 6 \ \pm 1 \cdot 2$	S3	$36 \cdot 58 \pm 1 \cdot 20$ (sub.)	49/12	$-99 \cdot 0 \ \pm 1 \cdot 7$
		$+38 \cdot 94 \pm 0 \cdot 98$	S4	$23 \cdot 9 \pm 0 \cdot 1^*$ (sub.)	59/36	$+62 \cdot 8 \ \pm 1 \cdot 0$
	F	$+40 \cdot 9 \ \pm 0 \cdot 4$				$+64 \cdot 8 \ \pm 0 \cdot 5$
Selected value		$+40 \cdot 9 \ \pm 0 \cdot 4$				$+64 \cdot 8 \ \pm 0 \cdot 5$
		$+43 \cdot 69 \pm 0 \cdot 24$	S4	$30 \cdot 0 \pm [1 \cdot 0]$ (sub.)	67/39	$+73 \cdot 7 \ \pm 1 \cdot 1$
	C	$+55 \cdot 78 \pm 0 \cdot 46$				
		$+25 \cdot 87 \pm 1 \cdot 30$				
		$+36 \cdot 92 \pm 0 \cdot 83$				
		$+18 \cdot 21 \pm 0 \cdot 53$	S4	$26 \cdot 9 \pm 0 \cdot 8$ (sub.)	65/19	$+45 \cdot 1 \ \pm 1 \cdot 0$
		$+33 \cdot 21 \pm 0 \cdot 53$	S4	$26 \cdot 9 \pm 0 \cdot 8$ (sub.)	65/19	$+60 \cdot 1 \ \pm 1 \cdot 0$
		$+34 \cdot 4 \ \pm 1 \cdot 0$	S4	$32 \cdot 02 \pm 0 \cdot 32$ (sub.)	66/20	$+66 \cdot 4 \ \pm 1 \cdot 1$
		$+31 \cdot 7 \ \pm 1 \cdot 5$	S4	$31 \cdot 06 \pm 0 \cdot 31$ (sub.)	66/20	$+62 \cdot 8 \ \pm 1 \cdot 6$
		$+37 \cdot 56 \pm 0 \cdot 53$				
		$+31 \cdot 11 \pm 0 \cdot 53$				
	CE	$+29 \cdot 9 \ \pm 2 \cdot 7$				

Determina

1 Formula	2 g.f.w.	3 Name	State	Purity mol %	Type	No. of expts.	Detn. of reactn.	$-\Delta H_r^\circ$ kcal/g.f.w.	R
$C_{20}H_{26}$	266·4302	Dihydrotetracyclo-pentadiene	c		SB	6	m	2766·0 ±2·8	3
$C_{21}H_{26}$	278·4414	[1, 8]-Paracyclophane	c		SB			2843·6 ±1·7	67
$C_{22}H_{38}$	302·5482	1, 2, 4, 5-Tetra-t-butyl-benzene	c		SB			3295·9 ±2·3	61
$C_{23}H_{22}$	298·4318	Tri-o-tolylethene	c		SB	5	m	2944·57±0·72	53
$C_{23}H_{22}$	298·4318	Tri-p-tolylethene	c		SB	5	m	2943·61±0·72	53
$C_{23}H_{24}$	300·4477	1, 1, 2-Tri-o-tolylethane	c		SB	5	m	2988·78±0·72	53
$C_{23}H_{24}$	300·4477	1, 1, 2-Tri-p-tolylethane	c		SB	5	m	2986·87±0·72	53
$C_{24}H_{18}$	306·4111	1, 3, 5-Triphenylbenzene	c	4s	SB	12	m	2925·75±1·27	39
$C_{24}H_{32}$	320·5226	[6, 6]-Paracyclophane	c		SB			3304·2 ±2·1	67
$C_{25}H_{20}$	320·4382	Tetraphenylmethane	c		SB	9	m	3093·5 ±0·6	46 47
$C_{26}H_{18}$	330·4334	9, 10-Diphenylanthracene	c		SB	4	m	3133·94±0·69	51
$C_{26}H_{20}$	332·4493	Tetraphenylethene	c	99·90	SB	5	m	3202·94±0·30	50
$C_{26}H_{22}$	334·4652	1, 1, 1, 2-Tetraphenyl-ethane	c		SB	10	m	3250·10±0·30	53
$C_{26}H_{22}$	334·4652	1, 1, 2, 2-Tetraphenyl-ethane	c		SB	10	m	3248·42±0·30	53
$C_{26}H_{46}$	358·6565	3-Phenyleicosane	c	98·2	SB	6	m	3885·4 ±1·5	44
$C_{26}H_{46}$	358·6565	9-Phenyleicosane	l	99·5	SB	7	m	3902·8 ±1·5	44
$C_{26}H_{52}$	364·7043	11-Cyclopentyl-heneicosane	l	97·1	SB	6	m	4066·5 ±1·6	44

1 kcal = 4·184 kJ

ΔH_r°	Remarks		5 ΔH_f° (l or c) kcal/g.f.w.	6 Determination of ΔH_v			7 ΔH_f° (g) kcal/g.f.w.
				Type	ΔH_v° kcal/g.f.w.	Ref.	
		CE	$-3\cdot1 \pm 2\cdot8$				
			$-19\cdot6 \pm 1\cdot7$		$26\cdot5\pm0\cdot5$ (sub.)	67/10	$+6\cdot9 \pm 1\cdot9$
			$-71\cdot2 \pm 2\cdot3$				
			$+29\cdot93\pm0\cdot77$				
			$+28\cdot97\pm0\cdot77$				
			$+5\cdot83\pm0\cdot77$				
			$+3\cdot92\pm0\cdot77$				
		A	$+53\cdot69\pm1\cdot30$	S4	$34\cdot3\pm[1\cdot0]$ (sub.)	58/39	$+88\cdot0 \pm1\cdot7$
			$-46\cdot1 \pm 2\cdot1$		$27\cdot5\pm0\cdot5$ (sub.)	67/10	$-18\cdot6 \pm2\cdot2$
		F	$+59\cdot1 \pm0\cdot7$				
		E	$+73\cdot78\pm0\cdot75$	S3	$37\cdot5\pm1\cdot0$ (sub.)	53/26	$+111\cdot3 \pm1\cdot2$
		C	$+74\cdot46\pm0\cdot41$				
		C	$+53\cdot31\pm0\cdot42$				
		C	$+51\cdot63\pm0\cdot42$				
			$-131\cdot2 \pm1\cdot6$				
			$-113\cdot8 \pm1\cdot6$				
			$-155\cdot0 \pm1\cdot7$				

C_aH_b

1 kcal $=4 \cdot 184$ kJ

1 Formula	2 g.f.w.	3 Name	State	Purity mol %	Type	No. of expts.	Detn. of reactn.	$-\Delta H_c^\circ$ kcal/g.f.w.	R
$C_{26}H_{52}$	364·7043	3-Cyclohexyleicosane	l	98·2	SB	7	m	4062·3 ±1·6	44
$C_{26}H_{52}$	364·7043	9-Cyclohexyleicosane	l	99·5	SB	6	m	4060·4 ±1·6	44
$C_{26}H_{54}$	366·7203	5-n-Butyldocosane	l	97·6	SB	6	m	4119·3 ±1·6	44
$C_{26}H_{54}$	366·7203	11-n-Butyldocosane	l		SB	6	m	4118·7 ±1·6	44
$C_{27}H_{48}$	372·6836	11-Phenylheneicosane	l	97·6	SB	6	m	4057·8 ±1·6	44
$C_{27}H_{48}$		11-Phenylheneicosane	l		SB	5	m	4058·5 ±1·6	46
$C_{27}H_{48}$		11-Phenylheneicosane							
$C_{27}H_{54}$	378·7314	11-Cyclohexylheneicosane	l	97·6	SB	13	m	4219·0 ±1·7	44
$C_{27}H_{54}$		11-Cyclohexylheneicosane	l		SB	4	m	4219·2 ±1·7	46
$C_{27}H_{54}$		11-Cyclohexylheneicosane							
$C_{28}H_{18}$	354·4557	9, 9'-Bianthryl	c		SB	5	m	3326·23±0·75	51
$C_{28}H_{18}$	354·4557	9, 9'-Biphenanthryl	c		SB	4	m	3299·12±0·66	51
$C_{28}H_{20}$	356·4716	Dianthracene	c		SB	6	m	3390·1 ±1·6	52
$C_{28}H_{22}$	358·4875	1, 1, 4, 4-Tetraphenyl- 1, 3-butadiene	c		SB	5	m	3463·74±0·72	53
$C_{28}H_{22}$	358·4875	1, 2, 3, 4-Tetraphenyl- 1, 3-butadiene	c		SB	5	m	3470·67±0·72	53
$C_{28}H_{26}$	362·5194	1, 1, 4, 4-Tetraphenyl- butane	c		SB	5	m	3560·54±0·96	53
$C_{30}H_{28}$	388·5577	Tetra-p-tolylethene	c		SB	5	m	3818·93±0·96	53
$C_{30}H_{30}$	390·5736	1, 1, 2, 2-Tetra-p- tolylethane	c		SB	5	m	3863·86±0·72	53
$C_{30}H_{60}$	420·8127	Cyclotriacontane	c		SB	3	m	4656·9 ±6·9	33
$C_{31}H_{56}$	428·7920	13-Phenylpentacosane	c		SB	5	m	4664·2 ±1·8	46
$C_{31}H_{62}$	434·8398	13-Cyclohexyl- pentacosane	l		SB	4	m	4843·9 ±1·9	46

1 kcal$=4\cdot184$ kJ

ΔH_r°		5 ΔH_f°(l or c) kcal/g.f.w.	6 Determination of ΔH_v			7 ΔH_f°(g) kcal/g.f.w.
Remarks			Type	ΔH_v° kcal/g.f.w.	Ref.	
		$-159\cdot2\ \pm1\cdot7$				
		$-161\cdot1\ \pm1\cdot7$				
		$-170\cdot5\ \pm1\cdot7$	E3	$30\cdot1\pm0\cdot2$		$-140\cdot4\ \pm1\cdot8$
		$-171\cdot1\ \pm1\cdot7$	E3	$29\cdot3\pm0\cdot2$		$-141\cdot8\ \pm1\cdot8$
		$-121\cdot1\ \pm1\cdot7$				
		$-120\cdot4\ \pm1\cdot7$				
Selected value		$-120\cdot8\ \pm1\cdot5$				
		$-164\cdot9\ \pm1\cdot7$				
		$-164\cdot7\ \pm1\cdot7$				
Selected value		$-164\cdot8\ \pm1\cdot5$				
	E	$+77\cdot97\pm0\cdot81$	S4	$30\cdot6\pm[1\cdot0]$ (sub.)	58/39	$+108\cdot6\ \pm[1\cdot3]$
	E	$+50\cdot86\pm0\cdot73$				
		$+73\cdot5\ \pm1\cdot7$				
		$+78\cdot85\pm0\cdot79$				
		$+85\cdot78\pm0\cdot79$				
		$+39\cdot02\pm1\cdot01$				
		$+40\cdot99\pm1\cdot03$				
		$+17\cdot60\pm0\cdot79$				
	BCE	$-214\cdot1\ \pm7\cdot0$				
		$-164\cdot2\ \pm1\cdot9$				
		$-189\cdot4\ \pm2\cdot0$				

THERMOCHEMISTRY OF ORGANIC AND ORGANOMETALLIC COMPOUNDS

C_aH_b 1 kcal = 4·184 kJ

Determina

1 Formula	2 g.f.w.	3 Name	State	Purity mol %	Type	No. of expts.	Detn. of reactn.	$-\Delta H_r^o$ kcal/g.f.w.	R
$C_{31}H_{64}$	436·8557	11-n-Decylheneicosane	l		SB	6	m	4900·4 ±2·0	46
$C_{31}H_{64}$		11-n-Decylheneicosane	l		SB	6	m	4897·7 ±1·9	44
$C_{31}H_{64}$		11-n-Decylheneicosane							
$C_{32}H_{26}$	410·5640	Pentaphenylethane	c	2s	SB	11	m	3988·85±0·72	53
$C_{32}H_{66}$	450·8828	n-Dotriacontane	c		SB	5	m	5032·7 ±2·0	46
$C_{32}H_{66}$		n-Dotriacontane	c		SB	3	m	5032·4 ±2·5	31
$C_{32}H_{66}$		n-Dotriacontane							
$C_{38}H_{30}$	486·6628	Hexaphenylethane	c		H	4	m	40·5 ±0·6	36
$C_{42}H_{28}$	532·6915	5, 6, 11, 12-Tetra- phenyltetracene	c		SB	4	m	5054·8 ±5·1	38
$C_{42}H_{28}$	532·6915	Pseudo-tetraphenyl- tetracene	c		SB	3	m	5035·6 ±5·0	38

$$1 \text{ kcal} = 4 \cdot 184 \text{ kJ}$$

$\Delta H°$			6 Determination of ΔH_v			
	Remarks	5 $\Delta H_f°$ (l or c) kcal/g.f.w.	Type	$\Delta H_v°$ kcal/g.f.w.	Ref.	7 $\Delta H_f°$ (g) kcal/g.f.w.
		$-201\cdot3$ $\pm2\cdot1$	E3	$34\cdot0\pm0\cdot2$		$-167\cdot3$ $\pm2\cdot2$
		$-204\cdot0$ $\pm2\cdot0$				$-170\cdot0$ $\pm2\cdot1$
	Selected value	$-202\cdot7$ $\pm2\cdot0$				$-168\cdot7$ $\pm2\cdot1$
		$+91\cdot12$ $\pm0\cdot80$				
$_m$ 26·67$\pm$0·54 (46/9)		$-231\cdot3$ $\pm2\cdot1$	E1$_m$	$64\cdot8\pm0\cdot6$		$-166\cdot5$ $\pm2\cdot2$
		$-231\cdot7$ $\pm2\cdot6$		(sub.)		$-166\cdot9$ $\pm2\cdot7$
	Selected value	$-231\cdot5$ $\pm1\cdot5$				$-166\cdot7$ $\pm1\cdot6$
$_8H_{30}$(c)+H$_2$ =2 Triphenylmethane(c){+40·9$\pm$0·4}		$+122\cdot3$ $\pm0\cdot8$				
	BCE	$+148\cdot2$ $\pm5\cdot2$	S4	$38\cdot4\pm[1\cdot0]$ (sub.)	58/39	$+186\cdot6$ $\pm5\cdot3$
	BCE	$+129\cdot0$ $\pm5\cdot1$				

THERMOCHEMISTRY OF ORGANIC AND ORGANOMETALLIC COMPOUNDS

C_aH_bO 1 kcal $=4 \cdot 184$ kJ

1 Formula	2 g.f.w.	3 Name	State	Purity mol %	Type	No. of expts.	Detn. of reactn.	$-\Delta H_r^\circ$ kcal/g.f.w.	R
CO	$28 \cdot 0106$	Carbon monoxide	g		FC	8	CO_2	$67 \cdot 63 \pm 0 \cdot 03$	31/5 39
CO		Carbon monoxide	g		SB		m	$67 \cdot 86 \pm 0 \cdot 10*$	32,
CO		Carbon monoxide							
$1/n\ [CH_2O]_n$	$30 \cdot 0265$	Polyoxymethylene	c		SB	10	m	$121 \cdot 44 \pm 0 \cdot 02$	63/
$1/n\ [CH_2O]_n$		Polyoxymethylene	c		SB	4	m	$120 \cdot 21 \pm 0 \cdot 12$	42
$1/n\ [CH_2O]_n$		Polyoxymethylene							
$1/8\ [8CH_2O. H_2O]$	$32 \cdot 2784$	Paraformaldehyde	c		SB	4	m	$119 \cdot 93 \pm 0 \cdot 12$	42,
CH_2O	$30 \cdot 0265$	Formaldehyde	aq		R		m	$-2 \cdot 4\ \pm 0 \cdot 5$	42,
CH_2O		Formaldehyde	g		R		m	$14 \cdot 9\ \pm 0 \cdot 1$	42/ 33
CH_2O		Formaldehyde	g		E		an	$2 \cdot 9 \pm [1 \cdot 0]$	33,
CH_2O		Formaldehyde	g		E		TP	$12 \cdot 24 \pm 0 \cdot 08$	59/
CH_2O		Formaldehyde	g	3s	FC	9	CO_2	$136 \cdot 42 \pm 0 \cdot 11$	M
CH_2O		Formaldehyde							
CH_4O	$32 \cdot 0424$	Methanol	g		FC	10	CO_2	$182 \cdot 61 \pm 0 \cdot 04$	32,
CH_4O		Methanol	l	$99 \cdot 9$	SB	5	CO_2	$173 \cdot 45 \pm 0 \cdot 03$	65/
CH_4O		Methanol							
C_2H_2O	$42 \cdot 0376$	Ketene	g		R	3	an	$47 \cdot 31 \pm 0 \cdot 65$	34,
C_2H_4O	$44 \cdot 0536$	Acetaldehyde	g		H	3	P/CO_2	$16 \cdot 51 \pm 0 \cdot 10$	38,
C_2H_4O	$44 \cdot 0536$	Ethylene oxide	g		FC	4	CO_2	$312 \cdot 15 \pm 0 \cdot 14$	65,
C_2H_4O		Ethylene oxide	g		FC		CO_2	$312 \cdot 55 \pm 0 \cdot 22$	42,
C_2H_4O		Ethylene oxide	l		SB	2	m	$301 \cdot 85 \pm 0 \cdot 30$	37,
C_2H_4O		Ethylene oxide							
C_2H_6O	$46 \cdot 0695$	Ethanol	g		FC	6	CO_2	$336 \cdot 81 \pm 0 \cdot 06$	32,
C_2H_6O		Ethanol	l	$99 \cdot 9$	SB	4	CO_2	$326 \cdot 86 \pm 0 \cdot 06$	65/
C_2H_6O		Ethanol							
C_2H_6O	$46 \cdot 0695$	Dimethyl ether	g	$99 \cdot 97 \pm 0 \cdot 01$	FC	6	CO_2	$349 \cdot 06 \pm 0 \cdot 11$	64,
C_3H_6O	$58 \cdot 0807$	Allyl alcohol	g		H	2	P/CO_2	$31 \cdot 22 \pm 0 \cdot 30$	38,

1 kcal $=4\cdot184$ kJ

ΔH_r°		5 ΔH_f° (l or c) kcal/g.f.w.	6 Determination of ΔH_v			7 ΔH_f° (g) kcal/g.f.w.
Remarks			Type	ΔH_v° kcal/g.f.w.	Ref.	
						$-26\cdot42\pm0\cdot04$
ss of CO from PVT measurements	E					$-26\cdot19\pm0\cdot11$*
Selected value						$-26\cdot42\pm0\cdot04$
		$-40\cdot93\pm0\cdot03$				
	CE	$-42\cdot16\pm0\cdot12$				
Selected value		$-40\cdot93\pm0\cdot03$				
	CE	$-42\cdot44\pm0\cdot13$				
[8CH$_2$O.H$_2$O] $\{-42\cdot44\pm0\cdot13\}+$aq $=$CH$_2$O(aq)		$-40\cdot0\ \ \pm0\cdot6$ (aq)				
$_2$O(g)$+$aq$=$CH$_2$O(aq) $\{-40\cdot0\pm0\cdot6\}$						$-25\cdot1\ \ \pm0\cdot7$
g) $\{-26\cdot42\pm0\cdot04\}+H_2$ $\rightleftharpoons$CH$_2$O(g). 2nd Law	E					$-29\cdot3\ \ \pm[1\cdot0]$
$_2$O(g) $\rightleftharpoons1/n$[CH$_2$O] $_n$(c)$\{-40\cdot93\pm0\cdot03\}$. 2nd Law						$-28\cdot69\pm0\cdot07$
I An $0\cdot99933\pm0\cdot00016$						$-25\cdot95\pm0\cdot12$
Selected value						$-25\cdot95\pm0\cdot12$
I An $1\cdot00016\pm0\cdot00036$		$-57\cdot01\pm0\cdot05$	V1	$8\cdot94\pm0\cdot01$	60/20	$-48\cdot07\pm0\cdot05$
		$-57\cdot23\pm0\cdot04$				$-48\cdot29\pm0\cdot05$
Selected value		$-57\cdot01\pm0\cdot05$				$-48\cdot07\pm0\cdot05$
$_2$O(g)$+$NaOH(aq) $\{-112\cdot36\}$ $=$CH$_3$CO$_2$Na(aq) $\{-173\cdot90\}$	E					$-14\cdot23\pm0\cdot70$
$_4$O(g)$+$H$_2$ $=$Ethanol(g) $\{-56\cdot24\pm0\cdot07\}$	E	$-45\cdot88\pm0\cdot13$	C1	$6\cdot15\pm0\cdot03$	49/13	$-39\cdot73\pm0\cdot12$
I An $1\cdot00001\pm0\cdot00022$		$-18\cdot54\pm0\cdot15$	C1	$5\cdot96\pm0\cdot01$	49/14	$-12\cdot58\pm0\cdot15$
		$-18\cdot14\pm0\cdot23$				$-12\cdot18\pm0\cdot23$
	BCE	$-22\cdot88\pm0\cdot31$				$-16\cdot92\pm0\cdot32$
Selected value		$-18\cdot54\pm0\cdot15$				$-12\cdot58\pm0\cdot15$
		$-66\cdot42\pm0\cdot08$	C1	$10\cdot18\pm0\cdot04$	31/8,	$-56\cdot24\pm0\cdot07$
		$-66\cdot19\pm0\cdot07$			61/20	$-56\cdot01\pm0\cdot08$
Selected value		$-66\cdot42\pm0\cdot08$				$-56\cdot24\pm0\cdot07$
An $0\cdot99963\pm0\cdot00027$						$-43\cdot99\pm0\cdot12$
opanol(g) $\{-61\cdot17\pm0\cdot16\}$	E	$-40\cdot9\pm0\cdot4$	E7	$11\cdot3\pm0\cdot3$		$-29\cdot55\pm0\cdot35$

C_aH_bO 1 kcal = 4·184 kJ

1 Formula	2 g.f.w.	3 Name	State	Purity mol %	Type	No. of expts.	Detn. of reactn.	$-\Delta H_r^\circ$ kcal/g.f.w.	R
C_3H_6O	58·0807	Propanal	l		SB	6	m	434·16±0·18	62
C_3H_6O		Propanal	g	99·75	E		glc	15·72±0·16	67
C_3H_6O		Propanal							
C_3H_6O	58·0807	Acetone	g		FC	3	CO_2	435·32±0·20	41
C_3H_6O		Acetone	g		H	4	P/CO_2	13·24±0·10	38
C_3H_6O		Acetone	g		E		glc	13·20±0·10	65
C_3H_6O		Acetone							
C_3H_6O	58·0807	Propylene oxide	l	99·86	SB	5	CO_2	458·26±0·26	62
C_3H_6O		Propylene oxide	l		SB	3	m	452·54±0·45	37
C_3H_6O		Propylene oxide	g	99·97±0·01	FC	6	CO_2	464·47±0·14	61
C_3H_6O		Propylene oxide							
C_3H_6O	58·0807	Trimethylene oxide	g	99·90±0·02	FC	4	CO_2	467·85±0·14	6
C_3H_8O	60·0966	n-Propanol	l	glc	SB	6	CO_2	483·12±0·18	61
C_3H_8O		n-Propanol	l	99·8	SB	4	CO_2	482·64±0·07	65
C_3H_8O		n-Propanol	l	99·75	SB	5	CO_2	482·90±0·30	6
C_3H_8O		n-Propanol							
C_3H_8O	60·0966	s-Propanol	l	glc	SB	5	CO_2	479·39±0·10	61
C_3H_8O		s-Propanol	l		SB	4	m	479·26±0·30	39
									50
C_3H_8O		s-Propanol	l	99·8	SB	4	CO_2	479·66±0·05	65
C_3H_8O		s-Propanol							
C_3H_8O	60·0966	Methyl ethyl ether	g	99·89±0·04	FC	4	CO_2	503·69±0·15	6
C_4H_4O	68·0759	Furan	l	99·98±0·01	SB	5	m	497·99±0·15	5
C_4H_4O		Furan	l		SB	2	m	499·1 ±1·0	2
C_4H_4O		Furan	g		H	4	P/CO_2	36·12±0·12	3
C_4H_4O		Furan	g	99·94±0·02	FC	4	CO_2	504·45±0·15	64
C_4H_4O		Furan							
C_4H_6O	70·0918	2-Butenal	l	99·95, glc	SB	6	m	546·71±0·18	6
C_4H_6O		2-Butenal	g		H	4	P/CO_2	24·91±0·10	3
C_4H_6O		2-Butenal							

1 kcal = $4 \cdot 184$ kJ

ΔH_r° Remarks	5 ΔH_f° (l or c) kcal/g.f.w.	6 Determination of ΔH_v Type	ΔH_v° kcal/g.f.w.	Ref.	7 ΔH_f° (g) kcal/g.f.w.
	$-52 \cdot 94 \pm 0 \cdot 19$	V1	$7 \cdot 09 \pm 0 \cdot 10$	67/1	$-45 \cdot 85 \pm 0 \cdot 22$
₆O(g)+H₂	$-52 \cdot 54 \pm 0 \cdot 22$				$-45 \cdot 45 \pm 0 \cdot 21$
⇌n-Propanol(g){$-61 \cdot 17 \pm 0 \cdot 16$}.					
2nd Law					
Selected value	$-52 \cdot 54 \pm 0 \cdot 22$				$-45 \cdot 45 \pm 0 \cdot 21$
	$-59 \cdot 15 \pm 0 \cdot 21$	C1	$7 \cdot 37 \pm 0 \cdot 01$	49/15,	$-51 \cdot 78 \pm 0 \cdot 21$
ropanol(g) {$-65 \cdot 12 \pm 0 \cdot 13$} E	$-59 \cdot 25 \pm 0 \cdot 17$			57/14	$-51 \cdot 88 \pm 0 \cdot 17$
₆O(g)+H₂	$-59 \cdot 29 \pm 0 \cdot 16$				$-51 \cdot 92 \pm 0 \cdot 16$
⇌2-Propanol(g){$-65 \cdot 12 \pm 0 \cdot 13$}.					
2nd Law					
Selected value	$-59 \cdot 27 \pm 0 \cdot 12$				$-51 \cdot 90 \pm 0 \cdot 12$
	$-28 \cdot 84 \pm 0 \cdot 26$	C1	$6 \cdot 67 \pm 0 \cdot 01$	62/19	$-22 \cdot 17 \pm 0 \cdot 26$
BCE	$-34 \cdot 56 \pm 0 \cdot 46$				$-27 \cdot 89 \pm 0 \cdot 46$
I An $1 \cdot 00015 \pm 0 \cdot 00016$	$-29 \cdot 30 \pm 0 \cdot 15$				$-22 \cdot 63 \pm 0 \cdot 15$
Selected value	$-29 \cdot 30 \pm 0 \cdot 15$				$-22 \cdot 63 \pm 0 \cdot 15$
I An $1 \cdot 00027 \pm 0 \cdot 00009$					$-19 \cdot 25 \pm 0 \cdot 15$
lysis ratio gave evidence of impurity	$-72 \cdot 29 \pm 1 \cdot 0$	C1	$11 \cdot 34 \pm 0 \cdot 02$	66/14	$-60 \cdot 95 \pm 1 \cdot 0$
	$-72 \cdot 77 \pm 0 \cdot 09$				$-61 \cdot 43 \pm 0 \cdot 10$
	$-72 \cdot 51 \pm 0 \cdot 30$				$-61 \cdot 17 \pm 0 \cdot 30$
Selected value	$-72 \cdot 51 \pm 0 \cdot 30$				$-61 \cdot 17 \pm 0 \cdot 30$
	$-76 \cdot 02 \pm 0 \cdot 12$	C1	$10 \cdot 90 \pm 0 \cdot 03$	63/5	$-65 \cdot 12 \pm 0 \cdot 13$
	$-76 \cdot 15 \pm 0 \cdot 31$				$-65 \cdot 25 \pm 0 \cdot 31$
	$-75 \cdot 75 \pm 0 \cdot 07$				$-64 \cdot 85 \pm 0 \cdot 08$
Selected value	$-76 \cdot 02 \pm 0 \cdot 12$				$-65 \cdot 12 \pm 0 \cdot 13$
I An $1 \cdot 00008 \pm 0 \cdot 00030$					$-51 \cdot 72 \pm 0 \cdot 16$
	$-14 \cdot 84 \pm 0 \cdot 16$	C1	$6 \cdot 61 \pm 0 \cdot 01$	52/7	$-8 \cdot 23 \pm 0 \cdot 16$
ABCE	$-13 \cdot 7 \pm 1 \cdot 0$				$-7 \cdot 1 \pm 1 \cdot 0$
ahydrofuran(g) {$-44 \cdot 02 \pm 0 \cdot 17$} E	$-14 \cdot 51 \pm 0 \cdot 21$				$-7 \cdot 90 \pm 0 \cdot 21$
	$-14 \cdot 99 \pm 0 \cdot 16$				$-8 \cdot 38 \pm 0 \cdot 16$
Selected value	$-14 \cdot 91 \pm 0 \cdot 15$				$-8 \cdot 30 \pm 0 \cdot 15$
	$-34 \cdot 44 \pm 0 \cdot 19$	E7	$9 \cdot 1 \pm 0 \cdot 2$		$-25 \cdot 34 \pm 0 \cdot 34$
anal(g) {$-48 \cdot 94 \pm 0 \cdot 34$} E	$-33 \cdot 13 \pm 0 \cdot 45$				$-24 \cdot 03 \pm 0 \cdot 37$
Selected value	$-33 \cdot 13 \pm 0 \cdot 45$				$-24 \cdot 03 \pm 0 \cdot 37$

THERMOCHEMISTRY OF ORGANIC AND ORGANOMETALLIC COMPOUNDS

C_aH_bO 1 kcal$=4\cdot184$ kJ

1 Formula	2 g.f.w.	3 Name	State	Purity mol %	Detn. Type	No. of expts.	Detn. of reactn.	$-\Delta H_r^{\circ}$ kcal/g.f.w.	R
C_4H_6O	$70\cdot0918$	Divinyl ether	g	$99\cdot93\pm0\cdot02$	FC	6	CO_2	$578\cdot14\pm0\cdot19$	63
C_4H_6O		Divinyl ether	g		H	4	P/CO_2	$56\cdot74\pm0\cdot10$	38
C_4H_6O		Divinyl ether							
C_4H_8O	$72\cdot1078$	Butanal	l		SB	10	m	$592\cdot05\pm0\cdot34$	60,
C_4H_8O		Butanal	l	$99\cdot92$, glc	SB	3	m	$592\cdot42\pm0\cdot34$	60,
C_4H_8O		Butanal	g	$99\cdot94$	E		glc	$16\cdot85\pm0\cdot30$	67
C_4H_8O		Butanal							
C_4H_8O	$72\cdot1078$	2-Methylpropanal	l		SB	5	m	$589\cdot67\pm0\cdot18$	62,
C_4H_8O	$72\cdot1078$	Methyl ethyl ketone	l	$99\cdot85$	SB	6	m	$584\cdot17\pm0\cdot27$	64
C_4H_8O		Methyl ethyl ketone	l	$99\cdot74$	SB	9	m	$582\cdot83\pm0\cdot20$	40 50
C_4H_8O		Methyl ethyl ketone	g		FC		CO_2		42
C_4H_8O		Methyl ethyl ketone	g		H	3	P/CO_2	$12\cdot97\pm0\cdot10$	38
C_4H_8O		Methyl ethyl ketone	g		E		glc	$12\cdot95\pm0\cdot16$	65
C_4H_8O		Methyl ethyl ketone							
C_4H_8O	$72\cdot1078$	Ethyl vinyl ether	g	$99\cdot99\pm0\cdot01$	FC	7	CO_2	$615\cdot98\pm0\cdot22$	63
C_4H_8O		Ethyl vinyl ether	g		H	3	P/CO_2	$26\cdot48\pm0\cdot06$	38
C_4H_8O		Ethyl vinyl ether							
C_4H_8O	$72\cdot1078$	1-Butene oxide	l		SB	2	m	$609\cdot10\pm0\cdot61$	37
C_4H_8O	$72\cdot1078$	Tetrahydrofuran	l		SB	4	m	$598\cdot9\ \pm0\cdot5$	58
C_4H_8O		Tetrahydrofuran	l		SB		m	$597\cdot8\ \pm0\cdot2$	57
C_4H_8O		Tetrahydrofuran	g	$99\cdot94\pm0\cdot02$	FC	5	CO_2	$605\cdot44\pm0\cdot16$	6
C_4H_8O		Tetrahydrofuran							
$C_4H_{10}O$	$74\cdot1237$	n-Butanol	l	$99\cdot92\pm0\cdot01$	SB	8	m	$639\cdot49\pm0\cdot11$	62 6
$C_4H_{10}O$		n-Butanol	l	glc	SB	3	m	$638\cdot25\pm0\cdot20$	60
$C_4H_{10}O$		n-Butanol	l	$>99\cdot75$	SB	9	CO_2	$639\cdot31\pm0\cdot20$	60
$C_4H_{10}O$		n-Butanol	l	$99\cdot7$	SB	4	CO_2	$639\cdot92\pm0\cdot15$	65
$C_4H_{10}O$		n-Butanol							
$C_4H_{10}O$	$74\cdot1237$	Isobutanol	l	glc	SB	5	CO_2	$637\cdot79\pm0\cdot20$	60
$C_4H_{10}O$		Isobutanol	l	$99\cdot0$	SB	4	CO_2	$638\cdot06\pm0\cdot14$	65
$C_4H_{10}O$		Isobutanol							

1 kcal=4·184 kJ

ΔH_r° Remarks	5 Δ_f (l or c) kcal/g.f.w.	6 Determination of ΔH_v			7 ΔH_f° (g) kcal/g.f.w.
		Type	ΔH_v° kcal/g.f.w.	Ref.	
H An 1·00008±0·00054	−9·27±0·28	V3	6·26±0·20	33/15	−3·01±0·20
ethyl ether(g) {−60·26±0·19} E	−9·78±0·30				−3·52±0·24
Selected value	−9·53±0·31				−3·27±0·25
rrection for butyric acid and water	−57·41±0·35	V1	8·05±0·10	67/1	−49·36±0·37
impurities					
	−57·04±0·35				−48·99±0·37
$_8$O(g)+H$_2$	−56·99±0·36				−48·94±0·34
⇌ n-Butanol(g){−65·79±0·14}.					
2nd Law					
Selected value	−56·99±0·36				−48·94±0·34
	−59·79±0·19	E7	7·54±0·30		−52·25±0·37
	−65·29±0·28	C1	8·34±0·01	61/22	−56·95±0·28
	−66·63±0·21				−58·29±0·21
sult converted to liq: ΔH_v used not given	−67·18±0·38				−58·84±0·38
Butanol(g) {−69·98±0·23}	−65·35±0·26				−57·01±0·26
H$_8$O(g)+H$_2$	−65·37±0·29				−57·03±0·29
⇌ s-Butanol(g){−69·98±0·23}.					
2nd Law	−65·36±0·20				−57·02±0·20
Selected value					
H An 0·99957±0·00020	−39·83±0·39	E1	6·35±0·30		−33·48±0·23
ethyl ether(g) {−60·26±0·19} E	−40·13±0·39				−33·78±0·22
Selected value	−39·98±0·37				−33·63±0·20
BCE	−40·36±0·62				
	−50·6 ±0·5	C1	7·65±0·01	M8	−42·9 ±0·5
E	−51·7 ±0·2				−44·0 ±0·2
H An 1·00024±0·00054	−51·67±0·17				−44·02±0·17
Selected value	−51·67±0·17				−44·02±0·17
An 0·9999±0·0002	−78·29±0·13	C1	12·50±0·02	66/14	−65·79±0·14
	−79·53±0·21				−67·03±0·21
An 0·99895±0·00032	−78·47±0·21				−65·97±0·21
	−77·86±0·16				−65·36±0·17
Selected value	−78·29±0·13				−65·79±0·14
An 0·99917±0·00012	−79·99±0·21	C1	12·15±0·02	66/14	−67·84±0·21
	−79·72±0·15				−67·57±0·16
Selected value	−79·99±0·21				−67·84±0·21

THERMOCHEMISTRY OF ORGANIC AND ORGANOMETALLIC COMPOUNDS

C_aH_bO

$$1 \text{ kcal} = 4 \cdot 184 \text{ kJ}$$

4
Determinati

1 Formula	2 g.f.w.	3 Name	State	Purity mol %	Type	No. of expts.	Detn. of reactn.	$-\Delta H_r^\circ$ kcal/g.f.w.	Re
$C_4H_{10}O$	$74 \cdot 1237$	s-Butanol	l	glc	SB	5	CO_2	$635 \cdot 91 \pm 0 \cdot 22$	60/
$C_4H_{10}O$		s-Butanol	l	$99 \cdot 7$	SB	4	CO_2	$635 \cdot 89 \pm 0 \cdot 13$	65/
$C_4H_{10}O$		s-Butanol							
$C_4H_{10}O$	$74 \cdot 1237$	t-Butanol	l	$99 \cdot 92$	SB	5	CO_2	$631 \cdot 92 \pm 0 \cdot 19$	60/
$C_4H_{10}O$		t-Butanol	l		E		an	$12 \cdot 8 \pm 0 \cdot 7$	34/
$C_4H_{10}O$		t-Butanol	l		E		an	$12 \cdot 6 \pm 0 \cdot 7$	55/
$C_4H_{10}O$		t-Butanol							
$C_4H_{10}O$	$74 \cdot 1237$	Diethyl ether	g	$99 \cdot 96 \pm 0 \cdot 01$	FC	9	CO_2	$657 \cdot 52 \pm 0 \cdot 18$	63/
$C_4H_{10}O$	$74 \cdot 1237$	Methyl n-propyl ether	g	glc	FC	5	CO_2	$660 \cdot 96 \pm 0 \cdot 25$	64/
$C_4H_{10}O$	$74 \cdot 1237$	Methyl isopropyl ether	g	glc	FC	8	CO_2	$657 \cdot 54 \pm 0 \cdot 22$	64/
C_5H_8O	$84 \cdot 1189$	Cyclopentanone	l		SB	6	m	$686 \cdot 97 \pm 0 \cdot 44$	62/
C_5H_8O		Cyclopentanone	g		H	3	P/CO_2	$12 \cdot 25 \pm 0 \cdot 15$	39/
C_5H_8O		Cyclopentanone							
C_5H_8O	$84 \cdot 1189$	Dihydropyran	l		SB	5	m	$705 \cdot 9 \pm 0 \cdot 3$	58/
$C_5H_{10}O$	$86 \cdot 1349$	Diethyl ketone	l	$99 \cdot 95$	SB	5	CO_2	$740 \cdot 96 \pm 0 \cdot 18$	M7
$C_5H_{10}O$	$86 \cdot 1349$	Methyl n-propyl ketone	l	$99 \cdot 93$	SB	7	CO_2	$740 \cdot 77 \pm 0 \cdot 23$	M7
$C_5H_{10}O$	$86 \cdot 1349$	Methyl isopropyl ketone	l	$99 \cdot 9$, glc	SB	5	CO_2	$740 \cdot 25 \pm 0 \cdot 18$	M7
$C_5H_{10}O$	$86 \cdot 1349$	Cyclopentanol	l	glc	SB	6	m	$740 \cdot 12 \pm 0 \cdot 40$	62/
$C_5H_{10}O$		Cyclopentanol	l		SB	6	m	$740 \cdot 11 \pm 0 \cdot 37$	50/
$C_5H_{10}O$		Cyclopentanol							
$C_5H_{10}O$	$86 \cdot 1349$	Tetrahydropyran	l		SB	4	m	$752 \cdot 8 \pm 1 \cdot 5$	58/
$C_5H_{10}O$		Tetrahydropyran	l	$99 \cdot 97 \pm 0 \cdot 01$	SB	3	CO_2	$750 \cdot 94 \pm 0 \cdot 31$	61/
$C_5H_{10}O$		Tetrahydropyran	l		SB		m	$749 \cdot 9 \pm 0 \cdot 2$	57/1
$C_5H_{10}O$		Tetrahydropyran	l		SB	9	m	$750 \cdot 28 \pm 0 \cdot 12$	58/1
$C_5H_{10}O$		Tetrahydropyran	g	$99 \cdot 98 \pm 0 \cdot 01$	FC	4	CO_2	$758 \cdot 44 \pm 0 \cdot 23$	65/
$C_5H_{10}O$		Tetrahydropyran							
$C_5H_{12}O$	$88 \cdot 1508$	1-Pentanol	l	$99 \cdot 5$	SB	5	CO_2	$794 \cdot 61 \pm 0 \cdot 09$	65/
$C_5H_{12}O$		1-Pentanol	l	$99 \cdot 87$	SB	7	CO_2	$795 \cdot 88 \pm 0 \cdot 16$	69/4
$C_5H_{12}O$		1-Pentanol							
$C_5H_{12}O$	$88 \cdot 1508$	2-Pentanol	l	$99 \cdot 5$	SB	4	CO_2	$792 \cdot 41 \pm 0 \cdot 16$	65/2

1 kcal = 4·184 kJ

ΔH_r° Remarks	5 ΔH_f° (l or c) kcal/g.f.w.	6 Determination of ΔH_v Type	ΔH_v° kcal/g.f.w.	Ref.	7 ΔH_f° (g) kcal/g.f.w.
'An 0·99904±0·00015	−81·87±0·23	C1	11·89±0·02	66/14	−69·98±0·23
	−81·89±0·14				−70·00±0·15
Selected value	−81·87±0·23				−69·98±0·23
'An 0·99977±0·00018	−85·86±0·20	C1	11·14±0·02	66/14	−74·72±0·21
ɔbutene(g) {−4·26±0·15}+H_2O(l)	−85·4 ±0·8				−74·3 ±0·8
$\rightleftharpoons C_4H_{10}O$ (l). 2nd Law					
ɔbutene(g) {−4·26±0·15}+H_2O(l)					−74·1 ±0·8
$\rightleftharpoons C_4H_{10}O$(l). 2nd Law					
Selected value	−85·86±0·20				−74·72±0·21
/H An 1·00004±0·00017	−66·75±0·21	C2	6·49±0·06	26/2	−60·26±0·19
/H An 0·99976±0·00020	−63·38±0·40	E7	6·56±0·30		−56·82±0·26
/H An 1·00038±0·00030	−66·51±0·39	E7	6·27±0·30		−60·24±0·23
	−56·55±0·45	C1	10·21±0·05	62/20	−46·34±0·46
	−55·93±0·36				−45·72±0·35
$_5H_8O$(g)+H_2					
=Cyclopentanol(g) {−57·97±0·30} E					
Selected value	−56·24±0·40				−46·03±0·40
	−37·6 ±0·3	V3	7·7±0·2	58/15	−29·9 ±0·4
	−70·87±0·20	C1	9·22±0·03	67/48	−61·65±0·21
	−71·06±0·25	C1	9·14±0·05	61/22	−61·92±0·26
	−71·58±0·20	C1	8·82±0·03	67/48	−62·76±0·21
	−71·71±0·41	C1	13·74±0·07	62/20	−57·97±0·42
	−71·72±0·38				−57·98±0·39
Selected value	−71·71±0·30				−57·97±0·31
/An 1·0013±0·0007	−59·0 ±1·5	V4	8·35±0·20	58/15	−50·6 ±1·6
	−60·89±0·33				−52·54±0·39
E	−61·9 ±0·2				−53·5 ±0·3
/An 1·0000±0·0001 E	−61·55±0·14				−53·20±0·25
/H An 1·00007±0·00008	−61·74±0·30				−53·39±0·24
Selected value	−61·74±0·30				−53·39±0·24
	−85·54±0·12	C1	13·61±0·04	66/14	−71·93±0·14
	−84·27±0·17				−70·66±0·18
Selected value	−84·27±0·17				−70·66±0·18
	−87·74±0·18	E7	12·56±0·30		−75·18±0·36

THERMOCHEMISTRY OF ORGANIC AND ORGANOMETALLIC COMPOUNDS

C_aH_oO

1 kcal = 4·184 kJ

| | | | | | No. | Detn. | | |
1 Formula	2 g.f.w.	3 Name	State	Purity mol %	Type	of expts.	Detn. of reactn.	$-\Delta H_r^\circ$ kcal/g.f.w.	Re
$C_5H_{12}O$	88·1508	3-Pentanol	l	99·5	SB	5	CO_2	791·65±0·11	65/2
$C_5H_{12}O$		3-Pentanol	g		E		glc	−13·56±0·16	65/
$C_5H_{12}O$		3-Pentanol							
$C_5H_{12}O$	88·1508	2-Methyl-1-butanol	l	99·5	SB	4	CO_2	794·92±0·13	65/2
$C_5H_{12}O$	88·1508	3-Methyl-1-butanol	l	99·5	SB	5	CO_2	794·98±0·12	65/2
$C_5H_{12}O$	88·1508	2-Methyl-2-butanol	l	99·5	SB	4	CO_2	789·45±0·11	65/2
$C_5H_{12}O$	88·1508	3-Methyl-2-butanol	l	99·0	SB	4	CO_2	792·53±0·15	65/2
$C_5H_{12}O$	88·1508	Methyl t-butyl ether	l		SB		m	803·0 ±1·2	61/2
C_6H_6O	94·1141	Phenol	c	99·93±0·01	SB	6	m	729·80±0·16	60/
C_6H_6O		Phenol	c		SB	9	m	730·41±0·24	54/
C_6H_6O		Phenol	c		SB	3	m	729·90±0·73	41/1
C_6H_6O		Phenol							
C_6H_6O	94·1141	Furylethylene	l		SB	3	m	766·8 ±0·8	29/
$C_6H_{10}O$	98·1460	1-Cyclopentenyl methanol	l		SB			914·1 ±0·8	63/2
$C_6H_{10}O$	98·1460	2-Methylene-cyclopentanol	l		SB			917·1 ±0·9	63/2
$C_6H_{10}O$	98·1460	2-Methylcyclopentanone	l		SB			842·5 ±1·3	63/2
$C_5H_{10}O$	98·1460	Cyclohexanone	l		SB	7	m	841·07±0·50	62/2
$C_6H_{10}O$		Cyclohexanone	l		SB			845·2 ±0·5	62/2
$H_6H_{10}O$		Cyclohexanone	g		H	3	P/CO_2	15·18±0·15	39/
$C_6H_{10}O$		Cyclohexanone							
$C_6H_{10}O$	98·1460	7-Oxabicyclo [2, 2, 1]-heptane	l		SB	5	CO_2	852·37±0·37	63/2
$C_6H_{12}O$	100·1619	Methyl n-butyl ketone	l	99·8	SB	5	CO_2	897·23±0·23	M7
$C_6H_{12}O$	100·1619	Methyl t-butyl ketone	l	99·97	SB	5	CO_2	895·67±0·20	M7
$C_6H_{12}O$	100·1619	Ethyl n-propyl ketone	l	99·96	SB	4	CO_2	897·68±0·19	M7

$$1 \text{ kcal} = 4\cdot184 \text{ kJ}$$

$\Delta H_r^?$ Remarks	5 ΔH_f° (l or c) kcal/g.f.w.	6 Determination of ΔH_v Type	ΔH_v° kcal/g.f.w.	Ref.	7 ΔH_f°(.g) kcal/g f.w.
	$-88\cdot50\pm0\cdot14$	E7	$12\cdot36\pm0\cdot30$		$-76\cdot14\pm0\cdot35$
$_5H_{12}O(g)$	$-87\cdot57\pm0\cdot40$				$-75\cdot21\pm0\cdot27$
$\rightleftharpoons H_2+$3-Pentanone(g) $\{-61\cdot65\pm0\cdot21\}$ 2nd Law					
Selected value	$-87\cdot57\pm0\cdot40$				$-75\cdot21\pm0\cdot27$
	$-85\cdot23\pm0\cdot15$	E7	$13\cdot04\pm0\cdot30$		$-72\cdot19\pm0\cdot35$
	$-85\cdot17\pm0\cdot14$	E7	$13\cdot15\pm0\cdot30$		$-72\cdot02\pm0\cdot35$
	$-90\cdot70\pm0\cdot13$	E7	$11\cdot63\pm0\cdot30$		$-79\cdot07\pm0\cdot35$
	$-87\cdot62\pm0\cdot17$	E7	$12\cdot27\pm0\cdot30$		$-75\cdot35\pm0\cdot36$
	$-77\cdot2\pm1\cdot2$		$7\cdot5\pm0\cdot5$	61/23	$-69\cdot7\pm1\cdot3$
	$-39\cdot45\pm0\cdot17$	S2	$16\cdot41\pm0\cdot12$ (sub.)	60/5	$-23\cdot04\pm0\cdot21$
	$-38\cdot84\pm0\cdot25$				$-22\cdot43\pm0\cdot29$
CE	$-39\cdot35\pm0\cdot74$				$-22\cdot94\pm0\cdot77$
Selected value	$-39\cdot45\pm0\cdot17$				$-23\cdot04\pm0\cdot21$
ABCE	$-2\cdot5\pm0\cdot8$	E7	$9\cdot1\pm0\cdot3$		$+6\cdot6\pm0\cdot9$
	$+8\cdot2\pm0\cdot8$				
	$+11\cdot2\pm0\cdot8$				
	$-63\cdot4\pm1\cdot3$				
	$-64\cdot81\pm0\cdot51$	C1	$10\cdot77\pm0\cdot05$	62/20	$-54\cdot04\pm0\cdot52$
	$-60\cdot7\pm0\cdot5$				$-49\cdot9\pm0\cdot5$
$_6H_{10}O(g)+H_2$	$-63\cdot97\pm0\cdot49$				$-53\cdot20\pm0\cdot48$
$=$Cyclohexanol(g) $\{-68\cdot38\pm0\cdot42\}$ E Selected value	$-64\cdot81\pm0\cdot51$				$-54\cdot04\pm0\cdot52$
$^:$/An $0\cdot99825\pm0\cdot00096$	$-53\cdot51\pm0\cdot40$				
	$-76\cdot97\pm0\cdot24$	E1	$10\cdot1\pm0\cdot1$		$-66\cdot87\pm0\cdot28$
	$-78\cdot53\pm0\cdot21$	E1	$9\cdot25\pm0\cdot10$		$-69\cdot28\pm0\cdot25$
	$-76\cdot52\pm0\cdot20$	C1	$10\cdot01\pm0\cdot05$	67/48	$-66\cdot51\pm0\cdot22$

C_aH_bO 1 kcal = 4·184 kJ

1 Formula	2 g.f.w.	3 Name	State	Purity mol %	Type	No. of expts.	Detn. of reactn.	$-\Delta H_r^\circ$ kcal/g.f.w.	Ref
$C_6H_{12}O$	100·1619	Ethyl isopropyl ketone	l	glc	SB	6	m	896·31±0·25	67/1
$C_6H_{12}O$	100·1619	1-Methylcyclopentanol	l		E		TP	13·34±0·34	52/
$C_6H_{12}O$	100·1619	Cyclohexanol	l		SB	6	m	891·17±0·52	62/2
$C_6H_{12}O$		Cyclohexanol	l		SB	6	m	890·82±0·45	50/1
$C_6H_{12}O$		Cyclohexanol	l		SB			890·6 ±0·5	62/2
$C_6H_{12}O$		Cyclohexanol							
$C_6H_{14}O$	102·1779	1-Hexanol	l	99·3	SB	4	CO_2	951·86±0·22	65/2
$C_6H_{14}O$	102·1779	Di-n-propyl ether	l	glc	SB	6	CO_2	963·93±0·19	65/2
$C_6H_{14}O$	102·1779	Di-isopropyl ether	l	glc	SB	4	CO_2	958·51±0·32	65/2
C_7H_6O	106·1253	Benzaldehyde	l		SB	2	m	842·7 ±1·7	29/
C_7H_8O	108·1412	Benzyl alcohol	l		SB	9	m	893·21±0·28	54/
C_7H_8O		Benzyl alcohol	l		SB	2	m	889·9 ±1·9	29/
C_7H_8O		Benzyl alcohol							
C_7H_8O	108·1412	o-Cresol	c	99·96±0·01	SB	5	m	882·72±0·24	60/
C_7H_8O	108·1412	m-Cresol	l	99·92±0·02	SB	4	m	885·25±0·14	60/
C_7H_8O		m-Cresol	l		SB	4	m	885·21±0·89	41/1
C_7H_8O		m-Cresol							
C_7H_8O	108·1412	p-Cresol	c	99·96±0·02	SB	5	m	883·99±0·16	60/
C_7H_8O	108·1412	Anisole (methyl phenyl ether)	l		SB	4	m	903·17±0·90	41/1
$C_7H_{10}O$	110·1572	Norcamphor	c		SB	6	m	945·8 ±1·0	34/
$C_7H_{12}O$	112·1731	2-Methylcyclohexanone	l		SB			999·4 ±0·8	63/2
$C_7H_{12}O$	112·1731	Cycloheptanone	l		SB			996·7 ±0·3	58/1
$C_7H_{12}O$	112·1731	1-Cyclohexenylmethanol	l		SB			976·8 ±0·5	63/2
$C_7H_{12}O$	112·1731	2-Methylenecyclohexanol	l		SB			1001·9 ±0·8	63/2

1 kcal $= 4 \cdot 184$ kJ

$\Delta H°$			6 Determination of ΔH_v			7
Remarks		5 $\Delta H_f°$ (l or c) kcal/g.f.w.	Type	$\Delta H_v°$ kcal/g.f.w.	Ref.	$\Delta H_f°$ (g) kcal/g.f.w.
		$-77 \cdot 89 \pm 0 \cdot 26$	C1	$9 \cdot 51 \pm 0 \cdot 03$	67/12	$-68 \cdot 38 \pm 0 \cdot 27$
Methylcyclopentene(g) $\{-0 \cdot 6 \pm 0 \cdot 5\} + H_2O(l)$ $\rightleftharpoons C_6H_{12}O(l)$. 2nd Law		$-82 \cdot 3 \pm 0 \cdot 6$				
		$-83 \cdot 03 \pm 0 \cdot 53$	C1	$14 \cdot 82 \pm 0 \cdot 07$	66/14	$-68 \cdot 21 \pm 0 \cdot 54$
		$-83 \cdot 38 \pm 0 \cdot 47$				$-68 \cdot 56 \pm 0 \cdot 50$
		$-83 \cdot 6 \pm 0 \cdot 5$				$-68 \cdot 8 \pm 0 \cdot 5$
	Selected value	$-83 \cdot 20 \pm 0 \cdot 40$				$-68 \cdot 38 \pm 0 \cdot 42$
		$-90 \cdot 65 \pm 0 \cdot 24$	V1	$15 \cdot 00 \pm 0 \cdot 30$	60/20	$-75 \cdot 65 \pm 0 \cdot 40$
		$-78 \cdot 58 \pm 0 \cdot 22$	E7	$8 \cdot 73 \pm 0 \cdot 30$		$-69 \cdot 85 \pm 0 \cdot 40$
		$-84 \cdot 00 \pm 0 \cdot 34$	E7	$7 \cdot 80 \pm 0 \cdot 30$		$-76 \cdot 20 \pm 0 \cdot 54$
	ABCE	$-20 \cdot 6 \pm 1 \cdot 7$	V4	$11 \cdot 8 \pm 1 \cdot 0$	57/46	$-8 \cdot 8 \pm 2 \cdot 0$
		$-38 \cdot 41 \pm 0 \cdot 30$	C2	$14 \cdot 41 \pm 0 \cdot 10$	26/2	$-24 \cdot 00 \pm 0 \cdot 33$
	ABCE	$-41 \cdot 7 \pm 1 \cdot 9$				$-27 \cdot 3 \pm 1 \cdot 9$
	Selected value	$-38 \cdot 41 \pm 0 \cdot 30$				$-24 \cdot 00 \pm 0 \cdot 33$
		$-48 \cdot 90 \pm 0 \cdot 25$	S2	$18 \cdot 17 \pm 0 \cdot 18$ (sub.)	60/5	$-30 \cdot 73 \pm 0 \cdot 31$
		$-46 \cdot 37 \pm 0 \cdot 16$	V2	$14 \cdot 75 \pm 0 \cdot 25$	60/5	$-31 \cdot 62 \pm 0 \cdot 30$
	CE	$-46 \cdot 41 \pm 0 \cdot 90$				$-31 \cdot 66 \pm 0 \cdot 92$
	Selected value	$-46 \cdot 37 \pm 0 \cdot 16$				$-31 \cdot 62 \pm 0 \cdot 30$
		$-47 \cdot 63 \pm 0 \cdot 18$	S2	$17 \cdot 67 \pm 0 \cdot 35$ (sub.)	60/5	$-29 \cdot 96 \pm 0 \cdot 40$
	CE	$-28 \cdot 45 \pm 0 \cdot 91$	C1	$11 \cdot 18 \pm [0 \cdot 1]$	67/48	$-17 \cdot 27 \pm 0 \cdot 93$
	CE	$-54 \cdot 1 \pm 1 \cdot 0$				
		$-68 \cdot 8 \pm 0 \cdot 8$				
		$-71 \cdot 5 \pm 0 \cdot 3$	E7	$12 \cdot 4 \pm 0 \cdot 3$		$-59 \cdot 1 \pm 0 \cdot 4$
		$-91 \cdot 4 \pm 0 \cdot 5$				
		$-66 \cdot 3 \pm 0 \cdot 8$				

THERMOCHEMISTRY OF ORGANIC AND ORGANOMETALLIC COMPOUNDS

C_aH_bO

1 kcal$=4\cdot184$ kJ

								4 Determinati
1 Formula	2 g.f.w.	3 Name	State	Purity mol %	No. of Type expts.	Detn. of reactn.	$-\Delta H_r^{\circ}$ kcal/g.f.w.	Re
$C_7H_{12}O$	$112\cdot1731$	*Endo*-2-Methyl-7-oxa-bicyclo [2, 2, 1] heptane	l		SB 5	CO_2	$1008\cdot43\pm0\cdot38$	63/2
$C_7H_{12}O$	$112\cdot1731$	*Exo*-2-Methyl-7-oxa-bicyclo [2, 2, 1] heptane	l		SB 5	CO_2	$1007\cdot58\pm0\cdot40$	63/2
$C_7H_{14}O$	$114\cdot1890$	Heptanal	l		SB 6	m	$1062\cdot1\ \pm0\cdot9$	60/2
$C_7H_{14}O$	$114\cdot1890$	Ethyl t-butyl ketone	l	glc	SB 6	m	$1051\cdot45\pm0\cdot32$	67/1
$C_7H_{14}O$	$114\cdot1890$	Di-isopropyl ketone	l	glc	SB 6	m	$1052\cdot23\pm0\cdot27$	67/1
$C_7H_{14}O$	$114\cdot1890$	Hexahydrobenzyl alcohol	l		SB 2	m	$1046\cdot2\ \pm2\cdot0$	29/.
$C_7H_{14}O$	$114\cdot1890$	*cis*-2-Methylcyclohexanol	l		SB 3	m	$1043\cdot3\ \pm1\cdot2$	31/9 58/1
$C_7H_{14}O$	$114\cdot1890$	*trans*-2-Methylcyclo-hexanol	l		SB 3	m	$1037\cdot2\ \pm2\cdot1$	31/9 58/1
$C_7H_{14}O$	$114\cdot1890$	*cis*-3-Methylcyclohexanol	l		SB 3	m	$1037\cdot1\ \pm2\cdot1$	31/9 58/1
$C_7H_{14}O$	$114\cdot1890$	*trans*-3-Methylcyclo-hexanol	l		SB 3	m	$1042\cdot3\ \pm2\cdot1$	31/9 58/1
$C_7H_{14}O$	$114\cdot1890$	*cis*-4-Methylcyclohexanol	l		SB 3	m	$1037\cdot8\ \pm1\cdot2$	31/9 58/1
$C_7H_{14}O$	$114\cdot1890$	*trans*-4-Methylcyclo-hexanol	l		SB 3	m	$1033\cdot0\ \pm2\cdot5$	31/9 58/1
$C_7H_{16}O$	$116\cdot2050$	1-Heptanol	l	$99\cdot3$	SB 4	CO_2	$1109\cdot59\pm0\cdot17$	65/2
$C_7H_{16}O$	$116\cdot2050$	Isopropyl t-butyl ether	l		SB	m	$1111\cdot0\ \pm0\cdot7$	61/2
C_8H_8O	$120\cdot1524$	Acetophenone (methyl phenyl ketone)	l		SB 4	CO_2	$991\cdot60\pm0\cdot21$	61/24
$C_8H_{10}O$	$122\cdot1683$	*o*-Ethylphenol	l	$99\cdot94\pm0\cdot01$	SB 6	CO_2	$1044\cdot07\pm0\cdot42$	63/25
$C_8H_{10}O$	$122\cdot1683$	*m*-Ethylphenol	l	$99\cdot94\pm0\cdot02$	SB 6	CO_2	$1042\cdot77\pm0\cdot37$	63/25
$C_8H_{10}O$	$122\cdot1683$	*p*-Ethylphenol	c	$99\cdot97\pm0\cdot01$	SB 5	CO_2	$1040\cdot35\pm0\cdot22$	63/25

1 kcal$=4\cdot184$ kJ

ΔH_r° Remarks		ΔH_f° (l or c) kcal/g.f.w.	6 Determination of ΔH_v			7 ΔH_f° (g) kcal/g.f.w.
		5	Type	ΔH_v° kcal/g.f.w.	Ref.	
An $0\cdot99928\pm0\cdot00062$		$-59\cdot82\pm0\cdot42$				
An $0\cdot99974\pm0\cdot00021$		$-60\cdot67\pm0\cdot43$				
rrection for heptanoic acid and water impurities		$-74\cdot5\ \pm0\cdot9$	E7	$11\cdot4\ \pm0\cdot3$		$-63\cdot1\ \pm1\cdot0$
		$-85\cdot11\pm0\cdot33$	C1	$10\cdot12\pm0\cdot01$	67/12	$-74\cdot99\pm0\cdot33$
		$-84\cdot33\pm0\cdot28$	C1	$9\cdot93\pm0\cdot01$	67/12	$-74\cdot40\pm0\cdot28$
	ABCE	$-90\cdot4\ \pm2\cdot0$				
	CE	$-93\cdot3\ \pm1\cdot2$	E7	$15\cdot1\ \pm0\cdot5$		$-78\cdot2\ \pm1\cdot3$
	CE	$-99\cdot4\ \pm2\cdot1$	E7	$15\cdot1\ \pm0\cdot5$		$-84\cdot3\ \pm2\cdot2$
	CE	$-99\cdot5\ \pm2\cdot1$	E7	$15\cdot6\ \pm0\cdot5$		$-83\cdot9\ \pm2\cdot2$
	CE	$-94\cdot3\ \pm2\cdot1$	E7	$15\cdot6\ \pm0\cdot5$		$-78\cdot7\ \pm2\cdot2$
	CE	$-98\cdot8\ \pm1\cdot2$	E7	$15\cdot7\ \pm0\cdot5$		$-83\cdot1\ \pm1\cdot3$
	CE	$-103\cdot6\ \pm2\cdot5$	E7	$15\cdot8\ \pm0\cdot5$		$-87\cdot8\ \pm2\cdot6$
		$-95\cdot29\pm0\cdot20$	V1	$16\cdot20\pm0\cdot35$	60/20	$-79\cdot09\pm0\cdot42$
		$-93\cdot9\ \pm0\cdot7$	E1	$8\cdot4\ \pm1\cdot0$		$-85\cdot5\ \pm1\cdot2$
		$-34\cdot07\pm0\cdot24$	E7	$13\cdot36\pm0\cdot30$		$-20\cdot71\pm0\cdot40$
An $0\cdot99904$		$-49\cdot91\pm0\cdot43$	V2	$15\cdot20\pm0\cdot24$	63/25	$-34\cdot71\pm0\cdot48$
An $0\cdot99950$		$-51\cdot21\pm0\cdot38$	V2	$16\cdot30\pm0\cdot12$	63/25	$-34\cdot91\pm0\cdot40$
An $0\cdot9998$		$-53\cdot63\pm0\cdot24$	S2	$19\cdot20\pm0\cdot12$ (sub.)	63/25	$-34\cdot43\pm0\cdot27$

C_aH_bO

1 kcal $= 4\cdot184$ kJ

1 Formula	2 g.f.w.	3 Name	State	Purity mol %	Type	No. of expts.	Detn. of reactn.	$-\Delta H_r^\circ$ kcal/g.f.w.	Re
$C_8H_{10}O$	$122\cdot1683$	2, 3-Xylenol	c	$99\cdot93\pm0\cdot01$	SB	4	m	$1036\cdot33\pm0\cdot22$	60/
$C_8H_{10}O$	$122\cdot1683$	2, 4-Xylenol	l	$99\cdot97\pm0\cdot01$	SB	4	m	$1039\cdot31\pm0\cdot22$	60/
$C_8H_{10}O$	$122\cdot1683$	2, 5-Xylenol	c	$99\cdot90\pm0\cdot02$	SB	5	m	$1035\cdot04\pm0\cdot20$	60/
$C_8H_{10}O$	$122\cdot1683$	2, 6-Xylenol	c	$99\cdot89\pm0\cdot01$	SB	4	m	$1037\cdot25\pm0\cdot24$	60/
$C_8H_{10}O$	$122\cdot1683$	3, 4-Xylenol	c	$99\cdot97\pm0\cdot01$	SB	4	m	$1036\cdot07\pm0\cdot26$	60/
$C_8H_{10}O$	$122\cdot1683$	3, 5-Xylenol	c	$99\cdot96\pm0\cdot01$	SB	4	m	$1035\cdot57\pm0\cdot26$	60/
$C_8H_{10}O$	$122\cdot1683$	Phenetole (ethyl phenyl ether)	l		SB	5	m	$1055\cdot8\ \pm1\cdot1$	41/
$C_8H_{10}O$	$122\cdot1683$	3-Methoxy-1-methylbenzene	l		SB	4	m	$1056\cdot8\ \pm1\cdot1$	41/
$C_8H_{12}O$	$124\cdot1842$	cis-β-Bicyclo-octanone	l		SB	4	m	$1094\cdot3\ \pm1\cdot1$	35/
$C_8H_{12}O$	$124\cdot1842$	trans-β-Bicyclo-octanone	l		SB	4	m	$1100\cdot1\ \pm1\cdot1$	35/
$C_8H_{12}O$	$124\cdot1842$	Endo-ethylenecyclo-hexanone	l		SB	7	m	$1096\cdot8\ \pm1\cdot1$	34/
$C_8H_{14}O$	$126\cdot2002$	2-Ethyl-2-hexenal	l	glc	SB	6	m	$1168\cdot17\pm0\cdot34$	60/2
$C_8H_{14}O$	$126\cdot2002$	Cyclo-octanone	l		SB	3	m	$1152\cdot7\ \pm1\cdot2$	33/
$C_8H_{14}O$	$126\cdot2002$	3-Oxa-bicyclo [3, 2, 2]-nonane	c		SB			$1164\cdot74\pm0\cdot10$	67/1
$C_8H_{16}O$	$128\cdot2161$	2-Ethyl-1-hexanal	l	glc	SB	9	m	$1215\cdot63\pm0\cdot36$	60/2
$C_8H_{16}O$	$128\cdot2161$	Isopropyl t-butyl ketone	l	glc	SB	6	m	$1207\cdot74\pm0\cdot36$	67/1
$C_8H_{16}O$	$128\cdot2161$	cis-3, cis-5,-Dimethyl-cyclohexanol	l		SB	3	m	$1197\cdot5\ \pm2\cdot7$	39/1
$C_8H_{16}O$	$128\cdot2161$	trans-3, trans-5,-Dimethylcyclohexanol	l		SB	3	m	$1183\cdot0\ \pm2\cdot7$	39/1

1 kcal $= 4 \cdot 184$ kJ

ΔH_r°			6 Determination of ΔH_v			7
Remarks		5 ΔH_f° (l or c) kcal/g.f.w.	Type	ΔH_v° kcal/g.f.w.	Ref.	ΔH_f° (g) kcal/g.f.w.
		$-57 \cdot 65 \pm 0 \cdot 24$	S2	$20 \cdot 08 \pm 0 \cdot 24$ (sub.)	60/5	$-37 \cdot 57 \pm 0 \cdot 33$
		$-54 \cdot 67 \pm 0 \cdot 24$	V2	$15 \cdot 74 \pm 0 \cdot 05$	60/5	$-38 \cdot 93 \pm 0 \cdot 25$
		$-58 \cdot 94 \pm 0 \cdot 22$	S2	$20 \cdot 31 \pm 0 \cdot 06$ (sub.)	60/5	$-38 \cdot 63 \pm 0 \cdot 24$
		$-56 \cdot 73 \pm 0 \cdot 26$	S2	$18 \cdot 07 \pm 0 \cdot 04$ (sub.)	60/5	$-38 \cdot 66 \pm 0 \cdot 27$
		$-57 \cdot 91 \pm 0 \cdot 28$	S2	$20 \cdot 49 \pm 0 \cdot 03$ (sub.)	60/5	$-37 \cdot 42 \pm 0 \cdot 29$
		$-58 \cdot 41 \pm 0 \cdot 28$	S2	$19 \cdot 80 \pm 0 \cdot 07$ (sub.)	60/5	$-38 \cdot 61 \pm 0 \cdot 30$
	CE	$-38 \cdot 2 \pm 1 \cdot 1$	V3	$11 \cdot 9 \pm 0 \cdot 4$	47/15	$-26 \cdot 3 \pm 1 \cdot 2$
	CE	$-37 \cdot 2 \pm 1 \cdot 1$	E7	$12 \cdot 3 \pm 0 \cdot 4$		$-24 \cdot 9 \pm 1 \cdot 2$
	BCE	$-68 \cdot 0 \pm 1 \cdot 1$	E7	$13 \cdot 0 \pm 0 \cdot 5$		$-55 \cdot 0 \pm 1 \cdot 3$
	BCE	$-62 \cdot 2 \pm 1 \cdot 1$	E7	$12 \cdot 8 \pm 0 \cdot 5$		$-49 \cdot 4 \pm 1 \cdot 3$
	CE	$-65 \cdot 5 \pm 1 \cdot 1$	E7	$12 \cdot 4 \pm 0 \cdot 5$		$-53 \cdot 1 \pm 1 \cdot 3$
		$-62 \cdot 44 \pm 0 \cdot 36$				
		$-77 \cdot 9 \pm 1 \cdot 2$	E7	$13 \cdot 0 \pm 0 \cdot 5$		$-64 \cdot 9 \pm 1 \cdot 3$
		$-65 \cdot 87 \pm 0 \cdot 14$				
		$-83 \cdot 30 \pm 0 \cdot 38$	E7	$11 \cdot 70 \pm 0 \cdot 30$		$-71 \cdot 60 \pm 0 \cdot 46$
		$-91 \cdot 19 \pm 0 \cdot 38$	C1	$10 \cdot 35 \pm 0 \cdot 02$	67/12	$-80 \cdot 84 \pm 0 \cdot 38$
percooled liquid examined	CE	$-101 \cdot 4 \pm 2 \cdot 7$				
	CE	$-115 \cdot 9 \pm 2 \cdot 7$				

THERMOCHEMISTRY OF ORGANIC AND ORGANOMETALLIC COMPOUNDS

C_aH_bO

1 kcal = 4·184 kJ

Determina▮

1 Formula	2 g.f.w.	3 Name	State	Purity mol %	Type	No. of expts.	Detn. of reactn.	$-\Delta H_r^\circ$ kcal/g.f.w.	R▮
$C_8H_{16}O$	128·2161	*cis*-3, *trans*-5,-Dimethyl- cyclohexanol	l		SB	3	m	1177·1 ±2·7	39/
$C_8H_{18}O$	130·2321	1-Octanol	l	99	SB	4	CO_2	1265·65±0·24	65/
$C_8H_{18}O$		1-Octanol	l	99·95, glc	SB	7	CO_2	1264·94±0·23	69/
$C_8H_{18}O$		1-Octanol							
$C_8H_{18}O$	130·2321	2-Ethyl-1-hexanol	l	glc	SB	7	m	1263·81±0·38	60/
$C_8H_{18}O$	130·2321	Di-n-butyl ether	l	glc	SB	5	CO_2	1276·92±0·22	65/
$C_8H_{18}O$		Di-n-butyl ether	l		SB		m	1277·1 ±0·8	57/
$C_8H_{18}O$		Di-n-butyl ether							
$C_8H_{18}O$	130·2321	Di-s-butyl ether	l	glc	SB	5	CO_2	1271·28±0·25	65/
$C_8H_{18}O$	130·2321	Di-t-butyl ether	l		SB		m	1271·1 ±0·3	61/
$C_9H_{10}O$	134·1795	Benzyl methyl ketone	l		SB		m	1151·73±0·40	54/
$C_9H_{10}O$		Benzyl methyl ketone	l		SB	3	m	1151·7 ±0·6	54/
$C_9H_{10}O$		Benzyl methyl ketone	l		SB	9	m	1149·18±0·28	54/
$C_9H_{10}O$		Benzyl methyl ketone							
$C_9H_{10}O$	134·1795	Ethyl phenyl ketone	l		SB	4	CO_2	1148·07±0·25	61/
$C_9H_{18}O$	142·2432	Di-n-butyl ketone	l	99·85	SB	5	CO_2	1366·11±0·31	M▮
$C_9H_{18}O$	142·2432	Di-t-butyl ketone	l		SB	5	m	1367·81±0·27	M▮
$C_9H_{20}O$	144·2592	1-Nonanol	l	98·6	SB	4	CO_2	1419·89±0·25	65/▮
$C_9H_{20}O$	144·2592	3, 5, 5-Trimethylhexanol	l		SB	5	m	1420·5 ±0·7	60/▮
$C_{10}H_8O$	144·1747	1-Naphthol	c		SB		m	1186·81±0·50	48/
$C_{10}H_8O$	144·1747	2-Naphthol	c		SB		m	1183·95±0·56	48/
$C_{10}H_8O$		2-Naphthol	c		SB	3	m	1183·81±0·56	46/▮
$C_{10}H_8O$		2-Naphthol							
$C_{10}H_{10}O$	146·1906	1-Tetralone	c		SB	3	m	1232·0 ±5·0	51/
$C_{10}H_{12}O$	148·2065	n-Propyl phenyl ketone	l		SB	3	CO_2	1305·26±0·40	61/▮
$C_{10}H_{12}O$	148·2065	1-Tetralol	c		SB	3	m	1292·7 ±6·0	51/
$C_{10}H_{12}O$	148·2065	5-Tetralol	c		SB	2	m	1282·2 ±1·3	50/▮ 51/▮

1 kcal = 4·184 kJ

ΔH_r°	Remarks		ΔH_f° (l or c) kcal/g.f.w.	6 Determination of ΔH_v			7 ΔH_f° (g) kcal/g.f.w.
			5	Type	ΔH_v° kcal/g.f.w.	Ref.	
		CE	$-121\cdot8\ \pm2\cdot7$				
			$-101\cdot59\pm0\cdot27$	E1	$17\cdot0\pm0\cdot1$		$-84\cdot59\pm0\cdot31$
			$-102\cdot30\pm0\cdot26$				$-85\cdot30\pm0\cdot30$
	Selected value		$-102\cdot30\pm0\cdot26$				$-85\cdot30\pm0\cdot30$
			$-103\cdot43\pm0\cdot40$	E7	$16\cdot12\pm0\cdot40$		$-87\cdot31\pm0\cdot56$
			$-90\cdot32\pm0\cdot25$	C2	$10\cdot50\pm0\cdot10$	31/3	$-79\cdot82\pm0\cdot27$
			$-90\cdot1\ \pm0\cdot8$				$-79\cdot6\ \pm0\cdot8$
	Selected value		$-90\cdot32\pm0\cdot25$				$-79\cdot82\pm0\cdot27$
			$-95\cdot96\pm0\cdot28$	E1	$9\cdot7\pm0\cdot3$		$-86\cdot26\pm0\cdot41$
			$-96\cdot1\ \pm0\cdot3$	V2	$9\cdot0\pm0\cdot2$	61/23	$-87\cdot1\ \pm0\cdot4$
			$-36\cdot30\pm0\cdot41$	V3	$12\cdot78\pm0\cdot06$	54/9	$-23\cdot52\pm0\cdot42$
			$-36\cdot3\ \pm0\cdot6$				$-23\cdot5\ \pm0\cdot6$
			$-38\cdot85\pm0\cdot30$				$-26\cdot07\pm0\cdot32$
	Selected value		$-36\cdot30\pm0\cdot41$				$-23\cdot52\pm0\cdot42$
			$-39\cdot96\pm0\cdot28$	E7	$13\cdot98\pm0\cdot40$		$-25\cdot98\pm0\cdot50$
An 1·0000			$-95\cdot18\pm0\cdot32$	C1	$12\cdot74\pm0\cdot02$	M6	$-82\cdot44\pm0\cdot32$
			$-93\cdot48\pm0\cdot29$	C1	$10\cdot84\pm0\cdot01$	M6	$-82\cdot64\pm0\cdot29$
			$-109\cdot72\pm0\cdot29$	V1	$18\cdot60\pm0\cdot45$	60/20	$-91\cdot12\pm0\cdot60$
			$-109\cdot1\ \pm0\cdot7$				
		CE	$-26\cdot96\pm0\cdot53$	V4$_m$	$21\cdot87\pm[0\cdot90]$ (sub.)	26/3, 27/4	$-5\cdot1\ \pm1\cdot1$
		CE	$-29\cdot82\pm0\cdot59$	V4$_m$	$19\cdot83\pm[0\cdot90]$ (sub.)	26/3, 27/4	$-10\cdot0\ \pm1\cdot1$
			$-29\cdot96\pm0\cdot59$				$-10\cdot1\ \pm1\cdot1$
	Selected value		$-29\cdot89\pm0\cdot40$				$-10\cdot1\ \pm1\cdot0$
			$-50\cdot1\ \pm5\cdot0$				
			$-45\cdot14\pm0\cdot44$	E7	$14\cdot51\pm0\cdot40$		$-30\cdot63\pm0\cdot57$
			$-57\cdot7\ \pm6\cdot0$				
			$-68\cdot2\ \pm1\cdot3$				

THERMOCHEMISTRY OF ORGANIC AND ORGANOMETALLIC COMPOUNDS

C_aH_bO

1 kcal = 4·184 kJ

Determina†

1 Formula	2 g.f.w.	3 Name	State	Purity mol %	Type	No. of expts.	Detn. of reactn.	$-\Delta H_r^\circ$ kcal/g.f.w.	R
$C_{10}H_{14}O$	150·2225	Thymol	c		SB		CO_2	1344·7 ±2·3	59/
$C_{10}H_{18}O$	154·2544	9-trans-Decalol	c		SB	1	m	1449·1 ±	50/ 51/
$C_{10}H_{20}O$	156·2703	t-Butyl neopentyl ketone	l	glc	SB	6	m	1517·85±0·50	67/
$C_{10}H_{22}O$	158·2862	1-Decanol	l	99	SB	5	CO_2	1577·35±0·18	65/
$C_{11}H_{14}O$	162·2336	Isobutyl phenyl ketone	l		SB	5	CO_2	1460·14±0·33	61/
$C_{11}H_{14}O$	162·2336	t-Butyl phenyl ketone	l		SB	4	CO_2	1462·85±0·54	61/
$C_{11}H_{14}O$	162·2336	2, 4, 5-Trimethyl-acetophenone	l		SB	3	m	1452·5 ±1·0	41/
$C_{11}H_{14}O$	162·2336	2, 4, 6-Trimethyl-acetophenone	l		SB	5	m	1448·9 ±0·7	41/
$C_{11}H_{14}O$	162·2336	α-Methyltetralol	c		SB	3	m	1454·4 ±4·4	51/
$C_{12}H_8O$	168·1970	Dibenzofuran	c		SB	3	m	1400·6 ±1·0	58/
$C_{12}H_{10}O$	170·2129	Diphenyl ether	c		SB	3	m	1461·2 ±0·9	58/
$C_{12}H_{10}O$		Diphenyl ether	l	99·999	SB	6	CO_2	1466·63±0·42	51/
$C_{12}H_{10}O$		Diphenyl ether							
$C_{13}H_{10}O$	182·2241	Dibenzopyran	c		SB	4	m	1549·0 ±0·9	58/
$C_{13}H_{10}O$	182·2241	Benzophenone (Diphenyl ketone)	c		SB	4	CO_2	1556·0 ±0·5	59/
$C_{13}H_{10}O$		Benzophenone	c		SB	5	m	1556·51±0·77	50/
$C_{13}H_{10}O$		Benzophenone	c		SB	3	m	1554·3 ±1·5	31/
$C_{13}H_{10}O$		Benzophenone							
$C_{13}H_{12}O$	184·2400	Diphenylcarbinol	c		SB	9	m	1607·51±0·46	54/
$C_{13}H_{12}O$	184·2400	2, 7-Dimethyl-4, 5-benzotropone	c		SB	6	m	1613·3 ±0·9	56/
$C_{14}H_{12}O$	196·2511	p-Methylbenzophenone	c		SB	4	CO_2	1708·0 ±0·5	59/
$C_{14}H_{12}O$	196·2511	Deoxybenzoin	c		SB	11	m	1709·64±0·68	62/2

$1 \text{ kcal} = 4 \cdot 184 \text{ kJ}$

ΔH_r°	Remarks	5 ΔH_f° (l or c) kcal/g.f.w.	6 Determination of ΔH_v			7 ΔH_f° (g) kcal/g.f.w.
			Type	ΔH_v° kcal/g.f.w.	Ref.	
		$-74 \cdot 0 \pm 2 \cdot 3$	S3	$21 \cdot 8 \pm 1 \cdot 0$ (sub.)	47/8	$-52 \cdot 2 \pm 2 \cdot 6$
		$-106 \cdot 3 \pm$				
		$-105 \cdot 81 \pm 0 \cdot 52$	Cl	$11 \cdot 66 \pm 0 \cdot 02$	67/12	$-94 \cdot 15 \pm 0 \cdot 52$
		$-114 \cdot 63 \pm 0 \cdot 24$	Vl	$19 \cdot 82 \pm 0 \cdot 50$	60/20	$-94 \cdot 81 \pm 0 \cdot 58$
		$-52 \cdot 63 \pm 0 \cdot 37$	E7	$14 \cdot 22 \pm 0 \cdot 40$		$-38 \cdot 41 \pm 0 \cdot 54$
		$-49 \cdot 92 \pm 0 \cdot 56$				
	C	$-60 \cdot 3 \pm 1 \cdot 1$	E7	$15 \cdot 1 \pm 0 \cdot 5$		$-45 \cdot 2 \pm 1 \cdot 2$
	C	$-63 \cdot 9 \pm 0 \cdot 8$	E7	$14 \cdot 9 \pm 0 \cdot 5$		$-49 \cdot 0 \pm 1 \cdot 0$
		$-58 \cdot 4 \pm 4 \cdot 4$				
		$-1 \cdot 3 \pm 1 \cdot 1$	V4$_{me}$	$21 \cdot 2 \pm [0 \cdot 5]$ (sub.)	58/15	$+19 \cdot 9 \pm 1 \cdot 2$
$_m 4 \cdot 11 \pm 0 \cdot 01$ (51/13)		$-9 \cdot 0 \pm 1 \cdot 0$	V4$_m$	$19 \cdot 61 \pm [0 \cdot 5]$ (sub.)	58/15	$+10 \cdot 61 \pm 1 \cdot 2$
	Selected value	$-3 \cdot 56 \pm 0 \cdot 44$ $-3 \cdot 56 \pm 0 \cdot 44$ (l)	V4	$15 \cdot 50 \pm [0 \cdot 5]$	48/6	$+11 \cdot 94 \pm 0 \cdot 67$ $+11 \cdot 94 \pm 0 \cdot 67$
		$-15 \cdot 2 \pm 1 \cdot 0$	V4$_{me}$	$26 \cdot 8 \pm [0 \cdot 5]$ (sub.)	58/15	$+11 \cdot 6 \pm 1 \cdot 2$
		$-8 \cdot 2 \pm 0 \cdot 6$	S3	$22 \cdot 7 \pm [0 \cdot 5]$ (sub.)	32/2	$+14 \cdot 5 \pm 0 \cdot 8$
		$-7 \cdot 73 \pm 0 \cdot 79$				$+15 \cdot 0 \pm 0 \cdot 9$
	CE	$-10 \cdot 0 \pm 1 \cdot 5$				$+12 \cdot 7 \pm 1 \cdot 6$
	Selected value	$-8 \cdot 11 \pm 0 \cdot 45$				$+14 \cdot 6 \pm 0 \cdot 7$
		$-25 \cdot 04 \pm 0 \cdot 50$				
	C	$-19 \cdot 3 \pm 0 \cdot 9$	E8$_{me}$	$18 \cdot 7 \pm [2 \cdot 5]$ (sub.)		$-0 \cdot 6 \pm [3 \cdot 0]$
		$-18 \cdot 6 \pm 0 \cdot 6$				
		$-16 \cdot 96 \pm 0 \cdot 70$	V4$_{me}$	$22 \cdot 3 \pm 1 \cdot 0$ (sub.)	47/15	$+5 \cdot 3 \pm 1 \cdot 3$

THERMOCHEMISTRY OF ORGANIC AND ORGANOMETALLIC COMPOUNDS

C_aH_bO

1 kcal = 4·184 kJ

1 Formula	2 g.f.w.	3 Name	State	Purity mol %	Type	No. of expts.	Detn. of reactn.	$-\Delta H_c^\circ$ kcal/g.f.w.	R
$C_{15}H_{14}O$	210·2782	p-Ethylbenzophenone	l		SB	4	CO_2	1873·6 ±0·5	59/
$C_{15}H_{14}O$	210·2782	Dibenzyl ketone	c		SB	3	m	1868·9 ±0·6	54/
$C_{15}H_{14}O$	210·2782	4, 5, 6-Trimethyl-benzoxalene	c		SB	4	m	1856·1 ±1·9	66/
$C_{15}H_{28}O$	224·3898	Cyclopentadecanone	c		SB	2	m	2249·6 ±2·2	33
$C_{16}H_{10}O$	218·2575	2, 3; 5, 6-Dibenzoxalene	c		SB	7	m	1863·2 ±1·9	66/
$C_{16}H_{16}O$	224·3053	p-Isopropylbenzophenone	l		SB	4	CO_2	2023·0 ±0·5	59/
$C_{16}H_{16}O$	224·3053	2, 7-Pentamethylene-4, 5-benzotropone	c		SB	6	m	2063·5 ±1·5	56/
$C_{16}H_{34}O$	242·4488	1-Hexadecanol	c		SB	7	m	2502·85±0·90	39/ 50/
$C_{16}H_{34}O$		1-Hexadecanol	c		SB	3	m	2501·5 ±1·4	31/
$C_{16}H_{34}O$		1-Hexadecanol	c	99·96	SB	6	CO_2	2501·97±0·43	69/
$C_{16}H_{34}O$		1-Hexadecanol							
$C_{17}H_{18}O$	238·3324	p-t-Butylbenzophenone	l		SB	4	CO_2	2181·3 ±0·5	59/
$C_{17}H_{30}O$	250·4281	Cycloheptadecenone	c		SB	2	m	2507·9 ±3·2	33/
$C_{17}H_{32}O$	252·4440	Cycloheptadecanone	c		SB	6	m	2563·8 ±2·6	33/
$C_{18}H_{12}O$	244·2957	6-Phenyl-2, 3-benzoxalene	c		SB	5	m	2133·4 ±2·1	66/
$C_{19}H_{16}O$	260·3388	Triphenylcarbinol	c		SB	9	m	2332·90±0·52	54/
$C_{23}H_{30}O$	322·4950	2, 7-Dodecamethylene-4, 5-benzotropone	c		SB	6	m	3125·0 ±2·7	56/
$C_{28}H_{44}O$	396·6629	Ergosterol	c		SB		m	3947·6 ±5·9	31/
$C_{28}H_{48}O$	400·6942	Cholesteryl methyl ether	c		SB			4117·2 ±1·2	56/
$C_{28}H_{48}O$	400·6942	3, 5-Cyclocholestan-6β-yl methyl ether	c		SB			4122·9 ±1·3	56/

$1 \text{ kcal} = 4 \cdot 184 \text{ kJ}$

ΔH_r°	Remarks	5 ΔH_f° (l or c) kcal/g.f.w.	6 Determination of ΔH_v			7 ΔH_f° (g) kcal/g.f.w.
			Type	ΔH_v° kcal/g.f.w.	Ref.	
		$-15\cdot4 \pm0\cdot6$				
		$-20\cdot1 \pm0\cdot7$	S5	$21\cdot3\pm[1\cdot2]$ (sub.)	54/10	$+1\cdot2 \pm[1\cdot5]$
		$-32\cdot9 \pm1\cdot9$	E4 $_{me}$	$[33\cdot4\pm0\cdot6]$ (sub.)	66/20	$+0\cdot5 \pm2\cdot2$
	BCE	$-117\cdot6 \pm2\cdot3$	S4	$18\cdot5\pm0\cdot2$ (sub.)	38/14	$-99\cdot1 \pm2\cdot4$
		$+16\cdot8 \pm2\cdot0$	S3	$30\cdot92\pm0\cdot31$ (sub.)	66/20	$+47\cdot7 \pm2\cdot1$
		$-28\cdot3 \pm0\cdot5$				
	C	$+12\cdot2 \pm1\cdot5$	E8 $_{me}$	$20\cdot6\pm[2\cdot5]$ (sub.)		$+32\cdot8 \pm[3\cdot0]$
		$-163\cdot32\pm0\cdot93$	S3	$40\cdot5\pm0\cdot5$ (sub.)	65/4	$-122\cdot82\pm1\cdot07$
√An $1\cdot0000$		$-164\cdot7 \pm1\cdot5$ $-164\cdot20\pm0\cdot46$				$-124\cdot2 \pm1\cdot6$ $-123\cdot70\pm0\cdot70$
	Selected value	$-164\cdot20\pm0\cdot46$				$-123\cdot70\pm0\cdot70$
		$-32\cdot4 \pm0\cdot6$				
	BCE	$-115\cdot7 \pm3\cdot3$	E1 $_{me}$	$18\cdot1\pm0\cdot2$ (sub.)		$-97\cdot6 \pm3\cdot4$
	BCE	$-128\cdot1 \pm2\cdot7$	S4	$18\cdot1\pm0\cdot2$ (sub.)	38/14	$-110\cdot0 \pm2\cdot8$
		$+30\cdot6 \pm2\cdot2$	S3	$34\cdot07\pm0\cdot35$ (sub.)		$+64\cdot7 \pm2\cdot3$
		$-0\cdot59\pm0\cdot60$				
		$-62\cdot9 \pm2\cdot7$	E8 $_{me}$	$24\cdot7\pm[2\cdot5]$ (sub.)		$-38\cdot2 \pm3\cdot7$
		$-188\cdot8 \pm6\cdot0$	S4	$28\cdot4\pm[1\cdot5]$ (sub.)	37/19	$-160\cdot4 \pm6\cdot3$
		$-155\cdot8 \pm1\cdot3$				
		$-150\cdot1 \pm1\cdot4$				

$C_aH_bO_2$

1 kcal = 4·184 kJ

1 Formula	2 g.f.w.	3 Name	State	Purity mol %	Type	No. of expts.	Detn. of reactn.	$-\Delta H_r^\circ$ kcal/g.f.w.	R
CO_2	44·0100	Carbon dioxide	g		SB	20	CO_2	94·050±0·024	38
CO_2		Carbon dioxide	g		SB	24	CO_2	94·037±0·013	38
CO_2		Carbon dioxide	g		SB	17	CO_2	94·065±0·013	44
CO_2		Carbon dioxide	g		SB	52	m	94·040±0·012	66,
CO_2		Carbon dioxide							44/
									65/
CH_2O_2	46·0259	Formic acid	l	>99·8	SB	4	CO_2	60·86±0·06	59/
CH_2O_2		Formic acid	l	99·8	SB	4	m	60·67±0·07	64/
CH_2O_2		Formic acid							
$C_2H_2O_2$	58·0370	Glyoxal	g		FC	3	CO_2	205·76±0·17	M9
$C_2H_4O_2$	60·0530	Acetic acid	l		SB	4	CO_2	209·02±0·06	59/
$C_2H_4O_2$		Acetic acid	l	99·8	SB	5	m	208·94±0·05	64/
$C_2H_4O_2$		Acetic acid							
$C_2H_6O_2$	62·0689	Ethylene glycol	l		SB	7	m	284·32±0·17	46/
$C_2H_6O_2$		Ethylene glycol	l		SB	4	m	284·75±0·29	37/
$C_2H_6O_2$		Ethylene glycol	l		SB	5	m	287·43±0·29	42/
$C_2H_6O_2$		Ethylene glycol							
$C_2H_6O_2$	62·0689	Ethyl hydroperoxide	l		SB		m	335·2 ±14·0	40/
$C_2H_6O_2$	62·0689	Dimethyl peroxide	g	>99, glc	SB		m	363·0 ±0·3*	65/2
C_3O_2	68·0323	Carbon suboxide	l	99·98	SB	4	m	254·13±0·24	65/2
$C_3H_2O_2$	70·0482	Propiolic acid	l		H	7	H_2	75·9 ±0·7	59/1
$C_3H_4O_2$	72·0641	Methylglyoxal	l		SB		m	344·9 ±1·0	38/
$C_3H_4O_2$	72·0641	β-Propiolactone	l		SB	7	m	339·94±0·20	66/2
$C_3H_4O_2$	72·0641	Acrylic acid	l		H	5	H_2	30·35±0·20	59/1
$C_3H_6O_2$	74·0801	2, 3-Epoxy-1-propanol	l		SB	3	m	415·84±0·20	54/3
$C_3H_6O_2$	74·0801	1, 3-Dioxolan	l		SB	4	CO_2	406·4 ±1·0	59/2
$C_3H_6O_2$		1, 3-Dioxolan	l		SB		m	407·5 ±0·1	57/1
$C_3H_6O_2$		1, 3-Dioxolan							
$C_3H_6O_2$	74·0801	Propionic acid	l	99·8	SB	5	m	365·03±0·04	64/1

5. THERMOCHEMICAL DATA

1 kcal$=4\cdot184$ kJ

ΔH_r°		5 ΔH_f° (l or c) kcal/g.f.w.	6 Determination of ΔH_v			
Remarks			Type	ΔH_v° kcal/g.f.w.	Ref.	ΔH_f° (g) kcal/g.f.w.
mbustion of graphite, 5s						$-94\cdot050\pm0\cdot024$
mbustion of graphite, 4s						$-94\cdot037\pm0\cdot013$
mbustion of graphite, 2s						$-94\cdot065\pm0\cdot013$
mbustion of graphite, 3s						$-94\cdot040\pm0\cdot012$
	Selected value					$-94\cdot051\pm0\cdot011$
An $0\cdot999\pm0\cdot001$		$-101\cdot51\pm0\cdot06$		$11\cdot03\pm[0\cdot10]$	68/4	$-90\cdot48\pm0\cdot12$
An $>0\cdot9997$		$-101\cdot70\pm0\cdot07$				$-90\cdot67\pm0\cdot13$
	Selected value	$-101\cdot60\pm0\cdot10$				$-90\cdot57\pm0\cdot14$
						$-50\cdot66\pm0\cdot18$
H An $0\cdot9966\pm0\cdot0016$						
An $>0\cdot9997$		$-115\cdot71\pm0\cdot10$		$12\cdot49\pm[0\cdot10]$	68/4	$-103\cdot22\pm0\cdot14$
		$-115\cdot79\pm0\cdot09$				$-103\cdot30\pm0\cdot13$
	Selected value	$-115\cdot75\pm0\cdot07$				$-103\cdot26\pm0\cdot12$
		$-108\cdot73\pm0\cdot18$	V4	$14\cdot8\pm1\cdot5$	37/10	$-93\cdot9\ \pm1\cdot5$
	BCE	$-108\cdot30\pm0\cdot30$				$-93\cdot5\ \pm1\cdot6$
	C	$-105\cdot62\pm0\cdot30$				$-90\cdot8\ \pm1\cdot6$
	Selected value	$-108\cdot73\pm0\cdot18$				$-93\cdot9\ \pm1\cdot5$
		$-57\cdot9\ \pm14\cdot0$	V3	$10\cdot3\pm1\cdot0$	51/16	$-47\cdot6\ \pm14\cdot0$
						$-30\cdot1\ \pm0\cdot3*$
	I	$-28\cdot02\pm0\cdot24$	C1	$5\cdot65\pm0\cdot35$	65/25	$-22\cdot37\pm0\cdot44$
$_3H_2O_2(l)+2H_2$ $=C_2H_5CO_2H(l)\ \{-122\cdot07\pm0\cdot08\}$		$-46\cdot2\ \pm0\cdot7$				
		$-73\cdot9\ \pm1\cdot0$	E8	$9\cdot1\pm0\cdot5$		$-64\cdot8\ \pm1\cdot2$
		$-78\cdot84\pm0\cdot21$	C1	$11\cdot24\pm0\cdot01$	66/22	$-67\cdot60\pm0\cdot21$
$_3H_4O_2(l)+H_2$ $=C_2H_5CO_2H(l)\ \{-122\cdot07\pm0\cdot08\}$		$-91\cdot72\pm0\cdot25$				
	E	$-71\cdot26\pm0\cdot21$				
/An $0\cdot9995\pm0\cdot0002$		$-80\cdot7\ \pm1\cdot0$	V5	$8\cdot5\pm0\cdot1$	59/20	$-72\cdot2\ \pm1\cdot0$
	E	$-79\cdot6\ \pm0\cdot1$				$-71\cdot1\ \pm0\cdot2$
	Selected value	$-79\cdot6\ \pm0\cdot1$				$-71\cdot1\ \pm0\cdot2$
/An $>0\cdot9997$		$-122\cdot07\pm0\cdot08$	E1	$13\cdot7\pm0\cdot5$	68/3	$-108\cdot4\pm0\cdot5$

$C_aH_bO_2$

1 kcal $= 4 \cdot 184$ kJ

1 Formula	2 g.f.w.	3 Name	State	Purity mol %	Type	No. of expts.	Detn. of reactn.	$-\Delta H_f^\circ$ kcal/g.f.w.	R
$C_3H_8O_2$	$76 \cdot 0960$	1, 2-Propanediol	l		SB	3	m	$435 \cdot 83 \pm 0 \cdot 44$	37
$C_3H_8O_2$	$76 \cdot 0960$	Dimethoxymethane	g	$99 \cdot 86$	FC	6	CO_2	$472 \cdot 14 \pm 0 \cdot 13$	M9
$C_3H_8O_2$	$76 \cdot 0960$	Propyl hydroperoxide	l	$98 \cdot 5$	SB		m	$479 \cdot 4 \pm 15 \cdot 0$	40
$C_4H_4O_2$	$84 \cdot 0753$	But-3-ynoic acid	c		H	3	H_2	$69 \cdot 8 \pm 0 \cdot 2$	58
$C_4H_6O_2$	$86 \cdot 0912$	Diacetyl	l		SB		m	$493 \cdot 83 \pm 0 \cdot 19$	54
$C_4H_6O_2$		Diacetyl	l		SB	3	m	$494 \cdot 8 \pm 0 \cdot 2$	54
$C_4H_6O_2$		Diacetyl	l		SB	9	m	$493 \cdot 58 \pm 0 \cdot 19$	54
$C_4H_6O_2$		Diacetyl							
$C_4H_6O_2$	$86 \cdot 0912$	Vinyl acetate	g		H	4	P/CO_2	$30 \cdot 88 \pm 0 \cdot 06$	38
$C_4H_8O_2$	$88 \cdot 1072$	n-Butyric acid	l	$99 \cdot 9$	SB	4	m	$521 \cdot 87 \pm 0 \cdot 14$	64
$C_4H_8O_2$	$88 \cdot 1072$	Ethyl acetate	l	$99 \cdot 9$	R	4	m	$-0 \cdot 89 \pm 0 \cdot 08$	58
$C_4H_8O_2$		Ethyl acetate	g		E		an	$3 \cdot 97 \pm 0 \cdot 20$	42
$C_4H_8O_2$		Ethyl acetate							
$C_4H_8O_2$	$88 \cdot 1072$	1, 4-Dioxan	l	$99 \cdot 97 \pm 0 \cdot 01$	SB	6	CO_2	$564 \cdot 99 \pm 0 \cdot 14$	61
$C_4H_8O_2$		1, 4-Dioxan	l		SB	4	m	$560 \cdot 6 \pm 0 \cdot 9$	33
$C_4H_8O_2$		1, 4-Dioxan							
$C_4H_8O_2$	$88 \cdot 1072$	1, 3-Dioxan	l	$99 \cdot 5$, glc	SB	3	CO_2	$559 \cdot 47 \pm 0 \cdot 14$	61/7
$C_4H_8O_2$		1, 3-Dioxan	l		SB	4	CO_2	$555 \cdot 0 \pm 1 \cdot 4$	59/2
$C_4H_8O_2$		1, 3-Dioxan	l		SB		m	$557 \cdot 2 \pm 0 \cdot 2$	57/1
$C_4H_8O_2$		1, 3-Dioxan	l	$99 \cdot 99$, glc	SB	6	m	$557 \cdot 34 \pm 0 \cdot 20$	67/1
$C_4H_8O_2$		1, 3-Dioxan							
$C_4H_{10}O_2$	$90 \cdot 1231$	1, 2-Butanediol	l		SB	2	m	$592 \cdot 63 \pm 0 \cdot 59$	37
$C_4H_{10}O_2$	$90 \cdot 1231$	1, 3-Butanediol	l		SB	3	m	$594 \cdot 70 \pm 0 \cdot 59$	37
$C_4H_{10}O_2$	$90 \cdot 1231$	2, 3-Butanediol	l		SB	2	m	$588 \cdot 35 \pm 0 \cdot 59$	37
$C_4H_{10}O_2$	$90 \cdot 1231$	2-Methyl-1, 2-propanediol	l		SB	2	m	$588 \cdot 79 \pm 0 \cdot 59$	37
$C_4H_{10}O_2$	$90 \cdot 1231$	1, 1-Dimethoxyethane	g		FC	6	CO_2	$624 \cdot 52 \pm 0 \cdot 30$	M9

$$1 \text{ kcal} = 4 \cdot 184 \text{ kJ}$$

$_\Delta H_r^\circ$ Remarks	5 ΔH_f° (l or c) kcal/g.f.w.	6 Determination of ΔH_v			7 ΔH_f° (g) kcal/g.f.w.
		Type	ΔH_v° kcal/g.f.w.	Ref.	
BCE	$-119 \cdot 58 \pm 0 \cdot 45$	V4	$14 \cdot 95 \pm 0 \cdot 50$	35/16	$-104 \cdot 63 \pm 0 \cdot 70$
An $0 \cdot 99932 \pm 0 \cdot 00030$	$-90 \cdot 17 \pm 0 \cdot 15$	C1	$6 \cdot 90 \pm 0 \cdot 05$	64/34	$-83 \cdot 27 \pm 0 \cdot 14$
	$-76 \cdot 0 \ \pm 15 \cdot 0$				
$_4O_2(c) + 2H_2$ $= n\text{-}C_3H_7CO_2H(l) \ \{-127 \cdot 59 \pm 0 \cdot 16\}$	$-57 \cdot 8 \ \pm 0 \cdot 3$				
	$-87 \cdot 32 \pm 0 \cdot 20$	V4	$9 \cdot 25 \pm 0 \cdot 22$	54/9	$-78 \cdot 07 \pm 0 \cdot 30$
	$-86 \cdot 4 \ \pm 0 \cdot 2$				$-77 \cdot 1 \ \pm 0 \cdot 3$
	$-87 \cdot 57 \pm 0 \cdot 20$				$-78 \cdot 32 \pm 0 \cdot 30$
Selected value	$-87 \cdot 45 \pm 0 \cdot 15$				$-78 \cdot 20 \pm 0 \cdot 26$
yl acetate (g)$\{-106 \cdot 34 \pm 0 \cdot 16\}$ E	$-83 \cdot 78 \pm 0 \cdot 35$	V3	$8 \cdot 32 \pm 0 \cdot 30$	47/15	$-75 \cdot 46 \pm 0 \cdot 18$
An $0 \cdot 9997$	$-127 \cdot 59 \pm 0 \cdot 16$	E1	$15 \cdot 2 \pm 0 \cdot 5$		$-112 \cdot 4 \ \pm 0 \cdot 6$
$)Ac(l) + H_2O(l)$ $= EtOH(l) \ \{-66 \cdot 42 \pm 0 \cdot 08\}$ $+ HOAc(l) \ \{-115 \cdot 75 \pm 0 \cdot 07\}$	$-114 \cdot 74 \pm 0 \cdot 15$	C1	$8 \cdot 40 \pm 0 \cdot 05$	66/14	$-106 \cdot 34 \pm 0 \cdot 16$
$)Ac(g) \ \{-103 \cdot 8\} + EtOH(g) \ \{-56 \cdot 24\}$ $\rightleftharpoons EtOAc(g) + H_2O(g)$. 2nd Law	$-113 \cdot 21 \pm 0 \cdot 34$				$-105 \cdot 81 \pm 0 \cdot 34$
Selected value	$-114 \cdot 74 \pm 0 \cdot 15$				$-106 \cdot 34 \pm 0 \cdot 16$
An $0 \cdot 99930 \pm 0 \cdot 00050$	$-84 \cdot 47 \pm 0 \cdot 14$	V2	$8 \cdot 96 \pm 0 \cdot 10$	38/15	$-75 \cdot 51 \pm 0 \cdot 17$
BCE	$-88 \cdot 9 \ \pm 0 \cdot 9$				$-79 \cdot 9 \ \pm 0 \cdot 9$
Selected value	$-84 \cdot 47 \pm 0 \cdot 14$				$-75 \cdot 51 \pm 0 \cdot 17$
An $0 \cdot 99893 \pm 0 \cdot 00065$	$-89 \cdot 99 \pm 0 \cdot 16$	V5	$8 \cdot 5 \pm 0 \cdot 2$	59/20	$-81 \cdot 49 \pm 0 \cdot 28$
An $0 \cdot 9965 \pm 0 \cdot 0016$	$-94 \cdot 5 \ \pm 1 \cdot 4$				$-86 \cdot 0 \ \pm 1 \cdot 4$
E	$-92 \cdot 3 \ \pm 0 \cdot 2$				$-83 \cdot 8 \ \pm 0 \cdot 3$
	$-92 \cdot 12 \pm 0 \cdot 21$				$-83 \cdot 62 \pm 0 \cdot 29$
Selected value	$-92 \cdot 21 \pm 0 \cdot 15$				$-83 \cdot 71 \pm 0 \cdot 27$
BCE	$-125 \cdot 15 \pm 0 \cdot 60$				
BCE	$-123 \cdot 08 \pm 0 \cdot 60$	V4	$15 \cdot 57 \pm 0 \cdot 50$	35/16	$-107 \cdot 51 \pm 0 \cdot 75$
BCE	$-129 \cdot 43 \pm 0 \cdot 60$	V4	$14 \cdot 16 \pm 0 \cdot 50$	46/16	$-115 \cdot 27 \pm 0 \cdot 75$
BCE	$-128 \cdot 99 \pm 0 \cdot 60$				
					$-93 \cdot 26 \pm 0 \cdot 31$

$C_4H_bO_2$

1 kcal = 4·184 kJ

1 Formula	2 g.f.w.	3 Name	State	Purity mol %	Type	No. of expts.	Detn. of reactn.	$-\Delta H_r^{\circ}$ kcal/g.f.w.	R
$C_4H_{10}O_2$	90·1231	Diethyl peroxide	l	>99, glc	SB		m	664·4 ±0·3*	65
$C_4H_{10}O_2$		Diethyl peroxide	l		SB		m	662·2	52
$C_4H_{10}O_2$		Diethyl peroxide							
$C_4H_{10}O_2$	90·1231	t-Butyl hydroperoxide	l		SB		m	654·2	51
$C_4H_{10}O_2$		t-Butyl hydroperoxide	l		SB		m	647·6 ±1·2	64
$C_4H_{10}O_2$		t-Butyl hydroperoxide							
$C_5H_4O_2$	96·0864	Furfural	l		SB			560·3	60
$C_5H_4O_2$		Furfural	l		SB	2	m	558·7 ±1·1	29
$C_5H_4O_2$		Furfural							
$C_5H_6O_2$	98·1024	Pent-3-ynoic acid	c		H	3	H_2	63·84±0·90	58
$C_5H_6O_2$	98·1024	Furfuryl alcohol	l		SB	6	m	609·18±0·30	50
$C_5H_6O_2$		Furfuryl alcohol	l		SB	3	m	607·0 ±1·2	29
$C_5H_6O_2$		Furfuryl alcohol							
$C_5H_8O_2$	100·1183	Acetylacetone	l		SB	8	m	642·23±0·36	57
$C_5H_8O_2$	100·1183	4-Pentenoic acid	l	2s	SB		m	640·6 ±0·6	37
$C_5H_8O_2$	100·1183	3-Pentenoic acid	l	2s	SB		m	639·6 ±0·5	37
$C_5H_8O_2$	100·1183	2-Pentenoic acid	l	2s	SB		m	636·8 ±0·5	37
$C_5H_8O_2$	100·1183	Methyl crotonate	l	2s	SB		m	652·0 ±0·4	36
$C_5H_8O_2$	100·1183	Isopropenyl acetate	l	99·93	R	4	m	14·39 ±0·12	57
$C_5H_{10}O_2$	102·1343	Pentanoic acid	l	99·85	SB	7	m	678·29±0·15	65
$C_5H_{10}O_2$		Pentanoic acid	l	99·9	SB	5	m	677·95±0·16	64
$C_5H_{10}O_2$		Pentanoic acid	l		SB	5	m	681·2 ±1·4	54
$C_5H_{10}O_2$		Pentanoic acid	l	2s	SB		m	677·2 ±0·3	37
$C_5H_{10}O_2$		Pentanoic acid							
$C_5H_{10}O_2$	102·1343	α-Methylbutyric acid	l		SB		m	679·3 ±1·4	54
$C_5H_{10}O_2$	102·1343	Isovaleric acid	l		SB		m	677·6 ±1·4	54
$C_5H_{10}O_2$	102·1343	Pivalic acid (trimethylacetic acid)	l		SB		m	676·9 ±1·4	54

Determina

$$1 \text{ kcal} = 4 \cdot 184 \text{ kJ}$$

H_r° / Remarks		5 ΔH_f° (l or c) kcal/g.f.w.	6 Determination of ΔH_v			7 ΔH_f° (g) kcal/g.f.w.
			Type	ΔH_v° kcal/g.f.w.	Ref.	
erimental details not given		$-53\cdot4$ $\pm0\cdot3$*		$7\cdot3\pm0\cdot5$	39/13	$-46\cdot1$ $\pm0\cdot6$*
		$-55\cdot6$				$-48\cdot3$
	Selected value	$-53\cdot4$ $\pm0\cdot3$*				$-46\cdot1$ $\pm0\cdot6$*
erimental details not given		$-63\cdot6$	V5	$11\cdot41\pm0\cdot04$	64/20	$-52\cdot2$
		$-70\cdot2$ $\pm1\cdot2$				$-58\cdot8$ $\pm1\cdot2$
	Selected value	$-70\cdot2$ $\pm1\cdot2$				$-58\cdot8$ $\pm1\cdot2$
		$-46\cdot6$	V5	$9\cdot2\pm[0\cdot6]$	60/25	$-34\cdot5$
	BCE	$-48\cdot2$ $\pm1\cdot1$	C2	$12\cdot10\pm0\cdot10$	26/2	$-36\cdot1$ $\pm1\cdot2$
	Selected value	$-48\cdot2$ $\pm1\cdot1$		$12\cdot10\pm0\cdot10$		$-36\cdot1$ $\pm1\cdot2$
$\mathrm{I_6O_2(c)+2H_2}$ $=\text{n-}\mathrm{C_4H_9CO_2H(l)}$ $\{-133\cdot71\pm0\cdot18\}$		$-69\cdot87\pm0\cdot95$				
		$-66\cdot02\pm0\cdot31$	E7	$15\cdot40\pm0\cdot40$		$-50\cdot62\pm0\cdot50$
	BCE	$-68\cdot2$ $\pm1\cdot2$				$-52\cdot8$ $\pm1\cdot3$
	Selected value	$-66\cdot02\pm0\cdot31$				$-50\cdot62\pm0\cdot50$
		$-101\cdot29\pm0\cdot37$	E7	$10\cdot82\pm0\cdot40$		$-90\cdot47\pm0\cdot54$
	BCE	$-102\cdot9$ $\pm0\cdot6$				
	BCE	$-103\cdot9$ $\pm0\cdot5$				
	BCE	$-106\cdot7$ $\pm0\cdot5$				
	BCE	$-91\cdot5$ $\pm0\cdot4$	E7	$9\cdot8\pm0\cdot3$		$-81\cdot7$ $\pm0\cdot5$
$\mathrm{H_8O_2(l)+H_2O(l)}$ $=\mathrm{(CH_3)_2CO(l)}$ $\{-59\cdot27\pm0\cdot12\}$ $+\mathrm{HOAc(l)}$ $\{-115\cdot75\pm0\cdot07\}$		$-92\cdot31\pm0\cdot23$				
An $>0\cdot9997$		$-133\cdot54\pm0\cdot18$	V1	$16\cdot56\pm0\cdot40$	65/27	$-116\cdot98\pm0\cdot45$
		$-133\cdot88\pm0\cdot19$				$-117\cdot32\pm0\cdot45$
	C	$-130\cdot6$ $\pm1\cdot4$				$-114\cdot0$ $\pm1\cdot5$
	BCE	$-134\cdot6$ $\pm0\cdot4$				$-118\cdot0$ $\pm0\cdot6$
	Selected value	$-133\cdot71\pm0\cdot18$				$-117\cdot15\pm0\cdot45$
	C	$-132\cdot5$ $\pm1\cdot4$				
	C	$-134\cdot2$ $\pm1\cdot4$				
	C	$-134\cdot9$ $\pm1\cdot4$				

$C_aH_bO_2$

1 kcal = 4·184 kJ

1 Formula	2 g.f.w.	3 Name	State	Purity mol %	Type	No. of expts.	Detn. of reactn.	$-\Delta H_r^\circ$ kcal/g.f.w.	F
$C_5H_{10}O_2$	102·1343	Ethyl propionate	g		E		an	5·40±[0·10]	38
$C_5H_{10}O_2$	102·1343	Isopropyl acetate	l	99·9	R	4	m	−0·54±0·10	58
$C_5H_{10}O_2$	102·1343	cis-Cyclopentanediol-1, 2	c		SB	2	m	695·9 ±1·0	42
$C_5H_{10}O_2$	102·1343	trans-Cyclopentanediol-1, 2	c		SB	5	m	694·7 ±0·7	42
$C_5H_{10}O_2$	102·1343	Tetrahydrofurfuryl alcohol	l		SB	2	m	707·7 ±1·4	29
$C_5H_{10}O_2$	102·1343	Ethylene glycol acetone	l		SB	5	m	710·7 ±0·7	42
$C_5H_{10}O_2$	102·1343	2-Methyl-1, 3-dioxan	l	99·99, glc	SB	7	m	707·23±0·25	67
$C_5H_{10}O_2$	102·1343	4-Methyl-1, 3-dioxan	l	99·99, glc	SB	5	m	712·03±0·08	67
$C_5H_{10}O_2$	102·1343	Tetramethylene formal	l		SB		m	719·2 ±0·2	57
$C_5H_{12}O_2$	104·1502	2, 2-Dimethoxypropane	l		R	3	m	−3·95±0·05	62
$C_6H_4O_2$	108·0976	p-Benzoquinone	c		SB	4	m	656·33±0·10	56
$C_6H_4O_2$		p-Benzoquinone	c		SB	9	m	656·88±0·13	54
$C_6H_4O_2$		p-Benzoquinone	c		SB		m	651·51±0·20	56
$C_6H_4O_2$		p-Benzoquinone							
$C_6H_6O_2$	110·1135	Hydroquinone	c		SB	6	m	681·79±0·21	56
$C_6H_6O_2$		Hydroquinone	c		SB	9	m	682·55±0·23	54
$C_6H_6O_2$		Hydroquinone	c		SB		m	675·16±0·27	56
$C_6H_6O_2$		Hydroquinone							
$C_6H_{10}O_2$	114·1454	Ethyl crotonate	l	2s	SB		m	805·5 ±0·5	36
$C_6H_{12}O_2$	116·1613	Hexanoic acid	l	99·9	SB	4	m	834·72±0·15	65
$C_6H_{12}O_2$		Hexanoic acid	l	99·9	SB			834·23±0·13	64
$C_6H_{12}O_2$		Hexanoic acid							

1 kcal = 4·184 kJ

H_r°		5 ΔH_f° (l or c) kcal/g.f.w.	6 Determination of ΔH_v			7 ΔH_f° (g) kcal/g.f.w.
Remarks			Type	ΔH_v° kcal/g.f.w.	Ref.	
H(g) {−56·24} ·EtCO₂H(g) {−108·4±0·5} ⇌ EtCO₂Et(g)+H₂O(g)		−110·7 ±0·6	C2	9·03±0·10	26/2	−101·7 ±0·6
OAc(l)+H₂O(l) = i-PrOH(l) {−76·02} +HOAc(l) {−115·75}		−124·01±0·18	C1	8·89±0·05	66/14	−115·12±0·19
	C	−115·9 ±1·0				
	C	−117·1 ±0·7				
	BCE	−104·1 ±1·4	E7	15·9 ±0·5		−88·2 ±1·5
	C	−101·1 ±0·7				
		−104·60±0·29	E7	9·23±0·30		−95·37±0·42
		−99·80±0·15	E7	9·36±0·30		−90·46±0·35
	E	−92·6 ±0·2	E7	9·8 ±0·3		−82·8 ±0·4
₁₂O₂(l)+H₂O(l) = (CH₃)₂CO(l) {−59·27} +2CH₃OH(l) {−57·01}		−108·92±0·40	E8	7·03±0·30		−101·89±0·50
		−44·61±0·12	S5	15·0 ±0·8 (sub.)	56/15	−29·61±0·84
		−44·06±0·15				−29·06±0·85
		−49·43±0·21				−34·43±0·87
	Selected value	−44·33±0·30				−29·33±0·90
		−87·46±0·23	S4	23·7 ±0·4 (sub.)	56/15	−63·76±0·45
		−86·70±0·25				−63·00±0·46
		−94·09±0·29				−70·39±0·49
	Selected value	−87·08±0·30				−63·38±0·50
	BCE	−100·4 ±0·5	E7	10·6 ±0·3		−89·8 ±0·6
		−139·48±0·18	E1	17·0 ±0·3		−122·5 ±0·4
An >0·9997		−139·97±0·17				−123·0 ±0·4
	Selected value	−139·73±0·25				−122·7 ±0·5

THERMOCHEMISTRY OF ORGANIC AND ORGANOMETALLIC COMPOUNDS

$C_aH_bO_2$ 1 kcal = 4·184 kJ

Determina

1 Formula	2 g.f.w.	3 Name	State	Purity mol %	Type	No. of expts.	Detn. of reactn.	$-\Delta H_r^\circ$ kcal/g.f.w.	R
$C_6H_{12}O_2$	116·1613	Methyl pentanoate	l	99·85	SB	4	m	851·30±0·14	65
$C_6H_{12}O_2$		Methyl pentanoate	l		SB	5	m	848·0 ±1·7	54
$C_6H_{12}O_2$		Methyl pentanoate							
$C_6H_{12}O_2$	116·1613	Methyl α-methylbutyrate	l		SB		m	846·5 ±1·7	54
$C_6H_{12}O_2$	116·1613	Methyl isovalerate	l		SB		m	845·4 ±1·7	54
$C_6H_{12}O_2$	116·1613	Methyl pivalate	l		SB		m	842·2 ±1·7	54
$C_6H_{12}O_2$	116·1613	n-Butyl acetate	l	>99·9	R	4	m	−0·80±0·10	5
$C_6H_{12}O_2$	116·1613	Cyclohexane hydroperoxide	l		SB	5	m	908·9 ±0·7	56
$C_6H_{12}O_2$	116·1613	2-Methoxytetrahydro-pyran	l		SB		m	868·5 ±0·3	57
$C_6H_{12}O_2$	116·1613	cis-2, 4-Dimethyl-1, 3-dioxan	l	99·99, glc	SB	5	m	862·41±0·34	67
$C_6H_{12}O_2$	116·1613	4, 5-Dimethyl-1, 3-dioxan	l	99·9, glc	SB	5	m	865·88±0·14	67
$C_6H_{12}O_2$	116·1613	5, 5-Dimethyl-1, 3-dioxan	l	99·99, glc	SB	7	m	863·67±0·26	67
$C_6H_{12}O_2$	116·1613	Pentamethylene formal	l		SB		m	883·7 ±0·3	57
$C_6H_{14}O_2$	118·1773	1-Hydroperoxyhexane	l		SB	5	m	970·9 ±1·2	56
$C_6H_{14}O_2$	118·1773	2-Hydroperoxyhexane	l		SB	5	m	968·4 ±0·7	56
$C_6H_{14}O_2$	118·1773	3-Hydroperoxyhexane	l		SB	5	m	969·6 ±1·8	56
$C_7H_6O_2$	122·1247	Benzoic acid	c	99·999	SB	19	CO_2	771·29±0·05	42
$C_7H_6O_2$		Benzoic acid	c	99·98	SB	38	m	771·32±0·13	56
$C_7H_6O_2$		Benzoic acid	c	99·977	SB	14	m	771·35±0·08	55
$C_7H_6O_2$		Benzoic acid	c	99·99	SB	10	m	771·32±0·08	44
$C_7H_6O_2$		Benzoic acid	c		SB	67	m	771·24±0·07	34
$C_7H_6O_2$		Benzoic acid	c	99·9996	SB	15	m	771·37±0·06	58
$C_7H_6O_2$		Benzoic acid							

1 kcal = 4·184 kJ

ΔH_r°			6 Determination of ΔH_v			
Remarks		5 ΔH_f° (l or c) kcal/g.f.w.	Type	ΔH_v° kcal/g.f.w.	Ref.	7 ΔH_f° (g) kcal/g.f.w.
		−122·90±0·16	E7	10·2 ±0·3		−112·7 ±0·4
	C	−126·2 ±1·7				−116·0 ±1·8
	Selected value	−122·90±0·16				−112·7 ±0·4
	C	−127·7 ±1·7	E7	10·0 ±0·3		−117·7 ±1·8
	C	−128·8 ±1·7	E7	9·8 ±0·3		−119·0 ±1·8
	C	−132·0 ±1·7	E7	9·2 ±0·3		−122·8 ±1·8
uOAc(l)+H₂O(l) = n-BuOH(l) {−78·29} +HOAc(l) {−115·75}		−126·52±0·19	C1	10·42±0·05	66/14	−116·10±0·20
		−65·3 ±0·7				
	E	−105·7 ±0·3	E8	10·2 ±0·3		−95·5 ±0·5
		−111·79±0·40	E7	9·53±0·30		−102·26±0·50
		−108·32±0·18	E7	10·16±0·30		−98·16±0·37
		−110·53±0·29	E7	9·86±0·30		−100·67±0·42
	E	−90·5 ±0·3	E8	10·0 ±0·3		−80·5 ±0·5
		−71·6 ±1·2				
		−74·1 ±0·7				
		−72·9 ±1·8				
		−92·01±0·09	S3	21·85±0·10	54/27	−70·16±0·14
		−91·98±0·14		(sub.)		−70·13±0·17
		−91·95±0·09				−70·10±0·14
		−91·98±0·09				−70·13±0·14
		−92·06±0·08				−70·21±0·13
		−91·93±0·07				−70·08±0·13
	Selected value	−91·99±0·05				−70·14±0·12

$C_aH_bO_2$

1 kcal = 4·184 kJ

1 Formula	2 g.f.w.	3 Name	State	Purity mol %	Type	No. of expts.	Detn. of reactn.	$-\Delta H_r^o$ kcal/g.f.w.	R
$C_7H_6O_2$	122·1247	1, 3-Dioxaindane	l		SB	3	m	819·3 ±0·5	58
$C_7H_6O_2$	122·1247	Tropolone	c		SB	2	m	806·12±0·20	52
$C_7H_6O_2$		Tropolone	c		SB		m	806·2 ±0·9	51
$C_7H_6O_2$		Tropolone	c		SB	3	m	806·0 ±0·6	56
$C_7H_6O_2$		Tropolone							
$C_7H_{10}O_2$	126·1566	Methyl hex-2-ynoate	l		H	3	H_2	71·1 ±1·2	58
$C_7H_{10}O_2$	126·1566	Ethyl pent-4-ynoate	l	2s	SB		m	932·6 ±0·6	38
$C_7H_{10}O_2$	126·1566	Ethyl pent-3-ynoate	l	2s	SB		m	931·2 ±0·6	38
$C_7H_{10}O_2$	126·1566	Ethyl pent-2-ynoate	l	2s	SB		m	927·8 ±0·5	38
$C_7H_{10}O_2$	126·1566	Ethyl β-vinylacrylate	l	2s	SB		m	919·1 ±0·6	38
$C_7H_{10}O_2$	126·1566	Ethyl α-furyl carbinol	l		SB			893·1 ±1·7	65
$C_7H_{12}O_2$	128·1725	Ethyl 4-pentenoate	l	2s	SB		m	965·1 ±0·5	37
$C_7H_{12}O_2$	128·1725	Ethyl cis-3-pentenoate	l		SB	8	m	964·9 ±0·8	38
$C_7H_{12}O_2$	128·1725	Ethyl trans-3-pentenoate	l	2s	SB	8	m	963·8 ±0·6	37 38
$C_7H_{12}O_2$	128·1725	Ethyl cis-2-pentenoate	l		SB	8	m	962·9 ±0·6	38
$C_7H_{12}O_2$	128·1725	Ethyl trans-2-pentenoate	l		SB	8	m	962·5 ±0·8	38
$C_7H_{12}O_2$	128·1725	n-Propyl crotonate	l	2s	SB		m	962·3 ±0·6	36
$C_7H_{12}O_2$	128·1725	Isopropyl crotonate	l	2s	SB		m	959·0 ±0·6	36
$C_7H_{14}O_2$	130·1884	Heptanoic acid	l	99·9	SB	4	m	991·18±0·19	65
$C_7H_{14}O_2$		Heptanoic acid	l	99·8	SB	5	m	990·43±0·14	64
$C_7H_{14}O_2$		Heptanoic acid							
$C_7H_{14}O_2$	130·1884	Methyl hexanoate	l	99·9	SB	4	m	1007·45±0·22	65
$C_7H_{14}O_2$	130·1884	Ethyl pentanoate	l		SB	5	m	1004·2 ±2·0	54
$C_7H_{14}O_2$		Ethyl pentanoate	l	2s	SB		m	1004·4 ±0·6	37
$C_7H_{14}O_2$		Ethyl pentanoate							
$C_7H_{14}O_2$	130·1884	Ethyl α-methylbutyrate	l		SB	5	m	1001·1 ±2·0	54

Determina

$$1 \text{ kcal} = 4 \cdot 184 \text{ kJ}$$

ΔH_r°		5 ΔH_f° (l or c) kcal/g.f.w.	6 Determination of ΔH_v			7 ΔH_f° (g) kcal/g.f.w.
	Remarks		Type	ΔH_v° kcal/g.f.w.	Ref.	
		$-44\cdot0\ \pm0\cdot5$	V5	$9\cdot9\ \pm0\cdot5$	58/20	$-34\cdot1\ \pm0\cdot8$
An $0\cdot99985\pm0\cdot00007$		$-57\cdot18\pm0\cdot21$	S3	$20\cdot0\ \pm0\cdot2$ (sub.)	51/15	$-37\cdot18\pm0\cdot28$
		$-57\cdot1\ \pm0\cdot9$				$-37\cdot1\ \pm1\cdot0$
		$-57\cdot3\ \pm0\cdot6$				$-37\cdot3\ \pm0\cdot7$
	Selected value	$-57\cdot18\pm0\cdot21$				$-37\cdot18\pm0\cdot28$
$H_{10}O_2(l)+2H_2$ $= n\text{-}C_5H_{11}CO_2CH_3(l)\ \{-129\cdot11\pm0\cdot25\}$		$-58\cdot0\ \pm1\cdot2$				
	BCE	$-67\cdot3\ \pm0\cdot6$	E7	$11\cdot6\ \pm0\cdot3$		$-55\cdot7\ \pm0\cdot7$
	BCE	$-68\cdot7\ \pm0\cdot6$	E7	$11\cdot9\ \pm0\cdot3$		$-56\cdot8\ \pm0\cdot7$
	BCE	$-72\cdot1\ \pm0\cdot5$	E7	$12\cdot3\ \pm0\cdot3$		$-59\cdot8\ \pm0\cdot6$
	BCE	$-80\cdot8\ \pm0\cdot6$	E7	$11\cdot6\ \pm0\cdot3$		$-69\cdot2\ \pm0\cdot7$
		$-106\cdot8\ \pm1\cdot7$				
	BCE	$-103\cdot1\ \pm0\cdot5$	E7	$11\cdot0\ \pm0\cdot3$		$-92\cdot1\ \pm0\cdot6$
	BCE	$-103\cdot3\ \pm0\cdot8$	E7	$10\cdot7\ \pm0\cdot3$		$-92\cdot6\ \pm0\cdot9$
	BCE	$-104\cdot4\ \pm0\cdot6$	E7	$11\cdot2\ \pm0\cdot3$		$-93\cdot2\ \pm0\cdot7$
	BCE	$-105\cdot3\ \pm0\cdot6$	E7	$11\cdot0\ \pm0\cdot3$		$-94\cdot3\ \pm0\cdot7$
	BCE	$-105\cdot7\ \pm0\cdot8$	E7	$11\cdot5\ \pm0\cdot3$		$-94\cdot2\ \pm0\cdot9$
	BCE	$-105\cdot9\ \pm0\cdot6$	E7	$11\cdot5\ \pm0\cdot3$		$-94\cdot4\ \pm0\cdot7$
	BCE	$-109\cdot2\ \pm0\cdot6$	E7	$11\cdot0\ \pm0\cdot3$		$-98\cdot2\ \pm0\cdot7$
		$-145\cdot38\pm0\cdot22$	E1	$18\cdot1\ \pm0\cdot3$	68/3	$-127\cdot3\ \pm0\cdot4$
An $>0\cdot9997$		$-146\cdot13\pm0\cdot18$				$-128\cdot0\ \pm0\cdot4$
	Selected value	$-145\cdot76\pm0\cdot35$				$-127\cdot7\ \pm0\cdot5$
		$-129\cdot11\pm0\cdot25$	E7	$11\cdot1\ \pm0\cdot3$		$-118\cdot0\ \pm0\cdot4$
	C	$-132\cdot4\ \pm2\cdot0$	E7	$11\cdot0\ \pm0\cdot3$		$-121\cdot4\ \pm2\cdot1$
	BCE	$-132\cdot2\ \pm0\cdot6$				$-121\cdot2\ \pm0\cdot7$
	Selected value	$-132\cdot2\ \pm0\cdot6$				$-121\cdot2\ \pm0\cdot7$
	C	$-135\cdot5\ \pm2\cdot0$	E7	$10\cdot6\ \pm0\cdot3$		$-124\cdot9\ \pm2\cdot1$

$C_aH_bO_2$ 1 kcal $= 4 \cdot 184$ kJ

1 Formula	2 g.f.w.	3 Name	State	Purity mol %	Type	No. of expts.	Detn. of reactn.	$-\Delta H_r^\circ$ kcal/g.f.w.	Re
$C_7H_{14}O_2$	$130 \cdot 1884$	Ethyl isovalerate	l		SB	5	m	$1000 \cdot 1 \pm 2 \cdot 0$	54/
$C_7H_{14}O_2$	$130 \cdot 1884$	Ethyl pivalate	l		SB	5	m	$998 \cdot 6 \pm 2 \cdot 0$	54/
$C_7H_{14}O_2$	$130 \cdot 1884$	2-*cis*-4-*trans*-6-Trimethyl-1, 3-dioxan	l		SB	3	m	$1019 \cdot 48 \pm 0 \cdot 24$	67/
$C_7H_{14}O_2$	$130 \cdot 1884$	Methylcyclohexane hydroperoxide	l		SB	5	m	$1057 \cdot 6 \pm 1 \cdot 5$	56/
$C_7H_{16}O_2$	$132 \cdot 2044$	2, 2-Diethoxypropane	l		R	3	m	$-5 \cdot 04 \pm 0 \cdot 05$	62/
$C_7H_{16}O_2$	$132 \cdot 2044$	1-Hydroperoxyheptane	l		SB	5	m	$1122 \cdot 9 \pm 1 \cdot 2$	56/
$C_7H_{16}O_2$	$132 \cdot 2044$	2-Hydroperoxyheptane	l		SB	5	m	$1122 \cdot 1 \pm 0 \cdot 5$	56/
$C_7H_{16}O_2$	$132 \cdot 2044$	3-Hydroperoxyheptane	l		SB	5	m	$1122 \cdot 0 \pm 1 \cdot 0$	56/
$C_7H_{16}O_2$	$132 \cdot 2044$	4-Hydroperoxyheptane	l		SB	5	m	$1125 \cdot 1 \pm 0 \cdot 5$	56/
$C_8H_8O_2$	$136 \cdot 1518$	2-Methoxybenzaldehyde	c		SB	4	m	$962 \cdot 0 \pm 1 \cdot 8$	40/
$C_8H_8O_2$	$136 \cdot 1518$	3-Methoxybenzaldehyde	l		SB	4	m	$959 \cdot 7 \pm 1 \cdot 8$	40/
$C_8H_8O_2$	$136 \cdot 1518$	4-Methoxybenzaldehyde	l		SB	3	m	$961 \cdot 8 \pm 1 \cdot 2$	40/
$C_8H_8O_2$	$136 \cdot 1518$	*o*-Hydroxyacetophenone	c		SB	5	m	$940 \cdot 2 \pm 0 \cdot 9$	37/
$C_8H_8O_2$	$136 \cdot 1518$	*m*-Hydroxyacetophenone	c		SB	5	m	$937 \cdot 1 \pm 1 \cdot 0$	37/
$C_8H_8O_2$	$136 \cdot 1518$	*p*-Hydroxyacetophenone	c		SB	5	m	$938 \cdot 6 \pm 1 \cdot 0$	37/
$C_8H_8O_2$		*o*-Toluic acid	c		SB	3	m	$923 \cdot 6 \pm 0 \cdot 9$	50/ 51/
$C_8H_8O_2$ $C_8H_8O_2$		*o*-Toluic acid *o*-Toluic acid	c		SB	5	CO_2	$926 \cdot 12 \pm 0 \cdot 18$	61/
$C_8H_8O_2$	$136 \cdot 1518$	*m*-Toluic acid	c		SB	5	CO_2	$923 \cdot 82 \pm 0 \cdot 19$	61/
$C_8H_8O_2$	$136 \cdot 1518$	*p*-Toluic acid	c		SB	3	m	$924 \cdot 5 \pm 0 \cdot 9$	50/ 51/
$C_8H_8O_2$ $C_8H_8O_2$		*p*-Toluic acid *p*-Toluic acid	c		SB	5	CO_2	$923 \cdot 08 \pm 0 \cdot 22$	61/

1 kcal = 4·184 kJ

| ΔH_f° | | | 6 Determination of ΔH_v | | | 7 |
Remarks		5 ΔH_f° (l or c) kcal/g.f.w.	ΔH_v° Type kcal/g.f.w.		Ref.	ΔH_f° (g) kcal/g.f.w.
	C	−136·5 ±2·0	E7 10·5 ±0·3			−126·0 ±2·1
	C	−138·0 ±2·0	C1 9·86±0·03		66/14	−128·1 ±2·0
		−117·08±0·31	E7 10·32±0·30			−106·76±0·43
		−79·0 ±1·5				
H₁₆O₂(l)+H₂O(l) = (CH₃)₂CO(l) {−59·27} +2EtOH(l) {−66·42}		−128·83±0·40	V3 7·61±0·20		62/23	−121·22±0·50
		−82·0 ±1·2				
		−82·8 ±0·5				
		−82·9 ±1·0				
		−79·8 ±0·5				
	BCE	−63·7 ±1·8				
	BCE	−66·0 ±1·8				
	BCE	−63·9 ±1·2	V4 15·42±[0·50]		47/15	−48·5 ±1·4
	BCE	−85·5 ±0·9				
	BCE	−88·6 ±1·0				
	BCE	−87·1 ±1·0				
		−102·1 ±0·9				
		−99·55±0·22				
Selected value		−99·55±0·22				
		−101·85±0·23				
		−101·2 ±0·9				
		−102·59±0·26				
Selected value		−102·59±0·26				

$H_{16}O_2(l)+H_2O(l)$
$= (CH_3)_2CO(l)$ {−59·27}
$+2EtOH(l)$ {−66·42}

$C_aH_bO_2$ 1 kcal $= 4\cdot184$ kJ

1 Formula	2 g.f.w.	3 Name	State	Purity mol %	Type	No. of expts.	Detn. of reactn.	$-\Delta H_r^\circ$ kcal/g.f.w.	Re
$C_8H_8O_2$	136·1518	Phenyl acetate	l		R	4	m	6·86±0·08	60/
$C_8H_8O_2$	136·1518	1, 4-Dioxatetralin	l		SB	3	m	964·8 ±0·3	58/
$C_8H_{10}O_2$	138·1677	o-Dimethoxybenzene	l		SB	4	m	1024·6 ±0·5	58/
$C_8H_{10}O_2$	138·1677	p-Xylene-αα′-diol	c		SB	11	m	999·83±0·40	62/
$C_8H_{14}O_2$	142·1996	n-Butyl crotonate	l	2s	SB		m	1118·8 ±0·7	36/
$C_8H_{14}O_2$	142·1996	Isobutyl crotonate	l	2s	SB		m	1117·3 ±0·7	36/
$C_8H_{14}O_2$	142·1996	s-Butyl crotonate	l	2s	SB		m	1117·2 ±0·4	36/
$C_8H_{14}O_2$	142·1996	n-Propyl 4-pentenoate	l	2s	SB		m	1123·7 ±0·7	37/
$C_8H_{14}O_2$	142·1996	Isopropyl 4-pentenoate	l	2s	SB		m	1119·6 ±0·4	37/
$C_8H_{14}O_2$	142·1996	n-Propyl 3-pentenoate	l	2s	SB		m	1122·0 ±0·7	37/
$C_8H_{14}O_2$	142·1996	Isopropyl 3-pentenoate	l	2s	SB		m	1117·5 ±0·5	37/
$C_8H_{14}O_2$	142·1996	n-Propyl 2-pentenoate	l	2s	SB		m	1119·5 ±0·8	37/
$C_8H_{14}O_2$	142·1996	Isopropyl 2-pentenoate	l	2s	SB		m	1116·2 ±0·8	37/
$C_8H_{14}O_2$	142·1996	Methyl α-t-butylacrylate	l		SB		m	1129·9 ±2·0	52/
$C_8H_{14}O_2$	142·1996	cis-Cyclopentanediol acetone	l		SB	4	m	1117·8 ±1·1	42/
$C_8H_{16}O_2$	144·2155	Octanoic acid	l	99·9	SB	3	m	1147·26±0·19	65/
$C_8H_{16}O_2$	144·2155	Octanoic acid	l	99·8	SB	4	m	1146·73±0·13	64/
$C_8H_{16}O_2$		Octanoic acid							
$C_8H_{16}O_2$	144·2155	Methyl heptanoate	l	99·9	SB	4	m	1163·39±0·19	65/
$C_8H_{16}O_2$	144·2155	n-Propyl pentanoate	l	2s	SB		m	1159·6 ±0·3	37/
$C_8H_{16}O_2$	144·2155	Isopropyl pentanoate	l	2s	SB		m	1157·4 ±0·7	37/
$C_8H_{16}O_2$	144·2155	s-Butyl butyrate	l	2s	SB		m	1157·3 ±0·9	36/

1 kcal = 4·184 kJ

ΔH_r°		5 ΔH_f° (l or c) kcal/g.f.w.	6 Determination of ΔH_v			7 ΔH_f° (g) kcal/g.f.w.
Remarks			Type	ΔH_v° kcal/g.f.w.	Ref.	
OAc(l)+H₂O(l) = PhOH(c) {−39·45±0·17} +HOAc(l) {−115·75}		−80·02±0·22	E7	13·0 ±0·4		−67·0 ±0·5
		−60·9 ±0·4	V5	12·1 ±0·6	58/20	−48·8 ±0·8
		−69·4 ±0·5	V5	16·0 ±0·5	58/20	−53·4 ±0·7
		−94·15±0·41				
	BCE	−111·8 ±0·7	E7	12·4 ±0·3		−99·4 ±0·8
	BCE	−113·3 ±0·7	E7	12·0 ±0·3		−101·3 ±0·8
	BCE	−113·4 ±0·4	E7	11·8 ±0·3		−101·6 ±0·5
	BCE	−106·9 ±0·7	E7	11·8 ±0·3		−95·1 ±0·8
	BCE	−111·0 ±0·4	E7	11·3 ±0·3		−99·7 ±0·5
	BCE	−108·6 ±0·7	E7	12·0 ±0·3		−96·6 ±0·8
	BCE	−113·1 ±0·5	E7	11·5 ±0·3		−101·6 ±0·6
	BCE	−111·1 ±0·8	E7	12·4 ±0·3		−98·7 ±0·9
	BCE	−114·4 ±0·8	E7	11·8 ±0·3		−102·6 ±0·9
		−100·7 ±2·0				
	C	−112·8 ±1·1				
/An >0·9997		−151·67±0·22	E1	19·2 ±0·3	68/3	−132·5 ±0·4
		−152·20±0·18				−133·0 ±0·4
	Selected value	−151·93±0·25				−132·7 ±0·4
		−135·54±0·22	E7	12·0 ±0·3		−123·5 ±0·4
	BCE	−139·3 ±0·4	E7	11·8 ±0·3		−127·5 ±0·5
	BCE	−141·5 ±0·7	E7	11·3 ±0·3		−130·2 ±0·8
	BCE	−141·6 ±0·9	E7	11·3 ±0·3		−130·3 ±1·0

$C_aH_bO_2$

1 Formula	2 g.f.w.	3 Name	State	Purity mol %	Type	No. of expts.	Detn. of reactn.	$-\Delta H_r^\circ$ kcal/g.f.w.	Ref
$C_8H_{16}O_2$	144·2155	Methyl α-t-butyl- propionate	l		SB		m	1177·9 ±2·0	52/1
$C_8H_{18}O_2$	146·2315	Di-t-butyl peroxide	l	>99, glc	SB		m	1276·2 ±0·2*	65/
$C_8H_{18}O_2$		Di-t-butyl peroxide	l		SB		m	1273	48/
$C_8H_{18}O_2$		Di-t-butyl peroxide	l		SB		m	1275·0	51/
$C_8H_{18}O_2$		Di-t-butyl peroxide							
$C_9H_8O_2$	148·1629	cis-Cinnamic acid	c	m.p. 42°C	SB	5	m	1043·5 ±0·4	35/
$C_9H_8O_2$		cis-Cinnamic acid	c	m.p. 42°C	SB	4	m	1043·97±0·41	37/
$C_9H_8O_2$		cis-Cinnamic acid	c	m.p. 58°C	SB	4	m	1045·35±0·41	37/
$C_9H_8O_2$		cis-Cinnamic acid	c	m.p. 68°C	SB	6	m	1047·75±0·41	37/
$C_9H_8O_2$	148·1629	trans-Cinnamic acid	c		SB	11	m	1039·19±0·41	62/2
$C_9H_8O_2$		trans-Cinnamic acid	c		SB	3	m	1038·4 ±1·0	29/
$C_9H_8O_2$		trans-Cinnamic acid							
$C_9H_{10}O_2$	150·1789	2, 3-Dimethylbenzoic acid	c		SB	5	CO_2	1080·38±0·19	61/2
$C_9H_{10}O_2$	150·1789	2, 4-Dimethylbenzoic acid	c		SB	5	CO_2	1078·45±0·20	61/2
$C_9H_{10}O_2$	150·1789	2, 5-Dimethylbenzoic acid	c		SB	5	CO_2	1079·01±0·19	61/2
$C_9H_{10}O_2$	150·1789	2, 6-Dimethylbenzoic acid	c		SB	5	CO_2	1082·70±0·19	61/2
$C_9H_{10}O_2$	150·1789	3, 4-Dimethylbenzoic acid	c		SB	4	CO_2	1075·99±0·25	61/2
$C_9H_{10}O_2$	150·1789	3, 5-Dimethylbenzoic acid	c		SB	2	CO_2	1076·55±0·17	61/2
$C_9H_{10}O_2$	150·1789	m-Cresyl acetate	l		R	3	m	4·39±0·14	57/1
$C_9H_{10}O_2$	150·1789	2, 3-Benzo- 1, 4-dioxacycloheptene	l		SB	3	m	1130·4 ±0·2	58/2
$C_9H_{12}O_2$	152·1948	Cumene hydroperoxide	l		SB			1220·9 ±1·6	64/2
$C_9H_{16}O_2$	156·2267	Isopentyl crotonate	l	2s	SB		m	1273·3 ±0·7	36/1
$C_9H_{16}O_2$	156·2267	n-Butyl 4-pentenoate	l	2s	SB		m	1278·3 ±0·4	37/
$C_9H_{16}O_2$	156·2267	Isobutyl 4-pentenoate	l	2s	SB		m	1276·3 ±0·8	37/

4
Determinati

1 kcal $= 4 \cdot 184$ kJ

ΔH_r°			6 Determination of ΔH_v			7
Remarks		5 ΔH_f° (l or c) kcal/g.f.w.	ΔH_v° Type kcal/g.f.w.		Ref.	ΔH_f° (g) kcal/g.f.w.
		$-121 \cdot 0 \pm 2 \cdot 0$				
		$-91 \cdot 0 \pm 0 \cdot 3$*	V4 $7 \cdot 6 \pm 0 \cdot 7$		51/16	$-83 \cdot 4 \pm 0 \cdot 8$
o experimental details		-94				-86
o experimental details		$-92 \cdot 2$				$-84 \cdot 6$
	Selected value	$-91 \cdot 0 \pm 1 \cdot 0$				$-83 \cdot 4 \pm 1 \cdot 2$
	C	$-76 \cdot 2 \pm 0 \cdot 4$				
	C	$-75 \cdot 75 \pm 0 \cdot 42$				
	C	$-74 \cdot 37 \pm 0 \cdot 42$				
	C	$-71 \cdot 97 \pm 0 \cdot 42$				
		$-80 \cdot 53 \pm 0 \cdot 42$				
	ABCE	$-81 \cdot 3 \pm 1 \cdot 0$				
	Selected value	$-80 \cdot 53 \pm 0 \cdot 42$				
		$-107 \cdot 65 \pm 0 \cdot 23$				
		$-109 \cdot 58 \pm 0 \cdot 24$				
		$-109 \cdot 02 \pm 0 \cdot 23$				
		$-105 \cdot 33 \pm 0 \cdot 23$				
		$-112 \cdot 04 \pm 0 \cdot 30$				
		$-111 \cdot 48 \pm 0 \cdot 21$				
m-CresylOAc(l)+H$_2$O(l) = m-Cresyl OH(l) {$-46 \cdot 37$} +HOAc(l) {$-115 \cdot 75$}		$-89 \cdot 41 \pm 0 \cdot 24$	V3 $14 \cdot 51 \pm 0 \cdot 50$		47/8	$-74 \cdot 90 \pm 0 \cdot 56$
		$-57 \cdot 6 \pm 0 \cdot 3$	V5 $13 \cdot 3 \pm 0 \cdot 6$		58/20	$-44 \cdot 3 \pm 0 \cdot 7$
		$-35 \cdot 5 \pm 1 \cdot 6$	V3 $16 \cdot 70 \pm 0 \cdot 04$		64/20	$-18 \cdot 8 \pm 1 \cdot 6$
	BCE	$-119 \cdot 7 \pm 0 \cdot 7$	E7 $12 \cdot 9 \pm 0 \cdot 3$			$-106 \cdot 8 \pm 0 \cdot 8$
	BCE	$-114 \cdot 7 \pm 0 \cdot 4$	E7 $12 \cdot 6 \pm 0 \cdot 3$			$-101 \cdot 2 \pm 0 \cdot 5$
	BCE	$-116 \cdot 7 \pm 0 \cdot 8$	E7 $12 \cdot 3 \pm 0 \cdot 3$			$-104 \cdot 4 \pm 0 \cdot 9$

$C_aH_bO_2$

1 kcal = 4·184 kJ

1 Formula	2 g.f.w.	3 Name	State	Purity mol %	Type	No. of expts.	Detn. of reactn.	$-\Delta H_r^\circ$ kcal/g.f.w.	Ref.
$C_9H_{16}O_2$	156·2267	s-Butyl 4-pentenoate	l	2s	SB		m	1274·2 ±0·4	37/6
$C_9H_{16}O_2$	156·2267	n-Butyl 3-pentenoate	l	2s	SB		m	1276·7 ±0·6	37/6
$C_9H_{16}O_2$	156·2267	Isobutyl 3-pentenoate	l	2s	SB		m	1272·2 ±0·6	37/6
$C_9H_{16}O_2$	156·2267	s-Butyl 3-pentenoate	l	2s	SB		m	1272·8 ±0·6	37/6
$C_9H_{16}O_2$	156·2267	n-Butyl 2-pentenoate	l	2s	SB		m	1274·7 ±0·6	37/6
$C_9H_{16}O_2$	156·2267	Isobutyl 2-pentenoate	l	2s	SB		m	1271·9 ±0·6	37/6
$C_9H_{16}O_2$	156·2267	s-Butyl 2-pentenoate	l	2s	SB		m	1271·0 ±0·5	37/6
$C_9H_{18}O_2$	158·2426	Nonanoic acid	l	99·75	SB	5	m	1304·09±0·22	65/26
$C_9H_{18}O_2$		Nonanoic acid	l		SB	5	m	1303·12±0·23	64/19
$C_9H_{18}O_2$		Nonanoic acid							
$C_9H_{18}O_2$	158·2426	Methyl octanoate	l	99·9	SB	6	m	1320·21±0·19	65/26
$C_9H_{18}O_2$	158·2426	n-Butyl pentanoate	l	2s	SB		m	1314·7 ±0·4	37/6
$C_9H_{18}O_2$	158·2426	Isobutyl pentanoate	l	2s	SB		m	1313·1 ±0·8	37/6
$C_9H_{18}O_2$	158·2426	s-Butyl pentanoate	l	2s	SB		m	1312·1 ±0·4	37/6
$C_9H_{20}O_2$	160·2586	Dibutyl formal	l		SB		m	1398·3 ±0·4	57/16
$C_{10}H_6O_2$	158·1581	1, 4-Naphthoquinone	c		SB		m	1101·63±0·44	56/15
$C_{10}H_{12}O_2$	164·2059	2, 4-Dimethylphenylacetic acid	c		SB	3	m	1231·9 ±1·5	36/11
$C_{10}H_{12}O_2$	164·2059	2, 3, 4-Trimethylbenzoic acid	c		SB	5	CO_2	1234·09±0·22	64/22
$C_{10}H_{12}O_2$	164·2059	2, 3, 5-Trimethylbenzoic acid	c		SB	5	CO_2	1233·60±0·20	64/22
$C_{10}H_{12}O_2$	164·2059	2, 3, 6-Trimethylbenzoic acid	c		SB	5	CO_2	1236·71±0·20	64/22
$C_{10}H_{12}O_2$	164·2059	2, 4, 5-Trimethylbenzoic acid	c		SB	5	CO_2	1231·93±0·30	64/22

4
Determinatio

1 kcal = 4·184 kJ

ΔH_r°		5 ΔH_f° (l or c) kcal/g.f.w.	6 Determination of ΔH_v			7 ΔH_f° (g) kcal/g.f.w.
	Remarks		Type kcal/g.f.w.	ΔH_v°	Ref.	
	BCE	$-118\cdot8 \pm0\cdot4$	E7	$12\cdot1 \pm0\cdot3$		$-106\cdot7 \pm0\cdot5$
	BCE	$-116\cdot3 \pm0\cdot6$	E7	$12\cdot8 \pm0\cdot3$		$-103\cdot5 \pm0\cdot7$
	BCE	$-120\cdot8 \pm0\cdot6$	E7	$12\cdot5 \pm0\cdot3$		$-108\cdot3 \pm0\cdot7$
	BCE	$-120\cdot2 \pm0\cdot6$	E7	$12\cdot3 \pm0\cdot3$		$-107\cdot9 \pm0\cdot7$
	BCE	$-118\cdot3 \pm0\cdot6$	E7	$13\cdot2 \pm0\cdot3$		$-105\cdot1 \pm0\cdot7$
	BCE	$-121\cdot1 \pm0\cdot6$	E7	$12\cdot8 \pm0\cdot3$		$-108\cdot3 \pm0\cdot7$
	BCE	$-122\cdot0 \pm0\cdot5$	E7	$12\cdot6 \pm0\cdot3$		$-109\cdot4 \pm0\cdot6$
An >0·9997		$-157\cdot20\pm0\cdot25$ $-158\cdot17\pm0\cdot26$	V3	$19\cdot7 \pm0\cdot1$	68/3	$-137\cdot50\pm0\cdot28$ $-138\cdot47\pm0\cdot29$
	Selected value	$-157\cdot68\pm0\cdot48$				$-137\cdot98\pm0\cdot50$
		$-141\cdot08\pm0\cdot23$	V3	$13\cdot6 \pm0\cdot4$	53/27	$-127\cdot48\pm0\cdot50$
	BCE	$-146\cdot6 \pm0\cdot4$	E7	$12\cdot7 \pm0\cdot3$		$-133\cdot9 \pm0\cdot5$
	BCE	$-148\cdot2 \pm0\cdot8$	E7	$12\cdot3 \pm0\cdot3$		$-135\cdot9 \pm0\cdot9$
	BCE	$-149\cdot2 \pm0\cdot4$	E7	$12\cdot2 \pm0\cdot3$		$-137\cdot0 \pm0\cdot5$
	E	$-131\cdot3 \pm0\cdot4$	E8	$11\cdot5 \pm0\cdot6$		$-119\cdot8 \pm0\cdot8$
		$-43\cdot83\pm0\cdot46$	S3 (sub.)	$17\cdot3 \pm0\cdot9$	56/15	$-26\cdot5 \pm1\cdot0$
	CE	$-118\cdot5 \pm1\cdot5$				
		$-116\cdot31\pm0\cdot25$				
		$-116\cdot80\pm0\cdot23$				
		$-113\cdot69\pm0\cdot23$				
		$-118\cdot47\pm0\cdot33$				

$C_aH_bO_2$

1 kcal = 4·184 kJ

1 Formula	2 g.f.w.	3 Name	State	Purity mol %	Type	No. of expts.	Detn. of reactn.	$-\Delta H_r^\circ$ kcal/g.f.w.	Re
								Determinat	
$C_{10}H_{12}O_2$	164·2059	2, 4, 6-Trimethylbenzoic acid	c		SB	5	CO_2	1236·19±0·22	64/
$C_{10}H_{12}O_2$	164·2059	3, 4, 5-Trimethylbenzoic acid	c		SB	5	CO_2	1230·68±0·22	64/
$C_{10}H_{12}O_2$	164·2059	Hinokitiol	c		SB	4	m	1269·0 ±0·9	56/
$C_{10}H_{12}O_2$	164·2059	5-Tetralin hydroperoxide	c		SB	3	m	1305·6 ±1·3	50/ 51/
$C_{10}H_{12}O_2$ $C_{10}H_{12}O_2$		5-Tetralin hydroperoxide 5-Tetralin hydroperoxide	c		SB	5	m	1307 ±2	56/
$C_{10}H_{18}O_2$	170·2538	9-trans- Decalin hydroperoxide	c		SB	3	m	1471·9 ±1·5	50/ 51/
$C_{10}H_{20}O_2$	172·2697	Decanoic acid	c	99·85	SB	5	m	1453·07±0·21	65/:
$C_{10}H_{20}O_2$	172·2697	Methyl nonanoate	l	99·75	SB	5	m	1476·38±0·36	65/2
$C_{10}H_{22}O_2$	174·2856	1, 10-Decanediol	c		SB	11	m	1526·24±0·60	62/:
$C_{11}H_{10}O_2$	174·2012	Cyclopentadiene benzoquinone	c		SB	4	m	1336·9 ±1·3	35/
$C_{11}H_{14}O_2$	178·2330	2, 3, 4, 5-Tetramethyl-benzoic acid	c		SB	5	CO_2	1389·82±0·31	64/2
$C_{11}H_{14}O_2$	178·2330	2, 3, 4, 6-Tetramethyl-benzoic acid	c		SB	5	CO_2	1391·43±0·24	64/2
$C_{11}H_{14}O_2$	178·2330	2, 3, 5, 6-Tetramethyl-benzoic acid	c		SB	5	CO_2	1391·81±0·25	64/2
$C_{11}H_{14}O_2$	178·2330	α-Methyltetralin hydroperoxide	c		SB	3	m	1475·2 ±4·4	51/
$C_{11}H_{22}O_2$	186·2968	Undecanoic acid	c	99·75	SB	6	m	1610·14±0·22	65/2
$C_{11}H_{22}O_2$	186·2968	Methyl decanoate	l	99·65	SB	5	m	1632·94±0·41	65/2
$C_{12}H_{10}O_2$	186·2123	1-Naphthyl acetate	c		SB	2	m	1401·30±0·32	48/.
$C_{12}H_{10}O_2$	186·2123	2-Naphthyl acetate	c		SB	3	m	1397·47±0·57	48/!

5. THERMOCHEMICAL DATA

1 kcal = 4·184 kJ

ΔH_r° Remarks	5 ΔH_f° (l or c) kcal/g.f.w.	6 Determination of ΔH_v			7 ΔH_f° (g) kcal/g.f.w.
		Type	ΔH_v° kcal/g.f.w.	Ref.	
	$-114·21\pm0·25$				
	$-119·72\pm0·25$				
	$-81·4\ \pm0·9$				
	$-44·8\ \pm1·3$				
	$-43\ \ \ \pm2$				
Selected value	$-44·0\ \pm1·0$				
	$-83·4\ \pm1·5$				
form: $\Delta H_m, 7·03\pm0·28$	$-170·59\pm0·24$	S3	$28·4\ \pm0·5$ (sub.)	68/3	$-151·19\pm0·56$
	$-147·28\pm0·38$	El	$15·1\ \pm0·5$	68/3	$-132·18\pm0·63$
	$-165·74\pm0·62$				
CE	$-39·2\ \pm1·3$				
	$-122·95\pm0·35$				
	$-121·34\pm0·28$				
	$-120·96\pm0·29$				
	$-37·6\ \pm4·4$				
C' form: $\Delta H_m, 6·16\pm0·25$	$-175·89\pm0·27$	S3	$29·0\ \pm0·3$ (sub.)	68/3	$-146·89\pm0·41$
	$-153·09\pm0·44$	V4	$16·5\ \pm0·5$	53/27	$-136·59\pm0·67$
CE	$-68·89\pm0·36$				
CE	$-72·72\pm0·60$				

$C_aH_bO_2$
1 kcal = 4·184 kJ

1 Formula	2 g.f.w.	3 Name	State	Purity mol %	Type	No. of expts.	Detn. of reactn.	$-\Delta H_r^\circ$ kcal/g.f.w.	Re
$C_{12}H_{16}O_2$	192·2601	Pentamethylbenzoic acid	c		SB	5	CO_2	1547·00±0·31	64/2
$C_{12}H_{24}O_2$	200·3239	Dodecanoic acid	c	99·7	SB	5	m	1763·26±0·22	65/2
$C_{12}H_{24}O_2$		Dodecanoic acid	c		SB	3	m	1774·3 ±1·8	44/9
$C_{12}H_{24}O_2$		Dodecanoic acid							
$C_{12}H_{24}O_2$	200·3239	Methyl undecanoate	l	99·7	SB	4	m	1789·41±0·31	65/2
$C_{12}H_{24}O_2$	200·3239	t-Butyl octanoate	l		SB	8	CO_2	1781·4 ±3·0	64/2
$C_{13}H_{10}O_2$	198·2235	Phenyl benzoate	c		SB	4	m	1506·5 ±0·5	67/1
$C_{13}H_{10}O_2$	198·2235	o-Phenylbenzoic acid	c		SB	3	m	1480·8 ±1·5	35/4
$C_{13}H_{26}O_2$	214·3510	Tridecanoic acid	c	99·7	SB	5	m	1917·97±0·31	65/2
$C_{13}H_{26}O_2$	214·3510	Methyl dodecanoate	l	99·7	SB	4	m	1945·12±0·38	65/2
$C_{14}H_8O_2$	208·2187	9, 10-Anthraquinone	c		SB		m	1540·38±0·75	56/1
$C_{14}H_8O_2$		9, 10-Anthraquinone	c		SB	5	m	1545·8 ±0·8	31/7
$C_{14}H_8O_2$		9, 10-Anthraquinone							
$C_{14}H_8O_2$	208·2187	9, 10-Phenanthraquinone	c		SB		m	1534·79±0·31	56/1
$C_{14}H_{10}O_2$	210·2346	Benzil	c		SB	11	m	1621·51±0·65	62/2
$C_{14}H_{10}O_2$		Benzil	c		SB	4	m	1615·7 ±0·8	54/1
$C_{14}H_{10}O_2$		Benzil							
$C_{14}H_{10}O_2$	210·2346	Anthracene transannular peroxide	c		SB	5	m	1678·1 ±0·8	52/9
$C_{14}H_{12}O_2$	212·2505	Benzoin	c		SB	11	m	1667·39±0·67	62/2
$C_{14}H_{28}O_2$	228·3781	Tetradecanoic acid	c	99·85	SB	7	m	2073·91±0·33	65/20
$C_{14}H_{28}O_2$		Tetradecanoic acid	c		SB	9	CO_2	2073·6 ±2·3	64/2
$C_{14}H_{28}O_2$		Tetradecanoic acid							
$C_{14}H_{28}O_2$	228·3781	Methyl tridecanoate	l	99·7	SB	5	m	2101·53±0·43	65/20
$C_{14}H_{28}O_2$	228·3781	t-Butyl decanoate	l		SB	7	CO_2	2093·2 ±1·9	64/24

1 kcal $= 4 \cdot 184$ kJ

ΔH_r°		6 Determination of ΔH_v			
		5 ΔH_f° (l or c) kcal/g.f.w.	Δ_v Type kcal/g.f.w.	Ref.	7 ΔH_f° (g) kcal/g.f.w.
Remarks					
		$-128 \cdot 13 \pm 0 \cdot 35$			
orm: $\Delta H_m, 8 \cdot 77 \pm 0 \cdot 35$		$-185 \cdot 13 \pm 0 \cdot 28$	S3 $28 \cdot 0 \ \pm 0 \cdot 7$ (sub.)	57/19	$-153 \cdot 43 \pm 0 \cdot 80$
		$-174 \cdot 1 \ \pm 1 \cdot 8$	S3 $31 \cdot 7 \ \pm 0 \cdot 4$ (sub.)	68/3	$-143 \cdot 4 \ \pm 2 \cdot 0$
	Selected value	$-185 \cdot 13 \pm 0 \cdot 28$	$31 \cdot 7 \ \pm 0 \cdot 4$ (sub.)		$-153 \cdot 43 \pm 0 \cdot 80$
		$-158 \cdot 98 \pm 0 \cdot 35$	E1 $17 \cdot 3 \ \pm 0 \cdot 5$	68/3	$-141 \cdot 68 \pm 0 \cdot 61$
		$-167 \cdot 0 \ \pm 3 \cdot 0$			
		$-57 \cdot 7 \ \pm 0 \cdot 5$	V4$_m$ $23 \cdot 0 \ \pm 0 \cdot 4$ (sub.)	47/15	$-34 \cdot 7 \ \pm 0 \cdot 8$
	BCE	$-83 \cdot 4 \ \pm 1 \cdot 5$			
form: $\Delta H_{tr} A' \to C, 2 \cdot 13 \pm 0 \cdot 09$; $\Delta H_m (C), 8 \cdot 17 \pm 0 \cdot 33$		$-192 \cdot 79 \pm 0 \cdot 36$	E1$_m$ $35 \cdot 0 \ \pm 0 \cdot 5$ (sub.)	68/3	$-157 \cdot 79 \pm 0 \cdot 62$
		$-165 \cdot 64 \pm 0 \cdot 42$	V3 $19 \cdot 3 \ \pm 0 \cdot 5$	56/41	$-146 \cdot 34 \pm 0 \cdot 66$
		$-49 \cdot 59 \pm 0 \cdot 77$	S3 $26 \cdot 8 \ \pm 1 \cdot 4$ (sub.)	56/15	$-22 \cdot 8 \ \pm 1 \cdot 6$
		$-44 \cdot 6 \ \pm 0 \cdot 8$			$-17 \cdot 8 \ \pm 1 \cdot 6$
	Selected value	$-49 \cdot 59 \pm 0 \cdot 77$			$-22 \cdot 8 \ \pm 1 \cdot 6$
		$-55 \cdot 18 \pm 0 \cdot 34$	S3 $21 \cdot 9 \ \pm 1 \cdot 1$ (sub.)	56/15	$-33 \cdot 3 \ \pm 1 \cdot 3$
		$-36 \cdot 78 \pm 0 \cdot 67$	S4 $23 \cdot 52 \pm 0 \cdot 27$ (sub.)	59/36	$-13 \cdot 26 \pm 0 \cdot 73$
		$-42 \cdot 6 \ \pm 0 \cdot 8$			$-19 \cdot 1 \ \pm 0 \cdot 9$
	Selected value	$-36 \cdot 78 \pm 0 \cdot 67$			$-13 \cdot 26 \pm 0 \cdot 73$
		$+19 \cdot 8 \ \pm 0 \cdot 8$			
		$-59 \cdot 21 \pm 0 \cdot 69$			
form: $\Delta H_m, 10 \cdot 68 \pm 0 \cdot 40$		$-199 \cdot 21 \pm 0 \cdot 36$	S4 $33 \cdot 4 \ \pm 0 \cdot 9$ (sub.)	61/ 55	$-165 \cdot 8 \ \pm 1 \cdot 0$
form		$-199 \cdot 5 \ \pm 2 \cdot 3$			$-166 \cdot 1 \ \pm 2 \cdot 5$
	Selected value	$-199 \cdot 21 \pm 0 \cdot 36$			$-165 \cdot 8 \ \pm 1 \cdot 0$
		$-171 \cdot 59 \pm 0 \cdot 45$	V3 $19 \cdot 5 \ \pm 1 \cdot 0$	68/3	$-142 \cdot 1 \ \pm 1 \cdot 1$
		$-179 \cdot 9 \ \pm 1 \cdot 9$			

$C_aH_bO_2$ 1 kcal = 4·184 kJ

1 Formula	2 g.f.w	3 Name	State	Purity mol %	Type	No. of expts.	Detn. of reactn.	$-\Delta H_r^\circ$ kcal/g.f.w.	Re
$C_{15}H_{12}O_2$	224·2617	Dibenzoylmethane	c		SB	11	m	1767·55±0·71	62/
$C_{15}H_{12}O_2$		Dibenzoylmethane	c	2s	SB	6	CO_2	1766·90±0·40	65/
$C_{15}H_{12}O_2$		Dibenzoylmethane							
$C_{15}H_{16}O_2$	228·2936	2, 2′ bis-(4-hydroxy-phenyl) propane	c		SB		m	1869·2 ±0·4	48/
$C_{15}H_{30}O_2$	242·4052	Pentadecanoic acid	c	99·75	SB	5	m	2229·53±0·36	65/
$C_{15}H_{30}O_2$	242·4052	Methyl tetradecanoate	l	99·85	SB	6	m	2257·69±0·50	65/
$C_{16}H_{12}O_2$	236·2728	Dibenzoylethylene	c		SB	9	m	1887·30±0·56	54/
$C_{16}H_{14}O_2$	238·2888	Dibenzoylethane	c		SB	9	m	1921·94±0·35	54/
$C_{16}H_{32}O_2$	256·4322	Hexadecanoic acid	c	99·8	SB	12	m	2384·79±0·38	65/
$C_{16}H_{32}O_2$		Hexadecanoic acid	c		SB	6	CO_2	2384·7 ±2·1	64/
$C_{16}H_{32}O_2$		Hexadecanoic acid	c	99·9	SB	5	m	2386·60±0·46	64/
$C_{16}H_{32}O_2$		Hexadecanoic acid							
$C_{16}H_{32}O_2$	256·4322	Methyl pentadecanoate	l	99·75	SB	5	m	2413·59±0·36	65/
$C_{16}H_{32}O_2$	256·4322	t-Butyl dodecanoate	l		SB	7	CO_2	2402·4 ±2·9	64/
$C_{17}H_{12}O_2$	248·2840	2-Naphthyl benzoate	c		SB		m	1964·6 ±0·8	48/
$C_{17}H_{16}O_2$	252·3159	1-Phenyl-1-ethoxy-2-benzoylethylene	c		SB	8	CO_2	2099·89±0·40	65/
$C_{17}H_{34}O_2$	270·4593	Heptadecanoic acid	c	99·7	SB	10	m	2539·29±0·41	65/
$C_{18}H_{10}O_2$	258·2792	5, 12-Tetracenequinone	c		SB		m	2000·37±0·80	56/
$C_{18}H_{10}O_2$	258·2792	1, 2-Benzanthra-9, 10-quinone	c		SB		m	1979·08±0·80	56/
$C_{18}H_{36}O_2$	284·4864	Octadecanoic acid	c	99·8	SB	11	m	2696·00±0·45	65/
$C_{18}H_{36}O_2$		Octadecanoic acid	c		SB	7	CO_2	2693·9 ±3·2	64/
$C_{18}H_{36}O_2$		Octadecanoic acid	c	99·8	SB	7	m	2696·21±0·51	64/
$C_{18}H_{36}O_2$		Octadecanoic acid	c		SB	7	m	2704·6 ±2·7	52/
$C_{18}H_{36}O_2$		Octadecanoic acid							

Determina

$$1 \text{ kcal} = 4\cdot184 \text{ kJ}$$

ΔH_f° Remarks	5 ΔH_f° (l or c) kcal/g.f.w.	6 Determination of ΔH_v — ΔH_v° Type kcal/g.f.w.	Ref.	7 ΔH_f° (g) kcal/g.f.w.
An $1\cdot0004\pm0\cdot0005$	$-53\cdot11\pm0\cdot73$ $-53\cdot76\pm0\cdot43$ Selected value $-53\cdot60\pm0\cdot40$			
An $0\cdot9998$	$-88\cdot1\ \pm0\cdot5$			
form: $\Delta H_{tr}\,B\to C,1\cdot74\pm0\cdot07$; $\Delta H_m\,(C)\,10\cdot21\pm0\cdot40$	$-205\cdot96\pm0\cdot41$	El$_m$ $38\cdot9\ \pm1\cdot0$ (sub.)	68/3	$-167\cdot1\ \pm1\cdot1$
	$-177\cdot80\pm0\cdot54$	V3 $21\cdot3\ \pm1\cdot0$	56/41	$-156\cdot5\ \pm1\cdot1$
	$-27\cdot41\pm0\cdot60$			
	$-61\cdot08\pm0\cdot40$			
form: ΔH_m, $12\cdot76\pm0\cdot48$ form An$>0\cdot9997$; $\Delta H_m\,10\cdot30\pm0\cdot10$	$-213\cdot07\pm0\cdot44$ $-213\cdot2\ \pm2\cdot1$ $-211\cdot26\pm0\cdot51$ Selected value $-213\cdot07\pm0\cdot44$ (C form)	S4 $36\cdot9\ \pm1\cdot0$ (sub.)	61/55	$-176\cdot2\ \pm1\cdot1$ $-176\cdot3\ \pm2\cdot3$ $-176\cdot2\ \pm1\cdot1$
	$-184\cdot27\pm0\cdot43$	V3 $21\cdot0\ \pm0\cdot3$	68/3	$-163\cdot3\ \pm0\cdot5$
	$-195\cdot5\ \pm2\cdot9$			
CE	$-44\cdot2\ \pm0\cdot8$			
An $0\cdot9993\pm0\cdot0007$	$-45\cdot50\pm0\cdot45$			
form: $\Delta H_{tr}\,B'\to C,1\cdot74\pm0\cdot07$; $\Delta H_m\,(C)\,12\cdot31\pm0\cdot48$	$-220\cdot93\pm0\cdot48$			
	$-34\cdot12\pm0\cdot82$	S3 $26\cdot0\ \pm1\cdot3$ (sub.)	56/15	$-8\cdot1\ \pm1\cdot5$
	$-55\cdot41\pm0\cdot82$	S3 $19\cdot8\ \pm1\cdot0$ (sub.)	56/15	$-35\cdot6\ \pm1\cdot3$
form: ΔH_m, $15\cdot06\pm0\cdot60$ form /An $>0\cdot9997$	$-226\cdot59\pm0\cdot52$ $-228\cdot7\ \pm3\cdot2$ $-226\cdot38\pm0\cdot57$ $-218\cdot0\ \pm2\cdot7$ Selected value $-226\cdot49\pm0\cdot46$	S4 $39\cdot8\ \pm1\cdot0$ (sub.)	61/55	$-186\cdot8\ \pm1\cdot2$ $-188\cdot9\ \pm3\cdot4$ $-186\cdot6\ \pm1\cdot2$ $-178\cdot2\ \pm2\cdot9$ $-186\cdot7\ \pm1\cdot2$

$C_aH_bO_2$ 1 kcal $= 4 \cdot 184$ kJ

1 Formula	2 g.f.w.	3 Name	State	Purity mol %	Type	No. of expts.	Detn. of reactn.	$-\Delta H_r^\circ$ kcal/g.f.w.	Re
$C_{18}H_{36}O_2$	284·4864	t-Butyl tetradecanoate	l		SB	4	CO_2	2716·4 $\pm 3\cdot 0$	64/
$C_{19}H_{36}O_2$	296·4976	Methyl cis-9-octadecenoate	l		SB	6	m	2841·1 $\pm 2\cdot 8$	37/
$C_{19}H_{36}O_2$	296·4976	Methyl trans-9-octadecenoate	l		SB	7	m	2840·5 $\pm 2\cdot 8$	37/
$C_{19}H_{38}O_2$	298·5135	Nonadecanoic acid	c	99·65	SB	9	m	2849·76 $\pm 0\cdot 55$	65/
$C_{20}H_{38}O_2$	310·5247	Ethyl cis-9-octadecenoate	l		SB	4	m	2993·6 $\pm 2\cdot 9$	37/
$C_{20}H_{38}O_2$	310·5247	Ethyl trans-9-octadecenoate	l		SB	4	m	2994·2 $\pm 2\cdot 9$	37/
$C_{20}H_{40}O_2$	312·5406	Eicosanoic acid	c	99·6	SB	7	m	3005·46 $\pm 0\cdot 36$	65/
$C_{21}H_{40}O_2$	324·5518	n-Propyl cis-9-octadecenoate	l		SB	4	m	3152·0 $\pm 3\cdot 1$	37/
$C_{21}H_{40}O_2$	324·5518	n-Propyl trans-9-octadecenoate	l		SB	3	m	3150·5 $\pm 3\cdot 1$	37/
$C_{22}H_{12}O_2$	308·3397	6, 13-Pentacenequinone	c		SB		m	2461·6 $\pm 1\cdot 5$	56/
$C_{22}H_{42}O_2$	338·5788	trans-13-Docosenoic acid	c		SB	11	m	3274·13 $\pm 0\cdot 65$	36/
$C_{22}H_{42}O_2$	338·5788	n-Butyl cis-9-octadecenoate	l		SB	5	m	3308·5 $\pm 3\cdot 3$	37/
$C_{22}H_{42}O_2$	338·5788	n-Butyl trans-9-octadecenoate	l		SB	4	m	3306·9 $\pm 3\cdot 3$	37/
$C_{23}H_{44}O_2$	352·6059	n-Pentyl trans-9-octadecenoate	l		SB		m	3462·8 $\pm 3\cdot 5$	37/
$C_{24}H_{12}O_2$	332·3620	3, 4: 9, 10-Dibenzpyrene-5, 8-quinone	c		SB		m	2607·1 $\pm 1\cdot 6$	56/1
$C_{38}H_{30}O_2$	518·6616	Di-triphenylmethyl peroxide	c		R	2	m	45·5 $\pm 0\cdot 5$	36/

1 kcal = 4·184 kJ

H_r°			6 Determination of ΔH_v			
Remarks		5 ΔH_f° (l or c) kcal/g.f.w.	ΔH_v° Type kcal/g.f.w.		Ref.	7 ΔH_f° (g) kcal/g.f.w.
		$-206\cdot2 \pm 3\cdot0$				
	CE	$-175\cdot5 \pm 2\cdot8$	V2 $20\cdot2 \pm[1\cdot0]$		52/34	$-155\cdot3 \pm 3\cdot0$
	CE	$-176\cdot1 \pm 2\cdot8$				
orm:$\Delta H_{tr} B' \to C, 2\cdot36\pm0\cdot10;$ ΔH_m (C) $13\cdot81\pm0\cdot55$		$-235\cdot19\pm0\cdot61$	El$_m$ $47\cdot5 \pm1\cdot2$ (sub.)		68/3	$-187\cdot7 \pm1\cdot4$
	CE	$-185\cdot4 \pm2\cdot9$				
	CE	$-184\cdot8 \pm2\cdot9$				
orm: $\Delta H_m, 17\cdot20\pm0\cdot68$		$-241\cdot86\pm0\cdot46$	S4 $47\cdot7 \pm1\cdot8$ (sub.)		61/55	$-194\cdot2 \pm1\cdot9$
	CE	$-189\cdot4 \pm3\cdot1$				
	CE	$-190\cdot9 \pm3\cdot1$				
		$-17\cdot4 \pm1\cdot6$	S3 $27\cdot8 \pm1\cdot4$ (sub.)		56/15	$+10\cdot4 \pm2\cdot2$
	BCE	$-229\cdot61\pm0\cdot72$				
	CE	$-195\cdot2 \pm3\cdot3$				
	CE	$-196\cdot8 \pm3\cdot3$				
	CE	$-203\cdot3 \pm3\cdot5$				
		$-60\cdot0 \pm1\cdot7$	S3 $26\cdot9 \pm1\cdot3$ (sub.)		56/15	$-33\cdot1 \pm2\cdot2$
$_5C_2(c)\{+122\cdot3\pm0\cdot8\}+O_2(g)$ $=C_{38}H_{30}O_2(c)$		$+76\cdot8 \pm1\cdot5$				

$C_aH_bO_2$ 1 kcal $= 4 \cdot 184$ kJ

									Determina
1 Formula	2 g.f.w.	3 Name	State	Purity mol %	Type	No. of expts.	Detn. of reactn.	$-\Delta H_r^\circ$ kcal/g.f.w.	
$C_{42}H_{28}O_2$	$564 \cdot 6903$	5, 12-Dihydro-5, 12-oxo; 11, 6-dihydro-, 11, 6- oxo-5, 6, 11, 12-tetra- phenyltetracene	c		SB	3	m	$4973 \cdot 1 \pm 5 \cdot 0$	3
$C_{42}H_{28}O_2$	$564 \cdot 6903$	5, 12-Dihydro-5, 12- peroxy-5, 6, 11, 12- tetraphenyltetracene	c		SB	5	m	$5030 \cdot 0 \pm 5 \cdot 0$	3
$C_{42}H_{28}O_2$	$564 \cdot 6903$	Pseudo-5, 6, 11, 12- tetraphenyltetracene peroxide	c		SB	3	m	$5051 \cdot 8 \pm 5 \cdot 1$	3

$1 \text{ kcal} = 4 \cdot 184 \text{ kJ}$

H_r°	Remarks	5 ΔH_f° (l or c) kcal/g.f.w.	6 Determination of ΔH_v			7 ΔH_f° (g) kcal/g.f.w.
			Type	ΔH_v° kcal/g.f.w.	Ref.	
		BCE $+66 \cdot 5 \pm 5 \cdot 0$				
		BCE $+123 \cdot 4 \pm 5 \cdot 0$				
		BCE $+145 \cdot 2 \pm 5 \cdot 1$				

$C_aH_bO_3$

1 kcal = 4·184 kJ

1 Formula	2 g.f.w.	3 Name	State	Purity mol %	Type	No. of expts.	Detn. of reactn.	$-\Delta H_c^{\circ}$ kcal/g.f.w.	R
$C_3H_4O_3$	88·0635	Ethylene carbonate	c		SB	5	CO_2	279·95±0·30	61
$C_3H_6O_3$	90·0795	α-Trioxan	c		SB	4	m	362·30±0·36	42
$C_3H_6O_3$		α-Trioxan	c		SB		m	356·7 ±0·9	43
$C_3H_6O_3$		α-Trioxan							
$C_3H_6O_3$	90·0795	L-Lactic acid	c		SB	5	m	321·22±0·22	59
$C_3H_8O_3$	92·0954	Glycerol	l		SB	3	m	395·63±0·24	46
$C_4H_2O_3$	98·0587	Maleic anhydride	c		SB	5	m	332·30±0·17	50
$C_4H_2O_3$		Maleic anhydride	c	99·85	SB	5	m	332·09±0·16	64
$C_4H_2O_3$		Maleic anhydride	c		R	2	m	8·33±0·22	42
$C_4H_2O_3$		Maleic anhydride							
$C_4H_4O_3$	100·0747	Succinic anhydride	c		SB	4	m	369·00±0·15	33
$C_4H_4O_3$		Succinic anhydride	c		R	2	m	11·20±0·02	42
$C_4H_4O_3$		Succinic anhydride							
$C_4H_6O_3$	102·0906	Acetic anhydride	l		R		m	13·96±0·10	42
$C_4H_6O_3$		Acetic anhydride	l	99·9 glc	R	6	m	14·00±0·10	62
$C_4H_6O_3$		Acetic anhydride							
$C_4H_{10}O_3$	106·1225	Diethylene glycol	l		SB	2	m	567·56±0·57	37
$C_5H_4O_3$	112·0858	Furoic acid	c		SB	5	m	487·77±0·25	50
$C_5H_4O_3$		Furoic acid	c		SB	2	m	486·6 ±0·5	29
$C_5H_4O_3$		Furoic acid							
$C_5H_4O_3$	112·0858	Methylmaleic anhydride	l		R	2	m	8·13±0·05	42
$C_5H_6O_3$	114·1018	Methylsuccinic anhydride	c		SB	4	m	527·02±0·27	33
$C_5H_6O_3$		Methylsuccinic anhydride	l		SB	4	m	529·11±0·27	33
$C_5H_6O_3$		Methylsuccinic anhydride	l		R	2	m	13·11±0·04	42
$C_5H_6O_3$		Methylsuccinic anhydride							
$C_5H_{12}O_3$	120·1496	1, 1, 1-Tris (hydroxy-methyl) ethane	c		SB	4	m	702·19±0·70	54

Determina

1 kcal = 4·184 kJ

ΔH_r°		5 ΔH_f° (l or c) kcal/g.f.w.	6 Determination of ΔH_v			7 ΔH_f° (g) kcal/g.f.w.
Remarks			Type	ΔH_v° kcal/g.f.w.	Ref.	
An 0·99959±0·00047; ΔH_m, 2·41±0·20 (62/24)		−138·83±0·31	V3$_m$	17·5 ±0·6 (sub.)	58/41	−121·3 ±0·7
	CE	−124·80±0·38 −130·4 ±0·9	V4$_{me}$	11·6 ±0·6 (sub.)	43/1	−113·2 ±0·7 −118·8 ±1·1
	Selected value	−124·80±0·38				−113·2 ±0·7
		−165·88±0·23				
		−159·78±0·25	V3	20·5±[0·50]	62/73	−139·3 ±0·58
		−112·22±0·18 −112·43±0·17	V4$_{me}$	17·1 ±1·2 (sub.)	49/33	−95·1 ±1·2 −95·3 ±1·2
H$_2$O$_3$(c)+H$_2$O(l) =Maleic acid (c){−188·73±0·15}	E	−112·08±0·17				−95·0 ±1·2
	Selected value	−112·32±0·15				−95·2 ±1·2
	BCE	−143·83±0·16				
H$_4$O$_3$(c)+H$_2$O(l) =Succinic acid (c){−224·86±0·10}	E	−145·34±0·12				
	Selected value	−145·34±0·12				
$_2$O(l)+H$_2$O(l) =2HOAc(l){−115·75±0·07}	E	−149·22±0·14	V4	12·4 ±0·30	54/28	−137·08±0·35
$_2$O(l)+H$_2$O(l) =2HOAc(l){−115·75±0·07}		−149·18±0·14				−137·04±0·35
	Selected value	−149·20±0·10				−137·06±0·33
	BCE	−150·22±0·58	V4	13·7 ±1·4	37/10	−136·5 ±1·6
		−119·12±0·26	S4	25·92±0·50 (sub.)	53/30	−93·20±0·57 −94·4 ±0·7
	ABCE	−120·3 ±0·5				
	Selected value	−119·12±0·26				−93·20±0·57
H$_4$O$_3$(l)+H$_2$O(l) =Citraconic acid (c){−197·04±0·35}	E	−120·59±0·38	V4	13·71±[0·50]	47/15	−106·88±[0·63]
	BCE	−148·18±0·28				
	BCE	−146·09±0·28				
H$_6$O$_3$(l)+H$_2$O(l) =Methylsuccinic acid (c) {−229·02±0·26}	E	−147·59±0·28				
	Selected value (liquid)	−147·59±0·28				
	E	−177·96±0·71				

THERMOCHEMISTRY OF ORGANIC AND ORGANOMETALLIC COMPOUNDS

$C_aH_bO_3$ 1 kcal = 4·184 kJ

1 Formula	2 g.f.w.	3 Name	State	Purity mol %	Type	No. of expts.	Detn. of reactn.	$-\Delta H_r^\circ$ kcal/g.f.w.	R
$C_6H_6O_3$	126·1129	Dimethylmaleic anhydride	c		SB		m	630·3 ±0·6	33,
$C_6H_8O_3$	128·1289	1, 1-Dimethylsuccinic anhydride	l		R	2	m	13·48±0·05	42
$C_6H_8O_3$		1, 1-Dimethylsuccinic anhydride	c		SB	4	m	681·88±0·34	33,
$C_6H_8O_3$	128·1289	cis-1, 2-Dimethylsuccinic anhydride	c		SB	4	m	680·31±0·34	33
$C_6H_8O_3$	128·1289	trans-1, 2-Dimethyl-succinic anhydride	c		R	2	m	6·91±0·05	42
$C_6H_8O_3$		trans-1, 2-Dimethyl-succinic anhydride	c		SB	7	m	678·53±0·34	33,
$C_6H_8O_3$		trans-1, 2-Dimethyl-succinic anhydride							
$C_6H_8O_3$	128·1289	Ethylsuccinic anhydride	l		SB	4	m	684·03±0·34	33,
$C_6H_{10}O_3$	130·1448	Propionic anhydride	l		R	2	m	13·52±0·06	42,
$C_6H_{12}O_3$	132·1607	Paraldehyde	l		SB	3	m	810·0 ±0·3	59/
$C_6H_{12}O_3$		Paraldehyde	l		SB	3	m	810·2 ±0·5	55/
$C_6H_{12}O_3$		Paraldehyde							
$C_6H_{12}O_3$	132·1607	2, 3-Butanediol mono-acetate	l		E		an	−0·44±0·50	45/
$C_6H_{14}O_3$	134·1767	1, 1, 1-Tris (hydroxy-methyl) propane	c		SB	6	m	863·03±0·86	53/
$C_7H_6O_3$	138·1241	Perbenzoic acid	c		R	4	m	75·6 ±3·0	54/
$C_7H_6O_3$	138·1241	Salicylic acid	c	3s	SB	15	m	722·30±0·21	35/
$C_7H_6O_3$		Salicylic acid	c		SB	3	m	722·49±0·72	37/1
$C_7H_6O_3$		Salicylic acid	c		SB	5	m	721·2 ±0·7	32/
$C_7H_6O_3$		Salicylic acid	c		SB	54	m	722·32±0·22	31/
$C_7H_6O_3$		Salicylic acid	c		SB	29	m	722·55±0·35	31/
$C_7H_6O_3$		Salicylic acid							
$C_7H_6O_3$	138·1241	Furylacrylic acid	c		SB	2	m	753·6 ±1·5	29/

$$1 \text{ kcal} = 4 \cdot 184 \text{ kJ}$$

H_r° Remarks	5 ΔH_f° (l or c) kcal/g.f.w.	6 Determination of ΔH_v			7 ΔH_f° (g) kcal/g.f.w.
		Type	ΔH_v° kcal/g.f.w.	Ref.	
	BCE $-139 \cdot 0 \pm 0 \cdot 6$				
$_8O_3(l)+H_2O(l)$ =1,1-Dimethylsuccinic acid (c) $\{-236 \cdot 08 \pm 0 \cdot 34\}$	E $-154 \cdot 28 \pm 0 \cdot 36$	V4	$15 \cdot 25 \pm [0 \cdot 50]$	47/15	$-139 \cdot 03 \pm [0 \cdot 63]$
	BCE $-155 \cdot 69 \pm 0 \cdot 35$				
	BCE $-157 \cdot 26 \pm 0 \cdot 35$				
$_8O_3(c)+H_2O(l)$ =1,2-Dimethylsuccinic acid (c) $\{-235 \cdot 13 \pm 0 \cdot 35\}$	E $-159 \cdot 90 \pm 0 \cdot 37$				
	BCE $-159 \cdot 04 \pm 0 \cdot 37$				
	Selected value $-159 \cdot 47 \pm 0 \cdot 45$				
	BCE $-153 \cdot 54 \pm 0 \cdot 35$				
$H_5CO)_2 O(l)+H_2O(l)$ =$2C_2H_5CO_2H(l)\{-122 \cdot 07 \pm 0 \cdot 08\}$	E $-162 \cdot 30 \pm 0 \cdot 18$	V4	$12 \cdot 56 \pm [0 \cdot 50]$	47/15	$-149 \cdot 74 \pm [0 \cdot 58]$
	$-164 \cdot 2 \pm 0 \cdot 3$	V5	$9 \cdot 9 \pm 0 \cdot 1$	59/20	$-154 \cdot 3 \pm 0 \cdot 4$
	$-164 \cdot 0 \pm 0 \cdot 5$				$-154 \cdot 1 \pm 0 \cdot 5$
	Selected value $-164 \cdot 1 \pm 0 \cdot 3$				$-154 \cdot 2 \pm 0 \cdot 4$
$_{12}O_3(l)+HOAc(l)\{-115 \cdot 75\}$ $\rightleftharpoons [CH_3CH(OAc)]_2$ $\{-218 \cdot 96 \pm 0 \cdot 80\}+H_2O(l)$	$-171 \cdot 97 \pm 0 \cdot 95$				
	$-179 \cdot 48 \pm 0 \cdot 87$				
$H_5CO_3H(c)+C_6H_5CHO(l)\{-20 \cdot 6 \pm 1 \cdot 7\}$ =$2C_6H_5CO_2H(c)\{-91 \cdot 99 \pm 0 \cdot 05\}$	$-87 \cdot 8 \pm 4 \cdot 0$				
	C $-141 \cdot 00 \pm 0 \cdot 22$	S3	$22 \cdot 74 \pm 0 \cdot 01$ (sub.)	54/27	$-118 \cdot 26 \pm 0 \cdot 25$
	CE $-140 \cdot 81 \pm 0 \cdot 73$				$-118 \cdot 07 \pm 0 \cdot 75$
	BCE $-142 \cdot 1 \pm 0 \cdot 7$				$-119 \cdot 4 \pm 0 \cdot 7$
	BCE $-140 \cdot 98 \pm 0 \cdot 23$				$-118 \cdot 24 \pm 0 \cdot 26$
	CE $-140 \cdot 75 \pm 0 \cdot 36$				$-118 \cdot 01 \pm 0 \cdot 38$
	Selected value $-141 \cdot 00 \pm 0 \cdot 15$				$-118 \cdot 26 \pm 0 \cdot 18$
	ABCE $-109 \cdot 7 \pm 1 \cdot 5$				

$C_aH_bO_3$

1 kcal = 4·184 kJ

1 Formula	2 g.f.w.	3 Name	State	Purity mol %	Type	No. of expts.	Detn. of reactn.	$-\Delta H_r^\circ$ kcal/g.f.w.	R
$C_7H_{10}O_3$	142·1560	Trimethylsuccinic anhydride	c		SB	6	m	835·40±0·42	33
$C_8H_4O_3$	148·1193	Phthalic anhydride	c 3s		SB	10	m	779·08±0·45	39, 50
$C_8H_8O_3$	152·1512	3-Hydroxy-4-methoxy-benzaldehyde	c		SB	4	m	917·3 ±1·4	40
$C_8H_8O_3$	152·1512	4-Acetylresorcinol	c		SB	5	m	888·6 ±0·9	37
$C_8H_8O_3$	152·1512	DL-Mandelic acid	c		SB		m	887·2 ±0·2	63,
$C_8H_8O_3$	152·1512	L-Mandelic acid	c		SB		m	886·9 ±0·2	63
$C_8H_{12}O_3$	156·1830	Tetramethylsuccinic anhydride	c		R	2	m	8·28±0·05	42,
$C_8H_{12}O_3$		Tetramethylsuccinic anhydride	c		SB	4	m	991·93±0·50	33
$C_8H_{12}O_3$		Tetramethylsuccinic anhydride							
$C_8H_{12}O_3$	156·1830	1, 1-Diethylsuccinic anhydride	c		SB	4	m	997·68±0·50	33,
$C_8H_{12}O_3$	156·1830	cis-1, 2-Diethylsuccinic anhydride	c		SB	4	m	996·58±0·50	33,
$C_8H_{12}O_3$	156·1830	trans-1, 2-Diethylsuccinic anhydride	l		SB	4	m	994·99±0·50	33,
$C_8H_{14}O_3$	158·1990	Ethyl α-ethylacetoacetate	l		SB	4	m	1059·35±0·50	35/
$C_8H_{14}O_3$	158·1990	Ethyl β-ethoxycrotonate	l		SB	3	m	1075·98±0·50	35/
$C_{10}H_8O_3$	176·1735	Phenylsuccinic anhydride	c		SB	4	m	1093·90±0·51	33/
$C_{10}H_{16}O_3$	184·2372	Triethylsuccinic anhydride	c		SB	4	m	1308·83±0·65	33/
$C_{10}H_{16}O_3$	184·2372	cis-Pinenonic racemic acid	c		SB	3	m	1309·9 ±1·6	36/:
$C_{10}H_{16}O_3$	184·2372	6-Oxoheptanoic acid γ-lactone	c		SB	3	m	1304·0 ±1·6	36/

Determina

1 kcal $= 4 \cdot 184$ kJ

ΔH_r° Remarks		5 ΔH_f° (l or c) kcal/g.f.w.	6 Determination of ΔH_ι			7 ΔH_f° (g) kcal/g.f.w.
			Type	ΔH_v° kcal/g.f.w.	Ref.	
	BCE	$-164 \cdot 53 \pm 0 \cdot 43$	$V4_{me}$	$17 \cdot 7 \pm 1 \cdot 0$ (sub.)	54/28	$-146 \cdot 8 \pm 1 \cdot 2$
		$-109 \cdot 96 \pm 0 \cdot 46$	S3	$21 \cdot 19 \pm 0 \cdot 07$ (sub.)	46/11	$-88 \cdot 77 \pm 0 \cdot 48$
	BCE	$-108 \cdot 4 \pm 1 \cdot 4$				
		$-137 \cdot 1 \pm 0 \cdot 9$				
		$-138 \cdot 5 \pm 0 \cdot 3$				
		$-138 \cdot 8 \pm 0 \cdot 3$				
$H_{12}O_3(c) + H_2O(l)$ = Tetramethylsuccinic acid (c) $\{-241 \cdot 98 \pm 0 \cdot 52\}$	**E**	$-165 \cdot 38 \pm 0 \cdot 54$				
	BCE	$-170 \cdot 37 \pm 0 \cdot 51$				
Selected value		$-165 \cdot 38 \pm 0 \cdot 54$				
	BCE	$-164 \cdot 62 \pm 0 \cdot 51$				
	BCE	$-165 \cdot 72 \pm 0 \cdot 51$				
	BCE	$-167 \cdot 31 \pm 0 \cdot 51$				
	E	$-171 \cdot 26 \pm 0 \cdot 51$	V4	$13 \cdot 59 \pm [0 \cdot 50]$	47/15	$-157 \cdot 67 \pm 0 \cdot 71$
	E	$-154 \cdot 63 \pm 0 \cdot 51$				
	BCE	$-119 \cdot 87 \pm 0 \cdot 52$				
	BCE	$-178 \cdot 20 \pm 0 \cdot 67$				
	CE	$-177 \cdot 1 \pm 1 \cdot 6$				
	CE	$-183 \cdot 0 \pm 1 \cdot 6$				

$C_aH_bO_3$

1 kcal = 4·184 kJ

1 Formula	2 g.f.w.	3 Name	State	Purity mol %	Type	No. of expts.	Detn. of reactn.	$-\Delta H_r^\circ$ kcal/g.f.w.	R
$C_{10}H_{18}O_3$	186·2532	Trimethylacetic anhydride	l		R	3	m	15·12±0·09	42
$C_{11}H_8O_3$	188·1846	3-Hydroxy-2-naphthoic acid	c		SB		m	1176·9 ±0·2	56/
$C_{12}H_{20}O_3$	212·2914	Tetraethylsuccinic anhydride	c		SB	5	m	1620·70±0·81	33
$C_{12}H_{24}O_3$	216·3233	Peroxydodecanoic acid	c		SB	6	CO_2	1785·8 ±2·0	64/
$C_{13}H_{10}O_3$	214·2229	Phenyl salicylate	c		SB	6	m	1459·9 ±1·1	31
$C_{13}H_{10}O_3$	214·2229	Diphenyl carbonate	c	99·9	SB	7	m	1468·36±0·45	58/
$C_{14}H_{10}O_3$	226·2340	Benzoic anhydride	c		SB	2	m	1555·3 ±1·6	50/ 51/
$C_{14}H_{12}O_3$	228·2499	trans-Stilbene ozonide	c		SB	3	m	1697 ±3	57/
$C_{14}H_{28}O_3$	244·3775	Peroxytetradecanoic acid	c		SB	7	CO_2	2093·9 ±2·3	64/
$C_{14}H_{28}O_3$	244·3775	t-Butyl peroxydecanoate	l		SB	7	CO_2	2108·6 ±2·1	64/
$C_{15}H_{18}O_3$	246·3089	Santonin	c		SB		m	1884·40±0·54	64/
$C_{15}H_{18}O_3$	246·3089	β-Santonin	c		SB		m	1885·23±0·43	64/
$C_{15}H_{18}O_3$	246·3089	6-α(H)-Santonin	c		SB		m	1885·80±0·36	64/
$C_{15}H_{18}O_3$	246·3089	6, 11-α(H)-Santonin	c		SB		m	1881·47±0·56	64/
$C_{16}H_{12}O_3$	252·2722	trans-1, 2-Diphenyl-succinic anhydride	c		SB	4	m	1814·35±0·91	33
$C_{16}H_{14}O_3$	254·2882	o-Toluic anhydride	c		SB	3	m	1855·5 ±1·9	50/ 51/
$C_{16}H_{14}O_3$	254·2882	p-Toluic anhydride	c		SB	2	m	1858·5 ±1·9	50/ 51/
$C_{16}H_{20}O_3$	260·3360	(−)α-Desmotropo-santonin methyl ether	c		SB		m	2031·24±0·52	64/
$C_{16}H_{20}O_3$	260·3360	(+)β-Desmotropo-santonin methyl ether	c		SB		m	2028·27±0·51	64/

Determina†

$$1 \text{ kcal} = 4 \cdot 184 \text{ kJ}$$

ΔH_r° Remarks		5 ΔH_f° (l or c) kcal/g.f.w.	6 Determination of ΔH_v			7 ΔH_f° (g) kcal/g.f.w.
			Type	ΔH_v° kcal/g.f.w.	Ref.	
$_3$CCO)$_2$O(l)+H$_2$O(l)		$-186 \cdot 4 \ \pm 1 \cdot 5$				
$=2$Me$_3$CCO$_2$H(l)$\{-134 \cdot 9 \pm 1 \cdot 4\}$	C	$-130 \cdot 9 \ \pm 0 \cdot 3$				
	BCE	$-191 \cdot 06 \pm 0 \cdot 82$				
		$-162 \cdot 6 \ \pm 2 \cdot 0$				
	CE	$-104 \cdot 3 \ \pm 1 \cdot 1$	V4$_{me}$	$22 \cdot 0 \ \pm 1 \cdot 0$ (sub.)	47/15	$-82 \cdot 3 \ \pm 1 \cdot 5$
		$-95 \cdot 88 \pm 0 \cdot 47$				
		$-103 \cdot 0 \ \pm 1 \cdot 7$	V4$_{me}$	$23 \cdot 1 \ \pm 1 \cdot 0$ (sub.)	47/15	$-79 \cdot 9 \ \pm 2 \cdot 0$
		$-30 \quad \pm 3$				
		$-179 \cdot 2 \ \pm 2 \cdot 3$				
		$-164 \cdot 5 \ \pm 2 \cdot 1$				
		$-141 \cdot 20 \pm 0 \cdot 57$				
		$-140 \cdot 37 \pm 0 \cdot 47$				
		$-139 \cdot 80 \pm 0 \cdot 43$				
		$-144 \cdot 31 \pm 0 \cdot 59$				
	BCE	$-100 \cdot 36 \pm 0 \cdot 93$				
		$-127 \cdot 5 \ \pm 1 \cdot 9$				
		$-124 \cdot 5 \ \pm 1 \cdot 9$				
		$-156 \cdot 73 \pm 0 \cdot 57$				
		$-159 \cdot 70 \pm 0 \cdot 56$				

1 kcal = 4·184 kJ

1 Formula	2 g.f.w.	3 Name	State	Purity mol %	Type	No. of expts.	Detn. of reactn.	$-\Delta_r$ kcal/g.f.w.	R
$C_{16}H_{32}O_3$	272·4316	Peroxyhexadecanoic acid	c		SB	6	CO_2	2406·2 ±2·2	64/
$C_{16}H_{32}O_3$	272·4316	t-Butyl peroxydodecanoate	l		SB	7	CO_2	2421·4 ±2·4	64/
$C_{18}H_{14}O_3$	278·3105	Cinnamic anhydride	c		SB	2	m	2088·0 ±2·0	50/ 51/
$C_{18}H_{36}O_3$	300·4858	Peroxyoctadecanoic acid	c		SB	9	CO_2	2717·7 ±2·7	64/
$C_{18}H_{36}O_3$	300·4858	t-Butyl peroxytetradecanoate	l		SB	4	CO_2	2732·4 ±2·5	64/

1 kcal $= 4 \cdot 184$ kJ

ΔH_r°	Remarks	5 $\Delta H_f^\bullet$ (l or c) kcal/g.f.w.	6 Determination of ΔH_v			7 ΔH_f° (g) kcal/g.f.w.
			Type	ΔH_v° kcal/g.f.w.	Ref.	
		$-191 \cdot 7 \ \pm 2 \cdot 2$				
		$-176 \cdot 5 \ \pm 2 \cdot 4$				
		$-83 \cdot 1 \ \pm 2 \cdot 0$				
		$-204 \cdot 9 \ \pm 2 \cdot 7$				
		$-190 \cdot 2 \ \pm 2 \cdot 5$				

$C_aH_bO_4$

1 kcal = 4·184 kJ

1 Formula	2 g.f.w.	5 Name	State	Purity mol %	Type	No. of expts.	Detn. of reactn.	$-\Delta H_r^o$ kcal/g.f.w.	R
$C_2H_2O_4$	90·0358	Oxalic acid	c		SB	4	m	60·59±0·11	26
$C_2H_2O_4$		Oxalic acid	c		SB	5	m	58·06±0·22	64
$C_2H_2O_4$		Oxalic acid							
$C_2H_6O_4$	94·0677	Bishydroxymethyl peroxide	c		R	4	m	70·1 ±1·2	53
$C_3H_4O_4$	104·0629	Malonic acid	c		SB	4	m	205·92±0·15	26
$C_3H_4O_4$		Malonic acid	c		SB	10	m	205·82±0·08	64
$C_3H_4O_4$		Malonic acid							
$C_4H_2O_4$	114·0581	Acetylene dicarboxylic acid	c		H	4	H_2	86·8 ±1·1	58
$C_4H_4O_4$	116·0741	Maleic acid	c		SB	6	m	324·15±0·09	38
$C_4H_4O_4$		Maleic acid	c		SB		m	324·56±0·46	58
$C_4H_4O_4$		Maleic acid	c		SB	5	m	323·89±0·16	64
$C_4H_4O_4$		Maleic acid	c		H	4	H_2	36·61±0·40	57
$C_4H_4O_4$		Maleic acid	c		H	5	H_2	36·34±0·20	59
$C_4H_4O_4$		Maleic acid							
$C_4H_4O_4$	116·0741	Fumaric acid	c		SB	7	m	318·72±0·09	38
$C_4H_4O_4$		Fumaric acid	c		SB		m	319·01±0·44	58
$C_4H_4O_4$		Fumaric acid	c		SB	5	m	319·00±0·20	64
$C_4H_4O_4$		Fumaric acid	c		H	4	H_2	31·15±0·30	57
$C_4H_4O_4$		Fumaric acid							
$C_4H_6O_4$	118·0900	Diacetyl peroxide	l		SB	4	m	453·2 ±2·2	57
$C_4H_6O_4$	118·0900	Succinic acid	c		SB	4	m	356·47±0·12	26
$C_4H_6O_4$		Succinic acid	c		SB	31	m	356·77±0·15	31
$C_4H_6O_4$		Succinic acid	c		SB	26	m	356·30±0·12	34
$C_4H_6O_4$		Succinic acid	c		SB	14	m	356·38±0·07	37
$C_4H_6O_4$		Succinic acid	c		SB	9	m	356·25±0·10	38
$C_4H_6O_4$		Succinic acid	c		SB	5	m	356·79±0·30	50/ 51
$C_4H_6O_4$		Succinic acid	c		SB	9	m	356·40±0·06	55
$C_4H_6O_4$		Succinic acid	c		SB	6	m	355·89±0·08	55
$C_4H_6O_4$		Succinic acid	c		SB		m	356·31±0·06	59
$C_4H_6O_4$		Succinic acid	c		SB	6	m	356·32±0·15	58
$C_4H_6O_4$		Succinic acid	c		SB	4	m	356·34±0·06	64/
$C_4H_6O_4$		Succinic acid	c	99·90	SB	9	m	356·28±0·10	64/
$C_4H_6O_4$		Succinic acid							

$$1 \text{ kcal} = 4 \cdot 184 \text{ kJ}$$

ΔH_r°		6 Determination of ΔH_v			7
Remarks	5 ΔH_f° (l or c) kcal/g.f.w.	Type	ΔH_v° kcal/g.f.w.	Ref.	ΔH_f° (g) kcal/g.f.w.
BCE	$-195 \cdot 83 \pm 0 \cdot 12$	S3	$23 \cdot 4 \pm 0 \cdot 5$ (sub.)	53/28	$-172 \cdot 43 \pm 0 \cdot 55$
	$-198 \cdot 36 \pm 0 \cdot 23$				$-174 \cdot 96 \pm 0 \cdot 59$
Selected value	$-198 \cdot 36 \pm 0 \cdot 23$				$-174 \cdot 96 \pm 0 \cdot 59$
ιOH(2N) $\{-112 \cdot 33\} + C_2H_6O_4$(c) $2H_2O(l) + 2HCO_2Na$(aq) $\{-159 \cdot 10\}$ $+H_2$	$-160 \cdot 1 \pm 1 \cdot 3$	S3	$22 \cdot 5 \pm 1 \cdot 0$ (sub.)	53/16	$-137 \cdot 6 \pm 2 \cdot 0$
BCE	$-212 \cdot 86 \pm 0 \cdot 16$				
	$-212 \cdot 96 \pm 0 \cdot 09$				
Selected value	$-212 \cdot 96 \pm 0 \cdot 09$				
4_2O_4(c)$+2H_2$ = Succinic acid(c) $\{-224 \cdot 86 \pm 0 \cdot 10\}$	$-138 \cdot 1 \pm 1 \cdot 1$				
	$-188 \cdot 68 \pm 0 \cdot 10$	S4	$26 \cdot 3 \pm 0 \cdot 6$ (sub.)	38/14	$-162 \cdot 4 \pm 0 \cdot 6$
	$-188 \cdot 27 \pm 0 \cdot 46$				$-162 \cdot 0 \pm 0 \cdot 7$
	$-188 \cdot 94 \pm 0 \cdot 17$				$-162 \cdot 6 \pm 0 \cdot 6$
4_4O_4(c)$+H_2$ = Succinic acid(c) $\{-224 \cdot 86 \pm 0 \cdot 10\}$	$-188 \cdot 25 \pm 0 \cdot 42$				$-162 \cdot 0 \pm 0 \cdot 7$
4_4O_4(c)$+H_2$ = Succinic acid(c) $\{-224 \cdot 86 \pm 0 \cdot 10\}$	$-188 \cdot 52 \pm 0 \cdot 23$				$-162 \cdot 2 \pm 0 \cdot 6$
Selected value	$-188 \cdot 73 \pm 0 \cdot 15$				$-162 \cdot 4 \pm 0 \cdot 6$
	$-194 \cdot 11 \pm 0 \cdot 10$	S4	$32 \cdot 5 \pm 1 \cdot 5$ (sub.)	38/14	$-161 \cdot 6 \pm 1 \cdot 5$
	$-193 \cdot 82 \pm 0 \cdot 44$				$-161 \cdot 3 \pm 1 \cdot 6$
	$-193 \cdot 83 \pm 0 \cdot 21$				$-161 \cdot 3 \pm 1 \cdot 5$
4_4O_4(c)$+H_2$ = Succinic acid(c) $\{-224 \cdot 86 \pm 0 \cdot 10\}$	$-193 \cdot 71 \pm 0 \cdot 32$				$-161 \cdot 2 \pm 1 \cdot 6$
Selected value	$-193 \cdot 83 \pm 0 \cdot 15$				$-161 \cdot 3 \pm 1 \cdot 5$
	$-127 \cdot 9 \pm 2 \cdot 2$				
BCE	$-224 \cdot 68 \pm 0 \cdot 13$	S4	$28 \cdot 1 \pm 0 \cdot 8$ (sub.)	60/47	$-196 \cdot 6 \pm 0 \cdot 9$
BCE	$-224 \cdot 38 \pm 0 \cdot 16$				$-196 \cdot 3 \pm 0 \cdot 9$
BCE	$-224 \cdot 85 \pm 0 \cdot 13$				$-196 \cdot 8 \pm 0 \cdot 9$
BCE	$-224 \cdot 77 \pm 0 \cdot 09$				$-196 \cdot 7 \pm 0 \cdot 8$
	$-224 \cdot 90 \pm 0 \cdot 12$				$-196 \cdot 8 \pm 0 \cdot 9$
C	$-224 \cdot 36 \pm 0 \cdot 31$				$-196 \cdot 3 \pm 0 \cdot 9$
	$-224 \cdot 75 \pm 0 \cdot 08$				$-196 \cdot 7 \pm 0 \cdot 8$
	$-225 \cdot 26 \pm 0 \cdot 10$				$-197 \cdot 2 \pm 0 \cdot 8$
	$-224 \cdot 84 \pm 0 \cdot 08$				$-196 \cdot 7 \pm 0 \cdot 8$
	$-224 \cdot 83 \pm 0 \cdot 16$				$-196 \cdot 7 \pm 0 \cdot 8$
	$-224 \cdot 81 \pm 0 \cdot 08$				$-196 \cdot 7 \pm 0 \cdot 8$
	$-224 \cdot 87 \pm 0 \cdot 12$				$-196 \cdot 8 \pm 0 \cdot 9$
Selected value	$-224 \cdot 86 \pm 0 \cdot 10$				$-196 \cdot 8 \pm 0 \cdot 8$

$C_4H_bO_4$
1 kcal = 4·184 kJ

1 Formula	2 g.f.w.	3 Name	State	Purity mol %	Type	No. of expts.	Detn. of reactn.	$-\Delta H_r^\circ$ kcal/g.f.w.	R
$C_4H_{10}O_4$	122·1219	i-Erythritol	c		SB	9	m	500·20±0·30	46
$C_5H_6O_4$	130·1012	Citraconic acid	c		SB	8	m	478·16±0·34	65,
$C_5H_6O_4$	130·1012	Itaconic acid	c		SB	8	m	474·17±0·12	65,
$C_5H_8O_4$	132·1171	Methylsuccinic acid	c		SB	4	m	514·50±0·25	33
$C_5H_8O_4$	132·1171	Glutaric acid	c		SB	4	m	514·24±0·12	26
$C_5H_8O_4$		Glutaric acid	c	99·80	SB	5	m	514·08±0·28	64,
$C_5H_8O_4$		Glutaric acid							
$C_5H_{10}O_4$	134·1331	Monoacetin (Glycerol monoacetate)	l		SB	6	m	594·53±0·89	56/
$C_5H_{12}O_4$	136·1490	1-(2-Hydroxyethoxy)-2, 3-propanediol	l		SB	5	m	673·8 ±1·2	55/
$C_5H_{12}O_4$	136·1490	Pentaerythritol	c		SB	5	m	660·12±0·66	54/
$C_6H_{10}O_4$	146·1442	1, 1-Dimethylsuccinic acid	c		SB	5	m	669·80±0·33	33,
$C_6H_{10}O_4$	146·1442	cis-1, 2-Dimethylsuccinic acid	c		SB	4	m	672·24±0·34	33,
$C_6H_{10}O_4$	146·1442	trans-1, 2-Dimethyl-succinic acid	c		SB	5	m	670·75±0·34	33,
$C_6H_{10}O_4$	146·1442	(−) 1, 2-Dimethylsuccinic acid	c		SB	2	m	671·05±0·34	33,
$C_6H_{10}O_4$	146·1442	Ethylsuccinic acid	c		SB	5	m	669·46±0·34	33/
$C_6H_{10}O_4$	146·1442	Adipic acid	c		SB	6	m	668·23±0·18	26/
$C_6H_{10}O_4$	146·1442	Diethyl oxalate	l	>99·5, glc	SB		m	713·37±	66/
$C_6H_{10}O_4$	146·1442	Dipropionyl peroxide	l		SB	3	CO_2	757·7 ±1·5	57/

Determina'

1 kcal = 4·184 kJ

ΔH_r°		5 ΔH_f° (l or c) kcal/g.f.w.	6 Determination of ΔH_v			7 ΔH_f° (g) kcal/g.f.w.
	Remarks		ΔH_v° Type kcal/g.f.w.		Ref.	
I$_m$, 5·57±0·01 (52/13)		−217·58±0·31	S4 32·3 ±[0·5] (sub.)		50/27	−185·28±0·61
		−197·04±0·35				
		−201·03±0·14				
	BCE	−229·02±0·26				
	BCE	−229·28±0·13				
		−229·44±0·29				
	Selected value	−229·44±0·29				
	BCE	−217·30±0·90				
	E	−206·3 ±1·2				
	E	−220·03±0·67	S4 34·4 ±0·2 (sub.)		53/28	−185·63±0·72
	BCE	−236·08±0·34				
	BCE	−233·64±0·35				
	BCE	−235·13±0·35				
	BCE	−234·83±0·35				
	BCE	−236·42±0·35				
	BCE	−237·65±0·20	S4 30·9 ±0·6 (sub.)		60/47	−206·75±0·65
	C	−192·51±	V4 15·17±[0·50]		47/15	−177·34±
		−148·2 ±1·5				

$C_aH_bO_4$ 1 kcal = 4·184 kJ

1 Formula	2 g.f.w.	3 Name	State	Purity mol %	Type	No. of expts.	Detn. of reactn.	$-\Delta H_r^\circ$ kcal/g.f.w.	Ref
$C_6H_{14}O_4$	150·1761	Triethylene glycol	l		SB	2	m	850·29±0·85	37/
$C_7H_{12}O_4$	160·1713	Trimethylsuccinic acid	c		SB	5	m	829·06±0·41	33/
$C_7H_{12}O_4$	160·1713	Pimelic acid	c		SB	5	m	827·00±0·25	26/
$C_7H_{12}O_4$	160·1713	5, 5'-Spirobis-1, 3-dioxan	c		SB	5	CO_2	900·4 ±1·3	59/2
$C_8H_6O_4$	166·1346	Phthalic acid	c		SB	5	m	770·52±0·32	39/1 50/1
$C_8H_6O_4$ $C_8H_6O_4$		Phthalic acid Phthalic acid	c		SB	7	m	770·43±0·24	61/2
$C_8H_6O_4$	166·1346	Isophthalic acid	c		SB	8	m	765·43±0·24	61/2
$C_8H_6O_4$	166·1346	Terephthalic acid	c		SB	8	m	762·29±0·20	61/2
$C_8H_{12}O_4$	172·1824	Cyclohexane-cis-1, 2-di-carboxylic acid	c		SB		m	932·6 ±1·8	30/
$C_8H_{12}O_4$	172·1824	Cyclohexane-trans-1, 2-dicarboxylic acid	c		SB		m	930·3 ±1·8	30/
$C_8H_{14}O_4$	174·1984	Tetramethylsuccinic acid	c		SB	4	m	988·63±0·50	33/9
$C_8H_{14}O_4$	174·1984	1, 1-Diethylsuccinic acid	c		SB	4	m	983·80±0·50	33/9
$C_8H_{14}O_4$	174·1984	cis-1, 2-Diethylsuccinic acid	c		SB	4	m	987·01±0·50	33/9
$C_8H_{14}O_4$	174·1984	trans-1, 2-Diethyl-succinic acid	c		SB	4	m	985·32±0·50	33/9
$C_8H_{14}O_4$	174·1984	Suberic acid	c		SB	4	m	982·53±0·30	26/
$C_8H_{14}O_4$	174·1984	2, 3-Butanediol diacetate	l		E		an	−5·33±0·50	45/1
$C_8H_{14}O_4$	174·1984	Dibutyryl peroxide	l		SB	6	CO_2	1069·7 ±1·0	57/2

$$1 \text{ kcal} = 4 \cdot 184 \text{ kJ}$$

ΔH_r°		ΔH_f° (l or c) kcal/g.f.w.	6 Determination of ΔH_v			7 ΔH_f° (g) kcal/g.f.w.
Remarks		5	ΔH_v° Type kcal/g.f.w.	Ref.		
	BCE	$-192 \cdot 22 \pm 0 \cdot 87$	V4 $18 \cdot 93 \pm 1 \cdot 89$	37/10		$-173 \cdot 3 \ \pm 2 \cdot 2$
	BCE	$-239 \cdot 19 \pm 0 \cdot 43$				
	BCE	$-241 \cdot 25 \pm 0 \cdot 28$				
An $0 \cdot 9968 \pm 0 \cdot 0012$		$-167 \cdot 8 \ \pm 1 \cdot 3$	V5$_{me}$ $17 \cdot 4 \ \pm 0 \cdot 4$ (sub.)	59/20		$-150 \cdot 4 \ \pm 1 \cdot 4$
		$-186 \cdot 83 \pm 0 \cdot 34$				
Selected value		$-186 \cdot 92 \pm 0 \cdot 26$ $-186 \cdot 88 \pm 0 \cdot 20$				
		$-191 \cdot 92 \pm 0 \cdot 60$	S4 $25 \cdot 5 \ \pm 0 \cdot 5$ (sub.)	62/58		$-166 \cdot 4 \ \pm 0 \cdot 6$
		$-195 \cdot 06 \pm 0 \cdot 23$	S3 $23 \cdot 48 \pm 0 \cdot 60$ (sub.)	34/9		$-171 \cdot 58 \pm 0 \cdot 70$
	CE	$-229 \cdot 7 \ \pm 1 \cdot 8$				
	CE	$-232 \cdot 0 \ \pm 1 \cdot 8$				
	BCE	$-241 \cdot 98 \pm 0 \cdot 52$				
	BCE	$-246 \cdot 81 \pm 0 \cdot 52$				
	BCE	$-243 \cdot 60 \pm 0 \cdot 52$				
	BCE	$-245 \cdot 29 \pm 0 \cdot 52$				
	BCE	$-248 \cdot 08 \pm 0 \cdot 35$	S4 $34 \cdot 2 \ \pm 0 \cdot 9$ (sub.)	60/47		$-213 \cdot 9 \ \pm 1 \cdot 0$
$H_3CHOH-]_2(l) \ \{-129 \cdot 43 \pm 0 \cdot 60\}$ $+2HOAc(l) \ \{-115 \cdot 75\}$ $\rightleftharpoons C_8H_{14}O_4(l) + 2H_2O(l)$		$-218 \cdot 96 \pm 0 \cdot 80$				
		$-160 \cdot 9 \ \pm 1 \cdot 1$				

THERMOCHEMISTRY OF ORGANIC AND ORGANOMETALLIC COMPOUNDS

$C_aH_bO_4$

1 kcal $= 4 \cdot 184$ kJ

							Determinat
1 Formula	2 g.f.w.	3 Name	State	Purity mol %	No. of Type expts.	Detn. of reactn.	$-\Delta H_r^\circ$ kcal/g.f.w.	Re
$C_9H_8O_4$	$180 \cdot 1617$	o-Acetylhydroxybenzoic acid	c		R 5	m	$-6 \cdot 50 \pm 0 \cdot 07$	64/
$C_9H_{16}O_4$	$188 \cdot 2255$	Azelaic acid	c		SB 4	m	$1141 \cdot 00 \pm 0 \cdot 45$	26,
$C_{10}H_{10}O_4$	$194 \cdot 1888$	Phenylsuccinic acid	c		SB 4	m	$1081 \cdot 09 \pm 0 \cdot 54$	33,
$C_{10}H_{10}O_4$	$194 \cdot 1888$	2, 4-Diacetylresorcinol	c		SB 6	m	$1102 \cdot 0 \ \pm 1 \cdot 5$	37,
$C_{10}H_{10}O_4$	$194 \cdot 1888$	4, 6-Diacetylresorcinol	c		SB 5	m	$1096 \cdot 5 \ \pm 1 \cdot 6$	37,
$C_{10}H_{12}O_4$	$196 \cdot 2047$	1-Monobenzoylglycerol	c		SB 7	m	$1164 \cdot 61 \pm 0 \cdot 26$	65/
$C_{10}H_{12}O_4$	$196 \cdot 2047$	2-Monobenzoylglycerol	c		SB 7	m	$1165 \cdot 69 \pm 0 \cdot 26$	65/
$C_{10}H_{16}O_4$	$200 \cdot 2366$	Dimethyl cis-hexahydro-isophthalate	l		SB 6	m	$1284 \cdot 7 \ \pm 2 \cdot 5$	39/
$C_{10}H_{16}O_4$	$200 \cdot 2366$	Dimethyl trans-hexa-hydroisophthalate	l		SB 6	m	$1282 \cdot 8 \ \pm 1 \cdot 9$	39/
$C_{10}H_{18}O_4$	$202 \cdot 2526$	Triethylsuccinic acid	c		SB 4	m	$1300 \cdot 49 \pm 0 \cdot 65$	33,
$C_{10}H_{18}O_4$	$202 \cdot 2526$	Sebacic acid	c		SB 5	m	$1296 \cdot 60 \pm 0 \cdot 50$	26,
$C_{11}H_{20}O_4$	$216 \cdot 2797$	Undecanedioic acid	c		SB 4	m	$1454 \cdot 95 \pm 0 \cdot 6$	26,
$C_{12}H_{14}O_4$	$222 \cdot 2430$	Diethyl phthalate	l		SB 6	m	$1421 \cdot 2 \ \pm 2 \cdot 8$	52/
$C_{12}H_{20}O_4$	$228 \cdot 2908$	Diethyl cis-hexahydro-isophthalate	l		SB 6	m	$1588 \cdot 6 \ \pm 3 \cdot 0$	39/
$C_{12}H_{20}O_4$	$228 \cdot 2908$	Diethyl trans-hexahydro-isophthalate	l		SB 6	m	$1586 \cdot 8 \ \pm 3 \cdot 0$	39/
$C_{12}H_{22}O_4$	$230 \cdot 3067$	Tetraethylsuccinic acid	c		SB 4	m	$1618 \cdot 02 \pm 0 \cdot 81$	33,
$C_{12}H_{22}O_4$	$230 \cdot 3067$	Dodecanedioic acid	c		SB 5	m	$1610 \cdot 00 \pm 0 \cdot 68$	26,
$C_{13}H_{24}O_4$	$244 \cdot 3338$	Tridecanedioic acid	c		SB 5	m	$1768 \cdot 0 \ \pm 0 \cdot 8$	26/
$C_{13}H_{26}O_4$	$246 \cdot 3498$	1-Monocaprin (Glycerol 1-decanoate)	c		SB 8	m	$1845 \cdot 71 \pm 0 \cdot 30$	65/

1 kcal = 4·184 kJ

ΔH_r°		6 Determination of ΔH_v			7
Remarks	5 ΔH_f° (l or c) kcal/g.f.w.	Type	ΔH_v° kcal/g.f.w.	Ref.	ΔH_f° (g) kcal/g.f.w.
$H_8O_4(c) + H_2O(l)$ = Salicylic acid(c) {$-141 \cdot 00 \pm 0 \cdot 15$} $+HOAc(l)$ {$-115 \cdot 75$}	$-194 \cdot 93 \pm 0 \cdot 20$				
BCE	$-251 \cdot 98 \pm 0 \cdot 50$				
BCE	$-201 \cdot 00 \pm 0 \cdot 56$				
	$-180 \cdot 1 \pm 1 \cdot 5$				
	$-185 \cdot 6 \pm 1 \cdot 6$				
	$-185 \cdot 79 \pm 0 \cdot 30$				
	$-184 \cdot 71 \pm 0 \cdot 30$				
CE	$-202 \cdot 3 \pm 2 \cdot 5$				
CE	$-204 \cdot 2 \pm 1 \cdot 9$				
BCE	$-254 \cdot 86 \pm 0 \cdot 67$				
BCE	$-258 \cdot 75 \pm 0 \cdot 56$	S4	$38 \cdot 4 \pm 0 \cdot 7$ (sub.)	60/47	$-220 \cdot 4 \pm 0 \cdot 9$
BCE	$-262 \cdot 76 \pm 0 \cdot 66$				
BCE	$-185 \cdot 6 \pm 2 \cdot 8$	V3	$21 \cdot 1 \pm [0 \cdot 5]$	58/39	$-164 \cdot 5 \pm 2 \cdot 9$
CE	$-223 \cdot 2 \pm 3 \cdot 0$				
CE	$-225 \cdot 0 \pm 3 \cdot 0$				
BCE	$-262 \cdot 06 \pm 0 \cdot 84$				
BCE	$-270 \cdot 08 \pm 0 \cdot 71$	S4	$36 \cdot 6 \pm 0 \cdot 7$ (sub.)	60/47	$-233 \ 5 \pm 1 \cdot 0$
BCE	$-274 \cdot 44 \pm 0 \cdot 88$				
	$-265 \cdot 05 \pm 0 \cdot 36$				

$C_aH_bO_4$ 1 kcal = 4·184 kJ

1 Formula	2 g.f.w.	3 Name	State	Purity mol %	Type	No. of expts.	Detn. of reactn.	$-\Delta H_r^\circ$ kcal/g.f.w.	Re
$C_{13}H_{26}O_4$	246·3498	2-Monocaprin (Glycerol 2-decanoate)	c		SB	6	m	1848·87±0·46	65/
$C_{14}H_{10}O_4$	242·2334	Dibenzoyl peroxide	c		SB	3	m	1564·6 ±1·5	50/ 51/
$C_{14}H_{10}O_4$		Dibenzoyl peroxide	c		R	3	m	116 ±5	61/
$C_{14}H_{10}O_4$		Dibenzoyl peroxide							
$C_{14}H_{14}O_4$	246·2653	Diallyl phthalate	l		SB	8	m	1663·9 ±3·2	55/
$C_{14}H_{14}O_4$		Diallyl phthalate	l		SB	13	m	1662·9 ±1·7	55/
$C_{14}H_{14}O_4$		Diallyl phthalate							
$C_{15}H_{30}O_4$	274·4040	1-Monolaurin (Glycerol 1-dodecanoate)	c		SB	7	m	2158·03±0·39	65/
$C_{15}H_{30}O_4$	274·4040	2-Monolaurin (Glycerol 2-dodecanoate)	c		SB	7	m	2160·02±0·39	65/
$C_{16}H_{14}O_4$	270·2876	cis-1, 2-Diphenylsuccinic acid	c		SB	5	m	1807·72±0·90	33/
$C_{16}H_{14}O_4$	270·2876	trans-1, 2-Diphenyl-succinic acid	c		SB	5	m	1806·13±0·90	33/
$C_{16}H_{14}O_4$	270·2876	Di-o-toluyl peroxide	c		SB	3	m	1863·4 ±1·9	50/ 51/
$C_{16}H_{14}O_4$	270·2876	Di-p-toluyl peroxide	c		SB	3	m	1875·1 ±1·9	50/ 51/
$C_{16}H_{22}O_4$	278·3513	Dibutyl phthalate	l	2s	SB	7	m	2054·9 ±3·0	36/
$C_{17}H_{20}O_4$	288·3466	(−) α-Desmotropo-santonin acetate	c		SB		m	2073·95±0·64	64/
$C_{17}H_{20}O_4$	288·3466	(+) β-Desmotropo-santonin acetate	c		SB		m	2071·88±0·43	64/
$C_{17}H_{34}O_4$	302·4581	1-Monomyristin (Glycerol 1-tetradecanoate)	c		SB		m	2468·30±0·59	40/
$C_{17}H_{34}O_4$		1-Monomyristin (Glycerol 1-tetradecanoate)	c		SB	10	m	2467·92±0·35	65/
$C_{17}H_{34}O_4$		1-Monomyristin (Glycerol 1-tetradecanoate)							

Determinati

4

$$1 \text{ kcal} = 4 \cdot 184 \text{ kJ}$$

of Δ_r		5 ΔH_f° (l or c) kcal/g.f.w.	6 Determination of ΔH_v			7 ΔH_f° (g) kcal/g.f.w.
	Remarks		$\Delta_v H_v^\circ$ Type kcal/g.f.w.		Ref.	
		$-261 \cdot 89 \pm 0 \cdot 50$				
		$-93 \cdot 7 \ \pm 1 \cdot 5$				
$(C_2H_5)_6Sn_2(l)$ $\{-52\pm 2\}+C_{14}H_{10}O_4(c)$ $= 2(C_2H_5)_3SnOCOC_6H_5(c)$ $\{-138\pm 1\}$		$-108 \ \pm 6$				
	Selected value	$-100 \ \pm 10$				
	E	$-131 \cdot 0 \ \pm 3 \cdot 2$				
	BCE	$-132 \cdot 0 \ \pm 1 \cdot 7$				
	Selected value	$-131 \cdot 6 \ \pm 1 \cdot 5$				
		$-277 \cdot 46 \pm 0 \cdot 45$				
		$-275 \cdot 47 \pm 0 \cdot 45$				
	BCE	$-175 \cdot 30 \pm 0 \cdot 92$				
	BCE	$-176 \cdot 89 \pm 0 \cdot 92$				
		$-119 \cdot 6 \ \pm 1 \cdot 9$				
		$-107 \cdot 9 \ \pm 1 \cdot 9$				
	BC	$-201 \cdot 4 \ \pm 3 \cdot 0$	V3 $21 \cdot 9 \ \pm 1 \cdot 0$		49/32	$-179 \cdot 5 \ \pm 3 \cdot 3$
		$-208 \cdot 07 \pm 0 \cdot 67$				
		$-210 \cdot 14 \pm 0 \cdot 50$				
	D	$-291 \cdot 92 \pm 0 \cdot 64$				
		$-292 \cdot 30 \pm 0 \cdot 43$				
	Selected value	$-292 \cdot 17 \pm 0 \cdot 38$				

$C_aH_bO_4$ 1 kcal = 4·184 kJ

								4
								Determination
					No.	Detn.		
1	2	3		Purity	of	of	$-\Delta H_r^\circ$	
Formula	g.f.w.	Name	State	mol %	Type	expts.	reactn.	kcal/g.f.w.	Ref
$C_{17}H_{34}O_4$	302·4581	2-Monomyristin (Glycerol 2-tetradecanoate)	c		SB	8	m	2470·34±0·43	65/3
$C_{18}H_{14}O_4$	294·3099	Cinnamoyl peroxide	c		SB	3	m	2086·0 ±2·0	50/1 51/1
$C_{18}H_{26}O_4$	306·4055	Di-n-pentyl phthalate	l		SB	5	m	2360·1 ±3·2	36/1
$C_{18}H_{26}O_4$	306·4055	Di-isopentyl phthalate	l		SB		m	2356·1 ±1·2	57/2
$C_{19}H_{38}O_4$	330·5123	1-Monopalmitin (Glycerol 1-hexadecanoate)	c		SB	7	m	2783·32±0·36	40/4
$C_{19}H_{38}O_4$		1-Monopalmitin (Glycerol 1-hexadecanoate)	c		SB	8	m	2778·67±0·51	65/3
$C_{19}H_{38}O_4$		1-Monopalmitin (Glycerol 1-hexadecanoate)							
$C_{19}H_{38}O_4$	330·5123	2-Monopalmitin (Glycerol 2-hexadecanoate)	c		SB	8	m	2792·85±0·67	40/4
$C_{19}H_{38}O_4$		2-Monopalmitin (Glycerol 2-hexadecanoate)	c		SB	8	m	2781·72±0·44	65/3
$C_{19}H_{38}O_4$		2-Monopalmitin (Glycerol 2-hexadecanoate)	c						
$C_{20}H_{14}O_4$	318·3322	Diphenyl phthalate	c		SB		m	2242·3 ±2·2	57/2
$C_{20}H_{26}O_4$	330·4278	Dicyclohexyl phthalate	c		SB		m	2546·5 ±2·6	57/2
$C_{21}H_{42}O_4$	358·5665	1-Monostearin (Glycerol 1-octadecanoate)	c		SB	8	m	3090·05±0·42	65/3
$C_{21}H_{42}O_4$	358·5665	2-Monostearin (Glycerol 2-octadecanoate)	c		SB	8	m	3093·89±0·55	65/3
$C_{24}H_{38}O_4$	390·5681	Di-(1-ethylhexyl) phthalate	c		SB	8	m	3296·1 ±3·3	55/2

1 kcal $= 4 \cdot 184$ kJ

ΔH_r° Remarks	5 $\Delta H_f'$ (l or c) kcal/g.f.w.	6 Determination of ΔH_v			7 ΔH_f° (g) kcal/g.f.w.
		Type	ΔH_v° kcal/g.f.w.	Ref.	
	$-289 \cdot 88 \pm 0 \cdot 49$				
	$-85 \cdot 1\ \pm 2 \cdot 0$				
BC	$-220 \cdot 9\ \pm 3 \cdot 2$	V4	$25 \cdot 5\ \pm 1 \cdot 0$	48/14	$-195 \cdot 4\ \pm 3 \cdot 4$
BCE	$-224 \cdot 9\ \pm 1 \cdot 2$				
D	$-301 \cdot 63 \pm 0 \cdot 45$				
	$-306 \cdot 28 \pm 0 \cdot 58$				
Selected value	$-306 \cdot 28 \pm 0 \cdot 58$				
D	$-292 \cdot 10 \pm 0 \cdot 73$				
	$-303 \cdot 23 \pm 0 \cdot 52$				
Selected value	$-303 \cdot 23 \pm 0 \cdot 52$				
BCE	$-116 \cdot 9\ \pm 2 \cdot 2$				
BCE	$-222 \cdot 6\ \pm 2 \cdot 6$				
	$-319 \cdot 64 \pm 0 \cdot 52$				
	$-315 \cdot 80 \pm 0 \cdot 63$				
BCE	$-259 \cdot 1\ \pm 3 \cdot 3$				

THERMOCHEMISTRY OF ORGANIC AND ORGANOMETALLIC COMPOUNDS

$C_aH_bO_5-C_aH_bO_{12}$ 　　　　　　　1 kcal = 4·184 kJ

						No. of	Detn. of		4 Determinatio
1 Formula	2 g.f.w.	3 Name	State	Purity mol %	Type	expts.	reactn.	$-\Delta H_r^\circ$ kcal/g.f.w.	Re
$C_4H_6O_5$	134·0894	DL-Malic acid	c		SB	4	m	316·89±0·14	64/
$C_4H_6O_5$	134·0894	L-Malic acid	c		SB	2	m	317·39±	64/
$C_5H_6O_5$	146·1006	α-Ketoglutaric acid	c		SB	8	m	429·94±0·20	65/3
$C_5H_{10}O_5$	150·1325	α-D-Xylose	c		SB		m	559·0 ±0·2	57/
$C_5H_{10}O_5$	150·1325	D-Ribose	c		SB	5	m	557·9 ±1·1	63/2
$C_5H_{12}O_5$	152·1484	Xylitol	c		SB		m	612·82±0·15	63/2
$C_7H_{12}O_5$	176·1707	Diacetin (Glycerol diacetate)	l		SB	9	m	800·4 ±1·6	56/
$C_8H_{18}O_5$	194·2297	Tetraethylene glycol	l		SB	2	m	1132·6 ±1·1	37/
$C_6H_6O_6$	174·1111	*trans*-Aconitic acid	c		SB	4	m	474·62±0·59	65/3
$C_6H_6O_6$	174·1111	*cis*-Aconitic acid	c		SB	4	m	476·6 ±1·8	65/3
$C_6H_{12}O_6$	180·1589	α-D-Glucose	c		SB	9	m	669·60±0·13	38/1
$C_6H_{12}O_6$		α-D-Glucose	c		SB		m	670·0 ±0·1	57/1
$C_6H_{12}O_6$		α-D-Glucose	c		SB		m	670·40±0·31	60/2
$C_6H_{12}O_6$		α-D-Glucose	c		SB	4	m	669·17±0·56	61/1
$C_6H_{12}O_6$		α-D-Glucose							
$C_6H_{12}O_6$	180·1589	L-Sorbose	c		SB	7	m	670·31±0·09	39/1
$C_6H_{12}O_6$	180·1589	β-D-Levulose	c		SB	6	m	671·71±0·08	39/1
$C_6H_{12}O_6$	180·1589	D-Mannose	c		SB	5	m	672·33±0·81	63/2
$C_6H_{12}O_6$	180·1589	α-D-Galactose	c		SB	5	m	666·77±0·10	39/1
$C_6H_{12}O_6$		α-D-Galactose	c		SB	5	m	670·1 ±2·6	63/2
$C_6H_{12}O_6$		α-D-Galactose							
$C_6H_{14}O_6$	182·1749	Mannitol	c		SB	6	m	722·94±0·43	46/8
$C_6H_{14}O_6$	182·1749	Dulcitol	c		SB	5	m	720·65±0·43	46/8
$C_7H_{14}O_6$	194·1860	α-Methylglucofuranoside	l		SB		m	849·1 ±0·2	57/1

$1 \text{ kcal} = 4 \cdot 184 \text{ kJ}$

ΔH_r°		6 Determination of ΔH_v				
	Remarks	5 ΔH_f° (l or c) kcal/g.f.w.	Type	ΔH_v° kcal/g.f.w.	Ref.	7 ΔH_f° (g) kcal/g.f.w.
		$-264 \cdot 26 \pm 0 \cdot 15$				
		$-263 \cdot 76 \pm$				
		$-245 \cdot 26 \pm 0 \cdot 21$				
	E	$-252 \cdot 8 \pm 0 \cdot 2$				
		$-253 \cdot 9 \pm 1 \cdot 1$				
		$-267 \cdot 33 \pm 0 \cdot 19$				
	BCE	$-267 \cdot 8 \pm 1 \cdot 6$				
	BCE	$-234 \cdot 6 \pm 1 \cdot 1$	V4	$23 \cdot 6 \pm 2 \cdot 4$	37/10	$-211 \cdot 0 \pm 2 \cdot 8$
		$-294 \cdot 63 \pm 0 \cdot 60$				
		$-292 \cdot 7 \pm 1 \cdot 8$				
	E	$-304 \cdot 60 \pm 0 \cdot 16$ $-304 \cdot 2 \pm 0 \cdot 1$ $-303 \cdot 80 \pm 0 \cdot 32$ $-305 \cdot 03 \pm 0 \cdot 57$				
	Selected value	$-304 \cdot 30 \pm 0 \cdot 15$				
		$-303 \cdot 89 \pm 0 \cdot 13$				
		$-302 \cdot 49 \pm 0 \cdot 12$				
		$-301 \cdot 87 \pm 0 \cdot 82$				
		$-307 \cdot 43 \pm 0 \cdot 13$ $-304 \cdot 1 \pm 2 \cdot 6$				
	Selected value	$-307 \cdot 43 \pm 0 \cdot 13$				
$H_m, 5 \cdot 39 \pm 0 \cdot 01$ (52/13)		$-319 \cdot 57 \pm 0 \cdot 45$				
$H_m, 7 \cdot 09 \pm 0 \cdot 04$ (52/13)		$-321 \cdot 86 \pm 0 \cdot 45$				
	E	$-287 \cdot 5 \pm 0 \cdot 2$				

THERMOCHEMISTRY OF ORGANIC AND ORGANOMETALLIC COMPOUNDS

$C_aH_bO_5-C_aH_bO_{12}$

1 kcal = 4·184 kJ

1 Formula	2 g.f.w.	3 Name	State	Purity mol %	Type	No. of expts.	Detn. of reactn.	$-\Delta H_r^\circ$ kcal/g.f.w.	Re
$C_7H_{14}O_6$	194·1860	α-D-Methylgluco-pyranoside	c		SB		m	841·8 ±0·2	57/
$C_7H_{14}O_6$		α-D-Methylgluco-pyranoside	c		SB	5	m	843·3 ±0·8	57/2
$C_7H_{14}O_6$		α-D-Methylgluco-pyranoside							
$C_7H_{14}O_6$	194·1860	β-D-Methylgluco-pyranoside	c		SB		m	840·8 ±0·1	57/
$C_9H_{14}O_6$	218·2083	Triacetin (Glycerol triacetate)	l		SB	7	m	1006·6 ±1·0	56/
$C_9H_{18}O_6$	222·2402	Tetrahydro-3, 3, 5, 5-tetrakis (hydroxy-methyl)-4-hydroxypyran	c		SB	6	m	1158·3 ±1·2	56/2
$C_{14}H_{22}O_6$	286·3278	Dicyclohexyl peroxidicarbonate	c		SB		m	1805±5	62/2
$C_6H_8O_7$	192·1265	Citric acid	c		SB	8	m	468·6 ±1·1	64/1
$C_6H_8O_7$	192·1265	D-Glucaric acid-1, 4-lactone	c		SB		m	494·4 ±1·0	64/2
$C_6H_8O_7$	192·1265	D-Glucaric acid-3, 6-lactone	c		SB		m	494·0 ±0·7	64/2
$C_6H_{14}O_7$	198·1743	α-D-Glucose hydrate	c		SB	3	m	666·75±0·18	38/1
$C_{10}H_{22}O_7$	254·2826	Dipentaerythritol	c		SB	9	m	1316·2 ±1·9	56/2
$C_6H_{10}O_8$	210·1418	Citric acid monohydrate	c		SB	12	m	466·7 ±0·1	58/2
$C_{12}H_{22}O_{11}$	342·3025	Sucrose	c		SB	4	m	1347·37±0·40	35/1
$C_{12}H_{22}O_{11}$		Sucrose	c		SB		m	1348·80±0·43	60/2
$C_{12}H_{22}O_{11}$		Sucrose							
$C_{12}H_{22}O_{11}$	342·3025	β-Lactose	c		SB	6	m	1345·49±0·12	39/1

1 kcal = 4·184 kJ

ΔH_r°	Remarks	5 ΔH_f° (l or c) kcal/g.f.w.	6 Determination of ΔH_v			7 ΔH_f° (g) kcal/g.f.w.
			Type	ΔH_v° kcal/g.f.w.	Ref.	
	E	$-294\cdot8 \pm 0\cdot2$				
		$-293\cdot3 \pm 0\cdot8$				
	Selected value	$-294\cdot8 \pm 0\cdot2$				
	E	$-295\cdot8 \pm 0\cdot1$				
	BCE	$-318\cdot1 \pm 1\cdot0$	V3	$19\cdot6 \pm 0\cdot2$	63/59	$-298\cdot5 \pm 1\cdot0$
	E	$-303\cdot0 \pm 1\cdot2$				
		-263 ± 5	S3	24 ± 2 (sub.)	62/21	-239 ± 5
		$-369\cdot0 \pm 1\cdot1$				
		$-343\cdot2 \pm 1\cdot0$				
		$-343\cdot6 \pm 0\cdot7$				
		$-375\cdot76 \pm 0\cdot21$				
	E	$-375\cdot8 \pm 1\cdot9$				
		$-439\cdot2 \pm 0\cdot1$				
	C	$-532\cdot71 \pm 0\cdot43$				
	E	$-531\cdot28 \pm 0\cdot46$				
	Selected value	$-532\cdot00 \pm 0\cdot70$				
		$-534\cdot59 \pm 0\cdot21$				

THERMOCHEMISTRY OF ORGANIC AND ORGANOMETALLIC COMPOUNDS

$C_aH_bO_5-C_aH_bO_{12}$ 1 kcal = 4·184 kJ

1 Formula	2 g.f.w.	3 Name	State	Purity mol %	Type	No. of expts.	Detn. of reactn.	$-\Delta H_r^\circ$ kcal/g.f.w.	Re
$C_{16}H_{22}O_{11}$	390·3471	α-D-Glucose pentaacetate	c		SB	6	m	1718·66±0·38	44/
$C_{16}H_{22}O_{11}$	390·3471	β-D-Glucose pentaacetate	c		SB	7	m	1722·67±0·46	44/
$C_{12}H_{24}O_{12}$	360·3179	β-Maltose monohydrate	c		SB	8	m	1360·52±0·13	39/
$C_{12}H_{24}O_{12}$	360·3179	α-Lactose monohydrate	c		SB	8	m	1354·68±0·23	39/

Determinati

4

1 kcal $= 4\cdot184$ kJ

ΔH_r°	Remarks	5 ΔH_f° (l or c) kcal/g.f.w.	6 Determination of ΔH_v			7 ΔH_f° (g) kcal/g.f.w.
			Type	ΔH_v° kcal/g.f.w.	Ref.	
		$-537\cdot62\pm0\cdot43$				
		$-533\cdot61\pm0\cdot50$				
		$-587\cdot87\pm0\cdot22$				
		$-593\cdot71\pm0\cdot29$				

THERMOCHEMISTRY OF ORGANIC AND ORGANOMETALLIC COMPOUNDS

1 kcal = 4·184 kJ

1 Formula	2 g.f.w.	3 Name	State	Purity mol %	Type	No. of expts.	Detn. of reactn.	$-\Delta H_r^\circ$ kcal/g.f.w.	Re
CH_5N	31·0577	Methylamine	l	>99	SB		CO_2	253·54±0·09	M/
C_2H_5N	43·0689	Ethyleneimine (Azirane)	l	99·9	SB	6	CO_2	380·86±0·14	52/
C_2H_7N	45·0848	Ethylamine	l	>99	SB		CO_2	409·50±0·12	M/
C_2H_7N	45·0848	Dimethylamine	g	>99·5, an	E		an	4·68±[0·40]	60/
C_2H_7N C_2H_7N		Dimethylamine Dimethylamine	l	>99	SB		CO_2	416·71±0·10	M/
C_3H_3N	53·0641	Acrylonitrile	l		SB		m	420·7 ±0·7	45/
C_3H_5N	55·0800	Propargylamine	l		SB	3	m	502·1 ±0·2	66/
C_3H_9N	59·1119	n-Propylamine	l	99·97±0·01	SB	7	m	565·31±0·07	67/
C_3H_9N	59·1119	Isopropylamine	l	99·99±0·01	SB	8	m	562·74±0·12	67/
C_3H_9N C_3H_9N	59·1119	Trimethylamine Trimethylamine	l g	>99	SB E		CO_2 an	578·64±0·15 3·08±[0·4]	M/ 60/
C_3H_9N		Trimethylamine	g		E		an	−7·82±[0·4]	60/
C_3H_9N		Trimethylamine							
C_4H_5N	67·0912	Pyrrole	l	99·99	SB	11	m	562·07±0·08	67/
C_4H_7N	69·1071	n-Propyl cyanide	l	glc	SB	3	CO_2	613·93±0·22	59/2
C_4H_7N	69·1071	Isopropyl cyanide	l		SB	3	CO_2	612·40±0·18	59/2
C_4H_9N C_4H_9N C_4H_9N	71·1230	Pyrrolidine Pyrrolidine Pyrrolidine	l l	99·85 99·85	SB SB	5 6	CO_2 m	673·83±0·20 673·84±0·18	59/2 59/2
$C_4H_{11}N$	73·1390	n-Butylamine	l	glc	SB	6	CO_2	721·43±0·27	59/2
$C_4H_{11}N$	73·1390	s-Butylamine	l	glc	SB	6	CO_2	719·08±0·22	59/2

Determinati

1 kcal $= 4 \cdot 184$ kJ

ΔH_r°		6 Determination of ΔH_v			
Remarks	5 ΔH_f° (l or c) kcal/g.f.w.	Type	ΔH_v° kcal/g.f.w.	Ref.	7 ΔH_f° (g) kcal/g.f.w.
	$-11 \cdot 30 \pm 0 \cdot 11$	C1	$5 \cdot 80 \pm 0 \cdot 04$	37/13	$-5 \cdot 50 \pm 0 \cdot 12$
	$+21 \cdot 97 \pm 0 \cdot 14$	V2	$8 \cdot 27 \pm [0 \cdot 15]$	56/21	$+30 \cdot 24 \pm 0 \cdot 20$
	$-17 \cdot 71 \pm 0 \cdot 13$	V1	$6 \cdot 36 \pm [0 \cdot 10]$	62/75	$-11 \cdot 35 \pm 0 \cdot 17$
$H_3NH_2(g) \{-5 \cdot 50 \pm 0 \cdot 12\}$ $\rightleftharpoons (CH_3)_2NH(g) + NH_3(g) \{-11 \cdot 02\}$ 3rd Law	$-10 \cdot 73 \pm [0 \cdot 5]$	C1	$6 \cdot 07 \pm 0 \cdot 01$	39/16	$-4 \cdot 66 \pm [0 \cdot 5]$
Selected value	$-10 \cdot 50 \pm 0 \cdot 12$ $-10 \cdot 50 \pm 0 \cdot 12$				$-4 \cdot 43 \pm 0 \cdot 12$ $-4 \cdot 43 \pm 0 \cdot 12$
	$+36 \cdot 1 \pm 0 \cdot 7$	V3	$8 \cdot 0 \pm 0 \cdot 4$	45/11	$+44 \cdot 1 \pm 0 \cdot 8$
E	$+49 \cdot 2 \pm 0 \cdot 2$				
	$-24 \cdot 26 \pm 0 \cdot 09$	V1	$7 \cdot 49 \pm 0 \cdot 05$	67/16	$-16 \cdot 77 \pm 0 \cdot 13$
	$-26 \cdot 83 \pm 0 \cdot 14$	V1	$6 \cdot 81 \pm 0 \cdot 05$	67/16	$-20 \cdot 02 \pm 0 \cdot 17$
$CH_3)_2NH(g) \{-4 \cdot 43 \pm 0 \cdot 12\}$ $\rightleftharpoons CH_3NH_2(g) \{-5 \cdot 50 \pm 0 \cdot 12\}$ $+ (CH_3)_3N(g)$. 3rd Law	$-10 \cdot 93 \pm 0 \cdot 17$ $-11 \cdot 70 \pm [0 \cdot 5]$	C1	$5 \cdot 26 \pm 0 \cdot 02$	44/11	$-5 \cdot 67 \pm 0 \cdot 18$ $-6 \cdot 44 \pm [0 \cdot 5]$
$H_3(g) + (CH_3)_3N(g)$ $\rightleftharpoons CH_3NH_2(g) \{-5 \cdot 50 \pm 0 \cdot 12\}$ $+ (CH_3)_2NH(g) \{-4 \cdot 43 \pm 0 \cdot 12\}$ 3rd Law	$-11 \cdot 99 \pm [0 \cdot 5]$				$-6 \cdot 73 \pm [0 \cdot 5]$
Selected value	$-10 \cdot 93 \pm 0 \cdot 17$				$-5 \cdot 67 \pm 0 \cdot 18$
An $0 \cdot 9998$	$+15 \cdot 08 \pm 0 \cdot 10$	C1	$10 \cdot 80 \pm 0 \cdot 03$	67/17	$+25 \cdot 88 \pm 0 \cdot 11$
	$-1 \cdot 38 \pm 0 \cdot 24$	V3	$9 \cdot 50 \pm 0 \cdot 30$	47/15	$+8 \cdot 12 \pm 0 \cdot 39$
	$-2 \cdot 91 \pm 0 \cdot 20$	E1	$9 \cdot 1 \pm 0 \cdot 3$		$+6 \cdot 2 \pm 0 \cdot 4$
Selected value	$-9 \cdot 79 \pm 0 \cdot 21$ $-9 \cdot 78 \pm 0 \cdot 19$ $-9 \cdot 78 \pm 0 \cdot 14$	C1	$8 \cdot 98 \pm 0 \cdot 02$	59/23	$-0 \cdot 81 \pm 0 \cdot 22$ $-0 \cdot 80 \pm 0 \cdot 20$ $-0 \cdot 80 \pm 0 \cdot 15$
	$-30 \cdot 51 \pm 0 \cdot 29$	E8	$7 \cdot 8 \pm 0 \cdot 3$		$-22 \cdot 7 \pm 0 \cdot 4$
	$-32 \cdot 86 \pm 0 \cdot 25$	E8	$7 \cdot 5 \pm 0 \cdot 3$		$-25 \cdot 4 \pm 0 \cdot 4$

C_aH_bN

1 kcal = 4·184 kJ

1 Formula	2 g.f.w.	3 Name	State	Purity mol %	Type	No. of expts.	Detn. of reactn.	$-\Delta H_c^\circ$ kcal/g.f.w.	R
$C_4H_{11}N$	73·1390	t-Butylamine	l	glc	SB	4	CO_2	716·01±0·27	59
$C_4H_{11}N$		t-Butylamine	l	99·90	SB	5	m	715·94±0·10	67
$C_4H_{11}N$		t-Butylamine							
$C_4H_{11}N$	73·1390	Diethylamine	l	>99	SB		CO_2	727·16±0·27	M
C_5H_5N	79·1023	Pyridine	l	99·85±0·07	SB	6	CO_2	665·03±0·72	54
C_5H_5N		Pyridine	l	99·92±0·02	SB	8	m	664·98±0·10	61
C_5H_5N		Pyridine							
C_5H_9N	83·1342	Dimethylpropargylamine	l		SB	3	m	824·9 ±0·4	66
C_5H_9N	83·1342	1, 2, 5, 6-Tetrahydro-pyridine	l		SB	5	CO_2	785·68±0·54	63
$C_5H_{11}N$	85·1501	Piperidine (Hexahydro-pyridine)	l		SB	5	CO_2	824·94±0·56	63
$C_5H_{11}N$		Piperidine (Hexahydro-pyridine)	g		E			46·12±[0·50]	35
$C_5H_{11}N$		Piperidine (Hexahydro-pyridine)	g		E		an	46·31±0·18	57
$C_5H_{11}N$		Piperidine (Hexahydro-pyridine)							
C_6H_7N	93·1294	Aniline	l		SB	5	m	810·83±0·24	42 51
C_6H_7N		Aniline	l	99·85	SB	6	m	810·95±0·24	62
C_6H_7N		Aniline							
C_6H_7N	93·1294	2-Methylpyridine	l	99·93±0·04	SB	5	CO_2	817·52±0·62	54
C_6H_7N		2-Methylpyridine	l	99·90	SB	6	m	816·96±0·16	63
C_6H_7N		2-Methylpyridine							
C_6H_7N	93·1294	3-Methylpyridine	l	99·97±0·02	SB	5	CO_2	819·74±0·48	54
C_6H_7N		3-Methylpyridine	l	99·88	SB	6	m	818·21±0·12	63
C_6H_7N		3-Methylpyridine							
C_6H_7N	93·1294	4-Methylpyridine	l	99·91±0·05	SB	5	CO_2	816·99±0·30	54
$C_6H_{15}N$	101·1932	Triethylamine	l	>99	SB		CO_2	1044·62±0·27	M
$C_6H_{15}N$		Triethylamine	l		SB	5	m	1046·15±0·13	66
$C_6H_{15}N$		Triethylamine							
C_7H_5N	103·1246	Phenyl cyanide	l	glc	SB	4	CO_2	868·15±0·30	59

$$1 \text{ kcal} = 4 \cdot 184 \text{ kJ}$$

ΔH_r° / Remarks	5 ΔH_f° (l or c) kcal/g.f.w.	6 Determination of ΔH_v			7 ΔH_f° (g) kcal/g.f.w.
		Type	ΔH_v° kcal/g.f.w.	Ref.	
An 0·9996	$-35 \cdot 93 \pm 0 \cdot 29$	V1	$7 \cdot 10 \pm 0 \cdot 05$	67/16	$-28 \cdot 83 \pm 0 \cdot 31$
	$-36 \cdot 00 \pm 0 \cdot 12$				$-28 \cdot 90 \pm 0 \cdot 15$
Selected value	$-36 \cdot 00 \pm 0 \cdot 12$				$-28 \cdot 90 \pm 0 \cdot 15$
	$-24 \cdot 78 \pm 0 \cdot 29$	V1	$7 \cdot 62 \pm [0 \cdot 10]$	62/75	$-17 \cdot 16 \pm 0 \cdot 31$
An 0·9999 ± 0·0001	$+23 \cdot 99 \pm 0 \cdot 73$	C1	$9 \cdot 61 \pm 0 \cdot 01$	57/24	$+34 \cdot 60 \pm 0 \cdot 73$
	$+23 \cdot 94 \pm 0 \cdot 12$				$+34 \cdot 55 \pm 0 \cdot 12$
Selected value	$+23 \cdot 94 \pm 0 \cdot 12$				$+34 \cdot 55 \pm 0 \cdot 12$
E	$+47 \cdot 2 \pm 0 \cdot 4$				
	$+8 \cdot 01 \pm 0 \cdot 55$				
	$-21 \cdot 05 \pm 0 \cdot 57$	V1	$9 \cdot 39 \pm 0 \cdot 05$	63/29	$-11 \cdot 66 \pm 0 \cdot 58$
ridine(g) $\{+34 \cdot 55 \pm 0 \cdot 12\} +3H_2$ E $\rightleftharpoons C_5H_{11}N(g)$. 2nd Law					$-11 \cdot 57 \pm [0 \cdot 55]$
ridine(g) $\{+34 \cdot 55 \pm 0 \cdot 12\} +3H_2$ E $\rightleftharpoons C_5H_{11}N(g)$. 2nd Law					$-11 \cdot 76 \pm 0 \cdot 24$
Selected value	$-21 \cdot 15 \pm 0 \cdot 25$				$-11 \cdot 76 \pm 0 \cdot 24$
	$+7 \cdot 42 \pm 0 \cdot 25$	C1	$13 \cdot 33 \pm 0 \cdot 01$	62/26	$+20 \cdot 75 \pm 0 \cdot 25$
	$+7 \cdot 54 \pm 0 \cdot 25$				$+20 \cdot 87 \pm 0 \cdot 25$
Selected value	$+7 \cdot 48 \pm 0 \cdot 18$				$+20 \cdot 81 \pm 0 \cdot 18$
An 1·0004 ± 0·0001	$+14 \cdot 11 \pm 0 \cdot 63$	C1	$10 \cdot 15 \pm 0 \cdot 01$	63/30	$+24 \cdot 26 \pm 0 \cdot 63$
	$+13 \cdot 55 \pm 0 \cdot 18$				$+23 \cdot 70 \pm 0 \cdot 18$
Selected value	$+13 \cdot 55 \pm 0 \cdot 18$				$+23 \cdot 70 \pm 0 \cdot 18$
An 1·0000 ± 0·0001	$+16 \cdot 33 \pm 0 \cdot 49$	C1	$10 \cdot 62 \pm 0 \cdot 01$	63/31	$+26 \cdot 95 \pm 0 \cdot 49$
	$+14 \cdot 80 \pm 0 \cdot 14$				$+25 \cdot 42 \pm 0 \cdot 14$
Selected value	$+14 \cdot 80 \pm 0 \cdot 14$				$+25 \cdot 42 \pm 0 \cdot 14$
	$+13 \cdot 58 \pm 0 \cdot 31$	V1	$10 \cdot 83 \pm 0 \cdot 10$	57/26	$+24 \cdot 41 \pm 0 \cdot 34$
	$-32 \cdot 05 \pm 0 \cdot 30$	V1	$8 \cdot 46 \pm [0 \cdot 10]$	62/75	$-23 \cdot 59 \pm 0 \cdot 32$
	$-30 \cdot 52 \pm 0 \cdot 16$				$-22 \cdot 06 \pm 0 \cdot 19$
Selected value	$-30 \cdot 52 \pm 0 \cdot 16$				$-22 \cdot 06 \pm 0 \cdot 19$
	$+39 \cdot 00 \pm 0 \cdot 32$	V4	$12 \cdot 54 \pm [0 \cdot 40]$	47/15	$+51 \cdot 54 \pm 0 \cdot 51$

C_aH_bN

1 kcal = 4·184 kJ

1 Formula	2 g.f.w.	3 Name	State	Purity mol %	Type	No. of expts.	Detn. of reactn.	$-\Delta H_c^\circ$ kcal/g.f.w.	R
C_7H_7N	105·1405	2-Vinylpyridine	l		SB		m	935·0 ±0·9	60/
C_7H_9N	107·1565	Methyldipropargylamine	l		SB	4	m	1077·2 ±0·3	66/
C_7H_9N	107·1565	2-Ethylpyridine	l		SB		m	966·1 ±1·0	60/
C_7H_9N	107·1565	2, 3-Dimethylpyridine	l	99·65±0·17	SB	5	m	970·40±0·30	58/
C_7H_9N	107·1565	2, 4-Dimethylpyridine	l	99·80±0·10	SB	6	m	969·63±0·18	58/
C_7H_9N	107·1565	2, 5-Dimethylpyridine	l	99·85±0·07	SB	5	m	970·23±0·22	58/
C_7H_9N	107·1565	2, 6-Dimethylpyridine	l	99·89±0·05	SB	7	m	968·80±0·36	58/
C_7H_9N	107·1565	3, 4-Dimethylpyridine	l	99·88±0·06	SB	6	m	970·14±0·24	58/
C_7H_9N	107·1565	3, 5-Dimethylpyridine	l	99·74±0·13	SB	6	m	971·14±0·20	58/
$C_7H_{11}N$	109·1724	1-Dimethylaminopent- 4-en-2-yne	l		SB	4	m	1092·8 ±0·4	66/
$C_7H_{13}N$	111·1884	1-Azabicyclo [2, 2, 2]- octane	c		SB			1089·25±0·14	67/
C_8H_7N	117·1517	Indole	c		SB		m	1019·4 ±1·0	32/
$C_8H_{11}N$	121·1836	N-Ethylaniline	l		SB		m	1129·1 ±1·0	52/
$C_8H_{11}N$	121·1836	Exo-2-cyanobicyclo- [2, 2, 1] heptane	l		SB		m	1132·69±0·31	64/
$C_8H_{11}N$	121·1836	Endo-2-cyanobicyclo- [2, 2, 1] heptane	c		SB		m	1132·98±0·35	64/
$C_8H_{15}N$	125·2155	3-Azabicyclo [3, 2, 2]- nonane	c		SB			1240·51±0·08	67/
$C_8H_{17}N$	127·2314	n-Butylisobutyraldimine	l		SB	5	m	1294·6 ±2·4	48/
$C_8H_{17}N$		n-Butylisobutyraldimine	l		SB	6	CO_2	1301·35±0·81	62/
$C_8H_{17}N$		n-Butylisobutyraldimine							
$C_8H_{19}N$	129·2473	n-Butylisobutylamine	l		SB	7	CO_2	1349·8 ±1·2	62/6
C_9H_9N	131·1788	Tripropargylamine	l		SB	3	m	1348·5 ±0·1	66/2

1 kcal $= 4 \cdot 184$ kJ

ΔH_r° / Remarks		ΔH_f° (l or c) kcal/g.f.w.	6 Determination of ΔH_v			7 ΔH_f° (g) kcal/g.f.w.
			Type	ΔH_v° kcal/g.f.w.	Ref.	
		$+37 \cdot 5 \ \pm 0 \cdot 9$				
	E	$+111 \cdot 4 \ \pm 0 \cdot 3$				
		$+0 \cdot 3 \ \pm 1 \cdot 0$				
		$+4 \cdot 62 \pm 0 \cdot 31$	V1	$11 \cdot 70 \pm 0 \cdot 10$	60/29	$+16 \cdot 32 \pm 0 \cdot 33$
		$+3 \cdot 85 \pm 0 \cdot 20$	V1	$11 \cdot 42 \pm 0 \cdot 10$	60/29	$+15 \cdot 27 \pm 0 \cdot 23$
		$+4 \cdot 45 \pm 0 \cdot 24$	V1	$11 \cdot 43 \pm 0 \cdot 10$	60/29	$+15 \cdot 88 \pm 0 \cdot 27$
		$+3 \cdot 02 \pm 0 \cdot 37$	V1	$11 \cdot 01 \pm 0 \cdot 10$	60/29	$+14 \cdot 03 \pm 0 \cdot 39$
		$+4 \cdot 36 \pm 0 \cdot 26$	V1	$12 \cdot 38 \pm 0 \cdot 10$	60/29	$+16 \cdot 74 \pm 0 \cdot 29$
		$+5 \cdot 36 \pm 0 \cdot 22$	V1	$12 \cdot 04 \pm 0 \cdot 10$	60/29	$+17 \cdot 40 \pm 0 \cdot 25$
	E	$+58 \cdot 7 \ \pm 0 \cdot 4$				
		$-13 \cdot 16 \pm 0 \cdot 17$				
	BCE	$+27 \cdot 9 \ \pm 1 \cdot 0$	S3	$16 \cdot 7 \ \pm 0 \cdot 2$ (sub.)	55/36	$+44 \cdot 6 \ \pm 1 \cdot 1$
		$+1 \cdot 0 \ \pm 1 \cdot 0$	V4	$12 \cdot 5 \ \pm 1 \cdot 0$	52/15	$+13 \cdot 5 \ \pm 1 \cdot 4$
		$+4 \cdot 55 \pm 0 \cdot 34$				
		$+4 \cdot 84 \pm 0 \cdot 38$				
		$-24 \cdot 26 \pm 0 \cdot 14$				
	E	$-38 \cdot 5 \ \pm 2 \cdot 4$				
An $0 \cdot 9980 \pm 0 \cdot 0010$		$-31 \cdot 74 \pm 0 \cdot 82$				
	Selected value	$-31 \cdot 74 \pm 0 \cdot 82$				
An $0 \cdot 9987 \pm 0 \cdot 0025$		$-51 \cdot 6 \ \pm 1 \cdot 2$	V4	$10 \cdot 73 \pm 0 \cdot 30$	62/27	$-40 \cdot 9 \ \pm 1 \cdot 3$
	E	$+194 \cdot 6 \ \pm 0 \cdot 2$				

C_aH_bN

1 kcal = 4·184 kJ

1 Formula	2 g.f.w.	3 Name	State	Purity mol %	Type	No. of expts.	Detn. of reactn.	$-\Delta H_r^\circ$ kcal/g.f.w.	R
$C_9H_{15}N$	137·2266	2, 4, 5-Trimethyl-3-ethylpyrrole	c		SB		m	1337·5 ±1·3	33/
$C_9H_{21}N$	143·2744	Tripropylamine	l		SB	4	m	1514·26±0·19	66/
$C_{10}H_9N$	143·1899	1-Naphthylamine	c		SB	5	m	1264·1 ±1·3	32
$C_{10}H_9N$	143·1899	2-Naphthylamine	c		SB	5	m	1262·9 ±1·3	32
$C_{10}H_9N$		2-Naphthylamine	c		SB	4	m	1261·7 ±1·1	47/
$C_{10}H_9N$		2-Naphthylamine							
$C_{10}H_9N$	143·1899	1-Phenylpyrrole	c		SB		m	1284·8 ±1·3	32
$C_{10}H_9N$	143·1899	2-Phenylpyrrole	c		SB		m	1281·2 ±1·3	32
$C_{12}H_9N$	167·2122	Carbazole	c		SB		m	1465·93±0·86	57/
$C_{12}H_{11}N$	169·2282	2-Aminobiphenyl	c		SB	3	m	1531·1 ±1·5	35
$C_{12}H_{11}N$	169·2282	4-Aminobiphenyl	c		SB	3	m	1523·7 ±1·5	35
$C_{12}H_{11}N$	169·2282	Diphenylamine	c		SB	7	m	1535·54±0·45	39/
$C_{12}H_{11}N$		Diphenylamine	c		SB	7	m	1532·4 ±0·5	42/
$C_{12}H_{11}N$		Diphenylamine	c	3s	SB	22	m	1535·42±0·34	51/
$C_{12}H_{11}N$		Diphenylamine							55/
$C_{12}H_{27}N$	185·3557	Tributylamine	l		SB	4	m	1983·55±0·26	66/
$C_{13}H_9N$	179·2234	Acridine	c		SB	8	m	1578·1 ±1·6	47/
$C_{13}H_9N$	179·2234	2-Cyanobiphenyl	c		SB	3	m	1585·7 ±1·6	35/
$C_{13}H_{11}N$	181·2393	Benzilidene anil	c		SB	5	m	1638·5 ±1·7	48/
$C_{13}H_{13}N$	183·2553	N-Methyldiphenylamine	l		SB	7	m	1695·5 ±1·7	56/
$C_{13}H_{13}N$	183·2553	4-Methyldiphenylamine	c		SB	5	m	1678·4 ±1·7	56/
$C_{14}H_{15}N$	197·2824	4, 4'-Dimethyldiphenyl-amine	c		SB	5	m	1826·3 ±1·8	56/

Determina

1 kcal = 4·184 kJ

H_r° / Remarks	5 ΔH_f° (l or c) kcal/g.f.w.	6 Determination of ΔH_v ΔH_v° Type kcal/g.f.w.	Ref.	7 ΔH_f° (g) kcal/g.f.w.
BCE	−21·3 ±1·3			
	−49·51±0·24			
BCE	+16·2 ±1·3	V4$_{me}$ 21·5 ±1·0 (sub.)	47/15	+37·7 ±1·7
BCE	+15·0 ±1·3	V4$_{me}$ 21·1 ±1·0 (sub.)	47/15	+36·1 ±1·7
BCE	+13·8 ±1·1			+34·9 ±1·5
Selected value	+14·4 ±0·8			+35·5 ±1·3
BCE	+36·9 ±1·3			
BCE	+33·3 ±1·3			
BCE	+29·90±0·88	S4 20·2 ±0·2 (sub.)	55/36	+50·1 ±0·9
BCE	+26·8 ±1·5			
BCE	+19·4 ±1·5			
E	+31·19±0·47	S4 23·1 ±[0·6] (sub.)	53/31	+54·3±0·8
	+28·0 ±0·5			+51·1 ±0·8
	+31·07±0·36			+54·2 ±0·7
Selected value	+31·07±0·36			+54·2 ±0·7
	−67·32±0·32			
BCE	+48·0 ±1·6			
	+55·6 ±1·6			
E	+40·1 ±1·7	E8$_{me}$ 20·5 ±0·5 (sub.)	48/4	+60·6 ±1·8
BCE	+28·8 ±1·7			
BCE	+11·7 ±1·7			
BCE	−2·8 ±1·8			

C_aH_bN

Determina

1 Formula	2 g.f.w.	3 Name	State	Purity mol %	Type	No. of expts.	Detn. of reactn.	$-\Delta H_r^o$ kcal/g.f.w.	R
$C_{16}H_{13}N$	219·2887	N-Phenyl- 2-naphthylamine	c		SB	4	CO_2	1987·07±0·43	48
$C_{16}H_{13}N$		N-Phenyl- 2-naphthylamine	c		SB	6	m	1987·4 ±1·2	54
$C_{16}H_{13}N$		N-Phenyl- 2-naphthylamine							
$C_{17}H_{13}N$	231·2999	N-Methyl-2, 3; 5, 6- dibenzazalene	c		SB	5	m	2092·0 ±3·0	66
$C_{18}H_{39}N$	269·5182	Trihexylamine	l		SB	4	m	2921·57±0·35	66
$C_{19}H_{17}N$	259·3540	N-Benzyldiphenylamine	c		SB	5	m	2411·8 ±2·4	56
$C_{21}H_{21}N$	287·4082	Tribenzylamine	c		SB	8	m	2726·0 ±2·7	56
$C_{24}H_{51}N$	353·6808	Trioctylamine	l		SB	4	m	3859·44±0·50	66
$C_{27}H_{57}N$	395·7620	Trinonylamine	l		SB	6	m	4328·24±0·40	66
$C_{30}H_{63}N$	437·8433	Tridecylamine	l		SB	5	m	4797·06±0·74	66

1 kcal $= 4 \cdot 184$ kJ

ΔH_r°	Remarks	5 ΔH_f° (l or c) kcal/g.f.w.	6 Determination of ΔH_v			7 ΔH_f° (g) kcal/g.f.w.
			Type	ΔH_v° kcal/g.f.w.	Ref.	
		$+38 \cdot 21 \pm 0 \cdot 46$				
	E	$+38 \cdot 5 \ \pm 1 \cdot 2$				
	Selected value	$+38 \cdot 21 \pm 0 \cdot 46$				
		$+49 \cdot 1 \ \pm 3 \cdot 0$	S4	$31 \cdot 54 \pm 0 \cdot 32$ (sub.)	66/20	$+80 \cdot 6 \ \pm 3 \cdot 1$
		$-103 \cdot 49 \pm 0 \cdot 44$				
	BCE	$+44 \cdot 2 \ \pm 2 \cdot 4$				
	BCE	$+33 \cdot 6 \ \pm 2 \cdot 7$				
		$-139 \cdot 82 \pm 0 \cdot 62$				
		$-158 \cdot 12 \pm 0 \cdot 56$				
		$-176 \cdot 39 \pm 0 \cdot 86$				

1 Formula	2 g.f.w.	3 Name	State	Purity mol %	Type	No. of expts.	Detn. of reactn.	$-\Delta H_f^\circ$ kcal/g.f.w.	Re
CH_2N_2	42·0405	Cyanamide	c		SB	6	m	176·42±0·13	48/ 51/
CH_6N_2	46·0724	Methylhydrazine	l	>99·75	SB	3	m	311·95±0·14	52/
C_2N_2 C_2N_2 C_2N_2	52·0357	Cyanogen Cyanogen Cyanogen	g g		FC FC	10 17	m CO_2	261·4 ±0·2 261·94±0·43	33/ 51,
$C_2H_8N_2$ $C_2H_8N_2$ $C_2H_8N_2$	60·0995	1, 1-Dimethylhydrazine 1, 1-Dimethylhydrazine 1, 1-Dimethylhydrazine	l l	>99·5	SB SB	4 3	m m	473·15±0·70 472·91±0·86	52/ 60/
$C_2H_8N_2$	60·0995	1, 2-Dimethylhydrazine	l		SB	1	m	473·95±	52/
$C_3H_2N_2$	66·0628	Malononitrile	c		SB	6	m	395·03±0·30	67/
$C_3H_4N_2$	68·0787	Pyrazole	c		SB	5	CO_2	446·5 ±1·1	62/
$C_3H_4N_2$	68·0787	Imidazole	c		SB	5	CO_2	432·75±0·78	62/
C_4N_2	76·0580	Dicyanoacetylene	l	2s	SB	5	CO_2	495·8 ±0·3	60/ 63/
$C_4H_2N_2$	78·0739	Fumaronitrile	c		SB	5	m	508·63±0·40	67/
$C_4H_4N_2$	80·0899	Pyridazine	l	2s	SB	8	m	566·57±0·22	62/
$C_4H_4N_2$	80·0899	Pyrimidine	l		SB	5	m	547·87±0·20	62/
$C_4H_4N_2$	80·0899	Pyrazine	c	>99·99, glc	SB	6	m	546·24±0·28	62/
$C_4H_{10}N_2$	86·1377	Piperazine	c		SB	5	CO_2	706·88±0·38	63/
$C_6H_4N_2$	104·1122	3-Cyanopyridine	c		SB	4	m	747·17±0·20	48/ 51/
$C_6H_4N_2$	104·1122	1, 4-Dicyanobutyne-2	c		SB	2	m	788·6 ±1·7	66/
$C_6H_8N_2$	108·1441	p-Phenylenediamine	c		SB		m	838·30±0·15	49/
$C_6H_8N_2$	108·1441	Phenylhydrazine	l		SB	6	m	871·60±0·17	42/ 51/

1 kcal $= 4 \cdot 184$ kJ

ΔH_r°	Remarks	ΔH_f° (l or c) kcal/g.f.w.	6 Determination of ΔH_v			7 ΔH_f° (g) kcal/g.f.w.
			Type	ΔH_v° kcal/g.f.w.	Ref.	
		$+14 \cdot 05 \pm 0 \cdot 14$				
	C	$+12 \cdot 95 \pm 0 \cdot 15$	Cl	$9 \cdot 65 \pm 0 \cdot 02$	51/27	$+22 \cdot 60 \pm 0 \cdot 15$
			Cl	$4 \cdot 96 \pm 0 \cdot 02$	39/18	$+73 \cdot 3 \ \pm 0 \cdot 2$
						$+73 \cdot 84 \pm 0 \cdot 43$
	Selected value	$+68 \cdot 88 \pm 0 \cdot 43$				$+73 \cdot 84 \pm 0 \cdot 43$
	C	$+11 \cdot 79 \pm 0 \cdot 71$	Cl	$8 \cdot 37 \pm 0 \cdot 02$	53/17	$+20 \cdot 08 \pm 0 \cdot 71$
	D	$+11 \cdot 55 \pm 0 \cdot 86$				$+19 \cdot 92 \pm 0 \cdot 86$
	Selected value	$+11 \cdot 67 \pm 0 \cdot 60$				$+20 \cdot 04 \pm 0 \cdot 60$
	C	$+12 \cdot 59 \pm$	Cl	$9 \cdot 40 \pm 0 \cdot 02$	51/18	$+21 \cdot 99 \pm$
		$+44 \cdot 56 \pm 0 \cdot 31$	S4	$18 \cdot 9 \ \pm 0 \cdot 2$ (sub.)	67/45	$+63 \cdot 5 \ \pm 0 \cdot 4$
	D	$+27 \cdot 7 \ \pm 1 \cdot 1$				
	D	$+13 \cdot 97 \pm 0 \cdot 79$				
		$+119 \cdot 6 \pm 0 \cdot 3$	V2	$6 \cdot 88 \pm [0 \cdot 15]$	57/43	$+126 \cdot 5 \ \pm 0 \cdot 4$
		$+64 \cdot 11 \pm 0 \cdot 41$	S4	$17 \cdot 2 \ \pm 0 \cdot 2$ (sub.)	67/45	$+81 \cdot 3 \ \pm 0 \cdot 5$
		$+53 \cdot 74 \pm 0 \cdot 23$	Cl	$12 \cdot 78 \pm 0 \cdot 10$	62/28	$+66 \cdot 52 \pm 0 \cdot 25$
		$+35 \cdot 04 \pm 0 \cdot 21$	Cl	$11 \cdot 95 \pm 0 \cdot 06$	62/28	$+46 \cdot 99 \pm 0 \cdot 23$
		$+33 \cdot 41 \pm 0 \cdot 29$	Cl	$13 \cdot 45 \pm 0 \cdot 11$ (sub.)	62/28	$+46 \cdot 86 \pm 0 \cdot 33$
		$-10 \cdot 90 \pm 0 \cdot 39$				
		$+46 \cdot 23 \pm 0 \cdot 21$				
		$+87 \cdot 6 \ \pm 1 \cdot 7$				
	C	$+0 \cdot 73 \pm 0 \cdot 17$				
		$+34 \cdot 03 \pm 0 \cdot 19$	V4	$14 \cdot 69 \pm 0 \cdot 20$	42/9	$+48 \cdot 72 \pm 0 \cdot 29$

THERMOCHEMISTRY OF ORGANIC AND ORGANOMETALLIC COMPOUNDS

$C_aH_bN_2$ 1 kcal = 4·184 kJ

1 Formula	2 g.f.w.	3 Name	State	Purity mol %	Type	No. of expts.	Detn. of reactn.	$-\Delta H_r^\circ$ kcal/g.f.w.	R
$C_6H_{12}N_2$	112·1759	Triethylenediamine	c		SB	4	m	970·8 ±2·0	64
$C_6H_{14}N_2$	114·1919	Azoisopropane	l		SB	5	m	1053·1 ±0·3	48
$C_7H_{14}N_2$	126·2030	Di-isopropylcyanamide	l		SB	7	m	1124·33±0·73	48 51
$C_7H_{14}N_2$	126·2030	Di-isopropylcarbodi-imide	l		SB	5	m	1131·66±0·56	48 51
$C_8H_4N_2$	128·1345	Phthalonitrile	c		SB	4	m	954·86±0·24	48 51
$C_8H_{16}N_2$	140·2301	1, 4-Bis (dimethylamino)-but-2-yne	l		SB	4	m	1338·3 ±0·3	66
$C_9H_{10}N_2$	146·1935	Di-α-pyrrylmethane	c		SB		m	1218·2 ±1·2	32
$C_{12}H_8N_2$	180·2110	Phenazine	c		SB	3	m	1461·0 ±1·5	46 47
$C_{12}H_{10}N_2$	182·2269	cis-Azobenzene	c		SB	7	m	1556·9 ±0·6	39 51
$C_{12}H_{10}N_2$	182·2269	trans-Azobenzene	c		SB	6	m	1546·8 ±0·4	39 51
$C_{12}H_{12}N_2$	184·2428	Hydrazobenzene	c		SB		m	1591·4 ±0·3	51
$C_{12}H_{12}N_2$		Hydrazobenzene	c		SB		m	1590·4 ±0·4	44
$C_{12}H_{12}N_2$		Hydrazobenzene							
$C_{12}H_{12}N_2$	184·2428	Benzidine	c		SB		m	1555·4 ±0·4	44
$C_{12}H_{12}N_2$		Benzidine	c		R	9	m	24·57±0·24	44
$C_{12}H_{12}N_2$		Benzidine							
$C_{12}H_{16}N_2$	188·2747	2, 5, 2', 5'-Tetramethyl-dipyrryl-(1, 1')	c		SB	3	CO_2	1706·74±0·20	66
$C_{13}H_{10}N_2$	194·2381	5-Aminoacridine	c		SB	6	m	1602·3 ±1·7	47
$C_{13}H_{10}N_2$	194·2381	3-Aminoacridine	c		SB	8	m	1604·0 ±3·2	47
$C_{14}H_{12}N_2$	208·2651	Dibenzilidene azine	c		SB	5	m	1807·7 ±3·1	48

1 kcal $=$ 4·184 kJ

Remarks		ΔH_f° (l or c) kcal/g.f.w.	6 Determination of ΔH_v			7 ΔH_f° (g) kcal/g.f.w.
		5	Type	ΔH_v° kcal/g.f.w.	Ref.	
		$-3\cdot4\ \pm2\cdot0$	S4	$14\cdot8\ \pm0\cdot8$ (sub.)	60/32	$+11\cdot4\ \pm2\cdot3$
	E	$+10\cdot6\ \pm0\cdot4$	E8	$8\cdot5\ \pm0\cdot5$	48/4	$+19\cdot1\ \pm0\cdot7$
		$-12\cdot23\pm0\cdot75$				
		$-4\cdot90\pm0\cdot58$				
		$+65\cdot82\pm0\cdot26$				
	E	$+39\cdot4\ \pm0\cdot3$				
	BCE	$+30\cdot2\ \pm1\cdot2$				
	BCE	$+59\cdot1\ \pm1\cdot5$				
		$+86\cdot7\ \pm0\cdot6$				
		$+76\cdot6\ \pm0\cdot4$				
		$+52\cdot9\ \pm0\cdot3$				
		$+51\cdot9\ \pm0\cdot4$				
Selected value		$+52\cdot9\ \pm0\cdot3$				
		$+16\cdot9\ \pm0\cdot4$				
drazobenzene(c) {$+52\cdot9\pm0\cdot3$}		$+28\cdot3\ \pm0\cdot5$				
= Benzidine(c)						
Selected value		$+16\cdot9\ \pm0\cdot4$				
An $0\cdot99944\pm0\cdot00012$		$+31\cdot61\pm0\cdot25$				
	BCE	$+38\cdot1\ \pm1\cdot7$				
	BCE	$+39\cdot8\ \pm3\cdot2$				
	E	$+81\cdot1\ \pm3\cdot1$	E8$_{me}$	$22\cdot3\ \pm0\cdot5$ (sub.)	48/4	$+103\cdot4\ \pm3\cdot3$

ΔH_r°

$C_aH_bN_2$ 1 kcal = 4·184 kJ

								Determinat
						No.	Detn.	
1	2	3		Purity		of	of	$-\Delta H_r^\circ$	
Formula	g.f.w.	Name	State	mol %	Type	expts.	reactn.	kcal/g.f.w.	R
$C_{16}H_{16}N_2$	236·3193	Dibenzilidene ethylene-diamine	c		SB	4	m	2116·2 ±2·0	48
$C_{17}H_{24}N_2$	256·3942	3, 3′, 5, 5′-Tetramethyl-4, 4′-diethyl-2, 2′-pyrromethene	c		SB		m	2412·1 ±2·4	33/
$C_{24}H_{20}N_2$	336·4404	Tetraphenylhydrazine	c		SB	5	m	3049·8 ±0·6	42/ 51

1 kcal = 4·184 kJ

ΔH_r°		5	6 Determination of ΔH_v			7
	Remarks	ΔH_f° (l or c) kcal/g.f.w.	Type	ΔH_v° kcal/g.f.w.	Ref.	ΔH_f° (g) kcal/g.f.w.
	E	+64·9 ±2·0	E8$_{me}$	22·0 ±0·5 (sub.)	48/4	+86·9 ±2·2
	BCE	−6·5 ±2·4				
		+109·4 ±0·7				

$C_aH_bN_3$—$C_aH_bN_{10}$ 1 kcal = 4·184 kJ

Determinati

1 Formula	2 g.f.w.	3 Name	State	Purity mol %	No. of Type expts.	Detn. of reactn.	$-\Delta H_f^\circ$ kcal/g.f.w.	Re
C_5HN_3	103·0838	Tricyanoethylene	c		SB 4	m	609·4 ±0·6	63/
$C_5H_3N_3$	105·0998	1, 1, 1-Tricyanoethane	c		SB 2	m	656·6 ±1·2	66/
$C_5H_7N_3$	109·1316	2, 6-Diaminopyridine	c		SB	m	707·80±0·11	56/
$C_5H_9N_3$	111·1476	Cyclopentyl azide	l		SB 4	m	820·46±0·38	54/
$C_6H_5N_3$	119·1269	Benzotriazole	c		SB 6	m	794·84±0·25	53/
$C_6H_{11}N_3$	125·1747	Cyclohexyl azide	l		SB 6	m	965·95±0·39	54/
$C_7H_3N_3$	129·1221	1, 1, 1-Tricyanobutyne-3	c		SB 2	m	905·6 ±1·8	66/
$C_7H_5N_3$	131·1380	1, 1, 1-Tricyanobutene-3	l		SB 2	m	939·8 ±2·5	66/
$C_8H_5N_3$	143·1492	Pyridinium dicyanomethylide	c		SB 6	m	1018·3 ±0·8	67/
$C_{10}H_7N_3$	169·1874	2, 3-Naphthotriazole	c		SB 5	m	1245·15±0·48	35
$C_{10}H_7N_3$	169·1874	1, 2-Naphthotriazole	c		SB 5	m	1241·87±0·48	35
$C_{12}H_{11}N_3$	197·2416	Diazoaminobenzene	c		SB 6	m	1583·60±0·31	53/
$C_{13}H_{11}N_3$	209·2527	2, 8-Diaminoacridine	c		SB 6	m	1628·8 ±1·6	47/
CH_2N_4	70·0539	Tetrazole	c		SB 4	m	219·02±0·21	51/
$C_2H_4N_4$	84·0810	1-Cyanoguanidine	c		SB 4	m	330·69±0·12	48/ 51/
$C_2H_4N_4$ $C_2H_4N_4$		1-Cyanoguanidine 1-Cyanoguanidine	c		SB 7	m	329·81±0·48	52/
$C_2H_4N_4$	84·0810	3-Amino-1, 2, 4-triazole	c		SB 3	m	343·09±0·94	57/
$C_3H_6N_4$	98·1081	1, 5-Dimethyltetrazole	c		SB 4	m	532·18±0·85	51/
$C_4H_{12}N_4$	116·1670	Tetramethyltetrazene	l		SB 5	m	840·24±0·42	60/3
C_6N_4 C_6N_4 C_6N_4	128·0937	Tetracyanoethylene Tetracyanoethylene Tetracyanoethylene	c c		SB 2 SB 4	m m	713·8 ±0·5 713·4 ±0·4	66/ 63/

$$1 \text{ kcal} = 4 \cdot 184 \text{ kJ}$$

H_c°		5 ΔH_f° (l or c) kcal/g.f.w.	6 Determination of ΔH_v			7 ΔH_f° (g) kcal/g.f.w.
	Remarks		Type	ΔH_v° kcal/g.f.w.	Ref.	
		$+105 \cdot 0 \pm 0 \cdot 6$	$V4_{me}$ (sub.)	$19 \cdot 4 \pm 1 \cdot 2$	63/33	$+124 \cdot 4 \pm 1 \cdot 5$
	D	$+83 \cdot 9 \pm 1 \cdot 2$				
		$-1 \cdot 56 \pm 0 \cdot 14$				
		$+42 \cdot 79 \pm 0 \cdot 40$	E	$10 \cdot 0 \pm 1 \cdot 0$	54/15	$+52 \cdot 8 \pm 1 \cdot 2$
	D	$+59 \cdot 75 \pm 0 \cdot 26$				
		$+25 \cdot 91 \pm 0 \cdot 42$	E	$11 \cdot 0 \pm 1 \cdot 0$	54/15	$+36 \cdot 9 \pm 1 \cdot 2$
	D	$+144 \cdot 8 \pm 1 \cdot 8$				
	D	$+110 \cdot 7 \pm 2 \cdot 5$				
		$+95 \cdot 1 \pm 0 \cdot 8$	S4 (sub.)	$30 \cdot 0 \pm 0 \cdot 3$	67/45	$+125 \cdot 1 \pm 0 \cdot 9$
	CE	$+65 \cdot 54 \pm 0 \cdot 53$				
	CE	$+62 \cdot 26 \pm 0 \cdot 53$				
	D	$+79 \cdot 25 \pm 0 \cdot 34$				
	BCE	$+30 \cdot 4 \pm 1 \cdot 6$				
		$+56 \cdot 65 \pm 0 \cdot 22$	S4 (sub.)	$23 \cdot 30 \pm 1 \cdot 00$	51/19	$+79 \cdot 95 \pm 1 \cdot 10$
		$+5 \cdot 96 \pm 0 \cdot 13$				
	Selected value	$+5 \cdot 08 \pm 0 \cdot 49$ $+5 \cdot 96 \pm 0 \cdot 13$				
		$+18 \cdot 36 \pm 0 \cdot 95$				
		$+45 \cdot 08 \pm 0 \cdot 86$				
	D	$+54 \cdot 15 \pm 0 \cdot 44$	V3	$10 \cdot 55 \pm 0 \cdot 50$	60/30	$+64 \cdot 70 \pm 0 \cdot 67$
	D Selected value	$+149 \cdot 5 \pm 0 \cdot 5$ $+149 \cdot 1 \pm 0 \cdot 4$ $+149 \cdot 1 \pm 0 \cdot 4$	S4 (sub.)	$19 \cdot 4 \pm 1 \cdot 4$	63/33	$+168 \cdot 9 \pm 1 \cdot 6$ $+168 \cdot 5 \pm 1 \cdot 5$ $+168 \cdot 5 \pm 1 \cdot 5$

$C_aH_bN_3-C_aH_bN_{10}$ 1 kcal $= 4 \cdot 184$ kJ

1 Formula	2 g.f.w.	3 Name	State	Purity mol %	Type	No. of expts.	Detn. of reactn.	$-\Delta H_r^{\circ}$ kcal/g.f.w.	R
$C_6H_{12}N_4$	140·1893	Hexamethylenetetramine	c		SB	4	m	1003·5 ±1·0	4?
$C_7H_2N_4$	142·1208	1, 1, 2, 2-Tetracyano-cyclopropane	c		SB	2	m	868·1 ±2·5	66
$C_7H_6N_4$	146·1527	1-Phenyltetrazole	c		SB	4	m	949·80±0·36	51
$C_7H_6N_4$	146·1527	5-Phenyltetrazole	c		SB	4	m	933·25±0·33	51
$C_8H_8N_4$	160·1798	1-Phenyl-5-methyl-tetrazole	c		SB	2	m	1094·86±0·48	57
$C_8H_8N_4$	160·1798	1-Methyl-5-phenyl-tetrazole	c		SB	2	m	1095·55±0·30	57
$C_8H_8N_4$	160·1798	2-Phenyl-5-methyl-tetrazole	c		SB	4	m	1091·34±0·60	51
$C_8H_{12}N_4$	164·2116	2, 2'-Azobis-isobutyro-nitrile	c		SB	4	m	1217·0 ±1·0	51
$C_8H_{14}N_4$	166·2276	2, 2'-Hydrazobis-isobutyronitrile	c		SB	4	m	1259·2 ±1·1	51
$C_{12}H_4N_4$	204·1925	7, 7, 8, 8-Tetracyano-quinodimethan	c		SB	5	m	1424·2 ±0·6	63
$C_{12}H_{12}N_4$	212·2562	1, 3-Diphenyltetrazene	c		SB	4	m	1636·99±0·90	51
$C_{13}H_{10}N_4$	222·2515	1, 5-Diphenyltetrazole	c		SB	4	m	1663·64±0·52	51
$C_{13}H_{10}N_4$	222·2515	2, 5-Diphenyltetrazole	c		SB	4	m	1658·72±0·51	51
$C_{13}H_{12}N_4$	224·2674	Formazane	c		SB	4	m	1741·99±0·90	51
$C_{19}H_{16}N_4$	300·3662	Formazylbenzene	c		SB	4	m	2463·3 ±1·0	51
$C_{32}H_{38}N_4$	478·6865	Aetioporphyrin I	c		SB	3	m	4301·6 ±4·3	33
$C_{32}H_{38}N_4$	478·6865	Aetioporphyrin II	c		SB	3	m	4308·0 ±4·3	33
$C_{36}H_{46}N_4$	534·7948	Octaethylporphyrin	c		SB	3	m	4913·3 ±4·9	33

Determina

$1 \text{ kcal} = 4 \cdot 184 \text{ kJ}$

			6 Determination of ΔH_v			
ΔH_r°	Remarks	5 ΔH_f° (l or c) kcal/g.f.w.	Type	ΔH_v° kcal/g.f.w.	Ref.	7 ΔH_f° (g) kcal/g.f.w.
	CE	$+29 \cdot 3 \ \pm 1 \cdot 0$				
	D	$+141 \cdot 4 \ \pm 2 \cdot 5$				
		$+86 \cdot 50 \pm 0 \cdot 37$				
		$+69 \cdot 95 \pm 0 \cdot 34$				
		$+69 \cdot 19 \pm 0 \cdot 50$				
		$+69 \cdot 88 \pm 0 \cdot 33$				
		$+65 \cdot 67 \pm 0 \cdot 61$				
		$+54 \cdot 7 \ \pm 1 \cdot 0$				
		$+28 \cdot 6 \ \pm 1 \cdot 1$				
		$+159 \cdot 0 \ \pm 0 \cdot 6$	S4	$25 \cdot 1 \ \pm 2 \cdot 2$ (sub.)	63/33	$+184 \cdot 1 \ \pm 2 \cdot 4$
		$+98 \cdot 49 \pm 0 \cdot 91$				
		$+99 \cdot 40 \pm 0 \cdot 54$	S4	$29 \cdot 05 \pm 1 \cdot 00$ (sub.)	51/19	$+128 \cdot 45 \pm 1 \cdot 14$
		$+94 \cdot 48 \pm 0 \cdot 53$	S4	$28 \cdot 60 \pm 1 \cdot 00$ (sub.)	51/19	$+123 \cdot 08 \pm 1 \cdot 14$
		$+109 \cdot 44 \pm 0 \cdot 91$				
		$+129 \cdot 8 \ \pm 1 \cdot 0$				
	BCE	$-6 \cdot 0 \ \pm 4 \cdot 3$				
	BCE	$+0 \cdot 4 \ \pm 4 \cdot 3$				
	BCE	$-43 \cdot 8 \ \pm 4 \cdot 9$				

Determinat|

1 Formula	2 g.f.w.	3 Name	State	Purity mol %	Type	No. of expts.	Detn. of reactn.	$-\Delta H_r^\circ$ kcal/g.f.w.	R∈
CH_3N_5	85·0686	5-Aminotetrazole	c		SB	4	m	246·18±0·56	51/
C_2HN_5	95·0638	5-Cyanotetrazole	c		SB	2	m	318·34±0·36	57/
$C_2H_5N_5$	99·0957	1-Methyl-5-amino- tetrazole	c		SB	4	m	405·14±0·22	57/
$C_2H_5N_5$	99·0957	2-Methyl-5-amino- tetrazole	c		SB	5	m	409·28±0·96	57/
$C_2H_5N_5$	99·0957	5-Methylaminotetrazole	c		SB	3	m	407·29±0·94	57/
$C_3H_7N_5$	113·1227	1-Methyl-5-methylamino- tetrazole	c		SB	3	m	569·1 ±1·4	57/
$C_3H_7N_5$	113·1227	5-Dimethylamino- tetrazole	c		SB	2	m	564·9 ±1·0	57/
$C_4H_7N_5$	125·1339	1-Allyl-5-aminotetrazole	c		SB	6	m	678·74±0·44	57/
$C_4H_7N_5$	125·1339	2-Allyl-5-aminotetrazole	c		SB	2	m	682·92±0·38	57/
$C_5H_5N_5$	135·1291	Adenine (6-aminopurine)	c	3s	SB	10	m	663·99±0·21	35
$C_7H_7N_5$	161·1673	1-Phenylaminotetrazole	c		SB	3	m	971·79±0·80	57/
$C_7H_7N_5$	161·1673	5-Phenylaminotetrazole	c		SB	2	m	970·38±0·60	57/
$C_7H_{11}N_5$	165·1992	1-Allyl-5-allylamino- tetrazole	c		SB	2	m	1117·8 ±1·0	57/
$C_7H_{11}N_5$	165·1992	5-Diallylaminotetrazole	c		SB	4	m	1118·0 ±1·0	57/
$C_{10}H_{11}N_5$	201·2327	Benzal-3-hydrazino- 5-methyl-1, 2, 4-triazole	c		SB	3	m	1377·91±0·50	57/
$C_3H_6N_6$	126·1215	Melamine (triamino- triazine)	c		SB	6	m	469·97±0·13	48/ 51/
$C_3H_6N_6$		Melamine (triamino- triazine)	c		SB	5	m	469·55±0·47	56/
$C_3H_6N_6$		Melamine (triamino- triazine)							

1 kcal = 4·184 kJ

ΔH_r°		5 ΔH_f° (l or c) kcal/g.f.w.	6 Determination of ΔH_v			7 ΔH_f° (g) kcal/g.f.w.
	Remarks		Type	ΔH_v° kcal/g.f.w.	Ref.	
		$+49\cdot66\pm0\cdot57$				
		$+96\cdot08\pm0\cdot37$				
		$+46\cdot25\pm0\cdot24$				
		$+50\cdot39\pm0\cdot97$				
		$+48\cdot40\pm0\cdot95$				
		$+47\cdot8\ \pm1\cdot4$				
		$+43\cdot6\ \pm1\cdot0$				
		$+63\cdot43\pm0\cdot45$				
		$+67\cdot61\pm0\cdot40$				
		$+22\cdot95\pm0\cdot22$	S4	$26\cdot0\ \pm2\cdot0$ (sub.)	65/32	$+49\cdot0\ \pm2\cdot1$
		$+74\cdot33\pm0\cdot81$				
		$+72\cdot92\pm0\cdot62$				
		$+83\cdot7\ \pm1\cdot0$				
		$+83\cdot9\ \pm1\cdot0$				
		$+61\cdot67\pm0\cdot55$				
		$-17\cdot13\pm0\cdot15$	S4	$29\cdot5\ \pm1\cdot0$ (sub.)	60/33	$+12\cdot4\ \pm1\cdot1$
		$-17\cdot55\pm0\cdot48$				$+11\cdot9\ \pm1\cdot2$
	Selected value	$-17\cdot13\pm0\cdot15$				$+12\cdot4\ \pm1\cdot1$

$C_aH_bN_3–C_aH_bN_{10}$ 1 kcal = 4·184 kJ

1 Formula	2 g.f.w.	3 Name	State	Purity mol %	Type	No. of expts.	Detn. of reactn.	$-\Delta H_r^\circ$ kcal/g.f.w.	Re
$C_8H_8N_6$	188·1932	Benzalhydrazone of 5-Hydrazinotetrazole	c		SB	4	m	1135·15±0·95	51/
$C_8H_8N_6$	188·1932	Benzal-5-hydrazino-tetrazole	c		SB	3	m	1131·07±0·34	57/
$C_{12}H_4N_6$	232·2059	1, 1, 1, 6, 6, 6-Hexa-cyanohexyne-3	c		SB	2	m	1504·6 ±1·2	66/
$C_{12}H_6N_6$	234·2218	1, 1, 1, 6, 6, 6-Hexa-cyanohexene-3	c		SB	2	m	1533·7 ±6·3	66/
$C_{14}H_4N_6$	256·2282	1, 1, 1, 8, 8, 8-Hexa-cyano-octadiyne-3, 5	c		SB	2	m	1744·6 ±3·0	66/
$C_2H_5N_7$	127·1091	5-Guanylaminotetrazole	c		SB	4	m	399·44±0·37	51/
$C_4H_6N_8$	166·1460	1, 2-Di-(5-tetrazolyl) ethane	c		SB	2	m	687·36±0·30	57/
$C_2H_4N_{10}$	168·1212	5, 5'-Hydrazotetrazole	c		SB	4	m	459·9 ±1·0	51/
$C_4H_6N_{10}$	194·1594	cis-1, 1'-Dimethyl-5, 5'-azotetrazole	c		SB	2	m	769·74±0·60	57/
$C_4H_6N_{10}$	194·1594	trans-1, 1'-Dimethyl-5, 5'-azotetrazole	c		SB	9	m	770·5 ±2·4	57/
$C_4H_6N_{10}$	194·1594	2, 2'-Dimethyl-5, 5'-azotetrazole	c		SB	2	m	761·5 ±1·1	57/
$C_6H_{10}N_{10}$	222·2136	2, 2'-Diethyl-5, 5'-azotetrazole	c		SB	3	m	1062·5 ±1·4	57/

Determinat

1 kcal = 4·184 kJ

ΔH_r°			6 Determination of ΔH_v			
	Remarks	5 ΔH_f° (l or c) kcal/g.f.w.	Type	ΔH_v° kcal/g.f.w.	Ref.	7 ΔH_f° (g) kcal/g.f.w.
		+109·48±0·96				
		+105·40±0·37				
			D	+239·3 ±1·2		
			D	+200·1 ±6·3		
			D	+291·2 ±3·0		
		+40·55±0·38				
		+106·21±0·31				
		+135·2 ±1·0				
		+188·59±0·62				
		+189·3 ±2·4				
		+180·3 ±1·1				
		+156·6 ±1·4				

C_aH_bON

$1 \text{ kcal} = 4 \cdot 184 \text{ kJ}$

1 Formula	2 g.f.w.	3 Name	State	Purity mol %	Type	No. of expts.	Detn. of reactn.	$-\Delta H_r^\circ$ kcal/g.f.w.	R
C_3H_7ON	73·0953	Dimethylformamide	l		SB	5	m	464·12±0·69	57
C_4H_5ON	83·0906	5-Methylisoxazole	l		SB	4	m	540·57±0·54	40
C_4H_5ON	83·0906	3-Methylisoxazole	l		SB	4	m	541·96±0·54	40
C_4H_7ON	85·1065	2-Pyrrolidone	l		SB		m	546·9 ±0·1	55
C_5H_5ON	95·1017	Pyrrole-2-aldehyde	c		SB		m	615·6 ±0·6	33
C_5H_7ON	97·1176	3, 5-Dimethylisoxazole	l		SB	4	m	694·28±0·69	40
C_5H_9ON	99·1336	N-Methylpyrrolidone	l		SB		CO_2	715·0 ±0·1	59
C_5H_9ON	99·1336	2-Piperidone	c		SB		m	704·4 ±0·1	55
$C_5H_{11}ON$	101·1495	Pentanamide	c		SB		m	755·28±0·25	56
C_6H_9ON	111·1447	Hex-2-ynamide	c		H	5	H_2	72·9 ±0·8	58
C_6H_9ON	111·1447	3, 4, 5-Trimethyl-isoxazole	l		SB	4	m	854·95±0·85	40
$C_6H_{11}ON$	113·1607	N-Methylpiperidone	l		SB		CO_2	870·0 ±0·1	59
$C_6H_{11}ON$	113·1607	ε-Caprolactam	c		SB		m	861·3 ±0·2	55
$C_6H_{13}ON$	115·1766	Hexanamide	c		SB		m	906·87±0·18	56
$C_6H_{13}ON$	115·1766	N-Ethyl-N-methyl-propionamide	l		SB		CO_2	928·1 ±0·1	59
$C_6H_{13}ON$	115·1766	N-Butylacetamide	l		R	4	m	27·06±0·16	62
C_7H_7ON	121·1399	Benzamide	c		SB	6	m	849·04±0·26	42 51
$C_7H_{13}ON$	127·1878	N-Methylcaprolactam	l		SB		CO_2	1029·1 ±0·1	59

Determina[

1 kcal = 4·184 kJ

f ΔH_r°			6 Determination of ΔH_v			
		5				7
Remarks		ΔH_f° (l or c) kcal/g.f.w.	Type	ΔH_v° kcal/g.f.w.	Ref.	ΔH_f° (g) kcal/g.f.w.
	E	−57·14±0·70	V4	11·37±[0·30]	61/63	−45·8 ±0·8
	BCE	−6·42±0·55	E7	10·0 ±0·5		+3·6 ±0·8
	BCE	−5·03±0·55	E7	9·8 ±0·5		+4·8 ±0·8
	E	−68·4 ±0·1				
	BCE	−25·4 ±0·6				
	BCE	−15·08±0·73	E7	10·8 ±0·4		−4·3 ±0·9
C/An 1·0015		−62·7 ±0·1				
	E	−73·3 ±0·1				
	C	−90·70±0·27	S3	21·34±0·10 (sub.)	59/37	−69·36±0·30
$C_6H_9ON(c)+2H_2$ $= CH_3(CH_2)_4CONH_2(c)$ $\{−101·48±0·21\}$		−28·6 ±0·9				
	BCE	−16·77±0·86	E7	12·0 ±0·5		−4·8 ±1·0
C/An 1·0000		−70·0 ±0·1				
	E	−78·7 ±0·2	S3	19·9 ±[0·2] (sub.)	53/32	−58·8 ±0·3
	C	−101·48±0·21	S3	22·72±0·10 (sub.)	59/37	−78·76±0·25
C/An 1·0004		−80·3 ±0·1				
$Ac_2O(l) \{−149·20±0·10\}$ $+BuNH_2(l) \{−30·51±0·29\}$ $= BuNHAc(l)$ $+HOAc(l) \{−115·75±0·07\}$		−91·02±0·39	C1	18·2 ±0·3	65/33	−72·82±0·49
		−48·42±0·28				
C/An 1·0005		−73·3 ±0·2				

THERMOCHEMISTRY OF ORGANIC AND ORGANOMETALLIC COMPOUNDS

C_aH_bON

1 kcal = 4·184 kJ

4
Determinatio

1 Formula	2 g.f.w.	3 Name	State	Purity mol %	Type	No. of expts.	Detn. of reactn.	$-\Delta H_c^\circ$ kcal/g.f.w.	Ref.
$C_7H_{13}ON$	127·1878	5-Methylcaprolactam	c		SB		m	1015·4 ±0·3	55/2
$C_7H_{13}ON$	127·1878	7-Methylcaprolactam	c		SB		m	1015·8 ±0·3	55/2
$C_7H_{13}ON$	127·1878	ζ-Enantholactam	c		SB		m	1019·1 ±0·3	55/2
C_8H_9ON	135·1670	Acetanilide	c		R	5	m	24·11±0·10	62/2
C_8H_9ON	135·1670	p-Aminoacetophenone	c		SB		m	1130·0 ±0·2	49/1
$C_8H_{15}ON$	141·2149	N-Methylenantholactam	l		SB		CO_2	1187·0 ±0·3	59/2
$C_8H_{17}ON$	143·2308	Octanamide	c		SB		m	1220·0 ±0·2	56/1
C_9H_7ON	145·1622	5-Phenylisoxazole	c		SB	4	m	1104·8 ±1·1	40/6
C_9H_7ON	145·1622	3-Phenylisoxazole	l		SB	4	m	1103·6 ±1·1	40/6
C_9H_7ON	145·1622	8-Hydroxyquinoline	c		SB		m	1064·7 ±0·2	49/1
$C_9H_{11}ON$	149·1941	p-Dimethylamino-benzaldehyde	c		SB		m	1189·3 ±0·2	56/19
$C_9H_{13}ON$	151·2101	2, 4-Dimethyl-3-ethyl-5-formylpyrrole	c		SB	6	m	1229·6 ±1·2	33/13
$C_9H_{17}ON$	155·2419	2, 2, 6, 6-Tetramethyl-4-oxopiperidine	c		SB		m	1347·26±0·82	66/27
$C_9H_{19}ON$	157·2579	N-Butylpentanamide	c		SB		m	1384·3 ±0·4	66/28
$C_{10}H_9ON$	159·1893	3-Methyl-5-phenyl-isoxazole	c		SB	4	m	1252·3 ±1·3	40/6
$C_{10}H_9ON$	159·1893	3-Phenyl-5-methyl-isoxazole	c		SB	4	m	1251·2 ±1·3	40/6
$C_{12}H_9ON$	183·2116	2-Methyl-2′, 1′-naphthoxazole	l		SB	5	m	1424·00±0·56	35/5

5. THERMOCHEMICAL DATA

1 kcal = 4·184 kJ

ΔH_r° Remarks	5 ΔH_f° (l or c) kcal/g.f.w.	6 Determination of ΔH_v			7 ΔH_f° (g) kcal/g.f.w.
		Type	ΔH_v° kcal/g.f.w.	Ref.	
	E −87·0 ±0·3				
	E −86·6 ±0·3				
	E −83·3 ±0·3				
c$_2$O(l) {−149·20±0·10} +PhNH$_2$(l) {+7·48±0·18} = PhNHAc(c) +HOAc(l) {−115·75±0·07}	−50·08±0·33	S4	19·3 ±0·2 (sub.)	55/36	−30·8 ±0·4
	C +70·2 ±0·2				
/An 0·9999	−77·8 ±0·3				
	C −113·1 ±0·3	S3	26·4 ±0·7 (sub.)	59/37	−86·7 ±0·8
	BCE +19·2 ±1·1	E7$_{me}$	19·1 ±1·0 (sub.)		+38·3 ±1·5
	BCE +18·0 ±1·1	E7	15·3 ±1·0		+33·3 ±1·5
	C −20·9 ±0·3	S4	26·0 ±0·4 (sub.)	63/60	+5·1 ±0·5
	C −32·9 ±0·3				
	BCE −60·9 ±1·2				
	−79·88±0·84	S3	14·53±0·65 (sub.)	66/27	−65·4 ±1·3
	E −111·2 ±0·4				
	BCE +4·4 ±1·3	E7$_{me}$	20·1 ±1·0 (sub.)		+25·5 ±1·8
	BCE +3·3 ±1·3	E7$_{me}$	20·3 ±1·0 (sub.)		+23·6 ±1·8
	CE −12·03±0·60				

C_aH_bON
1 kcal = 4·184 kJ

1 Formula	2 g.f.w.	3 Name	State	Purity mol %	Type	No. of expts.	Detn. of reactn.	$-\Delta H^\circ$ kcal/g.f.w.	Re
$C_{12}H_9ON$	183·2116	2-Methyl-1′, 2′-naphthoxazole	c		SB	5	m	1415·1 $\pm 2·8$	35
$C_{12}H_9ON$	183·2116	2-Methyl-2′, 3′-naphthoxazole	c		SB	4	m	1423·5 $\pm 2·1$	35
$C_{14}H_{13}ON$	211·2658	N-Acetyldiphenylamine	c		SB	6	m	1750·5 $\pm 1·7$	56/
$C_{14}H_{19}ON$	217·3136	N-Acetyl-N-cyclohexyl-phenylamine	c		SB	10	m	1894·0 $\pm 3·8$	56/
$C_{15}H_{11}ON$	221·2610	3, 5-Diphenylisoxazole	c		SB	4	m	1821·1 $\pm 1·8$	40

1 kcal = 4·184 kJ

ΔH_r°			6 Determination of ΔH_v			
	Remarks	5 ΔH_f° (l or c) kcal/g.f.w.	Type	ΔH_v° kcal/g.f.w.	Ref.	7 ΔH_f° (g) kcal/g.f.w.
	CE	−20·9 ±2·8				
	CE	−12·5 ±2·1				
	BCE	−10·3 ±1·7				
	BCE	−71·7 ±3·8				
	BCE	+34·6 ±1·8				

$C_aH_bO_2N$

1 kcal = 4·184 kJ

1 Formula	2 g.f.w.	3 Name	State	Purity mol %	Type	No. of expts.	Detn. of reactn.	$-\Delta H_f^\circ$ kcal/g.f.w.	Re
CH_3O_2N	61·0406	Methyl nitrite	g		FC	6	CO_2	179·7 ±0·8	61/
CH_3O_2N		Methyl nitrite	g		R	12	m	2·14±0·20	62/
CH_3O_2N		Methyl nitrite							
CH_3O_2N	61·0406	Nitromethane	l		SB	3	m	175·25±0·18	49/
CH_3O_2N		Nitromethane	l		SB	3	m	173·7 ±0·3	58/
CH_3O_2N		Nitromethane	l		SB			169·49±0·14	58/
CH_3O_2N		Nitromethane							
$C_2H_5O_2N$	75·0677	Nitroethane	l		SB	4	m	325·42±0·30	49/
$C_2H_5O_2N$		Nitroethane	l		SB	3	m	326·3 ±0·3	58/
$C_2H_5O_2N$		Nitroethane	l		SB			324·57±0·25	58/
$C_2H_5O_2N$		Nitroethane							
$C_2H_5O_2N$	75·0677	Glycine	c		SB	4	m	232·57±0·10	37/
$C_2H_5O_2N$		Glycine	c		SB		m	230·5 ±0·2	58/ 63/
$C_2H_5O_2N$		Glycine							
$C_3H_7O_2N$	89·0947	1-Nitropropane	l		SB	4	m	481·22±0·61	49/
$C_3H_7O_2N$		1-Nitropropane	l		SB	5	m	481·6 ±0·7	58/
$C_3H_7O_2N$		1-Nitropropane	l		SB			480·91±0·29	58/
$C_3H_7O_2N$		1-Nitropropane							
$C_3H_7O_2N$	89·0947	2-Nitropropane	l		SB	5	m	477·49±0·17	49/
$C_3H_7O_2N$		2-Nitropropane	l		SB	4	m	477·7 ±0·3	58/
$C_3H_7O_2N$		2-Nitropropane	l		SB			478·17±0·19	58/
$C_3H_7O_2N$		2-Nitropropane							
$C_3H_7O_2N$	89·0947	Sarcosine	c		SB	7	m	399·7 ±0·3	52/ 52/
$C_3H_7O_2N$	89·0947	L-Alanine	c		SB		m	376·9 ±0·5	58/
$C_3H_7O_2N$	89·0947	DL-Alanine	c		SB	7	m	386·56±0·14	37/
$C_3H_7O_2N$	89·0947	D-Alanine	c		SB	3	m	387·12±0·13	36/
$C_4H_5O_2N$	99·0900	Succinimide	c		SB	3	CO_2	437·28±0·05	66/
$C_4H_9O_2N$	103·1218	1-Nitrobutane	l		SB	4	m	637·62±0·32	49/

$$1 \text{ kcal} = 4\cdot184 \text{ kJ}$$

ΔH_r°	Remarks	5 ΔH_f° (l or c) kcal/g.f.w.	6 Determination of ΔH_v			7 ΔH_f° (g) kcal/g.f.w.
			Type	ΔH_v° kcal/g.f.w.	Ref.	
OH(g) {−48·07±0·05} +NOCl(g) {+12·36} = CH₃ONO(g)+HCl(g)						−16·8 ±0·8 −15·79±0·25
	Selected value					−15·79±0·25
		−21·27±0·19	C1	9·17±0·01	54/16	−12·10±0·19
	D	−22·8 ±0·3				−13·6 ±0·3
.S. measurement		−27·03±0·15				−17·86±0·15
	Selected value	−27·03±0·15				−17·86±0·15
		−33·47±0·31	V2	9·94±0·10	49/18	−23·53±0·35
	D	−32·6 ±0·3				−22·7 ±0·3
.S. measurement		−34·32±0·26				−24·38±0·30
	Selected value	−34·32±0·26				−24·38±0·30
		−126·32±0·12	S4	32·6 ±0·1 (sub.)	65/53	−93·72±0·15
	C	−128·4 ±0·2				−95·8 ±0·2
	Selected value	−126·32±0·12				−93·72±0·15
		−40·04±0·62	V2	10·37±0·10	49/18	−29·67±0·64
		−39·7 ±0·7				−29·3 ±0·7
.S. measurement		−40·35±0·30				−29·98±0·33
	Selected value	−40·35±0·30				−29·98±0·33
		−43·77±0·18	V2	9·88±0·10	49/18	−33·89±0·22
		−43·6 ±0·3				−33·7 ±0·3
.S. measurement		−43·09±0·20				−33·21±0·23
	Selected value	−43·09±0·20				−33·21±0·23
		−121·6 ±0·3				
	C	−144·4 ±0·5	S4	33·0 ±0·2 (sub.)	65/53	−111·4 ±0·6
		−134·70±0·16				
		−134·14±0·15				
An 0·99997±0·00004		−109·71±0·07				
		−46·00±0·33	V2	11·61±0·12	49/18	−34·39±0·36

$C_aH_bO_2N$ 1 kcal = 4·184 kJ

1 Formula	2 g.f.w.	3 Name	State	Purity mol %	Type	No. of expts.	Detn. of reactn.	$-\Delta H_f^\circ$ kcal/g.f.w.	R
$C_4H_9O_2N$	103·1218	2-Nitrobutane	l		SB	4	m	634·03±0·36	49
$C_4H_9O_2N$	103·1218	4-Aminobutanoic acid	c		SB		m	545·5 ±0·2	55
$C_4H_9O_2N$	103·1218	n-Propylcarbamate	c		SB		m	551·55±0·10	56
$C_5H_9O_2N$	115·1330	DL-Proline	c		SB	2	m	652·39±0·13	60
$C_5H_{11}O_2N$	117·1489	5-Aminopentanoic acid	c		SB		m	701·6 ±0·2	55
$C_5H_{11}O_2N$	117·1489	L-Valine	c		SB		m	698·31±0·13	57, 63
$C_6H_{13}O_2N$	131·1760	L-Leucine	c		SB	6	m	856·01±0·20	37
$C_6H_{13}O_2N$		L-Leucine	c		SB		m	853·72±0·06	57
$C_6H_{13}O_2N$		L-Leucine							
$C_6H_{13}O_2N$	131·1760	DL-Leucine	c		SB	6	m	855·25±0·21	37
$C_6H_{13}O_2N$	131·1760	D-Leucine	c		SB	3	m	856·03±0·21	37
$C_6H_{13}O_2N$	131·1760	L-Isoleucine	c		SB		m	855·9 ±0·2	57, 63
$C_6H_{13}O_2N$	131·1760	DL-Isoleucine	c		SB	3	m	856·52±0·43	60
$C_6H_{13}O_2N$	131·1760	2-Aminohexanoic acid	c		SB		m	855·6 ±0·3	55
$C_6H_{13}O_2N$	131·1760	4-Aminohexanoic acid	c		SB		m	853·9 ±0·3	55
$C_6H_{13}O_2N$	131·1760	5-Aminohexanoic acid	c		SB		m	854·6 ±0·3	55
$C_6H_{13}O_2N$	131·1760	6-Aminohexanoic acid	c		SB		m	855·6 ±0·3	55
$C_7H_7O_2N$	137·1393	Salicylaldoxime	c		SB		m	853·55±0·16	56
$C_7H_9O_2N$	139·1553	Methylethylmaleimide	c		SB		m	853·5 ±0·9	33
$C_7H_{15}O_2N$	145·2031	Ethyldiethylcarbamate	l		SB		m	1029·15±0·51	57
$C_7H_{15}O_2N$	145·2031	7-Aminoheptanoic acid	c		SB		m	1011·2 ±0·3	66
$C_8H_5O_2N$	147·1346	Isatin	c		SB		m	859·1 ±0·9	33
$C_8H_7O_2N$	149·1505	ω-Nitrostyrene	c		SB		m	998·8 ±0·5	56

Determina

$$1 \text{ kcal} = 4 \cdot 184 \text{ kJ}$$

ΔH_r° Remarks	5 ΔH_f° (l or c) kcal/g.f.w.	6 Determination of ΔH_v Type	ΔH_v° kcal/g.f.w.	Ref.	7 ΔH_f° (g) kcal/g.f.w.
	$-49 \cdot 59 \pm 0 \cdot 37$	V2	$10 \cdot 48 \pm 0 \cdot 10$	49/18	$-39 \cdot 11 \pm 0 \cdot 39$
	$-138 \cdot 1 \ \pm 0 \cdot 2$				
C	$-132 \cdot 07 \pm 0 \cdot 12$	V4$_{me}$	$19 \cdot 40 \pm 0 \cdot 50$ (sub.)	47/15	$-112 \cdot 67 \pm 0 \cdot 54$
E	$-125 \cdot 28 \pm 0 \cdot 15$				
E	$-144 \cdot 4 \ \pm 0 \cdot 2$				
C	$-147 \cdot 68 \pm 0 \cdot 15$	S4	$38 \cdot 9 \ \pm 0 \cdot 2$ (sub.)	65/53	$-108 \cdot 78 \pm 0 \cdot 26$
	$-152 \cdot 34 \pm 0 \cdot 24$	S4	$36 \cdot 0 \ \pm 0 \cdot 2$ (sub.)	65/53	$-116 \cdot 34 \pm 0 \cdot 30$
C	$-154 \cdot 63 \pm 0 \cdot 11$				$-118 \cdot 63 \pm 0 \cdot 25$
Selected value	$-152 \cdot 34 \pm 0 \cdot 24$				$-116 \cdot 34 \pm 0 \cdot 30$
	$-153 \cdot 10 \pm 0 \cdot 25$				
	$-152 \cdot 32 \pm 0 \cdot 25$				
C	$-152 \cdot 5 \ \pm 0 \cdot 2$				
E	$-151 \cdot 83 \pm 0 \cdot 45$				
E	$-152 \cdot 8 \ \pm 0 \cdot 3$				
E	$-154 \cdot 5 \ \pm 0 \cdot 3$				
E	$-153 \cdot 8 \ \pm 0 \cdot 3$				
E	$-152 \cdot 8 \ \pm 0 \cdot 3$				
C	$-43 \cdot 91 \pm 0 \cdot 18$				
BCE	$-112 \cdot 3 \ \pm 0 \cdot 9$				
BCE	$-141 \cdot 57 \pm 0 \cdot 54$				
E	$-159 \cdot 5 \ \pm 0 \cdot 3$				
BCE	$-64 \cdot 1 \ \pm 0 \cdot 9$				
C	$+7 \cdot 3 \ \pm 0 \cdot 5$				

THERMOCHEMISTRY OF ORGANIC AND ORGANOMETALLIC COMPOUNDS

$C_aH_bO_2N$ 1 kcal = 4·184 kJ

1 Formula	2 g.f.w.	3 Name	State	Purity mol %	Type	No. of expts.	Detn. of reactn.	$-\Delta H_r^\circ$ kcal/g.f.w.	Ref
$C_8H_9O_2N$	151·1664	o-Nitroethylbenzene	l		SB	7	m	1048·2 ±1·5	57/2
$C_8H_9O_2N$	151·1664	p-Nitroethylbenzene	l		SB	6	m	1046·6 ±1·5	57/2
$C_8H_9O_2N$	151·1664	Methyl phenyl-carbamate	c		SB		m	1015·2 ±0·5	65/3
$C_8H_{11}O_2N$	153·1824	2-Methyl-3-carbethoxy-pyrrole	c		SB		m	1029·9 ±1·0	33/1
$C_8H_{11}O_2N$	153·1824	2, 4-Dimethyl-3-carbo-methoxypyrrole	c		SB		m	1027·2 ±1·0	32/
$C_8H_{11}O_2N$	153·1824	2, 4-Dimethyl-5-carbo-methoxypyrrole	c		SB		m	1027·2 ±1·0	32/
$C_8H_{15}O_2N$	157·2143	N-Butyldiacetimide	l		R	4	m	9·68±0·12	65/3.
$C_9H_9O_2N$	163·1776	1 Phenyl-2-nitropropene	c		SB		m	1150·2 ±0·5	56/1
$C_9H_{11}O_2N$	165·1935	L-Phenylalanine	c		SB		m	1110·6 ±0·2	58/2 63/3
$C_9H_{11}O_2N$	165·1935	DL-Phenylalanine	c		SB	6	m	1112·1 ±0·6	52/1 52/18
$C_9H_{11}O_2N$	165·1935	Ethyl p-aminobenzoate	c		SB		m	1122·3 ±0·2	49/1
$C_9H_{13}O_2N$	167·2095	2, 5-Dimethyl-3-carbethoxypyrrole	c		SB	5	m	1176·1 ±1·2	33/13
$C_9H_{13}O_2N$	167·2095	2, 4-Dimethyl-5-carbethoxypyrrole	c		SB	4	m	1177·1 ±1·2	33/13
$C_9H_{13}O_2N$	167·2095	2, 3-Dimethyl-4-carbethoxypyrrole	c		SB	6	m	1178·1 ±1·2	33/1
$C_9H_{13}O_2N$	167·2095	2, 4-Dimethyl-3-carbethoxypyrrole	c		SB	4	m	1179·8 ±1·2	33/13
$C_9H_{13}O_2N$	167·2095	2, 3-Dimethyl-5-carbethoxypyrrole	c		SB	4	m	1183·1 ±1·2	33/13

1 kcal = 4·184 kJ

			6 Determination of ΔH_v			
ΔH_r°		5 ΔH_f° (l or c) kcal/g.f.w.	ΔH_v° Type kcal/g.f.w.		Ref.	7 ΔH_f° (g) kcal/g.f.w.
Remarks						
	E	−11·6 ±1·5	E7 14·3 ±0·5			+2·7 ±1·6
	E	−13·2 ±1·5	E7 15·0 ±0·5			+1·8 ±1·6
		−44·6 ±0·5				
	BCE	−98·2 ±1·0				
	BCE	−100·9 ±1·0				
	BCE	−100·9 ±1·0				
₁NAc₂(l)+H₂O(l) = BuNHAc(l) {−91·02±0·39} +HOAc(l) {−115·75±0·07}		−128·77±0·44	Cl 15·4 ±0·1		65/33	−113·37±0·47
	C	−3·7 ±0·5				
	C	−111·6 ±0·3	S4 36·8 ±0·2 (sub.)		65/53	−74·8 ±0·4
		−110·1 ±0·6				
	C	−99·9 ±0·3				
	BCE	−114·4 ±1·2				
	BCE	−113·4 ±1·2				
	BCE	−112·4 ±1·2				
	BCE	−110·7 ±1·2				
	BCE	−107·4 ±1·2				

THERMOCHEMISTRY OF ORGANIC AND ORGANOMETALLIC COMPOUNDS

$C_aH_bO_2N$

1 kcal = 4·184 kJ

1 Formula	2 g.f.w.	3 Name	State	Purity mol %	No. of Type expts.	Detn. of reactn.	$-\Delta H_c^o$ kcal/g.f.w.	Re
$C_9H_{16}O_2N$	170·2334	2, 2, 6, 6-Tetramethyl-4-oxopiperidine-1-oxyl	c		SB	m	1321·63±1·55	66/
$C_9H_{17}O_2N$	171·2413	2, 2, 6, 6-Tetramethyl-1-hydroxy-4-oxo-piperidine	c		SB	m	1336·77±0·10	66/
$C_9H_{18}O_2N$	172·2493	2, 2, 6, 6-Tetramethyl-4-hydroxypiperidine-1-oxyl	c		SB	m	1367·45±1·82	66/
$C_9H_{19}O_2N$	173·2573	2, 2, 6, 6-Tetramethyl-1,4-dihydroxypiperidine	c		SB 5	m	1388·98±0·33	66/
$C_9H_{19}O_2N$	173·2573	9-Aminononanoic acid	c		SB	m	1321·5 ±0·4	55/
$C_{10}H_7O_2N$	173·1728	1-Nitronaphthalene	c		SB 5	m	1189·8 ±1·2	37/
$C_{10}H_{11}O_2N$	177·2047	N-Phenyldiacetimide	c		R 4	m	10·88±0·14	65/
$C_{10}H_{13}O_2N$	179·2206	Ethyl NN-phenylmethyl-carbamate	l		SB 9	m	1292·8 ±1·2	54/
$C_{10}H_{13}O_2N$	179·2206	Isopropyl phenyl-carbamate	c		SB	m	1278·9 ±0·5	65/
$C_{10}H_{15}O_2N$	181·2366	2, 4, 5-Trimethyl-3-carbethoxypyrrole	c		SB	m	1335·8 ±1·3	32/
$C_{11}H_{15}O_2N$	193·2477	Ethyl NN-phenylethyl-carbamate	l		SB 8	m	1446·4 ±1·4	54/
$C_{11}H_{15}O_2N$	193·2477	Butyl phenylcarbamate	c		SB	m	1452·4 ±0·6	65/3
$C_{11}H_{15}O_2N$	193·2477	2, 4-Dimethyl-3-vinyl-5-carbethoxypyrrole	c		SB	m	1434·0 ±1·4	32/
$C_{11}H_{17}O_2N$	195·2636	2, 4-Dimethyl-3-ethyl-5-carbethoxypyrrole	c		SB 6	m	1487·6 ±1·5	33/
$C_{12}H_9O_2N$	199·2110	3-Nitrobiphenyl	c		SB 3	m	1451·6 ±1·5	35/
$C_{12}H_9O_2N$	199·2110	4-Nitrobiphenyl	c		SB 3	m	1445·7 ±1·5	35/

$$1 \text{ kcal} = 4 \cdot 184 \text{ kJ}$$

ΔH_r°			6 Determination of ΔH_v			7
Remarks		5 ΔH_f° (l or c) kcal/g.f.w.	Type	ΔH_v° kcal/g.f.w.	Ref.	ΔH_f° (g) kcal/g.f.w.
		$-71\cdot35\pm1\cdot56$	S3	$19\cdot91\pm0\cdot40$ (sub.)	65/35	$-51\cdot44\pm1\cdot60$
		$-90\cdot37\pm0\cdot18$	S3	$19\cdot15\pm1\cdot10$ (sub.)	65/35	$-71\cdot22\pm1\cdot12$
		$-93\cdot84\pm1\cdot83$	S3	$24\cdot26\pm1\cdot24$ (sub.)	66/27	$-69\cdot58\pm2\cdot21$
		$-106\cdot47\pm0\cdot38$	S3	$24\cdot00\pm0\cdot15$ (sub.)	66/27	$-82\cdot47\pm0\cdot45$
	E	$-174\cdot0\ \pm0\cdot4$				
	CE	$+10\cdot2\ \pm1\cdot2$	S4	$25\cdot6\ \pm[0\cdot5]$ (sub.)	50/28	$+35\cdot8\ \pm1\cdot3$
NAc$_2$(c)+H$_2$O(l) = PhNHAc(c) {$-50\cdot08\pm0\cdot33$} +HOAc(l) {$-115\cdot75\pm0\cdot07$}		$-86\cdot63\pm0\cdot37$	C1	$21\cdot5\ \pm0\cdot2$ (sub.)	65/33	$-65\cdot13\pm0\cdot43$
	E	$-91\cdot8\ \pm1\cdot2$				
		$-105\cdot7\ \pm0\cdot5$				
	BCE	$-117\cdot1\ \pm1\cdot3$				
	E	$-100\cdot5\ \pm1\cdot4$				
		$-94\cdot5\ \pm0\cdot6$				
	BCE	$-112\cdot9\ \pm1\cdot4$				
	BCE	$-127\cdot6\ \pm1\cdot5$				
	BCE	$+15\cdot6\ \pm1\cdot5$				
	BCE	$+9\cdot7\ \pm1\cdot5$				

$C_aH_bO_2N$ 1 kcal = 4·184 kJ

1 Formula	2 g.f.w.	3 Name	State	Purity mol %	Type	No. of expts.	Detn. of reactn.	$-\Delta H_c^\circ$ kcal/g.f.w.	Ref
$C_{12}H_{19}O_2N$	209·2907	2, 4-Dimethyl-3-propyl-5-carbethoxypyrrole	c		SB		m	1641·6 ±1·6	32/
$C_{14}H_{11}O_2N$	225·2493	*cis*-4-Nitrostilbene	c		SB	4	m	1724·0 ±0·7	50/1
$C_{14}H_{11}O_2N$	225·2493	*trans*-4-Nitrostilbene	c		SB	5	m	1717·0 ±0·7	50/1
$C_{14}H_{21}O_2N$	235·3290	2, 4-Diethyl-3, 5-dipropionylpyrrole	c		SB		m	1919·2 ±1·9	32/
$C_{15}H_{15}O_2N$	241·2923	Ethyl NN-diphenyl-carbamate	c		SB	5	m	1856·0 ±1·9	54/1
$C_{16}H_{15}O_2N$	253·3035	Acetoacetyldiphenylamine	c		SB	6	m	1962·7 ±2·0	56/1
$C_{17}H_{13}O_2N$	263·2987	3-N-Phenylamino-2-naphthoic acid	c		SB		m	1979·9 ±0·5	56/1

Determinati

4

$$1 \text{ kcal} = 4 \cdot 184 \text{ kJ}$$

ΔH_r°		5 ΔH_f° (l or c) kcal/g.f.w.	6 Determination of ΔH_v			7 ΔH_f° (g) kcal/g.f.w.
Remarks			Type	ΔH_v° kcal/g.f.w.	Ref.	
	BCE	$-136\cdot0$ $\pm1\cdot6$				
	A	$+31\cdot6$ $\pm0\cdot7$				
	A	$+24\cdot6$ $\pm0\cdot7$				
	BCE	$-114\cdot8$ $\pm1\cdot9$				
	E	$-67\cdot1$ $\pm1\cdot9$				
	BCE	$-54\cdot4$ $\pm2\cdot0$				
	C	$-63\cdot0$ $\pm0\cdot6$				

$C_8H_9O_3N$

$$1 \text{ kcal} = 4\cdot184 \text{ kJ}$$

1 Formula	2 g.f.w.	3 Name	State	Purity mol %	Type	No. of expts.	Detn. of reactn.	$-\Delta H_r^\circ$ kcal/g.f.w.	Re
CH_3O_3N	$77\cdot0400$	Methyl nitrate	g		R	4	m	$8\cdot66\pm0\cdot20$	59/
$C_2H_5O_3N$	$91\cdot0671$	Ethyl nitrate	l		SB	5	CO_2	$313\cdot39\pm0\cdot23$	57/
$C_2H_5O_3N$	$91\cdot0671$	2-Nitroethanol	l		SB	6	m	$275\cdot06\pm0\cdot54$	53/
$C_3H_7O_3N$	$105\cdot0941$	n-Propyl nitrate	l		SB	6	CO_2	$470\cdot00\pm0\cdot30$	57/
$C_3H_7O_3N$	$105\cdot0941$	Isopropyl nitrate	l		SB	5	CO_2	$466\cdot35\pm0\cdot30$	57/
$C_3H_7O_3N$		Isopropyl nitrate	l		SB		m	$467\cdot24\pm0\cdot47$	57/
$C_3H_7O_3N$		Isopropyl nitrate							
$C_4H_5O_3N$	$115\cdot0894$	Sarcosine-N-carboxylic anhydride	c		SB	4	m	$403\cdot5\ \pm0\cdot7$	52/ 52/1
$C_4H_9O_3N$	$119\cdot1212$	2-Nitro-2-methyl- propanol	c	$99\cdot6$	SB	5	m	$585\cdot60\pm0\cdot40$	52/
$C_4H_9O_3N$	$119\cdot1212$	3-Nitro-2-butanol	l		SB	4	m	$590\cdot4\ \pm1\cdot2$	52/1
$C_4H_9O_3N$	$119\cdot1212$	L-Threonine	c		SB		m	$490\cdot7\ \pm0\cdot2$	58/
$C_4H_9O_3N$	$119\cdot1212$	DL-Threonine	c		SB	2	m	$502\cdot27\pm0\cdot10$	60/
$C_5H_9O_3N$	$131\cdot1324$	L-Hydroxyproline	c		SB	4	m	$619\cdot67\pm0\cdot43$	60/
$C_8H_{11}O_3N$	$169\cdot1818$	3-Hydroxy-4-carbethoxy- 2-methylpyrrole	c		SB		m	$981\cdot9\ \pm1\cdot0$	33/
$C_9H_9O_3N$	$179\cdot1770$	Hippuric acid	c		SB	25	m	$1007\cdot78\pm0\cdot30$	38/
$C_9H_9O_3N$		Hippuric acid	c		SB		m	$1008\cdot3\ \pm0\cdot4$	51/
$C_9H_9O_3N$		Hippuric acid	c		SB		m	$1009\cdot4\ \pm0\cdot4$	54/1
$C_9H_9O_3N$		Hippuric acid	c		SB	6	m	$1008\cdot09\pm0\cdot38$	57/
$C_9H_9O_3N$		Hippuric acid	c		SB	11	CO_2	$1008\cdot35\pm0\cdot39$	61/1
$C_9H_9O_3N$		Hippuric acid							
$C_9H_{11}O_3N$	$181\cdot1929$	L-Tyrosine	c		SB	9	m	$1058\cdot45\pm0\cdot37$	37/1
$C_9H_{11}O_3N$		L-Tyrosine	c		SB			$1061\cdot7\ \pm0\cdot4$	44/1
$C_9H_{11}O_3N$		L-Tyrosine							
$C_9H_{13}O_3N$	$183\cdot2089$	2, 4-Dimethyl- 3-carbethoxyhydroxy- pyrrole	c		SB		m	$1126\cdot9\ \pm1\cdot1$	32/

1 kcal = 4·184 kJ

| ΔH_r° | | | | 6 Determination of ΔH_v | | 7 |
| | | 5 | | | | |
Remarks		ΔH_f° (l or c) kcal/g.f.w.	Type	ΔH_v° kcal/g.f.w.	Ref.	ΔH_f° (g) kcal/g.f.w.
$O_5(g)\{+2\cdot7\}+CH_3ONO(g)\{-15\cdot79\}$ $=0\cdot62N_2O_4(g)\{+2\cdot19\}$ $+0\cdot76NO_2(g)\{+7\cdot93\}+CH_3NO_3(g)$		$-37\cdot26\pm0\cdot32$	V2	$8\cdot15\pm[0\cdot10]$	57/44	$-29\cdot11\pm0\cdot30$
		$-45\cdot50\pm0\cdot25$	V2	$8\cdot67\pm[0\cdot10]$	57/44	$-36\cdot83\pm0\cdot28$
		$-83\cdot83\pm0\cdot56$				
		$-51\cdot26\pm0\cdot31$	V2	$9\cdot70\pm[0\cdot10]$	57/44	$-41\cdot56\pm0\cdot33$
		$-54\cdot91\pm0\cdot31$	V2	$9\cdot27\pm[0\cdot10]$	57/44	$-45\cdot64\pm0\cdot34$
		$-54\cdot02\pm0\cdot48$				$-44\cdot75\pm0\cdot50$
		$-54\cdot91\pm0\cdot31$				$-45\cdot64\pm0\cdot34$
	D	$-143\cdot5\ \pm0\cdot7$				
		$-98\cdot02\pm0\cdot41$				
		$-93\cdot2\ \pm1\cdot2$				
	C	$-192\cdot9\ \pm0\cdot2$				
	E	$-181\cdot35\pm0\cdot12$				
	E	$-158\cdot00\pm0\cdot44$				
	BCE	$-146\cdot2\ \pm1\cdot0$				
		$-146\cdot10\pm0\cdot32$				
		$-145\cdot6\ \pm0\cdot4$				
	C	$-144\cdot5\ \pm0\cdot4$				
		$-145\cdot79\pm0\cdot39$				
		$-145\cdot53\pm0\cdot40$				
Selected value		$-145\cdot61\pm0\cdot30$				
		$-163\cdot75\pm0\cdot40$				
		$-160\cdot5\ \pm0\cdot4$				
Selected value		$-163\cdot75\pm0\cdot40$				
	BCE	$-163\cdot6\ \pm1\cdot1$				

$C_aH_bO_3N$ 1 kcal = 4·184 kJ

1 Formula	2 g.f.w.	3 Name	State	Purity mol %	Type	No. of expts.	Detn. of reactn.	$-\Delta H_r^\circ$ kcal/g.f.w.	R
$C_{10}H_9O_3N$	191·1881	Phenylalanine-N-carboxylic anhydride	c		SB	4	m	1115·9 ±0·8	52 52
$C_{10}H_{13}O_3N$	195·2200	2-Nitro-2-methyl-1-phenylpropanol	c	99·8	SB	5	m	1308·8 ±1·4	52
$C_{10}H_{13}O_3N$	195·2200	2-Nitro-2-methyl-3-phenylpropanol	c	99·9	SB	4	m	1301·5 ±1·1	52
$C_{10}H_{13}O_3N$	195·2200	2, 4-Dimethyl-5-carbethoxy-3-formylpyrrole	c		SB		m	1231·4 ±1·2	3
$C_{10}H_{13}O_3N$	195·2200	2, 4-Dimethyl-3-carbethoxy-5-formylpyrrole	c		SB		m	1230·2 ±1·2	3
$C_{11}H_{15}O_3N$	209·2471	2, 4-Dimethyl-3-acetyl-5-carbethoxypyrrole	c		SB	6	m	1390·5 ±1·4	33
$C_{12}H_{17}O_3N$	223·2742	2, 4-Dimethyl-3-propionyl-5-carbethoxypyrrole	c		SB		m	1546·9 ±1·5	3
$C_{12}H_{17}O_3N$	223·2742	2, 4-Dimethyl-5-propionyl-3-carbethoxypyrrole	c		SB		m	1546·4 ±1·5	3

1 kcal = 4·184 kJ

ΔH_r°	Remarks	5 ΔH_f° (l or c) kcal/g.f.w.	6 Determination of ΔH_v			7 ΔH_f° (g) kcal/g.f.w.
			Type	ΔH_v° kcal/g.f.w.	Ref.	
	D	$-132\cdot0\ \pm0\cdot8$				
		$-75\cdot8\ \pm1\cdot4$				
		$-83\cdot1\ \pm1\cdot1$				
	BCE	$-153\cdot2\ \pm1\cdot2$				
	BCE	$-154\cdot4\ \pm1\cdot2$				
	BCE	$-156\cdot4\ \pm1\cdot4$				
	BCE	$-162\cdot4\ \pm1\cdot5$				
	BCE	$-162\cdot9\ \pm1\cdot5$				

$C_aH_bO_4N - C_aH_bO_6N$ 1 kcal = 4·184 kJ

1 Formula	2 g.f.w.	3 Name	State	Purity mol %	Type	No. of expts.	Detn. of reactn.	$-\Delta H_r^\circ$ kcal/g.f.w.	R
$C_3H_5O_4N$	119·0776	Methyl nitroacetate	l		SB		m	342·03±0·06	49/
$C_4H_7O_4N$	133·1047	L-Aspartic acid	c		SB	7	m	382·68±0·19	36/
$C_4H_7O_4N$		L-Aspartic acid	c		SB			385·0 ±1·0	44/
$C_4H_7O_4N$		L-Aspartic acid							
$C_4H_9O_4N$	135·1206	2-Nitro-2-methyl-1, 3-propanediol	c	99·6	SB	4	m	545·00±0·66	52/
$C_4H_9O_4N$		2-Nitro-2-methyl-1, 3-propanediol	c		SB	6	m	547·27±0·55	54/
$C_4H_9O_4N$		2-Nitro-2-methyl-1, 3-propanediol							
$C_5H_9O_4N$	147·1318	L-Glutamic acid	c		SB		m	536·35±0·18	57/
$C_5H_9O_4N$		L-Glutamic acid	c		SB			536·9 ±0·5	44/
$C_5H_9O_4N$		L-Glutamic acid							
$C_5H_9O_4N$	147·1318	D-Glutamic acid	c		SB	10	m	537·41±0·28	36/
$C_5H_{11}O_4N$	149·1477	2-Nitro-2-ethyl-1, 3-propanediol	c	99·7	SB	5	m	700·76±0·32	52/
$C_5H_{11}O_4N$		2-Nitro-2-ethyl-1, 3-propanediol	c		SB	8	m	702·34±0·70	53/
$C_5H_{11}O_4N$		2-Nitro-2-ethyl-1, 3-propanediol							
$C_6H_{13}O_4N$	163·1748	2-Nitro-2-n-propyl-1, 3-propanediol	c	99·5	SB	4	m	858·92±0·69	25/
$C_6H_{13}O_4N$	163·1748	2-Nitro-2-isopropyl-1, 3-propanediol	c	99·8	SB	5	m	859·4 ±1·0	52/
$C_9H_{11}O_4N$	197·1923	2, 4-Dimethyl-3-carbo-methoxy-5-carboxy-pyrrole	c		SB		m	1021·0 ±1·0	32,
$C_{10}H_{13}O_4N$	211·2194	2, 4-Dimethyl-3, 5-dicarbomethoxypyrrole	c		SB		m	1182·3 ±1·2	32,
$C_{11}H_{13}O_4N$	223·2306	2, 4-Dimethyl-5-carbo-methoxypyrrole-3-acrylic acid	c		SB		m	1285·4 ±1·3	32,
$C_{12}H_{15}O_4N$	237·2577	Phenylalanine-N-carboxylic acid-dimethyl ester-	c		SB	2	m	1457·8 ±1·4	52/ 52/

ΔH_r°		5 ΔH_f° (l or c) kcal/g.f.w.	6 Determination of ΔH_v			7 ΔH_f° (g) kcal/g.f.w.
	Remarks		Type	ΔH_v° kcal/g.f.w.	Ref.	
	C	$-110\cdot91\pm0\cdot08$				
		$-232\cdot63\pm0\cdot22$				
		$-230\cdot3\ \pm1\cdot0$				
	Selected value	$-232\cdot63\pm0\cdot22$				
		$-138\cdot62\pm0\cdot67$				
		$-136\cdot35\pm0\cdot57$				
	Selected value	$-137\cdot5\ \pm1\cdot1$				
		$-241\cdot32\pm0\cdot20$				
		$-240\cdot8\ \pm0\cdot5$				
	Selected value	$-241\cdot32\pm0\cdot20$				
		$-240\cdot26\pm0\cdot30$				
		$-145\cdot23\pm0\cdot33$				
	BCE	$-143\cdot65\pm0\cdot71$				
	Selected value	$-144\cdot4\ \pm0\cdot9$				
		$-149\cdot43\pm0\cdot70$				
		$-149\cdot0\ \pm1\cdot0$				
	BCE	$-201\cdot2\ \pm1\cdot0$				
	BCE	$-202\cdot3\ \pm1\cdot2$				
	BCE	$-193\cdot2\ \pm1\cdot3$				
	D	$-183\cdot2\ \pm1\cdot4$				

$C_aH_bO_4N$—$C_aH_bO_6N$ 1 kcal = 4·184 kJ

1 Formula	2 g.f.w.	3 Name	State	Purity mol %	No. of Type expts.	Detn. of reactn.	$-\Delta H_r^\circ$ kcal/g.f.w.	R
$C_{12}H_{17}O_4N$	239·2736	2, 4-Dimethyl-5-carbo-methoxy-pyrrole-3-propionic acid methyl ester	c		SB	m	1506·2 ±1·5	32
$C_{12}H_{17}O_4N$	239·2736	2, 4-Dimethyl-3, 5-dicarbethoxypyrrole	c		SB 5	m	1490·2 ±1·5	33
$C_{12}H_{17}O_4N$	239·2736	2, 4-Dimethyl-5-carb-ethoxypyrrole-3-propionic acid	c		SB 5	m	1485·6 ±1·5	33
$C_4H_9O_5N$	151·1200	Tris-(hydroxymethyl)-nitromethane	c		SB 4	m	509·32±0·73	52
$C_4H_9O_5N$		Tris-(hydroxymethyl)-nitromethane	c		SB 8	m	506·26±0·75	53
$C_4H_9O_5N$		Tris-(hydroxymethyl)-nitromethane						
$C_{12}H_{17}O_5N$	255·2730	4-Methyl-3, 5-dicarb-ethoxy-2-hydroxy-methylpyrrole	c		SB	m	1456·5 ±1·5	32
$C_{13}H_{17}O_5N$	267·2841	2, 4-Dimethyl-5-carb-ethoxypyrrole-3-glyoxylic acid ethyl ester	c		SB	m	1560·7 ±1·6	32
$C_{13}H_{19}O_5N$	269·3001	4-Methyl-3, 5-dicarb-ethoxy-2-α-hydroxy-ethylpyrrole	c		SB	m	1600·2 ±1·6	32
$C_{14}H_{21}O_5N$	283·3272	3, 5-Dicarbethoxy-2-α-hydroxypropyl-4-methylpyrrole	c		SB	m	1774·8 ±1·8	33
$C_{17}H_{25}O_6N$	339·3919	5-Carbethoxy-2, 4-dimethylpyrrole-3-methylmalonic acid diethyl ester	c		SB	m	2117·6 ±2·1	33

Determina

1 kcal = 4·184 kJ

ΔH_f°	Remarks		5 ΔH_f° (l or c) kcal/g.f.w.	6 Determination of ΔH_v			7 ΔH_f° (g) kcal/g.f.w.
				Type	ΔH_v° kcal/g.f.w.	Ref.	
		BCE	$-203 \cdot 1 \ \pm 1 \cdot 5$				
		BCE	$-219 \cdot 1 \ \pm 1 \cdot 5$				
		BCE	$-223 \cdot 7 \ \pm 1 \cdot 5$				
			$-174 \cdot 30 \pm 0 \cdot 74$				
			$-177 \cdot 36 \pm 0 \cdot 76$				
	Selected value		$-175 \cdot 8 \ \pm 1 \cdot 5$				
		BCE	$-252 \cdot 8 \ \pm 1 \cdot 5$				
		BCE	$-242 \cdot 6 \ \pm 1 \cdot 6$				
		BCE	$-271 \cdot 5 \ \pm 1 \cdot 6$				
		BCE	$-259 \cdot 2 \ \pm 1 \cdot 8$				
		BCE	$-335 \cdot 2 \ \pm 2 \cdot 1$				

$C_aH_bON_2$ 1 kcal = 4·184 kJ

1 Formula	2 g.f.w.	3 Name	State	Purity mol %	No. of Type	of expts.	Detn. of reactn.	$-\Delta H_r^\circ$ kcal/g.f.w.	Ref
CH_4ON_2	60·0558	Urea	c		SB	6	m	151·00±0·04	40/
CH_4ON_2		Urea	c		SB	3	m	151·50±0·40	46/
CH_4ON_2		Urea	c		SB	5	m	150·90±0·02	63/
CH_4ON_2		Urea							
$C_2H_4ON_2$	72·0670	Methylene-urea	c		SB	5	m	246·55±0·36	56/
C_3ON_2	80·0463	Carbonyl cyanide	l		SB	2	m	332·7 ±1·3	48/
$C_4H_{10}ON_2$	102·1371	Trimethylurea	c		SB	6	m	638·8 ±1·0	56/
$C_5H_6ON_2$	110·1164	N-Acetylimidazole	c		R	4	m	4·83±0·06	60/
$C_5H_6ON_2$		N-Acetylimidazole	c		R	4	m	18·07±0·12	62/
$C_5H_6ON_2$		N-Acetylimidazole							
$C_5H_6ON_2$	110·1164	Pyrrole-2-aldoxime	c		SB		m	678·1 ±0·7	33/
$C_8H_6ON_2$	146·1498	Phenylfurazan	c		SB	4	m	1014·2 ±1·0	31/
$C_9H_8ON_2$	160·1769	p-Tolylfurazan	c		SB	4	m	1164·6 ±1·2	31/
$C_9H_8ON_2$	160·1769	Methylphenyloxadiazole	c		SB	4	m	1128·6 ±2·2	31/
$C_9H_8ON_2$	160·1769	3-Methyl-5-phenyl-azoxime	c		SB	4	m	1144·4 ±1·3	31/
$C_9H_8ON_2$	160·1769	3-Phenyl-5-methyl-azoxime	c		SB	4	m	1143·0 ±1·3	31/
$C_9H_8ON_2$	160·1769	Di-α-pyrryl ketone	c		SB		m	1113·7 ±1·1	32/
$C_9H_{12}ON_2$	164·2088	NN-Phenylmethyl-N'-methylurea	c		SB	4	m	1204·3 ±1·2	56/
$C_9H_{20}ON_2$	172·2726	Tetraethylurea	l		SB	7	m	1397·2 ±1·4	56/
$C_{10}H_{14}ON_2$	178·2359	NN-Phenylmethyl-N'-ethylurea	c		SB	5	m	1345·9 ±1·4	56/
$C_{12}H_{10}ON_2$	198·2263	p-Nitrosodiphenylamine	c		SB	5	m	1521·12±0·76	56/

Determinati

$$1 \text{ kcal} = 4 \cdot 184 \text{ kJ}$$

ΔH_r°		5 ΔH_f° (l or c) kcal/g.f.w.	6 Determination of ΔH_v			7 ΔH_f° (g) kcal/g.f.w.
Remarks			ΔH_v° Type kcal/g.f.w.		Ref.	
		$-79 \cdot 68 \pm 0 \cdot 05$	S4	$21 \cdot 0 \pm [0 \cdot 5]$	56/43	$-58 \cdot 7 \pm [0 \cdot 5]$
		$-79 \cdot 18 \pm 0 \cdot 41$		(sub.)		$-58 \cdot 2 \pm [0 \cdot 7]$
		$-79 \cdot 78 \pm 0 \cdot 04$				$-58 \cdot 8 \pm [0 \cdot 5]$
	Selected value	$-79 \cdot 73 \pm 0 \cdot 05$				$-58 \cdot 7 \pm [0 \cdot 5]$
	BCE	$-78 \cdot 18 \pm 0 \cdot 37$				
	BCE	$+50 \cdot 5 \pm 1 \cdot 3$	V3	$8 \cdot 6 \pm 0 \cdot 8$	48/11	**+59·1** ± 1.5
	BCE	$-79 \cdot 0 \pm 1 \cdot 0$				
Ac-imidazole(c)+H$_2$O(l) = Imidazole(c) {+13·97±0·79} +HOAc(l) {−115·75±0·07}		$-28 \cdot 63 \pm 0 \cdot 82$				
Ac-imidazole(c) +nBuNH$_2$(l) {−30·51±0·29} = Imidazole(c) {+13·97±0·79} +BuNHAc(l) {−91·02±0·39}		$-28 \cdot 47 \pm 0 \cdot 95$				
	Selected value	$-28 \cdot 55 \pm 0 \cdot 60$				
	BCE	$+2 \cdot 9 \pm 0 \cdot 7$				
	BCE	$+56 \cdot 8 \pm 1 \cdot 0$				
	BCE	$+44 \cdot 9 \pm 1 \cdot 2$				
	BCE	$+8 \cdot 9 \pm 2 \cdot 2$				
	BCE	$+24 \cdot 7 \pm 1 \cdot 3$				
	BCE	$+23 \cdot 3 \pm 1 \cdot 3$				
	BCE	$-6 \cdot 0 \pm 1 \cdot 1$				
	BCE	$-52 \cdot 0 \pm 1 \cdot 2$				
	BCE	$-132 \cdot 4 \pm 1 \cdot 4$				
	BCE	$-72 \cdot 8 \pm 1 \cdot 4$				
	E	$+50 \cdot 93 \pm 0 \cdot 80$				

$C_aH_bON_2$

1 kcal = 4·184 kJ

| | | | | | | | | 4 |
| | | | | | | | | Determinatic |

1 Formula	2 g.f.w.	3 Name	State	Purity mol %	Type	No. of expts.	Detn. of reactn.	$-\Delta H_f^\circ$ kcal/g.f.w.	Ref
$C_{12}H_{10}ON_2$	198·2263	Diphenylnitrosamine	c		SB	5	m	1524·5 ±1·5	54/1
$C_{13}H_{12}ON_2$	212·2534	NN-Diphenylurea	c		SB	4	m	1603·22±0·80	52/1
$C_{13}H_{14}ON_2$	214·2693	Tetrahydrocarbazole-urea	c		SB	5	m	1443·5 ±1·4	56/1
$C_{14}H_{10}ON_2$	222·2486	Diphenylfurazan	c		SB	4	m	1736·9 ±2·0	31/1
$C_{14}H_{10}ON_2$	222·2486	Diphenylazoxime	c		SB	4	m	1707·8 ±2·7	31/1
$C_{14}H_{10}ON_2$	222·2486	Diphenyloxadiazole	c		SB	4	m	1697·8 ±2·0	31/1
$C_{14}H_{14}ON_2$	226·2805	NN-Diphenyl-N′-methylurea	c		SB	8	m	1769·4 ±1·7	52/1
$C_{14}H_{14}ON_2$	226·2805	N-Acetylhydrazobenzene	c		SB		m	1792·9 ±2·8	44/1:
$C_{14}H_{14}ON_2$	226·2805	N-Acetylbenzidine	c		SB		m	1756·1 ±2·4	44/1:
$C_{14}H_{14}ON_2$		N-Acetylbenzidine	c		R	8	m	35·19±0·25	44/1:
$C_{14}H_{14}ON_2$		N-Acetylbenzidine							
$C_{15}H_{16}ON_2$	240·3076	NN-Phenylmethyl-N′N′-phenylmethyl-urea	c		SB	8	m	1939·8 ±1·9	52/12
$C_{15}H_{16}ON_2$	240·3076	NN-Diphenyl-N′-ethylurea	c		SB	7	m	1920·8 ±1·9	52/12
$C_{15}H_{32}ON_2$	256·4351	NN′-Diheptylurea	c		SB	5	m	2353·9 ±1·3	64/31
$C_{16}H_{16}ON_2$	252·3187	NN-Diphenyl-N′-allylurea	c		SB	7	m	2042·1 ±1·0	56/11
$C_{16}H_{18}ON_2$	254·3347	NN-Phenylmethyl-N′N′-phenylethylurea	c		SB	5	m	2089·4 ±1·0	52/12
$C_{17}H_{20}ON_2$	268·3618	NN-Phenylethyl-N′N′-Phenylethylurea	c		SB		m	2256·9 ±2·3	39/17
$C_{17}H_{24}ON_2$	272·3936	2, 2′, 4, 4′-Tetramethyl-3, 3′-diethyldipyrryl ketone	c		SB		m	2358·0 ±2·4	32/7
$C_{17}H_{36}ON_2$	284·4893	NN′-Dioctylurea	c		SB	5	m	2657·6 ±1·1	64/31

1 kcal = 4·184 kJ

ΔH_r°		5 ΔH_f° (l or c) kcal/g.f.w.	6 Determination of ΔH_v			7 ΔH_f° (g) kcal/g.f.w.
Remarks			Type	ΔH_v° kcal/g.f.w.	Ref.	
	E	+54·3 ±1·5				
	BCE	−29·33±0·83				
	BCE	−257·4 ±1·4				
	BCE	+78·6 ±2·0				
	BCE	+49·5 ±2·7				
	BCE	+39·5 ±2·0				
	BCE	−25·5 ±1·7				
		−2·0 ±2·8				
		−38·8 ±2·4				
-Ac-hydrazobenzene(c) {−2·0±2·8} = N-Ac-benzidine(c)		−37·2 ±3·0				
	Selected value	−38·0 ±2·0				
	BCE	−17·5 ±1·9				
	BCE	−36·5 ±1·9				
		−149·9 ±1·3				
	BCE	−9·2 ±1·0				
	BCE	−30·3 ±1·0				
	CE	−25·1 ±2·3				
	BCE	−60·6 ±2·4				
		−170·9 ±1·2				

1 Formula	2 g.f.w.	3 Name	State	Purity mol %	4 Determinatio[n] Type	No. of expts.	Detn. of reactn.	$-\Delta H_r^\circ$ kcal/g.f.w.		Ref.
$C_{18}H_{20}ON_2$	280·3729	NN-Diphenyl-N′-piperidylurea	c		SB	12	m	2350·2	±2·4	56/1
$C_{21}H_{44}ON_2$	340·5976	NN′-Didecylurea	c		SB	5	m	3268·3	±1·7	64/3
$C_{23}H_{18}ON_2$	338·4127	N-(2-Naphthyl)-N′N′-diphenylurea	c		SB	7	m	2793·3	±2·8	55/2.
$C_{23}H_{18}ON_2$	338·4127	N-(1-Naphthyl)-N′N′-diphenylurea	c		SB	4	m	2796·7	±2·8	56/1

1 kcal $= 4 \cdot 184$ kJ

ΔH_r°	Remarks		5 ΔH_f° (l or c) kcal/g.f.w.	6 Determination of ΔH_v			7 ΔH_f° (g) kcal/g.f.w.
				Type	ΔH_v° kcal/g.f.w.	Ref.	
		BCE	$-25 \cdot 9 \ \pm 2 \cdot 4$				
			$-209 \cdot 7 \ \pm 1 \cdot 8$				
		BCE	$+15 \cdot 3 \ \pm 2 \cdot 8$				
		BCE	$+18 \cdot 7 \ \pm 2 \cdot 8$				

$C_aH_bO_2N_2$ 1 kcal = 4·184 kJ

1 Formula	2 g.f.w.	3 Name	State	Purity mol %	Type	No. of expts.	Detn. of reactn.	$-\Delta H_r^\circ$ kcal/g.f.w.	Ref.
$C_2H_4O_2N_2$	88·0664	Oxamide	c		SB		m	201·5 ±1·0	57/2
$C_2H_4O_2N_2$	88·0664	Glyoxime	c		SB		m	303·10±0·30	36/1
$C_2H_6O_2N_2$	90·0823	Dimethylnitramine	c		SB	4	CO_2	376·1 ±1·0	58/2
$C_3H_6O_2N_2$	102·0935	Malonic diamide	c		SB	9	m	356·58±0·36	55/2
$C_3H_6O_2N_2$	102·0935	Methylglyoxime	c		SB		m	456·81±0·46	36/1
$C_4H_6O_2N_2$	114·1046	5-Methylhydantoin	c		SB	7	m	464·85±0·25	55/22
$C_4H_8O_2N_2$	116·1206	Succinamide	c		SB		m	510·54±0·51	57/22
$C_4H_8O_2N_2$	116·1206	Dimethylglyoxime	c		SB		m	606·95±0·61	36/16
$C_4H_{10}O_2N_2$	118·1365	Diethylnitramine	l		SB	3	CO_2	692·4 ±2·0	58/25
$C_5H_8O_2N_2$	128·1317	5, 5-Dimethylhydantoin	c		SB	8	m	616·06±0·30	55/22
$C_5H_{12}O_2N_2$	132·1636	DL-Ornithine	c		SB	3	m	724·17±0·43	60/27
$C_6H_6O_2N_2$	138·1269	o-Nitroaniline	c		SB	7	m	762·96±0·76	57/23
$C_6H_6O_2N_2$	138·1269	m-Nitroaniline	c		SB	8	m	762·5 ±1·5	57/23
$C_6H_6O_2N_2$	138·1269	p-Nitroaniline	c		SB		m	759·34±0·15	51/5
$C_6H_6O_2N_2$		p-Nitroaniline	c		SB	5	m	760·05±0·76	54/12
$C_6H_6O_2N_2$		p-Nitroaniline							
$C_6H_{10}O_2N_2$	142·1588	5-Methyl-5-ethyl- hydantoin	c		SB	8	m	770·58±0·38	55/22
$C_6H_{14}O_2N_2$	146·1907	DL-Lysine	c		SB	4	m	880·30±0·35	60/27
$C_7H_8O_2N_2$	152·1540	5-Nitro-o-toluidine	c		SB		m	909·80±0·30	49/17
$C_7H_8O_2N_2$	152·1540	3-Nitro-p-toluidine	c		SB		m	914·48±0·12	49/17

1 kcal = 4·184 kJ

ΔH_r°		5 ΔH_f° (l or c) kcal/g.f.w.	6 Determination of ΔH_v			7 ΔH_f° (g) kcal/g.f.w.
Remarks			Type	ΔH_v° kcal/g.f.w.	Ref.	
	BCE	−123·2 ±1·0	S4	27·0 ±0·5 (sub.)	53/28	−96·2 ±1·1
	BCE	−21·63±0·31				
An 0·9939±0·0030		−16·9 ±1·0	S4	16·7 ±0·5 (sub.)	52/20	−0·2 ±1·2
	BCE	−130·52±0·37				
	BCE	−30·29±0·47				
	BCE	−116·30±0·27				
	BCE	−1′8·92±0·52				
	BCE	−42·51±0·62	S4	23·2 ±[0·5] (sub.)	56/42	−19·3 ±0·8
An 0·9966±0·0025		−25·4 ±2·0	V4	12·7 ±0·8	58/25	−12·7 ±2·3
	BCE	−127·46±0·33				
	E	−155·98±0·44				
	E	−6·29±0·77	S3	21·5 ±[1·0] (sub.)	58/39	+15·2 ±[1·3]
	E	−6·8 ±1·5	S3	23·3 ±[1·0] (sub.)	58/39	+16·5 ±1·8
		−9·91±0·17	S3	26·1 ±[1·0] (sub.)	58/39	+16·2 ±[1·1]
	E	−9·20±0·77				+16·9 ±[1·3]
Selected value		−9·91±0·17				+16·2 ±[1·1]
	BCE	−135·30±0·40				
	E	−162·21±0·37				
	C	−21·82±0·32				
	C	−17·14±0·16				

THERMOCHEMISTRY OF ORGANIC AND ORGANOMETALLIC COMPOUNDS

$C_aH_bO_2N_2$

1 kcal = 4·184 kJ

4
Determinati‹

1 Formula	2 g.f.w.	3 Name	State	Purity mol %	No. of Type	Detn. of expts. reactn.	$-\Delta H_r^\circ$ kcal/g.f.w.	Ref
$C_8H_6O_2N_2$	162·1492	3-Phenyl-5-hydroxy- azoxime	c		SB 4	m	931·2 ±1·4	31/1
$C_8H_6O_2N_2$	162·1492	3-Hydroxy-5-phenyl- azoxime	c		SB 4	m	956·3 ±1·4	31/1
$C_8H_8O_2N_2$	164·1652	Phenylglyoxime (α-form)	c		SB	m	1020·8 ±1·0	36/1
$C_8H_8O_2N_2$		Phenylglyoxime (β-form)	c		SB	m	1035·8 ±1·0	36/1
$C_8H_{18}O_2N_2$	174·2449	cis-Dimer of nitroso- isobutane	c		SB 5	CO_2	1321·1 ±0·9	59/2
$C_9H_{10}O_2N_2$	178·1923	p-Tolylglyoxime (α-form)	c		SB	m	1225·9 ±1·2	36/1
$C_9H_{10}O_2N_2$		p-Tolylglyoxime (β-form)	c		SB	m	1242·8 ±1·2	36/1
$C_9H_{12}O_2N_2$	180·2082	p-Phenetylurea	c		SB		1144·3 ±0·8	54/1
$C_9H_{12}O_2N_2$	180·2082	m-Ethoxyphenylurea	c		SB		1153·7 ±1·0	54/1
$C_{10}H_8O_2N_2$	188·1875	Linear-Dimethyl-2, 2'- benzo-4, 5, 4', 5'- bisoxazole	c		SB 8	m	1162·97±0·88	36/1
$C_{10}H_8O_2N_2$	188·1875	Angular-Dimethyl-2, 2'- benzo-4, 5, 4', 5'- bisoxazole	c		SB 7	m	1164·17±0·44	36/1
$C_{10}H_8O_2N_2$	188·1875	Methylbenzoylfurazan	c		SB 4	m	1242·3 ±1·4	31/1
$C_{11}H_{12}O_2N_2$	204·2305	L-Tryptophan	c		SB	m	1345·2 ±0·2	58/2
$C_{11}H_{14}O_2N_2$	206·2464	2, 4-Dimethyl-3-cyano- methyl-5-carbethoxy- pyrrole	c		SB	m	1420·2 ±1·4	33/1
$C_{12}H_{10}O_2N_2$	214·2257	o-Nitrodiphenylamine	c		SB	m	1485·6 ±1·5	57/2
$C_{14}H_{12}O_2N_2$	240·2639	Diphenylglyoxime (α-form)	c		SB	m	1736·6 ±1·7	36/1
$C_{14}H_{12}O_2N_2$		Diphenylglyoxime (β-form)	c		SB	m	1730·9 ±1·7	36/1
$C_{14}H_{12}O_2N_2$		Diphenylglyoxime (γ-form)	c		SB	m	1739·3 ±1·7	36/1
$C_{14}H_{12}O_2N_2$	240·2639	sym-Dibenzoylhydrazine	c		SB 4	m	1675·5 ±0·3	42/8 51/

$1 \text{ kcal} = 4 \cdot 184 \text{ kJ}$

ΔH_r°	Remarks		ΔH_f° (l or c) kcal/g.f.w.	6 Determination of ΔH_v			7 ΔH_f° (g) kcal/g.f.w.
			5	Type	ΔH_v° kcal/g.f.w.	Ref.	
	BCE		$-26 \cdot 2 \ \pm 1 \cdot 4$				
	BCE		$-1 \cdot 1 \ \pm 1 \cdot 4$				
	BCE		$-4 \cdot 9 \ \pm 1 \cdot 0$				
	BCE		$+10 \cdot 1 \ \pm 1 \cdot 0$				
			$-46 \cdot 1 \ \pm 0 \cdot 9$				
	BCE		$+37 \cdot 9 \ \pm 1 \cdot 2$				
	BCE		$+54 \cdot 8 \ \pm 1 \cdot 2$				
			$-112 \cdot 0 \ \pm 0 \cdot 8$	S5	20 ± 2 (sub.)	54/18	$-92 \cdot 0 \ \pm 2 \cdot 2$
			$-102 \cdot 6 \ \pm 1 \cdot 0$	S5	18 ± 2 (sub.)	54/18	$-84 \cdot 6 \ \pm 2 \cdot 2$
	CE		$-50 \cdot 80 \pm 0 \cdot 90$				
	CE		$-49 \cdot 60 \pm 0 \cdot 47$				
	BCE		$+28 \cdot 5 \ \pm 1 \cdot 4$				
	C		$-99 \cdot 3 \ \pm 0 \cdot 3$				
	BCE		$-92 \cdot 6 \ \pm 1 \cdot 4$				
	BCE		$+15 \cdot 4 \ \pm 1 \cdot 5$				
	BCE		$+10 \cdot 0 \ \pm 1 \cdot 8$				
	BCE		$+4 \cdot 3 \ \pm 1 \cdot 8$				
	BCE		$+12 \cdot 7 \ \pm 1 \cdot 8$				
			$-51 \cdot 1 \ \pm 0 \cdot 4$				

$C_aH_bO_2N_2$ 1 kcal = 4·184 kJ

							4 Determinatio
1 Formula	2 g.f.w.	3 Name	State	Purity mol %	No. of Type expts.	Detn. of reactn.	$-\Delta H_r^\circ$ kcal/g.f.w.	Ref.
$C_{15}H_{16}O_2N_2$	256·3070	NN-Diphenyl-N'-2-hydroxyethylurea	c		SB 9	m	1881·0 ±1·8	55/22
$C_{15}H_{34}O_2N_2$	274·4504	n-Heptylammonium n-heptylcarbamate	c		SB 5	m	2350·2 ±1·6	64/31
$C_{16}H_{16}O_2N_2$	268·3181	NN'-Diacetylhydrazo-benzene	c		SB	m	2006·3 ±1·1	44/12
$C_{16}H_{16}O_2N_2$	268·3181	NN'-Diacetylbenzidine	c		SB	m	1938·4 ±1·4	44/12
$C_{17}H_{18}O_2N_2$	282·3452	4-Diphenylamino-carbonylmorpholine	c		SB 9	m	2161·7 ±1·4	55/22
$C_{17}H_{22}O_2N_2$	286·3771	2-Anilinomethyl-3-methyl-4-ethyl 5-carbethoxypyrrole	c		SB	m	2259·3 ±2·3	33/10
$C_{17}H_{38}O_2N_2$	302·5046	n-Octylammonium n-octylcarbamate	c		SB 5	m	2653·8 ±1·2	64/31
$C_{21}H_{46}O_2N_2$	358·6130	n-Decylammonium n-decylcarbamate	c		SB 5	m	3264·8 ±1·8	64/31

1 kcal $= 4 \cdot 184$ kJ

| ΔH_r° | | 5 | 6 Determination of ΔH_v | | | 7 |
| | | ΔH_f° (l or c) | | ΔH_v° | | ΔH_f° (g) |
	Remarks	kcal/g.f.w.	Type	kcal/g.f.w.	Ref.	kcal/g.f.w.
	BCE	$-76 \cdot 3 \ \pm 1 \cdot 9$				
		$-221 \cdot 9 \ \pm 1 \cdot 7$				
		$-45 \cdot 0 \ \pm 1 \cdot 2$				
		$-112 \cdot 9 \ \pm 1 \cdot 5$				
	BCE	$-52 \cdot 0 \ \pm 1 \cdot 5$				
	BCE	$-91 \cdot 0 \ \pm 2 \cdot 3$				
		$-243 \cdot 1 \ \pm 1 \cdot 3$				
		$-281 \cdot 5 \ \pm 1 \cdot 9$				

$C_aH_bO_3N_2$—$C_aH_bO_8N_2$　　　　　1 kcal = 4·184 kJ

1 Formula	2 g.f.w.	3 Name	State	Purity mol %	Type	No. of expts.	Detn. of reactn.	$-\Delta H_r^\circ$ kcal/g.f.w.	Re
$C_3H_8O_3N_2$	120·1088	Dimethylolurea	c		SB	4	m	384·04±0·16	48/
$C_4H_8O_3N_2$	132·1200	L-Asparagine	c		SB	7	m	460·78±0·17	51/ 36/
$C_4H_8O_3N_2$	132·1200	Glycylglycine	c		SB	6	m	470·76±0·13	42/
$C_5H_{10}O_3N_2$	146·1471	L-Glutamine	c		SB		m	614·32±0·15	57/2
$C_5H_{10}O_3N_2$	146·1471	DL-Alanylglycine	c		SB	7	m	625·94±0·20	42/
$C_7H_{14}O_3N_2$	174·2012	Glycylvaline	c		SB		m	937·00±0·09	62/3
$C_7H_{14}O_3N_2$	174·2012	Glycine valine anhydride	c		SB	3	m	948·05±0·16	63/3
$C_8H_{16}O_3N_2$	188·2283	Valine alanine anhydride	c		SB		m	1107·7 ±0·2	62/3
$C_8H_{16}O_3N_2$	188·2283	DL-Leucylglycine	c		SB	5	m	1093·43±0·28	42/1
$C_9H_6O_3N_2$	190·1598	3-Hydroxy-5-Benzoyl-azoxime	c		SB	4	m	1041·2 ±2·0	31/1
$C_9H_6O_3N_2$	190·1598	3-Benzoyl-5-hydroxy-azoxime	c		SB	4	m	986·9 ±1·9	31/1
$C_{11}H_{14}O_3N_2$	222·2458	Glycylphenylalanine	c		SB		m	1349·20±0·43	62/3
$C_{11}H_{14}O_3N_2$	222·2458	Glycine phenylalanine anhydride	c		SB		m	1361·9 ±0·4	62/3
$C_{11}H_{22}O_3N_2$	230·3096	Valine leucine anhydride	c		SB		m	1568·1 ±0·2	62/3
$C_{12}H_{16}O_3N_2$	236·2729	Alanylphenylalanine	c		SB		m	1505·4 ±0·1	62/3
$C_{12}H_{16}O_3N_2$	236·2729	Alanine phenylalanine anhydride	c		SB		m	1517·9 ±0·6	62/3
$C_{14}H_{20}O_3N_2$	264·3271	Valylphenylalanine	c		SB		m	1816·84±0·36	63/3
$C_{14}H_{20}O_3N_2$	264·3271	Valine phenylalanine anhydride	c		SB		m	1837·7 ±0·9	62/3
$C_{16}H_{10}O_3N_2$	278·2697	Dibenzoylfurazan	c		SB	4	m	1871·3 ±2·2	31/1
$C_{18}H_{14}O_3N_2$	306·3239	Di-p-toluylfurazan	c		SB	4	m	2176·0 ±3·1	31/1
$C_{18}H_{20}O_3N_2$	312·3717	Phenylalanine anhydride	c		SB	3	m	2239·01±0·22	63/3

4
Determinati

1 kcal $= 4\cdot184$ kJ

ΔH_r°		5 ΔH_f° (l or c) kcal/g.f.w.	6 Determination of ΔH_v			7 ΔH_f° (g) kcal/g.f.w.
	Remarks		Type	ΔH_v° kcal/g.f.w.	Ref.	
		$-171\cdot37\pm0\cdot18$				
		$-188\cdot68\pm0\cdot19$				
	D	$-178\cdot70\pm0\cdot15$				
	C	$-197\cdot51\pm0\cdot17$				
		$-185\cdot89\pm0\cdot21$				
		$-199\cdot56\pm0\cdot14$				
		$-188\cdot51\pm0\cdot19$				
		$-191\cdot2\ \pm0\cdot3$				
		$-205\cdot50\pm0\cdot32$				
	BCE	$-10\cdot2\ \pm2\cdot0$				
	BCE	$-64\cdot5\ \pm1\cdot9$				
		$-163\cdot57\pm0\cdot48$				
		$-150\cdot9\ \pm0\cdot5$				
		$-217\cdot9\ \pm0\cdot3$				
		$-169\cdot7\ \pm0\cdot2$				
		$-157\cdot2\ \pm0\cdot6$				
		$-183\cdot02\pm0\cdot40$				
		$-162\cdot2\ \pm0\cdot9$				
	BCE	$+24\cdot9\ \pm2\cdot2$				
	BCE	$+4\cdot9\ \pm3\cdot1$				
		$-137\cdot06\pm0\cdot30$				

THERMOCHEMISTRY OF ORGANIC AND ORGANOMETALLIC COMPOUNDS

$C_aH_bO_3N_2$—$C_aH_bO_8N_2$ 1 kcal = 4·184 kJ

1 Formula	2 g.f.w.	3 Name	State	Purity mol %	Type	No. of expts.	Detn. of reactn.	$-\Delta H_r^\circ$ kcal/g.f.w.	Re
$C_2H_4O_4N_2$	120·0652	1, 2-Dinitroethane	c		SB	4	m	282·73±0·28	54/
$C_3H_6O_4N_2$	134·0923	1, 1-Dinitropropane	l		SB	4	m	446·33±0·33	49/
$C_3H_6O_4N_2$	134·0923	1, 3-Dinitropropane	l	98·2	SB	3	m	433·59±0·33	49/
$C_3H_6O_4N_2$	134·0923	2, 2-Dinitropropane	c	99·6	SB	5	m	442·23±0·64	49/
$C_4H_8O_4N_2$	148·1194	Tartramide	c		SB		m	364·2 ±1·4	57/
$C_4H_{10}O_4N_2$	150·1353	L-Asparagine hydrate	c		SB	4	m	458·08±0·19	36/
$C_5H_{10}O_4N_2$	162·1465	Dinitroneopentane	c		SB		m	746·08±0·77	45/
$C_6H_4O_4N_2$	168·1098	N-Nitrosuccinimide	c		SB		m	435·0 ±0·5	56/
$C_6H_4O_4N_2$ $C_6H_4O_4N_2$ $C_6H_4O_4N_2$	168·1098	m-Dinitrobenzene m-Dinitrobenzene m-Dinitrobenzene	c c		SB SB	7	m m	692·22±0·69 693·45±0·69	39/ 39/
$C_6H_{12}O_4N_2$	176·1735	2, 3-Dimethyl- 2, 3-dinitrobutane	c		SB		m	921·4 ±1·6	45/
$C_7H_6O_4N_2$ $C_7H_6O_4N_2$ $C_7H_6O_4N_2$	182·1369	2, 4-Dinitrotoluene 2, 4-Dinitrotoluene 2, 4-Dinitrotoluene	c c		SB SB	5 5	m m	846·14±0·85 846·25±0·85	39/ 39/
$C_7H_6O_4N_2$	182·1369	2, 6-Dinitrotoluene	c		SB	4	m	851·08±0·85	39/
$C_8H_8O_4N_2$	196·1640	Methyl p-nitrophenyl- carbamate	c		SB		m	923·4 ±1·1	57/
$C_8H_8O_4N_2$	196·1640	2, 4-Dinitro- 1-ethylbenzene	l		SB	7	m	1004·4 ±1·0	55/
$C_8H_8O_4N_2$	196·1640	2, 4-Dinitro-m-xylene	c		SB	4	m	1005 2 ±1·0	39/
$C_8H_8O_4N_2$	196·1640	4, 6-Dinitro-m-xylene	c		SB	4	m	1001·3 ±1·0	39/
$C_8H_8O_4N_2$	196·1640	NN'-Bisuccinimide	c		SB	3	CO_2	856·13±0·39	66/
$C_8H_{16}O_4N_2$	204·2277	NN'-Dicarbethoxy- ethylenediamine	c		SB	6	m	1064·5 ±1·0	52/

1 kcal = 4·184 kJ

H_r°	Remarks	5 ΔH_f° (l or c) kcal/g.f.w.	6 Determination of ΔH_v			7 ΔH_f° (g) kcal/g.f.w.
			Type	ΔH_v° kcal/g.f.w.	Ref.	
	E	$-42\cdot00\pm0\cdot29$				
		$-40\cdot77\pm0\cdot34$	V2	$14\cdot93\pm0\cdot15$	49/18	$-25\cdot84\pm0\cdot38$
		$-53\cdot51\pm0\cdot34$				
		$-44\cdot87\pm0\cdot65$				
	BCE	$-285\cdot3\ \pm1\cdot4$				
		$-259\cdot70\pm0\cdot23$				
	C	$-65\cdot75\pm0\cdot78$				
	C	$-265\cdot9\ \pm0\cdot5$				
	BCE	$-8\cdot72\pm0\cdot70$	S4	$19\cdot4\ \pm0\cdot4$	50/28	$+10\cdot7\ \pm0\cdot8$
	BCE	$-7\cdot49\pm0\cdot70$		(sub.)		$+11\cdot9\ \pm0\cdot8$
	Selected value	$-8\cdot10\pm0\cdot60$				$+11\cdot3\ \pm0\cdot7$
	C	$-52\cdot8\ \pm1\cdot6$				
	BCE	$-17\cdot16\pm0\cdot86$				
	BCE	$-17\cdot05\pm0\cdot86$				
	Selected value	$-17\cdot10\pm0\cdot65$				
	BCE	$-12\cdot22\pm0\cdot86$				
	BCE	$-102\cdot3\ \pm1\cdot1$				
	BCE	$-21\cdot3\ \pm1\cdot0$				
	BCE	$-20\cdot5\ \pm1\cdot0$				
	BCE	$-24\cdot4\ \pm1\cdot0$				
An $0\cdot99174\pm0\cdot00680$		$-169\cdot54\pm0\cdot40$				
	BCE	$-234\cdot4\ \pm1\cdot0$				

THERMOCHEMISTRY OF ORGANIC AND ORGANOMETALLIC COMPOUNDS

$C_aH_bO_3N_2$—$C_aH_bO_8N_2$ 1 kcal = 4·184 kJ

1 Formula	2 g.f.w.	3 Name	State	Purity mol %	Type	No. of expts.	Detn. of reactn.	$-\Delta H_r^\circ$ kcal/g.f.w.	P
$C_{10}H_6O_4N_2$	218·1703	1, 5-Dinitronaphthalene	c		SB	5	m	1152·1 ±1·2	37
$C_{10}H_6O_4N_2$		1, 5-Dinitronaphthalene	c		SB	7	m	1153·5 ±1·7	56
$C_{10}H_6O_4N_2$		1, 5-Dinitronaphthalene							
$C_{10}H_6O_4N_2$	218·1703	1, 8-Dinitronaphthalene	c		SB	4	m	1153·4 ±1·2	37
$C_{10}H_6O_4N_2$		1, 8-Dinitronaphthalene	c		SB	11	m	1156·5 ±1·2	56
$C_{10}H_6O_4N_2$		1, 8-Dinitronaphthalene							
$C_{11}H_{12}O_4N_2$	236·2293	Hippurylglycine	c		SB	12	m	1245·42±0·33	4
$C_{11}H_{14}O_4N_2$	238·2452	Glycine tyrosine anhydride	c		SB		m	1322·0 ±0·1	62
$C_{12}H_{18}O_4N_2$	254·2883	1-Amino-2, 5-dimethyl-3, 4-dicarbethoxypyrrole	c		SB		m	1561·8 ±1·6	33
$C_{14}H_{10}O_4N_2$	270·2468	cis-4, 4′-Dinitrostilbene	c		SB	6	m	1676·0 ±0·5	50
$C_{14}H_{10}O_4N_2$	270·2468	trans-4, 4′-Dinitrostilbene	c		SB	7	m	1671·4 ±0·3	50
$C_{19}H_{24}O_4N_2$	344·4141	2, 2′, 4, 4′-Tetramethyl-3, 3′-dicarbethoxy-dipyrrylmethene	c		SB		m	2409·8 ±2·4	3
$C_{19}H_{26}O_4N_2$	346·4301	4, 4′-Dicarbethoxy-2, 2′, 5, 5′-tetramethyl-3, 3′-pyrromethane	c		SB		m	2466·2 ±2·5	33
$C_{19}H_{26}O_4N_2$	346·4301	5, 5′-Dicarbethoxy-2, 2′, 4, 4′-tetramethyl 3, 3′-pyrromethane	c		SB		m	2471·1 ±2·5	33
$C_{19}H_{26}O_4N_2$	346·4301	2, 2′, 4, 4′-Tetramethyl-3, 3′-dicarbethoxy-dipyrrylmethane	c		SB		m	2467·5 ±2·5	3
$C_{20}H_{28}O_4N_2$	360·4572	2, 2′, 4, 4′-Tetramethyl-3, 3′-dicarbethoxy-dipyrrylmethylmethane	c		SB		m	2624·3 ±2·6	3
$C_4H_4O_5N_2$	160·0869	Alloxan	c		SB	8	m	273·65±0·11	3
$C_4H_8O_5N_2$	164·1188	Ethanolamine methyl-urethan nitrate	l		SB	7	m	400·96±0·24	55

Determina

$$1 \text{ kcal} = 4 \cdot 184 \text{ kJ}$$

ΔH°			6 Determination of ΔH_v			7
Remarks		5 ΔH_f° (l or c) kcal/g.f.w.	Type	ΔH_v° kcal/g.f.w.	Ref.	ΔH_f° (g) kcal/g.f.w.
	CE	$+6 \cdot 6 \ \pm 1 \cdot 2$				
	E	$+8 \cdot 0 \ \pm 1 \cdot 7$				
Selected value		$+7 \cdot 3 \ \pm 1 \cdot 1$				
	CE	$+7 \cdot 9 \ \pm 1 \cdot 2$				
	E	$+11 \cdot 0 \ \pm 1 \cdot 2$				
Selected value		$+9 \cdot 0 \ \pm 2 \cdot 0$				
		$-199 \cdot 03 \pm 0 \cdot 36$				
		$-190 \cdot 8 \ \pm 0 \cdot 2$				
	BCE	$-181 \cdot 6 \ \pm 1 \cdot 6$				
	A	$+17 \cdot 7 \ \pm 0 \cdot 5$				
	A	$+13 \cdot 1 \ \pm 0 \cdot 3$				
	BCE	$-196 \cdot 9 \ \pm 2 \cdot 4$				
	BCE	$-208 \cdot 9 \ \pm 2 \cdot 5$				
	BCE	$-204 \cdot 0 \ \pm 2 \cdot 5$				
	BCE	$-207 \cdot 6 \ \pm 2 \cdot 5$				
	BCE	$-213 \cdot 1 \ \pm 2 \cdot 6$				
		$-239 \cdot 18 \pm 0 \cdot 13$				
	BCE	$-248 \cdot 50 \pm 0 \cdot 26$				

$C_aH_bO_3N_2$—$C_aH_bO_8N_2$ 1 kcal = 4·184 kJ

Determina◆

1 Formula	2 g.f.w.	3 Name	State	Purity mol %	Type	No. of expts.	Detn. of reactn.	$-\Delta H_r^\circ$ kcal/g.f.w.	R◆
$C_6H_4O_5N_2$	184·1092	2, 4-Dinitrophenol	c		SB	5	m	644·58±0·64	42/
$C_6H_4O_5N_2$		2, 4-Dinitrophenol	c		SB	4	m	646·05±0·64	39/
$C_6H_4O_5N_2$		2, 4-Dinitrophenol							
$C_6H_4O_5N_2$	184·1092	2, 6·Dinitrophenol	c		SB	4	m	650·77±0·65	42/
$C_6H_{12}O_5N_2$	192·1729	Serylserine	c		SB		m	692·72±0·07	62/
$C_6H_{12}O_5N_2$	192·1729	Serine anhydride	c		SB		m	696·6 ±0·2	62/
$C_7H_6O_5N_2$	198·1363	2, 4-Dinitroanisole	c		SB		m	818·69±0·82	39/
$C_7H_6O_5N_2$	198·1363	2, 6-Dinitroanisole	c		SB	4	m	818·06±0·82	42/
$C_8H_8O_5N_2$	212·1634	2, 4-Dinitrophenetole	c		SB	6	m	971·34±0·97	42/
$C_8H_8O_5N_2$		2, 4-Dinitrophenetole	c		SB	5	m	971·77±0·97	39/
$C_8H_8O_5N_2$		2, 4-Dinitrophenetole							
$C_{10}H_6O_5N_2$	234·1697	2, 4-Dinitro-1-naphthol	c		SB	4	m	1102·1 ±1·1	46/
$C_{16}H_{20}O_5N_2$	320·3482	3-Hydroxy-4, 4'-di- carbethoxy-5, 5'- dimethyl-2, 3-bispyrrole	c		SB		m	1947·1 ±1·9	33/
$C_6H_4O_6N_2$	200·1086	2, 4-Dinitroresorcinol	c		SB	6	m	601·60±0·60	54/
$C_6H_4O_6N_2$	200·1086	4, 6-Dinitroresorcinol	c		SB	5	m	595·87±0·60	54/
$C_8H_{12}O_6N_2$	232·1946	4, 7-Diketo- 5,6- diazadecanedioic acid	c		SB		m	844·56±0·49	56/
$C_{10}H_{16}O_8N_2$	292·2475	Ethylenediamine- tetra-acetic acid	c		SB	3	m	1066·5 ±0·2	67/◆

1 kcal $= 4\cdot184$ kJ

H_r°		5	6 Determination of ΔH_v			7
	Remarks	ΔH_f° (l or c) kcal/g.f.w.	ΔH_v° Type kcal/g.f.w.		Ref.	ΔH_f° (g) kcal/g.f.w.
	CE	$-56\cdot36\pm0\cdot65$	S3	$25\cdot0 \pm[1\cdot0]$ (sub.)	58/39	$-31\cdot4 \pm1\cdot3$
	CE	$-54\cdot89\pm0\cdot65$				$-29\cdot9 \pm1\cdot3$
	Selected value	$-55\cdot63\pm0\cdot80$				$-30\cdot6 \pm1\cdot4$
	CE	$-50\cdot17\pm0\cdot66$	S3	$26\cdot8 \pm[1\cdot0]$ (sub.)	58/39	$-23\cdot4 \pm1\cdot3$
		$-281\cdot48\pm0\cdot11$				
		$-277\cdot6 \pm0\cdot3$				
	CE	$-44\cdot61\pm0\cdot83$				
	CE	$-45\cdot24\pm0\cdot83$				
	CE	$-54\cdot33\pm0\cdot98$				
	CE	$-53\cdot90\pm0\cdot98$				
	Selected value	$-54\cdot12\pm0\cdot75$				
	BCE	$-43\cdot4 \pm1\cdot1$				
	BCE	$-240\cdot9 \pm1\cdot9$				
	E	$-99\cdot34\pm0\cdot61$				
	E	$-105\cdot07\pm0\cdot61$				
	C	$-317\cdot74\pm0\cdot53$				
		$-420\cdot5 \pm0\cdot3$				

THERMOCHEMISTRY OF ORGANIC AND ORGANOMETALLIC COMPOUNDS

$C_aH_bON_3—C_aH_bO_{10}N_3$ 1 kcal = 4·184 kJ

1 Formula	2 g.f.w.	3 Name	State	Purity mol %	Type	No. of expts.	Detn. of reactn.	$-\Delta H_r^\circ$ kcal/g.f.w.	
$C_2H_5ON_3$	87·0817	2-Triazoethanol	l		SB	5	m	381·46±0·45	53
$C_4H_7ON_3$	113·1199	Creatinine	c		SB	9	m	558·31±0·10	36
$C_8H_7ON_3$	161·1645	Phenylaminofurazan	c		SB	4	m	1044·3 ±1·2	31
$C_4H_9O_2N_3$	131·1352	Creatine	c		SB	7	m	555·23±0·20	36
$C_9H_7O_2N_3$	189·1750	Benzoylaminofurazan	c		SB	4	m	1104·5 ±1·2	31
$C_{10}H_9O_2N_3$	203·2021	p-Toluylaminofurazan	c		SB	4	m	1258·0 ±1·3	31
$C_6H_5O_4N_3$	183·1245	2, 3-Dinitroaniline	c		SB		m	732·3 ±[0·7]	62
$C_6H_5O_4N_3$	183·1245	2, 4-Dinitroaniline	c		SB		m	719·4 ±[0·7]	62
$C_6H_5O_4N_3$		2, 4-Dinitroaniline	c		SB	6	m	718·33±0·72	54
$C_6H_5O_4N_3$		2, 4-Dinitroaniline							
$C_6H_5O_4N_3$	183·1245	2, 5-Dinitroaniline	c		SB		m	724·5 ±[0·7]	62
$C_6H_5O_4N_3$	183·1245	2, 6-Dinitroaniline	c		SB		m	723·0 ±[0·7]	62
$C_6H_5O_4N_3$	183·1245	3, 4-Dinitroaniline	c		SB		m	727·3 ±[0·7]	62
$C_6H_5O_4N_3$	183·1245	3, 5-Dinitroaniline	c		SB		m	725·8 ±[0·7]	62
$C_7H_7O_4N_3$	197·1515	N-Methyl-2, 6-dinitroaniline	c		SB	4	m	894·12±0·45	57
$C_7H_7O_4N_3$	197·1515	N-Methyl-2, 4-dinitroaniline	c		SB	6	m	881·32±0·88	54
$C_8H_9O_4N_3$	211·1786	2, 4-Dinitro-NN-dimethylaniline	c		SB	7	m	1052·5 ±1·6	56
$C_{12}H_9O_4N_3$	259·2232	2, 4-Dinitro-diphenylamine	c		SB		m	1441·4 ±1·4	57
$C_{14}H_{19}O_4N_3$	293·3252	Glycylalanylphenyl-alanine	c		SB		m	1744·2 ±0·1	62

1 kcal = 4·184 kJ

H_r°	Remarks	5 ΔH_f° (l or c) kcal/g.f.w.	6 Determination of ΔH_v			7 ΔH_f° (g) kcal/g.f.w.
			ΔH_v° Type kcal/g.f.w.		Ref.	
	D	+22·57±0·46				
		−57·00±0·12				
	BCE	+52·8 ±1·2				
		−128·39±0·22				
	BCE	+18·9 ±1·2				
	BCE	+10·1 ±1·3				
		−2·8 ±[0·7]				
	E	−15·7 ±[0·7] −16·76±0·73				
	Selected value	−16·25±0·75				
		−10·6 ±[0·7]				
		−12·1 ±[0·7]				
		−7·8 ±[0·7]				
		−9·3 ±[0·7]				
	E	−3·34±0·48				
	E	−16·14±0·89				
	E	−7·3 ±1·6				
	BCE	+5·4 ±1·4				
		−221·5 ±0·2				

$C_aH_bON_3$—$C_aH_bO_{10}N_3$ 1 kcal = 4·184 kJ

1 Formula	2 g.f.w.	3 Name	State	Purity mol %	Type	No. of expts.	Detn. of reactn.	$-\Delta H_r^\circ$ kcal/g.f.w.	F
$C_3H_7O_5N_3$	165·1063	Hydroxyethylmethyl-nitramine nitrate	c		SB	5	m	476·78±0·48	57
CHO_6N_3	151·0356	Trinitromethane	c		SB	8	m	116·71±0·48	67
CHO_6N_3		Trinitromethane	l		SB	6	m	120·36±0·38	67
CHO_6N_3		Trinitromethane	l		SB	1	m	109·58±	49
CHO_6N_3		Trinitromethane							
$C_5H_9O_6N_3$	207·1440	2-Methyl-2, 3, 3-trinitrobutane	c		SB		m	698·43±0·72	45
$C_6H_3O_6N_3$	213·1073	1, 3, 5-Trinitrobenzene	c		SB	3	m	656·29±0·66	39
$C_6H_3O_6N_3$		1, 3, 5-Trinitrobenzene	c		SB		m	656·46±0·66	39
$C_6H_3O_6N_3$		1, 3, 5-Trinitrobenzene							
$C_6H_{11}O_6N_3$	221·1711	2-Methyl-2, 3, 3-trinitropentane	c		SB		m	870·65±0·44	45
$C_7H_5O_6N_3$	227·1344	2, 4, 6-Trinitrotoluene	c		SB	4	m	813·06±0·81	39
$C_7H_5O_6N_3$		2, 4, 6-Trinitrotoluene	c		SB		m	813·17±0·81	39
$C_7H_5O_6N_3$		2, 4, 6-Trinitrotoluene	c		SB		m	813·7 ±[0·5]	56
$C_7H_5O_6N_3$		2, 4, 6-Trinitrotoluene							
$C_8H_7O_6N_3$	241·1615	2, 4, 6-Trinitroethyl-benzene	c		SB	9	m	974·01±0·97	54
$C_8H_7O_6N_3$	241·1615	2, 4, 6-Trinitro-m-xylene	c		SB	5	m	966·63±0·97	39
$C_8H_7O_6N_3$		2, 4, 6-Trinitro-m-xylene	c		SB		m	967·32±0·97	39
$C_8H_7O_6N_3$		2, 4, 6-Trinitro-m-xylene							
$C_{10}H_5O_6N_3$	263·1679	1, 3, 8- Trinitro-naphthalene	c		SB	5	m	1117·1 ±1·1	37
$C_{10}H_5O_6N_3$	263·1679	1, 4, 5-Trinitro-naphthalene	c		SB	2	m	1120·0 ±1·1	37
$C_6H_3O_7N_3$	229·1067	2, 4, 6-Trinitrophenol	c		SB		m	615·55±0·32	60
$C_6H_3O_7N_3$		2, 4, 6-Trinitrophenol	c		SB	4	m	614·03±0·61	42
$C_6H_3O_7N_3$		2, 4, 6-Trinitrophenol	c		SB	5	m	616·3 ±0·6	32
$C_6H_3O_7N_3$		2, 4, 6-Trinitrophenol	c		SB		m	613·68±0·61	39
$C_6H_3O_7N_3$		2, 4, 6-Trinitrophenol							

1 kcal = 4·184 kJ

		6 Determination of ΔH_v			
	5 ΔH_f° (l or c)		ΔH_v°		7 ΔH_f° (g)
Remarks	kcal/g.f.w.	Type	kcal/g.f.w.	Ref.	kcal/g.f.w.

ΔH_r°

Remarks	ΔH_f° (l or c) kcal/g.f.w.	Type	ΔH_v° kcal/g.f.w.	Ref.	ΔH_f° (g) kcal/g.f.w.
E	$-44·48\pm0·49$				
	$-11·50\pm0·49$	S4	$11·15\pm0·10$ (sub.)	67/18	$-0·35\pm0·51$
	$-7·85\pm0·40$	V3	$7·79\pm0·10$	67/18	$-0·06\pm0·43$
	$-18·63\pm$				
Selected value	$\Big\{\begin{array}{l}-8·19\pm0·50\text{(l)}\\-11·35\pm0·50\text{(c)}\end{array}$				$-0·20\pm0·50$
C	$-79·24\pm0·73$				
CE	$-10·49\pm0·67$	S4	$23·8\pm[0·5]$ (sub.)	50/28	$+13·3\pm0·9$
CE	$-10·32\pm0·67$				$+13·5\pm0·9$
Selected value	$-10·40\pm0·45$				$+13·4\pm0·7$
C	$-69·39\pm0·45$				
CE	$-16·09\pm0·82$	S4	$28·3\pm1·0$ (sub.)	50/17	$+12·2\pm1·3$
CE	$-15·98\pm0·82$				$+12·3\pm1·3$
C	$-15·4\pm[0·5]$				$+12·9\pm1·2$
Selected value	$-16·03\pm0·65$				$+12·3\pm1·2$
E	$-17·50\pm0·98$				
CE	$-24·88\pm0·98$				
CE	$-24·19\pm0·98$				
Selected value	$-24·53\pm0·70$				
CE	$+5·8\pm1·1$				
CE	$+8·7\pm1·1$				
	$-51·23\pm0·33$				
CE	$-52·75\pm0·62$				
BCE	$-50·5\pm0·6$				
CE	$-53·10\pm0·62$				
Selected value	$-51·23\pm0·33$				

THERMOCHEMISTRY OF ORGANIC AND ORGANOMETALLIC COMPOUNDS

$C_aH_bON_3$—$C_aH_bO_{10}N_3$　　　　　1 kcal = 4·184 kJ

					No. of	Detn. of		
1 Formula	2 g.f.w.	3 Name	State	Purity mol %	Type expts.	reactn.	$-\Delta H_r^\circ$ kcal/g.f.w.	R
$C_7H_5O_7N_3$	243·1338	2, 4, 6-Trinitro-*m*-cresol	c		SB	3	m	767·85±0·77	42
$C_7H_5O_7N_3$		2, 4, 6-Trinitro-*m*-cresol	c		SB	4	m	768·14±0·77	39
$C_7H_5O_7N_3$		2, 4, 6-Trinitro-*m*-cresol							
$C_7H_5O_7N_3$	243·1338	2, 4, 6-Trinitroanisole	c		SB	5	m	791·90±0·79	42
$C_7H_5O_7N_3$		2, 4, 6-Trinitroanisole	c		SB		m	791·14±0·79	39
$C_7H_5O_7N_3$		2, 4, 6-Trinitroanisole							
$C_8H_7O_7N_3$	257·1609	2, 4, 6-Trinitrophenetole	c		SB	4	m	942·58±0·94	42
$C_8H_7O_7N_3$		2, 4, 6-Trinitrophenetole	c		SB	5	m	942·66±0·94	39
$C_8H_7O_7N_3$		2, 4, 6-Trinitrophenetole							
$C_4H_7O_8N_3$	225·1157	2-Nitro-2-methyl-1, 3-propanediol dinitrate	c		SB	6	m	525·90±0·53	54
$C_5H_9O_8N_3$	239·1428	2-Nitro-2-ethyl-1, 3-propanediol-dinitrate	l		SB	7	m	689·9 ±1·0	53
$C_8H_7O_8N_3$	273·1603	2-(2, 4-Dinitrophenoxy)-ethyl nitrate	c		SB	5	m	922·96±0·92	53
$C_8H_7O_8N_3$		2-(2, 4-Dinitrophenoxy)-ethyl nitrate	c		SB		m	920·10±0·92	57
$C_8H_7O_8N_3$		2-(2, 4-Dinitrophenoxy)-ethyl nitrate							
$C_3H_5O_9N_3$	227·0880	Glycerol trinitrate	l		SB	4	m	364·61±0·90	39
$C_3H_5O_9N_3$		Glycerol trinitrate	l		SB	5	m	364·0 ±0·9	47
$C_3H_5O_9N_3$		Glycerol trinitrate							
$C_6H_{11}O_9N_3$	269·1693	1, 1, 1-Tris(hydroxy-methyl) propane trinitrate	c		SB	5	m	825·35±0·82	53
$C_5H_9O_{10}N_3$	271·1416	1-(2-Hydroxyethoxy)-2, 3-propanediol trinitrate	l		SB	5	m	649·50±0·65	55
$C_{14}H_{23}O_{10}N_3$	393·3535	Diethylenetriaminepenta-acetic acid	c		SB	4	m	1570·5 ±0·2	67

1 kcal = 4·184 kJ

H_r°		Remarks	5 ΔH_f° (l or c) kcal/g.f.w.	6 Determination of ΔH_v			7 ΔH_f° (g) kcal/g.f.w.
				Type ΔH_v° kcal/g.f.w.		Ref.	
		CE	$-61\cdot30\pm0\cdot78$				
		CE	$-61\cdot01\pm0\cdot78$				
		Selected value	$-61\cdot16\pm0\cdot55$				
		CE	$-37\cdot25\pm0\cdot80$	S4	$31\cdot8\ \pm[0\cdot5]$	50/28	$-5\cdot5\ \pm0\cdot9$
		CE	$-38\cdot01\pm0\cdot80$		(sub.)		$-6\cdot2\ \pm0\cdot9$
		Selected value	$-37\cdot63\pm0\cdot60$				$-5\cdot8\ \pm0\cdot8$
		CE	$-48\cdot93\pm0\cdot95$	S4	$28\cdot8\ \pm[0\cdot5]$	50/28	$-20\cdot1\ \pm1\cdot1$
		CE	$-48\cdot85\pm0\cdot95$		(sub.)		$-20\cdot1\ \pm1\cdot1$
		Selected value	$-48\cdot89\pm0\cdot75$				$-20\cdot1\ \pm1\cdot1$
		E	$-89\cdot41\pm0\cdot54$				
		BCE	$-87\cdot8\ \pm1\cdot0$				
		BCE	$-68\cdot55\pm0\cdot93$				
			$-71\cdot41\pm0\cdot93$				
		Selected value	$-70\cdot0\ \pm1\cdot5$				
		CE	$-88\cdot33\pm0\cdot91$	V4	$23\cdot9\ \pm1\cdot0$	59/38	$-64\cdot4\ \pm1\cdot4$
		E	$-88\cdot9\ \pm0\cdot9$				$-65\cdot0\ \pm1\cdot4$
		Selected value	$-88\cdot6\ \pm0\cdot7$				$-64\cdot7\ \pm1\cdot2$
		BCE	$-114\cdot69\pm0\cdot83$				
		E	$-128\cdot17\pm0\cdot66$				
			$-531\cdot8\ \pm0\cdot3$				

$C_aH_bON_4$—$C_aH_bO_{16}N_4$ 1 kcal = 4·184 kJ

1 Formula	2 g.f.w.	3 Name	State	Purity mol %	Type	No. of expts.	Detn. of reactn.	$-\Delta H_f^\circ$ kcal/g.f.w.	
CH_2ON_4	86·0533	5-Hydroxytetrazole	c		SB	5	m	163·87±0·46	5
$C_2H_4ON_4$	100·0804	5-Methoxytetrazole	c		SB	3	m	341·24±0·34	5
$C_3H_4ON_4$	112·0915	N-Acetyltetrazole	c		R	4	m	10·31±0·10	6
$C_3H_4ON_4$		N-Acetyltetrazole	c		R	3	m	20·22±0·16	6
$C_3H_4ON_4$		N-Acetyltetrazole							
$C_3H_6ON_4$	114·1075	1, 4-Dimethyl-5-tetrazolone	c		SB	2	m	480·5 ±1·0	5
$C_5H_4ON_4$	136·1138	Hypoxanthine (6-Hydroxypurine)	c		SB	8	m	580·42±0·17	3
$C_7H_6ON_4$	162·1521	1-Phenyl-5-hydroxy-tetrazole	c		SB	4	m	890·01±0·67	5
$CH_4O_2N_4$	104·0686	Nitroguanidine	c		SB			208·6 ±	5
$C_2H_4O_2N_4$	116·0798	Azodicarbamide	c		SB	2	m	254·82±0·50	57
$C_2H_4O_2N_4$	116·0798	Urazine	c		SB		m	262·3 ±0·3	56
$C_2H_6O_2N_4$	118·0957	Oxalyl dihydrazide	c		SB		m	322·50±0·13	49
$C_2H_6O_2N_4$	118·0957	Hydrazodicarbamide	c		SB	4	m	273·86±0·28	57
$C_3H_8O_2N_4$	132·1228	Malonyl dihydrazide	c		SB		m	475·92±0·13	49
$C_4H_6O_2N_4$	142·1180	Acetylene diurea	c		SB	6	m	463·37±0·46	56
$C_4H_{10}O_2N_4$	146·1499	Succinyl dihydrazide	c		SB		m	630·15±0·11	49
$C_5H_4O_2N_4$	152·1132	Xanthine	c		SB	8	m	516·17±0·21	3
$C_6H_{14}O_2N_4$	174·2041	D-Arginine	c		SB	9	m	893·49±0·30	37
$C_8H_6O_2N_4$	190·1626	2-Phenyl-5-carboxy-tetrazole	c		SB	4	m	947·48+0·80	51

Determin

1 kcal $= 4\cdot184$ kJ

ΔH_r° Remarks		5 ΔH_f° (l or c) kcal/g.f.w.	6 Determination of ΔH_v			7 ΔH_f° (g) kcal/g.f.w.
			Type	ΔH_v° kcal/g.f.w.	Ref.	
		$+1\cdot50\pm0\cdot47$				
		$+16\cdot51\pm0\cdot35$				
Ac-tetrazole(c)$+H_2O$(l)		$+19\cdot53\pm0\cdot27$				
= Tetrazole(c) $\{+56\cdot65\pm0\cdot22\}$						
$+$HOAc(l) $\{-115\cdot75\pm0\cdot07\}$						
Ac-tetrazole(c)$+$PhNH$_2$(l) $\{+7\cdot48\pm0\cdot18\}$		$+19\cdot31\pm0\cdot50$				
= Tetrazole(c) $\{+56\cdot65\pm0\cdot22\}$						
$+$PhNHAc(c) $\{-50\cdot08\pm0\cdot33\}$						
Selected value		$+19\cdot49\pm0\cdot24$				
		$-6\cdot6\ \pm1\cdot0$				
		$-26\cdot47\pm0\cdot18$				
		$+26\cdot71\pm0\cdot68$				
.S. measurement		$-22\cdot1\ \pm$				
		$-69\cdot91\pm0\cdot51$				
	C	$-62\cdot4\ \pm0\cdot3$				
	C	$-70\cdot55\pm0\cdot14$				
		$-119\cdot19\pm0\cdot29$				
	C	$-79\cdot49\pm0\cdot14$				
	BCE	$-117\cdot78\pm0\cdot47$				
	C	$-87\cdot63\pm0\cdot13$				
		$-90\cdot72\pm0\cdot22$				
		$-149\cdot02\pm0\cdot32$				
		$-9\cdot87\pm0\cdot81$				

$C_aH_bON_4$—$C_aH_bO_{16}N_4$ 1 kcal = 4·184 kJ

1 Formula	2 g.f.w.	3 Name	State	Purity mol %	Type	No. of expts.	Detn. of reactn.	$-\Delta H_r^\circ$ kcal/g.f.w.	R
$C_{26}H_{22}O_2N_4$	422·4908	Bis-diphenylurea	c		SB	5	m	3209·8 ±1·6	56
$C_{28}H_{26}O_2N_4$	450·5450	Ethylene bis-diphenylurea	c		SB	9	m	3474·8 ±3·5	56
$C_{32}H_{36}O_2N_4$	508·6693	Pyrroporphyrin (XV) monomethyl ester	c		SB	3	m	4146·6 ±4·1	33
$C_{33}H_{38}O_2N_4$	522·6964	γ-Phylloporphyrin monomethyl ester	c		SB	3	m	4310·4 ±4·3	33
$C_{33}H_{42}O_2N_4$	526·7283	2, 9-Diacetyl-1, 3, 5, 6, 8, 10-hexamethyl- 4, 7-diethyl-tetrapyrro- 14-ene	c		SB		m	4466·7 ±4·5	33
$C_{34}H_{38}O_2N_4$	534·7076	Desoxophylloerythrin monomethyl ester	c		SB	3	m	4452·3 ±4·4	33
$C_4H_6O_3N_4$	158·1174	Allantoin	c		SB	6	m	409·65±0·16	35
$C_5H_4O_3N_4$	168·1126	Uric acid	c		SB	6	m	458·98±0·20	35
$C_{34}H_{36}O_3N_4$	548·6910	Pyrophaeophorbid- a-monomethyl ester	c		SB	3	m	4337·8 ±4·3	33
$C_{34}H_{36}O_3N_4$	548·6910	Phylloerythrin monomethyl ester	c		SB	3	m	4339·4 ±4·3	33
$C_2H_6O_4N_4$	150·0945	Ethylenedinitramine	c		SB	6	m	368·44±0·37	55
$C_6H_6O_4N_4$	198·1391	2, 4-Dinitrophenyl- hydrazine	c		SB	8	m	781·19±0·78	54
$C_{12}H_{10}O_4N_4$	274·2379	4, 4′-Dinitrohydrazo- benzene	c		SB		m	1496·4 ±0·3	51
$C_{34}H_{34}O_4N_4$	562·6745	Protoporphyrin	c		SB	3	m	4234·4 ±4·2	33
$C_{34}H_{36}O_4N_4$	564·6904	Verdoporphyrin dimethyl ester	c		SB	3	m	4278·5 ±4·3	33
$C_{34}H_{38}O_4N_4$	566·7064	Rhodoporphyrin (XV) dimethyl ester	c		SB	3	m	4368·6 ±4·4	33
$C_{34}H_{38}O_4N_4$	566·7064	Rhodoporphyrin (XXI) dimethyl ester	c		SB	3	m	4366·9 ±4·4	33

1 kcal = 4·184 kJ

ΔH_r°			5	6 Determination of ΔH_v			7
	Remarks		ΔH_f° (l or c) kcal/g.f.w.	Type	ΔH_v° kcal/g.f.w.	Ref.	ΔH_f° (g) kcal/g.f.w.
		BCE	+13·0 ±1·7				
		BCE	−46·7 ±3·5				
		BCE	−92·7 ±4·2				
		BCE	−91·3 ±4·4				
		BCE	−71·6 ±4·6				
		BCE	−43·4 ±4·5				
			−171·50±0·17				
			−147·91±0·21				
		BCE	−89·6 ±4·4				
		BCE	−88·0 ±4·4				
		E	−24·61±0·38				
		E	+11·94±0·79				
			+26·2 ±0·3				
		BCE	−124·7 ±4·3				
		BCE	−148·9 ±4·4				
		BCE	−127·1 ±4·5				
		BCE	−128·8 ±4·5				

$C_aH_bON_4-C_aH_bO_{16}N_4$ 1 kcal = 4·184 kJ

1 Formula	2 g.f.w.	3 Name	State	Purity mol %	No. of Type expts.	Detn. of reactn.	$-\Delta H_c^\circ$ kcal/g.f.w.	R
$C_{35}H_{40}O_4N_4$	580·7335	Chloroporphyrin- e_4-dimethyl ester	c		SB 3	m	4499·0 ±4·5	33
$C_{35}H_{40}O_4N_4$	580·7335	Chlorin-e_4-dimethyl ester	c		SB 3	m	4508·3 ±4·5	33
$C_{36}H_{38}O_4N_4$	590·7287	Protoporphyrin dimethyl ester	c		SB 3	m	4556·6 ±4·6	33
$C_{36}H_{42}O_4N_4$	594·7605	Mesoporphyrin (IX) dimethyl ester	c		SB 3	m	4619·1 ±4·6	33
$C_{34}H_{36}O_5N_4$	580·6898	Phaeopurpurin-18-mono- methyl ester	c		SB 3	m	4192·5 ±4·2	33
$C_{36}H_{38}O_5N_4$	606·7281	Phaeoporphyrin- a_5-dimethyl ester	c		SB 3	m	4514·6 ±4·5	33
$C_{36}H_{38}O_5N_4$	606·7281	Methylphaeophorbide-a	c	2s	SB 6	m	4522·7 ±4·5	33
$C_4H_6O_6N_4$	206·1156	Dimethyldinitro- oxamide	c		SB 6	m	508·14±0·51	55
$C_7H_6O_6N_4$	242·1491	2, 4, 6-Trinitro- N-methylaniline	c	2s	SB 14	m	851·40±0·85	55
$C_7H_6O_6N_4$	242·1491	2, 4-Dinitrophenyl- methylnitramine	c		SB 7	m	867·16±0·88	56
$C_7H_6O_6N_4$	242·1491	2, 6-Dinitrophenyl- methylnitramine	c		SB 5	m	872·74±0·87	57
$C_{36}H_{36}O_6N_4$	620·7115	Methylphaeophorbide-b	c		SB 3	m	4409·5 ±4·4	33
$C_{36}H_{40}O_6N_4$	624·7434	Chlorin-p_6-trimethyl ester	c		SB 3	m	4455·1 ±4·4	33
$C_{37}H_{42}O_6N_4$	638·7705	Chloroporphyrin- e_6-trimethyl ester	c		SB 3	m	4679·4 ±4·7	33
$C_{37}H_{42}O_6N_4$	638·7705	Chlorin-e_6-trimethyl ester	c		SB 3	m	4688·3 ±4·7	33
$C_4H_6O_7N_4$	222·1150	N-(1, 1-Dinitroethyl)- 1-nitroacetaldoxime	c		SB	m	541·6 ±[0·5]	56
$C_{37}H_{40}O_7N_4$	652·7540	Dimethyl- phaeopurpurin-7	c		SB 3	m	4595·7 ±4·6	33

Determina

1 kcal = 4·184 kJ

ΔH_r° Remarks	5 ΔH_f° (l or c) kcal/g.f.w.	6 Determination of ΔH_v			7 ΔH_f° (g) kcal/g.f.w.
		Type	ΔH_v° kcal/g.f.w.	Ref.	
	BCE $-159\cdot1\ \pm4\cdot6$				
	BCE $-149\cdot8\ \pm4\cdot6$				
	BCE $-127\cdot2\ \pm4\cdot7$				
	BCE $-201\cdot4\ \pm4\cdot7$				
	BCE $-234\cdot9\ \pm4\cdot3$				
	BCE $-169\cdot2\ \pm4\cdot6$				
	BCE $-161\cdot1\ \pm4\cdot6$				
	E $-73\cdot01\pm0\cdot52$				
	E $-11\cdot90\pm0\cdot86$				
	E $+3\cdot86\pm0\cdot89$				
	E $+9\cdot44\pm0\cdot88$				
	BCE $-206\cdot0\ \pm4\cdot5$				
	BCE $-297\cdot0\ \pm4\cdot5$				
	BCE $-235\cdot1\ \pm4\cdot8$				
	BCE $-226\cdot3\ \pm4\cdot8$				
	C $-39\cdot5\ \pm[0\cdot5]$				
	BCE $-250\cdot5\ \pm4\cdot7$				

THERMOCHEMISTRY OF ORGANIC AND ORGANOMETALLIC COMPOUNDS

$C_aH_bON_4$—$C_aH_bO_{16}N_4$ $1 \text{ kcal} = 4 \cdot 184 \text{ kJ}$

1 Formula	2 g.f.w.	3 Name	State	Purity mol %	Type	No. of expts.	Detn. of reactn.	$-\Delta H_c^{\circ}$ kcal/g.f.w.	Re
CO_8N_4	196·0332	Tetranitromethane	l		SB	3	m	103·9 ±1·0	44/
CO_8N_4		Tetranitromethane	l	>99·9, glc	R	6	m	508·7 ±0·7	63/
CO_8N_4		Tetranitromethane							
$C_4H_6O_8N_4$	238·1144	2, 2, 3, 3-Tetranitro-butane	c		SB		m	582·59±0·24	45/
$C_4H_8O_8N_4$	240·1304	Bis-(2-hydroxyethyl-nitrate) nitramine	c		SB	7	m	575·09±0·58	54/
$C_4H_8O_8N_4$		Bis-(2-hydroxyethyl-nitrate) nitramine	c		SB	7	m	576·14±0·36	55/
$C_4H_8O_8N_4$		Bis-(2-hydroxyethyl-nitrate) nitramine							
$C_6H_{10}O_8N_4$	266·1686	Dinitrobutyleneglycol-diurethan	c		SB		m	706·32±0·71	57/
$C_{40}H_{46}O_8N_4$	710·8346	Coproporphyrin-(I) tetramethyl ester	c		SB	3	m	4979·5 ±4·9	33/
$C_8H_6O_{10}N_4$	318·1578	2-(2, 4, 6-Trinitro-phenoxy)ethyl nitrate	c		SB	3	m	888·63±0·88	53/
$C_8H_6O_{10}N_4$		2-(2, 4, 6-Trinitro-phenoxy)ethyl nitrate	c		SB		m	886·11±0·88	57/
$C_8H_6O_{10}N_4$		2-(2, 4, 6-Trinitro-phenoxy)ethyl nitrate							
$C_5H_8O_{12}N_4$	316·1391	Pentaerythritol tetranitrate	c		SB		m	614·2 ±0·6	39/
$C_5H_8O_{12}N_4$		Pentaerythritol tetranitrate	c		SB	2	m	614·8 ±0·3	66/
$C_5H_8O_{12}N_4$		Pentaerythritol tetranitrate							
$C_7H_{10}O_{14}N_4$	374·1762	α-D-Glucose tetranitrate	c		SB	6	m	802·55±0·80	57/
$C_{48}H_{54}O_{16}N_4$	942·9828	Isouroporphyrin (II) octamethyl ester	c		SB	3	m	5731·8 ±5·7	33/

1 kcal = 4·184 kJ

ΔH_r°		5 ΔH_f° (l or c) kcal/g.f.w.	6 Determination of ΔH_v			7 ΔH_f° (g) kcal/g.f.w.
Remarks			Type ΔH_v° kcal/g.f.w.		Ref.	
$NO_2)_4(l) + 6CO(g)$ {$-26\cdot42\pm0\cdot04$} $= 7CO_2 + 2N_2$		$+9\cdot8 \pm1\cdot0$ $+8\cdot9 \pm0\cdot7$	V4	$10\cdot3 \pm0\cdot5$	52/21	$+20\cdot1 \pm1\cdot2$ $+19\cdot2 \pm0\cdot8$
	Selected value	$+8\cdot9 \pm0\cdot7$				$+19\cdot2 \pm0\cdot8$
	C	$+1\cdot43\pm0\cdot25$				
	E	$-74\cdot37\pm0\cdot59$				
	BCE	$-73\cdot32\pm0\cdot38$				
	Selected value	$-73\cdot84\pm0\cdot50$				
	BCE	$-199\cdot56\pm0\cdot74$				
	BCE	$-353\cdot8 \pm5\cdot0$				
	BCE	$-68\cdot72\pm0\cdot90$				
	BCE	$-71\cdot24\pm0\cdot90$				
	Selected value	$-70\cdot0 \pm1\cdot5$				
	CE	$-129\cdot3 \pm0\cdot6$	S4	$36\cdot3 \pm0\cdot5$ (sub.)	53/33	$-93\cdot0 \pm0\cdot8$
		$-128\cdot7 \pm0\cdot3$				$-92\cdot4 \pm0\cdot7$
	Selected value	$-128\cdot8 \pm0\cdot3$				$-92\cdot5 \pm0\cdot7$
	E	$-197\cdot38\pm0\cdot83$				
	BCE	$-627\cdot2 \pm5\cdot8$				

$C_aH_bON_5-C_aH_bO_{16}N_5$ 1 kcal $= 4\cdot184$ kJ

1 Formula	2 g.f.w.	3 Name	State	Purity mol %	Type	No. of expts.	Detn. of reactn.	$-\Delta H_r^{\circ}$ kcal/g.f.w.	Re
$C_3H_5ON_5$	127·1062	5-Acetamidotetrazole	c		SB	4	m	451·13±0·44	51/
$C_5H_5ON_5$	151·1285	Guanine	c		SB	6	m	597·09±0·19	35/
$CH_5O_2N_5$	119·0833	Nitroaminoguanidine	c		SB	4	m	270·12±0·18	51/
$C_2H_3O_2N_5$	129·0785	3-Nitramino-1, 2, 4-triazole	c		SB	4	m	317·4 ±1·0	57/
$C_3H_5O_2N_5$	143·1056	3-Nitramino-5-methyl-1, 2, 4-triazole	c		SB	4	m	465·69±0·38	57/
$C_4H_7O_2N_5$	157·1327	5-Tetrazolylurethan	c		SB	4	m	562·74±0·65	51/
$C_2H_5O_3N_5$	147·0939	3-Amino-1, 2, 4-triazole nitrate	c		SB	3	m	318·0 ±1·2	57/
$C_2H_5O_3N_5$	147·0939	1-Formamido-2-nitroguanidine	c		SB	2	m	323·79±0·60	57/
$C_2H_5O_3N_5$	147·0939	N-Nitro-N′-guanidinourea	c		SB	8	m	284·02±0·28	56/
$C_2H_5O_3N_5$		N-Nitro-N′-guanidinourea	c		SB		m	286·6 ±1·4	57/
$C_2H_5O_3N_5$		N-Nitro-N′-guanidinourea							
$C_3H_7O_3N_5$	161·1209	1-Acetamido-2-nitroguanidine	c		SB	2	m	475·0 ±1·4	57/
$C_3H_7O_3N_5$	161·1209	3-Amino-5-methyl-1, 2, 4-triazole nitrate	c		SB	2	m	466·67±0·82	57/
$C_3H_9O_4N_5$	179·1363	Acetamidoguanidine nitrate	c		SB	2	m	471·5 ±1·1	57/
$C_6H_5O_6N_5$	243·1367	2, 4, 6-Trinitrophenyl-hydrazine	c		SB	6	m	743·85±0·74	54/

1 kcal $= 4 \cdot 184$ kJ

ΔH_r°			6 Determination of ΔH_v			7
Remarks		5 ΔH_f° (l or c) kcal/g.f.w.	ΔH_v° Type kcal/g.f.w.		Ref.	ΔH_f° (g) kcal/g.f.w.
		$-1 \cdot 81 \pm 0 \cdot 45$				
		$-43 \cdot 95 \pm 0 \cdot 20$				
		$+5 \cdot 28 \pm 0 \cdot 19$				
		$+26 \cdot 8 \pm 1 \cdot 0$				
		$+12 \cdot 75 \pm 0 \cdot 39$				
		$-52 \cdot 57 \pm 0 \cdot 66$				
		$-40 \cdot 9 \pm 1 \cdot 2$				
		$-35 \cdot 10 \pm 0 \cdot 61$				
	E	$-74 \cdot 87 \pm 0 \cdot 30$				
	BCE	$-72 \cdot 3 \pm 1 \cdot 4$				
Selected value		$-73 \cdot 6 \pm 1 \cdot 5$				
		$-46 \cdot 3 \pm 1 \cdot 4$				
		$-54 \cdot 59 \pm 0 \cdot 83$				
		$-118 \cdot 1 \pm 1 \cdot 1$				
		$+8 \cdot 76 \pm 0 \cdot 75$				

$C_4H_bON_5$—$C_aH_bO_{16}N_5$　　　　1 kcal = 4·184 kJ

					No.	Detn.		
					of	of		Determina
1 Formula	2 g.f.w.	3 Name	State	Purity mol %	Type	expts.	reactn.	$-\Delta H_f^\circ$ kcal/g.f.w. R
$C_4H_4O_8N_5$	250·1052	Bis (trinitroethyl)- nitrazine	c		SB		m	481·8 ±[0·5] 56
$C_7H_5O_8N_5$	287·1466	2, 4, 6-Trinitrophenyl- methylnitramide	c		SB	5	m	835·43±0·83 39
$C_7H_5O_8N_5$		2, 4, 6-Trinitrophenyl- methylnitramide	c		SB		m	832·1 ±[0·5] 56
$C_7H_3O_8N_5$		2, 4, 6-Trinitrophenyl- methylnitramide						
$C_{12}H_5O_8N_5$	347·2024	2, 4, 5, 7-Tetranitro- carbazole	c		SB		m	1303·9 ±1·3 57
$C_9H_{13}O_{16}N_5$	447·2279	2, 2, 4, 4-Tetrakis- (nitratomethyl)- 1-pyranol nitrate	c		SB	6	m	1101·0 ±2·2 54

1 kcal = 4·184 kJ

ΔH_r°	Remarks	5 ΔH_f° (l or c) kcal/g.f.w.	6 Determination of ΔH_v			7 ΔH_f° (g) kcal/g.f.w.
			Type	ΔH_v° kcal/g.f.w.	Ref.	
	C	$-31\cdot0$ $\pm[0\cdot5]$				
	CE	$+6\cdot28\pm0\cdot85$				
	C	$+2\cdot9$ $\pm[0\cdot5]$				
	Selected value	$+6\cdot28\pm0\cdot85$				
	BCE	$+4\cdot5$ $\pm1\cdot3$				
	E	$-189\cdot5$ $\pm2\cdot2$				

$C_aH_bO_cN_6$—$C_aH_bO_cN_{10}$ 1 kcal = 4·184 kJ

1 Formula	2 g.f.w.	3 Name	State	Purity mol %	Type	No. of expts.	Detn. of reactn.	$-\Delta H_r^\circ$ kcal/g.f.w.	Re
$C_5H_8ON_6$	168·1591	3-(5-Tetrazolylazo)- butanone-2	c		SB	3	m	783·52±0·60	57/
$CH_2O_2N_6$	130·0661	Nitroguanyl azide	c		SB	4	m	233·7 ±1·0	51/
$C_5H_{10}O_2N_6$	186·1745	Dinitrosopentamethylene- tetramine	c		SB	4	m	864·48±0·86	42
$C_5H_{10}O_2N_6$		Dinitrosopentamethylene- tetramine	c		SB	6	m	866·36±0·86	56/
$C_5H_{10}O_2N_6$		Dinitrosopentamethylene- tetramine							
$C_6H_8O_2N_6$	196·1697	3-(5-Tetrazolylazo)- pentandione-2, 4	c		SB	4	m	841·0 ±1·3	57/
$CH_4O_3N_6$	148·0814	5-Aminotetrazole nitrate	c		SB	4	m	224·09±0·45	51/
$CH_4O_3N_6$	148·0814	Guanyl azide nitrate	c		SB	4	m	234·47±0·35	51/1
$CH_8O_3N_6$	152·1133	Diaminoguanidine nitrate	c		SB	4	m	329·74±0·20	51/
$C_3H_6O_3N_6$	174·1197	Trinitrosotrimethylene triamine	c		SB	4	m	555·42±0·56	42/
$C_3H_6O_3N_6$		Trinitrosotrimethylene triamine	c		SB		m	552·9 ±0·5	56/
$C_3H_6O_3N_6$		Trinitrosotrimethylene triamine							
$C_3H_8O_3N_6$	176·1356	1, 3-Dimethyl-5-imino- tetrazole nitrate	c		SB	2	m	554·13±0·60	57/
$C_3H_{12}O_3N_6$	180·1675	Guanidine carbonate	c		SB	8	m	459·75±0·13	40/
$C_3H_{12}O_3N_6$		Guanidine carbonate	c		SB	9	m	461·25±0·92	55/
$C_3H_{12}O_3N_6$		Guanidine carbonate							
$C_3H_6O_6N_6$	222·1179	1, 3, 5-Trinitrohexa- hydro-s-triazine	c		SB	4	m	503·0 ±0·5	42/
$C_3H_6O_6N_6$		1, 3, 5-Trinitrohexa- hydro-s-triazine	c		SB		m	500·0 ±[0·5]	56/1
$C_3H_6O_6N_6$		1, 3, 5-Trinitrohexa- hydro-s-triazine							

1 kcal = 4·184 kJ

ΔH_r°		5 ΔH_f° (l or c) kcal/g.f.w.	6 Determination of ΔH_v			7 ΔH_f° (g) kcal/g.f.w.
	Remarks		ΔH_v° Type kcal/g.f.w.		Ref.	
		+40·00±0·61				
		+71·3 ±1·0				
	CE	+52·65±0·88				
	E	+54·53±0·88				
	Selected value	+53·6 ±1·0				
		+3·4 ±1·3				
		−6·59±0·46				
		+3·79±0·36				
		−37·57±0·21				
	CE	+68·32±0·58				
	C	+65·8 ±0·5				
	Selected value	+68·32±0·58				
		−1·28±0·61				
		−232·29±0·15				
	E	−230·79±0·93				
	Selected value	−232·29±0·15				
	CE	+15·9 ±0·5				
	C	+12·9 ±[0·5]				
	Selected value	+15·9 ±0·5				

THERMOCHEMISTRY OF ORGANIC AND ORGANOMETALLIC COMPOUNDS

$C_6H_9O_cN_6$—$C_6H_9O_cN_{10}$ 1 kcal = 4·184 kJ

						4 Determination			
1 Formula	2 g.f.w.	3 Name	State	Purity mol %	No. of Type expts.	Detn. of reactn.	$-\Delta H_r^\circ$ kcal/g.f.w.	Ref
$C_6H_{14}O_6N_6$	266·2151	Hexamethylene-tetramine dinitrate	c		SB 4	m	951·0 ±1·0	42/
$C_6H_8O_{12}N_6$	356·1637	NN'-Dinitro-NN'-di-(2-hydroxyethylnitrate) oxamide	c		SB 8	m	709·55±0·71	54/1
$C_{14}H_6O_{12}N_6$	450·2369	trans-2, 2', 4, 4', 6, 6'-Hexanitrostilbene	c		SB 8	m	1535·5 ±1·0	68/1
$C_8H_{12}O_{14}N_6$	416·2166	Ethylenedinitramine diurethanglycol dinitrate	c		SB 6	m	927·05±0·93	56/2
$C_{10}H_{16}O_{19}N_6$	524·2678	Dipentaerythritol hexanitrate	c		SB 7	m	1252·9 ±1·3	56/2
$C_{12}H_{14}O_{27}N_8$	702·2828	Saccharose octanitrate	c		SB 3	m	1280·8 ±2·6	57/2
$C_2H_7O_2N_9$	189·1372	5-Nitroaminotetrazole guanidinate	c		SB 4	m	453·77±0·88	51/1
$C_9H_9O_{12}N_9$	435·2252	2, 4, 6-Trinitro-1, 3, 5-tris-(methylnitramino)-benzene	c		SB 4	m	1182·1 ±2·3	53/1
$C_2H_8ON_{10}$	188·1525	1-(5-Tetrazolyl)-4-guanyl-tetrazene hydrate	c		SB 2	m	506·55±0·60	57/2

1 kcal $= 4\cdot184$ kJ

ΔH_r°	Remarks	5 ΔH_f° (l or c) kcal/g.f.w.	6 Determination of ΔH_v			7 ΔH_f° (g) kcal/g.f.w.
			Type	ΔH_v° kcal/g.f.w.	Ref.	
		CE $\quad -91\cdot5\ \pm1\cdot0$				
		E $\quad -128\cdot02\pm0\cdot72$				
		$+13\cdot8\ \pm1\cdot0$				
		E $\quad -235\cdot25\pm0\cdot95$				
		E $\quad -234\cdot1\ \pm1\cdot3$				
		E $\quad -326\cdot0\ \pm2\cdot6$				
		$+26\cdot56\pm0\cdot89$				
		BCE $\quad +28\cdot17\pm2\cdot3$				
		$+45\cdot19\pm0\cdot61$				

C_aH_bS

1 kcal = 4·184 kJ

Determina

1 Formula	2 g.f.w.	3 Name	State	Purity mol %	Type	No. of expts.	Detn. of reactn.	$-\Delta H_f^\circ$ kcal/g.f.w.	R
CH_4S	48·107	Methanethiol	l	99·94±0·06	RB	9	m	363·47±0·12	61
C_2H_4S	60·118	Thiacyclopropane	l		RB		m	481·02±0·28	63
C_2H_6S	62·134	Ethanethiol	l	99·98±0·01	RB	6	m	519·40±0·10	57
C_2H_6S	62·134	2-Thiapropane	l	99·99±0·01	RB	8	m	521·38±0·08	57
C_3H_6S	74·145	2-Methylthiacyclopropane	l		RB		m	633·75±0·28	63
C_3H_6S	74·145	Thiacyclobutane	l		RB		m	636·78±0·38	63
C_3H_6S		Thiacyclobutane	l	99·95	RB	3	m	637·07±0·27	54
C_3H_6S		Thiacyclobutane							
C_3H_8S	76·161	1-Propanethiol	l	99·98±0·01	RB	7	m	675·50±0·11	54
C_3H_8S	76·161	2-Propanethiol	l	99·98±0·01	RB	7	m	674·07±0·11	54
C_5H_8S	76·161	2-Thiabutane	l	99·99±0·01	RB	6	m	677·48±0·25	54
C_4H_4S	84·140	Thiophene	l		SB	3	m	676·23±0·60	40
C_4H_4S		Thiophene	l		SB	6	m	675·77±0·20	49
C_4H_4S		Thiophene	l		RB	6	m	675·81±0·22	55
C_4H_4S		Thiophene	l	99·96±0·02	RB	5	m	676·09±0·11	55
C_4H_4S		Thiophene							63
C_4H_6S	86·156	2, 3-Dihydrothiophene	l	99·86, glc	RB	5	m	737·76±0·25	62
C_4H_6S	86·156	2, 5-Dihydrothiophene	l	99·88, glc	RB	9	m	736·34±0·25	62
C_4H_8S	88·172	2, 2-Dimethylthiacyclopropane	l		RB		m	787·56±0·28	63
C_4H_8S	88·172	cis-2, 3-Dimethylthiacyclopropane	l		RB		m	787·56±0·28	63
C_4H_8S	88·172	trans-2, 3-Dimethylthiacyclopropane	l		RB		m	786·32±0·28	63
C_4H_8S	88·172	Thiacyclopentane	l	99·95	RB	8	m	775·95±0·33	54
C_4H_8S		Thiacyclopentane	l	99·95, glc	RB	5	m	776·01±0·25	62
C_4H_8S		Thiacyclopentane							
$C_4H_{10}S$	90·188	1-Butanethiol	l	99·99±0·01	RB	5	m	831·94±0·26	58

$$1 \text{ kcal} = 4 \cdot 184 \text{ kJ}$$

ΔH_r°		5 ΔH_f° (l or c) kcal/g.f.w.	6 Determination of ΔH_v			7 ΔH_f° (g) kcal/g.f.w.
Remarks			Type	ΔH_v° kcal/g.f.w.	Ref.	
	G	$-11 \cdot 09 \pm 0 \cdot 13$		$5 \cdot 69 \pm 0 \cdot 02$	61/31	$-5 \cdot 40 \pm 0 \cdot 14$
		$+12 \cdot 41 \pm 0 \cdot 29$	V1	$7 \cdot 28 \pm 0 \cdot 05$	52/22	$+19 \cdot 69 \pm 0 \cdot 30$
	G	$-17 \cdot 53 \pm 0 \cdot 14$	C1	$6 \cdot 53 \pm 0 \cdot 01$	52/23	$-11 \cdot 00 \pm 0 \cdot 14$
	G	$-15 \cdot 55 \pm 0 \cdot 13$	C1	$6 \cdot 66 \pm 0 \cdot 01$	57/30	$-8 \cdot 89 \pm 0 \cdot 13$
		$+2 \cdot 77 \pm 0 \cdot 30$		$8 \cdot 25 \pm 0 \cdot 40$	63/37	$+11 \cdot 02 \pm 0 \cdot 50$
		$+5 \cdot 80 \pm 0 \cdot 40$	V1	$8 \cdot 58 \pm 0 \cdot 05$	53/6	$+14 \cdot 38 \pm 0 \cdot 41$
	G	$+6 \cdot 09 \pm 0 \cdot 30$				$+14 \cdot 67 \pm 0 \cdot 31$
	Selected value	$+6 \cdot 00 \pm 0 \cdot 26$				$+14 \cdot 58 \pm 0 \cdot 27$
	G	$-23 \cdot 79 \pm 0 \cdot 15$	C1	$7 \cdot 65 \pm 0 \cdot 02$	56/22	$-16 \cdot 14 \pm 0 \cdot 16$
	G	$-25 \cdot 22 \pm 0 \cdot 15$	C1	$7 \cdot 08 \pm 0 \cdot 02$	54/20	$-18 \cdot 14 \pm 0 \cdot 16$
	G	$-21 \cdot 81 \pm 0 \cdot 27$	C1	$7 \cdot 64 \pm 0 \cdot 02$	51/20	$-14 \cdot 17 \pm 0 \cdot 28$
uffman–Ellis method (35/1)	AG	$+19 \cdot 52 \pm 0 \cdot 61$	C1	$8 \cdot 30 \pm 0 \cdot 01$	49/19	$+27 \cdot 82 \pm 0 \cdot 61$
uffman–Ellis method (35/1)	G	$+19 \cdot 06 \pm 0 \cdot 23$				$+27 \cdot 36 \pm 0 \cdot 23$
	G	$+19 \cdot 10 \pm 0 \cdot 24$				$+27 \cdot 40 \pm 0 \cdot 24$
		$+19 \cdot 38 \pm 0 \cdot 15$				$+27 \cdot 68 \pm 0 \cdot 15$
	Selected value	$+19 \cdot 29 \pm 0 \cdot 15$				$+27 \cdot 59 \pm 0 \cdot 15$
		$+12 \cdot 73 \pm 0 \cdot 28$	C1	$9 \cdot 02 \pm 0 \cdot 10$	62/3	$+21 \cdot 75 \pm 0 \cdot 30$
		$+11 \cdot 31 \pm 0 \cdot 28$	C1	$9 \cdot 55 \pm 0 \cdot 06$	62/3	$+20 \cdot 86 \pm 0 \cdot 29$
		$-5 \cdot 78 \pm 0 \cdot 30$		$8 \cdot 55 \pm 0 \cdot 40$	63/37	$+2 \cdot 77 \pm 0 \cdot 50$
		$-5 \cdot 78 \pm 0 \cdot 30$		$8 \cdot 55 \pm 0 \cdot 40$	63/37	$+2 \cdot 77 \pm 0 \cdot 50$
		$-7 \cdot 02 \pm 0 \cdot 30$		$7 \cdot 95 \pm 0 \cdot 40$	63/37	$+0 \cdot 93 \pm 0 \cdot 50$
	G	$-17 \cdot 39 \pm 0 \cdot 35$	C1	$9 \cdot 29 \pm 0 \cdot 01$	52/24	$-8 \cdot 10 \pm 0 \cdot 35$
		$-17 \cdot 33 \pm 0 \cdot 28$				$-8 \cdot 04 \pm 0 \cdot 28$
	Selected value	$-17 \cdot 36 \pm 0 \cdot 25$				$-8 \cdot 07 \pm 0 \cdot 25$
	G	$-29 \cdot 72 \pm 0 \cdot 28$	C1	$8 \cdot 74 \pm 0 \cdot 02$	57/31	$-20 \cdot 98 \pm 0 \cdot 29$

C_aH_bS

1 Formula	2 g.f.w.	3 Name	State	Purity mol %	Type	No. of expts.	Detn. of reactn.	$-\Delta H_r^\circ$ kcal/g.f.w.	Ref
$C_4H_{10}S$	90·188	2-Butanethiol	l	99·9 ±0·1	RB	7	m	830·44±0·17	58/3
$C_4H_{10}S$	90·188	2-Methyl-1-propanethiol	l	99·99±0·01	RB	8	m	830·19±0·17	58/3
$C_4H_{10}S$	90·188	2-Methyl-2-propanethiol	l	99·99±0·01	RB	5	m	828·15±0·17	58/3
$C_4H_{10}S$	90·188	2-Thiapentane	l	99·98±0·01	RB	7	m	833·41±0·20	58/3
$C_4H_{10}S$	90·188	3-Thiapentane	l	99·99±0·01	RB	8	m	833·20±0·16	58/3
$C_4H_{10}S$	90·188	3-Methyl-2-thiabutane	l	99·98±0·01	RB	9	m	831·93±0·14	58/3
C_5H_6S	98·168	2-Methylthiophene	l	99·96±0·01	RB	9	m	829·82±0·18	56/2
C_5H_6S	98·168	3-Methylthiophene	l	99·99±0·01	RB	8	m	829·45±0·17	53/20
$C_5H_{10}S$	102·199	4-Thia-1-hexene	l		RB	5		950·64±0·59	62/3.
$C_5H_{10}S$	102·199	Cyclopentanethiol	l	99·99±0·01	RB	6	m	934·39±0·15	61/3.
$C_5H_{10}S$	102·199	Trimethylthiacyclopropane	l		RB		m	941·26±0·27	63/14
$C_5H_{10}S$	102·199	2-Methylthiacyclopentane							61/33
$C_5H_{10}S$	102·199	Thiacyclohexane	l	99·99±0·01	RB	8	m	930·41±0·17	54/21
$C_5H_{10}S$		Thiacyclohexane	l		RB		m	930·34±0·27	63/14
$C_5H_{10}S$		Thiacyclohexane							
$C_5H_{12}S$	104·215	1-Pentanethiol	l	99·92	RB	9	m	987·75±0·38	54/1
$C_5H_{12}S$		1-Pentanethiol	l		RB	4	m	988·01±0·16	55/29
$C_5H_{12}S$		1-Pentanethiol							63/14
$C_5H_{12}S$	104·215	3-Methyl-1-butanethiol							61/33
$C_5H_{12}S$	104·215	2-Methyl-2-butanethiol	l	99·89±0·05	RB	7	m	985·19±0·19	62/34
$C_5H_{12}S$	104·215	2-Thiahexane	l	99·96	RB	6	m	989·93±0·16	61/33
$C_5H_{12}S$		2-Thiahexane	l		RB	6	m	990·05±0·46	62/35
$C_5H_{12}S$		2-Thiahexane							
$C_5H_{12}S$	104·215	3-Thiahexane	l	99·97	RB	9	m	989·49±0·14	61/33
$C_5H_{12}S$	104·215	2-Methyl-3-thiapentane	l		RB	6	m	986·78±0·47	62/35

1 kcal = 4·184 kJ

H_r°		5 ΔH_f° (l or c) kcal/g.f.w.	6 Determination of ΔH_v			7 ΔH_f° (g) kcal/g.f.w.
Remarks			Type	ΔH_v° kcal/g.f.w.	Ref.	
	G	$-31\cdot22\pm0\cdot20$	Cl	$8\cdot13\pm0\cdot01$	58/27	$-23\cdot09\pm0\cdot20$
	G	$-31\cdot47\pm0\cdot20$	Cl	$8\cdot30\pm0\cdot01$	58/28	$-23\cdot17\pm0\cdot20$
	G	$-33\cdot51\pm0\cdot20$	Cl	$7\cdot39\pm0\cdot05$	53/19	$-26\cdot12\pm0\cdot21$
	G	$-28\cdot25\pm0\cdot23$	Cl	$8\cdot67\pm0\cdot02$	57/31	$-19\cdot58\pm0\cdot24$
	G	$-28\cdot46\pm0\cdot19$	Cl	$8\cdot57\pm0\cdot01$	52/25	$-19\cdot89\pm0\cdot19$
	G	$-29\cdot73\pm0\cdot18$	Cl	$8\cdot18\pm0\cdot01$	55/28	$-21\cdot55\pm0\cdot18$
	G	$+10\cdot74\pm0\cdot21$	Cl	$9\cdot30\pm0\cdot01$	56/23	$+20\cdot04\pm0\cdot21$
	G	$+10\cdot37\pm0\cdot20$	Cl	$9\cdot45\pm0\cdot05$	53/20	$+19\cdot82\pm0\cdot21$
	H	$-5\cdot09\pm0\cdot60$	V3	$9\cdot4\ \pm0\cdot3$	62/33	$+4\cdot3\ \pm0\cdot7$
	G	$-21\cdot34\pm0\cdot18$	Cl	$9\cdot93\pm0\cdot01$	61/32	$-11\cdot41\pm0\cdot18$
		$-14\cdot47\pm0\cdot30$		$9\cdot40\pm0\cdot30$	63/37	$-5\cdot07\pm0\cdot42$
published data (U.S. Bureau of Mines)						$-15\cdot12\pm0\cdot31$
	G	$-25\cdot32\pm0\cdot20$	Cl	$10\cdot22\pm0\cdot05$	54/21	$-15\cdot10\pm0\cdot21$
		$-25\cdot39\pm0\cdot30$				$-15\cdot17\pm0\cdot31$
Selected value		$-25\cdot34\pm0\cdot17$				$-15\cdot12\pm0\cdot18$
	G	$-36\cdot28\pm0\cdot41$	Cl	$9\cdot83\pm0\cdot02$	65/36	$-26\cdot45\pm0\cdot42$
		$-36\cdot02\pm0\cdot20$				$-26\cdot19\pm0\cdot21$
Selected value		$-36\cdot07\pm0\cdot18$				$-26\cdot24\pm0\cdot18$
published data (U.S. Bureau of Mines)						$-27\cdot41\pm0\cdot30$
	G	$-38\cdot84\pm0\cdot22$	Cl	$8\cdot54\pm0\cdot01$	62/34	$-30\cdot30\pm0\cdot22$
	G	$-34\cdot10\pm0\cdot20$	V1	$9\cdot73\pm0\cdot01$	61/33	$-24\cdot37\pm0\cdot20$
	H	$-33\cdot98\pm0\cdot49$				$-24\cdot25\pm0\cdot49$
Selected value		$-34\cdot08\pm0\cdot18$				$-24\cdot35\pm0\cdot18$
	G	$-34\cdot54\pm0\cdot18$	V1	$9\cdot58\pm0\cdot01$	61/33	$-24\cdot96\pm0\cdot18$
		$-37\cdot25\pm0\cdot50$	V4	$9\cdot3\ \pm0\cdot3$	62/35	$-28\cdot0\ \pm0\cdot6$

C_aH_bS

1 kcal = 4·184 kJ

							Determinat		
1 Formula	2 g.f.w.	3 Name	State	Purity mol %	Type	No. of expts.	Detn. of reactn.	$-\Delta H_r^\circ$ kcal/g.f.w.	Re
$C_5H_{12}S$	104·215	3, 3-Dimethyl-2-thiabutane	l	99·98	RB	9	m	986·56±0·15	62/
C_6H_6S	110·179	Benzenethiol	l	99·98±0·01	RB	9	m	928·43±0·15	56/
$C_6H_{12}S$	116·227	Cyclopentyl-1-thiaethane							61/
$C_6H_{12}S$	116·227	Cyclohexanethiol							61/ 67/
$C_6H_{12}S$	116·227	Tetramethylthiacyclo-propane	c		RB		m	1098·24±0·37	63/
$C_6H_{12}S$	116·227	Thiacycloheptane	l		RB		m	1091·12±0·37	63/
$C_6H_{14}S$	118·242	1-Hexanethiol	l	99·97±0·01	RB	9	m	1144·56±0·18	66,
$C_6H_{14}S$	118·242	2-Thiaheptane	l		RB	6	m	1146·53±0·47	62/
$C_6H_{14}S$	118·242	3-Thiaheptane	l		RB	9	m	1145·29±0·46	62/
$C_6H_{14}S$	118·242	4-Thiaheptane	l	99·96	RB	6	m	1145·86±0·18	61/
$C_6H_{14}S$	118·242	2, 2-Dimethyl-3-thiapentane	l		RB	6	m	1141·70±0·49	62/
$C_6H_{14}S$	118·242	2, 4-Dimethyl-3-thiapentane	l		RB	7	m	1142·90±0·48	62/.
$C_6H_{14}S$		2, 4-Dimethyl-3-thiapentane							61/.
$C_6H_{14}S$		2, 4-Dimethyl-3-thiapentane							67/
C_7H_8S	124·206	α-Toluenethiol	l	glc	RB	6	m	1086·07±0·48	62/
C_7H_8S	124·206	Phenyl methyl sulphide	l		RB	7	m	1087·00±0·47	62/3
$C_7H_{14}S$	130·254	4-Thia-5, 5-dimethyl-1-hexene	l		RB	6	m	1258·76±0·38	62/.
$C_7H_{16}S$	132·270	1-Heptanethiol	l	99·97±0·01	RB	8	m	1300·94±0·18	66/
$C_8H_{10}S$	138·233	Benzyl methyl sulphide	l		RB	6	m	1244·13±0·48	62/3
$C_8H_{10}S$	138·233	Phenyl ethyl sulphide	l		RB	7	m	1243·15±0·36	62/3

1 kcal = 4·184 kJ

ΔH_r° Remarks		5 ΔH_f° (l or c) kcal/g.f.w.	6 Determination of ΔH_v			7 ΔH_f° (g) kcal/g.f.w.
			Type	ΔH_v° kcal/g.f.w.	Ref.	
	G	−37·47±0·18	C1	8·56±0·01	62/36	−28·91±0·18
	G	+15·30±0·19	C1	11·64±0·05	56/24	+26·94±0·20
published data (U.S. Bureau of Mines)						−15·41±0·25
published data (U.S. Bureau of Mines)						−22·80±0·21
		−19·84±0·40				
		−26·96±0·40		11·30±0·30	63/37	−15·66±0·50
	G	−41·83±0·22	V1	10·94±0·05	66/5	−30·89±0·23
		−39·86±0·50	V4	10·8 ±0·3	62/35	−29·1 ±0·6
	H	−41·10±0·50	V4	10·8 ±0·3	62/35	−30·3 ±0·6
	G	−40·53±0·22	V1	10·66±0·01	61/33	−29·87±0·22
		−44·69±0·52	V4	9·4 ±0·3	62/35	−35·3 ±0·6
		−43·49±0·51	V4	9·4 ±0·3	62/35	−34·1 ±0·6
npublished data (U.S. Bureau of Mines)						−33·73±0·24
Selected value		−43·1 ±0·4				−33·73±0·24
		+10·57±0·50	E5	12·3 ±0·4	62/37	+22·9 ±0·7
		+11·50±0·50	V4	12·1 ±0·5	62/38	+23·6 ±0·7
	H	−21·68±0·42	V4	10·6 ±0·3	62/33	−11·1 ±0·5
	G	−47·82±0·22	V1	12·09±0·05	66/5	−35·73±0·23
		+6·27±0·50	V4	12·8 ±0·5	62/38	+19·1 ±0·7
		+5·29±0·39	V4	13·2 ±0·5	62/38	+18·5 ±0·7

C_eH_bS

1 kcal $= 4 \cdot 184$ kJ

1 Formula	2 g.f.w.	3 Name	State	Purity mol %	Type	No. of expts.	Detn. of reactn.	$-\Delta H_r^\circ$ kcal/g.f.w.	Re
$C_8H_{18}S$	$146 \cdot 297$	5-Thianonane	l	$99 \cdot 97$	RB	6	m	$1458 \cdot 47 \pm 0 \cdot 32$	61/
$C_8H_{18}S$		5-Thianonane	l		RB	5	m	$1458 \cdot 42 \pm 0 \cdot 42$	55/
$C_8H_{18}S$		5-Thianonane							63/
$C_8H_{18}S$	$146 \cdot 297$	2, 6-Dimethyl-4-thiaheptane	l		RB	7	m	$1456 \cdot 41 \pm 0 \cdot 37$	62/
$C_8H_{18}S$	$146 \cdot 297$	2, 2, 4, 4-Tetramethyl-3-thiapentane	l		RB	6	m	$1455 \cdot 58 \pm 0 \cdot 18$	62/
$C_9H_{12}S$	$152 \cdot 260$	Benzyl ethyl sulphide	l		RB	6	m	$1399 \cdot 06 \pm 0 \cdot 51$	62/
$C_9H_{14}S$	$154 \cdot 276$	Thia-adamantane	c		RB	6	m	$1434 \cdot 32 \pm 0 \cdot 23$	61/
$C_{10}H_{22}S$	$174 \cdot 351$	1-Decanethiol	l	$99 \cdot 88 \pm 0 \cdot 02$	RB	10	m	$1769 \cdot 79 \pm 0 \cdot 29$	66/
$C_{10}H_{22}S$	$174 \cdot 351$	6-Thiaundecane	l		RB	6	m	$1772 \cdot 25 \pm 0 \cdot 40$	62/
$C_{10}H_{22}S$	$174 \cdot 351$	2, 8-Dimethyl-5-thianonane	l		RB	6	m	$1768 \cdot 58 \pm 0 \cdot 49$	62/3
$C_{12}H_{10}S$	$186 \cdot 278$	Diphenyl sulphide	l		RB	7	m	$1653 \cdot 19 \pm 0 \cdot 47$	62/3
$C_{14}H_{14}S$	$214 \cdot 332$	Dibenzyl sulphide	c		RB	6	m	$1962 \cdot 54 \pm 0 \cdot 48$	62/3

1 kcal = 4·184 kJ

H_r° Remarks		5 ΔH_f° (l or c) kcal/g.f.w.	6 Determination of ΔH_v			7 ΔH_f° (g) kcal/g.f.w.
			Type	ΔH_v° kcal/g.f.w.	Ref.	
	G	$-52·65\pm0·35$ $-52·70\pm0·44$	V1	$12·75\pm0·01$	61/33	$-39·90\pm0·35$ $-39·95\pm0·44$
Selected value		$-52·67\pm0·30$				$-39·92\pm0·30$
		$-54·71\pm0·42$	V4	$11·9\ \pm0·4$	62/35	$-42·8\ \pm0·6$
	H	$-55·54\pm0·20$	V4	$10·9\ \pm0·6$	62/35	$-44·6\ \pm0·7$
		$-1·17\pm0·52$	V4	$13·6\ \pm0·5$	62/38	$+12·4\ \pm0·7$
	G	$-34·22\pm0·27$				
	G	$-66·07\pm0·34$	E1	$15·42\pm0·10$	66/2	$-50·65\pm0·35$
		$-63·61\pm0·44$	V4	$14·7\ \pm0·5$	62/35	$-48·9\ \pm0·7$
		$-67·28\pm0·52$	V4	$14·4\ \pm0·4$	62/35	$-52·9\ \pm0·7$
	H	$+39·12\pm0·50$	V4	$16·2\ \pm0·5$	62/33	$+55·3\ \pm0·7$
		$+23·74\pm0·50$	$V4_m$	$22·3\ \pm0·8$ (sub.)	62/33	$+46·0\ \pm1·0$

THERMOCHEMISTRY OF ORGANIC AND ORGANOMETALLIC COMPOUNDS

$C_aH_bS_2$—$C_aH_bS_3$ 1 kcal = 4·184 kJ

Determina

1 Formula	2 g.f.w.	3 Name	State	Purity mol %	Type	No. of expts.	Detn. of reactn.	$-\Delta H_r^\circ$ kcal/g.f.w.	F
CS_2	76·139	Carbon disulphide	l	99·98	RB	10	m	403·24±0·12	61
CS_2		Carbon disulphide	l		SB	9	m		63
CS_2		Carbon disulphide	g		SB		m	265·8 ±2·0	49
CS_2		Carbon disulphide							
$C_2H_6S_2$	94·198	Ethanedithiol-1, 2	l	>99·8	RB	6	m	667·97±0·22	62
$C_2H_6S_2$	94·198	2, 3-Dithiabutane	l	99·97	RB	6	m	665·99±0·17	58
$C_3H_8S_2$	108·225	Propanedithiol-2, 3	l	>99·8	RB	7	m	824·35±0·26	62
$C_4H_{10}S_2$	122·252	Butanedithiol-1, 4	l	>99·8	RB	9	m	980·42±0·40	62
$C_4H_{10}S_2$	122·252	3, 4-Dithiahexane	l	99·92	RB	6	m	976·98±0·19	58
$C_5H_{12}S_2$	136·279	Pentanedithiol-1, 5	l	>99·9	RB	6	m	1136·92±0·28	62
$C_6H_{14}S_2$	150·306	4, 5-Dithiaoctane	l	99·97	RB	6	m	1289·45±0·17	58
$C_8H_{18}S_2$	178·361	5, 6-Dithiadecane	l	>99·8, glc	RB	6	m	1601·89±0·44	64
$C_8H_{18}S_2$	178·361	2, 7-Dimethyl-4, 5-dithiaoctane	l	>99·8, glc	RB	6	m	1599·52±0·28	64
$C_8H_{18}S_2$	178·361	2, 2, 5, 5-Tetramethyl-3, 4-dithiahexane	l	>99·8, glc	RB	7	m	1595·16±0·38	64
$C_{12}H_8S_2$	216·326	Thianthrene	c	99·8	RB	10	m	1733·15±0·53	54
$C_{12}H_8S_2$		Thianthrene	c		RB	5	m	1732·87±0·69	5?
$C_{12}H_8S_2$		Thianthrene	c	99·8	RB	10	m	1733·29±0·28	53
									63
$C_{12}H_8S_2$		Thianthrene	c		RB		m	1733·40±0·52	66
$C_{12}H_8S_2$		Thianthrene							
$C_{12}H_{10}S_2$	218·342	Diphenyl disulphide	c		RB	7	m	1793·60±0·65	62
$C_{16}H_{16}S_2$	272·434	cis-1, 2-Dibenzyl-dithioethylene	c		RB	5	m	2375·00±0·44	67
$C_{16}H_{16}S_2$	272·434	trans-1, 2-Dibenzyl-dithioethylene	c		RB	5	m	2371·88±0·28	67

$$1 \text{ kcal} = 4 \cdot 184 \text{ kJ}$$

ΔH_r° Remarks		5 ΔH_f° (l or c) kcal/g.f.w.	6 Determination of ΔH_v			7 ΔH_f° (g) kcal/g.f.w.
			Type	ΔH_v° kcal/g.f.w.	Ref.	
h method (34/3): ΔH_f worked out for	G	$+21\cdot43\pm0\cdot17$ $+21\cdot37\pm0\cdot37$	Cl	$6\cdot61\pm0\cdot02$	62/39	$+28\cdot04\pm0\cdot18$ $+27\cdot98\pm0\cdot38$
each run						
$_8$(g)$+3O_2$		$+23\cdot3\ \pm2\cdot0$				$+29\cdot9\ \pm2\cdot0$
$= CO_2+2SO_2(g)\ \{-70\cdot94\pm0\cdot05\}$						
Selected value		$+21\cdot43\pm0\cdot17$				$+28\cdot04\pm0\cdot18$
		$-12\cdot84\pm0\cdot28$	Cl	$10\cdot68\pm0\cdot03$	62/40	$-2\cdot16\pm0\cdot29$
	G	$-14\cdot82\pm0\cdot22$	Vl	$9\cdot18\pm0\cdot03$	50/18	$-5\cdot64\pm0\cdot23$
		$-18\cdot82\pm0\cdot32$	Cl	$11\cdot87\pm0\cdot03$	62/40	$-6\cdot95\pm0\cdot33$
		$-25\cdot12\pm0\cdot44$	Cl	$13\cdot22\pm0\cdot09$	62/40	$-11\cdot90\pm0\cdot45$
	G	$-28\cdot56\pm0\cdot24$	Cl	$10\cdot86\pm0\cdot05$	52/26	$-17\cdot70\pm0\cdot25$
		$-30\cdot99\pm0\cdot34$	Cl	$14\cdot17\pm0\cdot10$	62/40	$-16\cdot82\pm0\cdot36$
	G	$-40\cdot82\pm0\cdot23$	Cl	$12\cdot94\pm0\cdot10$	58/29	$-27\cdot88\pm0\cdot25$
		$-53\cdot11\pm0\cdot47$	E6	$15\cdot42\pm0\cdot40$	64/32	$-37\cdot69\pm0\cdot62$
		$-55\cdot48\pm0\cdot33$	E6	$14\cdot79\pm0\cdot40$	64/32	$-40\cdot69\pm0\cdot52$
		$-59\cdot84\pm0\cdot42$	E6	$12\cdot71\pm0\cdot40$	64/32	$-47\cdot13\pm0\cdot58$
	G	$+43\cdot52\pm0\cdot56$ $+43\cdot24\pm0\cdot71$ $+43\cdot66\pm0\cdot32$				
Selected value		$+43\cdot77\pm0\cdot55$ $+43\cdot62\pm0\cdot25$				
		$+35\cdot65\pm0\cdot67$	V4$_m$	$22\cdot7\ \pm0\cdot7$ (sub.)	62/33	$+58\cdot4\ \pm1\cdot0$
		$+35\cdot90\pm0\cdot49$				
		$+32\cdot78\pm0\cdot38$				

THERMOCHEMISTRY OF ORGANIC AND ORGANOMETALLIC COMPOUNDS

$C_aH_bS_2-C_aH_bS_3$

1 kcal = 4·184 kJ

1 Formula	2 g.f.w.	3 Name	State	Purity mol %	Determination Type	No. of expts.	Detn. of reactn.	$-\Delta H_r^\circ$ kcal/g.f.w.	Re
CH_2S_3	110·219	Trithiocarbonic acid	l		SB	10	m		63/.
CH_2S_3		Trithiocarbonic acid	l		E		TP	$-10·6 \pm 0·3$	63/.
CH_2S_3		Trithiocarbonic acid							
$C_3H_4S_3$	136·257	1, 3-Dithiolan-2-thione	c		RB	10	m	$853·5 \pm 0·4$	67/
$C_4H_6S_3$	150·284	1, 3-Dithian-2-thione	c		RB	10	m	$1009·7 \pm 0·1$	67/
$C_7H_8S_3$	188·334	4, 5-Tetramethylene-1, 3-dithiole-2-thione	c		RB	9	m	$1368·6 \pm 0·3$	67/
$C_7H_{10}S_3$	190·350	4, 5-Tetramethylene-1, 3-dithiolan-2-thione	c		RB	8	m	$1412·9 \pm 0·2$	67/

$$1 \text{ kcal} = 4 \cdot 184 \text{ kJ}$$

ΔH_r° Remarks	5 ΔH_f° (l or c) kcal/g.f.w.	6 Determination of ΔH_v Type	ΔH_v° kcal/g.f.w.	Ref.	7 ΔH_f° (g) kcal/g.f.w.
oth method (34/3): ΔH_f worked out for each run	$+6 \cdot 1 \ \pm 1 \cdot 2$				
C(SH)$_2$(l) $\rightleftharpoons$ H$_2$S(g) $\{-4 \cdot 93 \pm 0 \cdot 10\}$ +CS$_2$(l) $\{+21 \cdot 43 \pm 0 \cdot 17\}$. 2nd Law	$+5 \cdot 9 \ \pm 0 \cdot 5$				
Selected value	$+6 \cdot 0 \ \pm 0 \cdot 5$				
	$+3 \cdot 1 \ \pm 0 \cdot 4$	S4	$19 \cdot 56 \pm 0 \cdot 20$ (sub.)	67/47	$+22 \cdot 7 \ \pm 0 \cdot 5$
	$-3 \cdot 1 \ \pm 0 \cdot 2$	S4	$21 \cdot 85 \pm 0 \cdot 60$ (sub.)	67/47	$+18 \cdot 8 \ \pm 0 \cdot 7$
	$+5 \cdot 3 \ \pm 0 \cdot 4$	S4	$24 \cdot 40 \pm 0 \cdot 70$ (sub.)	67/47	$+29 \cdot 7 \ \pm 0 \cdot 8$
	$-18 \cdot 7 \ \pm 0 \cdot 3$	S4	$24 \cdot 83 \pm 0 \cdot 70$ (sub.)	67/47	$+6 \cdot 1 \ \pm 0 \cdot 8$

THERMOCHEMISTRY OF ORGANIC AND ORGANOMETALLIC COMPOUNDS

C_aH_bOS—$C_aH_bO_4S_2$ 1 kcal $= 4 \cdot 184$ kJ

1 Formula	2 g.f.w.	3 Name	State	Purity mol %	Type	No. of expts.	Detn. of reactn.	$-\Delta H_r^\circ$ kcal/g.f.w.	Re
COS	60·075	Carbon oxysulphide (carbonyl sulphide)	g		E		an	$-7\cdot98\pm0\cdot23$	32,
COS		Carbon oxysulphide (carbonyl sulphide)	g		E		e.m.f.	$7\cdot77\pm0\cdot06$	65/
COS		Carbon oxysulphide (carbonyl sulphide)							
C_2H_4OS	76·118	Thiolacetic acid	l		RB	3	m	$416\cdot22\pm0\cdot32$	55/
C_2H_4OS		Thiolacetic acid	l	99·8	R	7	an	$0\cdot64\pm0\cdot07$	63/ 57/
C_2H_4OS		Thiolacetic acid							
C_2H_6OS	78·134	Dimethyl sulphoxide	l		R		m	$33\cdot18\pm0\cdot10$	46/1
C_4H_8OS	104·172	Ethyl thiolacetate	l		R	4	m	$0\cdot95\pm0\cdot12$	57/:
$C_4H_{10}OS$	106·188	Diethyl sulphoxide	l		RB	7	m	$797\cdot69\pm0\cdot17$	61/3
$C_5H_{10}OS$	118·199	n-Propyl thiolacetate	l		R	4	m	$0\cdot93\pm0\cdot12$	57/3
$C_5H_{10}OS$	118·199	Isopropyl thiolacetate	l		R	4	m	$1\cdot39\pm0\cdot14$	57/3
$C_5H_{10}OS$	118·199	Allyl ethyl sulphoxide	l		RB	8	m	$913\cdot90\pm0\cdot16$	61/3
$C_6H_{12}OS$	132·226	n-Butyl thiolacetate	l		R	4	m	$1\cdot09\pm0\cdot12$	57/3
$C_6H_{12}OS$		n-Butyl thiolacetate	l		R	5	m	$13\cdot73\pm0\cdot20$	58/3
$C_6H_{12}OS$		n-Butyl thiolacetate							
$C_6H_{12}OS$	132·226	t-Butyl thiolacetate	l		R	4	m	$3\cdot00\pm0\cdot14$	57/3

Determinat

1 kcal = $4 \cdot 184$ kJ

H_r° Remarks	5 ΔH_f° (l or c) kcal/g.f.w.	6 Determination of ΔH_v			7 ΔH_f° (g) kcal/g.f.w.
		Type	ΔH_v° kcal/g.f.w.	Ref.	
$_2+H_2S(g)$ $\{-4\cdot93\}$ $\rightleftharpoons COS(g)+H_2O(g)$ $\{-58\cdot70\}$. 2nd Law E					$-33\cdot20\pm0\cdot25$
(g) $\{-26\cdot42\pm0\cdot04\}$ $+S(l)$ $\{+0\cdot34\}$ $\rightleftharpoons COS(g)$. 3rd Law					$-33\cdot85\pm0\cdot10$
: Pt/Ag/AgI/Ag$_{2+\delta}$S/Pt, CO(g), COS(g) Selected value					$-33\cdot85\pm0\cdot10$
	$-52\cdot39\pm0\cdot35$				
$_3COSH(l)+H_2O(l)$ $= CH_3CO_2H(l)$ $\{-115\cdot75\pm0\cdot07\}$ $+H_2S(g)$ $\{-4\cdot93\}$ Selected value	$-51\cdot72\pm0\cdot18$ $-52\cdot06\pm0\cdot40$				
$H_3)_2S(l)$ $\{-15\cdot55\pm0\cdot13\}$ $+0\cdot5O_2$ $= (CH_3)_2SO(l)$	$-48\cdot73\pm0\cdot17$	V2	$12\cdot64\pm0\cdot10$	48/12	$-36\cdot09\pm0\cdot20$
$SAc(l)+H_2O(l)$ $= HOAc(l)$ $\{-115\cdot75\pm0\cdot07\}$ $+EtSH(l)$ $\{-17\cdot53\pm0\cdot14\}$	$-64\cdot01\pm0\cdot21$				
	$-63\cdot97\pm0\cdot20$	E6	$14\cdot9\pm0\cdot3$	61/35	$-49\cdot1\pm0\cdot4$
$PrSAc(l)+H_2O(l)$ $= HOAc(l)$ $\{-115\cdot75\pm0\cdot07\}$ $+n\text{-}PrSH(l)$ $\{-23\cdot79\pm0\cdot15\}$	$-70\cdot29\pm0\cdot22$				
$rSAc(l)+H_2O(l)$ $= HOAc(l)$ $\{-115\cdot75\pm0\cdot07\}$ $+i\text{-}PrSH(l)$ $\{-25\cdot22\pm0\cdot15\}$	$-71\cdot26\pm0\cdot23$				
	$-41\cdot83\pm0\cdot20$	E6	$17\cdot1\pm0\cdot4$	61/35	$-24\cdot7\pm0\cdot5$
$BuSAc(l)+H_2O(l)$ $= HOAc(l)$ $\{-115\cdot75\pm0\cdot07\}$ $+n\text{-}BuSH(l)$ $\{-29\cdot72\pm0\cdot28\}$	$-76\cdot06\pm0\cdot32$				
$BuSAc(l)+n\text{-}BuNH_2(l)$ $\{-30\cdot51\pm0\cdot29\}$ $= n\text{-}BuSH(l)$ $\{-29\cdot72\pm0\cdot28\}$ $+n\text{-}BuNHAc(l)$ $\{-91\cdot02\pm0\cdot39\}$ Selected value	$-76\cdot50\pm0\cdot60$ $-76\cdot15\pm0\cdot30$				
$BuSAc(l)+H_2O(l)$ $= HOAc(l)$ $\{-115\cdot75\pm0\cdot07\}$ $+t\text{-}BuSH(l)$ $\{-33\cdot51\pm0\cdot20\}$	$-77\cdot94\pm0\cdot27$				

C_aH_bOS—$C_aH_bO_4S_2$ 1 kcal = 4·184 kJ

1 Formula	2 g.f.w.	3 Name	State	Purity mol %	Type	No. of expts.	Detn. of reactn.	$-\Delta H_r^{\circ}$ kcal/g.f.w.	R
$C_6H_{14}OS$	134·242	Dipropyl sulphoxide	l		RB	9	m	1107·74±0·16	61
C_8H_8OS	152·216	Phenyl thiolacetate	l		R	4	m	2·97±0·12	60
$C_{12}H_{10}OS$	202·277	Diphenyl sulphoxide	c		RB	10	m	1616·47±0·19	61
$C_2H_6O_2S$	94·133	Dimethyl sulphone	c		RB	6	m	429·42±0·12	61
$C_2H_6O_2S$		Dimethyl sulphone	c		R		m	58·16±0·20	46
$C_2H_6O_2S$		Dimethyl sulphone							
$C_3H_4O_2S$	104·128	Thiete sulphone	c		RB	6	m	512·97±0·20	M
$C_3H_6O_2S$	106·144	β-Thiolactic acid	l		SB	3	m	519·1 ±0·8	35
$C_3H_8O_2S$	108·160	Methyl ethyl sulphone	c		RB	8	m	583·12±0·12	61
$C_4H_6O_2S$	118·155	Divinyl sulphone	l		RB	6	m	675·55±0·83	M
$C_4H_6O_2S$	118·155	Butadiene sulphone	c		RB	7	m	648·88±0·30	61/3
$C_4H_6O_2S$	118·155	α-Butadiene sulphone	c		RB	4	m	647·80±0·39	M
$C_4H_8O_2S$	120·171	Methyl allyl sulphone	l		RB	6	m	701·39±0·24	61/3
$C_4H_{10}O_2S$	122·187	Diethyl sulphone	c		RB	11	m	738·53±0·10	61/3
$C_4H_{10}O_2S$	122·187	Methyl isopropyl sulphone	l		RB	6	m	741·22±0·26	61/3
$C_5H_8O_2S$	132·182	α-Isoprene sulphone	c		RB	5	m	801·46±0·45	M/
$C_5H_8O_2S$	132·182	β-Isoprene sulphone	c		RB	5	m	802·41±0·49	M/

1 kcal = 4·184 kJ

		6 Determination of ΔH_v			
ΔH_r°					7
Remarks	5 ΔH_f° (l or c) kcal/g.f.w.	ΔH_v° Type kcal/g.f.w.	Ref.		ΔH_f° (g) kcal/g.f.w.
	$-78\cdot65\pm0\cdot20$	E6 $17\cdot8$ $\pm0\cdot3$	$61/35$		$-60\cdot9$ $\pm0\cdot4$
Ac(l)+H$_2$O(l) = HOAc(l) $\{-115\cdot75\pm0\cdot07\}$ +PhSH(l) $\{+15\cdot30\pm0\cdot19\}$	$-29\cdot16\pm0\cdot25$				
	$+2\cdot40\pm0\cdot25$	E6$_m$ $23\cdot2$ $\pm0\cdot7$ (sub.)	$61/35$		$+25\cdot6$ $\pm0\cdot8$
H	$-107\cdot51\pm0\cdot14$	V4$_m$ $18\cdot4$ $\pm0\cdot7$ (sub.)	M/2		$-89\cdot1$ $\pm0\cdot8$
$_2$SO(l) $\{-48\cdot73\pm0\cdot17\}$+0·5O$_2$ = Me$_2$SO$_2$(c)	$-106\cdot89\pm0\cdot30$				$-88\cdot5$ $\pm0\cdot9$
Selected value	$-107\cdot51\pm0\cdot14$				$-89\cdot1$ $\pm0\cdot8$
	$-49\cdot69\pm0\cdot23$	V3$_m$ $20\cdot0$ $\pm0\cdot6$ (sub.)	M/2		$-29\cdot7$ $\pm0\cdot7$
ffman–Ellis method (35/1)	$-111\cdot9$ $\pm0\cdot8$				
H	$-116\cdot17\pm0\cdot17$	V4$_m$ $18\cdot6$ $\pm0\cdot7$ (sub.)	M/2		$-97\cdot6$ $\pm0\cdot8$
	$-49\cdot48\pm0\cdot85$	V4 $13\cdot5$ $\pm0\cdot2$	M/2		$-36\cdot0$ $\pm0\cdot9$
	$-76\cdot15\pm0\cdot34$	V4$_m$ $15\cdot0$ $\pm0\cdot6$ (sub.)	M/2		$-61\cdot2$ $\pm0\cdot7$
	$-77\cdot23\pm0\cdot42$	V4$_m$ $14\cdot7$ $\pm0\cdot6$ (sub.)	M/2		$-62\cdot5$ $\pm0\cdot7$
H	$-91\cdot95\pm0\cdot26$	V4 $19\cdot0$ $\pm0\cdot6$	M/2		$-73\cdot0$ $\pm0\cdot8$
	$-123\cdot13\pm0\cdot20$	V4$_m$ $20\cdot6$ $\pm0\cdot6$ (sub.)	M/2		$-102\cdot5$ $\pm0\cdot7$
	$-120\cdot44\pm0\cdot30$	V4 $16\cdot8$ $\pm0\cdot6$	M/2		$-103\cdot6$ $\pm0\cdot7$
	$-85\cdot94\pm0\cdot48$	V4$_m$ $14\cdot5$ $\pm0\cdot6$ (sub.)	M/2		$-71\cdot4$ $\pm0\cdot8$
	$-84\cdot99\pm0\cdot52$	V4$_m$ $15\cdot3$ $\pm0\cdot6$ (sub.)	M/2		$-69\cdot7$ $\pm0\cdot8$

$C_aH_bOS—C_aH_bO_4S_2$ 1 kcal = 4·184 kJ

1 Formula	2 g.f.w.	3 Name	State	Purity mol %	Type	No. of expts.	Detn. of reactn.	$-\Delta H_r^\circ$ kcal/g.f.w.	R
$C_5H_{10}O_2S$	134·198	Ethyl allyl sulphone	l		RB	5	m	858·76±0·20	61/
$C_5H_{12}O_2S$	136·214	Methyl n-butyl sulphone	l		RB	8	m	896·03±0·16	61/
$C_5H_{12}O_2S$	136·214	Methyl t-butyl sulphone	c		RB	6	m	891·22±0·66	61/
$C_6H_{14}O_2S$	150·241	Di-n-propyl sulphone	l		RB	10	m	1055·45±0·15	61/
$C_6H_{14}O_2S$	150·241	Ethyl t-butyl sulphone	c		RB	6	m	1048·33±0·30	61/
$C_7H_8O_2S$	156·205	Methyl phenyl sulphone	c		RB	10	m	993·01±0·16	61/
$C_8H_8O_2S$	168·216	Phenyl vinyl sulphone	c		RB	6	m	1119·18±0·40	M
$C_8H_{10}O_2S$	170·232	Methyl p-tolyl sulphone	c		RB	6	m	1148·77±0·13	61/
$C_8H_{10}O_2S$	170·232	Methyl benzyl sulphone	c		RB	6	m	1149·21±0·36	61/
$C_8H_{18}O_2S$	178·295	Di-n-butyl sulphone	c		RB	9	m	1365·36±0·45	61/
$C_8H_{18}O_2S$	178·295	Di-isobutyl sulphone	l		RB	9	m	1361·78±0·14	61/
$C_8H_{18}O_2S$	178·295	Di-t-butyl sulphone	c		RB	6	m	1358·14±0·53	61/
$C_9H_8O_2S$	180·227	Phenyl prop-1-ynyl sulphone	c		RB	5	m	1251·21±0·97	M/
$C_9H_8O_2S$	180·227	Phenyl prop-2-ynyl sulphone	c		RB	5	m	1247·24±0·60	M/
$C_9H_8O_2S$	180·227	Phenyl propadiene sulphone	c		RB	5	m	1238·9 ±1·2	M/
$C_9H_{10}O_2S$	182·243	p-Tolyl vinyl sulphone	c		RB	6	m	1273·51±0·72	M/
$C_{10}H_{10}O_2S$	194·254	p-Tolyl prop-1-ynyl sulphone	c		RB	6	m	1403·76±0·76	M/

$$1 \text{ kcal} = 4 \cdot 184 \text{ kJ}$$

H_r°	Remarks	5 ΔH_f° (l or c) kcal/g.f.w.	6 Determination of ΔH_v			7 ΔH_f° (g) kcal/g.f.w.
			Type	ΔH_v° kcal/g.f.w.	Ref.	
	H	$-96 \cdot 95 \pm 0 \cdot 24$	V4	$20 \cdot 0 \ \pm 0 \cdot 6$	M/2	$-77 \cdot 0 \ \pm 0 \cdot 7$
		$-128 \cdot 00 \pm 0 \cdot 20$	V4	$18 \cdot 2 \ \pm 0 \cdot 6$	M/2	$-109 \cdot 8 \ \pm 0 \cdot 7$
		$-132 \cdot 81 \pm 0 \cdot 70$	V4$_m$	$19 \cdot 7 \ \pm 0 \cdot 6$ (sub.)	M/2	$-113 \cdot 1 \ \pm 0 \cdot 9$
		$-130 \cdot 94 \pm 0 \cdot 20$	V4	$19 \cdot 1 \ \pm 0 \cdot 6$	M/2	$-111 \cdot 8 \ \pm 0 \cdot 7$
	H	$-138 \cdot 06 \pm 0 \cdot 32$	V4$_m$	$20 \cdot 7 \ \pm 0 \cdot 6$ (sub.)	M/2	$-117 \cdot 4 \ \pm 0 \cdot 7$
		$-82 \cdot 49 \pm 0 \cdot 20$	V4$_m$	$22 \cdot 0 \ \pm 0 \cdot 7$ (sub.)	M/2	$-60 \cdot 5 \ \pm 0 \cdot 9$
		$-50 \cdot 37 \pm 0 \cdot 44$	V4$_m$	$19 \cdot 6 \ \pm 0 \cdot 6$ (sub.)	M/2	$-30 \cdot 8 \ \pm 0 \cdot 7$
	H	$-89 \cdot 09 \pm 0 \cdot 18$	V4$_m$	$23 \cdot 9 \ \pm 0 \cdot 8$ (sub.)	M/2	$-65 \cdot 2 \ \pm 0 \cdot 9$
	H	$-88 \cdot 65 \pm 0 \cdot 40$	V4$_m$	$23 \cdot 7 \ \pm 0 \cdot 7$ (sub.)	M/2	$-65 \cdot 0 \ \pm 0 \cdot 9$
		$-145 \cdot 76 \pm 0 \cdot 50$	V4$_m$	$24 \cdot 0 \ \pm 0 \cdot 6$ (sub.)	M/2	$-121 \cdot 8 \ \pm 0 \cdot 9$
		$-149 \cdot 34 \pm 0 \cdot 20$	V4	$21 \cdot 4 \ \pm 0 \cdot 6$	M/2	$-127 \cdot 9 \ \pm 0 \cdot 7$
	H	$-152 \cdot 98 \pm 0 \cdot 57$	V4$_m$	$22 \cdot 5 \ \pm 0 \cdot 7$ (sub.)	M/2	$-130 \cdot 5 \ \pm 0 \cdot 9$
		$-12 \cdot 39 \pm 0 \cdot 99$	V4$_m$	$22 \cdot 8 \ \pm 0 \cdot 6$ (sub.)	M/2	$+10 \cdot 4 \ \pm 1 \cdot 2$
		$-16 \cdot 36 \pm 0 \cdot 63$	V4$_m$	$25 \cdot 1 \ \pm 0 \cdot 6$ (sub.)	M/2	$+8 \cdot 7 \ \pm 0 \cdot 8$
		$-24 \cdot 7 \ \pm 1 \cdot 2$	V4$_m$	$25 \cdot 2 \ \pm 0 \cdot 6$ (sub.)	M/2	$+0 \cdot 5 \ \pm 1 \cdot 4$
		$-58 \cdot 40 \pm 0 \cdot 75$	V4$_m$	$19 \cdot 7 \ \pm 0 \cdot 6$ (sub.)	M/2	$-38 \cdot 7 \ \pm 1 \cdot 0$
		$-22 \cdot 21 \pm 0 \cdot 80$	V4$_m$	$24 \cdot 7 \ \pm 0 \cdot 6$ (sub.)	M/2	$+2 \cdot 5 \ \pm 1 \cdot 0$

$C_aH_bOS—C_aH_bO_4S_2$ 1 kcal = 4·184 kJ

1 Formula	2 g.f.w.	3 Name	State	Purity mol %	No. of Type expts.	Detn. of reactn.	$-\Delta H_r^\circ$ kcal/g.f.w.	R
$C_{10}H_{10}O_2S$	194·254	p-Tolyl prop-2-ynyl sulphone	c		RB 5	m	1400·51±0·75	M
$C_{10}H_{10}O_2S$	194·254	p-Tolyl propadiene sulphone	c		RB 5	m	1391·25±0·91	M
$C_{10}H_{12}O_2S$	196·270	p-Tolyl prop-2-ene sulphone	c		RB 6	m	1422·87±0·56	M
$C_{10}H_{12}O_2S$	196·270	p-Tolyl trans-prop-1-ene sulphone	c		RB 5	m	1424·44±0·37	M
$C_{10}H_{12}O_2S$	196·270	p-Tolyl isopropenyl sulphone	c		RB 5	m	1426·14±0·45	M
$C_{11}H_{14}O_2S$	210·297	p-Tolyl but-3-ene sulphone	c		RB 5	m	1575·60±0·30	M
$C_{11}H_{14}O_2S$	210·297	p-Tolyl but-2-ene sulphone	c		RB 4	m	1573·47±0·40	M
$C_{11}H_{14}O_2S$	210·297	p-Tolyl but-1-ene sulphone	c		RB 5	m	1576·40±0·24	M
$C_{11}H_{14}O_2S$	210·297	p-Tolyl isobutenyl sulphone	c		RB 5	m	1575·00±0·24	M
$C_{11}H_{14}O_2S$	210·297	p-Tolyl 2-methyl-prop-2-ene sulphone	c		RB 6	m	1573·49±0·28	M
$C_{12}H_{10}O_2S$	218·276	Diphenyl sulphone	c		RB 6	m	1560·36±0·36	61/
$C_{14}H_{12}O_2S$	244·315	Phenyl trans-β-styryl sulphone	c		RB 7	m	1837·10±0·68	M
$C_{14}H_{14}O_2S$	246·330	Di-p-tolyl sulphone	c		RB 7	m	1864·48±0·13	61/
$C_{14}H_{14}O_2S$	246·330	Dibenzyl sulphone	c		RB 5	m	1871·33±0·24	61/
$C_{15}H_{14}O_2S$	258·342	p-Tolyl cis-β-styryl sulphone	c		RB 6	m	1990·80±0·80	M

1 kcal $= 4\cdot184$ kJ

ΔH_r°		5 ΔH_f° (l or c) kcal/g.f.w.	6 Determination of ΔH_v			7 ΔH_f° (g) kcal/g.f.w.
Remarks			Type	ΔH_v° kcal/g.f.w.	Ref.	
		$-25\cdot46\pm0\cdot79$	V4$_m$	$25\cdot7\ \pm0\cdot6$ (sub.)	M/2	$+0\cdot2\ \pm1\cdot0$
		$-34\cdot72\pm0\cdot93$	V4$_m$	$27\cdot0\ \pm0\cdot6$ (sub.)	M/2	$-7\cdot7\ \pm1\cdot2$
		$-71\cdot41\pm0\cdot59$	V4$_m$	$22\cdot9\ \pm0\cdot7$ (sub.)	M/2	$-48\cdot5\ \pm0\cdot9$
		$-69\cdot84\pm0\cdot42$	V4$_m$	$20\cdot0\ \pm0\cdot5$ (sub.)	M/2	$-49\cdot8\ \pm0\cdot7$
		$-68\cdot14\pm0\cdot48$	V4$_m$	$21\cdot2\ \pm0\cdot6$ (sub.)	M/2	$-46\cdot9\ \pm0\cdot8$
		$-81\cdot05\pm0\cdot34$	V4$_m$	$27\cdot1\ \pm0\cdot7$ (sub.)	M/2	$-54\cdot0\ \pm0\cdot8$
		$-83\cdot18\pm0\cdot43$	V4$_m$	$25\cdot7\ \pm0\cdot6$ (sub.)	M/2	$-57\cdot5\ \pm0\cdot8$
		$-80\cdot25\pm0\cdot28$	V4$_m$	$25\cdot4\ \pm0\cdot6$ (sub.)	M/2	$-54\cdot9\ \pm0\cdot7$
		$-81\cdot65\pm0\cdot28$	V4$_m$	$24\cdot4\ \pm0\cdot6$ (sub.)	M/2	$-57\cdot3\ \pm0\cdot7$
		$-83\cdot16\pm0\cdot32$	V4$_m$	$25\cdot5\ \pm0\cdot7$ (sub.)	M/2	$-57\cdot7\ \pm0\cdot8$
H		$-53\cdot71\pm0\cdot39$	V4$_m$	$25\cdot4\ \pm0\cdot7$ (sub.)	M/2	$-28\cdot3\ \pm0\cdot9$
		$-33\cdot38\pm0\cdot71$	V4$_m$	$25\cdot1\ \pm0\cdot9$ (sub.)	M/2	$-8\cdot3\ \pm1\cdot1$
		$-74\cdot32\pm0\cdot20$	V4$_m$	$26\cdot2\ \pm0\cdot7$ (sub.)	M/2	$-48\cdot1\ \pm0\cdot8$
H		$-67\cdot47\pm0\cdot27$	V4$_m$	$30\cdot0\ \pm0\cdot7$ (sub.)	M/2	$-37\cdot5\ \pm0\cdot8$
		$-42\cdot05\pm0\cdot83$	V4$_m$	$27\cdot8\ \pm0\cdot9$ (sub.)	M/2	$-14\cdot3\ \pm1\cdot2$

$C_aH_bOS–C_aH_bO_4S_2$ 1 kcal = 4·184 kJ

								Determinat	
1 Formula	2 g.f.w.	3 Name	State	Purity mol %	Type	No. of expts.	Detn. of reactn.	$-\Delta H_r^\circ$ kcal/g.f.w.	Re
$C_{15}H_{14}O_2S$	258·342	p-Tolyl trans-β-styryl sulphone	c		RB	7	m	1990·4 ±1·1	M
$C_2H_6O_3S$	110·132	Dimethyl sulphite	l		RB	6	m	411·86±0·23	M
$C_3H_8O_3S$	124·159	Methyl ethyl sulphite	l		RB	5	m	563·74±0·26	M
$C_4H_{10}O_3S$	138·187	Diethyl sulphite	l		RB	5	m	718·16±0·19	M
$C_6H_{14}O_3S$	166·241	Di-n-propyl sulphite	l		RB	6	m	1031·87±0·25	M
$C_8H_{18}O_3S$	194·295	Di-n-butyl sulphite	l		RB	6	m	1345·55±0·95	M
$C_2H_6O_4S$	126·132	Dimethyl sulphate	l		RB	6	m	361·20±0·20	M
$C_3H_8O_4S$	140·159	Isopropyl hydrogen sulphate	l		E		an	9·2 ±0·2	60/
$C_4H_{10}O_4S$	154·186	Diethyl sulphate	l		RB	6	m	667·38±0·23	M
$C_6H_{14}O_4S$	182·240	Di-n-propyl sulphate	l		RB	6	m	981·17±0·27	M
$C_8H_{18}O_4S$	210·294	Di-n-butyl sulphate	l		RB	6	m	1294·99±0·60	M
$C_6H_{10}O_2S_2$	178·273	Dithioadipic acid	c		RB	6	m	1091·0 ±0·5	46/ 50/
$C_8H_{14}O_2S_2$	206·328	Dithiosuberic acid	l		RB		m	1418·3 ±0·5	50/
$C_9H_{16}O_2S_2$	220·355	Dithioazelaic acid	c		RB	5	m	1564·5 ±0·5	46/ 50/
$C_{10}H_{18}O_2S_2$	234·382	Dithiosebacic acid	c		RB		m	1719·1 ±0·5	50/
$C_6H_{10}O_4S_2$	210·272	ββ′-Dithiodilactic acid	c		SB	3	m	962·8 ±0·9	35/
$C_7H_{16}O_4S_2$	228·331	2, 2-Bis(ethylsulphonyl)-propane	c		RB	3	m	1282·1 ±0·9	46/
$C_{12}H_{10}O_4S_2$	282·339	Diphenyl disulphone	c		RB	8	m	1604·36±0·34	64/

1 kcal = 4·184 kJ

ΔH_r° Remarks	5 ΔH_f° (l or c) kcal/g.f.w.	6 Determination of ΔH_v			7 ΔH_f° (g) kcal/g.f.w.
		ΔH_v° Type kcal/g.f.w.		Ref.	
	$-42\cdot5\ \pm1\cdot1$	V4$_m$ 25·9 ±0·6 (sub.)		M/2	$-16\cdot6\ \pm1\cdot3$
	$-125\cdot07\pm0\cdot26$	V4 9·6 ±0·4		M/2	$-115\cdot5\ \pm0\cdot5$
	$-135\cdot55\pm0\cdot30$	V4 10·4 ±0·4		M/2	$-125\cdot2\ \pm0\cdot5$
	$-143\cdot50\pm0\cdot24$	V4 11·6 ±0·4		M/2	$-131\cdot9\ \pm0\cdot5$
	$-154\cdot52\pm0\cdot28$	V4 14·0 ±0·4		M/2	$-140\cdot5\ \pm0\cdot5$
	$-165\cdot57\pm1\cdot00$	V4 16·2 ±0·4		M/2	$-149\cdot4\ \pm1\cdot1$
	$-175\cdot23\pm0\cdot23$	V4 11·6 ±0·4		M/2	$-164\cdot1\ \pm0\cdot5$
H$_6$(l) {+0·5±0·1} +[H$_2$SO$_4$. 2·68 H$_2$O](l) {−205·6} ⇌ i-C$_3$H$_7$SO$_4$H(l). 2nd Law	$-214\cdot3\ \pm0\cdot4$*				
	$-194\cdot28\pm0\cdot26$	V4 13·6 ±0·4		M/2	$-180\cdot7\ \pm0\cdot5$
	$-205\cdot22\pm0\cdot30$	V4 16·0 ±0·4		M/2	$-189\cdot2\ \pm0\cdot5$
	$-216\cdot13\pm0\cdot62$	V4 18·1 ±0·4		M/2	$-198\cdot0\ \pm0\cdot8$
GE	$-102\cdot6\ \pm0\cdot6$				
GE	$-100\cdot1\ \pm0\cdot6$				
GE	$-116\cdot2\ \pm0\cdot6$				
GE	$-124\cdot0\ \pm0\cdot6$				
ffman–Ellis method (35/1)	$-230\cdot8\ \pm1\cdot0$				
CGE	$-210\cdot5\ \pm1\cdot0$				
	$-153\cdot59\pm0\cdot39$	E6$_m$ 38·7 ±1·0 (sub.)		64/33	$-114\cdot9\ \pm1\cdot2$

C_6H_5NS—$C_6H_5N_2S_4$ 1 kcal = 4·184 kJ

1 Formula	2 g.f.w.	3 Name	State	Purity mol %	Type	No. of expts.	Detn. of reactn.	$-\Delta H_r^\circ$ kcal/g.f.w.	Re
C_2H_3NS	73·117	Methyl isothiocyanate	c		RB	3	m	453·50±0·28	55/ 63/
C_4H_5NS	99·155	4-Methylthiazole	l	glc	RB	5	m	707·18±0·14	66/
CH_4N_2S	76·120	Thiourea	c		RB	4	m	352·40±0·50	46/
$C_4H_2N_2S$	110·138	4-Cyanothiazole	c	glc	RB	6	m	641·03±0·19	66/
$C_6H_8N_2S$	140·208	3, 3′-Thiodipropionitrile	l		R		m, an	30·2 ± 0·6	66/
CH_5N_3S	91·135	Hydrazinium thiocyanate	c		RB	4	m	418·3 ±1·2	57/
CH_5N_3S	91·135	Thiosemicarbazide	c		RB	4	m	414·7 ±1·2	57/
$C_3H_7N_3S$	117·173	Acetaldehyde thiosemicarbazone	c		RB	3	m	680·5 ±2·0	57/
$C_4H_9N_3S$	131·200	Acetone thiosemi- carbazone	c		RB	4	m	832·5 ±2·4	57/
$C_2N_2S_2$	116·164	Dithiocyanogen	l		R	2	m	10·50±0·15	66/
$C_4H_{10}N_2S_2$	150·266	NN′-Dithiodiethylamine	l		R	6	m	87·3 ±0·7	62/
$C_6H_{10}N_4S_2$	202·301	Tetrahydro-3, 7-dimethyl- s-triazolo(a)-s-triazole- 1, 5-dithione	c		RB	4	m	1227·0 ±3·6	57/
$C_6H_{12}N_2S_3$	208·368	Tetramethylthiuram monosulphide	c		RB	7	m	1417·42±0·24	61/
$C_6H_{12}N_2S_4$	240·432	Tetramethylthiuram disulphide	c		RB	7	m	1559·62±0·34	61/

1 kcal $= 4 \cdot 184$ kJ

ΔH_r°			6 Determination of ΔH_v			7
Remarks		5 ΔH_f° (l or c) kcal/g.f.w.	Type	ΔH_v° kcal/g.f.w.	Ref.	ΔH_f° (g) kcal/g.f.w.
		$+19 \cdot 04 \pm 0 \cdot 30$				
		$+16 \cdot 31 \pm 0 \cdot 17$	C1	$10 \cdot 48 \pm 0 \cdot 05$	66/31	$+26 \cdot 79 \pm 0 \cdot 18$
	CGE	$-22 \cdot 16 \pm 0 \cdot 52$				
		$+52 \cdot 63 \pm 0 \cdot 21$	C1	$17 \cdot 67 \pm 0 \cdot 10$ (sub.)	66/31	$+70 \cdot 30 \pm 0 \cdot 24$
$H_2 = CH\ CN(l)\ \{+36 \cdot 1 \pm 0 \cdot 7\}$ $+(NH_4)_2S\ (aq.)\ \{-55 \cdot 4\}$ $= S(CH_2CH_2CN)_2(l)$ $+2NH_3\ (aq.)\ \{-19 \cdot 2\}$		$+25 \cdot 0\ \pm 1 \cdot 0$				
	GE	$+9 \cdot 6\ \pm 1 \cdot 2$				
	GE	$+6 \cdot 0\ \pm 1 \cdot 2$				
	GE	$+15 \cdot 4\ \pm 2 \cdot 0$				
	GE	$+5 \cdot 0\ \pm 2 \cdot 4$				
$N)_2(l)+3I^-\ (aq)\ \{-13 \cdot 19\}$ $= 2CNS^-\ (aq)\ \{+18 \cdot 27\}$ $+I_3^-\ (aq)\ \{-12 \cdot 3\}$		$+74 \cdot 3\ \pm 0 \cdot 3$	E8	$8 \cdot 0\ \pm 0 \cdot 5$	66/49	$+82 \cdot 3\ \pm 0 \cdot 6$
$_2H_5)_2NH(l)\ \{-24 \cdot 78 \pm 0 \cdot 29\}$ $+S_2Cl_2(l)\ \{-14 \cdot 2\}$ $= [(C_2H_5)_2N]_2S_2(l)$ $+2(C_2H_5)_2NH_2Cl(c)\ \{-85 \cdot 76 \pm 0 \cdot 33\}$		$-29 \cdot 1\ \pm 0 \cdot 9$	E4	$12 \cdot 6\ \pm [1 \cdot 0]$	62/59	$-16 \cdot 5\ \pm [1 \cdot 5]$
	GE	$+33 \cdot 4\ \pm 3 \cdot 6$				
	G	$+11 \cdot 58 \pm 0 \cdot 37$				
	G	$+9 \cdot 90 \pm 0 \cdot 50$				

$C_aH_bO_cN_dS_e$ 　　　　　 1 kcal = 4·184 kJ

1 Formula	2 g.f.w.	3 Name	State	Purity mol %	Type	No. of expts.	Detn. of reactn.	$-\Delta H_r^\circ$ kcal/g.f.w.	R
$C_4H_{10}ON_2S$	134·201	NN'-Thionylbis- (diethylamine)	l		R	6	m	90·7 ±1·0	62
$C_3H_7O_2NS$	121·159	L-Cysteine	c		SB	3	m	537·5 ±0·5	35
$C_3H_7O_2NS$		L-Cysteine	c		RB	2	m	542·0 ±0·5	46
$C_3H_7O_2NS$		L-Cysteine							
$C_4H_{10}O_2N_2S$	150·201	NN'-Sulphurylbis- (diethylamine)	l		R	6	m	130·0 ±1·7	62
$C_6H_{12}O_4N_2S_2$	240·302	L-Cystine	c		SB	3	m	1011·0 ±0·9	35
$C_6H_{12}O_4N_2S_2$		L-Cystine	c		RB	3	m	1015·3 ±0·9	46
$C_6H_{12}O_4N_2S_2$		L-Cystine							

Determina

$$1 \text{ kcal} = 4 \cdot 184 \text{ kJ}$$

H_r°		5 ΔH_f° (l or c) kcal/g.f.w.	6 Determination of ΔH_v			7 ΔH_f° (g) kcal/g.f.w.
	Remarks		Type	ΔH_v° kcal/g.f.w.	Ref.	
$H_5)_2NH(l)$ $\{-24 \cdot 78 \pm 0 \cdot 29\}$ $+SOCl_2(l)$ $\{-58 \cdot 7\}$ $= [(C_2H_5)_2N]_2SO(l)$ $+2(C_2H_5)_2NH_2Cl(c)$ $\{-85 \cdot 76 \pm 0 \cdot 33\}$		$-77 \cdot 0 \pm 1 \cdot 4$				
man–Ellis method (35/1)			CG	$-127 \cdot 6 \pm 0 \cdot 6$		
			CGE	$-123 \cdot 1 \pm 0 \cdot 6$		
	Selected value			$-123 \cdot 1 \pm 0 \cdot 6$		
$H_5)_2NH(l)$ $\{-24 \cdot 78 \pm 0 \cdot 29\}$ $SO_2Cl_2(l)$ $\{-94 \cdot 2\}$ $= [(C_2H_5)_2N]_2SO_2(l)$ $+2(C_2H_5)_2NH_2Cl(c)$ $\{-85 \cdot 76 \pm 0 \cdot 33\}$		$-151 \cdot 9 \pm 2 \cdot 2$				
man–Ellis method (35/1)			CG	$-251 \cdot 0 \pm 1 \cdot 0$		
			CGE	$-246 \cdot 7 \pm 1 \cdot 0$		
	Selected value			$-246 \cdot 7 \pm 1 \cdot 0$		

$C_aH_bHal_j$ 1 kcal = 4·184 kJ

1 Formula	2 g.f.w.	3 Name	State	Purity mol %	Type	No. of expts.	Detn. of reactn.	$-\Delta H_r^\circ$ kcal/g.f.w.	R
CF_4	88·005	Carbon tetrafluoride	g		R	4	an, ms	63·5 ±0·4	56
CF_4		Carbon tetrafluoride	g		RB		m, an	41·5 ±1·0	5(
CF_4		Carbon tetrafluoride	g		R	5	an	325·5 ±2·2	60
CF_4		Carbon tetrafluoride	g		RB	8	m, an	41·38±0·32	6!
CF_4		Carbon tetrafluoride	g		SB	7	m	222·87±0·38	6`
CF_4		Carbon tetrafluoride	g		SB	10	m	247·92±0·07	6`
CF_4		Carbon tetrafluoride	g		SB	5	m	246·84±0·14	67
CF_4		Carbon tetrafluoride	g		SB		m	223·05±0·20	68
CF_4		Carbon tetrafluoride	g		R			218·13±0·20	M
CF_4		Carbon tetrafluoride							
CF_3Cl	104·459	Trifluorochloromethane	g		R	8	an	345·9 ±1·4	55
CF_3Cl		Trifluorochloromethane	g	99·7, ms	R	10	an	331·0 ±1·1	62 63
CF_3Cl		Trifluorochloromethane	g		E	16	glc	−17·27±0·26	67
CF_3Cl		Trifluorochloromethane	g	ir, gsc	E	17	ms	10·69±0·30	67
CF_3Cl		Trifluorochloromethane	g	ir, gsc	E	12	gsc	−10·49±0·40	67
CF_3Cl		Trifluorochloromethane							
CF_2Cl_2	120·914	Difluorodichloromethane	g	ms	R	10	an	354·1 ±1·3	62 63
CF_2Cl_2		Difluorodichloromethane	g		R	8	an	367±3	55
CF_2Cl_2		Difluorodichloromethane							
$CFCl_3$	137·369	Fluorotrichloromethane	g		R	8	an	385±2	55

$$1 \text{ kcal} = 4\cdot184 \text{ kJ}$$

| H_r° | | 6 Determination of ΔH_v | | | |
	Remarks	5 ΔH_f° (l or c) kcal/g.f.w.	ΔH_v° Type kcal/g.f.w.	Ref.	7 ΔH_f° (g) kcal/g.f.w.
$_4$(g) $\{-157\cdot9\pm0\cdot8\}$ $= CF_4(g)+C(am)$ $\{+1\cdot85\pm0\cdot2\}$					$-223\cdot3 \pm0\cdot9$
$_{(g)}+42H_2O(l)$ $= CO_2+4$ [HF.10H$_2$O](l)					$-223\cdot5 \pm1\cdot0*$
$_{(g)}+4Na(c)$ $= 4NaF(c)\ \{-137\cdot9\}+C(c, gr.)$					$-226\cdot1 \pm2\cdot2*$
$_{(g)}+82H_2O(l) = CO_2+4$ [HF.20H$_2$O](l)					$-223\cdot84\pm0\cdot32*$
gr.)$+2F_2(g) = CF_4(g)$					$-222\cdot87\pm0\cdot38$
$)\ (C_2F_4)_n$ (polymer) $\{-198\cdot4\pm0\cdot5\}$ $-2F_2(g) = 2CF_4(g)$					$-223\cdot16\pm0\cdot50$
$)\ (C_2F_4)_n$ (polymer) $\{-198\cdot8\pm1\cdot2\}$ $-2F_2(g) = 2CF_4(g)$					$-222\cdot8 \pm0\cdot6$
, gr.)$+2F_2(g) = CF_4(g)$					$-223\cdot05\pm0\cdot20$
$C_2N_2(g)\ \{+73\cdot84\pm0\cdot43\}$ $-1\cdot33NF_3(g)\ \{-31\cdot6\pm0\cdot2\}$ $= CF_4(g)+1\cdot166N_2$					$-223\cdot3 \pm0\cdot5$
	Selected value				$-223\cdot3 \pm0\cdot4$
$_8Cl(g)+4K(c)$ $= C(c, gr.)+3KF(c)\ \{-135\cdot9\}$ $+KCl(c)\ \{-104\cdot4\}$					$-166\cdot2 \pm1\cdot4*$
$_8Cl(g)+4Na(c)$ $= C(am)\ \{+3\cdot95\}+3NaF(c)\ \{-137\cdot9\}$ $+NaCl(c)\ \{-98\cdot5\}$					$-177\cdot3 \pm1\cdot1*$
$_8Cl(g)+I_2(g)$ $\rightleftharpoons CF_3I(g)\ \{-140\cdot6\pm0\cdot8\}$ $+ICl(g)\ \{+4\cdot25\}.$ 3rd Law					$-168\cdot5 \pm0\cdot9*$
$_8Br(g)\ \{-155\cdot0\pm0\cdot6\}+Cl_2(g)$ $\rightleftharpoons CF_3Cl(g)+BrCl(g)\ \{+3\cdot50\}.$ 3rd Law					$-169\cdot2 \pm0\cdot7*$
$_8Cl(g)+Br_2(g)$ $\rightleftharpoons CF_3Br(g)\ \{-155\cdot0\pm0\cdot6\}$ $+BrCl(g)\ \{+3\cdot50\}.$ 3rd Law					$-169\cdot4 \pm0\cdot7*$
	Selected value				$-169\cdot0 \pm1\cdot0*$
$_2Cl_2(g)+4Na(c)$ $= C(am)\ \{+3\cdot95\}+2NaF(c)\ \{-137\cdot9\}$ $+2NaCl(c)\ \{-98\cdot5\}$					$-114\cdot8 \pm1\cdot3*$
$_2Cl_2(g)+4K(c)$ $= C(c, gr.)+2KF(c)\ \{-135\cdot9\}$ $+2KCl\ (c)\{-104\cdot4\}$					$-114\pm3*$
	Selected value				$-114\cdot8 \pm1\cdot3*$
$Cl_3(g)+4K(c)$ $= C(c, gr.)+KF(c)\ \{-135\cdot9\}$ $+3KCl(c)\ \{-104\cdot4\}$					-64 ± 2

$C_aH_bHal_f$ 1 kcal = 4·184 kJ

1 Formula	2 g.f.w.	3 Name	State	Purity mol %	Type	No. of expts.	Detn. of reactn.	$-\Delta H_f^\circ$ kcal/g.f.w.	R
CCl₄	153·823	Carbon tetrachloride	g		H	6		61·9 ±0·6	6:
CCl₄		Carbon tetrachloride	l		SB		m	87·4 ±[2·0]	5:
CCl₄		Carbon tetrachloride	l		R	2	an	22·3 ±1·0	56
CCl₄		Carbon tetrachloride							
CF₃Br	148·915	Trifluorobromomethane	g	glc	E	12	an	3·3 ±0·6	63
CF₃Br		Trifluorobromomethane	g	glc	E	13	gsc	4·59±0·28	67
CF₃Br		Trifluorobromomethane	g		E	10	glc	−9·55±0·06*	67
CF₃Br		Trifluorobromomethane							
CCl₃Br	198·279	Trichlorobromomethane	g		E		an	0·80±0·13	65
									51
CF₃I	195·911	Trifluoroiodomethane	g	>99·9, ms	E	16	glc	−17·10±0·34	67
CHF₃	70·014	Trifluoromethane	g		SB	5	an	90·33±0·65 (for z= 0)	58
CHF₂Cl	86·469	Difluorochloromethane	g		E		TP	29·1 ±[1·0]	64
CHCl₃	119·378	Chloroform (Trichloromethane)	l		SB		m	113·3 ±[2·0]	53
CH₂F₂	52·024	Difluoromethane	g	99·7, ms	SB	6	an	139·83±0·22	58
CH₂Cl₂	84·933	Dichloromethane	l		SB		m	144·8 ±[2·0]	53
CH₂Cl₂		Dichloromethane	g	glc	H	6	an	39·05±0·30	67
CH₂Cl₂		Dichloromethane							
CH₃Cl	50·488	Methyl chloride	g	>99	H	4	an	19·32±0·10	56
CH₃Cl		Methyl chloride	g		H	4	an	19·48±[0·10]	65
CH₃Cl		Methyl chloride							

Determina

1 kcal = 4·184 kJ

ΔH_r° Remarks	5 ΔH_f° (l or c) kcal/g.f.w.	6 Determination of ΔH_v Type	ΔH_v° kcal/g.f.w.	Ref.	7 ΔH_f° (g) kcal/g.f.w.
$_4(g)+2H_2 = C(c, gr.)+4HCl(g)$	$-34\cdot1\ \pm0\cdot6^*$	C1	$7\cdot75\pm0\cdot02$	59/28	$-26\cdot3\ \pm0\cdot6^*$
$:$ As$_2$O$_3$ solution EJ	$-29\cdot4\ \pm[2\cdot0]$				$-21\cdot\dot{o}\ \pm[2\cdot0]$
$Cl_3(l) \{-31\cdot9\pm2\cdot0\}+Cl_2(g) = CCl_4(l)+HCl(g)$	$-32\cdot1\ \pm[2\cdot5]$				$-24\cdot3\ \pm[2\cdot5]$
Selected value	$-33\cdot0\ \pm1\cdot5$				$-25\cdot2\ \pm1\cdot5$
$F_3(g) \{-166\cdot3\pm0\cdot7\}+Br_2(g) \rightleftharpoons CF_3Br(g)+HBr(g).$ 2nd Law					$-153\cdot5\ \pm1\cdot0^*$
$F_3(g) \{-166\cdot3\pm0\cdot7\}+Br_2(g) \rightleftharpoons CF_3Br(g)+HBr(g).$ 3rd Law					$-154\cdot8\ \pm0\cdot8^*$
$_3Br(g)+I_2(g) \rightleftharpoons CF_3I(g) \{-140\cdot6\pm0\cdot8\} +IBr(g) \{+9\cdot76\}.$ 3rd Law					$-155\cdot3\ \pm0\cdot8^*$
Selected value					$-155\cdot0\ \pm0\cdot6^*$
$Cl_3(g) \{-24\cdot6\pm2\cdot0\}+Br_2(g) \rightleftharpoons CCl_3Br(g)+HBr(g).$ 3rd Law					$-9\cdot3\ \pm[2\cdot0]$
$F_3(g) \{-166\cdot3\pm0\cdot7\}+I_2(g) \rightleftharpoons CF_3I(g)+HI(g).$ 3rd Law					$-140\cdot6\ \pm0\cdot8^*$
$F_3(g)+0\cdot5O_2+0\cdot5[45(3\text{-}4z)-1]H_2O(l) = (1-z)CO_2+zCF_4(g) \{-223\cdot3\pm0\cdot4\} +(3-4z) [HF.22H_2O](l)$					$-166\cdot3\ \pm0\cdot7^*$
$F_4(g) \{-157\cdot9\pm0\cdot8\}+2HCl(g) \rightleftharpoons 2CHF_2Cl(g).$ 3rd Law					$-115\cdot6\ \pm[1\cdot4]$
d: As$_2$O$_3$ solution EJ	$-31\cdot9\ \pm[2\cdot0]$	C2	$7\cdot3\ \pm[0\cdot1]$	26/2	$-24\cdot6\ \pm[2\cdot0]$
$_2F_2(g)+O_2+46H_2O(l) = CO_2+2[HF.23H_2O](l)$					$-108\cdot13\pm0\cdot22^*$
d:As$_2$O$_3$ solution EJ	$-28\cdot9\ \pm[2\cdot0]$	C2	$6\cdot8\ \pm[0\cdot1]$	26/2	$-22\cdot1\ \pm[2\cdot0]$
$_2Cl_2(g)+2H_2 = CH_4(g) \{-17\cdot89\pm0\cdot07\}+2HCl(g)$	$-29\cdot8\ \pm0\cdot4$				$-22\cdot96\pm0\cdot35$
Selected value	$-29\cdot8\ \pm0\cdot4$				$-22\cdot96\pm0\cdot35$
$_3Cl(g)+H_2 = CH_4(g) \{-17\cdot89\pm0\cdot07\} +HCl(g)$ E					$-20\cdot63\pm0\cdot12$
$_3Cl(g)+H_2 = CH_4(g) \{-17\cdot89\pm0\cdot07\} +HCl(g)$ E					$-20\cdot47\pm[0\cdot12]$
Selected value					$-20\cdot55\pm0\cdot10$

$C_aH_bHal_j$ 1 kcal = 4·184 kJ

1 Formula	2 g.f.w.	3 Name	State	Purity mol %	Type	No. of expts.	Detn. of reactn.	$-\Delta H_r^\circ$ kcal/g.f.w.	R
CH_3Br	94·944	Methyl bromide	g		E	12		2·41±[0·10]	48
CH_3Br		Methyl bromide	g		E	12		2·32±[0·10]	48
CH_3Br		Methyl bromide	g		H	6	an	17·59±0·33	65/
CH_3Br		Methyl bromide	l	>99·7, glc	R	8	m	3·30±0·30	66/
CH_3Br		Methyl bromide							
CH_3I	141·939	Methyl iodide	l		R	6	m	72·4 ±0·3	49/
CH_3I		Methyl iodide	g		R	10	an	11·0 ±2·5	52/
CH_3I		Methyl iodide	l		H	10	m	15·0 ±0·3	61/
CH_3I		Methyl iodide	g	99·98	E	12	uv	12·59±0·26	65/
CH_3I		Methyl iodide	g		E	19	glc	12·67±0·10	65/
CH_3I		Methyl iodide							
C_2F_4	100·016	Tetrafluoroethylene	g		H	8	an	147·8 ±1·1	56/
C_2F_4		Tetrafluoroethylene	g	99·5	R	9	an	385·0 ±1·1	62/
C_2F_4		Tetrafluoroethylene							
$(1/n)(C_2F_4)_n$	100·016	Polytetrafluoroethylene	c		RB	11	m, an	160·3 ±0·9 ($z = 0$) 118·8 ±0·5 ($z = 1$)	56/
C_2F_3Cl	116·471	Trifluorochloroethylene	g		R	7	an	391±3	55/
C_2F_3Cl		Trifluorochloroethylene	g		R	9	an	368·8 ±1·3	63/4
C_2F_3Cl		Trifluorochloroethylene							
C_2Cl_4	165·834	Tetrachloroethylene	l		SB		m	198·6 ±[2·0]	53/

1 kcal = 4·184 kJ

ΔH_r° Remarks	5 ΔH_f° (l or c) kcal/g.f.w.	6 Determination of ΔH_v			7 ΔH_f° (g) kcal/g.f.w.
		Type	ΔH_v° kcal/g.f.w.	Ref.	
$I_3Cl(g)$ {$-20\cdot55\pm0\cdot10$} $+HBr(g)$ $\rightleftharpoons CH_3Br(g)+HCl(g)$. 2nd Law	$-15\cdot10\pm[0\cdot15]$	C2	$5\cdot50\pm[0\cdot05]$	38/12	$-9\cdot60\pm[0\cdot14]$
$I_3Cl(g)$ {$-20\cdot55\pm0\cdot10$} $+HBr(g)$ $\rightleftharpoons CH_3Br(g)+HCl(g)$. 3rd Law	$-15\cdot01\pm[0\cdot15]$				$-9\cdot51\pm[0\cdot14]$
$I_3Br(g)+H_2$ $= CH_4(g)$ {$-17\cdot89\pm0\cdot07$} $+HBr(g)$	$-14\cdot50\pm0\cdot37$				$-9\cdot00\pm0\cdot36$
$I_3Br(l)+0\cdot5H_2$ $= CH_4(g)$ {$-17\cdot89\pm0\cdot07$} $+0\cdot5Br_2(l)$	$-14\cdot59\pm0\cdot32$				$-9\cdot09\pm0\cdot35$
Selected value	$-14\cdot6\ \pm0\cdot3$				$-9\cdot1\ \pm0\cdot3$
$H_3)_2Cd(l)$ {$+16\cdot7\pm0\cdot2$} $+2I_2(c)$ $= CdI_2(c)$ {$-48\cdot6\pm1\cdot0$} $+2CH_3I(l)$	$-3\cdot6\ \pm1\cdot2$	V3	$6\cdot7\ \pm[0\cdot3]$	36/18	$+3\cdot1\ \pm1\cdot3$
$I_3I(g)+HI(g)$ $= CH_4(g)$ {$-17\cdot89\pm0\cdot07$} $+I_2(g)$	$-5\cdot0\ \pm2\cdot6$				$+1\cdot7\ \pm2\cdot5$
$I_3I(l)+0\cdot5H_2$ $= CH_4(g)$ {$-17\cdot89\pm0\cdot07$} $+0\cdot5I_2(c)$	$-2\cdot9\ \pm0\cdot3$				$+3\cdot8\ \pm0\cdot5$
$I_3I(g)+HI(g)$ $\rightleftharpoons CH_4(g)$ {$-17\cdot89\pm0\cdot07$} $+I_2(g)$. 3rd Law	$-3\cdot40\pm0\cdot40$				$+3\cdot30\pm0\cdot28$
$I_3I(g)+HI(g)$ $\rightleftharpoons CH_4(g)$ {$-17\cdot89\pm0\cdot07$} $+I_2(g)$. 3rd Law	$-3\cdot32\pm0\cdot35$				$+3\cdot38\pm0\cdot12$
Selected value	$-3\cdot3\ \pm0\cdot3$				$+3\cdot4\ \pm0\cdot3$
$_2F_4(g)+2H_2+72H_2O(l)$ $= 2C(am)$ {$+1\cdot5\pm0\cdot2$} $+4[HF.18H_2O](l)$					$-157\cdot0\ \pm1\cdot1$*
$_2F_4(g)+4Na(c)$ $= 2C(am)$ {$+3\cdot98\pm0\cdot08$} $+4NaF(c)$ {$-137\cdot9$}					$-158\cdot8\ \pm1\cdot1$*
Selected value					$-157\cdot9\ \pm0\cdot8$*
$/n)\ (C_2F_4)_n+O_2+42(1-z)H_2O(l)$ $= (2-z)CO_2+zCF_4(g)$ $+4(1-z)\ [HF.10H_2O](l)$	$-198\cdot8\ \pm0\cdot9$*				
$_2F_3Cl(g)+4K(c)$ $= 2C(c, gr.)+KCl(c)$ {$-104\cdot4$} $+3KF(c)$ {$-135\cdot9$}					$-121\ \ \pm3$
$_2F_3Cl(g)+4Na(c)$ $= 2C(am)$ {$+3\cdot95$} $+NaCl(c)$ {$-98\cdot5$} $+3NaF(c)$ {$-137\cdot9$}					$-135\cdot5\ \pm1\cdot3$*
Selected value					-125 ± 4
ed: As_2O_3 solution	EJ $\quad -12\cdot2\ \pm[2\cdot0]$	C2	$9\cdot5\ \pm[0\cdot2]$	26/2	$-2\cdot7\ \pm[2\cdot0]$

$C_aH_bHal_j$

1 kcal = 4·184 kJ

1 Formula	2 g.f.w.	3 Name	State	Purity mol %	Type	No. of expts.	Detn. of reactn.	$-\Delta H_r^\circ$ kcal/g.f.w.	Re
C_2F_6	138·013	Hexafluoroethane	g	ms, glc	R	5	m	103·9 ±1·0	66/
C_2F_6		Hexafluoroethane	g	gsc	E	20	gsc	−3·66±0·30	67/
C_2F_6		Hexafluoroethane							
$C_2F_4Cl_2$	170·922	1, 2-Dichloro-tetrafluoroethane	g		R	4	an	57·3 ±0·2	49/
$C_2F_3Cl_3$	187·377	1, 1, 2-Trifluoro-trichloroethane	g		R	3	an	48·8 ±0·5	49/
C_2Cl_6	236·740	Hexachloroethane	c		SB		m	173·8 ±[2·0]	53/
C_2Cl_6		Hexachloroethane	c		R	4	an	36·7 ±0·6	56/
C_2Cl_6		Hexachloroethane	g		E		glc	31·7 ±1·0	63/4
C_2Cl_6		Hexachloroethane							
$C_2F_4Br_2$	259·834	1, 2-Dibromo-tetrafluoroethane	g		R	16	an	38·5 ±0·7	56/
$C_2F_3ClBr_2$	276·289	1, 2-Dibromotrifluoro-chloroethane	g		R	16	an	31·6 ±0·3	56/
C_2HF_3	82·025	Trifluoroethylene	g	99·5, ms	SB		an	233·4 ±2·0	62/4
C_2HCl_3	131·389	Trichloroethylene	l		SB		m	228·6 ±[2·0]	53/
C_2HCl_3		Trichloroethylene	l		R			18·6 ±[1·0]	56/
C_2HCl_3		Trichloroethylene							
C_2HCl_5	202·295	Pentachloroethane	l		SB		m	206·1 ±[2·0]	53/4
C_2HCl_5		Pentachloroethane	l		R			21·7 ±[1·0]	56/2
C_2HCl_5		Pentachloroethane							
C_2HF_4Br	180·933	1, 1, 2, 2-Tetrafluoro-bromoethane	g	ms	R	4	an	33·00±0·13	50/2
C_2HF_3ClBr	197·387	1, 1, 2-Trifluoro-2-chloro-1-bromoethane	g		R	5	an	26·10±0·14	50/2

$$1\ kcal = 4\cdot184\ kJ$$

ΔH_r° Remarks	5 ΔH_f° (l or c) kcal/g.f.w.	6 Determination of ΔH_v Type	ΔH_v° kcal/g.f.w.	Ref.	7 ΔH_f° (g) kcal/g.f.w.
$_2F_6(g)+0\cdot66NF_3(g)$ $\{-31\cdot6\pm0\cdot2\}$ $= 2CF_4(g)$ $\{-223\cdot3\pm0\cdot4\}$ $+0\cdot33N_2$					$-321\cdot6\ \pm1\cdot3*$
$_2F_6(g)+Br_2(g)\rightleftharpoons 2CF_3Br(g)$ $\{-155\cdot0\pm0\cdot6\}$					$-321\cdot1\ \pm1\cdot3*$
Selected value					$-321\cdot3\ \pm0\cdot9*$
$_2F_4(g)$ $\{-157\cdot9\pm0\cdot8\}+Cl_2(g)$ $= C_2F_4Cl_2(g)$	E $-220\cdot8\ \pm1\cdot0*$	C1	$5\cdot6\ \pm[0\cdot1]$	37/15	$-215\cdot2\ \pm0\cdot9*$
$_2F_3Cl(g)$ $\{-125\pm4\}$ $+Cl_2(g)$ $= C_2F_3Cl_3(g)$	E -181 ± 4	C2, V3	$6\cdot8\ \pm0\cdot1$	39/21, 63/42	$-174\ \pm4$
Red: As_2O_3 solution	EJ $-48\cdot3\ \pm[2\cdot0]$	S3	$14\cdot1\ \pm0\cdot4$ (sub.)	47/18	$-34\cdot2\ \pm[2\cdot1]$
$_2Cl_4(l)$ $\{-12\cdot2\pm2\cdot0\}$ $+Cl_2(g) = C_2Cl_6(c)$	$-48\cdot9\ \pm[2\cdot1]$				$-34\cdot8\ \pm[2\cdot2]$
$_2Cl_4(g)$ $\{-2\cdot7\pm2\cdot0\}$ $+Cl_2(g)$ $\rightleftharpoons C_2Cl_6(g)$. 3rd Law	$-48\cdot5\ \pm[2\cdot6]$				$-34\cdot4\ \pm[2\cdot5]$
Selected value	$-48\cdot6\ \pm1\cdot3$				$-34\cdot5\ \pm1\cdot4$
$_2F_4(g)$ $\{-157\cdot9\pm0\cdot8\}$ $+Br_2(g)$ $= C_2F_4Br_2(g)$	E				$-189\cdot0\ \pm1\cdot1*$
$_2F_3Cl(g)$ $\{-125\pm4\}$ $+Br_2(g)$ $= C_2F_3ClBr_2(g)$	E				$-149\ \pm4$
$_2HF_3(g)+1\cdot5O_2+91H_2O(l)$ $= 2CO_2+3[HF.30H_2O](l)$					$-117\cdot3\ \pm2\cdot0*$
Red: As_2O_3 solution	EJ $-10\cdot7\ \pm[2\cdot0]$	C	$8\cdot3\pm[0\cdot1]$	26/2	$-2\cdot2\ \pm[2\cdot0]$
$_2H_2Cl_4(l)$ $\{-46\cdot7\pm1\cdot4\}$ $+0\cdot5Ca(OH)_2(c)$ $\{-235\cdot8\}$ $+199H_2O(l)$ $= C_2HCl_3(l)+0\cdot5$ $[CaCl_2.400H_2O](l)$ $\{-209\cdot3\}$	$-10\cdot3\ \pm[1\cdot8]$	V3	$8\cdot7\ \pm[0\cdot2]$	44/3	$-1\cdot8\ \pm[1\cdot8]$
Selected value	$-10\cdot5\ \pm1\cdot6$		$8\cdot5\ \pm0\cdot1$		$-2\cdot0\ \pm1\cdot6$
Red: As_2O_3 solution	EJ $-44\cdot5\ \pm[2\cdot0]$	V4	$10\cdot9\ \pm[0\cdot5]$	56/26	$-33\cdot6\ \pm[2\cdot1]$
$_2HCl_5(l)+0\cdot5Ca(OH)_2(c)$ $\{-235\cdot8\}$ $+199H_2O(l)$ $= C_2Cl_4(l)$ $\{-12\cdot2\pm2\cdot0\}$ $+0\cdot5$ $[CaCl_2.400H_2O](l)$ $\{-209\cdot3\}$	$-45\cdot6\ \pm[2\cdot5]$				$-34\cdot7\ \pm[2\cdot6]$
Selected value	$-45\cdot0\ \pm[1\cdot6]$				$-34\cdot1\ \pm[1\cdot7]$
$C_2F_4(g)$ $\{-157\cdot9\pm0\cdot8\}$ $+HBr(g)$ $= C_2HF_4Br(g)$	E				$-199\cdot6\ \pm0\cdot9*$
$C_2F_3Cl(g)$ $\{-125\pm4\}$ $+HBr(g)$ $= C_2HF_3ClBr(g)$	E				$-160\ \pm4$

$C_aH_bHal_j$

1 kcal = 4·184 kJ

1 Formula	2 g.f.w.	3 Name	State	Purity mol %	Type	No. of expts.	Detn. of reactn.	$-\Delta H_r^\circ$ kcal/g.f.w.	Rer
$C_2H_2F_2$	64·035	1, 1-Difluoroethylene	g		SB	5	an	262·2 ±0·8	56/2
$C_2H_2F_2$		1, 1-Difluoroethylene	g	99·5, ms	SB		an	259·8 ±2·4	62/4
$C_2H_2F_2$		1, 1-Difluoroethylene							
$C_2H_2Cl_2$	96·944	1, 1-Dichloroethylene	l	99·8	SB	5	m	262·03±0·30	58/3
$C_2H_2Cl_2$	96·944	cis-1, 2-Dichloroethylene	l		SB		m	261·4 :[2·0]	53/
$C_2H_2Cl_2$	96·944	trans-1, 2-Dichloro-ethylene	l		SB		m	261·9 ±[2·0]	53/
$C_2H_2Cl_4$	167·850	1, 1, 2, 2-Tetrachloro-ethane	l		SB		m	232·5 ±[2·0]	53/
$C_2H_2Cl_4$		1, 1, 2, 2-Tetrachloro-ethane	l		R	3	an	40·4 ±0·2	56/2
$C_2H_2Cl_4$		1, 1, 2, 2-Tetrachloro-ethane							
$C_2H_2Cl_2Br_2$	256·762	1, 2-Dichloro-1, 2-dibromoethane	g		E	8	TP	17·3 ±0·2	39/2
C_2H_3Cl	62·499	Vinyl chloride	g	99·8	H	4	an	51·19±0·18	56/5
C_2H_3Cl		Vinyl chloride	g		R	8	an	24·06±0·28	62/4
C_2H_3Cl		Vinyl chloride							
C_2H_3Br	106·955	Vinyl bromide	g	>99, ir	H	7	an	47·65±0·45	57/3
$C_2H_3F_3$	84·041	1, 1, 1-Trifluoroethane	g		SB	6	an	241·0 ±0·4	65/4
$C_2H_3F_3$	84·041	1, 1, 2-Trifluoroethane	g		H	3	an	63·98±0·50	56/28
$C_2H_3Cl_3$	133·405	1, 1, 2-Trichloroethane	l		R	2	an	27·8 ±0·2	56/26
$C_2H_4F_2$	66·0510	1, 1-Difluoroethane	g	>99·9, glc	SB	8	an, CO$_2$	291·5 ±2·0	68/18
$C_2H_4Cl_2$	98·960	1, 1-Dichloroethane	l		SB		m	298·0 ±[2·0]	53/
$C_2H_4Cl_2$		1, 1-Dichloroethane	g	glc	H	8	an	33·66±0·25	67/25
$C_2H_4Cl_2$		1, 1-Dichloroethane							

1 kcal = 4·184 kJ

ΔH_r° Remarks	5 ΔH_f° (l or c) kcal/g.f.w.	6 Determination of ΔH_v			7 ΔH_f° (g) kcal/g.f.w.
		Type	ΔH_v° kcal/g.f.w.	Ref.	
$_2F_2(g)+2O_2+110H_2O(l)$ $=2CO_2+2\,[HF.55H_2O](l)$					$-79\cdot9\ \pm0\cdot8*$
$_2F_2(g)+2O_2+60H_2O(l)$ $=2CO_2+2\,[HF.30H_2O](l)$					$-82\cdot2\ \pm2\cdot4*$
Selected value					$-80\cdot5\ \pm1\cdot0*$
d: As_2O_3 solution J	$-5\cdot72\pm0\cdot35$	C1	$6\cdot33\pm0\cdot04$	59/29	$+0\cdot61\pm0\cdot36$
d: As_2O_3 solution EJ	$-6\cdot4\ \pm[2\cdot0]$	V5	$7\cdot4\ \pm[0\cdot3]$	47/19	$+1\cdot0\ \pm[2\cdot1]$
d: As_2O_3 solution EJ	$-5\cdot8\ \pm[2\cdot0]$	V5	$7\cdot0\ \pm[0\cdot3]$	47/19	$+1\cdot2\ \pm[2\cdot1]$
d: As_2O_3 solution EJ	$-46\cdot6\ \pm[2\cdot0]$	C2	$10\cdot8\ \pm[0\cdot3]$	26/2	$-35\cdot8\ \pm[2\cdot1]$
$_2Cl_2(l)\ \{-6\cdot4\pm2\cdot0\}+Cl_2(g)$ $=C_2H_2Cl_4(l)$	$-46\cdot8\ \pm[2\cdot0]$				$-36\cdot0\ \pm[2\cdot1]$
Selected value	$-46\cdot7\ \pm1\cdot4$				$-35\cdot9\ \pm1\cdot5$
$_2Cl_2(g)\ \{+1\cdot1\pm2\cdot1\}+Br_2(g)$ $\rightleftharpoons C_2H_2Cl_2Br_2(g).$ 2nd Law					$-8\cdot8\ \pm[2\cdot1]$
$_3Cl(g)+2H_2$ $=C_2H_6(g)\ \{-20\cdot24\pm0\cdot12\}+HCl(g)$					$+8\cdot89\pm0\cdot22$
$_2(g)\ \{+54\cdot34\pm0\cdot19\}+HCl(g)$ $=C_2H_3Cl(g)$					$+8\cdot22\pm0\cdot34$
Selected value					$+8\cdot6\ \pm0\cdot3$
$_3Br(g)+2H_2$ $=C_2H_6(g)\ \{-20\cdot24\pm0\cdot12\}+HBr(g)$					$+18\cdot71\pm0\cdot52$
$_3F_3(g)+2O_2+135H_2O(l)$ $=2CO_2+3\,[HF.45H_2O](l)$					$-178\cdot0\ \pm0\cdot4*$
$F_3Cl(g)\ \{-125\pm4\}+2H_2$ E $=C_2H_3F_3(g)+HCl(g)$					$-167\ \pm4$
$-C_2H_4Cl_2(l)\ \{-39\cdot2\pm0\cdot4\}+Cl_2(g)$ $=C_2H_3Cl_3(l)+HCl(g)$	$-44\cdot9\ \pm0\cdot5$	C1	$9\cdot4\ \pm[0\cdot2]$	57/36	$-35\cdot5\ \pm0\cdot6$
$_4F_2(g)+2\cdot5O_2+39H_2O(l)$ $=2CO_2+2[HF.20H_2O]\,(l)$					$-118\cdot8\ \pm2\cdot0*$
d: As_2O_3 solution EJ	$-38\cdot0\ \pm[2\cdot0]$	C2	$7\cdot35\pm0\cdot02$	56/29	$-30\cdot6\ \pm[2\cdot0]$
$_4Cl_2(g)+2H_2$ $=C_2H_6(g)\ \{-20\cdot24\pm0\cdot12\}+2HCl(g)$	$-38\cdot05\pm0\cdot30$				$-30\cdot70\pm0\cdot30$
Selected value	$-38\cdot0\ \pm0\cdot3$				$-30\cdot65\pm0\cdot30$

$C_aH_bHal_j$

1 kcal = 4·184 kJ

1 Formula	2 g.f.w.	3 Name	State	Purity mol %	Type	No. of expts.	Detn. of reactn.	$-\Delta H_r^\circ$ kcal/g.f.w.	R
$C_2H_4Cl_2$	98·960	1, 2-Dichloroethane	g	99·88	R	3	an	43·68±0·15	38
$C_2H_4Cl_2$		1, 2-Dichloroethane	l		SB		m	297·9 ±[2·0]	53
$C_2H_4Cl_2$		1, 2-Dichloroethane	l	99·9	SB	8	m	296·72±0·40	58
$C_2H_4Cl_2$		1, 2-Dichloroethane	g	glc	H	5	an	34·18±0·23	67
$C_2H_4Cl_2$		1, 2-Dichloroethane							
$C_2H_4Br_2$	187·872	1, 2-Dibromoethane	g	99·88	R	14	an	28·90±0·30	38
$C_2H_4I_2$	281·863	1, 2-Di-iodoethane	g		E	11	uv	13·4 ±[2·0]	35
$C_2H_4I_2$		1, 2-Di-iodoethane	g		E	18		11·5 ±0·4	54
$C_2H_4I_2$		1, 2-Di-iodoethane	g	99·5	E	17	TP, an	11·5 ±0·4	62
$C_2H_4I_2$		1, 2-Di-iodoethane							
C_2H_5Cl	64·515	Ethyl chloride	g		FC	4		341·1 ±2·5	51
C_2H_5Cl		Ethyl chloride	g		E	16	TP, an	17·16±[0·50]	55
C_2H_5Cl		Ethyl chloride	g		E	16	TP, an	17·35±[0·50]	55
C_2H_5Cl		Ethyl chloride	g	99·97	E	5	an	17·1 ±[0·5]	53
C_2H_5Cl		Ethyl chloride	g		H	4	an	16·56±0·10	56
C_2H_5Cl		Ethyl chloride							
C_2H_5Br	108·971	Ethyl bromide	g	99·97	E	5	TP	19·20±[0·50]	53
C_2H_5Br		Ethyl bromide	l	glc	H	6	m	−2·8 ±1·5	65
C_2H_5Br		Ethyl bromide	g		H	4	an	14·13±0·27	65
C_2H_5Br		Ethyl bromide							
C_2H_5I	155·967	Ethyl iodide	l		SB		m	350·5 ±1·0	49
C_2H_5I		Ethyl iodide	l		H	6	m	10·6 ±0·4	65
C_2H_5I		Ethyl iodide							

Determina

1 kcal = $4 \cdot 184$ kJ

ΔH_r° Remarks		5 ΔH_f° (l or c) kcal/g.f.w.	6 Determination of ΔH_v			7 ΔH_f° (g) kcal/g.f.w.
			Type	ΔH_v° kcal/g.f.w.	Ref.	
$I_4(g) \{+12\cdot45\pm0\cdot10\} +Cl_2(g)$ $= C_2H_4Cl_2(g)$	E	$-39\cdot70\pm0\cdot19$	Cl	$8\cdot47\pm0\cdot02$	58/32	$-31\cdot23\pm0\cdot18$
l: As_2O_3 solution	EJ	$-38\cdot2\ \pm[2\cdot0]$				$-29\cdot7\ \pm[2\cdot0]$
l: As_2O_3 solution	J	$-39\cdot34\pm0\cdot44$				$-30\cdot87\pm0\cdot45$
$I_4Cl_2(g)+2H_2$ $= C_2H_6(g) \{-20\cdot24\pm0\cdot12\} +2HCl(g)$		$-38\cdot65\pm0\cdot26$				$-30\cdot18\pm0\cdot26$
Selected value		$-39\cdot2\ \pm0\cdot4$				$-30\cdot7\ \pm0\cdot4$
$I_4(g) \{+12\cdot45\pm0\cdot10\} +Br_2(g)$ $= C_2H_4Br_2(g)$		$-18\cdot92\pm0\cdot35$	Cl	$9\cdot86\pm0\cdot02$	68/6	$-9\cdot06\pm0\cdot35$
$I_4(g) \{+12\cdot45\pm0\cdot10\} +I_2(g)$ $\rightleftharpoons C_2H_4I_2(g)$. 2nd Law			S5	$15\cdot7\ \pm[1\cdot0]$ (sub.)	54/22	$+14\cdot0\ \pm[2\cdot0]$
$I_4(g) \{+12\cdot45\pm0\cdot10\} +I_2(g)$ $\rightleftharpoons C_2H_4I_2(g)$. 2nd Law						$+15\cdot9\ \pm0\cdot4$
$I_4(g) \{+12\cdot45\pm0\cdot10\} +I_2(g)$ $\rightleftharpoons C_2H_4I_2(g)$. 2nd Law						$+15\cdot9\ \pm0\cdot4$
Selected value		$-0\cdot2\ \pm1\cdot1$ (c)				$+15\cdot5\ \pm0\cdot4$
$H_5Cl(g)+3O_2+138H_2O(l)$ $= 2CO_2+HCl(140H_2O)(l)$	E		Cl	$5\cdot91\pm[0\cdot1]$	26/5	$-23\cdot3\ \pm2\cdot5$
$H_4(g) \{+12\cdot45\pm0\cdot10\} +HCl(g)$ $\rightleftharpoons C_2H_5Cl(g)$. 2nd Law						$-26\cdot77\pm[0\cdot50]$
$H_4(g) \{+12\cdot45\pm0\cdot10\} +HCl(g)$ $\rightleftharpoons C_2H_5Cl(g)$. 3rd Law						$-26\cdot96\pm[0\cdot50]$
$H_4(g) \{+12\cdot45\pm0\cdot10\} +HCl(g)$ $\rightleftharpoons C_2H_5Cl(g)$. 3rd Law						$-26\cdot7\ \pm0\cdot5$
$H_5Cl(g)+H_2$ $= C_2H_6(g) \{-20\cdot24\pm0\cdot12\} +HCl(g)$						$-25\cdot74\pm0\cdot16$
Selected value		$-32\cdot0\ \pm0\cdot4$				$-26\cdot1\ \pm0\cdot4$
$H_4(g) \{+12\cdot45\pm0\cdot10\} +HBr(g)$ $\rightleftharpoons C_2H_5Br(g)$. 3rd Law		$-22\cdot1\ \pm[0\cdot60]$	E1	$6\cdot6\ \pm0\cdot3$		$-15\cdot45\pm[0\cdot50]$
$H_5Br(l)+0\cdot5H_2$ $= C_2H_6(g) \{-20\cdot24\pm0\cdot12\} +0\cdot5Br_2(l)$		$-23\cdot0\ \pm1\cdot5$				$-16\cdot4\ \pm1\cdot6$
$H_5Br(g)+H_2$ $= C_2H_6(g) \{-20\cdot24\pm0\cdot12\} +HBr(g)$		$-21\cdot4\ \pm0\cdot4$				$-14\cdot8\ \pm0\cdot3$
Selected value		$-21\cdot8\ \pm0\cdot5$				$-15\cdot2\ \pm0\cdot5$
		$-8\cdot4\ \pm1\cdot0^*$	Cl	$7\cdot58\pm0\cdot02$	68/6	$-0\cdot8\ \pm1\cdot0^*$
$H_5I(l)+0\cdot5H_2$ $= C_2H_6(g) \{-20\cdot24\pm0\cdot12\} +0\cdot5I_2(c)$		$-9\cdot6\ \pm0\cdot4$				$-2\cdot0\ \pm0\cdot4$
Selected value		$-9\cdot6\ \pm0\cdot4$				$-2\cdot0\ \pm0\cdot4$

THERMOCHEMISTRY OF ORGANIC AND ORGANOMETALLIC COMPOUNDS

$C_aH_bHal_j$

1 kcal = 4·184 kJ

								Determinat	
						No. of	Detn. of	$-\Delta H_r^\circ$	
1 Formula	2 g.f.w.	3 Name	State	Purity mol %	Type	expts.	reactn.	kcal/g.f.w.	Re
C_3F_8	188·021	Octafluoropropane	g	97·8, glc, ms	R	5	m, an	660·0 ±1·2	67/
$C_3H_3F_3$	96·053	1, 1, 1-Trifluoropropene	g		SB	12	an	366·2 ±1·6	67/
C_3H_5Br	120·982	Allyl bromide	l		R	4	an	3·7 ±0·5	49/
C_3H_5I	167·978	Allyl iodide	l		R	4	an	−2·1 ±0·2	49/
C_3H_5I		Allyl iodide	g		E		uv	8·0 ±0·6	66/
C_3H_5I		Allyl iodide	g		E		uv	9·5 ±1·0	66/
C_3H_5I		Allyl iodide							
$C_3H_5Cl_3$	147·432	1, 2, 3-Trichloropropane	l		RB	3	m	414·77±0·30	54/
$C_3H_6Cl_2$	112·987	1, 2-Dichloropropane	l		SB		m	450·1 ±[2·0]	53/
$C_3H_6Cl_2$		1, 2-Dichloropropane	g	glc	H	8	an	30·24±0·26	67/2
$C_3H_6Cl_2$		1, 2-Dichloropropane							
$C_3H_6Cl_2$	112·987	1, 3-Dichloropropane	l		SB		m	450·6 ±[2·0]	53/
$C_3H_6Cl_2$	112·987	2, 2-Dichloropropane	l		SB		m	449·2 ±[2·0]	53/
$C_3H_6Br_2$	201·899	1, 2-Dibromopropane	g	99·89	R	6	an	29·27±0·20	38/1
$C_3H_6I_2$	295·890	1, 2-Di-iodopropane	g	99·5	E	7	TP, an	11·2 ±0·8	62/4
C_3H_7F	62·088	1-Fluoropropane	g		H	7	an	22·0 ±[0·5]	56/3
C_3H_7F	62·088	2-Fluoropropane	g		H	5	an	20·2 ±[0·3]	56/3

1 kcal = 4·184 kJ

H_r° Remarks	5 ΔH_f° (l or c) kcal/g.f.w.	6 Determination of ΔH_v Type	ΔH_v° kcal/g.f.w.	Ref.	7 ΔH_f° (g) kcal/g.f.w.
$_8(g)+8Na(c)$ = $8NaF(c)$ {−137·9} +3C(am) {+5·4}					−427·0 ±1·2*
$_3F_3(g)+3O_2+90H_2O(l)$ =$3CO_2$+3 [HF.30H$_2$O](l)					−146·9 ±1·6*
$_5Br(l)+12H_2O(l)$ = [C$_3$H$_5$OH+HBr] (11H$_2$O) {−69·0±1·0}	+3·0 ±1·4	E1	7·9 ±0·5		+10·9 ±1·6
$_5I(l)+8H_2O(l)$ = [C$_3$H$_5$OH+HI] (7H$_2$O) {−53·0±1·0}	+13·2 ±1·3	E1	9·1 ±0·5		+22·3± 1·5
$_5I(g)+HI(g)$ ⇌ C$_3$H$_6$(g) {+4·88±0·16} +I$_2$(g). 2nd Law					+21·5 ±0·7
$_5I(g)+HI(g)$ ⇌ C$_3$H$_6$(g) {+4·88±0·16} +I$_2$(g). 3rd Law					+23·0 ±1·0
Selected value	+13·7 ±1·1				+22·8 ±1·0
d: As$_2$O$_3$ solution EJ	−55·17±0·34	E7	11·4 ±0·3		−43·8 ±0·5
d: As$_2$O$_3$ solution EJ $_6Cl_2(g)+2H_2$ = C$_3$H$_8$(g) {−24·83±0·14} +2HCl(g)	−48·3 ±[2·0] −47·3 ±[0·5]	V2	8·6 ±[0·4]	49/25	−39·7 ±[2·1] −38·71±0·30
Selected value	−47·3 ±0·5				−38·71±0·30
d: As$_2$O$_3$ solution EJ	−47·9 ±[2·0]	C1	9·67±0·03	68/6	−38·2 ±[2·0]
d: As$_2$O$_3$ solution EJ	−49·2 ±[2·0]	E7	7·8 ±0·3		−41·4 ±[2·1]
$_6(g)$ {+4·88±0·16} +Br$_2$(g) E = C$_3$H$_6$Br$_2$(g)					−17·00±0·26
$_6(g)$ {+4·88±0·16} +I$_2$(g) ⇌ C$_3$H$_6$I$_2$(g) 2nd Law					+8·5 ±0·9
$_7F(g)+H_2$ E = C$_3$H$_8$(g) {−24·83±0·14} +HF(g)					−67·6 ±[0·6]
$_7F(g)+H_2$ E = C$_3$H$_8$(g) {−24·83±0·14} +HF(g)					−69·4 ±[0·4]

$C_aH_bHal_j$

1 kcal = 4·184 kJ

1 Formula	2 g.f.w.	3 Name	State	Purity mol %	No. of Type expts.	Detn. of reactn.	$-\Delta H_r^\circ$ kcal/g.f.w.	R
C_3H_7Cl	78·542	1-Chloropropane	g glc		H 5	an	15·73±0·16	65/
C_3H_7Cl		1-Chloropropane	g glc		E	glc	3·25±[0·30]	65/
C_3H_7Cl		1-Chloropropane	g glc		E	glc	2·74±[0·30]	65/
C_3H_7Cl		1-Chloropropane						
C_3H_7Cl	78·542	2-Chloropropane	l		SB	m	484·8 ±[2·0]	53,
C_3H_7Cl		2-Chloropropane	g		E	TP, an	17·45±[0·50]	55/
C_3H_7Cl		2-Chloropropane	g glc		E	an	17·51±[0·50]	63/
C_3H_7Cl		2-Chloropropane	g glc		H 6	an	13·94±0·17	65/
C_3H_7Cl		2-Chloropropane						
C_3H_7Br	122·998	1-Bromopropane	l		E		2·2 ±[0·6]	44/
C_3H_7Br		1-Bromopropane	g		R 6	an	25·83±0·19	57/
C_3H_7Br		1-Bromopropane	g ir		E 2	ir	3·3 ±[0·6]	60/3
C_3H_7Br		1-Bromopropane	l		RB 5	m	491·58±0·34	61/
C_3H_7Br		1-Bromopropane	g glc		H 6	an	13·58±0·14	65/4
C_3H_7Br		1-Bromopropane						
C_3H_7Br	122·998	2-Bromopropane	g		R 6	an	20·10±0·14	50/2
C_3H_7Br		2-Bromopropane	g		R 4	an	20·39±0·14	57/
C_3H_7Br		2-Bromopropane	l		RB 3	m	490·45±0·40	61/
C_3H_7Br		2-Bromopropane	g ir		E	an, ir	19·28±[0·50]	62/4
C_3H_7Br		2-Bromopropane	g glc		H 5	an	10·85±0·22	65/4
C_3H_7Br		2-Bromopropane						
C_3H_7I	169·994	1-Iodopropane	l		R 4	m	51·6 ±0·6	52/2

1 kcal $= 4 \cdot 184$ kJ

ΔH_r° Remarks	ΔH_f° (l or c) kcal/g.f.w.	Determination of ΔH_v			ΔH_f° (g) kcal/g.f.w.
		Type	ΔH_v° kcal/g.f.w.	Ref.	
$H_7Cl(g)+H_2$	$-38 \cdot 1 \ \pm 0 \cdot 3$	C2	$6 \cdot 9 \ \pm[0 \cdot 2]$	31/3	$-31 \cdot 16 \pm 0 \cdot 21$
$= C_3H_8(g) \{-24 \cdot 83 \pm 0 \cdot 14\} +HCl(g)$	$-37 \cdot 2 \ \pm 1 \cdot 1$				$-30 \cdot 3 \ \pm 1 \cdot 1$
$C_3H_7Cl(g) \rightleftharpoons i\text{-}C_3H_7Cl(g) \{-33 \cdot 6 \pm 1 \cdot 0\}$ 2nd Law					
$C_3H_7Cl(g) \rightleftharpoons i\text{-}C_3H_7Cl(g) \{-33 \cdot 6 \pm 1 \cdot 0\}$ 3rd Law	$-37 \cdot 8 \ \pm 1 \cdot 1$				$-30 \cdot 9 \ \pm 1 \cdot 1$
Selected value	$-37 \cdot 9 \ \pm 0 \cdot 3$				$-31 \cdot 0 \ \pm 0 \cdot 2$
ed: As_2O_3 solution **EJ**	$-42 \cdot 1 \ \pm[2 \cdot 0]$	C2	$6 \cdot 5 \ \pm[0 \cdot 2]$	31/3	$-35 \cdot 6 \ \pm[2 \cdot 0]$
$_3H_6(g) \{+4 \cdot 88 \pm 0 \cdot 16\} +HCl(g)$	$-41 \cdot 1 \ \pm[0 \cdot 6]$				$-34 \cdot 63 \pm[0 \cdot 60]$
$\rightleftharpoons i\text{-}C_3H_7Cl.$ 2nd Law					
$_3H_6(g) \{+4 \cdot 88 \pm 0 \cdot 16\} +HCl(g)$	$-41 \cdot 2 \ \pm[0 \cdot 6]$				$-34 \cdot 69 \pm[0 \cdot 60]$
$\rightleftharpoons i\text{-}C_3H_7Cl.$ 2nd Law					
$C_3H_7Cl(g)+H_2$	$-39 \cdot 5 \ \pm 0 \cdot 3$				$-32 \cdot 95 \pm 0 \cdot 22$
$= C_3H_8(g) \{-24 \cdot 83 \pm 0 \cdot 14\} +HCl(g)$ Selected value	$-40 \cdot 1 \ \pm 1 \cdot 0$				$-33 \cdot 6 \ \pm 1 \cdot 0$
$C_3H_7Br(l) \rightleftharpoons i\text{-}C_3H_7Br(l) \{-30 \cdot 7 \pm 0 \cdot 6\}$	$-28 \cdot 5 \ \pm[0 \cdot 8]$	C1	$7 \cdot 62 \pm 0 \cdot 02$	66/14	$-20 \cdot 9 \ \pm[0 \cdot 8]$
2nd Law					
$C_3H_6(g) \{+12 \cdot 73 \pm 0 \cdot 14\} +HBr(g)$ **E**	$-29 \cdot 42 \pm 0 \cdot 24$				$-21 \cdot 80 \pm 0 \cdot 23$
$= n\text{-}C_3H_7Br(g)$					
$C_3H_7Br(g) \rightleftharpoons i\text{-}C_3H_7Br(g) \{-23 \cdot 5 \pm 0 \cdot 6\}$	$-27 \cdot 8 \ \pm[0 \cdot 8]$				$-20 \cdot 2 \ \pm[0 \cdot 8]$
2nd Law					
ed: As_2O_3 solution **J**	$-29 \cdot 68 \pm 0 \cdot 34$				$-22 \cdot 06 \pm 0 \cdot 35$
$_3H_7Br(g)+H_2$	$-27 \cdot 57 \pm 0 \cdot 21$				$-19 \cdot 95 \pm 0 \cdot 20$
$= C_3H_8(g) \{-24 \cdot 83 \pm 0 \cdot 14\} +HBr(g)$ Selected value	$-28 \cdot 1 \ \pm 1 \cdot 0$				$-20 \cdot 5 \ \pm 1 \cdot 0$
$_3H_6(g) \{+4 \cdot 88 \pm 0 \cdot 16\} +HBr(g)$ **E**	$-31 \cdot 13 \pm 0 \cdot 21$	C1	$7 \cdot 21 \pm 0 \cdot 02$	66/14	$-23 \cdot 92 \pm 0 \cdot 21$
$= i\text{-}C_3H_7Br(g)$					
$_3H_6(g) \{+4 \cdot 88 \pm 0 \cdot 16\} +HBr(g)$ **E**	$-31 \cdot 42 \pm 0 \cdot 21$				$-24 \cdot 21 \pm 0 \cdot 21$
$= i\text{-}C_3H_7Br(g)$					
ed: As_2O_3 solution **J**	$-30 \cdot 81 \pm 0 \cdot 44$				$-23 \cdot 60 \pm 0 \cdot 44$
$_3H_6(g) \{+4 \cdot 88 \pm 0 \cdot 16\} +HBr(g)$ **E**	$-30 \cdot 31 \pm[0 \cdot 55]$				$-23 \cdot 10 \pm[0 \cdot 55]$
$\rightleftharpoons i\text{-}C_3H_7Br(g).$ 2nd Law					
$C_3H_7Br(g)+H_2$	$-29 \cdot 89 \pm 0 \cdot 26$				$-22 \cdot 68 \pm 0 \cdot 26$
$= C_3H_8(g) \{-24 \cdot 83 \pm 0 \cdot 14\} +HBr(g)$ Selected value	$-30 \cdot 7 \ \pm 0 \cdot 6$				$-23 \cdot 5 \ \pm 0 \cdot 6$
$n\text{-}C_3H_7)_2Hg(l) \{-5 \cdot 0 \pm 1 \cdot 3\} +2I_2(c)$	$-15 \cdot 7 \ \pm 1 \cdot 8$	C1	$8 \cdot 57 \pm 0 \cdot 03$	68/6	$-7 \cdot 1 \ \pm 1 \cdot 8$
$= HgI_2(c) \{-25 \cdot 2 \pm 0 \cdot 2\} +2n\text{-}C_3H_7I(l)$					

THERMOCHEMISTRY OF ORGANIC AND ORGANOMETALLIC COMPOUNDS

$C_aH_bHal_j$

$$1 \text{ kcal} = 4 \cdot 184 \text{ kJ}$$

1 Formula	2 g.f.w.	3 Name	State	Purity mol %	Type	No. of expts.	Detn. of reactn.	$-\Delta H_r^\circ$ kcal/g.f.w.	R
C_3H_7I	169·994	2-Iodopropane	l		R	4	m	57·9 ±0·5	52/2
C_3H_7I		2-Iodopropane	g	ir	E		ir	2·9 ±1·0	62/4
C_3H_7I		2-Iodopropane							
C_4F_8	200·032	Octafluorocyclobutane	g	ms	R	6	an, m	714·5 ±2·2	64/3 68/2
$C_4H_8Br_2$	215·926	1, 2-Dibromobutane	g	99·9	R	4	an	29·44±0·20	38/1
$C_4H_8Br_2$		1, 2-Dibromobutane	g	99·9	R	3	an	28·90±0·30	41/1
$C_4H_8Br_2$ $C_4H_8Br_2$		1, 2-Dibromobutane 1, 2-Dibromobutane	l		RB	7	m	614·34±0·50	61/8
$C_4H_8Br_2$	215·926	2, 3-Dibromobutane	g	99·9	R	2	an	28·95±0·20	38/1.
$C_4H_8Br_2$		2, 3-Dibromobutane	g	99·9	R	3	an	30·17±0·20	38/1.
$C_4H_8Br_2$		2, 3-Dibromobutane							
$C_4H_8I_2$	309·917	1, 2-Di-iodobutane	g		E	8	uv	12·0 ±1·5	37/1
C_4H_9Cl	92·569	1-Chlorobutane	l		SB		m	646·3 ±[2·0]	53/4
C_4H_9Cl	92·569	1-Chloro-2-methyl-propane	l		SB		m	643·6 ±[2·0]	53/4
C_4H_9Cl	92·569	(±)-2-Chlorobutane	l		SB		m	643·2 ±[2·0]	53/4
C_4H_9Cl	92·569	2-Chloro-2-methyl-propane	g		E	11	TP	17·1 ±[0·5]	37/1(
C_4H_9Cl		2-Chloro-2-methyl-propane	g		E		TP	17·7 ±[0·5]	51/22
C_4H_9Cl		2-Chloro-2-methyl-propane	g	ir	E		an	17·5 ±[0·5]	64/37
C_4H_9Cl		2-Chloro-2-methyl-propane	l		SB		m	639·4 ±[2·0]	53/4
C_4H_9Cl		2-Chloro-2-methyl-propane							

$$1 \text{ kcal} = 4\cdot184 \text{ kJ}$$

ΔH_r° Remarks	5 ΔH_f° (l or c) kcal/g.f.w.	6 Determination of ΔH_v			7 ΔH_f° (g) kcal/g.f.w.
		Type	ΔH_v° kcal/g.f.w.	Ref.	
$C_3H_7)_2Hg(l)$ $\{-3\cdot1\pm1\cdot0\}$ $+2I_2(c)$	$-17\cdot9$ $\pm1\cdot5$	Cl	$8\cdot14\pm0\cdot02$	68/6	$-9\cdot8$ $\pm1\cdot5$
$= HgI_2(c)$ $\{-25\cdot2\pm0\cdot2\}$ $+2i\text{-}C_3H_7I(l)$	$-18\cdot1$ $\pm2\cdot2$				$-10\cdot0$ $\pm2\cdot2$
$C_3H_7I(g)$ $\{-7\cdot1\pm1\cdot8\}\rightleftharpoons i\text{-}C_3H_7I(g)$ 2nd Law					
Selected value	$-17\cdot9$ $\pm1\cdot5$				$-9\cdot8$ $\pm1\cdot5$
					$-369\cdot5$ $\pm2\cdot6$*
$F_8(g)+8Na(c)$					
$= 4C(am)$ $\{+4\cdot8\pm0\cdot4\}$ $+8NaF(c)$ $\{-137\cdot9\}$					
Butene(g) $\{-0\cdot20\pm0\cdot13\}$ $+Br_2(g)$ **E**	$-33\cdot05\pm0\cdot38$	V3	$10\cdot8$ $\pm0\cdot3$	41/12	$-22\cdot25\pm0\cdot24$
$= C_4H_8Br_2(g)$					
Butene(g) $\{-0\cdot20\pm0\cdot13\}$ $+Br_2(g)$	$-32\cdot43\pm0\cdot44$				$-21\cdot63\pm0\cdot35$
$= C_4H_8Br_2(g)$					
ed: As_2O_3 solution **J**	$-35\cdot12\pm0\cdot54$				$-24\cdot3$ $\pm0\cdot60$
Selected value	$-34\cdot3$ $\pm1\cdot0$				$-23\cdot5$ $\pm1\cdot0$
$ns\text{-}2\text{-Butene}(g)$ $\{-2\cdot99\pm0\cdot18\}$ $+Br_2(g)$	$-33\cdot6$ $\pm[0\cdot6]$	V4	$9\cdot0$ $\pm[0\cdot5]$	36/19	$-24\cdot55\pm0\cdot27$
$= C_4H_8Br_2(g)$					
$\text{-2-Butene}(g)$ $\{-1\cdot86\pm0\cdot20\}$ $+Br_2(g)$	$-33\cdot6$ $\pm[0\cdot6]$				$-24\cdot64\pm0\cdot20$
$= C_4H_8Br_2(g)$					
Selected value	$-33\cdot60\pm0\cdot55$				$-24\cdot60\pm0\cdot20$
Butene(g) $\{-0\cdot20\pm0\cdot13\}$ $+I_2(g)$ $\rightleftharpoons C_4H_3I_2(g)$. 2nd Law					$+2\cdot7$ $\pm1\cdot5$
ed: As_2O_3 solution **EJ**	$-43\cdot0$ $\pm[2\cdot0]$	Cl	$7\cdot93\pm0\cdot02$	68/6	$-35\cdot1$ $\pm[2\cdot0]$
ed: As_2O_3 solution **EJ**	$-45\cdot7$ $\pm[2\cdot0]$	Cl	$7\cdot57\pm0\cdot02$	68/6	$-38\cdot1$ $\pm[2\cdot0]$
ed: As_2O_3 solution **EJ**	$-46\cdot1$ $\pm[2\cdot0]$	Cl	$7\cdot54\pm0\cdot02$	68/6	$-38\cdot6$ $\pm[2\cdot0]$
Methylpropene(g) $\{-4\cdot26\pm0\cdot15\}$ $+HCl(g)\rightleftharpoons C_4H_9Cl(g)$. 2nd Law **E**	$-50\cdot3$ $\pm[0\cdot6]$	Cl	$6\cdot93\pm0\cdot02$	68/6	$-43\cdot4$ $\pm[0\cdot6]$
Methylpropene(g) $\{-4\cdot26\pm0\cdot15\}$ $+HCl(g)\rightleftharpoons C_4H_9Cl(g)$. 2nd Law **E**	$-50\cdot9$ $\pm[0\cdot6]$				$-44\cdot0$ $\pm[0\cdot6]$
Methylpropene(g) $\{-4\cdot26\pm0\cdot15\}$ $+HCl(g)\rightleftharpoons C_4H_9Cl(g)$. 2nd Law **E**	$-50\cdot7$ $\pm[0\cdot6]$				$-43\cdot8$ $\pm[0\cdot6]$
ed: As_2O_3 solution **EJ**	$-49\cdot9$ $\pm[2\cdot0]$				$-43\cdot0$ $\pm[2\cdot0]$
Selected value	$-50\cdot6$ $\pm0\cdot6$				$-43\cdot7$ $\pm0\cdot6$

1 kcal = 4·184 kJ

1 Formula	2 g.f.w.	3 Name	State	Purity mol %	Type	No. of expts.	Detn. of reactn.	$-\Delta H_f^{\circ}$ kcal/g.f.w.	Re
C_4H_9Br	137·025	1-Bromobutane	l		RB	3	m	649·26±0·30	61/4
C_4H_9Br C_4H_9Br	137·025	(±)-2-Bromobutane (±)-2-Bromobutane	l g	ir	RB R	4 4	m an	646·56±0·31 20·04±0·12	61/ 52/2
C_4H_9Br		(±)-2-Bromobutane	g	ir	R	5	an	18·42±0·12	52/2
C_4H_9Br		(±)-2-Bromobutane	g	ir	R	5	an	17·26±0·12	52/2
C_4H_9Br		(±)-2-Bromobutane							
C_4H_9Br	137·025	2-Bromo-2-methyl-propane	g		E	10	TP	18·90±[0·30]	57/3
C_4H_9Br		2-Bromo-2-methyl-propane	g	99·9	E		TP	18·95±[0·50]	37/1
C_4H_9Br		2-Bromo-2-methyl-propane							
C_4H_9I	184·021	2-Iodo-2-methylpropane	g		E		TP	19·2 ±2·0	37/1
C_4H_9I		2-Iodo-2-methylpropane	g		E			19·5 ±0·5	62/4
C_4H_9I		2-Iodo-2-methylpropane							
$C_5H_8Br_2$	227·938	1, 2-Dibromocyclo-pentane	g		R	2	an	28·61±0·30	41/12
$C_5H_{10}Br_2$	229·953	2, 3-Dibromo-2-methylbutane	g		R	2	an	30·38±0·20	38/13
$C_5H_{11}Cl$ $C_5H_{11}Cl$ $C_5H_{11}Cl$	106·596	1-Chloropentane 1-Chloropentane 1-Chloropentane	l l		SB RB	3	m m	800·8 ±[2·0] 801·0 ±0·9	53/4 54/3
$C_5H_{11}Cl$	106·596	1-Chloro-3-methylbutane	l		SB		m	800·0 ±[2·0]	53/4
$C_5H_{11}Cl$	106·596	2-Chloro-2-methylbutane	l		SB		m	795·3 ±[2·0]	53/4

$$1 \text{ kcal} = 4 \cdot 184 \text{ kJ}$$

ΔH_r° Remarks		5 ΔH_f° (l or c) kcal/g.f.w.	6 Determination of ΔH_v			7 ΔH_f° (g) kcal/g.f.w.
			Type	ΔH_v° kcal/g.f.w.	Ref.	
l: As_2O_3 solution	J	$-34\cdot36\pm0\cdot30$	Cl	$8\cdot76\pm0\cdot03$	66/14	$-25\cdot60\pm0\cdot30$
l: As_2O_3 solution	J	$-37\cdot06\pm0\cdot31$	Cl	$8\cdot21\pm0\cdot02$	66/14	$-28\cdot85\pm0\cdot31$
utene(g) $\{-0\cdot20\pm0\cdot13\}$ +HBr(g)	E	$-37\cdot15\pm0\cdot18$				$-28\cdot94\pm0\cdot18$
$= C_4H_9Br(g)$. 2nd Law						
2-Butene(g) $\{-1\cdot86\pm0\cdot20\}$ +HBr(g)	E	$-37\cdot19\pm0\cdot24$				$-28\cdot98\pm0\cdot24$
$= C_4H_9Br(g)$. 2nd Law						
ıs-2-Butene(g) $\{-2\cdot99\pm0\cdot18\}$ +HBr(g)	E	$-37\cdot16\pm0\cdot23$				$-28\cdot95\pm0\cdot23$
$= C_4H_9Br(g)$. 2nd Law						
Selected value		$-37\cdot15\pm0\cdot12$				$-28\cdot94\pm0\cdot12$
Methylpropene(g) $\{-4\cdot26\pm0\cdot15\}$ +HBr(g)	E	$-39\cdot3\pm[0\cdot4]$	V3	$7\cdot4\pm0\cdot2$	51/28	$-31\cdot86\pm[0\cdot35]$
$\rightleftharpoons C_4H_9Br(g)$. 2nd Law						
Methylpropene(g) $\{-4\cdot26\pm0\cdot15\}$ +HBr(g)	E	$-39\cdot3\pm[0\cdot6]$				$-31\cdot91\pm[0\cdot55]$
$\rightleftharpoons C_4H_9Br(g)$. 2nd Law						
Selected value		$-39\cdot3\pm0\cdot4$				$-31\cdot88\pm0\cdot30$
Methylpropene(g) $\{-4\cdot26\pm0\cdot15\}$ +HI(g)		$-25\cdot6\pm2\cdot0$	Cl	$8\cdot46\pm0\cdot02$	68/6	$-17\cdot1\pm2\cdot0$
$\rightleftharpoons C_4H_9I(g)$. 2nd Law						
Methylpropene(g) $\{-4\cdot26\pm0\cdot15\}$ +HI(g)		$-25\cdot9\pm0\cdot6$				$-17\cdot4\pm0\cdot6$
$\rightleftharpoons C_4H_9I(g)$. 3rd Law						
Selected value		$-25\cdot9\pm0\cdot6$				$-17\cdot4\pm0\cdot6$
$C_5H_8(g)$ $\{+8\cdot23\pm0\cdot22\}$ +$Br_2(g)$		$-24\cdot42\pm0\cdot48$	V3	$11\cdot43\pm0\cdot30$	41/12	$-12\cdot99\pm0\cdot37$
$= C_5H_8Br_2(g)$						
Methyl-2-butene(g) $\{-10\cdot12\pm0\cdot18\}$ +Br_2 (g) $= C_5H_{10}Br_2(g)$	E					$-33\cdot11\pm0\cdot27$
ed: As_2O_3 solution	EJ	$-50\cdot9\pm[2\cdot0]$	Cl	$9\cdot06\pm0\cdot03$	68/6	$-41\cdot8\pm[2\cdot0]$
ed: As_2O_3 solution	EJ	$-50\cdot7\pm0\cdot9$				$-41\cdot6\pm0\cdot9$
Selected value		$-50\cdot7\pm0\cdot9$				$-41\cdot6\pm0\cdot9$
ed: As_2O_3 solution	EJ	$-51\cdot7\pm[2\cdot0]$	E2	$8\cdot8\pm0\cdot3$		$-42\cdot9\pm[2\cdot1]$
ed: As_2O_3 solution	EJ	$-56\cdot4\pm[2\cdot0]$	V2	$8\cdot0\pm[0\cdot2]$	31/3	$-48\cdot4+[2\cdot0]$

$C_aH_bHal_j$

$$1 \text{ kcal} = 4 \cdot 184 \text{ kJ}$$

1 Formula	2 g.f.w.	3 Name	State	Purity mol %	Type	No. of expts.	Detn. of reactn.	$-\Delta H_r^c$ kcal/g.f.w.	R
$C_5H_{11}Br$	151·052	1-Bromopentane	l		RB	3	m	805·31±0·34	61
C_6F_6	186·057	Hexafluorobenzene	l	99·97	RB	5	m	584·03±0·29	64 69
C_6F_5Cl	202·512	Pentafluorochloro-benzene	l	99·94	RB	5	m	580·47±0·48	67 69
C_6Cl_6	284·785	Hexachlorobenzene	c	99·9	SB	9	m	567·7 ±1·0	58
C_6F_{10}	262·051	Decafluorocyclohexene	l	glc	RB	7	m	522·26±0·50	64
C_6HF_5	168·067	Pentafluorobenzene	l	99·87	RB	6	m	611·14±0·34	67 69
$C_6H_4F_2$	114·096	1, 2-Difluorobenzene	l	99·998	RB	6	m	707·65±0·13	62/
$C_6H_4F_2$	114·096	1, 3-Difluorobenzene	l	99·999	RB	7	m	704·31±0·17	62/
$C_6H_4F_2$	114·096	1, 4-Difluorobenzene	l		RB	6	m	704·72±0·17	62/
$C_6H_4Cl_2$ $C_6H_4Cl_2$ $C_6H_4Cl_2$	147·005	1, 2-Dichlorobenzene 1, 2-Dichlorobenzene 1, 2-Dichlorobenzene	l l l	99·9 99·88	RB SB SB	3 6 10	m m m	708·4 ±0·6 708·07±0·30 707·13±0·17	54/ 58/ 54/
$C_6H_4Cl_2$		1, 2-Dichlorobenzene							
$C_6H_4Cl_2$	147·005	1, 3-Dichlorobenzene	l	99·87	SB	6	m	706·44±0·25	54/
$C_6H_4Cl_2$	147·005	1, 4-Dichlorobenzene	c	99·55	SB	7	m	701·27±0·25	54/
$C_6H_4I_2$ $C_6H_4I_2$	329·908	1, 2-Di-iodobenzene 1, 2-Di-iodobenzene	l c		SB SB		m m	745·6 ±[1·0] 742·2 ±[1·0]	56/3 56/3

Determina

1 kcal = 4·184 kJ

ΔH_r° Remarks		5 ΔH_f° (l or c) kcal/g.f.w.	6 Determination of ΔH_v			7 ΔH_f° (g) kcal/g.f.w.
			Type	ΔH_v° kcal/g.f.w.	Ref.	
l: As_2O_3 solution	J	$-40\cdot68\pm0\cdot34$	Cl	$9\cdot83\pm0\cdot03$	66/14	$-30\cdot85\pm0\cdot35$
$_6(l)+4\cdot5O_2+123H_2O(l)$ $=6CO_2+6$ [HF.20H$_2$O](l)	J	$-237\cdot02\pm0\cdot29*$	Cl	$8\cdot53\pm0\cdot02$	65/44	$-228\cdot49\pm0\cdot29*$
$_5Cl(l)+4\cdot5O_2+253H_2O(l)$ $=6CO_2+[HCl.150H_2O](l)$ $+5[HF.20H_2O](l)$: Red: As_2O_3 solution		$-203\cdot34\pm0\cdot50*$	Cl	$9\cdot75\pm0\cdot10$	69/1	$-193\cdot59\pm0\cdot51*$
l: As_2O_3 solution	J	$-30\cdot6\pm1\cdot0$	S4	$22\pm[2]$ (sub.)	49/26	$-8\cdot6\pm[2\cdot3]$
$_{10}(l)+3\cdot5O_2+205H_2O(l)$ $=6CO_2+10[HF.20H_2O](l)$	J	$-470\cdot0\pm0\cdot5*$	V3	$7\cdot3\pm0\cdot2$	M/5	$-462\cdot7\pm0\cdot6*$
$_5H(l)+5O_2+102H_2O(l)$ $=6CO_2+5[HF.20H_2O](l)$		$-201\cdot30\pm0\cdot34*$	Cl	$8\cdot65\pm0\cdot05$	69/1	$-192\cdot65\pm0\cdot35*$
$H_4F_2(l)+6\cdot5O_2+59H_2O(l)$ $=6CO_2+2[HF.30H_2O](l)$		$-78\cdot91\pm0\cdot13*$	Cl	$8\cdot65\pm0\cdot02$	63/45	$-70\cdot26\pm0\cdot13*$
$H_4F_2(l)+6\cdot5O_2+59H_2O(l)$ $=6CO_2+2[HF.30H_2O](l)$		$-82\cdot25\pm0\cdot17*$	Cl	$8\cdot29\pm0\cdot02$	62/48	$-73\cdot96\pm0\cdot17*$
$H_4F_2(l)+6\cdot5O_2+59H_2O(l)$ $=6CO_2+2[HF.30H_2O](l)$		$-81\cdot84\pm0\cdot17*$	Cl	$8\cdot51\pm0\cdot02$	62/48	$-73\cdot33\pm0\cdot17*$
d: As_2O_3 solution	J	$-3\cdot9\pm0\cdot6$	V2	$11\cdot4\pm[0\cdot4]$	49/25	$+7\cdot5\pm[0\cdot7]$
d: As_2O_3 solution	J	$-4\cdot20\pm0\cdot34$				$+7\cdot2\pm[0\cdot5]$
$H_4Cl_2(l)+6\cdot5O_2+70H_2O(l)$ $=6CO_2+2[HCl.35\cdot5H_2O](l)$		$-4\cdot32\pm0\cdot20$				$+7\cdot1\pm[0\cdot4]$
d: $N_2H_4.2HCl$ solution Selected value		$-4\cdot28\pm0\cdot15$				$+7\cdot1\pm0\cdot4$
$H_4Cl_2(l)+6\cdot5O_2+70H_2O(l)$ $=6CO_2+2[HCl.35\cdot5H_2O](l)$ d: $N_2H_4.2HCl$ solution		$-5\cdot00\pm0\cdot23$	V2	$11\cdot1\pm[0\cdot4]$	49/25	$+6\cdot1\pm[0\cdot5]$
$H_4Cl_2(c)+6\cdot5O_2+70H_2O(l)$ $=6CO_2+2[HCl.35\cdot5H_2O](l)$ d: $N_2H_4.2HCl$ solution		$-10\cdot24\pm0\cdot25$	S3	$15\cdot5\pm0\cdot2$ (sub.)	61/43	$+5\cdot3\pm0\cdot3$
	E	$+44\cdot7\pm[1\cdot0]$	E1	$15\cdot5\pm1\cdot0$		$+60\cdot2\pm[1\cdot4]$
	E	$+41\cdot2\pm[1\cdot0]$				

$C_aH_bHal_j$

1 kcal = 4·184 kJ

1 Formula	2 g.f.w.	3 Name	State	Purity mol %	Type	No. of expts.	Detn. of reactn.	$-\Delta H_r^{\circ}$ kcal/g.f.w.	F
$C_6H_4I_2$	329·908	1, 3-Di-iodobenzene	c		SB		m	745·6 ±[1·0]	56
$C_6H_4I_2$	329·908	1, 4-Di-iodobenzene	c		SB		m	739·3 ±[1·0]	56
C_6H_5F	96·105	Fluorobenzene	l	99·95	RB	7	m	741·90±0·29	56
C_6H_5Cl	112·560	Chlorobenzene	l		SB		m	743·7 ±[2·0]	5
C_6H_5Cl		Chlorobenzene	l	99·99	SB	13	m	743·04±0·19	5
C_6H_5Cl		Chlorobenzene	l		R	6	an	32·0 ±0·9	56
C_6H_5Cl		Chlorobenzene	l	99·9, glc	RB	6	m	743·53±0·24	67
C_6H_5Cl		Chlorobenzene							
C_6H_5Br	157·016	Bromobenzene	l		R	3	an	78·5 ±0·8	51 56
C_6H_5I	204·011	Iodobenzene	l		R	3	m	37·2 ±0·7	51 56
C_6H_5I		Iodobenzene	l		SB		m	763·1 ±1·0	56
C_6H_5I		Iodobenzene							
$C_6H_{10}Br_2$	241·965	1, 2-Dibromocyclohexane	g		R	6	an	33·63±0·30	41
$C_6H_{10}Br_2$		1, 2-Dibromocyclohexane	l		RB	5	m	867·6 ±0·6	61
$C_6H_{10}Br_2$		1, 2-Dibromocyclohexane							
$C_6H_{11}Cl$	118·608	Chlorocyclohexane	l		R	4	an	34·2 ±0·3	56
$C_6H_{11}I$	210·059	Iodocyclohexane	g		R	9	an	7·8 ±2·0	56
$C_6H_{11}I$		Iodocyclohexane	l		SB		m	916·8 ±[1·0]	56
$C_6H_{11}I$		Iodocyclohexane							
$C_6H_{13}Cl$	120·624	2-Chlorohexane	l		R	3	an	33·4 ±0·3	56
$C_6H_{13}Br$	165·080	1-Bromohexane	l		RB	10	m	961·93±0·38	61

1 kcal = 4·184 kJ

H_r° Remarks		ΔH_f° (l or c) kcal/g.f.w.	6 Determination of ΔH_v			7 ΔH_f° (g) kcal/g.f.w.
		5	Type	ΔH_v° kcal/g.f.w.	Ref.	
	E	+44·7 ±[1·0]				
	E	+38·4 ±[1·0]				
$_5$F(l)+7O$_2$+58H$_2$O(l) = 6CO$_2$+[HF.60H$_2$O](l)		−36·03±0·29*	C1	8·27±0·02	56/32	−27·76±0·29*
As$_2$O$_3$ solution $_5$Cl(l)+7O$_2$+72H$_2$O(l) = 6CO$_2$+[HCl.74H$_2$O](l)	EJ	+3·0 ±[2·0] +2·50±0·19	C1	9·63±0·01	68/6	+12·6 ±[2·0] +12·13±0·19
N$_2$H$_4$.2HCl solution $_6$(l) {+11·72±0·13} +Cl$_2$(g) = C$_6$H$_5$Cl(l)+HCl(g)		+1·8 ±1·0				+11·4 ±1·0
As$_2$O$_3$ solution	J	+2·77±0·24				+12·40±0·24
Selected value		+2·58±0·16				+12·21±0·16
I$_5$)$_2$Hg(c) {+66·8±1·5} +2Br$_2$(l) = HgBr$_2$(c) {−40·6±0·2} +2C$_6$H$_5$Br(l)		+14·5 ±1·0	C1	10·47±0·03	68/6	+25·2 ±1·0
I$_5$)$_2$Hg(c) {+66·8±1·5} +2I$_2$(c) = HgI$_2$(c) {−25·2±0·2} +2C$_6$H$_5$I(l)		+27·4 ±1·0	E8	11·4 ±1·0		+38·8 ±1·4
	E	+28·0 ±1·0				+39·4 ±1·4
Selected value		+27·7 ±0·7				+39·1 ±1·3
H$_{10}$(g) {−1·08±0·18} +Br$_2$(g) = c-C$_6$H$_{10}$Br$_2$(g)		−39·39±0·46	V3	12·07±0·30	41/12	−27·32±0·35
As$_2$O$_3$ solution	J	−38·3 ±0·6				−26·2 ±0·7
Selected value		−38·8 ±0·4				−26·7 ±0·3
H$_{12}$(l) {−37·40±0·15} +Cl$_2$(g) = c-C$_6$H$_{11}$Cl(l)+HCl(g)		−49·54±0·34	E1	10·4 ±0·8		−39·1 ±[0·8]
$_6$H$_{11}$I(g)+HI(g) = c-C$_6$H$_{12}$(g) {−29·50±0·15} +I$_2$(g)		−24·4 ±2·1	V4	11·3 ±0·4	56/34	−13·1 ±2·0
	E	−23·2 ±[1·0]				−11·9 ±[1·1]
Selected value		−23·5 ±0·9				−12·2 ±1·0
$_5$H$_{14}$(l) {−47·46±0·18} +Cl$_2$(g) = C$_6$H$_{13}$Cl(l)+HCl(g) ...ture of 2- and 3-chlorohexanes formed; H_f° of these isomers expected to be similar		−58·8 ±0·4*	E1	9·6 ±0·3		−49·2 ±0·5*
: As$_2$O$_3$ solution	J	−46·42±0·38	C1	10·91±0·04	66/14	−35·51±0·39

$C_aH_bHal_j$ 1 kcal = 4·184 kJ

1 Formula	2 g.f.w.	3 Name	State	Purity mol %	Type	No. of expts.	Detn. of reactn.	$-\Delta H_r^\circ$ kcal/g.f.w.	
C_7F_{14}	350·056	Tetradecafluoromethyl- cyclohexane	l	glc	RB	10	m	440·93±0·11	5
C_7F_{16}	388·052	Hexadecafluoroheptane	l	99·8	RB	11	m	420·00±0·21	5
$C_7H_3F_5$	182·094	2, 3, 4, 5, 6-Pentafluoro- toluene	l	99·95	RB	5	m	763·47±0·36	6 6
$C_7H_4F_4$	164·104	Trifluoromethyl- 3-fluorobenzene	l	99·91	RB	10	m	767·61±0·14	5
$C_7H_5F_3$	146·113	Trifluoromethylbenzene	l	99·999	RB	6	m	805·75±0·39	5
$C_7H_5F_3$		Trifluoromethylbenzene	l	99·999	RB	6	m	805·29±0·12	6
$C_7H_5F_3$		Trifluoromethylbenzene							
C_7H_7F	110·132	4-Fluorotoluene	l	99·92	RB	6	m	895·60±0·17	62
C_7H_7Cl	126·587	Benzyl chloride	l		R	2	an	32·8 ±0·6	56
C_7H_7Br	171·043	Benzyl bromide	l		R	4	an	1·9 ±0·3	49
C_7H_7Br		Benzyl bromide	l		R	2	m	80·8 ±2·6	61
C_7H_7Br		Benzyl bromide	g		E	5	an	8·1 ±1·0	57
C_7H_7Br		Benzyl bromide	l		H	7	m	0·9 ±0·5	63
C_7H_7Br		Benzyl bromide							
C_7H_7I	218·038	Benzyl iodide	l		R	5	an	−3·0 ±0·4	49
C_7H_7I		Benzyl iodide	l		H	8	m	9·7 ±0·8	63
C_7H_7I		Benzyl iodide							
C_7H_7I	218·038	2-Iodotoluene	l		SB		m	916·2 ±[1·0]	56

1 kcal = 4·184 kJ

$\Delta H_r°$ Remarks	5 $\Delta H_f°$ (l or c) kcal/g.f.w.	6 Determination of ΔH_v			7 $\Delta H_f°$ (g) kcal/g.f.w.
		Type	$\Delta H_v°$ kcal/g.f.w.	Ref.	
$_{14}$(l)$+3\cdot5O_2+31\cdot5H_2O$(l) $= 4\cdot25CO_2+2\cdot75CF_4$(g) $\{-223\cdot3\pm0\cdot4\}$ $+3[HF.10H_2O]$(l)	$-701\cdot1 \pm1\cdot2$*	V1	$8\cdot11\pm0\cdot05$	59/9	$-693\cdot0 \pm1\cdot2$*
$_{16}$(l)$+3O_2+63H_2O$(l) $= 4\cdot5CO_2+2\cdot5CF_4 \{-223\cdot3\pm0\cdot4\}$ $+6[HF.10H_2O]$(l)	$-818\cdot0 \pm1\cdot0$*	V1	$8\cdot69\pm[0\cdot10]$	51/24	$-809\cdot3 \pm1\cdot1$*
$_3F_5$(l)$+6\cdot5O_2+101H_2O$(l) $= 7CO_2+5[HF.20H_2O]$(l)	$-211\cdot32\pm0\cdot36$*	Cl	$9\cdot78\pm0\cdot10$	69/1	$-201\cdot54\pm0\cdot38$*
$_4F_4$(l)$+7O_2+60H_2O$(l) $= 7CO_2+4[HF.15H_2O]$(l)	$-198\cdot51\pm0\cdot14$*	V1	$9\cdot07\pm0\cdot05$	59/9	$-189\cdot44\pm0\cdot15$*
$_5F_3$(l)$+7\cdot5O_2+59H_2O$(l) $= 7CO_2+3[HF.20H_2O]$(l)	$-151\cdot77\pm0\cdot39$*	Cl	$8\cdot98\pm0\cdot03$	59/30	$-142\cdot79\pm0\cdot40$*
$_5F_3$(l)$+7\cdot5O_2+59H_2O$(l) $= 7CO_2+3[HF.20H_2O]$(l)	$-152\cdot23\pm0\cdot12$*				$-143\cdot25\pm0\cdot13$*
Selected value	$-152\cdot18\pm0\cdot11$*				$-143\cdot20\pm0\cdot12$*
$_7F$(l)$+8\cdot5O_2+67H_2O$(l) $= 7CO_2+[HF.70H_2O]$(l)	$-44\cdot70\pm0\cdot17$*	Cl	$9\cdot42\pm0\cdot02$	62/49	$-35\cdot28\pm0\cdot17$*
$_5CH_3$(l) $\{+2\cdot91\pm0\cdot10\} +Cl_2$(g) $= C_7H_7Cl$(l)$+HCl$(g)	$-7\cdot8 \pm0\cdot6$	E7	$12\cdot3 \pm0\cdot4$		$+4\cdot5 \pm0\cdot8$
$_7Br$(l)$+12H_2O$(l) $= [C_7H_7OH+HBr] (11H_2O)$ (l) $\{-66\cdot5\pm0\cdot5\}$	$+3\cdot7 \pm0\cdot6$*	V5	$11\cdot3 \pm[1\cdot0]$	57/38	$+15\cdot0 \pm[1\cdot2]$
$_9Li$(l) $\{-31\cdot6\pm0\cdot8\} +C_7H_7Br$(l) $= LiBr$(c) $\{-83\cdot7\}$ $+C_{11}H_{16}$(l) $\{[-21\cdot4\pm0\cdot5]\}$	$+7\cdot3 \pm3\cdot0$				$+18\cdot6 \pm3\cdot2$
$_5CH_3$(g) $\{+11\cdot99\pm0\cdot10\} +Br_2$(g) $\rightleftharpoons C_7H_7Br(g)+HBr$(g). 3rd Law	$+8\cdot7 \pm[1\cdot4]$				$+20\cdot0 \pm1\cdot1$
$_7Br$(l)$+0\cdot5H_2$ $= C_7H_8$(l) $\{+2\cdot91\pm0\cdot10\} +0\cdot5Br_2$(l)	$+3\cdot8 \pm0\cdot5$				$+15\cdot1 \pm[1\cdot1]$
Selected value	$+5\cdot6 \pm1\cdot8$				$+16\cdot9 \pm1\cdot8$
$_7I$(l)$+8H_2O$(l) $= [C_7H_7OH+HI] (7H_2O)$ (l) $\{-50\cdot3\pm0\cdot6\}$	$+15\cdot0 \pm1\cdot0$*	V5	$11\cdot3 \pm[1\cdot0]$	57/38	$+26\cdot3 \pm[1\cdot4]$
$_7I$(l)$+0\cdot5H_2$ $= C_7H_8$(l) $\{+2\cdot91\pm0\cdot10\} +0\cdot5I_2$(c)	$+12\cdot6 \pm0\cdot9$				$+23\cdot9 \pm[1\cdot4]$
Selected value	$+13\cdot8 \pm1\cdot2$				$+25\cdot1 \pm1\cdot5$
E	$+18\cdot7 \pm[1\cdot0]$	E7	$13\cdot0 \pm1\cdot0$		$+31\cdot7 \pm[1\cdot4]$

THERMOCHEMISTRY OF ORGANIC AND ORGANOMETALLIC COMPOUNDS

$C_aH_bHal_j$ 1 kcal = 4·184 kJ

Determina

1 Formula	2 g.f.w.	3 Name	State	Purity mol %	Type	No. of expts	Detn. of reactn.	$-\Delta H_r^\circ$ kcal/g.f.w.	
C_7H_7I	218·038	3-Iodotoluene	l		SB		m	916·4 ±[1·0]	56
C_7H_7I	218·038	4-Iodotoluene	l		SB		m	913·6 ±[1·0]	56
$C_7H_{12}Br_2$	255·992	1, 2-Dibromocyclo-heptane	g		R	4	an	30·44±0·30	41
$C_7H_{14}Br_2$	258·008	1, 2-Dibromoheptane	g		R	4	an	30·24±0·30	41
$C_7H_{15}Br$	179·107	1-Bromoheptane	l		RB	4	m	1118·51±0·38	61
C_8F_{16}	400·064	Hexadecafluoroethyl-cyclohexane	l	glc	RB	10	m	501·68±0·23	59
$C_8H_6Cl_2$	173·043	2, 5-Dichlorostyrene	l	99·5	SB	6	m	977·23±0·44	58
$C_8H_6Cl_4$	243·949	1, 2, 4, 5-Tetrachloro-dimethylbenzene	c		RB	8	m	936·87±0·21	64
$C_8H_6Cl_4$		1, 2, 4, 5-Tetrachloro-dimethylbenzene	c		RB	8	m	937·84±0·14	64
$C_8H_6Cl_4$		1, 2, 4, 5-Tetrachloro-dimethylbenzene							
C_8H_9Cl	140·614	1-Chloro-2-ethylbenzene	l	99·78	SB	8	m	1052·40±0·30	54
C_8H_9Cl	140·614	1-Chloro-4-ethylbenzene	l	99·84	SB	7	m	1052·96±0·31	54
$C_8H_{14}Br_2$	270·019	1, 2-Dibromocyclo-octane	g		R	2	an	29·31±0·30	41
$C_8H_{17}Br$	193·134	1-Bromo-octane	l		RB	7	m	1274·52±0·54	61
$C_{10}H_7Cl$	162·620	1-Chloronaphthalene	l		SB		m	1198·3 ±[2·0]	53
$C_{10}H_7Cl$	162·620	2-Chloronaphthalene	c		SB		m	1198·5 ±[2·0]	53
$C_{10}H_7I$	254·072	1-Iodonaphthalene	l		SB		m	1218·2 ±[1·5]	56

1 kcal = 4·184 kJ

ΔH_r° Remarks		5 ΔH_f° (l or c) kcal/g.f.w.	6 Determination of ΔH_v			7 ΔH_f° (g) kcal/g.f.w.
			Type	ΔH_v° kcal/g.f.w.	Ref.	
	E	+18·9 ±[1·0]	E7	13·0 ±1·0		+31·9 ±[1·4]
	E	+16·1 ±[1·0]	E7	13·0 ±1·0		+29·1 ±[1·4]
$C_7H_{12}(g)\ \{-2\cdot19\pm0\cdot24\}\ +Br_2(g)$ $= C_7H_{12}Br_2(g)$		−37·67±0·49	V3	12·43±0·30	41/12	−25·24±0·38
$C_7H_{14}(g)\ \{-14\cdot81\pm0\cdot25\}\ +Br_2(g)$ $= C_7H_{14}Br_2(g)$		−50·67±0·52	V3	13·01±0·30	41/12	−37·66±0·43
d: As_2O_3 solution	J	−52·21±0·38	C1	12·05±0·04	66/14	−40·16±0·39
$F_{16}(l)+4O_2+31\cdot5H_2O(l)$ $= 4\cdot75CO_2+3\cdot25CF_4(g)\ \{-223\cdot3\pm0\cdot4\}$ $+3[HF.10H_2O](l)$		−799·1 ±1·3*	V1	9·20±0·05	59/9	−789·9 ±1·3*
d: As_2O_3 solution	J	+8·55±0·44				
$H_6Cl_4(c)+8\cdot5O_2+339H_2O(l)$ $= 8CO_2+4[HCl.85H_2O](l)$ d: $N_2H_4.2HCl$ solution d: As_2O_3 solution	J	−42·37±0·21* −41·44±0·14				
Selected value		−41·44±0·14				
$H_9Cl(l)+10O_2+101H_2O(l)$ $= 8CO_2+[HCl.105H_2O](l)$ d: $N_2H_4.2HCl$ solution		−12·93±0·30*	V2	11·3 ±[0·5]	49/25	−1·6 ±[0·6]
$H_9Cl(l)+10O_2+101H_2O(l)$ $= 8CO_2+[HCl.105H_2O](l)$ d: $N_2H_4.2HCl$ solution		−12·37±0·31*	V2	11·5 ±[0·5]	49/25	−0·9 ±[0·6]
$C_8H_{14}(g)\ \{-6\cdot45\pm0\cdot30\}\ +Br_2(g)$ $= C_8H_{14}Br_2(g)$		−41·41±0·52	V3	13·04±0·30	41/12	−28·37±0·43
d: As_2O_3 solution	J	−58·57±0·54	E1	13·14±0·10		−45·43±0·55
d: As_2O_3 solution	EJ	+13·0 ±[2·0]	E7	15·6 ±1·2		+28·6 ±[2·3]
d: As_2O_3 solution	EJ	+13·2 ±[2·0]	E7$_{me}$	19·6 ±1·4 (sub.)		+32·8 ±[2·4]
	E	+38·6 ±[1·5]	E7	17·3 ±1·4		+55·9 ±[2·1]

$C_aH_bHal_j$ 1 kcal = 4·184 kJ

1 Formula	2 g.f.w.	3 Name	State	Purity mol %	Type	No. of expts.	Detn. of reactn.	$-\Delta H_r^\circ$ kcal/g.f.w.	Re
$C_{10}H_7I$	254·072	2-Iodonaphthalene	c		SB		m	1214·1 ±[1·5]	56/
$C_{12}F_{22}$	562·099	Dodecafluorobicyclo-hexyl	c	99·98	RB	3	m	832·46±0·40	65/
$C_{12}F_{22}$		Dodecafluorobicyclo-hexyl	c	99·98	RB	5	m	969·00±0·20	65/
$C_{12}F_{22}$		Dodecafluorobicyclohexyl							
$C_{12}H_8F_2$	190·194	2, 2'-Difluorobiphenyl	c	glc	RB	8	m	1416·79±0·15	64/
$C_{12}H_8F_2$	190·194	4, 4'-Difluorobiphenyl	c	glc	RB	6	m	1416·61±0·22	64/
$C_{12}H_8Cl_2$	223·104	2, 2'-Dichlorobiphenyl	c	99·5, glc	RB	8	m	1420·05±0·41	64/
$C_{12}H_8Cl_2$	223·104	4 4'-Dichlorobiphenyl	c	glc	RB	8	m	1416·61±0·15	64/

Determinati

1 kcal = 4·184 kJ

ΔH_r°		5 ΔH_f° (l or c) kcal/g.f.w.	6 Determination of ΔH_v			7 ΔH_f° (g) kcal/g.f.w.
Remarks			Type	ΔH_v° kcal/g.f.w.	Ref.	
	E	+34·5 ±[1·5]	E7$_{me}$	21·7 ±1·6 (sub.)		+56·2 ±[2·2]
$_2F_{22}(c)+6·5O_2+131·2H_2O(l)$ $= 8·1CO_2+3·9CF_4(g) \{-223·3\pm0·4\}$ $+6·4[HF.20H_2O](l)$	J	−1074·1 ±1·6*				
$_2F_{22}(c)+6·5O_2+401·8H_2O(l)$ $= 11·4CO_2+0·6CF_4(g) \{-223·3\pm0·4\}$ $+19·6[HF.20H_2O](l)$	J	−1075·9 ±0·3*				
	Selected value	−1075·8 ±1·0*				
$_2H_8F_2(c)+13·5O_2+97H_2O(l)$ $= 12CO_2+2[HF.50H_2O](l)$		−70·73±0·15*	S3	22·7 ±1·0 (sub.)	64/6	−48·0 ±1·0*
$_2H_8F_2(c)+13·5O_2+97H_2O(l)$ $= 12CO_2+2[HF.50H_2O](l)$		−70·91±0·22*	S4	21·8 ±1·0 (sub.)	64/6	−49·1 ±1·1*
$_2H_8Cl_2(c)+13·5O_2+87H_2O(l)$ $= 12CO_2+2[HCl.45H_2O](l)$ d: $N_2H_4.2HCl$ solution		+7·49±0·41*	S4	23·0 ±1·0 (sub.)	64/6	+30·5 ±1·1*
$_2H_8Cl_2(c)+13·5O_2+87H_2O(l)$ $= 12CO_2+2[HCl.45H_2O](l)$ d: $N_2H_4.2HCl$ solution		+4·05±0·15*	S4	24·8 ±1·0 (sub.)	64/6	+28·9 ±1·0*

$C_aH_bOHal_j$ 1 kcal = 4·184 kJ

1 Formula	2 g.f.w.	3 Name	State	Purity mol %	Type	No. of expts.	Detn. of reactn.	$-\Delta H_r^{\circ}$ kcal/g.f.w.	Re
COF_2	66·007	Carbonyl fluoride	g	95·5	R	9	an	26·73±0·25	49/
$COCl_2$	98·917	Carbonyl chloride	g		E			25·9 ±0·1	61/
$COBr_2$	187·829	Carbonyl bromide	g		E	13	an	1·04±0·10	31/
C_2OCl_4	181·834	Trichloroacetyl chloride	l		R	5	m	28·1 ±0·1	50/
C_2HOCl_3	147·389	Dichloroacetyl chloride	l		R	6	m	25·7 ±0·1	50/
C_2HOCl_3	147·389	Trichloroacetaldehyde	l		R	4	m	24·58±0·05	50/
$C_2H_2OCl_2$	112·944	Chloroacetyl chloride	l		R	11	m	22·5 ±0·1	50/
C_2H_3OF	62·044	Acetyl fluoride	l		R	4	m	43·2 ±0·1	50/
C_2H_3OCl	78·499	Acetyl chloride	l	99·8, an	R	9	m	22·06±0·10	50/
C_2H_3OBr	122·955	Acetyl bromide	l	>99·8, an	R	6	m	23·03±0·14	49/
C_2H_3OI	169·950	Acetyl iodide	l	99·5, an	R	5	m	21·79±0·07	49/
C_2H_3OI		Acetyl iodide	g		E	2	uv	-0·74±0·50	66/
C_2H_3OI		Acetyl iodide							
$C_2H_3OF_3$	100·041	2, 2, 2-Trifluoroethanol	l	>98·5, an	SB	6	an	211·6 ±0·5	65/
C_3H_5OCl	92·526	Chloromethyloxirane	l		RB	2	m	423·13±0·10	54/
$C_3H_6OCl_2$	128·987	1, 3-Dichloropropan-2-ol	l		RB	3	m	406·3 ±0·5	54/

1 kcal = 4·184 kJ

ΔH_r° Remarks	5 ΔH_f° (l or c) kcal/g.f.w.	6 Determination of ΔH_v			7 ΔH_f° (g) kcal/g.f.w.
		Type	ΔH_v° kcal/g.f.w.	Ref.	
$OF_2(g)+71H_2O(l) = CO_2+2[HF.35H_2O](l)$					$-152\cdot95\pm0\cdot25$*
$O(g)\ \{-26\cdot42\pm0\cdot04\} +Cl_2(g) \rightleftharpoons COCl_2(g)$ 2nd Law					$-52\cdot3\ \pm0\cdot1$*
$O(g)\ \{-26\cdot42\pm0\cdot04\} +Br_2(g) \rightleftharpoons COBr_2(g)$ 2nd Law					$-20\cdot1\ \pm0\cdot1$*
$Cl_3COCl(l)+1301H_2O(l)$ $= [CCl_3CO_2H\{-120\cdot4\pm2\cdot0\}+HCl]$ $(1300H_2O)$ (l)	$-66\cdot4\ \pm2\cdot0$	E7	$9\cdot8\ \pm0\cdot5$		$-56\cdot6\ \pm2\cdot1$
$HCl_2COCl(l)+1301H_2O(l)$ $= [CHCl_2CO_2H\{-118\cdot6\pm2\cdot0\}+HCl]$ $(1300H_2O)$ (l)	$-67\cdot1\ \pm2\cdot0$	E7	$9\cdot4\ \pm0\cdot5$		$-57\cdot7\ \pm2\cdot1$
$Cl_3CHO(l)+[NaOH.275H_2O](l)\ \{-112\cdot2\}$ $= [CHCl_3\{-34\cdot1\pm2\cdot0\}$ $+HCO_2Na\{-158\cdot8\}]\ (275H_2O)$ (l)	$-56\cdot1\ \pm2\cdot0$				
$H_2ClCOCl(l)+1301H_2O(l)$ $= [CH_2ClCO_2H\{-122\cdot1\pm2\cdot0\}+HCl]$ $(1300H_2O)$ (l)	$-68\cdot0\ \pm2\cdot0$	E7	$9\cdot3\ \pm0\cdot5$		$-58\cdot7\ \pm2\cdot1$
$H_3COF(l)+15[NaOH.275H_2O](l)\ \{-112\cdot2\}$ $= [CH_3CO_2Na\{-173\cdot9\}+NaF\{-137\cdot8\}$ $+13NaOH]\ (4126H_2O)$ (l)	$-112\cdot4\ \pm0\cdot1$*	E7	$6\cdot0\ \pm0\cdot5$		$-106\cdot4\ \pm0\cdot5$*
$H_3COCl(l)+1301H_2O(l)$ $= [CH_3CO_2H\{-116\cdot05\pm0\cdot10\}+HCl]$ $(1300H_2O)$ (l)	$-65\cdot6\ \pm0\cdot14$	C2	$7\cdot2\ \pm[0\cdot1]$	31/3	$-58\cdot4\ \pm0\cdot2$
$H_3COBr(l)+121H_2O(l)$ $= [CH_3CO_2H+HBr]\ (120H_2O)$ (l)	$-53\cdot5\ \pm0\cdot2$*	C2	$7\cdot9\ \pm[0\cdot1]$	26/2	$-45\cdot6\ \pm0\cdot3$*
$H_3COI(l)+141H_2O(l)$ $= [CH_3CO_2H+HI]\ (140H_2O)$ (l)	$-39\cdot0\ \pm0\cdot13$*	E7	$9\cdot2\ \pm0\cdot8$		$-29\cdot8\ \pm0\cdot9$
$H_3CHO(g)\ \{-39\cdot73\pm0\cdot12\} +I_2(g)$ $\rightleftharpoons CH_3COI(g)+HI(g)$. 3rd Law	$-39\cdot6\ \pm1\cdot0$				$-30\cdot40\pm0\cdot52$
Selected value	$-39\cdot3\ \pm0\cdot9$				$-30\cdot1\ \pm0\cdot4$
$H_3OF_3(l)+1\cdot5O_2+90H_2O(l)$ $= 2CO_2+3[HF.30H_2O]$ (l)	$-207\cdot4\ \pm0\cdot5$*				
d: As_2O_3 solution EJ	$-35\cdot48\pm0\cdot14$	E7	$9\cdot7\ \pm1\cdot0$		$-25\cdot8\ \pm[1\cdot0]$
d: As_2O_3 solution EJ	$-92\cdot1\ \pm0\cdot5$	E7	$16\cdot0\ \pm1\cdot0$		$-76\cdot1\ \pm[1\cdot1]$

$C_aH_bOHal_j$ 1 kcal = 4·184 kJ

1 Formula	2 g.f.w.	3 Name	State	Purity mol %	Type	No. of expts.	Detn. of reactn.	$-\Delta H_r^\circ$ kcal/g.f.w.	R
$C_3H_6OCl_2$	128·987	2, 3-Dichloropropan-2-ol	l		RB	3	m	407·2 ±0·5	54/
C_6HOF_5	184·066	Pentafluorophenol	c	99·97	RB	6	m	567·57±0·47	67/ 69/
C_6HOCl_5	266·339	Pentachlorophenol	c	99·7	SB	5	m	556·8 ±0·7	58/
C_6H_5OCl	128·559	3-Chlorophenol	c		SB		m	691·4 ±[2·0]	53/
C_6H_5OCl		3-Chlorophenol	l		SB		m	695·5 ±[2·0]	53/
C_6H_5OCl	128·559	4-Chlorophenol	c		SB		m	693·5 ±[2·0]	53/
C_6H_5OCl		4-Chlorophenol	l		SB		m	697·4 ±[2·0]	53/
C_6H_5OI	220·011	2-Iodophenol	c		SB		m	712·2 ±[1·0]	56/
C_6H_5OI	220·011	3-Iodophenol	c		SB		m	712·5 ±[1·0]	56/
C_6H_5OI	220·011	4-Iodophenol	c		SB		m	712·3 ±[1·0]	56/
C_7H_5OCl	140·570	2-Chlorobenzaldehyde	l		SB		m	806·5 ±[2·0]	53/
C_7H_5OCl	140·570	3-Chlorobenzaldehyde	l		SB		m	804·7 ±[2·0]	53/
C_7H_5OCl	140·570	4-Chlorobenzaldehyde	c		SB		m	799·8 ±[2·0]	53/
C_7H_5OCl	140·570	Benzoyl chloride	l		R	5	m	24·35±0·10	50/
C_7H_5OBr	185·026	Benzoyl bromide	l		R	5	m	27·04±0·10	50/
C_7H_5OI	232·022	Benzoyl iodide	l		R	6	m	24·47±0·09	50/

1 kcal = 4·184 kJ

ΔH_r°		5 ΔH_f° (l or c) kcal/g.f.w.	6 Determination of ΔH_v			7 ΔH_f° (g) kcal/g.f.w.
Remarks			Type	ΔH_v° kcal/g.f.w.	Ref.	
ed: As$_2$O$_3$ solution	EJ	$-91\cdot2\ \pm0\cdot5$	E7	$15\cdot6\ \pm1\cdot0$		$-75\cdot6\ \pm[1\cdot1]$
HOF$_5$(c)$+4\cdot5$O$_2+102$H$_2$O(l) $=6$CO$_2+5$[HF.20H$_2$O] (l)		$-244\cdot86\pm0\cdot47*$	S4	$16\cdot1\ \pm0\cdot4$ (sub.)	69/1	$-228\cdot8\ \pm0\cdot6*$
ed: As$_2$O$_3$ solution	J	$-70\cdot0\ \pm0\cdot7$	S4	$16\cdot1\ \pm[0\cdot5]$ (sub.)	M/4	$-53\cdot9\ \pm0\cdot9$
ed: As$_2$O$_3$ solution	EJ	$-49\cdot4\ \pm[2\cdot0]$	S4	$12\cdot7\ \pm0\cdot6*$ (sub.)	38/14	$-36\cdot7\ \pm[2\cdot1]$
ed: As$_2$O$_3$ solution	EJ	$-45\cdot3\ \pm[2\cdot0]$				
ed: As$_2$O$_3$ solution	EJ	$-47\cdot3\ \pm[2\cdot0]$	S4	$12\cdot4\ \pm0\cdot6*$ (sub.)	38/14	$-34\cdot9\ \pm[2\cdot1]$
ed: As$_2$O$_3$ solution	EJ	$-43\cdot4\ \pm[2\cdot0]$				
	E	$-22\cdot9\ \pm[1\cdot0]$				
	E	$-22\cdot6\ \pm[1\cdot0]$				
	E	$-22\cdot8\ \pm[1\cdot0]$				
ed: As$_2$O$_3$ solution	EJ	$-28\cdot4\ \pm[2\cdot0]$	V2	$13\cdot3\ \pm0\cdot6$	49/25	$-15\cdot1\ \pm[2\cdot1]$
ed: As$_2$O$_3$ solution	EJ	$-30\cdot2\ \pm[2\cdot0]$				
ed: As$_2$O$_3$ solution	EJ	$-35\cdot1\ \pm[2\cdot0]$				
H$_5$COCl(l)$+1001$H$_2$O(l) $=$ C$_6$H$_5$CO$_2$H(c) $\{-91\cdot99\pm0\cdot05\}$ $+$[HCl.1000H$_2$O] (l)		$-39\cdot17\pm0\cdot11$	E7	$13\cdot1\ \pm1\cdot0$		$-26\cdot1\ \pm1\cdot0$
H$_5$COBr(l)$+1001$H$_2$O(l) $=$ C$_6$H$_5$CO$_2$H(c) $\{-91\cdot99\pm0\cdot05\}$ $+$[HBr.1000H$_2$O] (l)		$-25\cdot58\pm0\cdot11$	E1	$14\cdot0\ \pm1\cdot5$		$-11\cdot6\ \pm1\cdot5$
H$_5$COI(l)$+1001$H$_2$O(l) $=$ C$_6$H$_5$CO$_2$H(c) $\{-91\cdot99\pm0\cdot05\}$ $+$[HI.1000H$_2$O] (l)		$-12\cdot31\pm0\cdot10$	E1	$14\cdot8\ \pm1\cdot5$		$+2\cdot5\ \pm1\cdot5$

$C_aH_bO_2Hal_j$ 1 kcal = 4·184 kJ

1 Formula	2 g.f.w.	3 Name	State	Purity mol %	No. of Type expts.	Detn. of reactn.	$-\Delta H_r^\circ$ kcal/g.f.w.	Ref.
$C_2O_2Cl_2$	126·927	Oxalyl chloride	l	99·8	R 3	m, an	99·7 ±0·5	67/3
$C_2HO_2F_3$	114·024	Trifluoroacetic acid	l		SB 7	an	95·4 ±0·8	66/3
$C_2HO_2Cl_3$	163·388	Trichloroacetic acid	c		SB	m	118·9 ±[2·0]	53/
$C_2H_2O_2Cl_2$	128·943	Dichloroacetic acid	l		SB	m	149·1 ±[2·0]	53/
$C_2H_3O_2Cl$	94·498	Chloroacetic acid	c		SB	m	174·2 ±[2·0]	53/
$C_3H_5O_2Cl$	108·525	2-Chloropropionic acid	l		SB	m	333·7 ±[2·0]	53/
$C_3H_5O_2Cl$	108·525	3-Chloropropionic acid	c		SB	m	327·3 ±[2·0]	53/
$C_3H_5O_2I$	199·977	3-Iodopropionic acid	c		SB 4	m	343·0 ±1·2	44/
$C_3H_7O_2Cl$	110·541	1-Chloropropan-2, 3-diol	l		RB 3	m	401·34±0·20	54/
$C_3H_7O_2Cl$	110·541	2-Chloropropan-1, 3-diol	l		RB 3	m	403·21±0·26	54/
$C_4H_7O_2Cl$	122·552	2-Chlorobutyric acid	l		SB	m	483·4 ±[2·0]	53/
$C_4H_7O_2Cl$	122·552	3-Chlorobutyric acid	l		SB	m	488·0 ±[2·0]	53/4
$C_4H_7O_2Cl$	122·552	4-Chlorobutyric acid	l		SB	m	485·6 ±[2·0]	53/4
$C_5H_5O_2Cl_3$	203·453	Allyl trichloroacetate	l		SB	m	563·5 ±[2·0]	53/4
$C_5H_7O_2Cl_3$	205·469	n-Propyl trichloroacetate	l		SB	m	603·7 ±[2·0]	53/4
$C_5H_7O_2Cl_3$	205·469	Isopropyl trichloroacetate	l		SB	m	598·2 ±[2·0]	53/4
$C_5H_9O_2Cl$	136·579	n-Propyl chloroacetate	l		RB 3	m	660·1 ±0·5	54/3
$C_6O_2Cl_4$	245·878	Tetrachlorobenzoquinone	c		SB	m	518·9 ±[2·0]	53/4
$C_6HO_2Cl_3$	211·433	Trichlorobenzoquinone	c		SB	m	551·0 ±[2·0]	53/4
$C_6H_2O_2Cl_2$	176·988	2, 3-Dichlorobenzo- quinone	c		SB	m	584·6 ±[2·0]	53/4

$$1 \text{ kcal} = 4 \cdot 184 \text{ kJ}$$

ΔH_r°		5 ΔH_f° (l or c) kcal/g.f.w.	6 Determination of ΔH_v			7 ΔH_f° (g) kcal/g.f.w.
Remarks			Type	ΔH_v° kcal/g.f.w.	Ref.	
$_2O_2Cl_2(l)+4[KOH.30H_2O]$ $= [K_2CO_3\{-121\cdot0\}+2KCl\{-100\cdot1\}]$ $(122H_2O)$ $(l)+CO(g)$ $\{-26\cdot42\pm0\cdot04\}$		$-85\cdot6 \pm0\cdot6$	E4	$7\cdot6 \pm1\cdot0$		$-78\cdot0 \pm1\cdot2$
$_2HO_2F_3(l)+0\cdot5O_2+361H_2O(l)$ $= 2CO_2+3[HF.120H_2O]$ (l)		$-255\cdot4 \pm0\cdot8*$				
.ed: As_2O_3 solution	BEJ	$-120\cdot4 \pm[2\cdot0]$				
.ed: As_2O_3 solution	EJ	$-118\cdot6 \pm[2\cdot0]$				
.ed: As_2O_3 solution	EJ	$-122\cdot1 \pm[2\cdot0]$	$V2_m$	$18\pm[1]$ (sub.)	49/25, 28/1	$-104\cdot1 \pm[2\cdot3]$
.ed: As_2O_3 solution	EJ	$-125\cdot0 \pm[2\cdot0]$				
.ed: As_2O_3 solution	EJ	$-131\cdot4 \pm[2\cdot0]$				
	E	$-109\cdot9 \pm1\cdot2$				
.ed: As_2O_3 solution	EJ	$-125\cdot58\pm0\cdot24$				
.ed: As_2O_3 solution	EJ	$-123\cdot71\pm0\cdot30$				
.ed: As_2O_3 solution	EJ	$-137\cdot6 \pm[2\cdot0]$				
.ed: As_2O_3 solution	EJ	$-133\cdot0 \pm[2\cdot0]$				
.ed: As_2O_3 solution	EJ	$-135\cdot4 \pm[2\cdot0]$				
.ed: As_2O_3 solution	EJ	$-94\cdot5 \pm[2\cdot0]$	E7	$12\cdot5 \pm1\cdot0$		$-82\cdot0 \pm[2\cdot3]$
.ed: As_2O_3 solution	EJ	$-122\cdot7 \pm[2\cdot0]$	E7	$12\cdot7 \pm1\cdot0$		$-110\cdot0 \pm[2\cdot3]$
.ed: As_2O_3 solution	BEJ	$-128\cdot2 \pm[2\cdot0]$	E7	$12\cdot4 \pm1\cdot0$		$-115\cdot8 \pm[2\cdot3]$
.ed: As_2O_3 solution	EJ	$-123\cdot3 \pm0\cdot6$	E7	$11\cdot6 \pm1\cdot0$		$-111\cdot7 \pm1\cdot1$
.ed: As_2O_3 solution	BEJ	$-69\cdot0 \pm[2\cdot0]$	S4	$23\cdot6 \pm[2\cdot0]$ (sub.)	27/3	$-45\cdot4 \pm[2\cdot8]$
.ed: As_2O_3 solution	BEJ	$-64\cdot5 \pm[2\cdot0]$	S4	$21\cdot2 \pm[2\cdot0]$ (sub.)	27/3	$-43\cdot3 \pm[2\cdot8]$
.ed: As_2O_3 solution	BEJ	$-59\cdot3 \pm[2\cdot0]$				

THERMOCHEMISTRY OF ORGANIC AND ORGANOMETALLIC COMPOUNDS

$C_6H_bO_2Hal_j$

1 kcal = 4·184 kJ

Determination

1 Formula	2 g.f.w.	3 Name	State	Purity mol %	Type	No. of expts.	Detn. of reactn.	$-\Delta H_r^\circ$ kcal/g.f.w.	Ref
$C_6H_2O_2Cl_2$	176·988	2, 5-Dichlorobenzo-quinone	c		SB		m	585·3 ±[2·0]	53/
$C_6H_2O_2Cl_2$	176·988	2, 6-Dichlorobenzo-quinone	c		SB		m	585·5 ±[2·0]	53/
$C_6H_2O_2Cl_4$	247·894	1, 4-Dihydroxytetra-chlorobenzene	c		SB		m	546·8 ±[2·0]	53/
$C_6H_3O_2Cl$	142·543	Chlorobenzoquinone	c		SB		m	619·7 ±[2·0]	53/
$C_6H_3O_2Cl_3$	213·449	1, 4-Dihydroxy-trichlorobenzene	c		SB		m	578·4 ±[2·0]	53/
$C_6H_4O_2Cl_2$	179·004	1, 4-Dihydroxy-2, 3-dichlorobenzene	c		SB		m	612·8 ±[2·0]	53/
$C_6H_4O_2Cl_2$	179·004	1, 4-Dihydroxy-2, 5-dichlorobenzene	c		SB		m	610·1 ±[2·0]	53/
$C_6H_4O_2Cl_2$	179·004	1, 4-Dihydroxy-2, 6-dichlorobenzene	c		SB		m	611·0 ±[2·0]	53/
$C_6H_5O_2Cl$	144·559	1, 4-Dihydroxy-2-chlorobenzene	c		SB		m	649·2 ±[2·0]	53/
$C_6H_9O_2Cl_3$	219·496	n-Butyl trichloroacetate	l		SB		m	758·2 ±[2·0]	35/1 53/4
$C_6H_9O_2Cl_3$	219·496	Isobutyl trichloroacetate	l		SB		m	756·4 ±[2·0]	35/1 53/4
$C_6H_{10}O_2Cl_2$	185·051	n-Butyl dichloroacetate	l		SB		m	785·7 ±[2·0]	35/1 53/4
$C_6H_{10}O_2Cl_2$	185·051	Isobutyl dichloroacetate	l		SB		m	784·8 ±[2·0]	35/1 53/4
$C_6H_{11}O_2Cl$	150·605	n-Butyl chloroacetate	l		SB	3	m	817·0 ±0·5	54/3
$C_6H_{11}O_2Cl$	150·605	n-Propyl 3-chloro-propionate	l		SB		m	817·2 ±[2·0]	35/13 53/4
$C_6H_{11}O_2Cl$	150·605	Ethyl 4-chlorobutyrate	l		SB		m	810·3 ±[2·0]	35/13 53/4

H$_b$O$_2$Hal$_j$

1 kcal = 4·184 kJ

		6 Determination of ΔH_v			
ΔH°.					
Remarks	5 ΔH_f° (l or c) kcal/g.f.w.	Type	ΔH_v° kcal/g.f.w.	Ref.	7 ΔH_f° (g) kcal/g.f.w.
ed: As$_2$O$_3$ solution	**BEJ** −58·6 ±[2·0]				
ed: As$_2$O$_3$ solution	**BEJ** −58·4 ±[2·0]	S3	16·7 ±[2·0] (sub.)	27/3	−41·7 ±[2·8]
ed: As$_2$O$_3$ solution	**BEJ** −108·5 ±[2·0]				
ed: As$_2$O$_3$ solution	**BEJ** −52·7 ±[2·0]	S4	16·5 ±[2·0] (sub.)	27/3	−36·2 ±[2·8]
ed: As$_2$O$_3$ solution	**BEJ** −105·4 ±[2·0]	S4	24·2 ±[2·0] (sub.)	27/3	−81·2 ±[2·8]
ed: As$_2$O$_3$ solution	**BEJ** −99·5 ±[2·0]				
ed: As$_2$O$_3$ solution	**BEJ** −102·2 ±[2·0]				
ed: As$_2$O$_3$ solution	**BEJ** −101·3 ±[2·0]	S4	22·0 ±[2·0] (sub.)	27/3	−79·3 ±[2·8]
ed: **As$_2$O$_3$** solution	**BEJ** −91·5 ±[2·0]	S4	16·5 ±[2·0] (sub.)	27/3	−75·0 ±[2·8]
ed: As$_2$O$_3$ solution	**EJ** −130·6 ±[2·0]	E7	12·8 ±1·0		−117·8 ±[2·3]
ed: As$_2$O$_3$ solution	**EJ** −132·4 ±[2·0]	E7	12·7 ±1·0		−119·7 ±[2·3]
ed: As$_2$O$_3$ solution	**EJ** −131·5 ±[2·0]	E7	12·5 ±1·0		−119·0 ±[2·3]
ed: As$_2$O$_3$ solution	**EJ** −132·4 ±[2·0]	E7	12·5 ±1·0		−119·9 ±[2·3]
ed: As$_2$O$_3$ solution	**EJ** −128·7 ±0·5	E7	12·2 ±1·0		−116·5 ±1·1
ed: As$_2$O$_3$ solution	**EJ** −128·5 ±[2·0]	E7	12·4 ±1·0		−116·1 ±[2·3]
ed: As$_2$O$_3$ solution	**EJ** −135·4 ±[2·0]	E7	12·6 ±1·0		−122·8 ±[2·3]

$C_aH_bO_2Hal_f$

Determinat

1 Formula	2 g.f.w.	3 Name	State	Purity mol %	Type	No. of expts.	Detn. of reactn.	$-\Delta H_c^\circ$ kcal/g.f.w.	Re
$C_7HO_2F_5$	212·077	Pentafluorobenzoic acid	c	99·96	RB	10	m	610·13±0·24	64/ 69/
$C_7H_5O_2F$	140·115	2-Fluorobenzoic acid	c	99·89	RB	8	m	736·30±0·36	56/
$C_7H_5O_2F$	140·115	3-Fluorobenzoic acid	c	99·91	RB	8	m	732·84±0·20	56/
$C_7H_5O_2F$	140·115	4-Fluorobenzoic acid	c	99·93	RB	8	m	731·74±0·29	56/
$C_7H_5O_2F$		4-Fluorobenzoic acid	c	99·95	RB	5	m	732·13±0·21	64/
$C_7H_5O_2F$		4-Fluorobenzoic acid							
$C_7H_5O_2Cl$	156·570	2-Chlorobenzoic acid	c		SB		m	739·5 ±[2·0]	53/
$C_7H_5O_2Cl$	156·570	3-Chlorobenzoic acid	c		SB		m	733·6 ±[2·0]	53/
$C_7H_5O_2Cl$	156·570	4-Chlorobenzoic acid	c		RB	4	m	732·70±0·40	54/
$C_7H_5O_2Cl$		4-Chlorobenzoic acid	c	99·9, an	SB		m	732·50±0·75	68/
$C_7H_5O_2Cl$		4-Chlorobenzoic acid							
$C_7H_5O_2Br$	201·026	4-Bromobenzoic acid	c		RB	7	m	738·73±0·50	59/
$C_7H_5O_2I$	248·021	2-Iodobenzoic acid	c		SB		m	756·9 ±[1·0]	56/3
$C_7H_5O_2I$	248·021	3-Iodobenzoic acid	c		SB		m	753·4 ±[1·0]	56/3
$C_7H_5O_2I$	248·021	4-Iodobenzoic acid	c		SB		m	753·6 ±[1·0]	56/3
$C_7H_{11}O_2Cl_3$	233·524	3-Methylbutyl trichloroacetate	l		SB		m	912·2 ±[2·0]	35/1 53/4
$C_7H_{12}O_2Cl_2$	199·078	3-Methylbutyl dichloroacetate	l		SB		m	942·1 ±[2·0]	35/1 53/4
$C_7H_{13}O_2Cl$	164·633	3-Methylbutyl chloroacetate	l		SB		m	971·5 ±[2·0]	35/13 53/4
$C_7H_{13}O_2Cl$	164·633	n-Butyl 2-chloro-propionate	l		SB		m	971·4 ±[2·0]	35/13 53/4

1 kcal = $4\cdot184$ kJ

ΔH_r° Remarks	5 ΔH_f° (l or c) kcal/g.f.w.	6 Determination of ΔH_v			7 ΔH_f° (g) kcal/g.f.w.
		Type	ΔH_v° kcal/g.f.w.	Ref.	
$F_5CO_2H(c)+5O_2+102H_2O(l)$ $=7CO_2+5$ [HF.$20H_2O$](l)	J $-296\cdot34\pm0\cdot24$*	S4	$21\cdot9\ \pm1\cdot0$ (sub.)	69/1	$-274\cdot4\ \pm1\cdot1$*
$H_4FCO_2H(c)+7O_2+48H_2O(l)$ $=7CO_2+$ [HF.$50H_2O$ [(l)	$-135\cdot67\pm0\cdot36$*				
$_4FCO_2H(c)+7O_2+48H_2O(l)$ $=7CO_2+$ [HF.$50H_2O$](l)	$-139\cdot13\pm0\cdot20$*				
$_4FCO_2H(c)+7O_2+48H_2O(l)$ $=7CO_2+$ [HF.$50H_2O$](l)	$-140\cdot23\pm0\cdot29$*	S4	$21\cdot8\ \pm0\cdot3$ (sub.)	69/1	$-118\cdot4\ \pm0\cdot4$*
$_4FCO_2H(c)+7O_2+18H_2O(l)$ $=7CO_2+$ [HF.$20H_2O$](l)	$-139\cdot81\pm0\cdot21$*				$-118\cdot0\ \pm0\cdot4$*
Selected value	$-140\cdot00\pm0\cdot17$*				$-118\cdot2\ \pm0\cdot3$*
ed: As_2O_3 solution	EJ $-95\cdot3\ \pm[2\cdot0]$	S4	$19\cdot0\ \pm0\cdot8$ (sub.)	38/14	$-76\cdot3\ \pm[2\cdot2]$
ed: As_2O_3 solution	EJ $-101\cdot2\ \pm[2\cdot0]$	S4	$19\cdot6\ \pm0\cdot8$ (sub.)	38/14	$-81\cdot6\ \pm[2\cdot2]$
ed: As_2O_3 solution ed: N_2H_4.2HCl solution	EJ $-102\cdot11\pm0\cdot44$ $-102\cdot31\pm0\cdot77$	S4	$21\cdot0\ \pm0\cdot8$ (sub.)	38/14	$-81\cdot1\ \pm0\cdot9$ $-81\cdot3\ \pm1\cdot1$
Selected value	$-102\cdot19\pm0\cdot36$				$-81\cdot2\ \pm0\cdot9$
ed: As_2O_3 solution	J $-90\cdot4\ \pm0\cdot5$	El$_{me}$	$21\cdot0\ \pm1\cdot0$ (sub.)		$-69\cdot4\ \pm1\cdot1$
	E $-72\cdot2\ \pm[1\cdot0]$				
	E $-75\cdot7\ \pm[1\cdot0]$				
	E $-75\cdot5\ \pm[1\cdot0]$	El$_{me}$	$21\cdot0\ \pm1\cdot0$ (sub.)		$-54\cdot5\ \pm[1\cdot4]$
ed: As_2O_3 solution	EJ $-138\cdot9\ \pm[2\cdot0]$	E7	$13\cdot8\ \pm1\cdot0$		$-125\cdot1\ \pm[2\cdot3]$
ed: As_2O_3 solution	EJ $-137\cdot5\ \pm[2\cdot0]$	E7	$13\cdot3\ \pm1\cdot0$		$-124\cdot2\ \pm[2\cdot3]$
ed: As_2O_3 solution	EJ $-136\cdot6\ \pm[2\cdot0]$	E7	$13\cdot1\ \pm1\cdot0$		$-123\cdot5\ \pm[2\cdot3]$
ed: As_2O_3 solution	EJ $-136\cdot7\ \pm[2\cdot0]$	E7	$13\cdot0\ \pm1\cdot0$		$-123\cdot7\ \pm[2\cdot3]$

THERMOCHEMISTRY OF ORGANIC AND ORGANOMETALLIC COMPOUNDS

$C_aH_bO_2Hal_f$

1 kcal = 4·184 kJ

1 Formula	2 g.f.w.	3 Name	State	Purity mol %	Type	No. of expts.	Detn. of reactn.	$-\Delta H_r^\circ$ kcal/g.f.w.	Re
								Determinati	
$C_7H_{13}O_2Cl$	164·633	Isobutyl 2-chloro-propionate	l		SB		m	963·9 ±[2·0]	35/ 53/
$C_7H_{13}O_2Cl$	164·633	n-Butyl 3-chloro-propionate	l		SB		m	974·7 ±[2·0]	35/ 53/
$C_7H_{13}O_2Cl$	164·633	Isobutyl 3-chloro-propionate	l		SB		m	971·2 ±[2·0]	35/ 53/
$C_7H_{13}O_2Cl$	164·633	n-Propyl 2-chlorobutyrate	l		SB		m	957·3 ±[2·0]	35/ 53/
$C_7H_{13}O_2Cl$	164·633	n-Propyl 3-chlorobutyrate	l		SB		m	967·0 ±[2·0]	35/ 53/
$C_7H_{13}O_2Cl$	164·633	n-Propyl 4-chlorobutyrate	l		SB		m	966·7 ±[2·0]	35/ 53/
$C_8H_5O_2F_3$	190·123	3-Trifluoromethylbenzoic acid	c		RB	6	m	795·86±0·14	62/4
$C_8H_7O_2I$	262·048	Methyl 2-iodobenzoate	l		SB		m	933·4 ±[1·0]	56/3
$C_8H_7O_2I$	262·048	Methyl 3-iodobenzoate	c		SB		m	925·0 ±[1·0]	56/3
$C_8H_7O_2I$	262·048	Methyl 4-iodobenzoate	c		SB		m	923·0 ±[1·0]	56/3
$C_8H_{15}O_2Cl$	178·661	3-Methylbutyl 2-chloropropionate	l		SB		m	1120·5 ±[2·0]	35/1 53/4
$C_8H_{15}O_2Cl$	178·661	3-Methylbutyl 3-chloropropionate	l		SB		m	1128·6 ±[2·0]	35/1 53/4
$C_8H_{15}O_2Cl$	178·661	n-Butyl 2-chlorobutyrate	l		SB		m	1113·8 ±[2·0]	35/1 53/4
$C_8H_{15}O_2Cl$	178·661	Isobutyl 2-chlorobutyrate	l		SB		m	1112·3 ±[2·0]	35/13 53/4
$C_8H_{15}O_2Cl$	178·661	n-Butyl 3-chlorobutyrate	l		SB		m	1124·4 ±[2·0]	35/13 53/4
$C_8H_{15}O_2Cl$	178·661	Isobutyl 3-chlorobutyrate	l		SB		m	1121·4 ±[2·0]	35/13 53/4

1 kcal = 4·184 kJ

ΔH_r° Remarks		5 ΔH_f° (l or c) kcal/g.f.w.	6 Determination of ΔH_v			7 ΔH_f° (g) kcal/g.f.w.
			Type	ΔH_v° kcal/g.f.w.	Ref.	
: As_2O_3 solution	EJ	$-144\cdot2 \pm[2\cdot0]$	E7	$12\cdot8 \pm1\cdot0$		$-131\cdot4 \pm[2\cdot3]$
: As_2O_3 solution	EJ	$-133\cdot4 \pm[2\cdot0]$	E7	$13\cdot3 \pm1\cdot0$		$-120\cdot1 \pm[2\cdot3]$
: As_2O_3 solution	EJ	$-136\cdot9 \pm[2\cdot0]$	E7	$13\cdot2 \pm1\cdot0$		$-123\cdot7 \pm[2\cdot3]$
: As_2O_3 solution	EJ	$-150\cdot8 \pm[2\cdot0]$	E7	$12\cdot5 \pm1\cdot0$		$-138\cdot3 \pm[2\cdot3]$
: As_2O_3 solution	EJ	$-141\cdot1 \pm[2\cdot0]$	E7	$12\cdot5 \pm1\cdot0$		$-128\cdot6 \pm[2\cdot3]$
: As_2O_3 solution	EJ	$-141\cdot4 \pm[2\cdot0]$	E7	$12\cdot8 \pm1\cdot0$		$-128\cdot6 \pm[2\cdot3]$
$H_4F_3CO_2H(c)+7\cdot5O_2+59H_2O(l)$ $= 8CO_2+3[HF.20H_2O]$ (l)		$-225\cdot71\pm0\cdot14*$				
	E	$-58\cdot1 \pm[1\cdot0]$				
	E	$-66\cdot5 \pm[1\cdot0]$				
	E	$-68\cdot5 \pm[1\cdot0]$				
d: As_2O_3 solution	EJ	$-149\cdot9 \pm[2\cdot0]$	E7	$12\cdot5 \pm1\cdot0$		$-137\cdot4 \pm[2\cdot3]$
d: As_2O_3 solution	EJ	$-141\cdot8 \pm[2\cdot0]$	E7	$12\cdot9 \pm1\cdot0$		$-128\cdot9 \pm[2\cdot3]$
d: As_2O_3 solution	EJ	$-156\cdot6 \pm[2\cdot0]$	E7	$12\cdot6 \pm1\cdot0$		$-144\cdot0 \pm[2\cdot3]$
d: As_2O_3 solution	EJ	$-158\cdot1 \pm[2\cdot0]$	E7	$12\cdot5 \pm1\cdot0$		$-145\cdot6 \pm[2\cdot3]$
d: As_2O_3 solution	EJ	$-146\cdot0 \pm[2\cdot0]$	E7	$12\cdot7 \pm1\cdot0$		$-133\cdot3 \pm[2\cdot3]$
d: As_2O_3 solution	EJ	$-149\cdot0 \pm[2\cdot0]$	E7	$12\cdot5 \pm1\cdot0$		$-136\cdot5 \pm[2\cdot3]$

$C_aH_bO_2Hal_f$ 1 kcal = 4·184 kJ

1 Formula	2 g.f.w.	3 Name	State	Purity mol %	Type	No. of expts.	Detn. of reactn.	$-\Delta H_r^{\circ}$ kcal/g.f.w.	R
$C_8H_{15}O_2Cl$	178·661	n-Butyl 4-chlorobutyrate	l		SB		m	1122·7 ±[2·0]	35, 53
$C_8H_{15}O_2Cl$	178·661	Isobutyl 4-chlorobutyrate	l		SB		m	1119·7 ±[2·0]	35, 53
$C_9H_{17}O_2Cl$	192·688	3-Methylbutyl 2-chlorobutyrate	l		SB		m	1269·3 ±[2·0]	35, 53
$C_9H_{17}O_2Cl$	192·688	3-Methylbutyl 3-chlorobutyrate	l		SB		m	1278·8 ±[2·0]	35, 53
$C_9H_{17}O_2Cl$	192·688	3-Methylbutyl 4-chlorobutyrate	l		SB		m	1275·9 ±[2·0]	35, 53

1 kcal = 4·184 kJ

| H_r° | | | 6 Determination of ΔH_v | | | 7 |
| | | 5 | | | | |
Remarks		ΔH_f° (l or c) kcal/g.f.w.	Type	ΔH_v° kcal/g.f.w.	Ref.	ΔH_f° (g) kcal/g.f.w.
: As$_2$O$_3$ solution	EJ	−147·7 ±[2·0]	E7	13·0 ±1·0		−134·7 ±[2·3]
: As$_2$O$_3$ solution	EJ	−150·7 ±[2·0]	E7	12·9 ±1·0		−137·8 ±[2·3]
: As$_2$O$_3$ solution	EJ	−163·5 ±[2·0]	E7	12·9 ±1·0		−150·6 ±[2·3]
: As$_2$O$_3$ solution	EJ	−154·0 ±[2·0]	E7	13·0 ±1·0		−141·0 ±[2·3]
: As$_2$O$_3$ solution	EJ	−156·9 ±[2·0]	E7	13·3 ±1·0		−143·6 ±[2·3]

$C_aH_bN_dHal_j$

1 kcal = 4·184 kJ

1 Formula	2 g.f.w.	3 Name	State	Purity mol %	Type	No. of expts.	Detn. of reactn.	$-\Delta H_r^\circ$ kcal/g.f.w.	R
CNCl	61·471	Cyanogen chloride	g		R	5	an	66·3 ±0·2	54
CNBr	105·927	Cyanogen bromide	c		R	6	m	56·10±0·17	54
CNI	152·922	Cyanogen iodide	c		R	5	m, an	46·4 ±0·1	54
C_5NF_{11}	283·045	Undecafluoropiperidine	l	99·5	RB	7	m	349·8 ±0·8	63
CN_2F_6	154·015	Hexafluoromethane-diamine	g	ir, ms	R	5	m	114·3 ±0·9	67
$C_6H_{12}N_2F_4$	188·170	1, 2-Bisdifluoroamino-4-methylpentane	l	99·8	RB	7	m	1082·1 ±0·8	63
CN_3F_5	149·023	Pentafluoroguanidine	g	ir, ms	R	5	m	246·0 ±0·8	67
CN_3F_7	187·020	Heptafluoromethane-triamine	g	ir, ms	R	6	m	175·3 ±0·6	67
CN_4F_8	220·025	Octafluoromethane-tetramine	g	ir, ms	R	5	m	223·5 ±1·3	67
$C_2N_5F_{11}$	303·038	Undecafluorobiguanide	g	ir, ms	R	6	m	359·6 ±1·6	67

Determina

1 kcal = 4·184 kJ

H_r° Remarks	5 ΔH_f° (l or c) kcal/g.f.w.	6 Determination of ΔH_v			7 ΔH_f° (g) kcal/g.f.w.
		Type	ΔH_v° kcal/g.f.w.	Ref.	
Cl(g)+20[NaOH.110H$_2$O](l) = [NaCNO{−93·4}+NaCl{−97·2} +18NaOH] (122·3H$_2$O) (l)					+31·6 ±0·2*
Br(c)+150[NaOH.55H$_2$O] (l) = [NaCNO{−93·4}+NaBr{−86·1} +148NaOH] (55·75H$_2$O) (l)	+32·50±0·17*	S3	10·8 ±[1·0] (sub.)	54/23	+43·3 ±[1·1]
I(c)+150[NaOH.55H$_2$O] (l) = [NaCNO{−93·4}+NaI{−70·6} +148NaOH] (55·75H$_2$O) (l)	+38·3 ±0·1*	S4	14·3 ±0·1* (sub.)	43/2	+52·6 ±0·2*
F$_{11}$(l)+2·25O$_2$+21H$_2$O(l) = 2·75CO$_2$+2·25CF$_4$(g) {−223·3±0·4} +0·5N$_2$+2[HF.10H$_2$O] (l)	−496·8 ±1·2*	V1	7·16±0·05	63/47	−489·6 ±1·2*
$_2$F$_6$(g) = CF$_4$(g) {−223·3±0·4} +F$_2$(g)+N$_2$					−109·0 ±0·9*
$_{12}$N$_2$F$_4$(l)+8O$_2$+196H$_2$O(l) = 6CO$_2$+N$_2$+4[HF.50H$_2$O] (l)	−65·4 ±0·8*	V1	10·51±0·05	63/48	−54·9 ±0·8*
$_3$F$_5$(g) = CF$_4$(g) {−223·3±0·4} +0·5F$_2$(g)+1·5N$_2$					+22·7 ±0·8*
$_3$F$_7$(g) = CF$_4$(g) {−223·3±0·4} +1·5F$_2$(g)+1·5N$_2$					−48·0 ±0·6*
$_4$F$_8$(g) = CF$_4$(g) {−223·3±0·4} +2F$_2$(g)+2N$_2$					+0·2 ±1·3*
$_{11}$N$_5$(g) = 2CF$_4$(g) {−223·3±0·4} +1·5F$_2$(g)+2·5N$_2$ **I**					−87·0 ±1·6*

$C_aH_bO_cN_dHal_f$ 1 kcal = 4·184 kJ

1 Formula	2 g.f.w.	3 Name	State	Purity mol %	Type	No. of expts.	Detn. of reactn.	$-\Delta H_f^\circ$ kcal/g.f.w.	Determin
$C_2H_2ONCl_3$	162·403	Trichloroacetamide	c		SB		m	187·8 ±[2·0]	
C_2H_4ONCl	93·513	Chloroacetamide	c		SB		m	249·5 ±[2·0]	
$C_4H_4O_2NCl$	133·535	N-Chlorosuccinimide	c		R	4	m	67·78±0·07	6
$C_4H_4O_2NBr$	177·991	N-Bromosuccinimide	c		R	4	m	62·22±0·11	6
$C_4H_9O_2N_2F$	136·127	1-N-Fluoronitramino- butane	l	ir, glc	RB	7	m	686·6 ±0·7	6
$C_4H_9O_2N_2F$	136·127	2-N-Fluoronitramino- butane	l	ir, glc	RB	7	m	687·8 ±2·1	6
$C_4H_9O_2N_2F$	136·127	2-Methyl-2-N-fluoro- nitraminopropane	l	ir, glc	RB	7	m	695·2 ±1·8	6
CO_6N_3F	169·026	Fluorotrinitromethane	l	>99·5, glc	RB	5	m	797·5 ±0·5	66
CO_6N_3Cl	185·481	Chlorotrinitromethane	l	ir, glc	RB		m	864·37±0·35	64

1 kcal = 4·184 kJ

| H_r° | | 6 Determination of ΔH_v | | |
Remarks	5 ΔH_f° (l or c) kcal/g.f.w.	ΔH_v° Type kcal/g.f.w.	Ref.	7 ΔH_f° (g) kcal/g.f.w.
$_2ONCl_3(c)+1·25O_2+1800·5H_2O(l)$ $= 2CO_2+0·5N_2+3[HCl.600H_2O]$ (l) : As$_2$O$_3$ solution **BEJ**	$-85·6 \pm[2·0]$			
$_4ONCl(c)+2·25O_2+600H_2O(l)$ $= 2CO_2+0·5N_2+[HCl.600H_2O]$ (l) : As$_2$O$_3$ solution **BEJ**	$-80·9 \pm[2·0]$			
$_4O_2NCl(c)$ $-17[N_2H_4.300H_2O]$ (l) $\{+8·21\pm0·03\}$ $= C_4H_5O_2N(c)$ $\{-109·71\pm0·07\}$ $+0·5N_2+[HCl.53·5H_2O]$ (l) $+16·5[N_2H_4.305·9H_2O]$ (l)	$-85·58\pm0·10$			
$_4O_2NBr(c)$ $-17[N_2H_4.300H_2O]$ (l) $\{+8·21\pm0·03\}$ $= C_4H_5O_2N(c)$ $\{-109·71\pm0·07\}$ $+0·5N_2+[HBr.53·5H_2O]$ (l) $+16·5[N_2H_4.305·9H_2O]$ (l)	$-80·35\pm0·15$			
$_9O_2N_2F(l)+5O_2+146H_2O(l)$ $= 4CO_2+N_2+[HF.150H_2O]$ (l)	$-39·9 \pm0·7*$	V3 $13·0 \pm[0·5]$	66/37	$-26·9 \pm1·0*$
$_9O_2N_2F(l)+5O_2+146H_2O(l)$ $= 4CO_2+N_2+[HF.150H_2O]$ (l)	$-38·7 \pm2·1*$			
$_9O_2N_2F(l)+5O_2+146H_2O(l)$ $= 4CO_2+N_2+[HF.150H_2O]$ (l)	$-31·3 \pm1·8*$			
$(NO_2)_3(l)$ $+$Ethyl oxalate $C_6H_{10}O_4$ $\{-192·51\}$ $+4·25O_2+115·5H_2O(l)$ $= 7CO_2+1·5N_2+[HF.120H_2O]$ (l)	$-52·8 \pm0·5*$	V4 $8·3 \pm[0·2]$	66/23	$-44·5 \pm0·6*$
$(NO_2)_3(l)$ $+C_6H_5CO_2H(c)$ $\{-91·99\pm0·05\}$ $+5·25O_2+157·5H_2O(l)$ $= 8CO_2+1·5N_2+[HCl.160H_2O]$ (l) d: As$_2$O$_3$ solution	$-6·54 \pm0·35*$	V4 $10·86\pm[0·20]$	64/38	$+4·32\pm0·40*$

Group I organometallic compounds $1 \text{ kcal} = 4 \cdot 184 \text{ kJ}$

1 Formula	2 g.f.w.	3 Name	State	Purity mol %	Type	No. of expts.	Detn. of reactn.	$-\Delta H_r^\circ$ kcal/g.f.w.	R
GROUP IA C_2H_5Li	$36 \cdot 001$	Ethyl-lithium	c		SB	10	m, an	$416 \cdot 3 \pm 1 \cdot 3$	62
C_4H_9Li	$64 \cdot 055$	n-Butyl-lithium	l		SB	5	m	$723 \cdot 0 \pm 1 \cdot 7$	62
C_4H_9Li		n-Butyl-lithium	l		R	3	m	$57 \cdot 4 \pm 0 \cdot 7$	61
C_4H_9Li		n-Butyl-lithium							
CH_4NLi	$36 \cdot 989$	Methylaminolithium	c		R		m	$101 \cdot 86 \pm 0 \cdot 22$	53

Determina

1 kcal $= 4 \cdot 184$ kJ

| H_f° | | 6 Determination of ΔH_v | | | 7 |
| | 5 ΔH_f° (l or c) kcal/g.f.w. | ΔH_v° | | | ΔH_f° (g) kcal/g.f.w. |
Remarks		Type	kcal/g.f.w.	Ref.	
duct Li$_2$O $\{-142 \cdot 4\}$: correction made for Li$_2$O$_2$+Li$_2$CO$_3$. LiEt vapour shown to be polymeric (61/44)	$-14 \cdot 0 \pm 1 \cdot 3$	S3	$27 \cdot 9 \pm 0 \cdot 2$ (sub.)	62/51	$+13 \cdot 9 \pm 1 \cdot 4$*
duct Li$_2$O $\{-142 \cdot 4\}$: vapour may be polymeric	$-31 \cdot 8 \pm 1 \cdot 7$	V4	$25 \cdot 6 \pm 0 \cdot 7$	62/50	$-6 \cdot 2 \pm 1 \cdot 9$*
$_4$H$_9$(l)+H$_2$O(g) $\{-57 \cdot 80\}$ = LiOH(c) $\{-116 \cdot 45\}$ +n-C$_4$H$_{10}$(g) $\{-30 \cdot 36 \pm 0 \cdot 16\}$	$-31 \cdot 6 \pm 0 \cdot 8$				$-6 \cdot 0 \pm 1 \cdot 1$*
Selected value	$-31 \cdot 6 \pm 0 \cdot 8$				$-6 \cdot 0 \pm 1 \cdot 1$*
HCH$_3$(c)+2HCl(g) = LiCl $\{-97 \cdot 70\}$ +CH$_3$NH$_3$Cl $\{-71 \cdot 20\}$	$-22 \cdot 92 \pm 0 \cdot 40$				

Group II organometallic compounds   1 kcal = 4·184 kJ

1 Formula	2 g.f.w.	3 Name	State	Purity mol %	Type	No. of expts.	Detn. of reactn.	$-\Delta H_r^\circ$ kcal/g.f.w.	R
GROUP IIA									
$C_{10}H_{10}Mg$	154·503	Dicyclopentadienyl-magnesium	c		R	5	m	34·05±0·70	67
GROUP IIB									
C_2H_6Zn	95·44	Dimethylzinc	l		SB	3	m	482·8 ±1·4	49
C_2H_6Zn		Dimethylzinc	l		R	3	m	81·7 ±0·2	49
C_2H_6Zn		Dimethylzinc							
$C_4H_{10}Zn$	123·49	Diethylzinc	l		SB	11	m	805·1 ±3·2	49
$C_4H_{10}Zn$		Diethylzinc	l		SB		m	806·0 ±0·5	58/ 63/
$C_4H_{10}Zn$		Diethylzinc	l		R	5	m	62·6 ±0·5	49/
$C_4H_{10}Zn$		Diethylzinc	l		R	5	m	84·7 ±1·0	49/
$C_4H_{10}Zn$		Diethylzinc	l		R	6	m	72·9 ±1·0	49/
$C_4H_{10}Zn$		Diethylzinc							
$C_6H_{14}Zn$	151·55	Di-n-propylzinc	l		SB	3	m	1112·1 ±5·5	49
$C_8H_{18}Zn$	179·60	Di-n-butylzinc	l		SB	6	m	1425·7 ±5·6	49
C_2H_6Cd	142·47	Dimethylcadmium	l		SB	6	m	475·3 ±2·9	49
C_2H_6Cd		Dimethylcadmium	l		R	7	m	48·5 ±0·2	49/
C_2H_6Cd		Dimethylcadmium	l		R	6	m	73·3 ±0·2	49/
C_2H_6Cd		Dimethylcadmium							

1 kcal = $4 \cdot 184$ kJ

H_r°				6 Determination of ΔH_v			
			5 ΔH_f° (l or c) kcal/g.f.w.	ΔH_v° Type kcal/g.f.w.		Ref.	7 ΔH_f° (g) kcal/g.f.w.
Remarks							
c)+2C$_5$H$_6$(l) {+25·16±0·31} = Mg(C$_5$H$_5$)$_2$(c)+H$_2$(g) hydrolysis with H$_2$SO$_4$			+16·27±0·85	S4	16·3 ±0·3 (sub.)	67/32	+32·6 ±0·9
duct ZnO(c, hex.) {−83·36±0·21} (33/14) CH$_3$)$_2$(l)+[H$_2$SO$_4$.100H$_2$O] (l) = [ZnSO$_4$.100H$_2$O] (l) {−252·70} +2CH$_4$(g) {−17·89±0·07}			+6·4 ±1·5 +5·4 ±2·0	V2	7·06±[0·10]	61/45	+13·5 ±1·5 +12·5 ±2·0
Selected value			+6·0 ±2·0				+13·1 ±2·0
duct ZnO(c, hex.) {−83·36±0·21} (33/14) duct ZnO(c, hex.) {−83·36±0·21} (33/14)			+4·0 ±3·2 +4·9 ±0·7	V4	9·6 ±[0·5]	46/14	+13·6 ±3·3 +14·5 ±0·9
C$_2$H$_5$)$_2$(l)+2H$_2$O(l) = Zn(OH)$_2$(c?) {−153·5} +2C$_2$H$_6$(g) {−20·24±0·12}			+5·3 ±2·0				+14·9 ±2·1
C$_2$H$_5$)$_2$(l)+[H$_2$SO$_4$.100H$_2$O] (l) = [ZnSO$_4$.100H$_2$O] (l) {−252·70} +2C$_2$H$_6$(g) {−20·24±0·12}			+3·7 ±2·0				+13·3 ±2·1
C$_2$H$_5$)$_2$(l)+2I$_2$(c) = ZnI$_2$(c) {−50·0} +2C$_2$H$_5$I(l) {−9·6±0·4}			+3·7 ±1·8				+13·3 ±2·0
Selected value			+4·0 ±1·5				+13·6 ±1·7
oduct ZnO(c, hex.) {−83·36±0·21} (33/14)			−13·8 ±5·6	V4	10·9 ±[0·6]	49/30	−2·9 ±5·7
oduct ZnO(c, hex.) {−83·36±0·21} (33/14)			−24·9 ±5·7	V4	13·0 ±[0·8]	49/30	−11·9 ±5·8
oduct CdO(c) {−61·2±0·2} (54/24) (CH$_3$)$_2$(l)+2H$_2$O(l) = Cd(OH)$_2$(c?) {−133·3} +2CH$_4$(g) {−17·89±0·07}			+21·1 ±3·0 +16·1 ±1·0*	Cl	9·07±0·03	56/35	+30·2 ±3·0 +25·2 ±1·0*
(CH$_3$)$_2$(l)+[H$_2$SO$_4$.100H$_2$O] (l) = [CdSO$_4$.100H$_2$O] (l) {−232·97} +2CH$_4$(g) {−17·89±0·07}			+16·7 ±0·3				+25·8 ±0·3
Selected value			+16·7 ±0·3				+25·8 ±0·3

Group II organometallic compounds   1 kcal = 4·184 kJ

1 Formula	2 g.f.w.	3 Name	State	Purity mol %	Type	No. of expts	Detn. of reactn.	$-\Delta H_r^\circ$ kcal/g.f.w.	R
$C_4H_{10}Cd$	170·52	Diethylcadmium	l		SB		m	800·6 ±0·8	58
$C_4H_{10}Cd$		Diethylcadmium	l		R	7	m	75·8 ±0·4	63 49
$C_4H_{10}Cd$		Diethylcadmium	l		R	6	m	80·8 ±1·0	49
$C_4H_{10}Cd$		Diethylcadmium							
C_2H_6Hg	230·66	Dimethylmercury	l		SB		CO_2		52
C_2H_6Hg		Dimethylmercury	l		R	4	m	72·2 ±0·6	50
C_2H_6Hg		Dimethylmercury	l		R	3	m	44·1 ±0·2	50
C_2H_6Hg		Dimethylmercury							
$C_4H_{10}Hg$	258·71	Diethylmercury	l		SB		m	741·7 ±1·0	58
$C_4H_{10}Hg$		Diethylmercury	l		SB		CO_2		63 52
$C_4H_{10}Hg$		Diethylmercury	l		R	4	m	91·23±0·33	51
$C_4H_{10}Hg$		Diethylmercury	l		R	4	m	52·0 ±0·6	51
$C_4H_{10}Hg$		Diethylmercury							
$C_6H_{14}Hg$	286·77	Di-n-propylmercury	l		SB		m	1037·6 ±[1·0]*	35
$C_6H_{14}Hg$		Di-n-propylmercury	l		R	4	m	91·8 ±0·5	52
$C_6H_{14}Hg$		Di-n-propylmercury							
$C_6H_{14}Hg$	286·77	Di-isopropylmercury	l		SB		m	1051·5 ±[1·0]*	35
$C_6H_{14}Hg$		Di-isopropylmercury	l		R	4	m	98·9 ±0·5	52
$C_6H_{14}Hg$		Di-isopropylmercury							
$C_8H_{18}Hg$	314·82	Di-n-butylmercury	l		SB		m	1343·9 ±[1·5]*	35
$C_8H_{18}Hg$	314·82	Di-isobutylmercury	l		SB		m	1342·9 ±[1·5]*	35

$$1 \text{ kcal} = 4 \cdot 184 \text{ kJ}$$

H_r°		6 Determination of ΔH_v			7
Remarks	5 ΔH_f° (l or c) kcal/g.f.w.	Type	ΔH_v° kcal/g.f.w.	Ref.	ΔH_f° (g) kcal/g.f.w.
duct CdO(c) $\{-61 \cdot 2 \pm 0 \cdot 2\}$ (54/24}	$+21 \cdot 6 \pm 0 \cdot 9$	E1	$11 \cdot 0 \pm 0 \cdot 5$		$+32 \cdot 6 \pm 1 \cdot 0$
$C_2H_5)_2(l) + [H_2SO_4.100H_2O]$ (l)	$+14 \cdot 5 \pm 0 \cdot 4$				$+25 \cdot 5 \pm 0 \cdot 7$
$= (CdSO_4.100H_2O]$ (l) $\{-232 \cdot 97\}$ $+2C_2H_6(g)$ $\{-20 \cdot 24 \pm 0 \cdot 12\}$					
$C_2H_5)_2(l) + 2I_2(c)$	$+13 \cdot 6 \pm 1 \cdot 2$				$+24 \cdot 7 \pm 1 \cdot 3$
$= CdI_2(c)$ $\{-48 \cdot 0\}$ $+2C_2H_5I(l)$ $\{-9 \cdot 6 \pm 0 \cdot 4\}$					
Selected value	$+14 \cdot 5 \pm 0 \cdot 4$				$+25 \cdot 5 \pm 0 \cdot 7$
duct Hg(l); correction made for HgO	$+14 \cdot 3 \pm 0 \cdot 1$	V2	$8 \cdot 26 \pm [0 \cdot 20]$	61/45	$+22 \cdot 6 \pm 0 \cdot 3$
$CH_3)_2(l) + 2Br_2(l)$	$+13 \cdot 4 \pm 0 \cdot 8$				$+21 \cdot 7 \pm 0 \cdot 9$
$= HgBr_2(c)$ $\{-40 \cdot 6 \pm 0 \cdot 2\}$ $+2CH_3Br(g)$ $\{-9 \cdot 1 \pm 0 \cdot 3\}$					
$CH_3)_2(l) + 2I_2(c)$	$+12 \cdot 3 \pm 0 \cdot 4$				$+20 \cdot 6 \pm 0 \cdot 5$
$= HgI_2(c)$ $\{-25 \cdot 2 \pm 0 \cdot 2\}$ $+2CH_3I(l)$ $\{-3 \cdot 3 \pm 0 \cdot 3\}$					
Selected value	$+14 \cdot 0 \pm 1 \cdot 0$				$+22 \cdot 3 \pm 1 \cdot 0$
duct Hg(l); correction made for HgO	$+22 \cdot 1 \pm 1 \cdot 0$	V4	$10 \cdot 7 \pm [0 \cdot 4]$	36/18	$+32 \cdot 8 \pm 1 \cdot 1$
duct Hg(l); correction made for HgO	$+6 \cdot 5 \pm 0 \cdot 2$				$+17 \cdot 2 \pm 0 \cdot 5$
$C_2H_5)_2(l) + 2Br_2(l)$	$+7 \cdot 6 \pm 0 \cdot 8$				$+18 \cdot 3 \pm 0 \cdot 9$
$= HgBr_2(c)$ $\{-40 \cdot 6 \pm 0 \cdot 2\}$ $+2C_2H_5Br(l)$ $\{-21 \cdot 8 \pm 0 \cdot 5\}$					
$C_2H_5)_2(l) + 2I_2(c)$	$+7 \cdot 6 \pm 0 \cdot 9$				$+18 \cdot 3 \pm 1 \cdot 0$
$= HgI_2(c)$ $\{-25 \cdot 2 \pm 0 \cdot 2\}$ $+2C_2H_5I(l)$ $\{-9 \cdot 6 \pm 0 \cdot 4\}$					
Selected value	$+7 \cdot 1 \pm 0 \cdot 7$				$+17 \cdot 8 \pm 0 \cdot 8$
duct Hg(l)	$-4 \cdot 9 \pm [1 \cdot 0]^*$	V4	$13 \cdot 2 \pm 0 \cdot 3$	52/31	$+8 \cdot 3 \pm [1 \cdot 1]^*$
$n-C_3H_7)_2(l) + 2Br_2(l)$	$-5 \cdot 0 \pm 1 \cdot 3$				$+8 \cdot 2 \pm 1 \cdot 4$
$= HgBr_2(c)$ $\{-40 \cdot 6 \pm 0 \cdot 2\}$ $+2n-C_3H_7Br(l)$ $\{-28 \cdot 1 \pm 1 \cdot 0\}$					
Selected value	$-5 \cdot 0 \pm 1 \cdot 3$				$+8 \cdot 2 \pm 1 \cdot 4$
duct Hg(l)	$+9 \cdot 0 \pm [1 \cdot 0]^*$	V4	$12 \cdot 8 \pm 0 \cdot 4$	52/31	$+21 \cdot 8 \pm [1 \cdot 1]^*$
$i-C_3H_7)_2(l) + 2Br_2(l)$	$-3 \cdot 1 \pm 1 \cdot 0$				$+9 \cdot 7 \pm 1 \cdot 1$
$= HgBr_2(c)$ $\{-40 \cdot 6 \pm 0 \cdot 2\}$ $+2i-C_3H_7Br(l)$ $\{-30 \cdot 7 \pm 0 \cdot 6\}$					
Selected value	$-3 \cdot 1 \pm 1 \cdot 0$				$+9 \cdot 7 \pm 1 \cdot 1$
duct Hg(l)	$-23 \cdot 4 \pm [1 \cdot 5^*]$	E1	$15 \cdot 6 \pm 1 \cdot 0$		$-7 \cdot 8 \pm [1 \cdot 8]^*$
duct Hg(l)	$-24 \cdot 4 \pm [1 \cdot 5]^*$	E1	$15 \cdot 2 \pm 1 \cdot 0$		$-9 \cdot 2 \pm [1 \cdot 8]^*$

Group II organometallic compounds   1 kcal = 4·184 kJ

1 Formula	2 g.f.w.	3 Name	State	Purity mol %	Type	No. of expts.	Detn. of reactn.	$-\Delta H_r^\circ$ kcal/g.f.w.	R
$C_{10}H_{22}Hg$	342·88	Di-3-methylbutylmercury	l		SB		m	1655·2 ±[2·0]*	35
$C_{12}H_{10}Hg$	354·80	Diphenylmercury	c		SB	5	CO_2	1537·1 ±1·8	56
$C_{12}H_{10}Hg$		Diphenylmercury	c		SB		CO_2		52
$C_{12}H_{10}Hg$		Diphenylmercury	c		SB				63
$C_{12}H_{10}Hg$		Diphenylmercury	c		R	4	m	51·4 ±1·2	56
$C_{12}H_{10}Hg$		Diphenylmercury							
CH_3ClHg	251·08	Methylmercury chloride	c		R	4	m	16·17±0·55	50
CH_3BrHg	295·53	Methylmercury bromide	c		R	4	m	14·59±0·75	50
CH_3IHg	342·53	Methylmercury iodide	c		R	2	m	9·51±0·38	50
C_2H_5ClHg	265·11	Ethylmercury chloride	c		R	2	m	20·96±0·87	51
C_2H_5BrHg	309·56	Ethylmercury bromide	c		R	3	m	17·8 ±1·1	51
C_2H_5IHg	356·56	Ethylmercury iodide	c		R	2	m	13·2 ±1·3	51
C_3H_7ClHg	279·13	n-Propylmercury chloride	c		R	3	m	21·17±0·30	52
C_3H_7ClHg	279·13	Isopropylmercury chloride	c		R	3	m	22·53±0·50	52
C_3H_7BrHg	323·59	n-Propylmercury bromide	c		R	3	m	18·15±0·30	52

1 kcal = 4·184 kJ

| H_r° | | 6 Determination of ΔH_v | | | 7 |
| | | | | | |
Remarks	5 ΔH_f° (l or c) kcal/g.f.w.	Type	ΔH_v° kcal/g.f.w.	Ref.	ΔH_f° (g) kcal/g.f.w.
duct Hg(l)	$-36\cdot8 \pm[2\cdot0]^*$	E1	$17\cdot0 \pm1\cdot5$		$-19\cdot8 \pm[3\cdot0]^*$
an $0\cdot9998\pm0\cdot0033$; product Hg(l); correction made for HgO	$+66\cdot9 \pm1\cdot8$	S3	$26\cdot95\pm0\cdot20$ (sub.)	58/5	$+93\cdot9 \pm1\cdot8$
duct Hg(l); correction made for HgO	$+71\cdot6 \pm0\cdot3$				$+98\cdot6 \pm0\cdot4$
duct Hg(l)	$+68\cdot3 \pm1\cdot5$				$+95\cdot3 \pm1\cdot5$
$C_6H_5)_2(c)+2HCl(g)$ $= HgCl_2(c) \{-53\cdot5\pm0\cdot5\}$ $+2C_6H_6(l) \{+11\cdot72\pm0\cdot13\}$	$+65\cdot5 \pm2\cdot0$				$+92\cdot5 \pm2\cdot0$
Selected value	$+66\cdot8 \pm1\cdot5$				$+93\cdot8 \pm1\cdot5$
$CH_3)_2(l) \{+14\cdot0\pm1\cdot0\}$ $-HgCl_2(c) \{-53\cdot5\pm0\cdot5\}$ $= 2CH_3HgCl(c)$	$-27\cdot8 \pm0\cdot6$	S3	$15\cdot5 \pm0\cdot4$ (sub.)	51/25	$-12\cdot3 \pm0\cdot7$
$CH_3)_2(l) \{+14\cdot0\pm1\cdot0\}$ $-HgBr_2(c) \{-40\cdot6\pm0\cdot2\}$ $= 2CH_3HgBr(c)$	$-20\cdot6 \pm0\cdot6$	S3	$16\cdot2 \pm0\cdot4$ (sub.)	51/25	$-4\cdot4 \pm0\cdot7$
$CH_3)_2(l) \{+14\cdot0\pm1\cdot0\}$ $+HgI_2(c)\{-25\cdot2\pm0\cdot2\}$ $= 2CH_3HgI(c)$	$-10\cdot4 \pm0\cdot6$	S3	$15\cdot6 \pm0\cdot4$ (sub.)	51/25	$+5\cdot2 \pm0\cdot7$
$C_2H_5)_2(l) \{+7\cdot1\pm0\cdot7\}$ $-HgCl_2(c) \{-53\cdot5\pm0\cdot5\}$ $= 2C_2H_5HgCl(c)$	$-33\cdot7 \pm0\cdot7$	S3	$18\cdot2 \pm0\cdot7$ (sub.)	51/25	$-15\cdot5 \pm1\cdot0$
$C_2H_5)_2(l) \{+7\cdot1\pm0\cdot7\}$ $+HgBr_2(c) \{-40\cdot6\pm0\cdot2\}$ $= 2C_2H_5HgBr(c)$	$-25\cdot7 \pm0\cdot8$	S3	$18\cdot3 \pm0\cdot7$ (sub.)	51/25	$-7\cdot4 \pm1\cdot0$
$C_2H_5)_2(l) \{+7\cdot1\pm0\cdot7\}$ $+HgI_2(c) \{-25\cdot2\pm0\cdot2\}$ $= 2C_2H_5HgI(c)$	$-15\cdot7 \pm0\cdot8$	S3	$19\cdot0 \pm0\cdot7$ (sub.)	51/25	$+3\cdot3 \pm1\cdot0$
$n-C_3H_7)_2(l) \{-5\cdot0\pm1\cdot3\}$ $+HgCl_2(c) \{-53\cdot5\pm0\cdot5\}$ $= 2 \, n-C_3H_7HgCl(c)$	$-39\cdot8 \pm1\cdot0$				
$i-C_3H_7)_2(l) \{-3\cdot1\pm1\cdot0\}$ $+HgCl_2(c) \{-53\cdot5\pm0\cdot5\}$ $= 2 \, i-C_3H_7HgCl(c)$	$-39\cdot6 \pm0\cdot6$				
$n-C_3H_7)_2(l) \{-5\cdot0\pm1\cdot3\}$ $+HgBr_2(c) \{-40\cdot6\pm0\cdot2\}$ $= 2 \, n-C_3H_7HgBr(c)$	$-31\cdot9 \pm1\cdot0$				

Group II organometallic compounds   1 kcal = 4·184 kJ

1 Formula	2 g.f.w.	3 Name	State	Purity mol %	Type	No. of expts.	Detn. of reactn.	$-\Delta H_f^\circ$ kcal/g.f.w.	R
C_3H_7BrHg	323·59	Isopropylmercury bromide	c		R	3	m	21·22±0·35	52
C_3H_7IHg	370·58	n-Propylmercury iodide	c		R	3	m	12·64±0·40	52
C_3H_7IHg	370·58	Isopropylmercury iodide	c		R	3	m	15·60±0·40	52
C_6H_5ClHg	313·15	Phenylmercury chloride	c		R	3	m	13·9 ±0·5	56
C_6H_5ClHg		Phenylmercury chloride	c		R	4	m	32·65±0·60	56
C_6H_5ClHg		Phenylmercury chloride							
C_6H_5BrHg	357·61	Phenylmercury bromide	c		R	3	m	8·49±0·28	51
C_6H_5IHg	404·60	Phenylmercury iodide	c		R	2	m	5·22±0·18	51

$$1 \text{ kcal} = 4 \cdot 184 \text{ kJ}$$

ΔH_r°		5 ΔH_f°(l or c) kcal/g.f.w.	6 Determination of ΔH_v			7 ΔH_f° (g) kcal/g.f.w.
Remarks			Type	ΔH_v° kcal/g.f.w.	Ref.	
$(i\text{-}C_3H_7)_2$(l) $\{-3 \cdot 1 \pm 1 \cdot 0\}$ $+HgBr_2$(c) $\{-40 \cdot 6 \pm 0 \cdot 2\}$ $= 2 \text{ i-}C_3H_7HgBr$(c)		$-32 \cdot 5 \ \pm 0 \cdot 6$				
$(n\text{-}C_3H_7)_2$(l) $\{-5 \cdot 0 \pm 1 \cdot 3\}$ $+HgI_2$(c) $\{-25 \cdot 2 \pm 0 \cdot 2\}$ $= 2 \text{ n-}C_3H_7HgI$(c)		$-21 \cdot 4 \ \pm 1 \cdot 0$				
$(i\text{-}C_3H_7)_2$(l) $\{-3 \cdot 1 \pm 1 \cdot 0\}$ $+HgI_2$(c) $\{-25 \cdot 2 \pm 0 \cdot 2\}$ $= 2i\text{-}C_3H_7HgI$(c)		$-22 \cdot 0 \ \pm 0 \cdot 6$				
$(C_6H_5)_2$(c) $\{+66 \cdot 8 \pm 1 \cdot 5\}$ $+HgCl_2$(c) $\{-53 \cdot 5 \pm 0 \cdot 5\}$ $= 2C_6H_5HgCl$(c)		$-0 \cdot 3 \ \pm 0 \cdot 8$				
$(C_6H_5)_2$(c) $\{+66 \cdot 8 \pm 1 \cdot 5\}+HCl$(g) $= C_6H_5HgCl$(c) $+C_6H_6$(l) $\{+11 \cdot 72 \pm 0 \cdot 13\}$	Selected value	$+0 \cdot 4 \ \pm 1 \cdot 2$ $0 \cdot 0 \ \pm 1 \cdot 0$				
$(C_6H_5)_2$(c) $\{+66 \cdot 8 \pm 1 \cdot 5\}$ $+HgBr_2$(c) $\{-40 \cdot 6 \pm 0 \cdot 2\}$ $= 2C_6H_5HgBr$(c)		$+8 \cdot 9 \ \pm 1 \cdot 2$				
$(C_6H_5)_2$(c) $\{+66 \cdot 8 \pm 1 \cdot 5\}$ $+HgI_2$(c) $\{-25 \cdot 2 \pm 0 \cdot 2\}$ $= 2C_6H_5HgI$(c)		$+18 \cdot 2 \ \pm 1 \cdot 2$				

Group III organometallic compounds   1 kcal = 4·184 kJ

Determina*

1 Formula	2 g.f.w.	3 Name	State	Purity mol %	Type	No. of expts.	Detn. of reactn.	$-\Delta H_r^\circ$ kcal/g.f.w.	R
C_3H_9B	55·916	Trimethylboron	l		SB	6	m	705·4 ±2·9	49
C_3H_9B		Trimethylboron	l	99·9	SB	6	CO_2	714·5 ±5·4	61
C_3H_9B		Trimethylboron							
$C_6H_{15}B$	97·997	Triethylboron	l		SB		m, an	1175·8 ±5·5	60
$C_6H_{15}B$		Triethylboron	l		SB	2	CO_2	1189·2 ±3·6	61
$C_6H_{15}B$		Triethylboron	l		R	5	an	172·5 ±2·1	63
$C_6H_{15}B$		Triethylboron							
$C_8H_{17}B$	124·036	n-Butylboracyclopentane	l		SB	13	m	1445·7 ±1·0	66
$C_9H_{21}B$	140·079	Tri-n-propylboron	l		SB		m, an	1639·9 ±8·0	60
$C_9H_{21}B$		Tri-n-propylboron	l		SB	3	m	1649·4 ±3·1	63
$C_9H_{21}B$		Tri-n-propylboron							
$C_9H_{21}B$	140·079	Tri-isopropylboron	l		SB	5	m	1645·7 ±2·8	63
$C_{12}H_{27}B$	182·160	Tri-n-butylboron	l		SB	2	CO_2	2127·4 ±2·4	61
$C_{12}H_{27}B$		Tri-n-butylboron	l		SB	11	m	2125·6 ±0·5	66 67
$C_{12}H_{27}B$		Tri-n-butylboron							
$C_{12}H_{27}B$	182·160	Tri-isobutylboron	l		SB	6	m	2111·3 ±1·3	63
$C_{12}H_{27}B$		Tri-isobutylboron	l		R	6	an	146·2 ±1·5	63
$C_{12}H_{27}B$		Tri-isobutylboron							
$C_{12}H_{27}B$	182·160	Tri-s-butylboron	l		SB	6	m	2130±6	56
$C_{15}H_{33}B$	224·241	Tri-3-methylbutylboron	l		SB	9	m	2581·7 ±1·6	63
$C_{18}H_{15}B$	242·131	Triphenylboron	c		R	7	m	256·8 ±1·8	67
$C_{18}H_{33}B$	260·275	Tricyclohexylboron	c		R	7	m	261·9 ±2·3	67
$C_{18}H_{39}B$	266·323	Tri-n-hexylboron	l		R	4	an	136·6 ±0·8	61

1 kcal = 4·184 kJ

| ΔH_r° | | 6 Determination of ΔH_v | | | |
Remarks	5 ΔH_f° (l or c) kcal/g.f.w.	Type	ΔH_v° kcal/g.f.w.	Ref.	7 ΔH_f° (g) kcal/g.f.w.
oduct B_2O_3 (am)	−34·0 ±3·0	C1	4·83±[0·02]	61/46	−29·2 ±3·0
oduct H_3BO_3(c)	−34·1 ±5·5				−29·3 ±5·5
Selected value	−34·1 ±5·5				−29·3 ±5·5
oduct B_2O_3 (am)	−50·7 ±5·6	C1	8·8 ±[0·1]	61/46	−41·9 ±5·6
oduct H_3BO_3(c)	−46·5 ±3·7				−37·7 ±3·7
H_6(g) {+9·15±0·60} +6C_2H_4(g) {+12·45±0·10} = 2$C_6H_{15}B$(l)	−45·2 ±1·2				−36·4 ±1·2
Selected value	−45·3 ±1·2				−36·5 ±1·2
oduct H_3BO_3(c)	−46·4 ±1·1				
oduct B_2O_3 (am)	−73·7 ±8·0	V3	10·0 ±[0·3]	46/14	−63·7 ±8·0
oduct B_2O_3(c)	−66·4 ±3·3				−56·4 ±3·3
Selected value	−66·4 ±3·3				−56·4 ±3·3
oduct B_2O_3(c)	−70·1 ±3·0	V3	10·0 ±[0·3]	46/14	−60·1 ±3·0
oduct H_3BO_3(c)	−82·5 ±2·5	E1	14·8 ±0·5		−67·7 ±2·5
oduct H_3BO_3(c)	−84·3 ±0·7				−69·5 ±0·9
Selected value	−83·4 ±1·0				−68·6 ±1·1
oduct B_2O_3(c)	−91·6 ±1·5	E1	14·3 ±0·5		−77·3 ±1·6
H_6(g) {+9·15±0·60} +6i-C_4H_8(g) {−4·26±0·15} = 2$C_{12}H_{27}B$(l)	−81·3 ±1·2				−67·0 ±1·3
Selected value	−81·3 ±1·2				−67·0 ±1·3
oduct B_2O_3(c)	−73±6	E1	14·5 ±0·5		−58±6
oduct B_2O_3(c)	−108·3 ±1·9	E1	17·2 ±0·6		−91·1 ±2·0
$_{18}H_{15}B$(c)+3H_2O_2(l) {−44·88} = 3C_6H_5OH(c) {−39·45±0·08} +H_3BO_3(c)	+11·6 ±1·9	V4$_m$	19·5 ±0·5 (sub.)	67/49	+31·1 ±2·1
$_{18}H_{33}B$(c)+3H_2O_2(l) {−44·88} = 3$C_6H_{11}OH$(l) {−83·20±0·40} +H_3BO_3(c)	−114·6 ±2·5	E1$_{me}$	19·5 ±1·0 (sub.)	67/49	−95·1 ±3·0
H_6(g) {+9·15±0·60} +6(1-Hexene) (l) {−17·29±0·29} = 2$C_{18}H_{39}B$(l)	−115·6 ±2·2	E1	21·2 ±0·7		−94·4 ±2·4

Group III organometallic compounds   1 kcal = 4·184 kJ

								Determinat
1 Formula	2 g.f.w.	3 Name	State	Purity mol %	No. of Type expts.	Detn. of reactn.	$-\Delta H_r^\circ$ kcal/g.f.w.	Re
$C_{21}H_{45}B$	308·404	Tri-n-heptylboron	l		R 3	an	135·4 ±0·4	61
$C_{24}H_{51}B$	350·485	Tri-n-octylboron	l		R 3	an	135·7 ±0·5	61
$C_{24}H_{51}B$	350·485	Tri-s-octylboron	l		R 5	an	116·3 ±0·6	61
$C_5H_{11}OB$	97·954	Methoxyboracylco- pentane	l		SB 6	m	911·7 ±0·6	66/
$C_8H_{19}OB$	142·051	Di-n-butylborinic acid	l		R 5	m	21·96±0·75	53/
$C_8H_{19}OB$		Di-n-butylborinic acid	l		R 6	m	24·94±0·75	53/
$C_8H_{19}OB$		Di-n-butylborinic acid						
$C_9H_{21}OB$	156·078	Methyl di-n-butyl- borinate	l		SB 7	m	1590·9 ±0·8	66/
$C_{12}H_{27}OB$	198·159	n-Butyl di-n-butyl- borinate	l		SB 8	m	2045·7 ±1·0	62/
$C_2H_7O_2B$	73·888	Dimethoxyborane	l		R 7	m	24·24±0·24	64/
$C_6H_7O_2B$	121·932	Phenylboronic acid	c		R 4	m	84·0 ±0·4	66/
$C_6H_{15}O_2B$	129·996	Di-isopropoxyborane	l		R 4	an	138·4 ±0·8	62/
$C_6H_{15}O_2B$		Di-isopropoxyborane	l		R 2	an	69·9 ±0·4	62/
$C_6H_{15}O_2B$		Di-isopropoxyborane						
$C_{12}H_{27}O_2B$	214·159	Di-n-butyl n-butylboronate	l		SB 7	m	1976·7 ±2·4	62/

1 kcal $= 4\cdot184$ kJ

ΔH_r°		6 Determination of ΔH_v			
Remarks	5 ΔH_f° (l or c) kcal/g.f.w.	7 ΔH_v° Type kcal/g.f.w.		Ref.	ΔH_f° (g) kcal/g.f.w.

Remarks	ΔH_f° (l or c) kcal/g.f.w.	Type	ΔH_v° kcal/g.f.w.	Ref.	ΔH_f° (g) kcal/g.f.w.
$H_6(g)$ {$+9\cdot15\pm0\cdot60$} $+6$(1-Heptene) (l) {$-23\cdot33\pm0\cdot25$} $= 2C_{21}H_{45}B(l)$	$-133\cdot1\ \pm1\cdot8$	E1	$24\cdot4\ \pm0\cdot7$		$-108\cdot7\ \pm2\cdot0$
$H_6(g)$ {$+9\cdot15\pm0\cdot60$} $+6$(1-Octene) (l) {$-29\cdot11\pm0\cdot30$} $= 2C_{24}H_{51}B(l)$	$-150\cdot6\ \pm1\cdot8$	E1	$27\cdot6\ \pm0\cdot7$		$-123\cdot0\ \pm2\cdot0$
$H_6(g)$ {$+9\cdot15\pm0\cdot60$} $+6$(cis+trans-2-Octene) (l) {$-31\cdot6\pm0\cdot6$} $= 2C_{24}H_{51}B(l)$	$-148\cdot4\ \pm3\cdot5$	E1	$27\cdot0\ \pm1\cdot0$		$-121\cdot4\ \pm4\cdot0$
oduct $H_3BO_3(c)$	$-93\cdot3\ \pm0\cdot7$				
$C_4H_9)_2BBr(l)$ {$-84\cdot3\pm1\cdot5$}$+7501H_2O$ (l) $= (n-C_4H_9)_2BOH(l)+[HBr.7500H_2O]$ (l)	$-145\cdot6\ \pm2\cdot0$	E1	15 ± 2		$-130\cdot6\ \pm3\cdot4$
$C_4H_9)_2BI(l)$ {$-66\cdot9\pm1\cdot5$}$+6501H_2O$ (l) $= (n-C_4H_9)_2BOH(l)+[HI.6500H_2O]$ (l)	$-147\cdot0\ \pm2\cdot0$				$-132\cdot0\ \pm3\cdot4$
Selected value	$-146\cdot3\ \pm1\cdot6$				$-131\cdot3\ \pm2\cdot8$
duct $H_3BO_3(c)$	$-131\cdot9\ \pm0\cdot9$				
duct B_2O_3 (am)	$-155\cdot0\ \pm1\cdot0$				
$(OCH_3)_2(l)+\infty H_2O(l)$ $= H_2+H_3BO_3(aq)$ {$-256\cdot29\pm0\cdot25$} $+2CH_3OH(aq)$ {$-58\cdot75\pm0\cdot10$}	$-144\cdot5\ \pm0\cdot5$	V2	$6\cdot14\pm0\cdot30$	33/18	$-138\cdot4\ \pm0\cdot6$
$H_5B(OH)_2(c)+H_2O_2(l)$ {$-44\cdot88$} $= C_6H_5OH(c)$ {$-39\cdot45\pm0\cdot08$} $+H_3BO_3(c)$	$-172\cdot0\ \pm0\cdot5$				
$H_6(g)$ {$+9\cdot15\pm0\cdot60$} $+4(CH_3)_2CO(l)$ {$-59\cdot27\pm0\cdot12$} $= 2[(i-C_3H_7O)_2BH]$ (l)	$-183\cdot2\ \pm1\cdot1$				
$H_6(g)$ {$+9\cdot15\pm0\cdot60$} $+4i-C_3H_7OH(l)$ {$-76\cdot02\pm0\cdot12$} $= 4H_2+2[(i-C_3H_7O)_2BH]$ (l)	$-182\cdot9\ \pm0\cdot8$				
Selected value	$-183\cdot1\ \pm0\cdot5$				
duct B_2O_3 (am)	$-224\cdot0\ \pm2\cdot4$				

Group III organometallic compounds   1 kcal $= 4 \cdot 184$ kJ

						No. of	Detn. of	$-\Delta H_r^\circ$	
1 Formula	2 g.f.w.	3 Name	State	Purity mol %	Type	expts.	reactn.	kcal/g.f.w.	Re
$C_3H_9O_3B$	$103 \cdot 914$	Trimethyl borate	l		R	6	m	$4 \cdot 62 \pm 0 \cdot 10$	52/
$C_6H_{15}O_3B$	$145 \cdot 996$	Triethyl borate	l		R	6	m	$5 \cdot 38 \pm 0 \cdot 12$	52/
$C_9H_{21}O_3B$	$188 \cdot 077$	Tri-n-propyl borate	l		R	6	m	$4 \cdot 60 \pm 0 \cdot 12$	52/
$C_{12}H_{27}O_3B$	$230 \cdot 158$	Tri-n-butyl borate	l		R	5	m	$4 \cdot 66 \pm 0 \cdot 12$	52/
$C_{12}H_{27}O_3B$ $C_{12}H_{27}O_3B$		Tri-n-butyl borate Tri-n-butyl borate	l		SB		m	$1925 \cdot 2 \pm 1 \cdot 2$	65/
$C_{16}H_{36}OB_2$	$266 \cdot 087$	Di-n-butylborinic anhydride	l		SB		m	$2839 \cdot 7 \pm 1 \cdot 1$	65/
$C_{24}H_{20}OB_2$	$330 \cdot 049$	Diphenylborinic anhydride	c		R	5	m	$388 \cdot 9 \pm 1 \cdot 9$	66/
$C_{18}H_{15}O_3B_3$	$311 \cdot 751$	Phenylboronic anhydride	c		R	4	m	$262 \cdot 0 \pm 2 \cdot 0$	66/
$C_3H_{12}NB$	$72 \cdot 947$	Trimethylaminoborane	c	$99 \cdot 95 \pm 0 \cdot 03$	RB	8	m, an		66
$C_6H_{18}NB$	$115 \cdot 028$	Triethylaminoborane	l	$99 \cdot 995$	RB	8	m, an	$1306 \cdot 60 \pm 0 \cdot 28$	67/
$C_8H_{20}NB$	$141 \cdot 066$	Di-n-butylboronamine	l		SB		m	$1499 \cdot 4 \pm 0 \cdot 8$	65/
$C_{12}H_{28}NB$	$197 \cdot 175$	Di-n-butylboron-n-butylamine	l		SB		m	$2136 \cdot 0 \pm 1 \cdot 2$	65/

Determinat

1 kcal = 4·184 kJ

ΔH_r°		6 Determination of ΔH_v			
Remarks	5 ΔH_f° (l or c) kcal/g.f.w.	Type	ΔH_v° kcal/g.f.w.	Ref.	7 ΔH_f° (g) kcal/g.f.w.
$I_3O)_3B(l)+3H_2O(l)$ = $3CH_3OH(l)$ {$-57·01\pm0·05$} + $H_3BO_3(c)$	$-222·9\ \pm0·5$	V3	$8·3\ \pm0·5$	31/13	$-214·6\ \pm0·7$
$H_5O)_3B(l)+3H_2O(l)$ = $3C_2H_5OH(l)$ {$-66·42\pm0·08$} + $H_3BO_3(c)$	$-250·4\ \pm0·6$	V3	$10·5\ \pm0·5$	31/13	$-239·9\ \pm0·8$
$C_3H_7O)_3B(l)+3H_2O(l)$ = $3n\text{-}C_3H_7OH(l)$ {$-72·51\pm0·15$} + $H_3BO_3(c)$	$-269·4\ \pm0·6$	E1	$11·8\ \pm1·0$		$-257·6\ \pm1·2$
$C_4H_9O)_3B(l)+3H_2O(l)$ = $3n\text{-}C_4H_9OH(l)$ {$-78·29\pm0·13$} + $H_3BO_3(c)$	$-286·7\ \pm0·6$	E1	$12·5\ \pm1·0$		$-274·2\ \pm1·2$
duct $H_3BO_3(c)$	$-284·7\ \pm1·3$				$-272·2\ \pm1·7$
Selected value	$-286·7\ \pm0·6$				$-274·2\ \pm1·2$
duct $H_3BO_3(c)$	$-212·8\ \pm1·3$				
$_6H_5)_2B]_2O(c)+4H_2O_2(l)$ {$-44·88$}$+H_2O(l)$ = $4C_6H_5OH(c)$ {$-39·45\pm0·08$} + $2H_3BO_3(c)$	$-84·0\ \pm2·0$				
$H_5BO)_3(c)+3H_2O_2(l)$ {$-44·88$}$+3H_2O(l)$ = $3C_6H_5OH(c)$ {$-39·45\pm0·08$} + $3H_3BO_3(c)$	$-301·2\ \pm2·3$				
$H_{12}NB(c, III)+6·75O_2$ $+18·674[HF.51·219H_2O]$ (l) = $3CO_2+0·5N_2$ $+[HBF_4.14·67HF.58·72H_2O]$ (l)	$-34·04\pm0·55$	S4	$13·6\ \pm0·2$ (sub.)	59/31	$-20·4\ \pm0·6$
$H_{18}NB(l)+11·250O_2$ $+18·67[HF.48·21H_2O]$ (l) = $6CO_2+0·5N_2$ $+[HBF_4.14·67HF.58·72H_2O]$ (l)	$-47·47\pm0·37$	V1	$14·5\ \pm0·2$	67/34	$-33·0\ \pm0·4$
duct $H_3BO_3(c)$	$-95·2\ \pm0·9$				
duct $H_3BO_3(c)$	$-108·0\ \pm1·3$				

Group III organometallic compounds   1 kcal $= 4\cdot184$ kJ

1 Formula	2 g.f.w.	3 Name	State	Purity mol %	Type	No. of expts.	Detn. of reactn.	$-\Delta H_f^\circ$ kcal/g.f.w.	R
$C_6H_{18}N_3B$	$143\cdot041$	Tris(dimethylamino)-borane	l		R	5	m	$68\cdot8 \pm 0\cdot3$	53,
$C_2H_{11}NB$	$70\cdot739$	N-Dimethylamino-diborane	g	$99\cdot94$	R		m, an	$89\cdot62 \pm 0\cdot65$	61,
$C_8H_9S_2B$	$180\cdot100$	2-Phenyl-1, 3, 2-dithioborolan	l		R	2	m	$12\cdot9 \pm 0\cdot5$	66
$C_9H_{11}S_2B$	$194\cdot127$	2-Phenyl-1, 3, 2-dithioborinan	l		R	3	m	$14\cdot3 \pm 0\cdot6$	66,
$C_3H_9S_3B$	$152\cdot107$	Trimethylthioborane	l		R	7	m	$17\cdot6 \pm 0\cdot3$	67,
$C_6H_{15}S_3B$	$194\cdot188$	Triethylthioborane	l		R	4	m	$22\cdot4 \pm 0\cdot4$	66,
$C_9H_{21}S_3B$	$236\cdot270$	Tri-n-propylthioborane	l		R	7	m	$22\cdot3 \pm 0\cdot6$	67,
$C_{12}H_{27}S_3B$	$278\cdot351$	Tri-n-butylthioborane	l		R	8	m	$23\cdot7 \pm 0\cdot3$	67,
$C_{15}H_{33}S_3B$	$320\cdot432$	Tri-n-pentylthioborane	l		R	9	m	$24\cdot2 \pm 0\cdot2$	67,
$C_{18}H_{15}S_3B$	$338\cdot322$	Triphenylthioborane	c		R	9	m	$21\cdot6 \pm 0\cdot3$	67,
$C_8H_{18}ClB$	$160\cdot497$	Di-n-butylchloroborane	l		R	5	m	$18\cdot54 \pm 0\cdot10$	53,

1 kcal = 4·184 kJ

ΔH_r°		6 Determination of ΔH_v			7
Remarks	5 ΔH_f° (l or c) kcal/g.f.w.	Type	ΔH_v° kcal/g.f.w.	Ref.	ΔH_f° (g) kcal/g.f.w.
$H_3)_2N]_3B(l)+3H_2O(l)$ $+(n+3)$ [HCl.54H$_2$O] (l) $= [H_3BO_3+3(CH_3)_2NH_2Cl.54H_2O]$ $(n[HCl.54H_2O])$ (l)	$-77\cdot1 \pm 1\cdot0$	V3	$11\cdot2 \pm 0\cdot2$	51/29	$-65\cdot9 \pm 1\cdot1$
$I_3)_2NB_2H_5(g)+6H_2O(l)$ $= (CH_3)_2NH(g) \{-4\cdot43\pm0\cdot12\}$ $+2H_3BO_3(c)+5H_2$	$-34\cdot75\pm0\cdot81$	V2	$6\cdot89\pm0\cdot05$	61/47	$-27\cdot86\pm0\cdot80$
$H_2CH_2SBC_6H_5(l)+\infty H_2O(l)$ $= HS.CH_2CH_2SH(l) \{-12\cdot84\pm0\cdot28\}$ $+C_6H_5B(OH)_2(aq) \{-168\cdot8\pm0\cdot6\}$	$-32\cdot1 \pm 1\cdot0$				
$H_2CH_2CH_2SBC_6H_5(l)+\infty H_2O(l)$ $= HS.(CH_2)_3SH(l) \{-18\cdot82\pm0\cdot32\}$ $+C_6H_5B(OH)_2(aq) \{-168\cdot8\pm0\cdot6\}$	$-36\cdot7 \pm 1\cdot1$				
$I_3S)_3B(l)+\infty H_2O(l)$ $= 3CH_3SH(g) \{-5\cdot40\pm0\cdot14\}$ $+H_3BO_3(aq) \{-256\cdot29\pm0\cdot25\}$	$-49\cdot8 \pm 0\cdot6$	V3	$12\cdot9 \pm 0\cdot2$	67/50	$-36\cdot9 \pm 0\cdot7$
$H_5S)_3B(l)+\infty H_2O(l)$ $= 3C_2H_5SH(l) \{-17\cdot53\pm0\cdot14\}$ $+H_3BO_3(aq) \{-256\cdot29\pm0\cdot25\}$	$-81\cdot4 \pm 0\cdot7$	V4	$14\cdot7 \pm 0\cdot5$	66/51	$-66\cdot7 \pm 0\cdot9$
$C_3H_7S)_3B(l)+\infty H_2O(l)$ $= 3n-C_3H_7SH(l) \{-23\cdot79\pm0\cdot15\}$ $+H_3BO_3(aq) \{-256\cdot29\pm0\cdot25\}$	$-100\cdot3 \pm 0\cdot7$	V4	$20\cdot8 \pm 0\cdot5$	67/50	$-79\cdot5 \pm 0\cdot9$
$C_4H_9S)_3B(l)+\infty H_2O(l)$ $= 3n-C_4H_9SH(l) \{-29\cdot72\pm0\cdot28\}$ $+H_3BO_3(aq) \{-256\cdot29\pm0\cdot25\}$	$-116\cdot7 \pm 0\cdot6$	V4	$22\cdot9 \pm 0\cdot5$	67/50	$-93\cdot8 \pm 0\cdot8$
$C_5H_{11}S)_3B(l)+\infty H_2O(l)$ $= 3n-C_5H_{11}SH(l) \{-36\cdot07\pm0\cdot18\}$ $+H_3BO_3(aq) \{-256\cdot29\pm0\cdot25\}$	$-135\cdot3 \pm 0\cdot7$	V4	$25\cdot0 \pm 0\cdot5$	67/50	$-110\cdot3 \pm 0\cdot9$
$H_5S)_3B(c)+\infty H_2O(l)$ $= 3C_6H_5SH(l) \{+15\cdot30\pm0\cdot19\}$ $+H_3BO_3(aq) \{-256\cdot29\pm0\cdot25\}$	$+16\cdot3 \pm 0\cdot7$	E1$_{me}$	$30\cdot0 \pm 1\cdot6$ (sub.)	67/50	$+46\cdot3 \pm 1\cdot8$
$C_4H_9)_2BCl(l)+5501H_2O(l)$ $= (n-C_4H_9)_2BOH(l) \{-146\cdot3\pm1\cdot6\}$ $+[HCl.5500H_2O]$ (l)	$-99\cdot4 \pm 1\cdot8$	V3	$12\cdot0 \pm 0\cdot3$	53/23	$-87\cdot4 \pm 1\cdot9$

Group III organometallic compounds   1 kcal = 4·184 kJ

1 Formula	2 g.f.w.	3 Name	State	Purity mol %	Type	No. of expts.	Detn. of reactn.	$-\Delta H_c^\circ$ kcal/g.f.w.	Re
$C_8H_{18}BrB$	204·953	Di-n-butylbromoborane	l		R	5	m	22·56±0·50	53/
$C_8H_{18}IB$	251·948	Di-n-butyliodoborane	l		R	5	m	20·16±0·15	53/
$C_{12}H_{10}ClB$	200·488	Diphenylchloroborane	l		R	5	m	17·2 ±0·2	67/
$C_{12}H_{10}BrB$	244·934	Diphenylbromoborane	l		R	8	m	22·0 ±0·4	67/
$C_6H_5Cl_2B$	158·824	Phenylboron dichloride	l		R	7	m	40·3 ±0·3	67/
$C_6H_5Br_2B$	247·736	Phenylboron dibromide	l		R	6	m	48·9 ±0·3	67/
$C_2H_4O_2ClB$	106·317	2-Chloro-1, 3, 2-dioxaborolan	l		R	5	m	6·9 ±0·6	64/
$C_2H_6O_2ClB$	108·333	Dimethoxychloroborane	l		R	6	m, an	6·4 ±0·2	61/
$C_3H_6O_2ClB$	120·344	4-Methyl-2-chloro-1, 3, 2-dioxaborolan	l		R	2	m	7·2 ±0·3	64/
$C_4H_8O_2ClB$	134·371	4, 5-Dimethyl-2-chloro-1, 3, 2-dioxaborolan	l		R	2	m	8·3 ±0·5	64/
$C_4H_{10}O_2ClB$	136·387	Diethoxychloroborane	l		R	9	m, an	23·5 ±0·7	54/

1 kcal = 4·184 kJ

ΔH_r°		5 ΔH_f° (l or c) kcal/g.f.w.	6 Determination of ΔH_v			7 ΔH_f° (g) kcal/g.f.w.
Remarks			Type	ΔH_v° kcal/g.f.w.	Ref.	
-C$_4$H$_9$)$_3$B(l) {−83·4±1·0}+HBr(g) = (n-C$_4$H$_9$)$_2$BBr(l) +n-C$_4$H$_{10}$(g) {−30·36±0·16}		−84·3 ±1·5	V3	12·5 ±0·3	53/23	−71·8 ±1·6
-C$_4$H$_9$)$_3$B(l) {−83·4±1·0}+HI(g) = (n-C$_4$H$_9$)$_2$BI(l) +n-C$_4$H$_{10}$(g) {−30·36±0·16}		−66·9 ±1·5	E1	13·0 ±0·6	53/23	−53·9 ±1·7
$_6$H$_5$)$_2$BCl(l)+2001H$_2$O(l) = (C$_6$H$_5$)$_2$BOH(l) {−77·4±1·8} (67/51) +[HCl.2000H$_2$O] (l)		−31·9 ±2·0	V4	9·9 ±0·5	67/49	−22·0 ±2·1
$_6$H$_5$)$_2$BBr(l)+1701H$_2$O(l) = (C$_6$H$_5$)$_2$BOH(l) {−77·4±1·8} (67/51) +[HBr.1700H$_2$O] (l)		−16·1 ±2·0	V4	14·4 ±0·5	67/49	−1·7 ±2·1
$_6$H$_5$BCl$_2$(l)+3002H$_2$O(l) = C$_6$H$_5$B(OH)$_2$(aq) {−168·8±0·6} +2[HCl.1500H$_2$O] (l)		−71·7 ±0·6	V3	8·1 ±0·2	67/49	−63·6 ±0·7
$_6$H$_5$BBr$_2$(l)+4002H$_2$O(l) = C$_6$H$_5$B(OH)$_2$(aq) {−168·8±0·6} +2[HBr.2000H$_2$O] (l)		−41·4 ±0·6	V4	10·5 ±0·5	67/49	−30·9 ±0·8
$\overline{CH_2CH_2OBCl}$(l)+3H$_2$O(l) = H$_3$BO$_3$(c)+HCl(g) +HO.CH$_2$CH$_2$OH(l) {−108·73±0·18}		−180·4 ±0·8				
$\overline{CH_3O)_2BCl}$(l)+3H$_2$O(l) = H$_3$BO$_3$(c) +2CH$_3$OH(l) {−57·01±0·05}+HCl(g)		−186·2 ±0·4	V3	8·2 ±0·3	31/13	−178·0 ±0·5
CH(CH$_3$)CH$_2$OBCl(l)+3H$_2$O(l) = H$_3$BO$_3$(c)+HCl(g) +HOCH(CH$_3$)CH$_2$OH(l) {−119·58±0·45}		−190·9 ±1·0				
$\overline{CH(CH_3)CH(CH_3)OBCl}$(l)+3H$_2$O(l) = H$_3BO_3$(c)+HCl(g) +HOCH(CH$_3$)CH(CH$_3$)OH(l) {−129·43±0·60}		−199·7 ±1·2				
C$_2$H$_5$O)$_2$BCl(l)+2703H$_2$O(l) = [H$_3$BO$_3$+HCl+2C$_2$H$_5$OH] (2700H O) (l)		−205·2 ±1·0	V3	9·3 ±0·2	31/13	−195·9 +1·1

Group III organometallic compounds   1 kcal = 4·184 kJ

								Determinat
					No. of	Detn. of	$-\Delta H_r^\circ$	
1 Formula	2 g.f.w.	3 Name	State	Purity mol %	Type expts.	reactn.	kcal/g.f.w.	R
$C_2H_5OCl_2B$	126·779	Ethoxydichloroborane	l		R	4	m	42·5 ±0·8	54/
$C_4H_{12}N_2ClB$	134·418	Bis-(dimethylamino)- chloroborane	l		R	4	m	59·1 ±0·8	54/
$C_2H_6NCl_2B$	125·794	Dimethylamino- dichloroborane	l		R	5	m	55·1 ±0·8	54/
$C_2H_4S_2ClB$	138·446	2-Chloro- 1, 3, 2-dithioborolan	l		R	3	m	38·3 ±0·4	66/
$C_3H_6S_2ClB$	152·473	2-Chloro- 1, 3, 2-dithioborinan	l		R		m	37·3 ±0·4	66/
C_3H_9Al	72·087	Trimethylaluminium	l		SB	6	m	761·1 ±2·3	49/
C_3H_9Al		Trimethylaluminium	l	99·5, an	R	5	m	123·0 ±0·9	63/
C_3H_9Al		Trimethylaluminium							
$C_4H_{11}Al$	86·114	Diethylaluminium hydride	l		SB	1	m	878·7 ±4·4*	65/
$C_4H_{11}Al$		Diethylaluminium hydride	l		SB	4	m	903·4 ±1·6	67/
$C_4H_{11}Al$		Diethylaluminium hydride							
$C_6H_{15}Al$	114·168	Di-n-propylaluminium hydride	l		SB	3	m	1218·8 ±0·8	67/
$C_6H_{15}Al$	114·168	Triethylaluminium	l		SB	7	m	1225·0 ±2·1	65/4
$C_6H_{15}Al$		Triethylaluminium	l		SB	4	m	1220·3 ±0·7	67/3
$C_6H_{15}Al$		Triethylaluminium	l		R		an	154·8 ±2·0	61/4
$C_6H_{15}Al$		Triethylaluminium							

1 kcal = 4·184 kJ

ΔH_r°	5 ΔH_f° (l or c) kcal/g.f.w.	6 Determination of ΔH_v			7 ΔH_f° (g) kcal/g.f.w.
Remarks		Type	ΔH_v° kcal/g.f.w.	Ref.	
$_2H_5OBCl_2$(l)$+3703H_2O$(l) $= [H_3BO_3+2HCl+C_2H_5OH]$ $(3700H_2O)$ (l)	$-157\cdot2$ $\pm1\cdot1$	V3	$8\cdot4$ $\pm0\cdot2$	31/13	$-148\cdot8$ $\pm1\cdot2$
$CH_3)_2N]_2BCl$(l)$+3H_2O$(l)$+101[HCl.54H_2O]$ $= [H_3BO_3+2(CH_3)_2NH_2Cl.27H_2O]$ $(100[HCl.54H_2O])$ (l)	$-94\cdot1$ $\pm1\cdot1$	V3	$10\cdot0$ $\pm0\cdot5$	51/29	$-84\cdot1$ $\pm1\cdot2$
$CH_3)_2NBCl_2$(l)$+4003H_2O$(l) $= [H_3BO_3+(CH_3)_2NH_2Cl$ $+HCl]$ $(4000H_2O)$ (l)	$-106\cdot9$ $\pm1\cdot1$	V3	$8\cdot9$ $\pm0\cdot3$	51/29	$-98\cdot0$ $\pm1\cdot2$
$\overline{CH_2CH_2SBCl}$(l)$+\infty H_2O$(l) $= HSCH_2CH_2SH$(l) $\{-12\cdot84\pm0\cdot28\}$ $+[H_3BO_3+HCl]$ (aq)	$-66\cdot0$ $\pm0\cdot7$				
$\overline{(CH_2)_3SBCl}$(l)$+\infty H_2O$(l) $= HS(CH_2)_3SH$(l) $\{-18\cdot82\pm0\cdot32\}$ $+[H_3BO_3+HCl]$ (aq.)	$-72\cdot9$ $\pm0\cdot7$				
roduct α-Al_2O_3(c) $\{-400\cdot48\pm0\cdot25\}$ (57/39)	$-28\cdot7$ $\pm2\cdot4$*	V1	$15\cdot1$ $\pm0\cdot4$	63/52, 41/13	$-13\cdot6$ $\pm2\cdot5$*
$CH_3)_3Al$(l)$+3CH_3CO_2H$ (toluene solution) $\{-115\cdot5\pm0\cdot1\}$ $= (CH_3CO_2)_3Al$(c) $\{-451\cdot8\pm0\cdot8\}$ $+3CH_4$(g) $\{-17\cdot89\pm0\cdot07\}$	$-36\cdot0$ $\pm1\cdot6$*		ΔH_v to monomer		$-20\cdot9$ $\pm1\cdot7$*
Selected value	$-36\cdot0$ $\pm1\cdot6$*				$-20\cdot9$ $\pm1\cdot7$*
roduct α-Al_2O_3(c) $\{-400\cdot48\pm0\cdot25\}$ (57/39)	$-73\cdot5$ $\pm4\cdot5$*	V4	$13\cdot8$ $\pm0\cdot5$*	65/48	
roduct α-Al_2O_3(c) $\{-400\cdot48\pm0\cdot25\}$ (57/39); C/An 0·991	$-48\cdot8$ $\pm1\cdot7$*		Vapour association probable		
Selected value	$-48\cdot8$ $\pm1\cdot7$*				
roduct α-Al_2O_3(c) $\{-400\cdot48\pm0\cdot25\}$ (57/39); C/An 0·995	$-58\cdot1$ $\pm0\cdot9$*				
roduct α-Al_2O_3(c) $\{-400\cdot48\pm0\cdot25\}$ (57/39); C/An 0·97	$-51\cdot9$ $\pm2\cdot2$*	V3	$17\cdot5$ $\pm0\cdot5$	61/48	$-34\cdot4$ $\pm2\cdot3$*
roduct α-Al_2O_3(c) $\{-400\cdot48\pm0\cdot25\}$ (57/39); C/An 0·992	$-56\cdot6$ $\pm0\cdot9$*		ΔH_v to monomer		$-39\cdot1$ $\pm1\cdot0$*
$C_2H_5)_3Al$(l)$+3H_2O$(g) $= Al(OH)_3$ (am) $\{-305\pm4\}$ $+3C_2H_6$(g) $\{-20\cdot24\pm0\cdot12\}$	$-37\cdot5$ $\pm5\cdot0$*				$-20\cdot0$ $\pm5\cdot0$*
Selected value	$-56\cdot6$ $\pm0\cdot9$*				$-39\cdot1$ $\pm1\cdot0$*

Group III organometallic compounds   1 kcal $= 4 \cdot 184$ kJ

1 Formula	2 g.f.w.	3 Name	State	Purity mol %	Type	No. of expts.	Detn. of reactn.	$-\Delta H_r^\circ$ kcal/g.f.w.	Ref
$C_8H_{19}Al$	$142 \cdot 222$	Di-n-butylaluminium hydride	l		SB	3	m	$1534 \cdot 0 \pm 1 \cdot 3$	67/3
$C_8H_{19}Al$	$142 \cdot 222$	Di-isobutylaluminium hydride	l		SB	5	m	$1505 \cdot 5 \pm 3 \cdot 9$	65/4
$C_8H_{19}Al$		Di-isobutylaluminium hydride	l		SB	4	m	$1532 \cdot 5 \pm 1 \cdot 0$	67/3
$C_8H_{19}Al$		Di-isobutylaluminium hydride							
$C_9H_{21}Al$	$156 \cdot 249$	Tri-n-propylaluminium	l		SB	4	m	$1687 \cdot 0 \pm 1 \cdot 0$	67/3
$C_{12}H_{27}Al$	$198 \cdot 330$	Tri-n-butylaluminium	l		SB	3	m	$2162 \cdot 1 \pm 1 \cdot 3$	67/3
$C_{12}H_{27}Al$	$198 \cdot 330$	Tri-isobutylaluminium	l		SB	5	m	$2181 \cdot 2 \pm 7 \cdot 1$	64/3
$C_{12}H_{27}Al$		Tri-isobutylaluminium	l		SB	3	m	$2158 \cdot 3 \pm 1 \cdot 8$	67/3
$C_{12}H_{27}Al$		Tri-isobutylaluminium							

GROUP IIIB

1 Formula	2 g.f.w.	3 Name	State	Purity mol %	Type	No. of expts.	Detn. of reactn.	$-\Delta H_r^\circ$ kcal/g.f.w.	Ref
C_3H_9Ga	$114 \cdot 83$	Trimethylgallium	l		SB	6	m	$701 \cdot 0 \pm 1 \cdot 9$	58/3
C_3H_9Ga		Trimethylgallium	l		R	8	m, an	$47 \cdot 8 \pm 2 \cdot 0$	58/3
C_3H_9Ga		Trimethylgallium							
CH_3I_2Ga	$338 \cdot 56$	Methylgallium di-iodide	c		R	8	m	$37 \cdot 9 \pm 1 \cdot 0$	58/3
C_3H_9In	$159 \cdot 93$	Trimethylindium	c		R	7	m, an	$159 \cdot 0 \pm [1 \cdot 0]$	68/1
C_5H_5Tl	$269 \cdot 47$	Cyclopentadienylthallium	c		R	5	m	$21 \cdot 75 \pm 0 \cdot 50$	67/36

1 kcal = $4 \cdot 184$ kJ

ΔH_r°		5	6 Determination of ΔH_v			7
Remarks		ΔH_f° (l or c) kcal/g.f.w.	Type	ΔH_v° kcal/g.f.w.	Ref.	ΔH_f° (g) kcal/g.f.w.
roduct α-Al$_2$O$_3$(c) $\{-400 \cdot 48 \pm 0 \cdot 25\}$ (57/39); C/An 0·995		$-67 \cdot 6 \pm 1 \cdot 4^*$				
roduct α-Al$_2$O$_3$(c) $\{-400 \cdot 48 \pm 0 \cdot 25\}$ (57/39); C/An 0·97		$-96 \cdot 1 \pm 4 \cdot 0^*$	V4	$10 \cdot 1 \pm 0 \cdot 5$	65/48	
roduct α-Al$_2$O$_3$(c) $\{-400 \cdot 48 \pm 0 \cdot 25\}$ (57/39); C/An 0·989		$-69 \cdot 1 \pm 1 \cdot 1^*$	Vapour associ- ation probable			
	Selected value	$-69 \cdot 1 \pm 1 \cdot 1^*$				
roduct α-Al$_2$O$_3$(c) $\{-400 \cdot 48 \pm 0 \cdot 25\}$ (57/39}; C/An 0·994		$-77 \cdot 0 \pm 1 \cdot 1^*$				
roduct α-Al$_2$O$_3$(c) $\{-400 \cdot 48 \pm 0 \cdot 25\}$ (57/39); C/An 0·995		$-89 \cdot 0 \pm 1 \cdot 4^*$				
roduct α-Al$_2$O$_3$(c) $\{-400 \cdot 48 \pm 0 \cdot 25\}$ (57/39)		$-69 \cdot 9 \pm 7 \cdot 2^*$				
roduct α-Al$_2$O$_3$(c) $\{-400 \cdot 48 \pm 0 \cdot 25\}$ (57/39}; C/An 0·995		$-92 \cdot 8 \pm 1 \cdot 9^*$				
	Selected value	$-92 \cdot 8 \pm 1 \cdot 9^*$				
roduct Ga$_2$O$_3$(c) $\{-261 \cdot 05 \pm 0 \cdot 30\}$ (62/54) CH$_3$)$_3$Ga(l)+3I$_2$(c) $\quad$ = GaI$_3$(c) $\{-57 \cdot 1\}$ $\quad\quad$ +3CH$_3$I(l) $\{-3 \cdot 3 \pm 0 \cdot 3\}$		$-19 \cdot 1 \pm 2 \cdot 0$ $-19 \cdot 2 \pm 3 \cdot 0$	V2	$7 \cdot 9 \pm 0 \cdot 2$	58/34	$-11 \cdot 2 \pm 2 \cdot 0$ $-11 \cdot 3 \pm 3 \cdot 0$
	Selected value	$-19 \cdot 1 \pm 1 \cdot 5$				$-11 \cdot 2 \pm 1 \cdot 6$
CH$_3$)$_3$Ga(l) $\{-19 \cdot 1 \pm 1 \cdot 5\}$+2I$_2$(c) $\quad$ = CH$_3$GaI$_2$(c)+2CH$_3$I(l) $\{-3 \cdot 3 \pm 0 \cdot 3\}$		$-50 \cdot 4 \pm 2 \cdot 0$				
CH$_3$)$_3$In(c)+3Br$_2$(l) $\quad$ = InBr$_3$(c) $\{-102 \cdot 5\}$ $\quad\quad$ +3CH$_3$Br(g) $\{-9 \cdot 1 \pm 0 \cdot 3\}$		$+29 \cdot 2 \pm [1 \cdot 4]$	S3	$11 \cdot 6 \pm 0 \cdot 6$ (sub.)	34/13	$+40 \cdot 8 \pm [1 \cdot 5]$
TlOH.2050H$_2$O] (l) $\{-54 \cdot 5\}$ +c-C$_5$H$_6$(g) $\{+31 \cdot 94 \pm 0 \cdot 28\}$ $\quad$ = TlC$_5$H$_5$(c)+2051H$_2$O(l)		$+24 \cdot 0 \pm 0 \cdot 6^*$				

Group IV organometallic compounds   1 kcal = 4·184 kJ

1 Formula	2 g.f.w.	3 Name	State	Purity mol %	No. of Type expts.	Detn. of reactn.	$-\Delta H_r^\circ$ kcal/g.f.w.	4 Determination Ref.

GROUP IVB

ALL STATIC-BOMB MEASUREMENTS ON ORGANOSILICON

$C_3H_{10}OSi$	90·199	Trimethylsilanol	l		R	5	m	1·2 ±0·1	67/52
$C_6H_{18}OSi_2$	162·382	Hexamethyldisiloxane	l		RB	16	m		64/3
$C_4H_{13}NSi$	103·241	Methylamino- trimethylsilane	l		R	6	m	5·8 ±0·1	67/52
$C_5H_{15}NSi$	117·268	Dimethylamino- trimethylsilane	l		R	5	m	0·8 ±0·1	67/52
$C_7H_{19}NSi$	145·322	Diethylamino- trimethylsilane	l		R	4	m	32·1 ±0·4	62/59
$C_6H_{19}NSi_2$	161·397	Hexamethyldisilazane	l		R	6	m	34·49±0·04	66/40
$C_6H_{19}NSi_2$		Hexamethyldisilazane	l		R	6	m	13·5 ±0·2	67/52
$C_6H_{19}NSi_2$		Hexamethyldisilazane							
$C_7H_{21}NSi_2$	175·424	Bis(trimethylsilyl)- methylamine	l		R	6	m	15·4 ±0·2	67/52
$C_9H_{27}NSi_3$	233·580	Tris(trimethylsilyl)- amine	c		R	6	m	27·3 ±0·5	67/52
$C_7H_{18}SSi$	162·372	Trimethylsilyl n-butyl sulphide	l		R	5	m	1·8 ±0·1	67/52

1 kcal = 4·184 kJ

| H_r° | | | 6 Determination of ΔH_v | | |
	Remarks	5 ΔH_f° (l or c) kcal/g.f.w.	ΔH_v° Type kcal/g.f.w.	Ref.	7 ΔH_f° (g) kcal/g.f.w.

MPOUNDS HAVE BEEN EXCLUDED FROM THESE TABLES

$_3)_3$SiOH(l) = 0·5[(CH$_3$)$_3$Si]$_2$O(l) {−194·7±1·3} +0·5H$_2$O(l)		−130·3 ±0·8	V3 10·9 ±0·4	53/36	−119·4 ±0·9
equation (80)		−194·7 ±1·3	V3 8·9 ±0·4	47/15	−185·8 ±1·4
$_3)_3$SiNHCH$_3$(l)+0·5H$_2$O(l) = 0·5[(CH$_3$)$_3$Si]$_2$O(l) {−194·7±1·3} +CH$_3$NH$_2$(g) {−5·50±0·12}		−62·9 ±0·8	E8 8·6 ±0·5	67/52	−54·3 ±1·0
$_3)_3$SiN(CH$_3$)$_2$(l)+0·5H$_2$O(l) = 0·5[(CH$_3$)$_3$Si]$_2$O(l) {−194·7±1·3} +(CH$_3$)$_2$NH(g) {−4·43±0·12}		−66·8 ±0·8	V4 7·6 ±0·4	58/44	−59·2 ±0·9
$_3)_3$SiCl(l) {−91·8±0·5} 2(C$_2$H$_5$)$_2$NH(l) {−24·78±0·29} = (CH$_3$)$_3$SiN(C$_2$H$_5$)$_2$(l) +(C$_2$H$_5$)$_2$NH$_2$Cl(c) {−85·76±0·33}		−87·7 ±0·9			
$_3)_3$Si]$_2$NH(l)+[3(HCl.1600·33H$_2$O)] (l) = [(CH$_3$)$_3$Si]$_2$O(l) {−194·7±1·3} +[NH$_4$Cl+2HCl] (4800H$_2$O) (l)		−123·8 ±1·4	E8 9·9 ±0·5	67/52	−113·9 ±1·6
$_3)_3$Si]$_2$NH(l)+H$_2$O(l) = [(CH$_3$)$_3$Si]$_2$O(l) {−194·7±1·3} +NH$_3$(g)		−123·9 ±1·6			−114·0 ±1·8
	Selected value	−123·8 ±1·0			−113·9 ±1·2
$_3)_3$Si]$_2$NCH$_3$(l)+H$_2$O(l) = [(CH$_3$)$_3$Si]$_2$O(l) {−194·7±1·3} +CH$_3$NH$_2$(g) {−5·50±0·12}		−116·5 ±1·6	E8 9·3 ±0·5	67/52	−107·2 ±1·8
$_3)_3$Si]$_3$N(c)+1·5H$_2$O(l) = 1·5[(CH$_3$)$_3$Si]$_2$O(l) {−194·7±1·3} +NH$_3$(g)		−173·3 ±2·2	E8$_{me}$ 13·0 ±2·0 (sub.)	67/52	−160·3 ±2·9
$_3)_3$SiSC$_4$H$_9$(l)+0·5H$_2$O(l) = 0·5[(CH$_3$)$_3$Si]$_2$O(l) {−194·7±1·3} +n-C$_4$H$_9$SH(l) {−29·72±0·28}		−91·1 ±0·8	E8 9·7 ±0·5	67/52	−81·4 ±1·0

Group IV organometallic compounds   1 kcal = 4·184 kJ

1 Formula	2 g.f.w.	3 Name	State	Purity mol %	Type	No. of expts.	Detn. of reactn.	$-\Delta H_r^\circ$ kcal/g.f.w.	F
CH_4Cl_2Si	115·035	Methyldichlorosilane	l	99·60, glc	RB	5	m	353·6 ±0·8	6
C_2H_7ClSi	94·617	Dimethylchlorosilane	l	99·52, glc	RB	5	m	570·5 ±0·6	6
$C_2H_6Cl_2Si$	129·062	Dimethyldichlorosilane	l	99·50, glc	RB	5	m	503·4 ±0·5	68
$C_2H_6Cl_2Si$		Dimethyldichlorosilane	l		R	2	m	32·09±0·03	66
$C_2H_6Cl_2Si$		Dimethyldichlorosilane							
C_3H_9ClSi	108·644	Trimethylchlorosilane	l		R	5	m	11·18±0·16	66
C_3H_9ClSi		Trimethylchlorosilane	l		R	6	m	−6·5 ±0·1	67
C_3H_9ClSi		Trimethylchlorosilane							
C_3H_9BrSi	153·100	Trimethylbromosilane	l		R	6	m	−6·0 ±0·1	67
$C_{12}H_{10}Cl_2Si$	253·206	Diphenyldichlorosilane	l		R	4	m	31·64±0·46	66
$C_8H_{20}Ge$	188·84	Tetraethylgermanium	l	glc	SB	5	CO_2	1519·30±0·76	64
$C_8H_{20}Ge$		Tetraethylgermanium	l		SB		m	1515·6 ±1·7	63
$C_8H_{20}Ge$		Tetraethylgermanium	l		RB	5	m	1517·8 ±1·5	64
$C_8H_{20}Ge$		Tetraethylgermanium							
$C_{12}H_{28}Ge$	244·95	Tetra-n-propylgermanium	l		SB	5	CO_2	2144·51±0·47	64
$C_{21}H_{30}Ge_2$	319·55	Hexaethyldigermane	l		SB		m	2321·0 ±2·0	63
$C_{12}H_{30}OGe_2$	335·55	Bis-triethylgermanium oxide	l		SB		m	2264·0 ±2·0	66

1 kcal = 4·184 kJ

ΔH_r° Remarks	5 ΔH_f° (l or c) kcal/g.f.w.	6 Determination of ΔH_v			7 ΔH_f° (g) kcal/g.f.w.
		Type	ΔH_v° kcal/g.f.w.	Ref.	
duct SiO_2 (colloidal soln.) {−217·44} : N_2H_4.2HCl solution	−105·9 ±0·8*	V3	6·7 ±0·4	47/15	−99·2 ±1·0*
duct SiO_2 (colloidal soln.) {−217·44} : N_2H_4.2HCl solution	−79·8 ±0·6*				
duct SiO_2 (colloidal soln.) {−217·44} : N_2H_4.2HCl solution	−118·4 ±0·5*	V3	8·2 ±0·4	47/15	−110·2 ±0·7*
$_3)_2SiCl_2(l)+1002H_2O(l)$ = $(CH_3)_2Si(OH)_2(aq)$ {−205} +2[HCl.500H_2O] (l)	−116·0 ±[1·0]				−107·8 ±[1·2]
Selected value	−118·4 ±0·5				−110·2 ±0·7
$_3)_3SiCl(l)+800·5H_2O(l)$ = 0·5[$(CH_3)_3Si]_2O(l)$ {−194·7±1·3} +[HCl.800H_2O] (l)	−91·9 ±0·8	V3	7·2 ±0·4	46/18	−84·7 ±0·9
$_3)_3SiCl(l)+0·5H_2O(l)$ = 0·5[$(CH_3)_3Si]_2O(l)$ {−194·7±1·3} +HCl(g)	−91·8 ±0·8				−84·6 ±0·9
Selected value	−91·8 ±0·5				−84·6 ±0·7
$_3)_3SiBr(l)+0·5H_2O(l)$ = 0·5[$(CH_3)_3Si]_2O(l)$ {−194·7±1·3} +HBr(g)	−77·9 ±0·8	E8	7·8 ±0·5	67/52	−71·1 ±1·0
$H_5)_2SiCl_2(l)+1002H_2O(l)$ = $(C_6H_5)_2Si(OH)_2(aq)$ {−155} +2[HCl.500H_2O] (l)	−66·5 ±[1·0]	V4	16·6 ±1·0	47/15	−49·9 ±[1·5]
An 0·99759±0·00043; product GeO_2 (am) {−129·08±0·13}	−45·34±0·80	V4	10·7 ±0·3	63/54	−34·6 ±1·0
oduct GeO_2(am) {−129·08±0·13}	−49·0 ±1·7				−38·3 ±1·7
oduct GeO_2(c, hex.) {−132·58±0·24} (66/42)	−50·3 ±1·6				−39·6 ±1·6
Selected value	−50·3 ±1·6				−39·6 ±1·6
An 0·99744±0·00032; product GeO_2(am) {−129·08±0·13}	−69·59±0·53	E6	14·7 ±1·0	64/40	−54·9 ±1·2
oduct GeO_2(am) {−129·08±0·13}	−90·5 ±2·0	E7	15·0 ±0·5	63/54	−75·5 ±2·2
An 0·999; product GeO_2(am) {−129·08±0·13}	−147·5 ±2·0	E4	14·0 ±1·0	66/43	−133·5 ±2·5

Group IV organometallic compounds   1 kcal = 4·184 kJ

1 Formula	2 g.f.w.	3 Name	State	Purity mol %	Type	No. of expts.	Detn. of reactn.	$-\Delta H_r^\circ$ kcal/g.f.w.	F
$C_4H_{12}Sn$	178·83	Tetramethyltin	l	99·94	SB	4	CO_2	912·40±0·45	6
$C_4H_{12}Sn$		Tetramethyltin	l		SB		m	905·7 ±[1·0]	57
$C_4H_{12}Sn$		Tetramethyltin	l		SB		m	912·2 ±0·8	66
$C_4H_{12}Sn$		Tetramethyltin							
$C_5H_{12}Sn$	190·84	Trimethylvinyltin	l		R	6	m	45·75±0·21	59
$C_5H_{12}Sn$		Trimethylvinyltin	l		SB		m	1058·8 ±0·8	63
$C_5H_{12}Sn$		Trimethylvinyltin							
$C_5H_{14}Sn$	192·86	Trimethylethyltin	l		SB	5	CO_2	1071·22±0·55	6
$C_6H_{16}Sn$	206·88	Trimethylisopropyltin	l	99·9, glc	SB	4	CO_2	1228·74±1·00	6
$C_7H_{18}Sn$	220·91	Trimethyl-t-butyltin	c		SB	4	CO_2	1383·06±1·05	6
$C_8H_{12}Sn$	226·87	Tetravinyltin	l		SB		m	1373·0 ±0·9	63
$C_8H_{18}Sn$	232·92	Triethylvinyltin	l		SB		m	1514·7 ±0·7	63
$C_8H_{20}Sn$	234·94	Tetraethyltin	l	glc	SB	7	CO_2	1551·49±0·56	63
$C_8H_{20}Sn$		Tetraethyltin	l		SB		m	1545·9 ±[1·5]	57
$C_8H_{20}Sn$		Tetraethyltin	l		SB		m	1551 ±2	66
$C_8H_{20}Sn$		Tetraethyltin							
$C_9H_{14}Sn$	240·90	Trimethylphenyltin	l		R	9	m	51·79±0·13	59
$C_{10}H_{16}Sn$	254·93	Trimethylbenzyltin	l		R	10	m	54·16±0·22	59
$C_{12}H_{20}Sn$	282·98	Tetra-allyltin	l		SB		m	1909·9 ±1·7	63
$C_{12}H_{28}Sn$	291·05	Tetra-n-propyltin	l		SB	5	CO_2	2173·3 ±1·2	63
$C_{12}H_{28}Sn$		Tetra-n-propyltin	l		SB		m	2176·4 ±1·4	63
$C_{12}H_{28}Sn$		Tetra-n-propyltin							

1 kcal = 4·184 kJ

ΔH_r° Remarks	5 ΔH_f° (l or c) kcal/g.f.w.	6 Determination of ΔH_v			7 ΔH_f° (g) kcal/g.f.w.
		Type	ΔH_v° kcal/g.f.w.	Ref.	
An 0·99994±0·00026; product SnO$_2$(c) {−138·81±0·08} (53/24)	−12·50±0·46	V4	7·9 ±0·3	30/2	−4·6 ±0·6
duct SnO$_2$(c) {−138·81±0·08} (53/24)	−19·2 ±[1·0]				−11·3 ±[1·1]
An 0·999; SnO$_2$(c) {−138·81±0·08} (53/24)	−12·7 ±0·8				−4·8 ±0·9
Selected value	−12·50±0·46				−4·6 ±0·6
H$_{12}$Sn(l)+Br$_2$(g) = (CH$_3$)$_3$SnBr(l) {−44·3±1·0}+ CH$_2$=CHBr(g) {+18·71±0·52}	+12·8 ±1·2	E1	8·9 ±0·5	63/1	+21·7 ±1·3
duct SnO$_2$(c) {−138·81±0·08} (53/24)	+39·8 ±0·8				+48·7 ±1·0
Selected value	+12·8 ±1·2				+21·7 ±1·3
duct SnO$_2$(c) {−138·81±0·08} (53/24)	−16·05±0·57	V4	9·0 ±0·4	30/2	−7·1 ±0·7
An 1·00089±0·00019; product SnO$_2$(c) {−138·81±0·08} (53/24)	−20·9 ±1·0	E1	9·7 ±0·5	66/3	−11·2 ±1·1
An 1·00086±0·00030; product SnO$_2$(c) {−138·81±0·08} (53/24)	−28·9 ±1·1	E1$_{me}$	12·9 ±1·0 (sub.)	66/3	−16·0 ±1·8
duct SnO$_2$(c) {−138·81±0·08} (53/24)	+71·9 ±0·9				
duct SnO$_2$(c) {−138·81±0·08} (53/24)	+8·6 ±0·7				
An 0·99923±0·00089; product SnO$_2$(c) {−138·81±0·08} (53/24)	−22·88±0·58	E1	12·2 ±0·5	63/1	−10·7 ±0·8
duct SnO$_2$(c) {−138·81±0·08} (53/24)	−28·5 ±[1·5]				−16·3 ±[1·6]
An 0·999; product SnO$_2$(c) {−138·81±0·08} (53/24)	−23 ±2				−11 ±2
Selected value	−22·88±0·58				−10·7 ±0·8
H$_3$)$_3$SnC$_6$H$_5$(l)+Br$_2$(g) = (CH$_3$)$_3$SnBr(l) {−44·3±1·0} +C$_6$H$_5$Br(l) {+14·5±1·0}	+14·6 ±1·7	E1	12·5 ±1·0	63/1	+27·1 ±2·0
H$_3$)$_3$SnC$_7$H$_7$(l)+Br$_2$(g) = (CH$_3$)$_3$SnBr(l) {−44·3±1·0} +C$_7$H$_7$Br(l) {+5·6±1·8}	+8·1 ±2·1	E1	13·5 ±1·0	63/1	+21·6 ±2·4
oduct SnO$_2$(c) {−138·81±0·08} (53/24)	−40·7 ±1·7				
An 1·00000±0·00086; product SnO$_2$(c) {−138·81±0·08} (53/24)	−50·5 ±1·2	E1	16·0 ±0·5	63/1	−34·5 ±1·3
oduct SnO$_2$(c) {−138·81±0·08} (53/24)	−47·4 ±1·4				−31·4 ±1·5
Selected value	−50·5 ±1·2				−34·5 ±1·3

Group IV organometallic compounds   1 kcal = 4·184 kJ

1 Formula	2 g.f.w.	3 Name	State	Purity mol %	Type	No. of expts.	Detn. of reactn.	$-\Delta H_r^\circ$ kcal/g.f.w.	R
$C_{12}H_{28}Sn$	291·05	Tetraisopropyltin	l		SB	5	CO_2	2179·0 ±1·3	66
$C_{12}H_{28}Sn$ $C_{12}H_{28}Sn$		Tetraisopropyltin Tetraisopropyltin	l		SB		m	2174·1 ±0·9	63
$C_{16}H_{36}Sn$	347·16	Tetra-n-butyltin	l	glc	SB	6	CO_2	2800·5 ±1·6	63
$C_{16}H_{36}Sn$ $C_{16}H_{36}Sn$		Tetra-n-butyltin Tetra-n-butyltin	l		SB		m	2801·1 ±0·7	63
$C_{16}H_{36}Sn$	347·16	Tetraisobutyltin	l		SB		m	2794·2 ±1·4	63
$C_{24}H_{20}Sn$	427·12	Tetraphenyltin	c		SB	3	CO_2	3177·72±0·82	64
$C_{24}H_{20}Sn$		Tetraphenyltin	c		SB	12	m	3195 ±3	65
$C_{24}H_{20}Sn$ $C_{24}H_{20}Sn$		Tetraphenyltin Tetraphenyltin	c		SB		m	3242·5 ±1·7	63
$C_{24}H_{44}Sn$	451·31	Tetracyclohexyltin	c		SB		m	3811·8 ±6·8	63
$C_6H_{18}Sn_2$	327·59	Hexamethyldistannane	l		R	4	m	70·24±0·50	57
$C_{12}H_{30}Sn_2$	411·75	Hexaethyldistannane	l		SB		m	2379 ±2	61/ 66/
$C_{36}H_{30}Sn_2$ $C_{36}H_{30}Sn_2$ $C_{36}H_{30}Sn_2$	700·02	Hexaphenyldistannane Hexaphenyldistannane Hexaphenyldistannane	c c		SB SB		m m	4870 ±4 4848 ±7	66/ 63/
$C_{10}H_{14}O_2Sn$	284·91	Trimethyltin benzoate	c		SB		m	1440 ±2	66/
$C_{13}H_{20}O_2Sn$	326·99	Triethyltin benzoate	c		SB		m	1907 ±1	61/ 66/
CH_3Cl_3Sn	240·08	Methyltin trichloride	l		R		glc	44·5 ±6·6	65/
$C_2H_6Cl_2Sn$	219·67	Dimethyltin dichloride	l		R		glc	23·2 ±3·5	65/

1 kcal = $4·184$ kJ

		5	6 Determination of ΔH_v			7
ΔH_r°		ΔH_f° (l or c) kcal/g.f.w.	Type	ΔH_v° kcal/g.f.w.	Ref.	ΔH_f° (g) kcal/g.f.w.
Remarks						
An $1·00055\pm0·00025$; product SnO_2(c) $\{-138·81\pm0·08\}$ (53/24)		$-44·8 \pm1·3$	E1	$15·5 \pm1·0$	66/3	$-29·3 \pm1·9$
duct SnO_2(c) $\{-138·81\pm0·08\}$ (53/24)		$-49·7 \pm0·9$				$-34·2 \pm1·4$
Selected value		$-44·8 \pm1·3$				$-29·3 \pm1·9$
An $0·99988\pm0·00016$; product SnO_2(c) C $\{-138·81\pm0·08\}$ (53/24)		$-72·8 \pm1·6$	E1	$19·8 \pm0·5$	63/1	$-53·0 \pm1·8$
duct SnO_2(c) $\{-138·81\pm0·08\}$ (53/24)		$-72·2 \pm0·7$				$-52·4 \pm0·9$
Selected value		$-72·8 \pm1·6$				$-53·0 \pm1·8$
product SnO_2(c) $\{-138·81\pm0·08\}$ (53/24)		$-79·1 \pm1·4$				
An $0·99998\pm0·00007$; product SnO_2(c) $\{-138·81\pm0·08\}$ (53/24)		$+98·54\pm0·87$	S3	$15·85\pm0·30$ (sub.)	62/5	$+114·39\pm0·95$
An $0·999$; product SnO_2(c) $\{-138·81\pm0·08\}$ (53/24)		$+116 \pm3$				$+132 \pm3$
duct SnO_2 $\{-138·81\pm0·08\}$ (53/24)		$+163·3 \pm1·7$				$+179·2 \pm1·8$
Selected value		$+98·54\pm0·87$				$+114·39\pm0·95$
product SnO_2(c) $\{-138·81\pm0·08\}$ (53/24)		$-87·1 \pm6·8$				
$H_3)_6Sn_2$(l)$+Br_2$(l) $= 2(CH_3)_3SnBr$(l) $\{-44·3\pm1·0\}$		$-18·4 \pm2·1$	E8	$12·0 \pm1·0$		$-6·4 \pm2·4$
duct SnO_2(c) $\{-138·81\pm0·08\}$ (53/24)		-52 ± 2	E4	15 ± 1		-37 ± 3
product SnO_2(c) $\{-138·81\pm0·08\}$ (53/24)		$+182 \pm4$	E4$_{me}$	28 ± 5 (sub.)	66/43	$+210 \pm7$
product SnO_2(c) $\{-138·81\pm0·08\}$ (53/24)		$+160 \pm7$				$+188 \pm9$
Selected value		$+182 \pm4$				$+210 \pm7$
An $0·999$; product SnO_2(c) $\{-138·81\pm0·08\}$ (53/24)		-118 ± 2				
An $0·999$; product SnO_2(c) $\{-138·81\pm0·08\}$ (53/24)		-138 ± 1				
$CH_3)_4Sn$(l) $\{-12·50\pm0·46\}$ $+3SnCl_4$(l) $\{-122·2\}$ $= 4CH_3SnCl_3$(l)		$-105·9 \pm2·5$				
$CH_3)_4Sn$(l) $\{-12·50\pm0·46\}$ $+SnCl_4$(l) $\{-122·2\}$ $= 2(CH_3)_2SnCl_2$(l)		$-79·0 \pm2·5$				

Group IV organometallic compounds   1 kcal = 4·184 kJ

1 Formula	2 g.f.w.	3 Name	State	Purity mol %	Type	No. of expts.	Detn. of reactn.	$-\Delta H_r^\circ$ kcal/g.f.w.	Re
C_3H_9ClSn	119·25	Trimethyltin chloride	l		R		glc	43·9 ±6·6	65/
C_3H_9BrSn	243·70	Trimethyltin bromide	l		R	7	m	48·3 ±0·7	57/
C_3H_9ISn	290·70	Trimethyltin iodide	l		R	7	m	44·0 ±0·7	57/
$C_{12}H_{27}BrSn$	369·95	Tri-n-butyltin bromide	l		R	6	m	54·69±0·09	59/
$C_{12}H_{10}Br_2Sn$	432·72	Diphenyltin dibromide	c		R	4	m	74·14±0·70	59/
$C_{18}H_{15}BrSn$	429·92	Triphenyltin bromide	c		R	4	m	35·15±0·35	59/
$C_4H_{12}Pb$	267·33	Tetramethyl-lead	l	99·95, glc	RB	6	m	887·0 ±0·3	59/
$C_8H_{20}Pb$	323·44	Tetraethyl-lead	l		RB	9	m	1525·7 ±0·6	56/

Determina

$$1\ \text{kcal} = 4 \cdot 184\ \text{kJ}$$

ΔH_r°		5 ΔH_f° (l or c) kcal/g.f.w.	6 Determination of ΔH_v			7 ΔH_f° (g) kcal/g.f.w.
Remarks			Type	ΔH_v° kcal/g.f.w.	Ref.	
$(CH_3)_4Sn(l)$ $\{-12 \cdot 50 \pm 0 \cdot 46\}$ $+SnCl_4(l)$ $\{-122 \cdot 2\}$ $= 4(CH_3)_3SnCl(l)$		$-50 \cdot 9 \pm 2 \cdot 5$				
$(CH_3)_4Sn(l)$ $\{-12 \cdot 50 \pm 0 \cdot 46\}+Br_2(g)$ $= (CH_3)_3SnBr(l)+CH_3Br(g)$ $\{-9 \cdot 1 \pm 0 \cdot 3\}$		$-44 \cdot 3 \pm 1 \cdot 0$	E4	$11 \cdot 3 \pm 1 \cdot 0$	63/1	$-33 \cdot 0 \pm 1 \cdot 4$
$(CH_3)_6Sn_2(l)$ $\{-18 \cdot 4 \pm 2 \cdot 1\}+I_2(c)$ $= 2(CH_3)_3SnI(l)$		$-31 \cdot 2 \pm 1 \cdot 1$	E4	$11 \cdot 5 \pm 1 \cdot 0$	63/1	$-19 \cdot 7 \pm 1 \cdot 5$
$(C_4H_9)_4Sn(l)$ $\{-72 \cdot 8 \pm 1 \cdot 6\}+Br_2(g)$ $= (n\text{-}C_4H_9)_3SnBr(l)$ $+n\text{-}C_4H_9Br(l)$ $\{-34 \cdot 36 \pm 0 \cdot 30\}$		$-85 \cdot 7 \pm 1 \cdot 7$	E4	20 ± 3	63/1	$-65 \cdot 7 \pm 4 \cdot 0$
$(C_6H_5)_4Sn(c)$ $\{+98 \cdot 54 \pm 0 \cdot 87\}+2Br_2(l)$ $= (C_6H_5)_2SnBr_2(c)$ $+2C_6H_5Br(l)$ $\{+14 \cdot 5 \pm 1 \cdot 0\}$		$-4 \cdot 6 \pm 1 \cdot 5$				
$(C_6H_5)_3SnBr(c)+Br_2(l)$ $= (C_6H_5)_2SnBr_2(c)$ $\{-4 \cdot 6 \pm 1 \cdot 5\}$ $+C_6H_5Br(l)$ $\{+14 \cdot 5 \pm 1 \cdot 0\}$		$+45 \cdot 1 \pm 2 \cdot 0$				
$(CH_3)_4Pb(l)+7 \cdot 5O_2$ $+2[HNO_3.40H_2O]$ (l) $\{-49 \cdot 44\}$ $= 4CO_2+7H_2O(l)$ $+Pb(NO_3)_2(c)$ $\{-108 \cdot 0\}$		$+23 \cdot 5 \pm 0 \cdot 6$	V1	$9 \cdot 1 \pm 0 \cdot 1$	59/2	$+32 \cdot 6 \pm 0 \cdot 7$
$(C_2H_5)_4Pb(l)+13 \cdot 5O_2$ $+2[HNO_3.30H_2O]$ (l) $\{-49 \cdot 43\}$ $= 8CO_2+11H_2O(l)$ $+Pb(NO_3)_2(c)$ $\{-108 \cdot 0\}$		$+12 \cdot 7 \pm 0 \cdot 9$	V3	$13 \cdot 6 \pm 0 \cdot 6$	36/21	$+26 \cdot 3 \pm 1 \cdot 1$

Group V organometallic compounds   1 kcal = 4·184 kJ

1 Formula	2 g.f.w.	3 Name	State	Purity mol %	Type	No. of expts.	Detn. of reactn.	$-\Delta H_r^\circ$ kcal/g.f.w.	R
GROUP VA $C_{12}H_{12}V$	207·171	Dibenzene vanadium	c		SB	5	m	1732·7 ±3·6	61
GROUP VB C_3H_9P	76·079	Trimethylphosphine	l		SB	7	m	763·6 ±1·1	57
$C_6H_{15}P$	118·160	Triethylphosphine	l		SB			1237·1 ±3·0	63
$C_{12}H_{27}P$	202·323	Tri-n-butylphosphine	l		R	6	m	105·7 ±2·0	56
$C_{18}H_{13}P$	260·278	9-Phenyl-9-phosphafluorene	c		SB	5	an	2385·8 ±4·0	62
$C_{18}H_{15}P$	262·294	Triphenylphosphine	c		SB	5	an	2464·0 ±4·6	60
$C_{18}H_{15}P$		Triphenylphosphine	c		SB	7	m	2461·5 ±3·0	60
$C_{18}H_{15}P$		Triphenylphosphine							
$C_{34}H_{25}P$	464·552	Pentaphenylphosphole	c		SB	4	an	4349·5 ±6·8	62
C_3H_9OP	92·078	Trimethylphosphine oxide	c		R	3	m	108·4 ±1·0	60
$C_{12}H_{27}OP$	218·322	Tri-n-butylphosphine oxide	c		SB	3	m	2143·0 ±7·8	66
$C_{18}H_{15}OP$	278·293	Triphenylphosphine oxide	c		SB	4	an	2394·1 ±6·0	60
CH_5O_3P	96·023	Methylphosphonic acid	c		R	4	m	47·0 ±0·3	55/3

1 kcal = 4·184 kJ

H_r°		5 ΔH_f° (l or c) kcal/g.f.w.	6 Determination of ΔH_v			7 ΔH_f° (g) kcal/g.f.w.
Remarks			Type	ΔH_v° kcal/g.f.w.	Ref.	

H_r° Remarks	ΔH_f° (l or c) kcal/g.f.w.	Type	ΔH_v° kcal/g.f.w.	Ref.	ΔH_f° (g) kcal/g.f.w.
duct V_2O_5(c) {$-370\cdot6\pm0\cdot5$} (60/40)	$+8\cdot9\ \pm3\cdot6$*				
$_9P$(l)$+6\cdot5O_2$ $= 3CO_2+3H_2O$(l)$+H_3PO_4$(c)	$-29\cdot2\ \pm1\cdot1$*	V3	$6\cdot7\ \pm[0\cdot5]$	40/8	$-22\cdot5\ \pm1\cdot2$*
$_{15}P$(l)$+11O_2$ $= 6CO_2+7\cdot5H_2O$(l)$+0\cdot25P_4O_{10}$; state of P_4O_{10} not given; assumed amorphous, P_4O_{10}(am) {-727}	$-21\cdot3\ \pm3\cdot0$*	V3	$9\cdot5\ \pm[0\cdot5]$	36/18	$-11\cdot8\ \pm3\cdot1$*
$H_9)_3P$(l)$+H_2O_2$(l) {$-44\cdot9$} $= (C_4H_9)_3PO$(c) {$-110\cdot4\pm7\cdot8$}$+H_2O$(l)	$-28\cdot1\ \pm8\cdot0$				
$_8H_{13}P$(c)$+22\cdot5O_2$ $= 18CO_2+5H_2O$(l)$+H_3PO_4$(c)	$+45\cdot6\ \pm4\cdot0$*				
$_8H_{15}P$(c)$+23O_2$ $= 18CO_2+6H_2O$(l)$+H_3PO_4$(c)	$+55\cdot5\ \pm4\cdot6$*	V4$_{me}$	23 ± 2 (sub.)	49/31, 64/1	$+78\cdot5\ \pm5\cdot0$*
$_8H_{15}P$(c)$+23O_2$ $= 18CO_2+6H_2O$(l)$+H_3PO_4$(c) D	$+53\cdot0\ \pm3\cdot0$*				$+76\cdot0\ \pm4\cdot0$*
Selected value	$+55\cdot5\ \pm4\cdot6$*				$+78\cdot5\ \pm5\cdot0$*
$_4H_{25}P$(c)$+41\cdot5O_2$ $= 34CO_2+11H_2O$(l)$+H_3PO_4$(c)	$+94\cdot6\ \pm6\cdot8$*				
$H_3)_3P$(l) {$-29\cdot2\pm1\cdot1$}$+H_2O_2$(l) {$-44\cdot9$} $= (CH_3)_3PO$(c)$+H_2O$(l)	$-114\cdot2\ \pm1\cdot5$*	E4$_{me}$	$12\cdot0\ \pm1\cdot0$ (sub.	60/43	$-102\cdot2\ \pm1\cdot8$*
$_4H_9)_3PO$(c)$+19\cdot5O_2$ $= 12CO_2+12\cdot097H_2O$(l) $+0\cdot806H_3PO_4$(c) $+0\cdot097H_4P_2O_7$(c) {$-535\cdot6$}	$-110\cdot4\ \pm7\cdot8$*				
$_6H_5)_3PO$(c)$+22\cdot5O_2$ $= 18CO_2+6H_2O$(l)$+H_3PO_4$(c)	$-14\cdot4\ \pm6\cdot0$				
H_3POCl_2(c) {$-148\cdot0\pm6\cdot2$}$+5002H_2O$(l) $= CH_3PO(OH)_2$(c)$+2[HCl.2500H_2O]$ (l)	$-252\cdot0\ \pm6\cdot2$	S5	$11\cdot5\ \pm[1\cdot0]$ (sub.)	55/32	$-240\cdot5\ \pm6\cdot3$

Group V organometallic compounds   1 kcal = 4·184 kJ

1 Formula	2 g.f.w.	3 Name	State	Purity mol %	Type	No. of expts.	Detn. of reactn.	$-\Delta H_r^\circ$ kcal/g.f.w.	R
$C_2H_7O_3P$	110·050	Ethylphosphonic acid	c		R	4	m	49·4 ±0·3	55
$C_2H_7O_3P$		Ethylphosphonic acid	c		R	10	m	45·90±0·13	61
$C_2H_7O_3P$		Ethylphosphonic acid							
$C_3H_9O_3P$	124·077	Trimethyl phosphite	l		R	6	m	73·7 ±1·0	55
$C_5H_{13}O_3P$	152·131	Diethyl methylphosphonate	l		R	4	m	84·9 ±2·0	56
$C_6H_{15}O_3P$	166·158	Triethyl phosphite	l		R	6	m	74·8 ±1·0	55
$C_6H_{15}O_3P$		Triethyl phosphite	l		R	3	m	72·7 ±2·0	56
$C_6H_{15}O_3P$		Triethyl phosphite							
$C_8H_{19}O_3P$	194·213	Di-isopropyl ethylphosphonate	l		R	4	m	88·7 ±2·0	56
$C_9H_{21}O_3P$	208·240	Tri-isopropyl phosphite	l		R	3	m	73·9 ±2·0	56

1 kcal = 4·184 kJ

H_r°		5 ΔH_f° (l or c) kcal/g.f.w.	6 Determination of ΔH_v			7 ΔH_f° (g) kcal/g.f.w.
Remarks			Type	ΔH_v° kcal/g.f.w.	Ref.	

Remarks	ΔH_f° (l or c) kcal/g.f.w.	Type	ΔH_v° kcal/g.f.w.	Ref.	ΔH_f° (g) kcal/g.f.w.
$_5POCl_2$(l) $\{-146\cdot8\pm3\cdot3\}+5002H_2O$(l) $= C_2H_5PO(OH)_2$(c)$+2[HCl.2500H_2O]$ (l)	$-253\cdot0\ \pm3\cdot4*$	S5	$12\cdot1\ \pm[1\cdot0]$ (sub.)	55/32	$-240\cdot9\ \pm3\cdot5$
$_5POCl_2$(l) $\{-146\cdot8\pm3\cdot3\}+2002H_2O$(l) $= C_2H_5PO(OH)_2$(c)$+2[HCl.1000H_2O]$ (l)	$-249\cdot5\ \pm3\cdot3*$				$-237\cdot4\ \pm3\cdot4$
Selected value	$-251\cdot3\ \pm2\cdot5$				$-239\cdot2\ \pm3\cdot0$
$_3$(l) $\{-76\cdot4\}+3CH_3OH$(l) $\{-57\cdot01\pm0\cdot05\}$ $\cdot3C_6H_5N(CH_3)_2$(l) $= (CH_3O)_3P$(l)$+3C_6H_5N(CH_3)_2.HCl$(c). $[C_6H_5N(CH_3)_2.HCl$(c) $-C_6H_5N(CH_3)_2$(l)]$\{-48\cdot01\pm0\cdot10\}$	$-177\cdot1\ \pm1\cdot1$	V5	$8\cdot8\ \pm[1\cdot0]$	56/38	$-168\cdot3\ \pm[1\cdot5]$
$_3POCl_2$(c) $\{-148\cdot0\pm6\cdot2\}$ $\cdot2C_2H_5OH$(l) $\{-66\cdot42\pm0\cdot08\}$ $\cdot2(C_2H_5)_3N$(l) $= CH_3PO(OC_2H_5)_2$(l)$+2(C_2H_5)_3N.HCl$(c) $[(C_2H_5)_3N.HCl$(c)$-(C_2H_5)_3N$(l)] $\{-60\cdot3\}$	$-245\cdot3\ \pm6\cdot5$	V5	$13\cdot5\ \pm[1\cdot0]$	56/38	$-231\cdot8\ \pm6\cdot6$
$_3$(l) $\{-76\cdot4\}$ $\cdot3C_2H_5OH$(l) $\{-66\cdot42\pm0\cdot08\}$ $\cdot3C_6H_5N(CH_3)_2$(l) $= (C_2H_5O)_3P$(l)$+3C_6H_5N(CH_3)_2.HCl$(c). $[C_6H_5N(CH_3)_2.HCl$(c) $-C_6H_5N(CH_3)_2$(l)]$\{-48\cdot01\pm0\cdot10\}$	$-206\cdot4\ \pm1\cdot1$	V5	$10\cdot0\ \pm[1\cdot0]$	56/38	$-196\cdot4\ \pm[1\cdot5]$
$_3$(l) $\{-76\cdot4\}$ $\cdot3C_2H_5OH$(l) $\{-66\cdot42\pm0\cdot08\}$ $\cdot3C_6H_5N(CH_3)_2$(l) $= (C_2H_5O)_3P$(l)$+3C_6H_5N(CH_3)_2.HCl$(c) $[C_6H_5N(CH_3)_2.HCl$(c) $-C_6H_5N(CH_3)_2$(l)]$\{-48\cdot01\pm0\cdot10\}$	$-204\cdot3\ \pm2\cdot0$				$-194\cdot3\ \pm2\cdot2$
Selected value	$-205\cdot9\ \pm1\cdot0$				$-195\cdot9\ \pm1\cdot3$
H_5POCl_2(l) $\{-146\cdot8\pm3\cdot3\}+2i\text{-}C_3H_7OH$(l) $-76\cdot02\pm0\cdot12\}+2(C_2H_5)_3N$(l) $= C_2H_5PO(Oi\text{-}C_3H_7)_3+2(C_2H_5)_3N.HCl$(c) $[(C_2H_5)_3N.HCl$(c)$-(C_2H_5)_3N$(l)] $\{-60\cdot3\}$	$-266\cdot9\ \pm3\cdot9$	V5	$14\cdot5\ \pm[1\cdot0]$	56/38	$-252\cdot4\ \pm4\cdot1$
l_3(l) $\{-76\cdot4\}+3i\text{-}C_3H_7OH$(l) $-76\cdot02\pm0\cdot12\}+3C_6H_5N(CH_3)_2$(l) $= (i\text{-}C_3H_7O)_3P$(l)$+3C_6H_5N(CH_3)_2.HCl$(c) $[C_6H_5N(CH_3)_2.HCl$(c) $-C_6H_5N(CH_3)_2$(l)]$\{-48\cdot01\pm0\cdot10\}$	$-234\cdot3\ \pm2\cdot0$	V5	$11\cdot0\ \pm[1\cdot0]$	56/38	$-223\cdot3\ \pm2\cdot2$

Group V organometallic compounds   1 kcal = 4·184 kJ

1 Formula	2 g.f.w.	3 Name	State	Purity mol %	Type	No. of expts.	Detn. of reactn.	$-\Delta H_r^\circ$ kcal/g.f.w.	F
$C_6H_{15}O_4P$	182·158	Triethyl phosphate	l		R	5	m	116·3 ±0·8	56
$C_6H_{15}O_4P$		Triethyl phosphate	l		SB	4	an	983·4 ±2·6	60
$C_6H_{15}O_4P$		Triethyl phosphate							
$C_9H_{21}O_4P$	224·239	Tri-n-propyl phosphate	l		SB	3	m	1452·0 ±5·1	66
$C_9H_{21}O_4P$	224·239	Tri-isopropyl phosphate	l		R	3	m	100·0 ±1·0	55
$C_{12}H_{27}O_4P$	266·320	Tri-n-butyl phosphate	l		SB	5	m	1905·9 ±2·7	66
$C_5H_{14}NP$	119·148	Trimethylphosphine- N-ethylimine	l		R	5	an	51·7 ±0·5	60
$C_{20}H_{20}NP$	305·363	Triphenylphosphine- N-ethylimine	c		R	5	an	11·7 ±0·5	60
$C_{12}H_{30}N_3P$	247·367	Tris(diethylamino)- phosphine	l		R	8	m	100·6 ±2·0	59
$C_6H_{18}N_3P_3$	225·152	Hexamethylcyclo- triphosphazatriene	c		SB	3	an	1665·2 ±5·4	60
$C_{13}H_{15}ON_2P$	246·251	Methylphosphonic dianilide	c		SB	6	m	1855·0 ±6·0	56
$C_{14}H_{17}ON_2P$	260·278	Ethylphosphonic dianilide	c		SB	7	m	2014·3 ±3·0	56

$$1 \text{ kcal} = 4\cdot184 \text{ kJ}$$

ΔH_r° — Remarks	5 ΔH_f° (l or c) kcal/g.f.w.	6 Determination of ΔH_v — Type	ΔH_v° kcal/g.f.w.	Ref.	7 ΔH_f° (g) kcal/g.f.w.
$_5O)_3P(l)$ {$-205\cdot9\pm1\cdot0$} $H_2O_2(l)$ {$-44\cdot9$} $= (C_2H_5O)_3PO(l)+H_2O(l)$	$-298\cdot8 \pm1\cdot3$	V4	$13\cdot7 \pm[1\cdot0]$	30/4	$-285\cdot1 \pm1\cdot6$
$_5O)_3PO(l)+9O_2$ $= 6CO_2+6H_2O(l)+H_3PO_4(c)$	$-296\cdot5 \pm2\cdot6^*$				$-282\cdot8 \pm2\cdot8$
Selected value	$-298\cdot2 \pm1\cdot2$				$-284\cdot5 \pm1\cdot5$
$_{21}O_4P(l)+13\cdot5O_2$ $= 9CO_2+9\cdot06H_2O(l)+0\cdot88H_3PO_4(c)$ $+0\cdot06H_4P_2O_7(c)$ {$-535\cdot6$}	$-314\cdot5 \pm5\cdot1$	V4	$15\cdot6 \pm[1\cdot0]$	30/4	$-298\cdot9 \pm5\cdot2$
$_3H_7O)_3P(l)$ {$-234\cdot3\pm2\cdot0$} $H_2O_2 (5000H_2O)(l)$ {$-45\cdot69$} $= (i\text{-}C_3H_7O)_3PO(l)+5001H_2O(l)$	$-311\cdot7 \pm2\cdot3$				
$_{27}O_4P(l)+18O_2$ $= 12CO_2+12\cdot06H_2O(l)+0\cdot88H_3PO_4(c)$ $+0\cdot06H_4P_2O_7(c)$ {$-535\cdot6$}	$-347\cdot7 \pm2\cdot7^*$	V4	$17\cdot2 \pm[1\cdot2]$	30/4	$-330\cdot5 \pm3\cdot0^*$
$_3)_3P = NC_2H_5(l)+[HCl.5001H_2O](l)$ $= [(CH_3)_3PO+C_2H_5NH_3Cl](5000H_2O)(l)$ $[(CH_3)_3PO.5000H_2O](l)$ {$-118\cdot3\pm1\cdot6$} $[C_2H_5.NH_3Cl.5000H_2O](l)$ {$-77\cdot4$}	$-35\cdot8 \pm1\cdot7^*$	V5	$14\cdot7 \pm[1\cdot0]$	60/43	$-21\cdot1 \pm2\cdot0$
$_5)_3P=NC_2H_5(c)+[HCl.5001H_2O](l)$ $= (C_6H_5)_3PO(c)$ {$-14\cdot4\pm6\cdot0$} $+[C_2H_5.NH_3Cl.5000H_2O](l)$ {$-77\cdot4$}	$+28\cdot1 \pm6\cdot0$	$E4_{me}$	$18\cdot0 \pm2\cdot0$ (sub.)	60/43	$+46\cdot1 \pm6\cdot4$
$_3(l)$ {$-76\cdot4$} $\cdot6(C_2H_5)_2NH(l)$ {$-24\cdot78\pm0\cdot29$} $= [N(C_2H_5)_2]_3P(l)$ $+3(C_2H_5)_2NH_2Cl(c)$ {$-85\cdot5$}	$-69\cdot2 \pm2\cdot4$	V5	$14\cdot5 \pm[1\cdot0]$	59/33	$-54\cdot7 \pm2\cdot6$
$_{18}N_3P_3(c)+14\cdot25O_2$ $= 6CO_2+4\cdot5H_2O(l)+1\cdot5N_2+3H_3PO_4(c)$	$-123\cdot6 \pm5\cdot4^*$				
$_3PO(NHC_6H_5)_2(c)+17\cdot5O_2$ $= 13CO_2+6H_2O(l)+N_2+H_3PO_4$ (soln.: 2M in crucible; $0\cdot5$M in bomb) {$-308\cdot2$ assumed}	$-85\cdot7 \pm6\cdot0^*$				
$_5PO(NHC_6H_5)_2(c)+19O_2$ $= 14CO_2+7H_2O(l)+N_2+H_3PO_4$ (soln.: 2M in crucible; $0\cdot5$M in bomb) {$-308\cdot2$ assumed}	$-88\cdot8 \pm3\cdot0^*$				

THERMOCHEMISTRY OF ORGANIC AND ORGANOMETALLIC COMPOUNDS

Group V organometallic compounds   1 kcal = 4·184 kJ

1 Formula	2 g.f.w.	3 Name	State	Purity mole %	Type	No. of expts.	Detn. of reactn.	$-\Delta H_r^\circ$ kcal/g.f.w.	Determir
$C_6H_{15}O_3SP$	198·222	Triethyl thionophosphate	l		R	5	m	26·6 ±0·4	5
$C_2H_5Cl_2P$	130·942	Ethyldichlorophosphine	l		R	4	m	40·0 ±0·4	5
CH_3OCl_2P	132·914	Methylphosphonic dichloride	c		R	4	m	53·0 ±0·2	5
$C_2H_5OCl_2P$	146·941	Ethylphosphonic dichloride	l		R	4	m	57·5 ±0·6	5
$C_4H_{10}NCl_2P$	174·011	Diethylamino-dichlorophosphine	l		R		m	27·0 ±[1·0]	64
$C_8H_{20}N_2ClP$	210·689	Bis(diethylamino)-chlorophosphine	l		R		m	25·5 ±[1·0]	64
C_3H_9As	120·027	Trimethylarsine	l		SB	7	m	664·2 ±2·4	56
$C_6H_{15}As$	162·108	Triethylarsine	l		SB			1158·3 ±4·0	63 58
$C_{18}H_{15}As$	306·242	Triphenylarsine	c		SB	8	m	2409·8 ±4·0	61
$C_{18}H_{15}As$		Triphenylarsine	c		RB	6	m	2361·5 ±1·5	64
$C_{18}H_{15}As$		Triphenylarsine							

$$1 \text{ kcal} = 4 \cdot 184 \text{ kJ}$$

ΔH_r° Remarks	5 ΔH_f° (l or c) kcal/g.f.w.	6 Determination of ΔH_v			7 ΔH_f° (g) kcal/g.f.w.
		Type	ΔH_v° kcal/g.f.w.	Ref.	
H₅O)₃P(l) {−205·9±1·0} +S (c, rhombic) = (C₂H₅O)₃PS(l)	−232·5 ±1·1				
H₅PCl₂(l)+SO₂Cl₂(l) {−94·2} = C₂H₅POCl₂(l) {−146·8±3·3} +SOCl₂(l) {−58·7}	−71·2 ±3·5				
₃POCl₂(c)+4C₆H₅NH₂(l) {+7·48±0·18} = 2C₆H₅NH₃Cl(c) {−42·70±0·40} +CH₃PO(NHC₆H₅)₂(c) {−85·7±6·0}	−148·0 ±6·2*	S5	14·9 ±[1·0] (sub).	55/32	−133·1 ±6·3*
H₅POCl₂(l)+4C₆H₅NH₂(l) {+7·48±0·18} = 2C₆H₅NH₃Cl(c) {−42·70±0·40} +C₂H₅PO(NHC₆H₅)₂(c) {−88·8±3·0}	−146·6 ±3·3*	V5	10·2 ±[1·0]	55/32	−136·4 ±3·5*
C₂H₅)₂]₃P(l) {−69·2±2·4} +2PCl₃(l) {−76·4} = 3(C₂H₅)₂NPCl₂(l)	−83·0 ±[1·3]				
N(C₂H₅)₂]₃P(l) {−69·2±2·4} +PCl₃(l) {−76·4} = 3[(C₂H₅)₂N]₂PCl(l)	−80·1 ±[1·8]				
H₉As(l)+6O₂ = 3CO₂+4·5H₂O(l) +0·5As₂O₃(c) {−157·0}	−3·9 ±2·4*	V3	6·9 ±[0·3]	55/33	+2·8 ±2·5*
H₁₅As(l)+10·5O₂ = 6CO₂+7·5H₂O(l) +0·5As₂O₃ (c) {−157·0}	+3·1 ±4·0*	V5	10·3 ±[1·0]	64/42	+13·4 ±4·2*
₃H₁₅As(c)+23O₂ = 18CO₂+7·5H₂O(l) +0·5As₂O₅(c) {−221·1}	**D, I** +94·0 ±4·0*	V4_me	23·5 ±[2·0] (sub.)	49/31, 64/1	+117·5 ±[4·5]*
₃H₁₅As(c)+22·5O₂ +43[NaOH.12·93H₂O] (l) {−112·45} = 18CO₂+[40NaOH.Na₃AsO₃.565H₂O] (l) Na₃AsO₃ (in NaOH solution) {−319·75±0·30}	+74·1 ±1·6*				+97·6 ±[2·5]*
Selected value	+74·1 ±1·6*				+97·6 ±[2·5]*

Group V organometallic compounds   1 kcal = 4·184 kJ

1 Formula	2 g.f.w.	3 Name	State	Purity mol %	Type	No. of expts.	Detn. of reactn.	$-\Delta H_r^\circ$ kcal/g.f.w.	Re
$C_3H_9O_3As$	168·025	Trimethyl arsenite	l		R	7	m	5·85±0·20	53/
$C_6H_{15}O_3As$	210·106	Triethyl arsenite	l		R	5	m	6·35±0·15	53/
$C_9H_{21}O_3As$	252·188	Tri-n-propyl arsenite	l		R	5	m	5·85±0·70	53/
C_3H_9Sb	166·86	Trimethylstibine	l		SB	6	m	698·2 ±6·0	55/
$C_6H_{15}Sb$	208·936	Triethylstibine	l		SB			1162·6 ±2·5	63/
$C_{18}H_{15}Sb$	353·070	Triphenylstibine	c		SB	7	m	2392·4 ±4·0	60/
C_3H_9Bi	254·085	Trimethylbismuth	l		SB	7	m	696·0 ±3·4	54/
$C_6H_{15}Bi$	296·166	Triethylbismuth	l		SB			1185·8 ±4·0	63/
$C_{18}H_{15}Bi$	440·300	Triphenylbismuth	c		SB	6	m	2386·0 ±4·0	60/

1 kcal = 4·184 kJ

ΔH_r°		5 ΔH_f° (l or c) kcal/g.f.w.	6 Determination of ΔH_v			7 ΔH_f° (g) kcal/g.f.w
	Remarks		Type ΔH_v° kcal/g.f.w.		Ref.	
$H_3O)_3As(l)+1\cdot5H_2O(l)$ $= 3CH_3OH(l)\ \{-57\cdot01\pm0\cdot05\}$ $+0\cdot5As_2O_3(c)\ \{-157\cdot0\}$		$-141\cdot2\ \pm0\cdot3$	V4 $10\cdot1\ \pm0\cdot3$		53/37	$-131\cdot1\ \pm0\cdot4$
$_2H_5O)_3As(l)+1\cdot5H_2O(l)$ $= 3C_2H_5OH(l)\ \{-66\cdot42\pm0\cdot08\}$ $+0\cdot5As_2O_3(c)\ \{-157\cdot0\}$		$-168\cdot9\ \pm0\cdot3$	V4 $12\cdot1\ \pm1\cdot0$		53/37	$-156\cdot8\ \pm1\cdot1$
$_3H_7O)_3As(l)+1\cdot5H_2O(l)$ $= 3n\text{-}C_3H_7OH(l)\ \{-72\cdot51\pm0\cdot30\}$ $+0\cdot5As_2O_3\ (c)\ \{-157\cdot0\}$		$-187\cdot7\ \pm1\cdot2$	E1 $14\cdot0\ \pm2\cdot0$		53/37	$-173\cdot7\ \pm2\cdot4$
$_3H_9Sb(l)+6\cdot25O_2$ $= 3CO_2+4\cdot5H_2O(l)$ $+0\cdot5Sb_2O_4(c)\ \{-216\cdot9\}$		$+0\cdot2\ \pm6\cdot0^*$	V4 $7\cdot5\ \pm[0\cdot3]$		55/33, 40/8, 46/14	$+7\cdot7\ \pm6\cdot0^*$
$_6H_{15}Sb(l)+10\cdot5O_2$ $= 6CO_2+7\cdot5H_2O(l)$ $+0\cdot5Sb_2O_3(c)\ \{-169\cdot4\}$		$+1\cdot2\ \pm2\cdot5^*$	V4 $10\cdot4\ \pm[1\cdot0]$		46/14	$+11\cdot6\ \pm2\cdot6^*$
$_8H_{15}Sb(c)+22\cdot75O_2$ $= 18CO_2+7\cdot5H_2O(l)$ $+0\cdot5Sb_2O_4(c)\ \{-216\cdot9\}$	D I	$+78\cdot7\ \pm4\cdot0^*$	$V4_{me}$ $25\cdot4\ \pm[2\cdot0]$ (sub.)		49/31, 64/1	$+104\cdot1\ \pm4\cdot5^*$
$_3H_9Bi(l)+6O_2$ $= 3CO_2+4\cdot5H_2O(l)$ $+0\cdot5Bi_2O_3(c)\ \{-137\cdot2\}$		$+37\cdot8\ \pm3\cdot4^*$	V2 $8\cdot6\ \pm[0\cdot3]$ $8\cdot0\ \pm[0\cdot3]$		61/53 55/33	$+46\cdot1\ \pm3\cdot4^*$
$_6H_{15}Bi(l)+10\cdot5O_2$ $= 6CO_2+7\cdot5H_2O(l)$ $+0\cdot5Bi_2O_3(c)\ \{-137\cdot2\}$		$+40\cdot6\ \pm4\cdot0^*$	V5 $11\cdot0\ \pm[1\cdot0]$		64/42	$+51\cdot6\ \pm4\cdot2^*$
$_8H_{15}Bi(c)+22\cdot5O_2$ $= 18CO_2+7\cdot5H_2O(l)$ $+0\cdot5Bi_2O_3(c)\ \{-137\cdot2\}$	D, I	$+112\cdot1\ \pm4\cdot0^*$	$E1_{me}$ $26\cdot5\ \pm2\cdot0$ (sub.)		64/1	$+138\cdot6\ \pm4\cdot5^*$

Group VI organometallic compounds   1 kcal = 4·184 kJ

1 Formula	2 g.f.w.	3 Name	State	Purity mol %	Type	No. of expts.	Detn. of reactn.	$-\Delta H_r^{\circ}$ kcal/g.f.w.	Re
GROUP VIA									
$C_{12}H_{12}Cr$	208·225	Dibenzene chromium	c		SB	5	m	1696·2 ±8·0	59/
$C_{12}H_{12}Cr$ $C_{12}H_{12}Cr$		Dibenzene chromium Dibenzene chromium	c		SB	5	m	1725·8 ±3·0	58/
$C_{18}H_{24}Cr$	292·388	Di-(1, 3, 5-trimethyl-benzene) chromium	c		SB	4	m	2638·1 ±6·0	61/
$C_{18}H_{24}Cr$	292·388	Di-(1, 2, 4-trimethyl-benzene) chromium	c		SB	4	m	2640·1 ±4·0	61/
C_6O_6Cr	220·059	Chromium hexacarbonyl	c		SB	5	m	443·1 ±1·0	56/
$C_{15}H_{21}O_6Cr$	349·327	Tris(acetylacetonato)-chromium	c		R	5	m	−11·06±0·08	67/
$C_{12}H_{12}Mo$	252·17	Dibenzene molybdenum	c		SB	4	m	1789·8 ±4·4	61/
C_6O_6Mo	264·00	Molybdenum hexacarbonyl	c		SB	6	m	507·5 ±0·6	56/
C_6O_6W	351·913	Tungsten hexacarbonyl	c		SB	6	m	537·8 ±0·6	56/
GROUP VIB									
CSe_2 CSe_2 CSe_2	169·93	Carbon diselenide Carbon diselenide Carbon diselenide	l l		SB SB	10 2	m m	254·3 ±2·8 241·1 ±2·0	66/ 36/
$C_4H_{10}Se$	137·08	Diethyl selenide	l		SB	4	m	748·6 ±0·8	36/
$C_{14}H_{14}Se$	261·23	Dibenzyl selenide	c		SB	6	m	1849·3 ±5·0	36/

$1 \text{ kcal} = 4\cdot184 \text{ kJ}$

ΔH_r° Remarks		5 ΔH_f° (l or c) kcal/g.f.w.	6 Determination of ΔH_v Type ΔH_v° kcal/g.f.w. Ref.	7 ΔH_f° (g) kcal/g.f.w.
oduct $Cr_2O_3(c)$ $\{-272\cdot7\}$	**D**	$+21\cdot3$ $\pm8\cdot0$	S4 $18\cdot7$ $\pm[1\cdot5]$ 58/36 (sub.)	$+40\cdot0$ $\pm8\cdot1$
oduct assumed to be $Cr_2O_3(c)$ $\{-272\cdot7\}$	**D**	$+50\cdot9$ $\pm3\cdot0$		$+69\cdot6$ $\pm3\cdot0$
Selected value		$+36\pm15$		$+55\pm15$
oduct $Cr_2O_3(c)$ $\{-272\cdot7\}$		$-10\cdot9$ $\pm6\cdot0$		
oduct $Cr_2O_3(c)$ $\{-272\cdot7\}$		$-8\cdot9$ $\pm4\cdot0$		
oduct $Cr_2O_3(c)$ $\{-272\cdot7\}$		$-257\cdot6$ $\pm1\cdot1$*	S4 $17\cdot2$ $\pm[1\cdot0]$ 35/15 (sub.) S4 $16\cdot6$ $\pm[1\cdot0]$ 52/32 (sub.)	$-240\cdot7$ $\pm[1\cdot4]$*
$(C_5H_7O_2)_3(c)+6H_2O(l)+3HCl(aq)$ $= 3C_5H_8O_2(l)$ $\{-101\cdot29\pm0\cdot37\}$ $+CrCl_3.6H_2O(c)$ $\{-581\cdot1\}$		$-366\cdot4$ $\pm1\cdot2$*		
oduct $MoO_3(c)$ $\{-178\cdot2\}$		$+73\cdot1$ $\pm4\cdot4$*	S5 $22\cdot6$ $\pm[2\cdot0]$ 61/50 (sub.)	$+95\cdot7$ $\pm4\cdot8$*
oduct $MoO_3(c)$ $\{-178\cdot2\}$		$-235\cdot0$ $\pm0\cdot7$*	S4 $16\cdot7$ $\pm[1\cdot0]$ 35/15 (sub.) S4 $17\cdot3$ $\pm[1\cdot0]$ 52/32 (sub.) S4 $16\cdot7$ $\pm[1\cdot0]$ 60/45 (sub.)	$-218\cdot0$ $\pm[1\cdot2]$*
oduct $WO_3(c)$ $\{-201\cdot5\pm0\cdot4\}$		$-228\cdot0$ $\pm0\cdot7$*	S4 $16\cdot7$ $\pm[1\cdot0]$ 52/32 (sub.) S4 $17\cdot7$ $\pm[1\cdot0]$ 35/15 (sub.)	$-210\cdot8$ $\pm[1\cdot2]$*
$e_2(l)+3O_2 = CO_2+2SeO_2(c)$ $\}-53\cdot86\}$		$+52\cdot5$ $\pm2\cdot8$*	V3 $8\cdot67\pm[0\cdot10]$ 66/45	$+61\cdot4$ $\pm2\cdot8$*
$e_2(l)+3O_2 = CO_2+2SeO_2(c)$ $\{-53\cdot86\}$		$+39\cdot3$ $\pm2\cdot0$*	V3 $9\cdot09\pm[0\cdot10]$ 47/20	$+48\cdot0$ $\pm2\cdot0$*
Selected value		$+52\cdot5$ $\pm2\cdot8$*		$+61\cdot4$ $\pm2\cdot8$*
$H_{10}Se(l)+7\cdot5O_2$ $= 4CO_2+5H_2O(l)+SeO_2(c)$ $\{-53\cdot86\}$		$-23\cdot0$ $\pm0\cdot8$*	E8 $9\cdot3$ $\pm[1\cdot0]$ 64/1	$-13\cdot7$ $\pm[1\cdot3]$*
$_4H_{14}Se(c)+18\cdot5O_2$ $= 14CO_2+7H_2O(l)+SeO_2(c)$ $\{-53\cdot86\}$		$+0\cdot5$ $\pm5\cdot0$		

Group VII organometallic compounds   1 kcal = 4·184 kJ

1 Formula	2 g.f.w.	3 Name	State	Purity mol %	No. of Type expts.	Detn. of reactn.	$-\Delta H_r^{\circ}$ kcal/g.f.w.	Re
GROUP VIIA								
$C_8H_5O_3Mn$	204·065	Tricarbonyl cyclo-pentadienylmanganese	c		SB 6	m	922·1 ±1·0	65/
$C_8H_5O_3Mn$		Tricarbonyl cyclo-pentadienylmanganese	c		SB		918·1 ±2·0	63/ 58/
$C_8H_5O_3Mn$		Tricarbonyl cyclo-pentadienylmanganese						
$C_{15}H_{21}O_6Mn$	352·269	Tris(acetylacetonato)-manganese (III)	c		R 5	m	−0·31±0·15	68/
$C_{10}O_{10}Mn_2$	335·044	Dimanganese decacarbonyl	c	99·92	RB 6	m	777·0 ±0·8	58/

Determinat

1 kcal $= 4 \cdot 184$ kJ

ΔH_r°		5 ΔH_f° (l or c) kcal/g.f.w.	6 Determination of ΔH_v			7 ΔH_f° (g) kcal/g.f.w
Remarks			Type	ΔH_v° kcal/g.f.w.	Ref.	
›duct MnO_2(c) $\{-124 \cdot 5\}$		$-125 \cdot 6 \pm 1 \cdot 0^*$				
›duct MnO_2(c) $\{-124 \cdot 5\}$	I	$-129 \cdot 6 \pm 2 \cdot 0^*$				
Selected value		$-126 \cdot 6 \pm 2 \cdot 0^*$				
$(C_5H_7O_2)_3$(c)$+4H_2O$(l) $+FeCl_2$(c) $\{-81 \cdot 86 \pm 0 \cdot 12\}+3HCl$(aq) $= 3C_5H_8O_2$(l) $\{-101 \cdot 29 \pm 0 \cdot 37\}$ $+MnCl_2.4H_2O$(c) $\{-407 \cdot 0 \pm 0 \cdot 2\}$ $+FeCl_3$(c) $\{-95 \cdot 70 \pm 0 \cdot 20\}$		$-331 \cdot 9 \pm 1 \cdot 2$	S4	$18 \cdot 6 \pm 0 \cdot 2^*$ (sub.)	64/52	$-313 \cdot 3 \pm 1 \cdot 2$
›$_2$(CO)$_{10}$(c)$+6O_2+4$[HNO$_3$.16H$_2$O] (l) $= 10CO_2+45 \cdot 4H_2O$(l) $+2$[Mn(NO$_3$)$_2$.10$\cdot$3H$_2$O] (l) $\{-148 \cdot 75\}$		$-400 \cdot 9 \pm 0 \cdot 8$	S4	$15 \cdot 0 \pm 1 \cdot 0$ (sub.)	60/46	$-385 \cdot 9 \pm 1 \cdot 3$

Group VIII organometallic compounds   1 kcal = 4·184 kJ

1 Formula	2 g.f.w.	3 Name	State	Purity mol %	4 Determinati Type	No. of expts.	Detn. of reactn.	$-\Delta H_r^\circ$ kcal/g.f.w.	Ref
$C_{10}H_{10}Fe$	186·038	Dicyclopentadienyliron	c		SB	4	m	1404·8 ±1·2	52/
C_5O_5Fe	195·900	Iron pentacarbonyl	l		SB	5	m	385·9 ±1·5	59/
$C_{15}H_{21}O_6Fe$	353·178	Tris(acetylacetonato)-iron (III)	c		R	5	m	−33·30±0·12	68/
C_4HO_4Co	171·983	Cobalt tetracarbonyl hydride	g		E		an	30·38±[0·50]	66/
$C_{10}H_{10}Ni$	188·90	Dicyclopentadienylnickel	c		SB	4	m	1345·1 ±0·6	54/
C_4O_4Ni	170·75	Nickel tetracarbonyl	g		E			36·5 ±1·0	55/
C_4O_4Ni		Nickel tetracarbonyl	l		SB	9	m	282·1 ±1·0	57/ 58/
C_4O_4Ni		Nickel tetracarbonyl							

1 kcal = 4·184 kJ

ΔH_r° Remarks		5 ΔH_f° (l or c) kcal/g.f.w.	6 Determination of ΔH_v			7 ΔH_f° (g) kcal/g.f.w.
			Type	ΔH_v° kcal/g.f.w.	Ref.	
$_5H_5)_2Fe(c)+13\cdot166O_2$ $= 10CO_2+5H_2O(l)$ $+0\cdot33Fe_3O_4(c)$ {$-267\cdot1\pm0\cdot4$}	D	$+33\cdot7\ \pm1\cdot3*$	S3	$17\cdot6\ \pm0\cdot1$ (sub.)	62/56	$+51\cdot3\ \pm1\cdot3*$
oduct $Fe_2O_3(c)$ {$-196\cdot5$}		$-182\cdot6\ \pm1\cdot5*$	V3	$9\cdot6\ \pm0\cdot2$	59/35	$-173\cdot0\ \pm1\cdot5*$
$_5H_7O_2)_3Fe(c)+3HCl(aq)$ $= FeCl_3(c)$ {$-95\cdot70\pm0\cdot20$}$+\infty H_2O(l)$ $+3C_5H_8O_2(l)$ {$-101\cdot29\pm0\cdot37$}		$-313\cdot0\ \pm1\cdot2$	S4	$15\cdot6\ \pm0\cdot8*$ (sub.)	64/52	$-297\cdot4\ \pm1\cdot5*$
$)(c)+4CO(g)$ {$-26\cdot42\pm0\cdot04$}$+0\cdot5H_2$ $= Co(CO)_4H(g)$ 2nd Law						$-136\cdot0\ \pm[0\cdot6]$
$_5H_5)_2Ni(c)+13O_2$ $= 10CO_2+5H_2O(l)+Ni(c)$	D	$+63\cdot0\ \pm0\cdot6*$	S5	$17\cdot3\ \pm0\cdot3$ (sub.)	67/38	$+80\cdot3\ \pm0\cdot7*$
$i(c)+4CO(g)$ {$-26\cdot42\pm0\cdot04$} $\rightleftharpoons Ni(CO)_4(g)$ 3rd Law		$-148\cdot8\ \pm1\cdot2*$	V4	$6\cdot6\ \pm0\cdot3$	62/57	$-142\cdot2\ \pm1\cdot1*$
$i(CO)_4(l)+2\cdot5O_2$ $= 4CO_2+NiO(c)$ {$-57\cdot3\pm0\cdot4$}		$-151\cdot4\ \pm1\cdot1*$				$-144\cdot8\ \pm1\cdot2*$
Selected value		$-150\cdot1\ \pm0\cdot9*$				$-143\cdot5\ \pm1\cdot0*$

Chapter 6

The Practical Application of Thermochemical Data for Organic and Organometallic Compounds

Applications of thermochemical data may be considered from two stand-points, the practical applications considered in this Chapter, and the applications to the more theoretical aspects of chemistry discussed in Chapter 7. This division is not a mutually exclusive one, because some practical applications have theoretical significance, while some theoretical applications, particularly the estimation of heats of formation and reaction, can have important practical consequences.

The practical application of thermochemical data is of importance to chemists, chemical engineers and technologists. We clearly cannot give an exhaustive discussion of possible uses of all the data tabulated in Chapter 5, but we will quote typical uses for some of the values in the hope that the reader himself will be tempted to exploit the rich store of data contained in Chapter 5. The reader will quickly discover that thermochemical data on some classes of organic compounds are plentiful and of excellent quality but that data on other classes are sparse, or poor quality, or totally lacking. In this situation it is important that reliable methods for correlating and predicting thermochemical data be developed, based on well-established experimental values for key compounds. The present status of methods for estimating heats of formation will be discussed in later Sections, but at this point we wish to emphasise that data used for establishing estimation methods should be experimental and not smoothed, interpolated or estimated values: in the past some proposed estimation methods have not been based on experimental data, resulting in schemes that were divorced from reality. The discussion of heats of reaction and their correlation that now follows is based directly on values tabulated in Chapter 5.

6.1. Calculation of the heats of reaction of organic compounds

The basis for the calculation of ΔH_r° for any chemical reaction is provided by equation (6), namely

$$\Delta H_r^\circ = \sum \Delta H_f^\circ \text{ (products)} - \sum \Delta H_f^\circ \text{ (reactants)}.$$

The calculation requires that the heats of formation of all reactants and products be known. Where the species concerned are organic the values should be sought in Chapter 5, and where they are inorganic in compilations such as the Landolt–Börnstein (61/1), National Bureau of Standards (50/1, 65/1, 68/4), or JANAF (61/3) tables. If a value for a heat of reaction is required for a temperature $T_2 \neq 298 \cdot 15°K$, then equation (9) should be invoked, with $T_1 = 298 \cdot 15°K$.

Examples of the use of equation (6) in calculating the heats of selected organic reactions are given below. The reactions have been chosen partly from those employed industrially, especially in the petrochemical industry, and partly from those employed in the laboratory by organic chemists. It must be kept in mind that the values of ΔH_r° discussed below apply to reactions in which all reactants and products are in their standard states, a situation that is unlikely to be found in practice. The difference, however, between ΔH_r° and ΔH_r, the heat of the reaction for real conditions, is unlikely to exceed 1 kcal, so that neglect of this difference will be acceptable to many users of the data. In cases where this neglect is unacceptable, thermodynamic methods (cf. Sections 2.2.2 and 3.2.5) may be used to calculate ΔH_r for the real conditions. For example, equation (4) should be used to calculate the heats of reactions involving real gases at moderate pressures.

In the Sections that follow we will apply the expressions "thermochemically favourable" and "thermochemically unfavourable" to reactions in which $\Delta H_r^{\circ} < 0$ and $\Delta H_r^{\circ} > 0$ respectively. The related expressions "thermodynamically favourable" and "thermodynamically unfavourable" will be applied to reactions in which $\Delta G_r^{\circ} < 0$ and $\Delta G_r^{\circ} > 0$ respectively. Now the driving force of a chemical reaction depends on the sign and magnitude of ΔG_r°, being greater the more negative is ΔG_r°, but consideration of the probable magnitudes of ΔH_r° and $T\Delta S_r^{\circ}$ [see equation (23)] shows that when $\Delta H_r^{\circ} \ll 0$, $\Delta G_r^{\circ} < 0$ in most instances. Thus for a reaction having $\Delta H_r^{\circ} = -10$ kcal, ΔS_r° would need to be more negative than -34 cal deg^{-1} before ΔG_r° became positive at $298 \cdot 15°K$. Such a large decrease in entropy is only likely to occur in reactions where there is a decrease in the number of moles of *gaseous* species, or in some polymerization reactions. With these exceptions in mind, we may aver that when $\Delta H_r^{\circ} < -10$ kcal a reaction is likely to be thermodynamically, as well as thermochemically, favourable at $298 \cdot 15°K$. It must always be remembered, however, that even a thermodynamically favourable reaction will not necessarily proceed spontaneously at a detectable rate. It may be necessary to use catalysts or to change the temperature or other reaction conditions to cause a reaction to proceed along a desired pathway at a useful rate.

6.1.1. HYDROGENATION AND HYDROGENOLYSIS REACTIONS

Table 11 gives values of ΔH_r° for the hydrogenation of some olefinic, diolefinic and benzenoid compounds. In practice, when hydrogen is used as the reducing agent, such reactions proceed at a useful rate only in the presence of catalysts, usually finely divided metals such as nickel, platinum or palladium. Nascent hydrogen or hydrogen sources such as the metal hydrides can sometimes be used to advantage in place of molecular hydrogen; the overall heat of reaction will then be equal to the sum of the heats for the [hypothetical] reaction in which molecular hydrogen is generated and for the reaction in which the unsaturated compound is reduced by hydrogen.

TABLE 11. Heats of some hydrogenation reactions at 25°C.

	Reaction	ΔH_r°, kcal.
I	$C_2H_4(g)+H_2(g) = C_2H_6(g)$	$-32 \cdot 69 \pm 0 \cdot 16$
II	$CH_2{=}CH.C_3H_7(g)+H_2(g) = n\text{-}C_5H_{12}(g)$	$-29 \cdot 77 \pm 0 \cdot 34$
III	$c\text{-}C_6H_{10}(l)+H_2(g) = c\text{-}C_6H_{12}(l)$	$-28 \cdot 32 \pm 0 \cdot 21$
IV	$c\text{-}C_6H_{10}(g)+H_2(g) = c\text{-}C_6H_{12}(g)$	$-28 \cdot 42 \pm 0 \cdot 24$
V	(c)$+H_2(g) = HO_2C.CH_2.CH_2.CO_2H(c)$	$-31 \cdot 03 \pm 0 \cdot 18$
VI	(g)$+2H_2(g) = n\text{-}C_5H_{12}(g)$	$-53 \cdot 22 \pm 0 \cdot 22$
VII	$CH_2{=}CH.CH_2.CH{=}CH_2(g)+2H_2(g) = n\text{-}C_5H_{12}(g)$	$-60 \cdot 35 \pm 0 \cdot 25$
VIII	(l)$+3H_2(g) = c\text{-}C_6H_{12}(l)$	$-49 \cdot 12 \pm 0 \cdot 19$
IX	(g)$+3H_2(g) = c\text{-}C_6H_{12}(g)$	$-49 \cdot 31 \pm 0 \cdot 20$
X	(c)$+5H_2(g) =$ (l)	$-73 \cdot 75 \pm 0 \cdot 34$

All the hydrogenation reactions in Table 11 are thermochemically favourable. The values of ΔH_r° for the hydrogenation of the mono-olefines (reactions I-V, with olefine and saturated product in the same physical state) are all seen to be in the region of -30 kcal. ΔH_r° for the hydrogenation of the unconjugated diolefine 1,4-pentadiene (reaction VII) is almost exactly double that for the corresponding mono-olefine 1-pentene (reaction II), but that for the conjugated diolefine *trans*-1, 3-pentadiene (reaction VI) is significantly less than double. Comment on the underlying reasons for the small, but significant, differences in the heats of hydrogenation of various mono-olefines and on the heats of hydrogenation of conjugated olefines and aromatic compounds is made in Chapter 7.

The heats of dehydrogenation of the saturated compounds in Table 11, corresponding to reactions I–X in reverse, must be equal in magnitude but opposite in sign to the heats of the respective hydrogenation reactions. Hence all these dehydrogenation reactions are thermochemically unfavourable; the question whether reaction I is also thermodynamically unfavourable is discussed in Section 6.2.1.

Values of ΔH_r° for the hydrogenation, or hydrogenolysis, of some compounds containing functional groups other than C = C are given in Table 12.

TABLE 12. Heats of some hydrogenation and hydrogenolysis reactions at 25°C.

	Reaction	ΔH_r°, kcal.
XI	$CH_3.CHO(g)+H_2(g) = CH_3.CH_2OH(g)$	$-16\cdot51\pm0\cdot14$
XII	$CH_3.COCH_3(g)+H_2(g) = CH_3.CH(OH)CH_3(g)$	$-13\cdot22\pm0\cdot18$
XIII	$CH_3.CO_2H(l)+2H_2(g) = CH_3.CH_2OH(l)+H_2O(l)$	$-18\cdot99\pm0\cdot11$
XIV	$CH_3.CO_2C_2H_5(l)+2H_2(g) = 2CH_3.CH_2OH(l)$	$-18\cdot10\pm0\cdot22$
XV	$C_2H_5Br(g)+H_2(g) = C_2H_6(g)+HBr(g)$	$-13\cdot7\ \pm0\cdot6$
XVI	$CH_3.COCl(l)+2H_2(g) = CH_3.CH_2OH(l)+HCl(g)$	$-22\cdot9\ \pm0\cdot2$
XVII	*trans* ⟨O⟩ - N=N - ⟨O⟩ (c) + H₂(g) = ⟨O⟩ - NH.NH - ⟨O⟩ (c)	$-23\cdot7\ \pm0\cdot5$
XVIII	$CH_3S.SCH_3(g)+H_2(g) = 2CH_3SH(g)$	$-5\cdot16\pm0\cdot37$
XIX	$CH_3NH.NHCH_3(g)+H_2(g) = 2CH_3NH_2(g)$	$-33\cdot0\ \pm[0\cdot5]$
XX	$CH_3O.OCH_3(g)+H_2(g) = 2CH_3OH(g)$	$-66\cdot0\ \pm0\cdot4*$

The hydrogenation of carbonyl compounds (exemplified by reactions XI and XII) is used industrially for the synthesis of higher alcohols. Hydrogenolysis reactions (exemplified by reactions XIII–XVI) are frequently employed in laboratory-scale syntheses. All these reactions are thermochemically favourable, though less so than the reactions involving hydrogenation of the C=C bond (Table 11); this is also the case for the hydrogenation of the N=N bond, exemplified by reaction XVII.

Reactions XVIII to XX, involving reductive bond fission, show an interesting gradation in the magnitude of ΔH_r°.

6.1.2. OXIDATION REACTIONS

In this Section we discuss reactions involving the addition of oxygen to organic molecules, sometimes accompanied by partial scission of the molecule; the complete oxidative degradation of organic compounds to carbon dioxide and water is discussed in Section 6.1.6. Some examples of oxidation reactions of organic compounds are given in Table 13.

TABLE 13. Heats of some oxidation reactions at 25°C.

	Reaction	ΔH_r°, kcal.
XXI	$C_2H_5OH(l) + 0\cdot 5 O_2(g) = CH_3.CHO(l) + H_2O(l)$	$-47\cdot 78 \pm 0\cdot 16$
XXII	$CH_3.CHO(l) + 0\cdot 5 O_2(g) = CH_3.CO_2H(l)$	$-69\cdot 87 \pm 0\cdot 15$
XXIII		$-163\cdot 22 \pm 0\cdot 12$
XXIV	$n\text{-}C_6H_{14}(l) + 3\cdot 8333 O_2(g)$ $= 2\cdot 66 CH_3.CO_2H(l) + 0\cdot 66 HCO_2H(l) + H_2O(l)$	$-397\cdot 26 \pm 0\cdot 27$
XXV	$c\text{-}C_6H_{12}(l) + 3\cdot 33 O_2(g) = 2\cdot 66 CH_3.CO_2H(l)$ $+ 0\cdot 66 HCO_2H(l)$	$-339\cdot 00 \pm 0\cdot 25$
XXVI	$C_2H_4(g) + 0\cdot 5 O_2(g) = CH_3.CHO(g)$	$-52\cdot 18 \pm 0\cdot 16$
XXVII		$-25\cdot 03 \pm 0\cdot 18$
XXVIII		$-453\cdot 32 \pm 0\cdot 51$
XXIX		$-309\cdot 09 \pm 0\cdot 53$
XXX	$n\text{-}C_4H_9.CH=CH_2(l) + CO(g) + H_2(g) = n\text{-}C_6H_{13}.CHO(l)$	$-30\cdot 8 \pm 1\cdot 0$

All the oxidation reactions listed in Table 13 are thermochemically favourable, particularly those involving scission of $C-C$ and $C-H$ bonds. When oxidations are carried out in the laboratory (reactions such as XXI–XXIII), reagents containing peroxy- or oxyanion-groups are frequently used, but in industrial practice molecular oxygen (of which air is the cheapest source) is commonly used as oxidant. Thus the D.C.L. process for the large-scale production of carboxylic acids from naphtha (exemplified by reactions XXIV and XXV) is believed to employ air under pressure at *ca.* 200°C in the presence of oil-soluble metal-containing catalysts (67/53).

A process used in Germany for the manufacture of acetaldehyde, reaction XXVI, is based on the use of compressed air as the oxidant, with an aqueous solution of cupric chloride and palladium chloride as catalyst (67/54). The left-hand sides of reactions XXVI and XXVII are identical but the products are isomeric. In this example of alternative reaction pathways there is a marked difference in the reaction heats, and the practical application of the less thermodynamically favourable reaction XXVII will depend on the availability of a highly specific catalyst; metallic silver fulfills this role in industrial processes (67/53).

Reactions XXVIII and XXIX form the bases of commercial processes for the manufacture of phthalic anhydride, using naphthalene (reaction XXVIII) or *o*-xylene (reaction XXIV) as raw materials. In the von Heyden process the organic vapours are oxidised at *ca.* 400°C in a stream of air over granulated vanadium pentoxide catalyst (67/54). The very large amount of heat evolved is transferred to a surrounding molten-salt bath, which serves to maintain the reactor at constant temperature.

Reaction XXX is quite different in character from the other reactions listed in Table 13. Two reducing agents, carbon monoxide and hydrogen, are employed, yet the end result of this hydroformylation, or OXO, reaction is the addition of oxygen to an organic molecule, hence in this discussion the reaction may be regarded as an oxidation. On an industrial scale the exothermic reaction XXX is carried out using $C_3 - C_{12}$ olefines as feedstock, hydrogen and carbon monoxide at high pressure, and cobalt carbonyl compounds as catalysts (67/54).

6.1.3. HYDROLYSIS AND HYDRATION REACTIONS

Table 14 contains examples of the types of hydrolysis reaction encountered in synthetic organic chemistry, together with some examples of hydration reactions that have technological importance.

The heat of hydrolysis of ethyl acetate (reaction XXXI) is close to zero. It would be expected that ΔS_r° for a reaction involving four liquids only would be relatively small and hence that ΔG_r° for reaction XXXI would be close to zero also. In consequence the equilibrium constant for this reaction, in terms of mole fractions, will be near unity—it is well known that the hydrolysis of ethyl acetate, and of most esters, proceeds under normal conditions to an equilibrium state in which all four components are present in analysable

TABLE 14. Heats of some hydrolysis and hydration reactions at 25°C.

	Reaction	ΔH_r°, kcal.
XXXI	$CH_3.CO_2C_2H_5(l) + H_2O(l) = CH_3.CO_2H(l) + C_2H_5OH(l)$	$+0.89 \pm 0.18$
XXXII	$(CH_3CO)_2O(l) + H_2O(l) = 2CH_3.CO_2H(l)$	-13.98 ± 0.18
XXXIII	$CH_3.COBr(l) + \infty H_2O(l) = [CH_3.CO_2H + HBr](aq.)$	-23.34 ± 0.25
XXXIV	$n\text{-}C_4H_9.CONH_2(c) + \infty H_2O(l)$ $= n\text{-}C_4H_9.CO_2H(l) + NH_3 \text{ (aq.)}$	$+6.12 \pm 0.35$
XXXV	$C_2H_5Br(l) + \infty H_2O(l) = [C_2H_5OH(l) + HBr] \text{ (aq.)}$	-7.9 ± 0.5
XXXVI	$n\text{-}C_4H_9Li(l) + H_2O(l) = n\text{-}C_4H_{10}(g) + LiOH(c)$	-46.9 ± 1.0
XXXVII	$C_2H_4(g) + H_2O(g) = C_2H_5OH(g)$	-10.89 ± 0.14
XXXVIII	$2C_2H_4(g) + H_2O(g) = C_2H_5OC_2H_5(g)$	-27.36 ± 0.22
XXXIX	$C_2H_2(g) + H_2O(l) = CH_3.CHO(l)$	-31.90 ± 0.23
XL	$\underset{\diagdown \text{ } O \text{ } \diagup}{CH_2 \!-\! CH_2} \text{ (g)} + H_2O(l) = C_2H_4(OH)_2(l)$	-27.83 ± 0.24

quantities. By contrast, the hydrolysis of an acid anhydride (reaction XXXII) is quite favourable thermochemically and the hydrolysis of an acyl halide (reaction XXXIII) even more so; as both of these reactions involve liquids only, ΔS_r° is small, so that $\Delta G_r^{\circ} \ll 0$ and hence these reactions will proceed essentially to completion. The hydrolysis of an amide is thermochemically unfavourable (e.g. for reaction XXXIV, $\Delta H_r^{\circ} = +6.12 \pm 0.35$) under the standard state conditions, but equilibrium in this hydrolysis reaction can be displaced and the reaction caused to proceed to completion by removal of NH_3, either by boiling or by adding acid. The hydrolysis of an alkyl halide (reaction XXXV) is thermochemically favourable, though much less so than the hydrolysis of the corresponding acyl halide, and one might suspect that the direction of this reaction could be reversed by altering the conditions. Indeed by use of gaseous hydrogen bromide, with continuous removal of water from the reaction mixture, higher alcohols can be converted smoothly into bromides. Reaction XXXVI is an example of the hydrolysis of an organometallic compound of the type often obtained as an intermediate in an organic synthesis; such reactions are usually highly exothermic, as in the given example.

The hydration of ethylene to ethanol (reaction XXXVII) is an important method of producing industrial (i.e. non-potable) alcohol. Most commercial processes employ a gas-phase reaction at about 300°C with phosphoric acid on an inert support as catalyst (67/53). This reaction is somewhat exothermic ($\Delta H_{r,\,600°K}^{\circ} = -11\cdot3$ kcal), but there is a decrease in the number of moles of gas as a result of reaction and $\Delta S_{r,\,600°K}^{\circ} = -31\cdot0$ cal deg^{-1}, giving $\Delta G_{r,\,600°K}^{\circ} = +7\cdot2$ kcal; hence the direction of this reaction can easily be reversed (cf. Section 6.2.2). Production of ethylene by this reversed reaction was once important commercially but has diminished to insignificance with the rise of the petrochemical industry. Diethyl ether is a byproduct of the ethylene–ethanol reaction, whichever its direction (see reaction XXXVIII).

The hydration of acetylene (reaction XXXIX), catalysed by mercuric salts, is the basis of one commercial process for the synthesis of acetaldehyde; the reaction is very exothermic. Commercial production of ethylene glycol is effected by hydration of ethylene oxide (reaction XL); this exothermic reaction is carried out at about 150°C under pressure (67/53).

6.1.4. REACTIONS OF HALOGENS AND ORGANOHALOGEN COMPOUNDS

TABLE 15. Heats of some reactions involving halogen compounds at 25°C.

	Reaction	ΔH_r°, kcal.
XLI	$C_2H_4(g)+Cl_2(g) = C_2H_4Cl_2(g)$	$-43\cdot3\ \pm0\cdot4$
XLII	$C_2H_4(g)+Br_2(g) = C_2H_4Br_2(g)$	$-28\cdot9\ \pm0\cdot4$
XLIII	$C_2H_4(g)+I_2(g) = C_2H_4I_2(g)$	$-11\cdot9\ \pm0\cdot4$
XLIV	$c\text{-}C_6H_{12}(l)+Cl_2(g) = c\text{-}C_6H_{11}Cl(l)+HCl(g)$	$-34\cdot20\pm0\cdot37$
XLV	$C_6H_6(l)+Cl_2(g) = C_6H_5Cl(l)+HCl(g)$	$-31\cdot20\pm0\cdot22$
XLVI	$C_2H_6(g)+Cl_2(g) = C_2H_5Cl(g)+HCl(g)$	$-27\cdot91\pm0\cdot60$
XLVII	$C_2H_6(g)+Br_2(g) = C_2H_5Br(g)+HBr(g)$	$-11\cdot1\ \pm0\cdot5$
XLVIII	$C_2H_6(g)+I_2(g) = C_2H_5I(g)+HI(g)$	$+9\cdot7\ \pm0\cdot4$
XLIX	$CCl_4(l)+2HF(g) = CCl_2F_2(g)+2HCl(g)$	$+3\cdot7\ \pm1\cdot5$
L	$4C_2H_5Cl(g)+4[\text{Pb-Na alloy}]$ $=Pb(C_2H_5)_4(l)+3Pb+4NaCl(c)$	$-277\cdot0\ \pm1\cdot8$

Reactions XLI, XLII and XLIII in Table 15 refer to the addition to ethylene of gaseous chlorine, bromine, and iodine respectively. The gaseous state was chosen for the halogens in these reactions to reveal the trend in the values of ΔH_r°, without the need to consider the differing states of matter of the halogens in their reference states at 25°C. The addition of the halogens to ethylene is thermochemically favourable in each case, least so for iodine. In these reactions there is a decrease in the number of moles of gas as a result of reaction, so one would predict that $\Delta G_r^\circ > \Delta H_r^\circ$ (cf. Section 6.1), and that for the addition of iodine (reaction XLIII) ΔG_r° would be sufficiently close to zero for this reaction to go in either direction, depending upon the experimental conditions. This prediction is in accord with observation, since vicinal di-iodides readily undergo dehalogenation.

Substitutive chlorination, as exemplified by reactions XLIV–XLVI, is a highly exothermic process; there is no great difference in ΔH_r° for substitution of aliphatic *vis-à-vis* aromatic compounds.

Reactions XLVI–XLVIII show the heats of substitution of ethane by the halogens. The reaction with chlorine is very favourable thermochemically, that with bromine is less so, but the reaction with iodine is thermochemically unfavourable, and indeed direct iodination is rarely encountered in organic chemistry.

Examples of important industrial metathetical reactions of organohalogen compounds are the syntheses of fluorochloro-compounds such as refrigerants and propellants (reaction XLIX), and the production of tetraethyl-lead used as an anti-knock agent in petroleum fuels (reaction L). The halogen-exchange reaction between carbon tetrachloride and anhydrous hydrogen fluoride, catalysed by antimony trichloride, yields the products $CFCl_3$, CF_2Cl_2 and CF_3Cl. The reaction producing CF_2Cl_2, shown in Table 15, is slightly unfavourable thermochemically. The commercial process is conducted at about 100°C with continuous displacement of the equilibrium by fractionating off the required products (67/53). By contrast, the reaction used for the synthesis of tetraethyl-lead (reaction L) is very favourable thermochemically.

6.1.5. ISOMERIZATION AND POLYMERIZATION REACTIONS

We discuss first the isomerization reactions, or intramolecular rearrangements, shown in Table 16. When these involve skeletal rearrangements of saturated molecules, ΔH_r° is generally small. An example is the isomerization of n-octane to 2, 2, 4-trimethylpentane (reaction LI), which is one of the reactions occurring when petroleum distillates are submitted to catalytic reforming. Reaction LI is mildly exothermic, whilst the isomerization of the bicyclic hydrocarbon β-pinene to the open-chain hydrocarbon myrcene, (reaction LII) is mildly endothermic; this latter reaction, which is brought about on an industrial scale by passing β-pinene through a tube at about 600°C, is the first step in the synthesis of certain perfumes from a cheap constituent of turpentine (64/50). Quite large reaction heats sometimes occur in isomerization reactions involving changes in the type of chemical bonding. For example, the isomerization of ethylene oxide to acetaldehyde (reaction LIII) is so favourable thermochemically that this reaction has been known to occur with explosive violence. Another well known isomerization reaction

TABLE 16. Heats of some isomerization and polymerization reactions at 25°C.

	Reaction	ΔH_r°, kcal.
LI	$CH_3(CH_2)_6CH_3(l) = (CH_3)_3C.CH_2.CH(CH_3)_2(l)$	$-2\cdot16\pm0\cdot45$
LII	$(l) = (CH_3)_2C{=}CH(CH_2)_2\overset{\shortparallel}{C}.CH{=}CH_2(l)$	$+5\cdot3\ \pm0\cdot9$
LIII	$\overset{\displaystyle O}{\overset{\diagup\ \diagdown}{CH_2{-}CH_2}}(g) = CH_3.CHO(g)$	$-27\cdot15\pm0\cdot19$
LIV	$\langle\!\bigcirc\!\rangle\text{-}NH{-}NH\text{-}\langle\!\bigcirc\!\rangle(c) = NH_2\text{-}\langle\!\bigcirc\!\rangle\text{-}\langle\!\bigcirc\!\rangle\text{-}NH_2(c)$	$-36\cdot0\ \pm0\cdot5$
LV	$CH_2{=}CH_2(g) = (1/n)\,(C_2H_4)_n\ \text{(polymer)}$	$-25\cdot5\ \pm0\cdot2$
LVI	$\overset{\displaystyle CH_2{-}CH_2}{\underset{CH_2}{\diagdown\ \diagup}}(g) = (1/n)\,(C_3H_6)_n\ \text{(polymer)}$	$-32\cdot0\ \pm0\cdot3$
LVII	$CH_2{=}CHCl(g) = (1/n)\,(C_2H_3Cl)_n\ \text{(polymer)}$	$-31\cdot1\ \pm0\cdot5$
LVIII	$CF_2{=}CF_2(g) = (1/n)\,(C_2F_4)_n\ \text{(polymer)}$	$-41\ \ \pm1$
LIX	$\langle\!\bigcirc\!\rangle\text{-}CH = CH_2(l) = (1/n)\,(C_8H_8)_n\ \text{(polymer)}$	$-16\cdot7\ \pm0\cdot4$
LX	$\overset{\displaystyle CH_2{-}CO}{\underset{CH_2{-}O}{\mid\qquad\mid}}(l) = (1/n)\,(C_3H_4O_2)_n\ \text{(polymer)}$	$-19\cdot4\ \pm0\cdot5$

is the benzidine rearrangement (reaction LIV) and this is very favourable thermochemically.

Intermolecular bond rearrangement can give rise to dimers or to a series of oligomers up to polymers of high molecular weight. We consider here the addition polymerization of monomers to form high polymers, the type of reaction which underlies the manufacture of most plastics. Examples are the production of polyethylene (reaction LV), polyvinylchloride (reaction LVII), and polytetrafluoroethylene (reaction LVIII) from the gaseous monomers; each of these reactions is very favourable thermochemically. It is noteworthy that the polymerization of cyclopropane to form polyethylene (reaction LVI) is also thermochemically very favourable but a method for carrying out this particular reaction has not yet been discovered. The above examples are all of reactions in which gaseous monomers are converted to solid polymers. When a liquid monomer is polymerized, as in the production of polystyrene (reaction LIX), the value of ΔH_r° is less negative. Reaction LX, the polymerization of liquid β-propiolactone, is quite favourable thermochemically. Here, the heat of reaction derives largely from the release of strain energy when the 4-membered lactone ring opens to form a chain structure.

The thermodynamic aspects of polymerization reactions are discussed in Section 6.2.3.

6.1.6. COMBUSTION AND DECOMPOSITION REACTIONS

Combustion reactions form the basis for producing heat from fossil fuels and hence for generating much of the motive and electric power used at the present time. Decomposition reactions are the source of the energy release from explosives. Both these types of reaction involve the degradation of organic molecules into smaller ones, and for this reason it is convenient to discuss them in the same Section.

The use of combustion reactions in calorimetric measurement was discussed in Section 3.2. The reactions which occur in a flame calorimeter (Section 3.2.6) are similar to those occurring in domestic or industrial equipment in which gaseous or liquid fuel is burned at a jet in air: virtually complete combustion of the organic constituents of the fuel to carbon dioxide and water takes place in properly designed appliances. There is also a formal similarity between the reactions that occur in a bomb calorimeter (Section 3.2.2) and those that occur in internal combustion engines of both the electrical-ignition and the compression-ignition types, but whereas complete combustion of the fuel takes place in a bomb calorimeter, combustion in an internal combustion engine may be incomplete: some fuel may be unburned and some partial combustion products such as carbon monoxide and aldehydes may be formed.

The maximum amount of energy that can be generated by combustion of unit mass of fuel can in principle be precisely calculated *a priori* if there is foreknowledge of (a) the composition of the fuel, (b) the composition and mean temperature of the products of combustion. Foreknowledge of the composition of the fuel, or more usefully of its *calorific value*†, can be obtained by laboratory experiments. Standardised methods for analysing fuels and determining their calorific values are published by bodies such as the British Standards Institution, The Institute of Petroleum, the American Society for Testing and Materials, and the International Standards Organization. The questions of the compositions and temperatures of the combustion products in an engine are less easily answered because these variables depend on the design and condition of the engine and on the manner in which it is operated.

---

† The precise definition of *calorific value* depends on the physical state of the fuel and on the physical state of the water produced by combustion. Readers are referred to reference (61/58).

The technological aspects of fuel performance are outside the scope of this book [see the monograph by Brame and King (67/55) and the handbook "Technical data on fuel" (55/37)], but some general observations can be made. Firstly, the water produced by combustion in any appliance or engine that has reached its normal operating temperature will remain in the gaseous state. Hence, to obtain the calorific value appropriate for the combustion of benzene, for example, under practical conditions in an engine, the value of $-\Delta H_c^{\circ}$ for benzene must be reduced by $3 \times \Delta H_v$ for water. Secondly, the heat evolved on complete combustion of unit mass of fuel can be calculated from equation (6), if ΔH_f° of the fuel is known or can be estimated from its chemical composition. Thus for combustion of a fuel of composition $C_a H_b O_c$, equation (6) can be written in the form

$$
-\frac{\Delta H_c^{\circ}}{M} = \frac{-1}{M}\left\{a \times \Delta H_f^{\circ}[CO_2(g)] + (b/2) \times \Delta H_f^{\circ}[H_2O(g)] - \Delta H_f^{\circ}[C_a H_b O_c]\right\}
$$

$$
= \frac{1}{M}\left\{94{\cdot}05a + 28{\cdot}9b + \Delta H_f^{\circ}[C_a H_b O_c]\right\}, \tag{144}
$$

where M is the effective molecular weight of the fuel and $-\Delta H_c^{\circ}/M$ is the heat output (at constant pressure) per gramme at 25°C. A high heat output per gramme is provided by fuels having relatively high values of a/M and b/M compared with values of $\Delta H_f^{\circ}/M$ that are positive or at least not too negative. Representative values of $-\Delta H_c^{\circ}/M$ for hydrocarbons, both gaseous and liquid, are given in Table 17.

Of the hydrocarbons listed in Table 17, that of lowest molecular weight, methane, has the highest value of $-\Delta H_c^{\circ}/M$. It is of interest to note that the heat output per gramme for methane is greater than that for graphite, namely 7830 cal g^{-1}, but much lower than the value for its other constituent element, hydrogen, namely 28 670 cal g^{-1}. On a volumetric basis, however, the heat of combustion of methane is greater than that of hydrogen. For the homologous series of n-alkanes, $-\Delta H_c^{\circ}/M$ decreases slowly as the length of the hydrocarbon chain increases. Chain branching in alkanes gives rise to a slight reduction in the heat output per gramme (exemplified in Table 17 by 2, 2, 4-trimethylpentane compared with n-octane) but branched-chain alkanes have better antiknock properties than the straight-chain compounds and for this reason are preferred as engine fuels (67/55). When the C_6 non-alkanes in Table 17 are considered, it is seen that the non-alkanes have less heat output per unit mass than the alkanes, but the differences are not great

TABLE 17. Heat of combustion per unit mass of some hydrocarbon fuels*

Fuel	$-\Delta H_c^\circ/M$, cal g^{-1}.
Methane (g), CH_4	11 952
n-Butane (g), C_4H_{10}	10 922
n-Hexane (l), C_6H_{14}	10 692
n-Octane (l), C_8H_{18}	10 617
2, 2, 4-Trimethylpentane (l), C_8H_{18}	10 598
1-Hexene (l), C_6H_{12}	10 620
Cyclohexane (l), C_6H_{12}	10 381
Benzene (l), C_6H_6	9 594

* Combustion to gaseous CO_2 and gaseous H_2O at $298 \cdot 15^\circ K$, at constant pressure.

and to a first approximation we may say that all hydrocarbons in the gasoline range give *ca.* 10 000 cal g^{-1} on combustion. Oxygen derivatives of the hydrocarbons generally give much less heat than this. This arises because the derivative must have a higher molecular weight than the parent hydrocarbon, and may have a lower value of b and a more negative value of ΔH_f° also (*cf.* equation (144)). For example, ethyl alcohol, which finds high-tonnage use as a fuel, gives only 6 405 cal g^{-1} on combustion. However, the heat output per gramme is by no means the only criterion by which an engine fuel is judged, and from the standpoints of completeness of combustion and resistance to knock, ethyl alcohol rates high.

In the above discussion on fuel combustion it was assumed that the oxidant was gaseous molecular oxygen, since in the form of air this is the cheapest and most abundant oxidant. In certain specialised applications, however, notably in rocket propulsion, other oxidants may be used (60/50, 66/52). These include liquid oxygen, liquid ozone, hydrogen peroxide, nitric acid and dinitrogen tetroxide. When these liquid oxidants are used, calculation of the heat output per unit mass of fuel requires data on the heat of formation of the oxidants (61/3, 68/4). Solid oxidants such as potassium nitrate and ammonium perchlorate, premixed with an appropriate fuel, are also used for rocket propulsion. Gunpowder (charcoal plus sulphur plus potassium nitrate) is of course the oldest example of a system of this type. Modern solid fuels for rockets are generally organic polymers containing an inorganic oxidant dispersed throughout the polymer matrix (60/50, 66/52).

An important discovery of the nineteenth century was that the oxidant function of a fuel system need not reside in a substance physically distinct from the fuel but can be chemically incorporated into the fuel itself. From the first example of such a compound, glycerol trinitrate or "nitroglycerin", has been developed the range of high explosives and smokeless powders used in weapons and in mining and quarrying. When glycerol trinitrate is detonated, it decomposes according to the following equation:

$$C_3H_5(NO_3)_3(l) = 3CO_2(g) + 2{\cdot}5H_2O(g) + 1{\cdot}5N_2(g) + 0{\cdot}25O_2(g).$$

It will be noted that one mole of liquid decomposes to give 7·25 moles of gas and it is this sudden and >1000-fold increase in volume which is the main source of the explosive power of glycerol trinitrate; the amount of heat released, 1487 cal g^{-1}, is quite modest compared with the combustion heats of hydrocarbons. In glycerol trinitrate there is an excess of oxygen over that required to convert all the carbon in the molecule to carbon dioxide and all the hydrogen to water. Hence for glycerol trinitrate, the heat of combustion in excess oxygen (to gaseous water) will be equal to the heat of decomposition. Pentaerythritol tetranitrate is an example of an explosive in which there is a deficiency of oxygen compared with that needed for complete combustion. It decomposes on detonation (66/29) according to the equation:

ONO$_2$

|

CH$_2$

|

O$_2$NOCH$_2$-C-CH$_2$ONO$_2$(c) = 3·5CO$_2$(g) + 1·5CO(g) + 3·5H$_2$O(g)

| $\qquad\qquad\qquad\qquad\qquad\qquad\qquad\qquad$ +0·5H$_2$(g) + 2N$_2$(g).

CH$_2$

|

ONO$_2$

Gunn and coworkers (66/29) measured the energy of combustion of pentaerythritol tetranitrate in excess oxygen as 1956 cal g^{-1}, and the energy of decomposition as 1499 cal g^{-1}. These values were obtained using a bomb calorimeter of special construction, capable of withstanding repeated explosive shocks; the numerical values refer to a final state in which water is present as a liquid. The difference between ΔU_c° and $\Delta U^\circ_{\text{decomp}}$ is reconcilable with the value of ΔU_c° for oxidizing 1·5 moles of carbon monoxide to carbon dioxide and 0·5 moles of hydrogen to water. Very large differences between ΔU_c° and $\Delta U^\circ_{\text{decomp}}$ occur with explosives containing relatively low ratios of O : C (or H). Sasiadek (67/56) has reported a 3-fold variation between ΔU_c and ΔU_{decomp} for silver picrate, which in the absence of added oxygen decomposes according to the equation:

$$C_6H_2(NO_2)_3OAg(c) = 6CO(g) + H_2O(g) + 1{\cdot}5N_2(g) + Ag(c).$$

Although the majority of high explosive substances in use are organic nitro- or nitrato-compounds, other types of organic compound can undergo exothermic decomposition. In principle, any organic compound that has a positive value of ΔG_f° will revert, usually exothermally, to the elements of which it is composed. Benzene, though generally considered to be a very stable substance, belongs to this class of compounds but fortunately for organic chemistry its rate of decomposition is negligible at ordinary temperatures. By contrast, certain heavy-metal organometallic compounds with positive values of ΔG_f° decompose on percussion with explosive violence; these compounds serve as detonators.

6.2. Calculation of the thermodynamic feasibility of organic reactions

In Sections 2.5.2, 2.5.3 and 3.4 we discussed methods for deriving heats of reaction from measurements of chemical equilibria. We now discuss the converse procedure, namely the use of heat-of-reaction and entropy data to calculate equilibrium constants and hence to predict whether a particular reaction is likely to give a reasonable yield of desired product. The precision needed for the basic data in calculations of this type was considered in Section 2.4.

We have defined the term "thermodynamically favourable" as applying to reactions having $\Delta G_r^\circ < 0$, and "thermodynamically unfavourable" to those having $\Delta G_r^\circ > 0$ (Section 6.1). The special case $\Delta G_r^\circ = 0$ corresponds to a reaction for which the equilibrium constant is unity. According to thermodynamic considerations (i.e. neglecting kinetic considerations), reactions with $\Delta G_r^\circ < 0$ are more propitious than those with $\Delta G_r^\circ > 0$, but it is nevertheless sometimes possible to obtain an acceptable yield of product from reactions of the latter type. For purposes of convenience only we introduce now the term "thermodynamically feasible" to describe reactions that are calculated to give an acceptable yield of a desired product at equilibrium. This term is not precise because what is acceptable as a yield will vary greatly from case to case and will often be decided by economic rather than chemical considerations. In general, the equilibrium ratio [desired organic product X]/[organic reactant A] will need to be greater than 10^{-2} for a yield to be considered acceptable and the reaction in question to be considered feasible. Corresponding values of K_p are given by equation (25), and they will range over many powers of 10, depending on the magnitudes of the stoicheiometric coefficients and on the number and concentrations (or partial pressures) of other species, B, C..., Y, Z... participating in the reaction. In the majority of cases, however, $K_p > 10^{-10}$ for $[X]/[A] > 10^{-2}$, and from equation (24) this corresponds to $\Delta G_{r,\ 298\cdot15^\circ K}^\circ < +14$ kcal, or $\Delta G_{r,\ 900^\circ K}^\circ$ $< +42$ kcal. Hence a reaction having a value of $\Delta G_r^\circ > +14$ kcal at

298·15°K is very unlikely to give an acceptable yield of product and may reasonably be described as "thermodynamically unfeasible". At the highest temperatures at which organic reactions are carried out, acceptable yields may in some circumstances be obtained from reactions having $\Delta G^\circ_{r,T}$ up to $+40$ kcal.

We now consider some organic reactions from the point of view of their "thermodynamic feasibility", with particular reference to the effect of changing the reaction temperature. Values of $\Delta G^\circ_{r,\ T}$ needed for this discussion will be calculated from values of ΔH°_r given in Chapter 5 and from values of $(H^\circ_T - H^\circ_{ref})$ and S°_T from named sources.

6.2.1. THE DEHYDROGENATION OF ETHANE

$$C_2H_6(g) \rightleftharpoons C_2H_4(g) + H_2(g).$$

For this reaction at 298·15°K, $\Delta H^\circ_r = +32·69 \pm 0·16$ kcal (see Table 11, reaction I). As there is an increase in the number of moles of gaseous species, ΔS°_r is positive and $\Delta G^\circ_r < \Delta H^\circ_r$. Numerical values of the thermodynamic quantities are presented in Table 18, for various temperatures. Ideal-gas behaviour has been assumed in the calculations and experimental errors have been excluded. From the discussion on errors in Section 2.4, it is expected that the main source of uncertainty will arise from the experimental error in ΔH°_r, and that the uncertainty in K_p will be less than $\pm 20\%$ at 298·15°K and less than $\pm 10\%$ at 900°K, leading to uncertainties in $p_{C_2H_4}/p_{C_2H_6}$ at

TABLE 18. Thermodynamic quantities for the equilibrium
$$C_2H_6(g) \rightleftharpoons C_2H_4(g) + H_2(g)$$

Quantity	T, °K			
	298·15	500	700	900
$\Delta(H^\circ_T - H^\circ_{298})$, †kcal	0·00	$+0·84$	$+1·59$	$+1·68$
$\Delta H^\circ_{r,\ T}$, kcal	$+32·69$	$+33·53$	$+34·28$	$+34·37$
$\Delta S^\circ_{r,\ T}$, †cal deg^{-1}	$+28·80$	$+31·00$	$+31·92$	$+32·30$
$T\Delta S^\circ_{r\ T}$, kcal	$+8·59$	$+15·50$	$+22·34$	$+29·07$
$\Delta G^\circ_{r,\ T}$, kcal	$+24·10$	$+18·03$	$+11·94$	$+5·30$
K_p, atm	$10^{-17·66}$	$10^{-7·88}$	$10^{-3·727}$	$10^{-1·287}$
$p_{C_2H_4}/p_{C_2H_6}$ for $p_{C_2H_6} = 1$ atm.	$1·48 \times 10^{-9}$	$1·15 \times 10^{-4}$	$1·37 \times 10^{-2}$	$0·227$

† Values calculated from data in reference 53/1.

298·15°K of $< \pm 5\%$ and at 900°K of $< \pm 4\%$. Uncertainties of this magnitude do not influence the following discussion.

From Table 18 it is seen that for $p_{C_2H_6} = 1$ atm, $p_{C_2H_4}/p_{C_2H_6}$ reaches 10^{-2} at *ca.* 700°K and that the yield of ethylene improves as the temperature is increased further. We conclude that this reaction is thermodynamically feasible at temperatures of 700°K and above. At very high temperatures an upper limit to the yield will be set by the occurrence of other cracking reactions leading to the production of acetylene, methane, carbon and hydrogen. Industrially, the catalytic cracking of alkane feedstocks to olefines, of which the ethane–ethylene reaction is the simplest example, is conducted at temperatures in the range 900–1100°K to secure high yields. These temperatures are higher than those to which organic compounds are generally subjected, and to avoid coke formation it is necessary to raise the temperature of the reactants very rapidly and to quench the products quickly also (67/53).

If equation (26) is used to calculate K_p, it is perhaps more easily seen why a reaction that is so unfavourable at 298·15°K becomes at *ca.* 900°K a basic reaction of the petrochemical industry. At 298·15°K, with ΔH_r° in cal and ΔS_r° in cal deg^{-1},

$$K_{p,\ 298\cdot15°K} = \exp\left[\frac{-\Delta H_r^\circ}{1\cdot987 \times 298\cdot15} + \frac{\Delta S_r^\circ}{1\cdot987}\right] = \exp[-55\cdot2 + 14\cdot5],$$

and at 900°K

$$K_{p,\ 900°K} = \exp\left[\frac{-\Delta H_{r,900°K}^\circ}{1\cdot987 \times 900} + \frac{\Delta S_{r,900°K}^\circ}{1\cdot987}\right] = \exp[-19\cdot2 + 16\cdot3].$$

At 298·15°K, the magnitude of the exponential is determined mainly by the term deriving from ΔH_r°, whereas at 900°K the terms deriving from $\Delta H_{r,\ 900°K}^\circ$ and $\Delta S_{r,\ 900°K}^\circ$ are of comparable magnitude, but opposite sign.

For gaseous reactions in which $v_X + v_Y + \ldots - v_A - v_B \ldots \neq 0$, such as the ethane–ethylene reaction, the ratio p_X/p_A will depend on p_A. For the ethane–ethylene system it follows from equation (25) that provided the hydrogen and ethylene present in the equilibrium mixture are derived solely from the cracking reaction, then

$$\frac{p_{C_2H_4}}{p_{C_2H_6}} = \sqrt{\frac{K_p}{p_{C_2H_6}}}. \tag{145}$$

Thus at any temperature, the ratio $p_{C_2H_4}/p_{C_2H_6}$ depends inversely on the square root of the partial pressure of ethane; the ratios quoted in Table 18 apply to

$p_{C_2H_6} = 1$ atm and to a system to which no hydrogen has been added. Equation (145) indicates that as the partial pressure of ethane is increased so the percentage conversion to ethylene will diminish. In technological practice a relatively low partial pressure of ethane is obtained by dilution of the feed gas with steam (67/53); the amount of ethyl alcohol formed as a byproduct by interaction of steam and ethylene is shown in the next Section to be negligible at 900°K.

Thermodynamic data bearing on the dehydrogenation equilibria of numerous hydrocarbons have been summarised by Mauras (56/47).

6.2.2. THE HYDRATION OF ETHYLENE

$$C_2H_4(g) + H_2O(g) \rightleftharpoons C_2H_5OH(g)$$

For this reaction at 298·15°K, $\Delta H_r^\circ = -10·89 \pm 0·14$ kcal (see Table 14, reaction XXXVII). In this reaction there is a decrease in the number of moles of gaseous species, hence we would expect that ΔS_r° is negative and $\Delta G_r^\circ > \Delta H_r^\circ$. Numerical values of the thermodynamic quantities are presented in Table 19, for various temperatures; ideal-gas behaviour has been assumed. Experimental errors are not included in Table 19, but it can be shown by arguments akin to those in Section 6.2.1 that the uncertainties in the data do not influence the discussion.

TABLE 19. Thermodynamic quantities for the equilibrium

$$C_2H_4(g) + H_2O(g) \rightleftharpoons C_2H_5OH(g)$$

Quantity	T, °K			
	298·15	500	700	900
$\Delta(H_T^\circ - H_{298}^\circ)$, †kcal	0·00	−0·34	−0·59	−0·22
$\Delta H_{r,\ T}^\circ$, kcal	−10·89	−11·23	−11·48	−11·11
$\Delta S_{r,\ T}^\circ$, †cal deg^{-1}	−30·02	−30·98	−31·02	−30·81
$T\Delta S_{r,\ T}^\circ$, kcal	− 8·95	−15·49	−21·71	−27·73
$\Delta G_{r,\ T}^\circ$, kcal	− 1·94	+ 4·26	+10·23	+16·62
K_p, atm^{-1}	$10^{+1·422}$	$10^{-1·862}$	$10^{-3·193}$	$10^{-4·035}$
$p_{C_2H_5OH}/p_{C_2H_4}$ for $p_{H_2O} = 1$ atm.	[26·4]‡	0·0137	$6·4 \times 10^{-4}$	$9·2 \times 10^{-5}$

† Values calculated from data in references 53/1 and 61/20.
‡ This value is hypothetical, because $p_{H_2O} = 1$ atm cannot be realised at 298·15°K

The listed values of $p_{C_2H_5OH}/p_{C_2H_4}$, which all refer to $p_{H_2O} = 1$ atm, are seen to become progressively less favourable for the production of alcohol as the temperature is increased. By 700°K the ratio is considerably less than 10^{-2}, so that any process for the hydration of ethylene would be feasible only if conducted at a temperature lower than 700°K. Under industrial conditions the optimum temperature for the process will be determined by the interplay of various factors, of which the chief are the thermodynamics of the reaction, the kinetics of the reaction, and the economics of the plant and operating costs. Industrial processes are operated at *ca.* 600°K with partial pressures of ethylene and steam of *ca.* 30 atm. Since

$$\frac{p_{C_2H_5OH}}{p_{C_2H_4}} = K_p \times p_{H_2O}, \tag{146}$$

the use of steam at moderately high pressure ensures good conversion of ethylene to ethyl alcohol: at $p_{H_2O} = 1$ atm, a relatively low yield of ethyl alcohol would be obtained at 600°K. The reaction is exothermic at all temperatures; the heat evolved can be used to pre-heat the reactants.

The hydration of ethylene to ethanol has been the subject of several experimental studies of the chemical equilibrium involved. Earlier work, summarised by Barrow (52/36), and later experimental work by Cope and Dodge (59/39) are in broad agreement with the calculated values in Table 19. Cope (59/39, 64/51) has also studied experimentally the equilibrium

$$2C_2H_5OH(g) \rightleftharpoons (C_2H_5)_2O(g) + H_2O(g),$$

which is a side reaction in the industrial process for the production of ethyl alcohol. At the time of Cope's earlier publication (59/39) the equilibrium concentrations calculated from the extant thermodynamic data were in flagrant disagreement with the experimental data; in his calculations Cope used the value -58.62 kcal mol^{-1} for the heat of formation of diethyl ether (gas), derived from the work of Thomsen in 1886. A recent determination of this heat of formation by Pilcher and co-workers, $\Delta H_f^\circ(g) = -60.26 \pm 0.19$ kcal (63/2), has brought about reconciliation of the calculated and experimental equilibrium data (64/51). This case-history illustrates the point made in Section 2.4, namely that calculation of chemical equilibrium constants from thermal data is a worthwhile endeavour only when values of ΔH_f° reliable to at least ± 0.3 kcal mol^{-1} are available.

6.2.3. CEILING TEMPERATURES IN POLYMERIZATION REACTIONS

The thermodynamic and kinetic aspects of addition polymerization have been discussed in a detailed review by Dainton and Ivin (58/42). In this Section we merely discuss the basic thermodynamic conditions for favourable polymerization of a monomer.

Heats of polymerization can be obtained in several ways: (i) by direct measurement using reaction calorimetry (see Section 3.3.4), (ii) by deduction from the measured energies of combustion of the monomer and the polymer, (iii) by measurement of the equilibrium between the monomer and polymer as a function of temperature (see Section 3.4.2), or (iv) by use of estimation procedures such as those discussed in Chapter 7. An account of the experimental methods available for obtaining heats of polymerization has been given by Dainton and Ivin (Chapter 12 of reference 62/2), together with an extensive tabulation of polymerization heats.

In general, a polymer represents a more ordered state than a monomer, so that ΔS_r° for the polymerization of 1 mole of monomer to one repeating unit of polymer will be negative; values of ΔS_r° can be obtained experimentally by method (iii) above. For polymerization to be favourable, i.e. for $\Delta G_{poly}^\circ < 0$, the process must be exothermic.

If polymerization is considered as a potentially reversible reaction between monomer and polymer, then polymerization will be thermodynamically favoured when $\Delta G_{poly}^\circ < 0$ and the reverse process, depolymerization, will be thermodynamically favoured when $\Delta G_{poly}^\circ > 0$. If, for a particular monomer–polymer system ΔH_{poly}° and ΔS_{poly}° are both considered to be temperature independent, then ΔG_{poly}° will be linearly dependent on absolute temperature, because $\Delta G_{poly}^\circ = \Delta H_{poly}^\circ - T\Delta S_{poly}^\circ$. As ΔS_{poly}° is negative (see above), ΔG_{poly}° will become less negative as the temperature increases. The temperature for which ΔG_{poly}° is zero may be defined as the *ceiling temperature*. It follows that the ceiling temperature is the temperature at which the polymer is in equilibrium with monomer at unit activity. Below the ceiling temperature polymerization is favoured, whereas above the ceiling temperature depolymerization is favoured. Knowledge of the ceiling temperature for a particular system is therefore useful in planning the experimental conditions for polymerization—clearly it would be pointless to attempt to polymerize a monomer at temperatures much above the ceiling temperature.

Table 20 lists the heats and entropies of polymerization of some liquid monomers to form crystalline polymers, together with the ceiling temperatures, calculated on the assumption that ΔH_{poly}° and ΔS_{poly}° are temperature-independent. The data have been mainly taken from Dainton and Ivin's review (58/42).

The data in Table 20 show that the polymerization of 1-butene, styrene, isoprene and 1,3-butadiene should be favourable at temperatures up to *ca.* 700°K but for α-methylstyrene a relatively low temperature for polymerization is essential. The ceiling temperature is also a measure of the temperature to which a polymer can be heated before it will decompose to the monomer. For poly(methyl methacrylate) heating to about 200°C will cause extensive decomposition to the monomer, whereas for polytetrafluoroethylene heating to 1000°K should cause no depolymerization.

TABLE 20. Heats and entropies of polymerization at 25°C, and corresponding ceiling temperatures.

Monomer	$-\Delta H_{poly.}^{\circ}$ (kcal)	$-\Delta S_{po'y.}^{\circ}$ (cal deg^{-1})	Ceiling Temp. (°K)
1-Butene	20·0	26·9	740
Styrene	16·7	24·9	670
α-Methylstyrene	8·2	26·3	310
1, 3-Butadiene	17·4	21·2	820
Isoprene	17·9	24·2	740
Methyl methacrylate	13·5	28·0	480
Tetrafluoroethylene	39	26·8	1450
Cyclopropane	27·0	16·5	1630
Cyclobutane	25·1	13·2	1900
Cyclopentane	5·2	10·2	510

The polymerization of cyclopropane, cyclobutane and cyclopentane is seen from Table 20 to be thermodynamically favourable and the first two of these have very high ceiling temperatures. However, no catalyst has yet been found to promote these polymerizations, so these constitute examples of thermodynamically favourable reactions whose occurrence is hindered by kinetic factors.

6.3. Calculation of the heats of dissociation of chemical bonds.

An important use of values of ΔH_f° of organic compounds is in the calculation of the heats of formation of organic free radicals. By suitable combination of values of ΔH_f° for radicals with values of ΔH_f° for other organic compounds, the heats of dissociation of individual chemical bonds may be derived. Bond dissociation energies are important in discussion of chemical kinetics, especially of pyrolytic reactions and electron-impact processes. Although these topics are very important in modern physical chemistry we shall treat them in outline only and refer interested readers to some of the reviews and monographs that deal with the subject in detail.

The *bond dissociation energy*, $D(R-X)$, can in practice be equated with the value of ΔH_r° for the reaction†

$$R-X(g) = R\cdot(g) + X\cdot(g), \qquad (147)$$

where $R\cdot$ is an organic free radical and $X\cdot$ may also be an organic free radical, an inorganic free radical, or an atom. It follows that:

$$D(R-X) = \Delta H_f^\circ[R\cdot(g)] + \Delta H_f^\circ[X\cdot(g)] - \Delta H_f^\circ[R-X(g)]. \quad (148)$$

The use of equation (148) to calculate values of $D(R-X)$ depends on the availability of values of $\Delta H_f^\circ(R\cdot)$, $\Delta H_f^\circ(X\cdot)$ and $\Delta H_f^\circ(R-X)$, all in the gaseous state. A comprehensive list of values of the heats of formation of gaseous atoms in their ground states is given in Table 24 of Chapter 7. Derivation of values of $\Delta H_f^\circ[R\cdot(g)]$ for various types of R depends on the measurement of $D(R-X)$ for key compounds. For organic compounds these are mostly derived from studies of reaction kinetics; Kerr (66/7) has reviewed the field and has offered selected values of $D(R-X)$ and $\Delta H_f(R\cdot)$. Other discussions of bond dissociation energies are to be found in references 55/7, 58/6, 62/68, 65/56, 68/11.

Some examples of the extensive applicability of equation (148) are now discussed with respect to one of the simplest organic radicals, the methyl radical. The most probable value for the first (C−H) bond dissociation energy in methane, as selected by Kerr, is $104\cdot0 \pm 1\cdot0$ kcal, i.e. for the process

$$CH_4(g) = CH_3\cdot(g) + H\cdot(g),$$

---

† The bond dissociation energy $D(R-X)$ is more strictly defined as $\Delta U(0°K)$ for the reaction,

$$(R-X)(g) = R\cdot(g) + X\cdot(g),$$

where $(R-X)$, $R\cdot$ and $X\cdot$ are in their ground vibrational states, and it is this quantity, D_0° which is generally derived from spectroscopic measurements. Dissociation energies are also derived from equilibrium studies and from activation energies; in these cases it is generally $\Delta H(298\cdot15°K)$ which is derived. Fortunately, $\Delta H(298\cdot15°K)$ is not very different from $\Delta U(0°K)$; Cottrell (58/6) has shown the maximum difference would be $2\cdot4$ kcal, which is reached in the dissociation of a polyatomic molecule if none of the vibrational degrees of freedom lost on dissociation are excited at $298\cdot15°K$. Some examples of the differences (kcal) between $0°K$ and $298\cdot15°K$ are:

Bond	$\Delta U(0°K)$	$\Delta H(298\cdot15°K)$
H-H	$103\cdot26$	$104\cdot20$
Cl-Cl	$57\cdot36$	$58\cdot16$
CH_3-H	$102\cdot4$	$104\cdot0$
HO-H	$117\cdot97$	$119\cdot21$

For the dissociation of bonds in polyatomic molecules, the uncertainty of the determination is rarely less than ± 1 kcal. The use of data obtained at $298\cdot15°K$ for dissociation energies should not cause confusion.

$\Delta H_r^\circ = 104 \cdot 0$ kcal; whence, using $\Delta H_f^\circ[CH_4(g)]$ from Chapter 5 and $\Delta H_f^\circ(H \cdot)$ from Chapter 7, we derive $\Delta H_f^\circ[CH_3 \cdot (g)]$ as $34 \cdot 0 \pm 1 \cdot 0$ kcal. Then from the measured heats of formation of substituted methanes CH_3X and of the corresponding atomic species $X \cdot (g)$ the derivation of bond dissociation energies is straightforward, e.g.

$$D(CH_3 - Cl) = 83 \cdot 5 \pm 1 \cdot 1 \text{ kcal,}$$

$$D(CH_3 - Br) = 69 \cdot 8 \pm 1 \cdot 2 \text{ kcal,}$$

$$D(CH_3 - I) = 56 \cdot 1 \pm 1 \cdot 2 \text{ kcal.}$$

Moreover, from $\Delta H_f^\circ[C_2H_6(g)]$ we can derive for the process

$$C_2H_6(g) = 2CH_3 \cdot (g),$$

$$D(CH_3 - CH_3) = 88 \cdot 2 \pm 2 \cdot 0 \text{ kcal.}$$

Hence from thermochemical measurements coupled with key measured values of $\Delta H_f^\circ[R \cdot (g)]$ it is possible to derive a large number of bond dissociation energies.

The term *mean bond dissociation energy* is often useful for discussions of bond strengths and reactivities. For a compound MR_n, where n R-groups are attached to a central atom M, the mean bond dissociation energy $\bar{D}(M - R)$ is defined as $1/n$ of ΔH_r° for the process

$$MR_n(g) = M \cdot (g) + nR \cdot (g),$$

i.e. $\bar{D}(M - R) = 1/n \{\Delta H_f^\circ[M \cdot (g)] + n\Delta H_f^\circ[R \cdot (g)] - \Delta H_f^\circ[MR_n(g)]\}$.   (149)
For example, the mean bond dissociation energies for the tetramethyl derivatives of some of the group IV elements can be calculated from heats of formation given in Chapter 5. Using $\Delta H_f^\circ[CH_3 \cdot (g)] = 34 \cdot 0 \pm 1 \cdot 0$ kcal (see above) and the appropriate value of $\Delta H_f^\circ[M \cdot (g)]$ from Table 24, Chapter 7, we obtain the following results:

$$\bar{D}[C(CH_3)_4] = 86 \cdot 8 \text{ kcal,}$$

$$\bar{D}[Sn(CH_3)_4] = 53 \cdot 2 \text{ kcal,}$$

$$\bar{D}[Pb(CH_3)_4] = 37 \cdot 5 \text{ kcal.}$$

It is well known that the pyrolysis of lead tetra-alkyls results in ready production of free radicals, in accord with the fact that the mean bond dissociation energy for $Pb(CH_3)_4$ is by far the smallest value in this group.

6.4. Empirical methods of estimating heats of reaction

As stated in Chapter 1, one of the important aims of experimental thermo-chemistry is the assembly of sufficient data to permit the calculation of the heat of any chemical reaction. The present Section and much of Chapter 7 will be devoted to discussing methods for estimating values of ΔH_f° of organic compounds where experimental measurements have not been made; the estimated values can then in turn be used for estimating the heats of organic reactions. In this Section we consider empirical prediction rules that are applicable within a given homologous series, *intra-series rules*, or are applicable to compounds of similar molecular structure in different homologous series, *inter-series rules*. Here we approach the problem of estimation solely from an empirical standpoint, whereas in Chapter 7 we consider the relations between the heat of formation of an organic compound and its molecular structure.

6.4.1. ESTIMATION OF $\Delta H_f^\circ(g)$ FROM INTRA-SERIES INCREMENTS

It has been known for many years that the values of the thermodynamic properties of a homologous series $Y-(CH_2)_m-H$ show a monotonic, often linear, dependence on m. Parks and Huffman (32/9) in their classic book, "The Free Energies of Some Organic Compounds", discussed such relations with particular reference to Gibbs energies, and Rossini (34/12) discussed the heats of formation of hydrocarbons in a similar way. Later, Prosen, Johnson and Rossini (46/17) in a very important but strangely neglected publication proposed the following general relation,

$$\Delta H_f^\circ[Y-(CH_2)_m-H(g)] = A'+Bm+\Delta, \qquad (150)$$

where "A' is a constant peculiar to the end group Y, B is a constant for all normal alkyl series independent of Y, and Δ is a term which has a small finite value for the lower members, being largest for $m = 0$, and becomes zero for the higher members beginning near $m = 4$". The validity of equation (150) and of the quoted statement rests on experimental data for n-alkanes, n-alk-l-enes, n-alkylbenzenes, n-alkylcyclopentanes, and n-alkylcyclohexanes† Equation (150) has come to be regarded as basic for the construction o. bond-energy-term schemes (see Chapter 7) and the constant B, which Prosen and Rossini (45/1) found to be $-4\cdot926$ kcal, has come to be regarded as a constant of Nature. Prosen and Rossini obtained $B = -4\cdot926$ kcal by least-squares analysis of $\Delta H_f^\circ(g)$ data for the n-alkanes C_6 to C_{12} inclusive, with individual values weighted inversely as the squares of the experimental uncertainty intervals.

In view of the importance of equation (150) we now propose to examine its applicability to reliable experimental data other than for hydrocarbons. Such data are not particularly abundant, and the best examples would seem to be from the n-alkyl bromide, thiol and alcohol series, since reliable experimental data are available in Chapter 5 for at least seven members of each series. Application of the method of least squares to weighted values of $\Delta H_f^\circ(g)$ for compounds having four or more carbon atoms gave the following results:—

$$\Delta H_f^\circ[\text{Br} - (\text{CH}_2)_m - \text{H(g)}] = -6 \cdot 143 - 4 \cdot 891m + \Delta \text{ kcal,} \qquad (151)$$

$$\Delta H_f^\circ[\text{HS} - (\text{CH}_2)_m - \text{H(g)}] = -1 \cdot 327 - 4 \cdot 933m + \Delta \text{ kcal,} \qquad (152)$$

$$\Delta H_f^\circ[\text{HO} - (\text{CH}_2)_m - \text{H(g)}] = -46 \cdot 435 - 4 \cdot 844m + \Delta \text{ kcal.} \qquad (153)$$

The coefficients of m in equations (151) and (152) are close to Prosen and Rossini's value, $-4 \cdot 926$ kcal, and in fact the differences between the coefficients of m in equations (151) and (152) and $-4 \cdot 926$ are not statistically significant when the experimental uncertainties in the values of $\Delta H_f^\circ(g)$ are taken into account. For the alcohols, the difference between $-4 \cdot 926$ and the least-squares coefficient in equation (153) is on the borderline of statistical significance. Table 21 shows values of Δ for the bromide, thiol and alcohol series calculated (i) using the least-squares coefficients from equations (151), (152) and (153) respectively, (ii) using $B = -4 \cdot 926$ kcal and mean values of A' for $m \geqslant 4$†. Shown for comparison are values of Δ for n-alkanes calculated using $B = -4 \cdot 926$ kcal and the mean value of $A' = -10 \cdot 53$ kcal, which was derived from $\Delta H_f^\circ(g)$ data in Chapter 5 for the C_4, C_5, C_6, C_7, C_8, C_{10}, and C_{16} n-alkanes. The least-squares value of B from these data, calculated by using weighting factors inversely proportional to the squares of the experimental uncertainty intervals, was $-4 \cdot 913$, close to Prosen and Rossini's value.

From consideration of the data in Table 21, the following conclusions may be drawn.

(a) Equation (150) with $B = -4 \cdot 926$ kcal satisfactorily fits the experimental data for the higher members of the n-alkyl bromide and thiol series. In both series the largest values of Δ are for $m = 1$ and are positive. (Cf. the n-alkane series, where Δ for $m = 1$ is negative.) Values of Δ for $m \geqslant 3$ in the n-alkyl bromide series and for $m \geqslant 2$ in the n-alkyl thiol series are less than the experimental errors.

(b) Equation (150) with $B = -4 \cdot 926$ kcal fits the experimental data for the higher members of the alcohol series less well than the corresponding equation with $B = -4 \cdot 844$ kcal. In the alcohol series the largest value of Δ is for $m = 1$ and is positive. When $B = -4 \cdot 844$ kcal, the values of Δ for $m \geqslant 2$ are small compared with the experimental uncertainties.

† See the footnote to table 21, concerning the precise significance of m.

TABLE 21. Values of Δ in equation (150), for the n-alkane, n-alkyl bromide, n-alkyl thiol and n-alkyl alcohol series.

Series	Equation (150)			Δ, kcal, for $m =$									
$Y-(CH_2)_m-H$		A (kcal)	B (kcal)	1	2	3	4	5	6	7	8	10	16
$Y = H$†	(ii)	−10·530	−4·926	−2·43	+0·14	+0·48	−0·13	+0·06	+0·17	+0·16	+0·08	+0·15	−0·45
$Y = Br$	(i)	− 6·143	−4·891	+1·9	+0·7	+0·3	+0·11	−0·25	−0·02	+0·22	−0·16	—	—
	(ii)	− 5·954	−4·926	+1·8	+0·6	+0·2	+0·06	−0·27	0·00	+0·28	−0·07	—	—
$Y = SH$	(i)	− 1·327	−4·933	+0·86	+0·19	−0·01	+0·08	−0·25	+0·04	+0·13	—	+0·01	—
	(ii)	− 1·374	−4·926	+0·90	+0·23	+0·01	+0·10	−0·24	+0·04	+0·13	−0·11	+0·02	—
$Y = OH$	(i)	−46·435	−4·844	+3·21	−0·12	−0·20	+0·02	0·00	—	—	−0·17	—	+0·24
	(ii)	−45·725	−4·926	+2·58	−0·66	−0·67	−0·36	−0·31	—	—	—	—	+0·84

† Prosen, Johnson and Rossini (46/17) defined Y as CH_3 for the alkanes, so that by their definition $m = 0$ for methane. Here we define Y = H for the alkanes, so that by our definition $m = 1$ for methane. The present definition ensures that m has the same value for both an alkyl derivative and its parent alkane.

TABLE 22. Application of the method of intra-series increments to estimation of $\Delta H^\circ_f(g)$.

Series R—X,	A'	$\Delta H^\circ_f(g)$, kcal, for $m =$								
X =	kcal	2	3	4	5	6	7	8	9	10
—F	−52·8	−62·7	−67·6	−72·5	−77·4	−82·4	−87·3	−92·2	−97·1	−102·0
—Cl	−17·0	−26·8	−31·8	−36·7	−41·6	−46·5	−51·4	−56·4	−61·3	−66·2
—Br	−6·0	−15·8	−20·7	−25·7	−30·6	−35·5	−40·4	−45·3	−50·3	−55·2
—I	+7·7	−2·2	−7·1	−12·0	−16·9	−21·9	−26·8	−31·7	−36·6	−41·5
—NH$_2$	−3·0	−12·9	−17·8	−22·7	−27·6	−32·5	−37·5	−42·4	−47·3	−52·2
—NO$_2$	−15·2	−25·1	−30·0	−34·9	−39·8	−44·8	−49·7	−54·6	−59·5	−64·4
—CN	+22·9	+13·0	+8·1	+3·2	−1·7	−6·7	−11·6	−16·5	−21·4	−26·3
—OCH$_3$	−42·0	−51·9	−56·8	−61·7	−66·6	−71·6	−76·5	−81·4	−86·3	−91·2
—CO$_2$CH$_3$	−93·0	−102·9	−107·8	−112·7	−117·6	−122·5	−127·5	−132·4	−137·3	−142·2

Figures in bold type are those from which values of A' were determined.

In summary, we may say that equation (150) and the associated quotation apply convincingly to the n-alkyl bromide and n-alkyl thiol series but less convincingly to the n-alkyl alcohol series in the range C_4–C_{16}. It is reasonable to expect that when the alkyl group in a compound $R - X$ is augmented by a CH_2 group, the increase in total molecular binding energy (Section 7.2.1), and the consequent decrease in $\Delta H_f^\circ(g)$, should be constant for $m \geqslant 3$ and independent of the nature of X. Hence we may regard the situation in the n-alkanes, n-alkyl bromides and n-alkyl thiols as normal and that in the n-alkyl alcohols as slightly abnormal, at least for the homologues in the C_4–C_{16} range. To what extent abnormalities may exist in other homologous series cannot be considered until more experimental data accrue.

For the present it would be reasonable to assume that $B = -4 \cdot 92$ for most simple homologous series ($m \geqslant 3$). With this assumption it is easy to estimate $\Delta H_f^\circ(g)$ for any member ($m \geqslant 3$) of a homologous series from a single reliable value of $\Delta H_f^\circ(g)$ for one member ($m \geqslant 3$) of that series, since with Δ equal to zero, A' in equation (150) can be deduced for the series from a single value of $\Delta H_f^\circ(g)$. This method of estimation may be called *the method of intra-series increments*. Examples of its application are shown in Table 22. Since Δ for $m = 2$ is near zero in some series but not in others (Table 21), the method should be applied with caution to C_2 compounds. Application of the method to C_1 compounds is not recommended, because lack of knowledge of the value of Δ may give an erroneous result. Prosen, Johnson and Rossini (46/17) showed how these values of Δ in certain hydrocarbon series depended on the molecular structure of the end group, but in general the estimation of Δ for $m = 1$ requires the use of a bond-energy-term scheme as discussed in Chapter 7.

6.4.2. ESTIMATION OF $\Delta H_f^\circ(g)$ FROM INTER-SERIES INCREMENTS

If, as discussed in the previous Section, $\Delta H_f^\circ(g)$ for members of two homologous series, $[Y - (CH_2)_m - H]$ and $[Z - (CH_2)_m - H]$ are given by equations of the type (150), it then follows for a fixed value of m greater than 2, that

$$\Delta H_f^\circ[Y - (CH_2)_m - H(g)] - \Delta H_f^\circ[Z - (CH_2)_m - H(g)] = A'_Y - A'_Z. \quad (154)$$

When A'_Y is taken as the constant for the n–alkane series (i.e. $A'_Y = A'_H = -10 \cdot 5$ kcal), values of the quantity $A'_Y - A'_Z = A'_H - A'_Z$ may be called the *inter-series increments*, and these may be used to estimate heats of formation. For example, the heat of formation of n-$C_6H_{13}Z$ in the gaseous state can be estimated from $\Delta H_f^\circ(g)$ of n–hexane and the appropriate inter-series increment:

$$\Delta H_f^\circ[\text{n-}C_6H_{13}Z(g)] = \Delta H_f^\circ[\text{n-}C_6H_{14}(g)] - (A'_H - A'_Z).$$

Some values for inter-series increments are given in Table 23.

For application to n–alkyl derivatives the method of inter-series increments offers no advantage whatever over the method of intra-series increments, but the inter-series increment method is more widely applicable because it may be used to estimate heats of formation of branched-chain derivatives. For example, reference to the data of Chapter 5 shows that the difference in $\Delta H_f^\circ(g)$ between an alkyl chloride and the iso-structural alkyl bromide is approximately constant, whether the alkyl group is straight-chain or branched. This fact makes possible the estimation of values of $\Delta H_f^\circ(g)$ for branched-chain compounds using inter-series increments derived from straight-chain compounds. By way of example, we may estimate $\Delta H_f^\circ(g)$ of t-butyl bromide from the value of $\Delta H_f^\circ(g)$ for either t-butyl chloride or t-butyl iodide, together with the appropriate inter-series increments from Table 23:

$$\Delta H_f^\circ[\text{t-C}_4\text{H}_9\text{Br}(g)] = \Delta H_f^\circ[\text{t-C}_4\text{H}_9\text{Cl}(g)] - (A'_H - A'_{Br}) + (A'_H - A'_{Cl}) = -32 \cdot 7 \text{ kcal,} \tag{155}$$

$$\Delta H_f^\circ[\text{t-C}_4\text{H}_9\text{Br}] = \Delta H_f^\circ[\text{t-C}_4\text{H}_9\text{I}(g)] - (A'_H - A'_{Br}) + (A'_H - A'_I) = -31 \cdot 1 \text{ kcal.} \tag{156}$$

These two independent estimates agree to within the accuracy of the basic data, and give a mean value of $-31 \cdot 9$ kcal for $\Delta H_f^\circ[\text{t}-\text{C}_4\text{H}_9\text{Br}(g)]$ which is in good agreement with the experimental value of $-31 \cdot 9 \pm 0 \cdot 3$ kcal.

TABLE 23. Some values of inter-series increments, kcal mol^{-1}

Z	$A'_H - A'_Z$	Z	$A'_H - A'_Z$
F	$+42 \cdot 3$	NH$_2$	$- 7 \cdot 5$
Cl	$+ 6 \cdot 5$	NO$_2$	$+ 4 \cdot 7$
Br	$- 4 \cdot 5$	CN	$-33 \cdot 4$
I	$-18 \cdot 2$	OCH$_3$	$+31 \cdot 5$
SH	$- 9 \cdot 2$	CO$_2$CH$_3$	$+82 \cdot 5$

6.4.3. INCREMENTAL METHODS OF ESTIMATING $\Delta H_f^\circ(l)$ AND $\Delta H_f^\circ(c)$

The preferred method of estimating the heat of formation of a compound in the liquid or crystalline state is first to estimate $\Delta H_f^\circ(g)$ then to subtract the appropriate value of ΔH_v°, or ΔH_s°, as estimated by one of the methods discussed in Sections 4.4 and 4.5. An alternative, but less trustworthy, method is to estimate $\Delta H_f^\circ(l)$ or $\Delta H_f^\circ(c)$ by an incremental method. Thus it was shown in Section 4.4.3. paragraph (a) that values of ΔH_v° for liquid members of a homologous series $C_mH_{2m+1}Y$ vary approximately linearly with m, the normal increment in ΔH_v° per CH$_2$-group being $1 \cdot 1_2$ kcal. This linear relationship can be combined with that expressed by equation (150), with $B = -4 \cdot 92$ kcal and $\Delta = 0$, to yield the equation

$$\Delta H_f^\circ(l) = A'' - (4\cdot 92 + 1\cdot 12)m = A'' - 6\cdot 04\, m \text{ kcal}, \tag{157}$$

applicable to compounds having $m \geqslant 3$. Equation (157) can be used to estimate missing values of $\Delta H_f^\circ(l)$ for members of a homologous series from $\Delta H_f^\circ(l)$ for a single liquid member of that series with $m \geqslant 3$.

There is no precise relationship between ΔH_s at 25°C and m, although the CH_2-increment in $\Delta H_{s, T_{tr}}$ is known to be approximately constant at 2·0 kcal (Section 4.5.1). If ΔH_s is assumed to be equal to $\Delta H_{s, T_{tr}}$, the following approximate relation is obtained:

$$\Delta H_f^\circ(c) \approx A''' - (4\cdot 92 + 2\cdot 0)m \approx A''' - 6\cdot 9\, m \text{ kcal}. \tag{158}$$

Equation (158) can be used to estimate missing values of $\Delta H_f^\circ(c)$ for members of a homologous series from $\Delta H_f^\circ(c)$ for a single solid member of that series ($m \geqslant 3$). The value obtained will be subject to considerable uncertainty, so that estimates of $\Delta H_f^\circ(c)$ made in this way should be treated with caution.

6.4.4. ESTIMATION OF THE HEATS OF INTER-RELATED REACTIONS

It follows from equations (154) and (157) that ΔH_r° for the conversion of members of a homologous series of reactants, in a defined physical state, to a homologous series of products, in the same physical state, must be a constant equal to A' (product series) $- A'$ (reactant series) for gas-state reactions and A'' (product series) $- A''$ (reactant series) for liquid-state reactions. Thus, $\Delta H_r^\circ = -27\cdot 3 \pm 0\cdot 6$ kcal for the reaction

$$CH_2 = CH.C_5H_{11}(l) + Br_2(l) \rightarrow CH_2Br.CHBr.C_5H_{11}(l),$$

and we would expect the heat of conversion of any other liquid n-alk-l-ene to a liquid 1,2–dibromoalkane to be close to $-27\cdot 3 \pm 0\cdot 6$ kcal also.

As an extension of the above concept, one may state that when structurally related reactants, in a defined physical state, are converted to structurally related products, in a defined physical state, the heat of reaction will be approximately constant. Several experimental demonstrations of this statement are available. We cite the work of Entelis and his colleagues (67/57), who showed that the heat of the reaction

$$R - \underset{\underset{O}{\diagdown \diagup}}{CH - CH_2}(l) + HNO_3(nH_2O)(l) = R.CH(OH).CH_2NO_3(nH_2O)(l)$$

is nearly constant for $R = CH_3, CH_2Cl, CH_2Br, CH_2OCH_3$ and $CH_2OC_6H_5$. Obviously the rule of approximately constant reaction heats is useful for predictive purposes.

Chapter 7

Some Theoretical Applications of Thermochemical Data

The purpose of the present Chapter is to examine how the heat of formation of an organic compound may be related to its molecular structure. Clearly, if precise, reliable relations could be established for all types of molecular structure, further thermochemical measurements would scarcely be necessary, since heats of formation of compounds not studied experimentally could be estimated from their structures. The search for reliable thermochemical–structural relations is a joint endeavour of theoretical chemists and thermochemists, and in discussing the present state of this subject we shall emphasize those treatments which in our opinion provide the best methods for estimating heats of formation from molecular structure.

7.1. Heats of atomization

It is first necessary to consider what thermochemical quantity can most meaningfully be related to structure. The total energy content of a molecule in its ground state includes (i) intramolecular energy due to the chemical binding of the constituent atoms, (ii) translational, rotational, and vibrational energy, (iii) intermolecular energy due to interaction of external force fields. Energy from effect (iii) can be removed from consideration by specifying that the molecule is in the ideal-gas state. It is conventional to make no attempt to separate effects (i) and (ii) and to regard the measure of the total chemical binding energy to be the energy change of the process:

Molecule (ground state, ideal gas, T_1)→Atoms (ground state, ideal gas, T_1).

This energy change, known as the *energy of atomization* of the molecule, uses an energy base-line that is amenable to experimental study, namely the total energy of the constituent atoms at infinite separation. For most purposes, it is convenient to combine the $\Delta(pV)$ quantity for the reaction with the energy change and hence to deal with the *heat of atomization*, written as $\Delta H_{a, T_1}$. If we equate the heat of atomization to the total chemical binding energy, the problem then arises as to how this quantity can be subdivided in terms of the energies of individual bonds. For a diatomic molecule the heat of atomization may be regarded as a direct measure of the strength of the bond, but for polyatomic molecules additional effects must be considered, since interaction between bonds may occur. In general, the total chemical binding energy may be divided into three terms, (i) the sum of the chemical bond energies in the molecule, (ii) destabilizing effects which decrease the total binding energy (e.g. those due to steric hindrance or angular strain within the molecule), and (iii) stabilizing effects which increase the total binding energy (e.g. those due to electron delocalization). This may be expressed by the general relation,

$$\Delta H_{a, T_1} = \sum \text{Chemical bond energies}$$

$$- \sum \text{Destabilization energies} + \sum \text{Stabilization energies.} \quad (159)$$

For each compound thermochemical measurements provide one single piece of information, namely the heat of atomization, and there is no unique, unambiguous way of apportioning it amongst the three terms in equation (159). To make a reasoned apportionment one has to be guided by other considerations, not the least of which is chemical intuition. The normal approach is to select those compounds for which it is reasonable to suppose that special stabilizing or destabilizing energy effects are absent, and from the heats of atomization of these compounds to derive a set of energy terms for individual bonds. The heats of atomization of compounds with more complex molecular structures can then be examined to derive strain or stabilization energies and thence to develop comprehensive schemes for the estimation of unknown heats of atomization or unknown heats of formation

7.1.1. DERIVATION OF HEATS OF ATOMIZATION AT 298·15°K

The heat of atomization at 298·15°K, which we shall write as ΔH_a without a temperature descriptor, is derived by application of equation (6). Thus for a compound $K_k L_l M_m \ldots$, the atomization process is represented by the equation

$$K_k L_l M_m \ldots (g) = k K(g) + l L(g) + m M(g) + \ldots$$

where all the species are in their ground states. From equation (6):

$$\Delta H_a = k \Delta H_f^\circ [K(g)] + l \Delta H_f^\circ [L(g)] + m \Delta H_f^\circ [M(g)] + \ldots$$

$$- \Delta H_f^\circ [K_k L_l M_m \ldots (g)]. \quad (160)$$

To obtain ΔH_a, standard values at 298·15°K for the heat of formation of the compound in the gaseous state and the heats of formation of the gaseous atoms from the elements in their reference states are required. Reliable values are available for the heats of formation of the gaseous atoms of most of the common elements, and are listed in Table 24. The values are taken from compilations by Brewer (62/60), Kondratiev (62/61) and the JANAF compilers (61/3), and from the references specifically cited.

TABLE 24. Heats of formation of gaseous atoms in their ground states at 298·15°K, kcal.

Atom	$\Delta H_f^\circ(g)$	(Ref.)	Atom	$\Delta H_f^\circ(g)$	(Ref.)
H	52·10±0·06		Rb	19·5 ±1·0	
Li	38·4 ±0·4		Sr	39·1 ±1·0	
Be	78·25±0·50		Y	98 ±2	
B	132·6 ±4·0		Zr	145·4 ±0·4	
C	170·90±0·45		Nb	173 ±2	
N	113·0 ±0·5		Mo	157·7	
O	59·56±0·03		Ru	153 ±2	
F	18·86±0·20		Rh	133 ±1	
Na	25·8 ±0·1		Pd	91 ±1	
Mg	34·29±0·10	(65/55)	Ag	68·4	
Al	78·0 ±0·4		Cd	26·54±0·10	(65/55)
Si	108·4 ±3·0	(61/56)	In	58 ±2	
P	75·5	(63/56)	Sn	72·0 ±2·0	
S	65·65±0·60		Sb	63 ±2	
Cl	28·92±0·03		Te	46 ±2	
K	21·3 ±0·2		I	25·54±0·01	
Ca	42·81±0·02	(62/62)	Cs	18·7 ±0·3	
Sc	91·2 ±0·3	(63/57)	Ba	42·5	
Ti	112·5		Hf	148·0	(63/57)
V	123		Ta	186·8 ±1·0	
Cr	95 ±1		W	201·8 ±2·0	
Mn	66·7		Re	187 ±2	
Fe	99·5 ±1		Os	187 ±2	
Co	101·6		Ir	159 ±2	
Ni	102·8		Pt	135·2 ±1·0	
Cu	81·1		Au	88·3 ±0·9	(62/63)
Zn	30·85±0·10	(65/55)	Hg	14·65±0·02	
Ga	69·0		Tl	43·0 ±1·0	
Ge	90·2 ±3·0		Pb	46·75±0·13	
As	69·0 ±3·0		Bi	49·5 ±1·0	
Se	49·4 ±1·0		Th	136·6	
Br	26·74±0·07		U	115 ±3	(62/64)

7.1.2. HEATS OF ATOMIZATION AT 0°K, AND THEIR CORRECTION FOR ZERO POINT ENERGY.

The use of the heat of atomization at 298·15°K as a measure of chemical binding energy was criticized more than 30 years ago by Zahn (34/10), on the grounds that there are energy contributions to the heat of atomization at $298 \cdot 15°K$ that cannot fairly be regarded as resulting from chemical binding. These contributions are the thermal energy of translation, rotation, and vibration of the molecule and the zero point energy. Zahn's criticism was that although it may be reasonable to apportion the chemical binding energy into bond contributions, with corrections for destabilization and stabilization, it is not obvious that the thermal and zero point energies should be apportioned in like manner. This question was discussed by Cottrell (48/15) with special reference to hydrocarbons, and his clear conclusion was that for most applications it makes little difference whether ΔH_a or $\Delta H_{a, \, 0°K}$ is used. It is instructive to consider the relative sizes of the quantities involved. Shown in Table 25 are values of $H_{298}° - H_0°$ for some hydrocarbons in the gaseous state, together with molar zero point energies (Z.P.E.) calculated from equation (161):

$$Z.P.E. = N \sum_i \tfrac{1}{2} h v_i. \tag{161}$$

where h is Planck's constant, N is the Avogadro constant and v_i is the frequency of the fundamental vibration i.

TABLE 25

$H_{298}° - H_0°$ and molar zero point energies for some hydrocarbons, kcal.

	$H_{298}° - H_0°$	Z.P.E.		$H_{298}° - H_0°$	Z.P.E.
Methane	2·40	27·1	Ethylene	2·53	30·5
Ethane	2·86	45·2	Propene	3·26	48·4
Propane	3·51	63·0	1-Butene	4·22	66·1
n-Butane	4·65	80·6	Cyclopropane	2·74	49·1
Isobutane	4·28	80·2	Cyclobutane	3·24	67·3
n-Pentane	5·67	98·5	Cyclopentane	3·60	86·0
Neopentane	5·03	96·8	Cyclohexane	4·24	104·5
n-Hexane	6·99	116·2	Benzene	3·40	67·1

It is seen from Table 25 that $H_{298}° - H_0°$ is comparatively small, increasing with molecular size with a $-CH_2-$ increment (above C_2) of approximately 1 kcal. Hence to a first approximation $H_{298}° - H_0°$ may be regarded as a bond additive function of molecular structure. It can also be seen from Table 25 that molar zero point energies are much larger quantities than $H_{298}° - H_0°$, and the question of whether they are bond additive functions of molecular structure is crucial. Of the $9n$ vibrational modes available to an alkane C_nH_{2n+2}, the vibration of the C–H bonds provides the main contribution to the zero point energy and these vibrational frequencies are to a first approximation independent of the structure of the rest of the molecule. Hence the contribution of the C–H bonds to the Z.P.E. is approximately an additive function of their total number in the molecule. It can also be shown that the contribution of each type of C–C bond to the Z.P.E. in n-alkanes is approximately constant, so that the Z.P.E. is to a fair approximation a bond additive function in n-alkanes. The main corollary of the above findings is that if ΔH_a is an additive function of structure in alkanes so also will be $\Delta H_{a,\,0°K}$ and $\Delta H_{a,\,0°K}$ (corrected for Z.P.E.) to a first approximation.

It is to be noted however, that the difference in Z.P.E. between the isomers n-pentane and neopentane implies that the simple Z.P.E. additivity rule is not fully applicable to highly branched alkanes. Since Cottrell's discussion in 1948, several bond energy schemes that account for differences in the heats of formation of isomers have been developed; differences in Z.P.E. are automatically included in these schemes.

From a practical standpoint it is fortunate that non-removal of $H_{298}^\circ - H_0^\circ$ and Z.P.E. from ΔH_a does not complicate the correlation of chemical binding energies with structure, because the number of compounds for which $H_{298}^\circ - H_0^\circ$ and Z.P.E. are known with sufficient accuracy is very limited compared with the number of compounds for which ΔH_a can be derived. Further support for this conclusion is provided by calculations in Section 7.2.1 and by the success of the bond energy schemes described in Section 7.3 onwards.

7.1.3. THE VALENCE STATE, AND INTRINSIC BOND ENERGIES

Bond energies derived from heats of atomization with respect to atoms in their ground states have been described by some critics as unrealistic, because atoms in molecules do not in general exhibit the same valence characteristics as in their ground states. For example, the carbon atom in its ground state has the outer-shell electronic configuration s^2p^2, a bivalent state, whereas in most of its compounds carbon is tetravalent. Mecke (30/3) argued that for a realistic computation of bond energies involving carbon, the promotion energy from the ground state to the tetravalent state should be included.

Consider for example, methane. The valence state for the carbon atom as described by van Vleck (34/11) is that in which the four outer-shell electrons are distributed in tetrahedral orbitals, $t_1t_2t_3t_4$, each derived by sp^3 hybridization. The electron spins will be randomly orientated with respect to each other, hence this valence-state is not a stationary state and cannot be observed spectroscopically. The promotion energy to the valence state cannot be measured directly but can be calculated from atomic spectroscopic levels by use of the Slater–Condon theory of atomic spectra (35/17).

The most recent calculations (62/65, 62/66) give for the promotion energy of carbon from the ground state to the sp^3-hybridized tetrahedral state (as in alkanes) 152 kcal mol^{-1}, to the sp^2-hybridized trigonal state (as in alkenes) 157 kcal mol^{-1}, and to the sp-hybridized digonal state (as in alkynes) 166 kcal mol^{-1}. Heats of atomization with respect to valence states defined in this way can be calculated with reasonably small uncertainty. Bond energies derived using such valence states are known as *intrinsic bond energies*.

More careful analysis, however, shows that the above description of the valence state is too simple. Voge (36/22) examined methane in terms of the Heitler–London–Pauling–Slater electron-pairing theory and showed that the valence state involves not only the tetravalent configuration $t_1 t_2 t_3 t_4$ but also contributions from a bivalent configuration $t_1{}^2 t_2 t_3$ and a zerovalent configuration $t_1{}^2 t_2{}^2$. The admixture of these lower valent configurations with the tetravalent configuration reduces the promotion energy to the valence state and results in greater stability of the methane molecule. The precise definition of the valence state of an atom in a molecule is clearly crucial if intrinsic bond energies are to be calculated and applied; but for most thermochemical purposes calculation of heats of atomization with respect to ground states is adequate. Only in special situations need the role of the valence state be considered. For instance, the energies of the step-wise formation of radicals, shown in Table 26, require consideration of valence tastes for their proper interpretation.

TABLE 26. Stepwise bond dissociation energies and mean bond dissociation energies in MX_n compounds, kcal.

	$D(MX_{n-1}-X)$	$D(MX_{n-2}-X)$	$D(MX_{n-3}-X)$	$D(MX_{n-4}-X)$	$\bar{D}(M-X)$
$HgCl_2$	81	25	—	—	53
$HgBr_2$	72	16	—	—	44
HgI_2	61	8	—	—	35
$Hg(CH_3)_2$	51	8	—	—	30
CH_4	104	104	108	81	99
CF_4	126	96	107	141	178
CCl_4	70	63	?	?	18
$TiCl_4$	80	101	106	124	103
SiF_4	?	?	169	125	142
NF_3	58	?	?	—	66
BF_3	165	?	?	—	153
$AlCl_3$	91	95	119	—	102
OH_2	118	102	—	—	110

The values given in Table 26 for CH_4 are from the review by Kerr (66/7), and those for CF_4 from Zmbov, Uy and Margrave (68/16). The remaining values are from the review by Skinner and Pilcher (63/58), where references to the original literature are cited.

Thus the step-wise bond dissociation energies [see equation (148)] in compounds of the type MX_n are in every case different from the mean bond dissociation energies $\bar{D}(M-X)$ [see equation (149)] and any discussion of this fact would need to take account of changes in valence-state energy during the various steps. For example, the effect of changes in the valence state has been discussed by Skinner (49/34) for the dissociation of the mercury halides, where the valence-state promotion energy appears to be mainly recovered in the second dissociation step, thus reducing its value compared with the first dissociation step. For the dissociation of titanium tetrachloride, intrinsic bond energies were calculated with respect to assumed valence states by Pilcher and Skinner (58/43).

This book is not concerned with the thermochemistry of radicals, ions or excited states so further discussion of intrinsic bond energies is unnecessary.

7.2. Bond energy terms

Bond energy terms are quantities assigned to the bonds in a molecule so that equation (159) is obeyed, with the restriction that bond energy terms for bonds of the same formal type are identical. No such restriction applies to the definition of bond dissociation energies (Section 6.3) and indeed experimental determination of these quantities shows that they are often not identical for bonds of the same formal type. (*Cf.* Table 26).

Bond energy terms were introduced by Fajans about 1920 and were subsequently used to good effect by Sidgwick in "The Covalent Link in Chemistry" (33/17) and by Pauling in "The Nature of the Chemical Bond" (39/23) to illuminate problems of chemical binding. The original assumption made by Fajans was that the bond energy term for a bond of a given type has a characteristic value, independent of environment and factors such as the hybridization states of the bonded atoms, i.e. bond energy terms were considered to be additive and transferable from one molecular structure to another. The Fajans concept can be expressed by the following form of equation (159):

$$\Delta H_a = \sum N_{AB}[\text{Bond energy term for bond } A-B]$$
$$-\sum \text{Destabilization energies} + \sum \text{Stabilization energies}, \qquad (162)$$

where N_{AB} is the number of bonds of type $A-B$ per molecule. The second and third Σ terms in equation (162) were considered to be insignificant for many compounds but in instances where the experimental heat of atomization differed markedly from the sum of the bond energy terms transferred from other compounds, the deviation was ascribed to either steric strain (destabilization) or resonance (stabilization).

For compounds of the type MX_n, which contain only one type of bond, there is no ambiguity in deriving the bond energy term $E(M-X)$. Comparison of equations (149), (160) and (162) reveals that for such structures the bond energy term is identical with the mean bond dissociation energy $\bar{D}(M-X)$, provided X is atomic and not a radical; i.e. for MX_n compounds,

$$E(M-X) = \bar{D}(M-X) = (1/n)(\Delta H_a[MX_n]).$$

For molecules containing more than one type of bond, such as ethane C_2H_6, it is necessary to make an assumption on how the heat of atomization is to be apportioned amongst the bonds. The simplest assumption in this case is that the C–H bond energy term in ethane is the same as in methane, permitting $E(C–C)$ to be derived. This kind of assumption is the basis of a simple Fajans-type scheme.

If for alkanes in general it is assumed that constant bond energy terms can be ascribed to the C–H and C–C bonds, then

$$\Delta H_a(C_nH_{2n+2}) = (n-1)E(C-C)+(2n+2)E(C-H); \qquad (163)$$

values for $E(C-C)$ and $E(C-H)$ can be obtained by solution of pairs of equations of type (163) using experimental values of ΔH_a for alkanes of different carbon number.

7.2.1. EXAMINATION OF THE CONCEPT OF CONSTANT BOND ENERGY TERMS

It is important to scrutinize the evidence for the notion that bond energy terms are constant and transferable. It is pertinent to start by examining the situation in the n-alkane series. For these compounds sufficient thermodynamic data (53/1) and spectroscopic data (56/46) are available to permit derivation of the heats of atomization at 298·15°K, at 0°K, and at 0°K with correction for Z.P.E. The resulting heats of atomization for the first six n-alkanes are listed in Table 27.

TABLE 27. Heats of atomization of n-alkanes, kcal.

	ΔH_a	$\Delta H_{a, \, 0°K}$	$\Delta H_{a, \, 0°K}$ (with correction for Z.P.E.)
CH_4	397·19	392·10	418·95
C_2H_6	674·64	665·49	710·75
C_3H_8	954·33	941·31	1004·23
n-C_4H_{10}	1234·96	1218·55	1299·17
n-C_5H_{12}	1514·80	1494·87	1593·19
n-C_6H_{14}	1794·72	1771·27	1887·47

From the data for n-C_5H_{12} and n-C_6H_{14} in Table 27 values for $E(C-H)$ and $E(C-C)$ were obtained using equation (163). The values, given in Table

TABLE 28. Values for some bond energy terms, kcal.

	ΔH_a	$\Delta H_{a,0°K}$	$\Delta H_{a, 0°K}$ (with correction for Z.P.E.)
$E(C-H)$	98·78	97·32	104·02
$E(C-C)$	82·36	81·76	86·24

28, were then used to estimate the heats of atomization of CH_4, C_2H_6, C_3H_8 and n-C_4H_{10}. The differences between the estimated and observed heats of atomization are listed in Table 29.

TABLE 29. Estimated minus observed heats of atomization, kcal.

	ΔH_a	$\Delta H_{a, 0°K}$	$\Delta H_{a, 0°K}$ (with correction for Z.P.E.)
CH_4	−2·07	−2·82	−2·87
C_2H_6	+0·40	+0·19	−0·39
C_3H_8	+0·63	+0·77	+0·41
n-C_4H_{10}	−0·08	−0·07	−0·25

It is apparent from Table 29 that there is little to choose between the schemes based on ΔH_a, $\Delta H_{a, 0°K}$ and $\Delta H_{a, 0°K}$ (with correction for Z.P.E.) since each leads to similar estimation errors; the error for methane in each case is quite substantial. Similar examples showing the weakness of the concept of constant, transferable, bond energy terms could be selected from other homologous series. Essentially this is the same phenomenon as was discussed in Section 6.4.1, namely the deviation of the heats of formation for the lower members of a homologous series from values linearly extrapolated from those of the higher members. For that part of a homologous series for which the heats of formation vary linearly with carbon number (i.e. above C_3), the difference in binding energy between two neighbouring members is simply $E(C-C)+2E(C-H)$ and for such compounds the assumption of constant, transferable, bond energy terms appears to be quite accurate.

Next, we consider how the concept of constant bond energy terms applies to structural isomers. Since sets of such isomers have the same numbers and types of chemical bonds, the Fajans concept necessarily predicts that members of a set should have identical heats of atomization and hence identical gas-state heats of formation. Some examples of the heats of formation of C_4H_9X structural isomers, taken from the data of Chapter 5, are given in Table 30.

TABLE 30

Heats of formation of some structural isomers in the gaseous state, kcal.

X =	$\Delta H_f^\circ(g)$ for the isomers		
	$CH_3.CH_2.CH_2.CH_2X$	$CH_3.CH_2$ $\diagdown$ CHX CH_3 $\diagup$	CH_3 $\diagdown$ $CH_3 - CX$ CH_3 $\diagup$
H	$-30\cdot36\pm 0\cdot16$		$-32\cdot41\pm0\cdot13$
CH_3	$-35\cdot10\pm 0\cdot15$	$-36\cdot85\pm 0\cdot15$	$-40\cdot27\pm0\cdot25$
C_6H_5	$- 3\cdot28\pm 0\cdot30$	$- 4\cdot15\pm 0\cdot31$	$- 5\cdot40\pm0\cdot31$
OH	$-65\cdot79\pm 0\cdot14$	$-69\cdot98\pm 0\cdot23$	$-74\cdot72\pm0\cdot21$
NH_2	$-22\cdot7 \pm 0\cdot4$	$-25\cdot4 \pm 0\cdot4$	$-28\cdot90\pm0\cdot15$
SH	$-20\cdot98\pm 0\cdot29$	$-23\cdot09\pm 0\cdot20$	$-26\cdot12\pm0\cdot21$
Cl	$-35\cdot1 \pm[2\cdot0]$	$-38\cdot6 \pm[2\cdot0]$	$-43\cdot7 \pm0\cdot6$
Br	$-25\cdot60\pm 0\cdot30$	$-28\cdot94\pm 0\cdot12$	$-31\cdot88\pm0\cdot30$

The isobutyl isomers have been omitted from Table 30 but were these to be included they would merely add to the general conclusion that the heats of formation of structural isomers are *not* equal. It is evident from Table 30 that chain branching in organic compounds leads to more negative heats of formation and hence to higher heats of atomization, reflecting increased binding energy. The effect can be large, e.g. the increase in binding energy from n-butanol to t-butanol is $8\cdot93\pm0\cdot27$ kcal mol^{-1} and even for the non-polar compounds n-pentane and neopentane the increase is $5\cdot17\pm0\cdot32$ kcal mol^{-1}. The failure of the concept of constant bond energy terms to account for differing heats of formation of structural isomers is another example of the weakness of this concept.

Finally we consider *single-centre redistribution reactions*, defined by Skinner (54/29) as those reactions in which the chemical bonds change in relative position but not in number or formal character. According to the Fajans concept such reactions should have $\Delta H_r^\circ(g) = 0$, i.e. they should be *thermoneutral*. Some examples of the heats of single-centre redistribution reactions, calculated from the data of Chapter 5, are shown in Table 31.

The redistribution reactions shown in Table 31 are clearly not thermoneutral so that again the weakness of the concept of constant, transferable, bond energy terms has been demonstrated.

TABLE 31. Heats of some gaseous redistribution reactions at $298\cdot15^\circ K$, kcal.

Reaction	$\Delta H_r^\circ(g)$
$\frac{3}{4}CF_4+\frac{1}{4}CH_4 = CHF_3$	$+5\cdot6\pm0\cdot9$
$\frac{3}{4}CF_4+\frac{1}{4}CCl_4 = CF_3Cl$	$+4\cdot8\pm1\cdot5$
$\frac{1}{2}CH_4+\frac{1}{2}C(CH_3)_4 = CH_3CH_2CH_3$	$+4\cdot3\pm0\cdot3$
$\frac{1}{2}Hg(CH_3)_2+\frac{1}{2}HgCl_2 = CH_3HgCl$	$-6\cdot0\pm1\cdot0$
$\frac{3}{4}C(CH_3)_4+\frac{1}{4}CCl_4 = (CH_3)_3CCl$	$-7\cdot2\pm0\cdot7$

We have seen from the above examples that the simple assumption made by Fajans breaks down in some circumstances, but since the Fajans scheme has proved useful in broad interpretations of chemical bonding (33/17, 39/23), refinement of the scheme, rather than its abandonment, is called for.

7.2.2. ZAHN'S BOND ENERGY SCHEME

The additivity of bond energy terms was questioned in 1934 by Zahn (34/10), who proposed a "more general type of energy model" in which the total chemical binding energy was written as a sum of bond energy terms, plus a further sum associated with interactions between pairs of bonds attached to the same atom. This model leads to the following expressions for the heats of atomization of the first five members of the n-alkane series:

$$
\begin{aligned}
\text{for CH}_4 \qquad & \Delta H_a = 4E(C-H)+6P_1, \\
\text{for C}_2\text{H}_6 \qquad & \Delta H_a = 6E(C-H)+E(C-C)+6P_1+6P_2, \\
\text{for C}_3\text{H}_8 \qquad & \Delta H_a = 8E(C-H)+2E(C-C)+7P_1+10P_2+P_3, \\
\text{for n-C}_4\text{H}_{10} \qquad & \Delta H_a = 10E(C-H)+3E(C-C)+8P_1+14P_2+2P_3, \\
\text{for n-C}_5\text{H}_{12} \qquad & \Delta H_a = 12E(C-H)+4E(C-C)+9P_1+18P_2+3P_3,
\end{aligned} \tag{164}
$$

where P_1, P_2 and P_3 are the interaction energies associated with the pairs

of bonds $\overset{H}{\underset{H}{>}}C, \overset{H}{\underset{C}{>}}C, \overset{C}{\underset{C}{>}}C$, respectively.

By introduction of *effective bond energy terms*, $B(C-H)$ and $B(C-C)$, defined by

$$
\begin{aligned}
B(C-H) &= E(C-H)+\frac{3}{2}P_1 \\
B(C-C) &= E(C-C)-3P_1+6P_2,
\end{aligned} \tag{165}
$$

and a *composite interaction parameter*, X_{CCC}, defined by

$$
X_{CCC} = P_1 - 2P_2 + P_3, \tag{166}
$$

the heats of atomization of the first five n-alkanes may be written as follows:

$$
\begin{aligned}
\text{for CH}_4 \qquad & \Delta H_a = 4B(C-H), \\
\text{for C}_2\text{H}_6 \qquad & \Delta H_a = 6B(C-H)+B(C-C), \\
\text{for C}_3\text{H}_8 \qquad & \Delta H_a = 8B(C-H)+2B(C-C)+X_{CCC}, \\
\text{for n-C}_4\text{H}_{10} \qquad & \Delta H_a = 10B(C-H)+3B(C-C)+2X_{CCC}, \\
\text{for n-C}_5\text{H}_{12} \qquad & \Delta H_a = 12B(C-H)+4B(C-C)+3X_{CCC}.
\end{aligned} \tag{167}
$$

536 THERMOCHEMISTRY OF ORGANIC AND ORGANOMETALLIC COMPOUNDS

This scheme when applied to alkanes in general gives:

$$\Delta H_a(C_nH_{2n+2}) = (n-1)B(C-C) + (2n+2)B(C-H) + \left(N_p - 6 + \tfrac{1}{2}N_s\right)X_{CCC},$$

$$(168)$$

where N_p = the number of primary $C-H$ bonds, and N_s = the number of secondary $C-H$ bonds per molecule.

Equation (168) represents an advance on equation (163) because it can, when $X_{CCC} \neq 0$, accommodate differences in the heats of formation of structural isomers. This can be illustrated by the calculation of ΔH_a for the isomeric butanes and pentanes, firstly from equation (163) with parameters from Table 28, then from equation (168) with the following parameters (kcal): $B(C-C) = 79.22$, $B(C-H) = 99.33$ and $X_{CCC} = 2.00$. The difference between calculated and observed values by the two methods are listed in Table 32.

TABLE 32. [Calculated−observed] heats of atomization, kcal.

Compound	ΔH_a[Calc.−Obs.]	
	Equation (163)	Equation (168)
n-Butane	−0·08	0·00
Isobutane	−2·13	−0·05
n-Pentane	0·00	+0·04
Isopentane	−1·75	+0·29
Neopentane	−5·17	+0·87

It is seen that the introduction of a single interaction parameter X_{CCC} results in markedly better agreement between calculated and observed heats of atomization. (The values of the B and X_{CCC} parameters were chosen to illustrate this improvement and should not be applied to general calculations on alkanes.) The Zahn scheme has not found wide application, because it was introduced at a time when precise thermochemical data were rare, and this scheme has now been superseded by more modern ones described in the next section. The importance of the Zahn scheme was its demonstration that the simple Fajans scheme could readily be refined.

7.3. Modern bond energy schemes

Since the 1940's several bond energy schemes, which represent marked improvements over the Fajans and Zahn schemes, have been proposed. In this Section we discuss the basic principles of these schemes, mainly in respect of alkanes; in later Sections we will discuss the application of these schemes to other types of compound. Initially we shall avoid discussion of those compounds in which significant destabilization energy effects may occur, so that in some instances the schemes will be restricted to a smaller group of compounds than the original author intended. Our purpose here is to develop clearly the bases of these schemes and to improve their applicability: in so doing we shall present "best values" of parameters that have not been published elsewhere.

7.3.1. GROUP METHODS OF ESTIMATING ΔH_a.

In 1932, Parks and Huffman (32/9) showed that the Gibbs energies of formation and entropies of organic compounds could plausibly be expressed as additive functions of structural-group contributions. Since then, the application of similar methods to the calculation of heats of formation and heats of atomization has been much discussed and several useful schemes have been proposed.

The basis of the Group method has been clearly described by Benson and Buss (58/8). As previously remarked in Section 7.2.1, the Fajans assumption of constant, transferable, bond energy terms implies that single-centre redistribution reactions of the general type,

$$X - Y - X + Z - Y - Z = 2X - Y - Z,$$

where Y is a single atom, should be thermoneutral, i.e. $\Delta H_r^\circ(g) = 0$. If now the central structural group is diatomic, $Y - Y'$, then the corresponding two-centre redistribution reaction is

$$X - Y - Y' - X + Z - Y - Y' - Z = X - Y - Y' - Z + Z - Y - Y' - X.$$

There is a crucial difference between the two types of redistribution reaction: in the single-centre type the act of switching groups creates a new combination of next neighbours, viz $X - Y - Z$, but in the two-centre type the act of switching groups leaves the number and types of next-neighbour combinations unchanged†, viz $X - Y - Y'$, $Y - Y' - X$, $Z - Y - Y'$ and $Y - Y' - Z$. Hence if we postulate that two-centre redistribution reactions are thermoneutral, we must extend the Fajans concept as follows: "the energy ascribed to a chemical bond is constant and transferable from structure to structure provided its nearest neighbours remain the same". This statement can be shown to be a corollary of the principle of additivity of group contributions.

---

† The question of the statistical weights of the various species can be ignored in discussion of heats, but is important in discussion of entropy changes, as demonstrated in Section 3.4.3.

Illustration of the point follows from consideration of the set of alkanes $C_2 H_n (CH_3)_{6-n}$ (58/8). Ten different formulae are possible, and if we select the six distinct two-centre frameworks

$$-CH_2-CH_2- \qquad -CH(CH_3)-CH_2- \qquad -C(CH_3)_2-CH_2-$$
$$-CH(CH_3)-CH(CH_3)- \qquad -C(CH_3)_2-CH(CH_3)-$$
$$-C(CH_3)_2-C(CH_3)_2-$$

we can write six redistribution reactions involving the ten formulae, based on these frameworks:

$$CH_3-Y-Y'-CH_3+H-Y-Y'-H$$
$$=CH_3-Y-Y'-H+H-Y-Y'-CH_3.$$

If the principle of additivity holds for all six types of $Y-Y'$, and for reactions involving all ten compounds, then four independent parameters are deducible from the heats of atomization of the compounds C_2H_n $(CH_3)_{6-n}$. Inspection shows that the four independent parameters correspond to the groups CH_3-, $-CH_2-$, $>CH-$ and $>C<$; in Benson and Buss's notation the four parameters are:

$$[C-(C)(H)_3], \qquad [C-(C)_2 (H)_2], \qquad [C-(C)_3 (H)], \qquad [C-(C)_4].$$

Thus if we know the heats of atomization of any four compounds of the series containing these groups at least once, the heats of atomization of the six others can be deduced via the redistribution reactions. The applicability of this Group method to the estimation of the heats of atomization of other alkanes can be demonstrated by the data of Chapter 5. Only for very highly branched alkanes does the method yield poor results, and for these compounds Benson and Buss (58/8) have shown how a special allowance for steric strain in the molecules can be made (*Cf.* Section 7.4).

Many group methods have been described in the literature, though none is so rigorous as that of Benson and Buss. Bremner and Thomas (47/21) developed group equations for estimating the thermodynamic properties of some hydrocarbons, including alicyclic compounds. Souders, Matthews and Hurd (49/35) applied the group method extensively to hydrocarbons and provided a simple empirical method for dealing with the steric strains. Franklin (49/36) published a similar scheme with group parameters applicaable between 0° and 1500°K. Franklin's parameters were extended by van Krevelen and Chermin (52/37), who in addition gave group parameters for the estimation of ΔG_f.

In later Sections we describe in more detail how the Group method may be applied to the estimation of ΔH_a of various types of organic compound.

We now proceed to examine some other approaches to the estimation of ΔH_a of alkanes.

7.3.2. THE LAIDLER METHOD OF ESTIMATING ΔH_a.

The $C-H$ bonds in alkanes vary slightly in length according to whether they are in primary $(-CH_3)$, secondary $(-CH_2-)$ or tertiary $(>CH-)$ groups. The three types of bond may be symbolized as $(C-H)_p$, $(C-H)_s$ and $(C-H)_t$ respectively [Section 4.4.3., paragraph (c)]. Lide (60/51) determined the bond lengths, r, in propane and 2-methylpropane to be as follows: $r, (C-H)_p = 1\cdot091$ Å; $r,(C-H)_s = 1\cdot096$ Å; and $r,(C-H)_t = 1\cdot108$ Å.

Laidler (56/8) proposed a bond energy scheme in which different bond energy term values are ascribed to the three types of $C-H$ bond, i.e. three formal types of $C-H$ bond are recognised. According to Laidler's scheme, the heat of atomization of an alkane is given by

$$\Delta H_a(C_nH_{2n+2}) = (n-1)E(C-C)+N_pE(C-H)_p+N_sE(C-H)_s$$
$$+N_tE(C-H)_t, \quad (169)$$

where $E(C-C)$ is the $C-C$ bond energy term, presumed constant, $E(C-H)_p$, $E(C-H)_s$ and $E(C-H)_t$ are the bond energy terms for the primary, secondary and tertiary $C-H$ bonds respectively and N_p, N_s and N_t are the numbers of such bonds in the molecule.

In Laidler's original work no special allowance for steric hindrance in alkanes was made, yet no restriction on the applicability of the scheme was implied. It now appears more reasonable to restrict this scheme to those compounds which do not show steric hindrance effects (see Section 7.4). It then follows that the Laidler scheme and the Group scheme (with no corrections for steric hindrance) for alkanes are equivalent because they both employ four parameters. This equivalence is shown below:

Group Parameters		*Laidler Parameters*
$[C-(C)(H)_3]$	$\equiv$	$\frac{1}{2}E(C-C)+3E(C-H)_p$
$[C-(C)_2(H)_2]$	$\equiv$	$E(C-C)+2E(C-H)_s$
$[C-(C)_3(H)]$	$\equiv$	$\frac{3}{2}E(C-C)+E(C-H)_t$
$[C-(C)_4]$	$\equiv$	$2E(C-C).$

So long as numerical values for the parameters in the two schemes are chosen in accordance with the above equivalences, the estimated values for the heat of atomization of any alkane must be the same by either method.

7.3.3. THE ALLEN METHOD OF ESTIMATING ΔH_a.

The scheme proposed by Allen (59/40) may be regarded as having been developed from the Zahn scheme (Section 7.2.2.). With the neglect of steric terms in this initial discussion†, Allen's scheme leads to the following expression for the heat of atomization of an alkane:

$$\Delta H_a(C_nH_{2n+2}) = (n-1)B(C-C) + (2n+2)B(C-H) + (N_p - 6 + \tfrac{1}{2}N_s)\Gamma_{CCC}$$
$$+ [n_t + 4n_q]\Delta_{CCC}. \quad (170)$$

The parameters in equation (170) have the same significance as their counterparts in equations (165), (166) [but with Γ_{CCC} written in place of X_{CCC}], and (168); additionally n_t and n_q are respectively the numbers of tertiary and quaternary carbon atoms in the molecule, and Δ_{CCC} is the energy effect associated with a non-bonded trio of carbon atoms in the structure, $[n_t + 4n_q]$ being the number of such trios. For example, there is one non-bonded carbon trio in isobutane, and there are four in neopentane, as can be seen

$$\begin{matrix} & & CH_3 \\ & & | \\ \text{from the molecular formulae } CH_3-CH-CH_3 \text{ and } CH_3-C-CH_3. \\ & | & | \\ & CH_3 & CH_3 \end{matrix}$$

As the Allen scheme contains four parameters, $B(C-C)$, $B(C-H)$ Γ_{CCC} and Δ_{CCC}, it is possible to derive equivalence relationships between these parameters and those of the four-parameter Group or Laidler schemes:

Group Parameters	Allen Parameters
$[C-(C)(H)_3]$	$\equiv \tfrac{1}{2}B(C-C) + 3B(C-H)$
$[C-(C)_2(H)_2]$	$\equiv B(C-C) + 2B(C-H) + \Gamma_{CCC}$
$[C-(C)_3(H)]$	$\equiv \tfrac{3}{2}B(C-C) + B(C-H) + 3\Gamma_{CCC} + \Delta_{CCC}$
$[C-(C)_4]$	$\equiv 2B(C-C) + 6\Gamma_{CCC} + 4\Delta_{CCC}.$

As before, so long as the numerical values of the parameters in the Allen and Group schemes are chosen in accord with the above equivalences, then for a particular alkane the estimated values for the heat of atomization must be the same, and identical with the value estimated by Laidler parameters conforming to the equivalence relations.

We have now demonstrated that the Group, the Laidler, and the Allen schemes, although starting from different empirical bases, are equivalent

---

† The proper recognition of these terms was an important feature of the Allen scheme.

one to another. Clearly, the three schemes must be in equal accord with experimental data, so the choice of scheme for a given calculation must depend on the preference of the user for thinking in terms of groups, in terms of chemical bonds of differing types (Laidler), or in terms of bonds plus interactions (Allen). Each of these schemes has found wide application, and each can be developed to deal with compounds other than alkanes; this development is discussed in Section 7.6.

For alkanes, the three schemes discussed are each four-parameter schemes, so one would certainly expect improved fit of experimental data compared with the two-parameter scheme of Fajans. The Group, Laidler and Allen schemes can be regarded as having a common fundamental basis, namely, that bond energy terms are constant if the nearest neighbours of the bonds are the same. Though the schemes differ superficially, allowance for next-neighbour effects is made in each, overtly in the Allen scheme and covertly in the other two.

7.3.4. THE TATEVSKII METHOD OF ESTIMATING ΔH_a

We now consider some schemes that have been specifically developed for alkanes. The Tatevskii scheme (61/59) is based on an extension of the Laidler scheme by assuming that *all* bonds in alkanes, $C-C$ as well as $C-H$, should be classified according to their environment. According to this scheme the heat of atomization of an alkane $C_n H_{2n+2}$ $(n>2)$ is given by:

$$\Delta H_a (C_n H_{2n+2}) = \sum_{i=1}^{3} N_i E_i (C-H) + \sum_{i \leqslant j, \, j=1}^{4} N_{ij} E_{ij} (C-C), \qquad (171)$$

where N_i is the number of $C-H$ bonds of type C_i-H, $E_i (C-H)$ is the bond energy term for C_i-H, N_{ij} is the number of $C-C$ bonds of type C_i-C_j, and $E_{ij} (C-C)$ is the bond energy term for C_i-C_j.

Equation (171) involves twelve parameters $E_i (C-H)$ and $E_{ij} (C-C)$, but there are three dependent relations, since

$$\begin{aligned} N_1 &= 3N_{12} + 3N_{13} + 3N_{14} \\ N_2 &= 2N_{22} + N_{12} + N_{23} + N_{24} \\ N_3 &= \tfrac{2}{3} N_{33} + \tfrac{1}{3} N_{13} + \tfrac{1}{3} N_{23} + \tfrac{1}{3} N_{34}. \end{aligned} \right\} \qquad (172)$$

By substituting these relations into equation (171) and by introducing composite bond energy terms, B_{ij}, defined by

$$B_{ij} = E_{ij} (C-C) + \left(\frac{4-i}{i} \right) E_i (C-H) + \left(\frac{4-j}{j} \right) E_j (C-H), \qquad (173)$$

we may write

$$\Delta H_a (C_n H_{2n+2}) = \sum_{i \leqslant j, \, j=1}^{4} N_{ij} B_{ij}. \tag{174}$$

There are nine independent parameters for which Tatevskii gives the following values (kcal mol^{-1}) for use in the estimation of ΔH_a:

$$B_{12} = 477{\cdot}27; \quad B_{13} = 412{\cdot}31; \quad B_{14} = 379{\cdot}94;$$
$$B_{22} = 280{\cdot}06; \quad B_{23} = 214{\cdot}46; \quad B_{24} = 181{\cdot}63;$$
$$B_{33} = 147{\cdot}69; \quad B_{34} = 113{\cdot}92; \quad B_{44} = 79{\cdot}27.$$

To illustrate Tatevskii's method let us consider the estimation of ΔH_a for 3-ethyl-2-methylpentane. It is advantageous to sketch the carbon skeleton and to label the atoms according to whether they are primary (1), secondary (2), tertiary (3) or quaternary (4).

$$\begin{array}{c} C_1 - C_3 - C_3 - C_2 - C_1 \\ \;\;\;\; | \quad\; | \\ \;\;\;\; C_1 \quad C_2 \\ \;\;\;\;\quad\quad | \\ \;\;\;\;\quad\quad C_1 \end{array}$$

Then from equation (174):

$$\Delta H_a = 2B_{13} + 2B_{12} + 2B_{23} + B_{33} = 2355{\cdot}77 \text{ kcal.}$$

The experimental value is $2355{\cdot}44 \pm 0{\cdot}35$ kcal.

7.3.5. THE PLATT METHOD OF ESTIMATING ΔH_a

Platt (47/10, 52/2) devised an energy model on a more generalised basis than Zahn's. In Platt's scheme, the contribution of each $C-H$ and $C-C$ bond to the heat of atomization is influenced not only by its nearest bond neighbours, but by neighbours once, twice and further removed.

For the αth $C-C$ bond, the contribution E_α to the heat of atomization is given by

$$E_\alpha = E_{CC}^\circ + p_1 f_{\alpha_1} + p_2 f_{\alpha_2} + \ldots . q_1 g_{\alpha_1} + q_2 g_{\alpha_2} + \ldots, \tag{175}$$

where p_j is the energy effect of the jth neighbour $C-C$ bond on E_α, q_j is the energy effect of the jth neighbour $C-H$ bond on E_α, $f_{\alpha j}$ is the number of $C-C$ bonds j bonds removed from the αth $C-C$ bond and $g_{\alpha j}$ is the number of $C-H$ bonds j bonds removed from the αth $C-C$ bond. A summation of such contributions from all the bonds of the molecule gives an expression which can be written in the form:

$$\Delta H_a = A + a_0(n-1) + a_1 \sum_\alpha f_{\alpha_1} + a_2 \sum_\alpha f_{\alpha_2} + \dots , \tag{176}$$

where a_1, a_2 etc. are linear combinations of $p_1 p_2 \dots q_1 q_2$ etc., and n is the number of carbon atoms per molecule. Platt found that the addition of quadratic terms improved the agreement between estimated and experimental heats of atomization, and his modified formula is:

$$\Delta H_a = A + a_0(n-1) + a_1 \sum f_{\alpha_1} + a_2 \sum f_{\alpha_2} + a_3 \sum f_{\alpha_3}$$
$$+ a_{11} \sum f_{\alpha_1}^2 + a_{12} \sum f_{\alpha_1} f_{\alpha_2} + a_{13} \sum f_{\alpha_2}^2. \tag{177}$$

Comment on the utility of this quite complicated empirical scheme is made in Section 7.5.

7.3.6. THE GREENSHIELDS–ROSSINI METHOD OF ESTIMATING ΔH_a

Greenshields and Rossini (58/11) devised the following empirical formula relating the standard heat of formation of a normal alkane to that of its branched chain isomers:

$$\Delta H_f^\circ (\text{branched, g}) - \Delta H_f^\circ (\text{normal, g}) = -0\cdot469 C_3 - 1\cdot364 C_4$$

$$+ 1\cdot139 \Delta P_3 + \frac{12\cdot508 \Delta W}{n(n-1)} + 1\cdot978 P_4'' + 5\cdot19 P_4'. \tag{178}$$

Te pharameters have the same significance as their counterparts in Section 4.4.3, paragraph (b). Additionally, P_4' is the number of pairs of quaternary carbon atoms separated by one carbon atom, and P_4'' is the number of pairs of quaternary–tertiary carbon atoms separated by one carbon atom. Comment on the agreement between calculated and observed heats of atomization using the Greenshields–Rossini equation is given in Section 7.5.

7.3.7. THE SOMAYAJULU–ZWOLINSKI METHOD OF ESTIMATING ΔH_a

Somayajulu and Zwolinski (66/53) examined a generalised model developed from the Allen scheme by including further contributions to the heat of atomization from vicinal (V) pairs of bonds,

$$V_1 \equiv \overset{H}{\diagdown} C - C \overset{H}{\diagup} ; \quad V_2 \equiv \overset{C}{\diagdown} C - C \overset{H}{\diagup} ; \quad V_3 \equiv \overset{C}{\diagdown} C - C \overset{C}{\diagup} .$$

Two separate cases were considered. In the first case the interaction terms for the vicinal pairs of bonds were taken to be additive using the simple relation:

$$V_2 = \tfrac{1}{2}(V_1 + V_3). \tag{179}$$

The following expression for the heat of atomization of an alkane was then obtained:

$$\Delta H_a(C_n H_{2n+2}) = (2n+2)E(C-H) + (n-1)E(C-C) + 3(n_1 - n_3)\delta_1$$
$$+ (6n_3 + 4n_2)\delta_2, \tag{180}$$

where δ_1 and δ_2 are composite interaction parameters, and n_1, n_2 and n_3 are the numbers of primary, secondary and tertiary hydrogen atoms respectively. The scheme is a four-parameter one, and although Somayajula and Zwolinski give a contrary impression it can be made equivalent to the four-parameter schemes previously described: the equivalence with the Group method is shown below.

Group Parameters		Parameters in equation (180)
$[C-(C)(H)_3]$	$\equiv$	$3E(C-H) + \tfrac{1}{2}E(C-C) + 9\delta_1$
$[C-(C)_2(H)_2]$	$\equiv$	$2E(C-H) + E(C-C) + 8\delta_2$
$[C-(C)_3(H)]$	$\equiv$	$E(C-H) + \tfrac{3}{2}E(C-C) - 3\delta_1 + 6\delta_2$
$[C-(C)_4]$	$\equiv$	$2E(C-C)$.

In the second case discussed by Somayajulu and Zwolinski, the V terms were considered to be given by equation (181):

$$V_2 - \tfrac{1}{2}(V_1 + V_3) = \delta_4. \tag{181}$$

The following expression for the heat of atomization of an alkane was then obtained:

$$\Delta H_a(C_n H_{2n+2}) = (2n+2)E(C-H) + (n-1)E(C-C) + 3n_1\delta_1 + 4n_2\delta_2$$
$$+ 3n_3\delta_3 + 2(m - 2n_2 - 3n_3)\delta_4. \tag{182}$$

The parameter m is defined by the relation

$$m = \sum_{i \leqslant j,\, j=1}^{4} m_{ij}, \tag{183}$$

where $m_{ij} = \tfrac{3}{2}(2i + 2j - ij - 3) + \tfrac{1}{2}(i-1)(j-1)$. $\tag{184}$

Equation (182) although containing six constants is actually a five-parameter expression for ΔH_a, because of a relationship between the δ parameters;

additionally Somayajulu and Zwolinski made separate allowance for steric interaction. The fit of equation (182) to experimental data is discussed in Section 7.5.

7.4. Steric correction terms in bond energy schemes

In preceding Sections we have referred to the presence of a destabilizing energy effect in highly branched alkanes. This effect is a manifestation of the well known phenomenon of steric hindrance: we refer to special corrections for this effect as *steric correction terms*. This type of steric hindrance arises

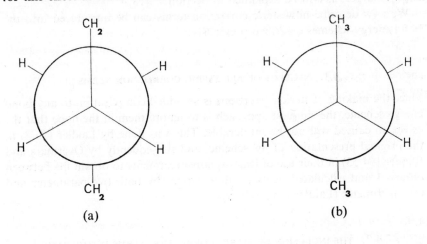

(a)

(b)

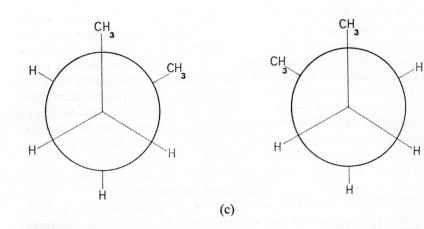

(c)

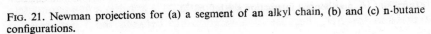

FIG. 21. Newman projections for (a) a segment of an alkyl chain, (b) and (c) n-butane configurations.

from interactions between hydrogen atoms, which according to the Lennard-Jones description of inter-atomic forces has an attractive component proportional to $1/r^6$ and a repulsive component proportional to $1/r^{12}$.† Clearly the repulsive forces have a significant effect only when atoms are close together, e.g. for hydrogen atoms when $r < 2$ Å. The lowest energy configuration of a molecule of an unbranched alkane in the gas state is that in which the bonds are fully staggered, as shown at (a) in the Newman projections (56/49) in Figure 21; the destabilizing effect of H ... H repulsions is minimized in this arrangement. In highly branched alkanes the destabilizing effect is larger, as will be explained in Section 7.4.3.

We now describe how steric correction terms can be introduced into the bond energy schemes described previously.

7.4.1. NEGLECT OF ALL STERIC CORRECTION TERMS

Since the making of steric corrections is an added complication to any bond energy scheme, the simplest approach is to ignore them in the hope that the errors so caused will not be intolerable. This was done by Laidler (56/8) in the original presentation of his scheme, and subsequently by Overmars and Blinder (64/53) in their use of least-squares treatments to obtain fits between observed and calculated heats of atomization, by both two-parameter and four-parameter (Laidler-type) schemes.

7.4.2. THE INCLUSION OF STERIC CORRECTION TERMS IN THE BASIC PARAMETERS

The Tatevskii, Platt, and Greenshields–Rossini schemes do not take special account of steric correction terms. The Tatevskii scheme, however, does include steric correction terms in a disguised form; Bernstein (62/69) has shown that five of the nine parameters, (B_{ij}, $i \geqslant 2$, $j \geqslant 3$) include steric terms. In Platt's scheme the inclusion of steric terms in the basic parameters is less obvious, but as the quadratic terms subtract from the total heat of atomization they resemble steric correction terms. In the Greenshields–Rossini scheme there are specific terms, P_4', P_4'' to allow for the large steric interaction between adjacent t-butyl and s-propyl groups. Moreover, as the ΔP_3 and ΔW terms are negative, they too correspond to steric correction terms.

Hence the parameters in these three schemes do make allowance for steric interactions and further corrections should be unnecessary.

---

† r is the distance between atomic centres. For a recent examination of various mathematical expressions for non-bonded interaction see ref. 68/22.

7.4.3. STERIC CORRECTION TERMS IN GROUP SCHEMES

Franklin (49/36) and Souders, Matthews and Hurd (49/35) introduced steric correction terms for specific groupings in alkanes. The following corrections (kcal mol^{-1}) to ΔH_a are taken from Franklin's paper:

for each C_2H_5 side chain, 1·5;
for three adjacent $\geqslant$ CH groups, 1·6;
for adjacent $>$C$<$ and $\geqslant$ CH groups, 2·5;
for adjacent $>$C$<$ and $>$C$<$ groups, 5·0;
for $>$C$<$ not adjacent to a terminal $-CH_3$ group, 2·1.

Such a scheme, though simple to apply, deals with the problem of steric correction terms by the rather primitive expedient of adding five extra parameters.

7.4.4. THE GAUCHE-1, 4-PAIR INTERACTION

The *gauche* configurations of n-butane [Figure 21(c)] are known (54/17) to be less stable than the *trans* configuration [Figure 21(b)] by *ca.* 0·5 kcal mol^{-1}, due to the greater interaction of the hydrogen atoms attached to the 1, 4-carbon atoms. In highly branched alkanes the presence of some *gauche*

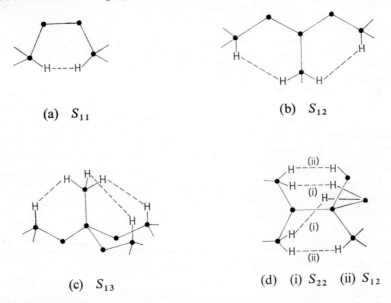

FIG. 22. Strain release by internal rotation. In S_{ij}, i and j represent the numbers of hydrogen atoms on the *gauche* 1, 4-carbon atoms involved in steric repulsion.

conformations in part of the structure is inevitable, and Allen (59/40) proposed that a correction of 0·5 kcal should be made for each pair of *gauche* 1, 4-(C—H) bonds. On some evidence the magnitude of this *gauche* 1, 4-steric correction term appears to be greater than 0·5 kcal, and Benson and Buss (58/8) suggested the value 0·7 kcal.

Skinner (62/70) suggested that the magnitude of the steric correction term depends on whether the H ... H repulsions in a given structure can be reduced by strain-release mechanisms, e.g. by internal rotation of groups or alteration of normal bond angles. Possibilities for strain release by internal rotation vary greatly between structures, as shown schematically in Figure 22.

In n-alkanes, the *trans* configuration is energetically preferred to the *gauche*, but in a branched structure, like that of 2-methylbutane, a *gauche* configuration cannot be avoided. However, the resulting 1, 4-interaction labelled S_{11} in figure 22(a) could be relieved by a slight rotation of one of the 1, 4-CH$_3$ groups, from the staggered towards the eclipsed configuration; only a small rotation is possible because of the comparatively large energy barrier for internal rotation (*ca.* 3 kcal mol^{-1}). The steric repulsions labelled S_{12}, S_{13} (Figure 22(b) and (c)) can also be partially released by internal rotation, but in each case only one of the H ... H repulsions can be reduced in this way. Steric repulsions of type S_{22} cannot be reduced by internal rotation because in a molecule like 2, 2, 3-trimethylbutane, rotation of a terminal

(a)

(b)

(c)

FIG. 23. Strain release by angle widening.

CH_3 group to reduce S_{22} simultaneously increases S_{12}. Based on these considerations Skinner suggested the following order of steric interactions:

$$S_{22} = S_{23} > S_{12} = S_{13} > S_{11}.$$

Skinner also considered the possibility of strain release by widening of the angles θ_i and θ_j [Figure 23(a)] from the normal tetrahedral angle, $109\frac{1}{2}°$.

Skinner suggested that angle rigidity decreases in the order $\theta_3 > \theta_2 > \theta_1$ (Figure 23(b)) and is greater when more than one CCC angle of a particular carbon atom is involved in steric repulsions, as exemplified in figure 23(c). Hence, steric correction terms can be classified according to both internal-rotation and angle-widening mechanisms. The values (kcal mol^{-1}) chosen by Skinner (62/70) for the steric correction terms are listed in Table 33.

TABLE 33. Values of steric correction terms (kcal mol^{-1}) according to Skinner.

	S_{11}	S_{12}	S_{22}
$(\theta_1\theta_2)$	0·33	0·38	0·43
$(\theta_1\theta_2{}^n)$	0·48	0·53	0·58
$(\theta_1\theta_3{}^m)$	0·52	0·57	0·62
$(\theta_2\theta_2)$	0·45	0·50	0·70
$(\theta_2\theta_2{}^n)$	0·75	0·80	1·00
$(\theta_2{}^n\theta_3{}^m)$	0·80	0·85	1·10
$(\theta_3{}^m\theta_3{}^m)$	0·85	0·90	1·25

Skinner also pointed out that for some molecules, e.g. 2, 2, 4-trimethyl-pentane, the staggered configuration about each C—C bond results in the very close approach of two hydrogen atoms on 1, 5 carbon atoms but it was not possible to provide a quantitative discussion of these 1, 5-steric interactions.

Somayajulu and Zwolinski (66/53) have approached the problem of steric interaction and its amelioration from the standpoint of angle release only. They assumed that to a first approximation the steric repulsion energy of an alkane arises from interactions associated with pairs of bonds, two C—C bonds apart. The three possible types of interaction are then:

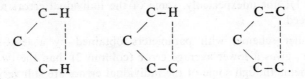

They expressed the steric repulsion energy of an alkane as the sum of the strain energies of all the CCC angles:

$$\sum S = \sum S_{ijk} = \sum l_{ijk}S = lS, \qquad (i, k = 1, 2, 3, 4; \; j = 2, 3, 4)$$

where S_{ijk} is the strain energy for the $C_i - C_j - C_k$ angle, S is the general strain parameter, and l_{ijk} is the strain number for the angle $C_iC_jC_k$; hence l is the strain number for the complete molecule. For angles other than $C_4\,C_2\,C_3$, $C_4C_3C_3$, $C_4C_4C_3$, $C_4C_2C_4$, $C_4C_3C_4$ and $C_4C_4C_4$, a general expression for l_{ijk} is:

$$l_{ijk} = (j-2)(i+k-2).$$

For angles $C_4C_2C_3$, $C_4C_3C_3$, $C_4C_4C_3$, $C_4C_2C_4$, $C_4C_3C_4$, $C_4C_4C_4$, steric strain is enhanced by 1, 5 and 1, 6 interactions and for these,

$$l_{ijk} = j(i+k-2) = (j-2)(i+k-2)+2(i+k-2), \quad i = 4, k = 3, j = 2, 3, 4$$

$$l_{ijk} = (j+2)(i+k-2) = (j-2)(i+k-2)+4(i+k-2), \quad i, k = 4, j = 2, 3, 4.$$

Then, $$l = l_1 + 10l_2 + 24l_3$$

where, $l_1 = l_{ijk} = (j-2)(i+k-2), \quad i, k = 1, 2, 3, 4; j = 2, 3, 4$

l_2 = number of $C_4 - C - C_3$ angles

l_3 = number of $C_4 - C - C_4$ angles

According to Somayajulu and Zwolinski, the value of the parameter S is $0.22\,\text{kcal mol}^{-1}$.

7.5. Comparison of observed and estimated heats of atomization of alkanes

The comparitive merits of the bond energy schemes described in Sections 7.2–7.4 are now tested by application of the schemes to all alkanes for which experimental data have been given in Chapter 5. The results of this test are shown in Table 34 in terms of ΔH_a (estimated) $- \Delta H_a$ (observed).

From inspection of Table 34 the following conclusions can be drawn:

(i)   The simple two-parameter scheme, employing only $E(C-C)$ and $E(C-H)$, gives quite a large average error of estimation, $1.5\,\text{kcal}$, (column 1); not unexpectedly, some of the individual errors are very large indeed.

(ii) The Laidler scheme, with parameters obtained by a least-squares technique, gives a lower average error (column 2) than the two-parameter scheme though some of the individual errors are still very large.

(iii) The results for the specialised schemes developed by Platt, Greenshields and Rossini, and Somayajulu and Zwolinski (columns 4–6) are impressive. Very few of the estimation errors are larger than the experimental errors in ΔH_a (observed) so one cannot expect a better correlation than these treatments provide. Unfortunately these methods have the disadvantage that the equations are cumbersome to apply and cannot readily be modified to treat compounds other than alkanes. The low estimation errors provided by the Tatevskii method (column 3) are also impressive, but again there are difficulties in extending this method to other types of compound, because of the large number of parameters required.

(iv) The remaining schemes tested in Table 34 are four-parameter schemes, with parameters selected to accord with the equivalences given in Sections 7.3.1–7.3.3. In the absence of other factors all such schemes would yield identical results but three columns have been devoted to this approach to demonstrate the effectiveness of the various ways of assessing the steric correction terms. The Skinner method (column 8) is seen to be the best; although it involves a large number of parameters it must be remembered that their values are selected on a reasoned basis. The Franklin method of estimating the steric correction terms (column 9) is seen to be quite good, but leads to large errors in highly branched alkanes such as 2, 2, 3- and 2, 3, 4-trimethylpentane.

The simplest method of applying steric correction terms is to use a constant value for each 1, 4-steric interaction, and in column 7 of Table 34 the constant value of 0·6 kcal has been used. This method is seen to provide a satisfactory estimate of ΔH_a for all but the highly branched alkanes: we suggest that for compounds containing the structure

$$
\begin{array}{ccc}
\text{C} & \text{C} & \\
| & | & \\
\text{C} - \text{C} - \text{C} - \text{C} & \\
| & & \\
\text{C} & &
\end{array}
$$

the calculated steric correction term should be increased by 2 kcal and for compounds containing the structure

$$
\begin{array}{ccc}
\text{C} & \text{C} & \\
| & | & \\
\text{C} - \text{C} - \text{C} - \text{C} & \\
| & | & \\
\text{C} & \text{C} &
\end{array}
$$

the calculated steric correction term should be increased by 4 kcal. When these additional corrections are applied, the average value of

ΔH_a (estimated) $- \Delta H_a$ (observed) falls to 0·40 kcal, and seven large discrepancies are removed from the list. Because of its excellence and simplicity, this is the scheme we shall extend to compounds other than alkanes. The selection of a particular form of four-parameter scheme to be used in conjunction with our recommended way of handling steric correction terms is arbitrary: as we have stressed already, the choice of the Group, the Laidler or the Allen method is a matter of personal preference, now that these schemes have been unified and the parameters made mutually compatible.

7.6. Application of bond energy schemes to substituted alkanes

In this Section we consider how bond energy schemes for alkanes can be adapted to substituted alkanes; discussion of olefines, acetylenes, aromatic compounds, other ring compounds, and compounds having electrostatically interacting groups is postponed to later Sections.

7.6.1. APPLICATION OF THE GROUP METHOD TO SUBSTITUTED ALKANES

Application of the group method presents no difficulty. In the compound alkyl–X, for example, the carbon atom of the $C-X$ bond can be primary, secondary or tertiary corresponding to the structures (i) $R_1.CH_2-X$, (ii) $R_1 R_2.CH-X$ and (iii) $R_1 R_2 R_3.C-X$. There will then be three group energies, which in Benson and Buss's notation may be written

$$[C-(C)(H)_2(X)], \quad [C-(C)_2(H)(X)], \quad [C-(C)_3 X],$$

and the heat of atomization of the compound will be given by the sum of the appropriate group energies. Hence, from the heats of atomization of three compounds, respectively containing the structures (i), (ii) and (iii), it is possible to derive the group parameters, and to apply them to any other member of the same class of compounds.

To apply the group method to methyl compounds CH_3-X, where X is monovalent, an extra group energy $[C-(H)_3(X)]$ would be required; this is a trivial matter because the extra parameter applies only to this compound, and is not needed for calculations on other compounds.

The group method was applied to substituted alkanes by Benson and Buss (58/8), and has since been applied by Lehmann and Ruschitzky (67/9) to organic oxygen and sulphur compounds to yield values of ΔH_f° (g) at various temperatures between 0° and 1000°K.

TABLE 34. Comparison of observed and estimated heats of atomization of alkanes at 298.15°K, kcal.

Alkane skeleton	ΔH_a(obs.)[a]	ΔH_a(est.) − ΔH_a(obs.)								
		1[b]	2[c]	3[d]	4[e]	5[f]	6[g]	7[hi]	8[hj]	9[hk]
CCC	954·33	+2·58	+0·81	+0·21	-0·13	—	+0·38	+0·33	+0·33	+0·33
CCCC	1234·96	+1·82	+0·14	-0·36	-0·42	—	-0·23	-0·28	-0·28	-0·28
C / CCC	1237·01	-0·23	-1·06	-0·08	-0·56	+0·36	-0·18	-0·30	-0·30	-0·30
CCCCC	1514·80	+1·84	+0·26	-0·14	-0·17	—	-0·18	-0·10	-0·10	-0·10
C / CCCC	1516·55	+0·09	-0·64	-0·20	-0·16	-0·27	-0·29	-0·42	-0·15	+0·18
C / CCC / C	1519·97	-3·33	-2·35	-0·21	+0·19	+0·97	+0·47	+0·27	+0·27	+0·27
CCCCCC	1794·72	+1·78	+0·31	-0·00	-0·11	—	+0·03	0·00	0·00	0·00
C / CCCCC	1796·57	-0·07	-0·69	-0·16	-0·25	-0·37	-0·18	-0·42	-0·15	+0·18
C / CCCCC	1795·93	+0·57	-0·05	-0·16	+0·07	-0·45	-0·03	-0·38	-0·14	+0·82
CC / CCCC	1797·41	-0·91	-0·68	-0·48	-0·12	-0·86	-0·3	+0·17	-0·13	+1·37

Alkane skeleton	$\Delta H_a(\text{obs.})^a$	$\Delta H_a(\text{est.}) - \Delta H_a(\text{obs.})$								
		1^b	2^c	3^d	4^e	5^f	6^g	7^{hi}	8^{hj}	9^{hk}
C CCCC C	1799·28	−2·78	−1·70	−0·56	−0·16	−0·28	−0·10	−0·22	−0·16	+0·98
CCCCCC	2074·75	+1·62	+0·24	+0·03	−0·16	—	+0·01	−0·01	−0·01	−0·01
C CCCCC	2076·42	−0·05	−0·58	+0·05	−0·26	−0·25	−0·01	−0·25	+0·02	+0·35
C CCCCC	2075·63	+0·74	+0·21	+0·20	+0·16	+0·00	+0·25	−0·06	+0·18	+1·14
C C CCCC	2075·15	+1·22	+0·69	+0·04	+0·13	−0·06	+0·29	−0·18	+0·18	+0·12
C CCCC C	2079·10	−2·73	−1·55	−0·32	−0·24	−0·01	+0·01	−0·02	+0·04	+1·18
CC CCCC	2077·23	−0·86	−0·54	−0·88	−0·66	−1·32	−0·60	−0·23	−0·51	+1·57
C C CCCC	2078·11	−1·74	−1·42	+0·05	−0·29	−0·51	−0·08	−0·51	+0·03	+0·69
C CCCC C	2077·98	−1·61	−0·43	−0·30	−0·37	−0·57	−0·32	−0·10	+0·02	+0·20

Alkane skeleton	ΔH_a(obs.)[a]	ΔH_a(est.) $-$ ΔH_a(obs.)								
		1[b]	2[c]	3[d]	4[e]	5[f]	6[g]	7[hi]	8[hj]	9[hk]
C CCCC CC	2078·77	−2·40	−0·37	−0·41	−0·34	+0·60	−0·36	+1·14	−0·36	+1·04
CCCCCCC	2354·86	+1·37	+0·10	−0·02	−0·29	—	−0·05	−0·10	−0·10	−0·10
C CCCCCC	2356·47	−0·24	−0·66	+0·06	−0·33	−0·26	−0·14	−0·28	−0·01	+0·32
C CCCCCC	2355·79	+0·44	+0·02	+0·10	−0·16	−0·06	−0·17	−0·20	+0·04	+1·00
C CCCCCC	2355·66	+0·57	+0·15	+0·23	−0·08	+0·30	−0·31	−0·07	+0·17	+1·13
C C CCCCC	2355·36	+0·87	+0·45	−0·11	−0·43	+0·14	+0·12	−0·37	−0·01	−0·07
C CCCCC C	2358·68	−2·45	−1·17	+0·16	−0·12	+0·45	+0·46	+0·42	+0·48	+1·62
CC CCCCC	2356·10	+0·13	+0·56	+0·31	+0·12	+0·07	+0·56	+0·92	+0·44	+2·72
C C CCCCC	2357·40	−1·17	−0·74	+0·18	−0·25	+0·31	+0·18	−0·38	+0·03	+1·42

Alkane skeleton	$\Delta H_a(\text{obs.})^a$	$\Delta H_a(\text{est.}) - \Delta H_a(\text{obs.})$								
		1[b]	2[c]	3[d]	4[e]	5[f]	6[g]	7[hi]	8[hj]	9[hk]
`C   C` `CCCCC`	2358·18	−1·95	−1·52	+0·04	−0·61	−0·62	−0·12	−0·56	−0·02	+0·64

`C`
`CCCCC`
`C`
	2357·58	−1·35	−0·07	+0·16	−0·37	+0·16	+0·15	+0·32	+0·44	+0·62

`CC` `CCCCC`	2355·87	+0·36	+0·79	−0·10	−0·06	+0·39	+0·31	+0·55	+0·29	+2·95

`C`
`CC`
`CCCC`
	2355·44	+0·79	+1·22	+0·33	+0·08	+0·27	+0·74	+0·38	+0·14	+1·88

`C`
`C`
`CCCC`
`C`
	2356·35	−0·12	+1·16	+0·29	−0·72	−0·22	+0·07	+0·35	+0·38	+0·35

`C`
`CCCC`
`CC`
	2357·58	−1·35	+0·78	+0·20	−0·20	+0·15	+0·42	+1·75	+0·32	+2·25

`C`
`CCCC`
`C C`
	2358·54	−2·31	−0·18	—		+0·85	+0·02	—	—	—

`CC`
`CCCC`
`C`
	2356·69	−0·46	+1·67	+0·63	−0·24	+0·12	+0·39	+2·04	+0·25	+1·04

Alkane skeleton	$\Delta H_a(\text{obs.})^a$	$\Delta H_a(\text{est.}) - \Delta H_a(\text{obs.})$								
		1^b	2^c	3^d	4^e	5^f	6^g	7^{hi}	8^{hj}	9^{hk}
CCC / CCCC	2356·94	−0·71	+0·57	−0·01	−0·13	−0·55	+0·42	+0·91	−0·24	+2·31
CC / CCCC / CC	2358·83	−2·60	+1·24	−0·08	−1·46	+0·01	−0·84	+3·41	−0·49	+2·01
CCCCCCCC	2634·76	+1·33	+0·16	+0·14		—	+0·11	+0·02	+0·02	+0·02
C / C / CCCC / C / C	2635·51	+0·58	+1·97	+0·09		−0·05	−0·56	+0·01	−0·15	−0·29
CC / CCCC / CC	2637·77	−0·68	+3·26	+1·10		−0·05	−0·14	+4·29	−0·65	+0·99
CCC / CCCC / C	2636·71	−0·62	+2·47	/		−0·04	+0·08	/	/	/
C C / CCCC / C C	2637·90	−1·81	+2·13	/		−0·05	+0·29	/	/	/
CCC / CCCC / C	2636·53	−0·44	+2·65	+0·43		−0·04	−0·10	+3·05	−0·45	+0·75

Alkane skeleton	$\Delta H_a(\text{obs.})^a$	$\Delta H_a(\text{est.}) - \Delta H_a(\text{obs.})$								
		1^b	2^c	3^d	4^e	5^f	6^g	7^{hi}	8^{hj}	9^{hk}
n-C_{10}	2914·84	+1·12	+0·04	+0·12		—		−0·04	−0·04	−0·04
C CCCCCCCC	2917·32	−1·36	−1·59	−0·67		−0·77		−1·09	−0·82	−0·49
C CCCCCCCC	2917·00	−1·04	−1·27	−0·99		−1·13		−1·37	−1·13	−0·17
n-C_{11}	3195·05	+0·77	−0·20	−0·03		—		−0·23	−0·23	−0·23
C C CCCCCC C C	3202·5	−6·7	−2·5	−0·6				−0·2	0·0	+1·3
C C CCCCCC C C	3196·4	−0·6	+3·6	—				—	—	—
C C CCCCC C CC	3197·4	−1·6	+3·4	—				—	—	—
n-C_{12}	3474·64	+1·04	+0·17	−0·44		—		+0·20	+0·20	+0·20
C C CCCCCC C C	3481·4	−5·7	−1·5	−0·5				−0·3	0·0	+0·3

Alkane skeleton	ΔH_a(obs.)[a]	ΔH_a(est.) − ΔH_a(obs.)								
		1[b]	2[c]	3[d]	4[e]	5[f]	6[g]	7[hi]	8[hj]	9[hk]
C C CCCCCCCC C C	3755·4	+0·1	+4·5	/				—	—	—
CCCCCC C C C C	3752·5	+3·0	+7·4	/				—	—	—
C C CCCCCCCCC C C	4318·3	−3·0	+1·5	/				—	—	—
C C CCCCCCCC C C	4313·6	+1·7	+6·2	/				—	—	—
n-C₁₆	4595·60	−0·47	−0·93	−0·28		—		−0·68	−0·68	−0·68
n-C₁₈	5155·0	−0·1	−0·4	+0·4		—		0·0	0·0	0·0
C CCCCC-C₁₉	7397·2	−3·4	−2·0	−0·2				−1·2	−1·0	−0·0
C C₉-CCC-C₁₃	7398·6	−4·8	−3·4	−1·6				−2·6	−2·4	−1·4
n-C₃₂	9074·1	−1·2	0·0	+2·2		—		+1·1	+1·1	+1·1
Average Error		1·47	1·30	0·25	0·29	0·28	0·25	0·66	0·30	0·78

a. Values of ΔH_a (observed) were calculated from the heats of formation of gaseous atoms (Table 24) and the heats of formation of gaseous alkanes given in Chapter 5.

b. Column 1: the values of ΔH_a (estimated) were calculated by means of the 2-parameter scheme of Overmars and Blinder (64/53) with correction of their parameters to accord with the data of Table 24, namely $E(C-H) = 99\cdot297$, $E(C-C) = 81\cdot269$ kcal.

c. Column 2: the values of ΔH_a (estimated) were calculated by Laidler's method, without steric corrections, and with parameters taken from Overmars and Blinder (64/53) modified to be compatible with Table 24, namely $E(C-C) = 83\cdot638$, $E(C-H)_p = 98\cdot589$, $E(C-H)_s = 98\cdot163$, $E(C-H)_t = 97\cdot735$ kcal. (Cf. note h below.)

d. Column 3: the values of ΔH_a (estimated) were obtained by Tatevskii's scheme, using parameters given in Section 7.3.4.

e. Column 4: the values of ΔH_a (estimated) were obtained by Platt's scheme using the equation,
$$\Delta H_a = 119\cdot79 + 277\cdot17n + 1\cdot57 \ \Sigma f_1 + 0\cdot545 \ \Sigma f_2 + 0\cdot14 \ \Sigma f_3 - 0\cdot12 \ \Sigma f_{11} - 0\cdot235 \ \Sigma f_{12} - 0\cdot07 \ \Sigma f_{22}. \tag{185}$$
The values of Σf_i and Σf_{ij} were taken from Platt (47/10, 52/2).

f. Column 5: the values of ΔH_a (estimated) were obtained by Greenshields and Rossini's scheme (Section 7.3.6.), using their parameters.

g. Column 6: the values of ΔH_a (estimated) were obtained by Somayajulu and Zwolinski's scheme (Section 7.3.7 and 7.4.4) using the parameters $E(C-H) = 99\cdot297$, $E(C-C) = 85\cdot22$, $\delta_1 = -1\cdot06$, $\delta_2 = -0\cdot95$, $\delta_3 = -0\cdot84$, $\delta_4 = 0\cdot02$, $S = 0\cdot22$ kcal.

h. Columns 7–9: the values of ΔH_a (estimated) were obtained by the unified 4-parameter scheme, employing either the group parameters $[C-((C)(H)_3] = 337\cdot32$, $[C-((C)_2(H)_2] = 280\cdot02$, $[C-(C_3)(H)] = 224\cdot75$, $[C-(C)_4] = 170\cdot96$ kcal, or the Laidler parameters $E(C-C) = 85\cdot48$, $E(C-H)_p = 98\cdot19$, $E(C-H)_s = 97\cdot27$, $E(C-H)_t = 96\cdot53$ kcal, or the Allen parameters $B(C-C) = 78\cdot84$, $B(C-H) = 99\cdot30$, $\Gamma_{CCC} = 2\cdot58$. $\Delta_{CCC} = -0\cdot55$ kcal.

i. Steric correction terms were calculated using the constant value of $0\cdot6$ kcal for each 1, 4-gauche interaction.

j. Steric correction terms were calculated using Skinner's values, (Table 33).

k. Steric correction terms were calculated using Franklin's group values (Section 7.4.3).

l. Results are omitted for alkanes having 1, 5-steric interactions (Section 7.7.1).

7.6.2. APPLICATION OF THE LAIDLER METHOD TO SUBSTITUTED ALKANES

The Laidler method was extended to substituted alkanes by Lovering and Laidler (60/10), and was later applied to a wide range of aliphatic and benzenoid compounds by Cox (62/71). There are two equivalent procedures for applying the Laidler method to substituted alkanes.

(a) One can consider that the bond-energy term $E(C-X)$ is constant and that values of $E(C-H)$ for the $C-H$ bonds attached to the carbon atom of $C-X$ depend on the degree of substitution of that carbon atom. To apply this principle to substituted alkanes, three extra parameters are needed, namely

$$E(C-X), \quad E(C-H)_s^X, \quad E(C-H)_t^X,$$

where $E(C-H)_s^X$ is the $C-H$ bond energy term in $R_1.CH_2-X$ and $E(C-H)_t^X$ is the $C-H$ bond energy term in $R_1 R_2.CH-X$. For methyl compounds CH_3X, an extra $C-H$ bond energy term $E(C-H)_p^X$ would be required, but as discussed in Section 7.6.1 this parameter is not needed for calculations on compounds other than CH_3X itself.

(b) One could equally well consider that the $C-H$ bond energy terms had the same values in alkyl$-X$ compounds as in alkanes but that there are three $C-X$ bond energy terms,

$$E(C-X)_s, \quad E(C-X)_t, \quad \text{and} \quad E(C-X)_q,$$

to be derived from $R_1.CH_2-X$, $R_1 R_2.CH-X$ and $R_1 R_2 R_3.C-X$ respectively. Again, the methyl derivative would require a special parameter, $E(C-X)_p$, that would not be needed for calculations on other compounds.

Clearly approaches (a) and (b) are equivalent, as they contain the same number of parameters. To avoid confusion, we will consistently choose approach (a), as this is more in keeping with the spirit of Laidler's scheme.

As was the case for alkanes, the Laidler and Group methods can be made equivalent for $R-X$ compounds. The equivalence relations are shown below:

Group Parameters	Laidler Parameters
$[C-(C)(H)_2(X)]$	$\equiv$ $E(C-X)+2E(C-H)_s^X+\frac{1}{2}E(C-C)$
$[C-(C)_2(H)(X)]$	$\equiv$ $E(C-X)+E(C-H)_t^X+E(C-C)$
$[C-(C)_3(X)]$	$\equiv$ $E(C-X)+\frac{3}{2}E(C-C).$

7.6.3. APPLICATION OF THE ALLEN METHOD TO SUBSTITUTED ALKANES

The Allen scheme has been extended to substituted alkanes by Skinner (62/70), Skinner and Pilcher (63/58), and McCullough and Good (61/60).

The necessary equation is

$$\Delta H_a (R-X) = N_X B(C-X) + N_1 B(C-H) + N_2 B(C-C) + b_3 \Gamma_{ccc} + c_4 \Delta_{ccc}$$
$$+ b_3' \Gamma_{ccx} + c_4' \Delta_{ccx} + b_3'' \Gamma_{cxc} + c_4'' \Delta_{ccc}^X - [S], \quad (186)$$

where the symbols have the same meanings as their counterparts in equation (170); additionally N_1 and N_2 are the numbers of $C-H$ and $C-C$ bonds, N_X is the number of $C-X$ bonds, b_3 is the number of $C-C-C$ interactions, b_3' is the number of $C-C-X$ interactions, b_3'' is the number of $C-X-C$ interactions, c_4 is the number of C_3 trios attached to carbon, c_4' is the number of C_2X trios attached to carbon, c_4'' is the number of C_3 trios attached to (multivalent) X, and $[S]$ is the total steric correction term.

To apply the Allen scheme to structures $R_1.CH_2-X$, $R_1 R_2.CH-X$ and $R_1 R_2 R_3.C-X$, three new parameters are needed, namely $B(C-X)$, Γ_{ccx}. and Δ_{ccx}. Again equivalence between the Allen scheme and the Group method exists, as shown below:

Group Parameters   *Allen Parameters*

$[C-(C)(H)_2(X)] \equiv B(C-X) + 2B(C-H) + \tfrac{1}{2}B(C-C) + \Gamma_{ccx}$

$[C-(C)_2(H)(X)] \equiv B(C-X) + B(C-H) + B(C-C) + \Gamma_{ccc} + 2\Gamma_{ccx} + \Delta_{ccx}$

$[C-(C)_3(X)] \equiv B(C-X) + \tfrac{3}{2}B(C-C) + 3\Gamma_{ccc} + 3\Gamma_{ccx} + \Delta_{ccc} + 3\Delta_{ccx}.$

The extension of these schemes to substituted alkanes where the substituent Y is a divalent atom or group is now discussed. For compounds of the general types $R_1.CH_2-Y-H$, $R_1R_2.CH-Y-H$ and $R_1R_2R_3.C-Y-H$ (e.g. the alcohol series if $Y =$ oxygen), the following equivalences can be derived:

Group Parameters   *Laidler Parameters*

$[C-(C)(H)_2(Y)] + [Y-(C)(H)] \equiv [E(C-Y) + E(Y-H)]$
$\qquad\qquad\qquad\qquad\qquad + 2E(C-H)_s^Y + \tfrac{1}{2}E(C-C)$

$[C-(C)_2(H)(Y)] + [Y-(C)(H)] \equiv [E(C-Y) + E(Y-H)]$
$\qquad\qquad\qquad\qquad\qquad + E(C-H)_t^Y + E(C-C)$

$[C-(C)_3(Y)] + [Y-(C)(H)] \equiv [E(C-Y) + E(Y-H)] + \tfrac{3}{2}E(C-C)$

Group Parameters   *Allen Parameters*

$[C-(C)(H)_2(Y)] + [Y-(C)(H)] \equiv [B(C-Y) + B(Y-H)] + 2B(C-H)$
$\qquad\qquad\qquad\qquad\qquad + \tfrac{1}{2}B(C-C) + \Gamma_{ccy}$

$[C-(C)_2(H)(Y)] + [Y-(C)(H)] \equiv [B(C-Y) + B(Y-H)] + B(C-H)$
$\qquad\qquad\qquad\qquad\qquad + B(C-C) + 2\Gamma_{ccy} + \Delta_{ccy}$

$[C-(C)_3(Y)] + [Y-(C)(H)] \equiv [B(C-Y) + B(Y-H)] + \tfrac{3}{2}B(C-C)$
$\qquad\qquad\qquad\qquad\qquad + 3\Gamma_{ccc} + 3\Gamma_{ccy} + \Delta_{ccc} + 3\Delta_{ccy}.$

Or for compounds of the types $R_1.CH_2-Y-CH_2.R_1'$, $R_1R_2.CH-Y-CH.R_1'R_2'$ and $R_1R_2R_3.C-Y-C.R_1'R_2'R_3'$ (e.g. the ether series if Y = oxygen) the following equivalences can be derived:

Group Parameters	*Laidler Parameters*
$2[C-(C)(H)_2(Y)]+[Y-(C)_2]$	$\equiv 2E(C-Y)+E(C-C)+4E(C-H)_s^Y$
$2[C-(C)_2(H)(Y)]+[Y-(C)_2]$	$\equiv 2E(C-Y)+2E(C-C)+2E(C-H)_t^Y$
$2[C-(C)_3(Y)]+[Y-(C)_2]$	$\equiv 2E(C-Y)+3E(C-C)$

Group Parameters	*Allen Parameters*
$2[C-(C)(H)_2(Y)]+[Y-(C)_2]$	$\equiv 2B(C-Y)+B(C-C)+4B(C-H)$ $+2\Gamma_{CCY}+\Gamma_{CYC}$
$2[C-(C)_2(H)(Y)]+[Y-(C)_2]$	$\equiv 2B(C-Y)+2B(C-C)+2B(C-H)$ $+2\Gamma_{CCC}+4\Gamma_{CCY}+\Gamma_{CYC}+2\Delta_{CCY}$
$2[C-(C)_3(Y)]+[Y-(C)_2]$	$\equiv 2B(C-Y)+3B(C-C)+6\Gamma_{CCC}$ $+6\Gamma_{CCY}+\Gamma_{CYC}+2\Delta_{CCC}+6\Delta_{CCY}.$

Similar equivalences can be derived for cases where Y has a valence higher than two. The relations have in common with those shown above that for a given series of compounds, some of the parameters occur in sets. For practical calculations it is unnecessary to separate these sets of parameters into individual terms for each bond. Thus, for alcohols the term $[C-(C)_n(H)_{3-n}(O)]$ will always occur with the term $[O-(C)(H)]$ in the Group method, the terms $E(C-O)$ and $E(O-H)$ will always occur together in the Laidler method and the terms $B(C-O)$ and $B(O-H)$ will always occur together in the Allen method. It is often convenient to work with composite bond energy terms in the Laidler and Allen schemes as well as in the Group scheme. If separation of the composite terms is required, an extra assumption must be made, since data from organic compounds alone provide an underdetermined set of equations. For example, it is not unreasonable to take $B(O-H)$ equal to the $O-H$ mean bond energy term in water and then by subtraction to deduce $B(C-O)$.

7.6.4. 1, 4-GAUCHE STERIC CORRECTION TERMS IN SUBSTITUTED ALKANES

One would expect these steric correction terms within the unsubstituted part of the molecule to have the same values as in the corresponding alkanes but the steric correction terms which involve the substituent are likely to be different for geometric and possibly electronic reasons. For example, in

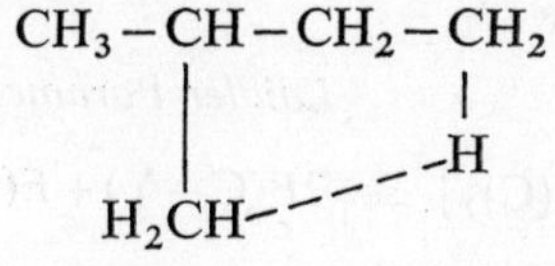

$$CH_3-CH-CH_2-CH_2$$
$$\quad | \qquad\qquad |$$
$$\quad | \qquad\qquad -H$$
$$H_2CH$$

there is one 1, 4-steric correction term, the magnitude of which is hardly likely to be the same as that of the corresponding terms in

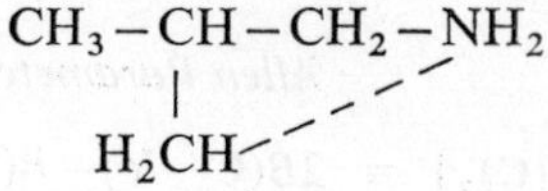

$$CH_3-CH-CH_2-NH_2$$
$$\quad |$$
$$H_2CH$$

or
$$CH_3-CH-CH_2-OH.$$
$$\quad |$$
$$H_2CH$$

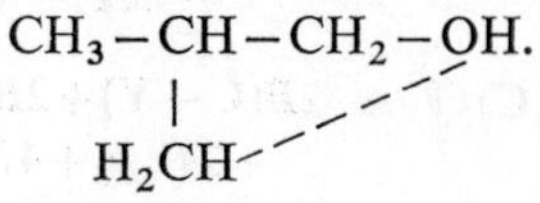

In compounds of the type R_1-Y-R_2, the presence of a significant 1, 4-*gauche* interaction across the Y-atom will depend critically on the $C-Y$ bond length and to some extent on the CYC bond angle also. For example, in ethers $r(C-O) \approx 1.43$ Å compared with $r(C-C) \approx 1.54$ Å, so one would expect 1, 4-steric interaction across the O-atom in ethers to be greater than that across a C-atom in alkanes (*cf.* Section 7.6.5). On the other hand, in thioethers $r(C-S) \approx 1.82$ Å so one would expect 1, 4-steric interaction across the S-atom in thioethers to be less than that across a C-atom in alkanes: according to McCullough and Good (61/60) this interaction across $-S-$ is zero, to within experimental error. One would therefore expect the 1, 4-steric repulsion term to decrease in the order

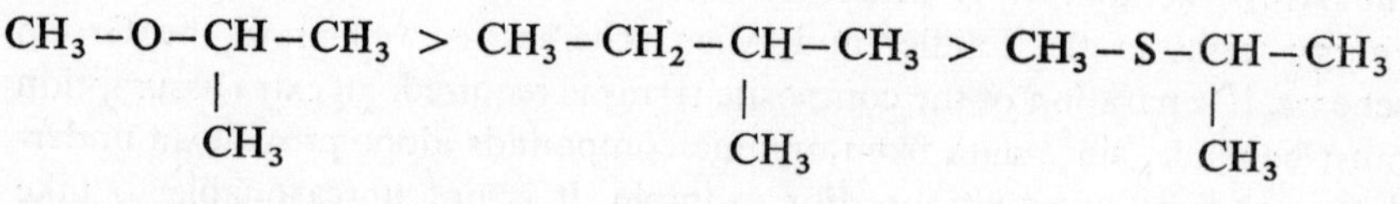

$$CH_3-O-CH-CH_3 > CH_3-CH_2-CH-CH_3 > CH_3-S-CH-CH_3$$
$$\qquad\quad | \qquad\qquad\qquad\qquad | \qquad\qquad\qquad\qquad |$$
$$\qquad\quad CH_3 \qquad\qquad\qquad\qquad CH_3 \qquad\qquad\qquad\qquad CH_3$$

as will be discussed further in the two following Sections.

7.6.5. APPLICATION OF BOND ENERGY SCHEMES TO ALCOHOLS AND ETHERS

Table 35 lists observed and estimated heats of atomization of alcohols. The observed values were calculated from the heats of formation of gaseous alcohols given in Chapter 5 and heats of formation of gaseous elements given in Table 24. The estimated values were calculated using the equivalent sets of parameters (kcal) given in paragraphs (a), (b) and (c) below.

(a) *The Group method*

$$[C-(C)(H)_2(O)]+[O-(C)(H)] = 432 \cdot 61,$$

$$[C-(C)_2(H)(O)]+[O-(C)(H)] = 379 \cdot 54,$$

$$[C-(C)_3(O)]+[O-(C)(H)] = 327 \cdot 07.$$

(b) *The Laidler method*

$$[E(C-O)+E(O-H)] = 198 \cdot 85, E(C-H)_s^O = 95 \cdot 51, E(C-H)_t^O = 95 \cdot 21.$$

(c) *The Allen method*

$$B(C-O) = 78 \cdot 15, B(O-H) = 110 \cdot 78, \Gamma_{CCO} = 5 \cdot 66, \Delta_{CCO} = -1 \cdot 43.$$

The constant value $0 \cdot 6$ kcal (Section 7.5) was used in correcting for each 1, 4-steric interaction, except that steric corrections involving the $O-H$ group were assumed to be zero.

It can be seen from Table 35 that the agreement between ΔH_a (observed) and ΔH_a (estimated) for the alcohol series is in most instances excellent.

TABLE 35

Estimated and observed heats of atomization of aliphatic alcohols at $298 \cdot 15°$K, kcal.

	ΔH_a(observed)	ΔH_a(estimated)	ΔH_a(est.)$-\Delta H_a$(obs.)
Methanol	486·93	486·83	−0·10
Ethanol	770·20	769·93	−0·27
n-Propanol	1050·23	1049·95	−0·28
s-Propanol	1054·18	1054·18	0·00
n-Butanol	1329·95	1329·97	+0·02
Isobutanol	1332·00	1332·00	0·00
s-Butanol	1334·14	1334·20	+0·06
t-Butanol	1338·88	1339·03	+0·15
n-Pentanol	1609·92	1609·99	+0·07
2-Pentanol	1614·44	1614·22	−0·22
3-Pentanol	1614·47	1614·22	−0·25
2-Methyl-l-butanol	1611·45	1611·41	−0·04
3-Methyl-l-butanol	1611·23	1611·41	+0·18
2-Methyl-2-butanol	1618·33	1618·45	+0·12
3-Methyl-2-butanol	1614·61	1615·65	+0·04
l-Hexanol	1890·01	1890·01	0·00
l-Heptanol	2168·55	2170·03	+1·48
l-Octanol	2449·86	2450·05	+0·19
2-Ethyl-l-hexanol	2451·87	2450·88	−0·99
l-Nonanol	2730·78	2730·07	−0·71
l-Decanol	3009·57	3010·09	+0·52
l-Hexadecanol	4489·06	4690·21	+1·15

Table 36 lists observed and estimated heats of atomization of aliphatic ethers. The additional parameters (kcal) needed for the calculations are given below.

(a) *The Group method*

$$2[C-(C)(H)_2(O)] + [O-(C)_2] = 649.56,$$

$$2[C-(C)_2(H)(O)] + [O-(C)_2] = 543.42,$$

$$2[C-(C)_3(O)] + [O-(C)_2] = 438.48.$$

(b) *The Laidler method* $E(C-O) = 91.02.$

(c) *The Allen method* $\Gamma_{COC} = 5.90,$

In the calculation of ΔH_a (estimated) *no* steric correction terms were included. The column of Table 36 headed $[S]$ lists the corresponding steric correction terms, according to Skinner (62/70), when the $-O-$ atom in the ether is replaced by $-CH_2-$. It is seen from Table 36, that in cases where $[S] = 0$ for the corresponding alkane there is excellent agreement between the observed and estimated heats of atomization of ethers, but in cases where the alkane shows a steric effect, so does the ether. We cannot at present devise precise methods for estimating steric correction terms in ethers, but from data in the last two columns of Table 36 it should be possible to guess values appropriate to other ethers.

TABLE 36

Estimated and observed heats of atomization of aliphatic ethers at $298.15°K$, kcal.

	ΔH_a(obs.)	ΔH_a(est.)	ΔH_a(est.)$-\Delta H_a$(obs.)	$[S]$
Dimethyl ether	757.95	758.00	+0.05	0
Methyl ethyl ether	1040.78	1041.10	+0.32	0
Diethyl ether	1324.42	1324.20	−0.22	0
Methyl n-propyl ether	1320.98	1321.12	+0.14	0
Methyl s-propyl ether	1324.40	1325.35	+0.95	0.33
Methyl t-butyl ether	1608.86	1610.20	+1.34	1.14
Di-n-propyl ether	1884.11	1884.24	+0.13	0
Di-s-propyl ether	1890.46	1892.70	+1.24	1.14
Isopropyl t-butyl ether	2174.86	2177.55	+2.69	3.79
Di-n-butyl ether	2444.28	2444.28	−0.00	0
Di-s-butyl ether	2450.72	2452.74	+2.02	2.12
Di-t-butyl ether	2451.56	2462.40	+10.84	8.0

7.6.6. APPLICATION OF BOND ENERGY SCHEMES TO THIOLS AND THIOETHERS

Table 37 lists observed and estimated heats of atomization of thiols. The additional parameters (kcal) needed for the calculations are given below.

(a) *The Group method*

$$[C-(C)(H)_2(S)]+[S-(C)(H)] = 393 \cdot 96,$$

$$[C-(C)_2(H)(S)]+[S-(C)(H)] = 338 \cdot 76$$

$$[C-(C)_3(S)]+[S-(C)(H)] \quad = 284 \cdot 39.$$

(b) *The Laidler method*

$$[E(C-S)+E(S-H)] = 156 \cdot 17, \quad E(C-H)_s^S = 97 \cdot 52, \quad E(C-H)_t^S = 97 \cdot 11.$$

(c) *The Allen method*

$$B(C-S) = 65 \cdot 29, \quad B(S-H) = 87 \cdot 35, \quad \Gamma_{CCS} = 3 \cdot 30, \quad \Delta_{CCS} = -1 \cdot 20.$$

Corrections for 1, 4-steric interactions were made using the constant value of 0·6 kcal (Section 7.5), except those involving the SH group which were assumed to be zero. Excellent agreement between observed and estimated heats of atomization is evident.

TABLE 37

Estimated and observed heats of atomization of aliphatic thiols at $298 \cdot 15°K$, kcal.

	ΔH_a(obs.)	ΔH_a(est.)	ΔH_a(est.)$-\Delta H_a$(obs.)
Methanethiol	450·35	450·54	+0·19
Ethanethiol	731·05	731·28	+0·23
1-Propanethiol	1011·29	1011·30	+0·01
2-Propanethiol	1013·29	1013·40	+0·11
1-Butanethiol	1291·23	1291·32	+0·09
2-Butanethiol	1293·34	1293·42	−0·08
2-Methyl-l-propanethiol	1293·42	1293·35	−0·07
2-Methyl-2-propanethiol	1296·37	1296·35	−0·02
1-Pentanethiol	1571·59	1571·34	−0·25
3-Methyl-1-butanethiol	1572·76	1572·77	+0·01
2-Methyl-2-butanethiol	1575·65	1575·77	+0·12
1-Hexanethiol	1851·34	1851·36	+0·02
1-Heptanethiol	2131·28	2131·38	+0·10
1-Decanethiol	2971·50	2971·44	−0·26

Table 38 lists observed and estimated heats of atomization of aliphatic thioethers. The additional parameters needed for the calculations are given below.

(a) *The Group method*

$$2[C-(C)(H)_2(S)]+[S-(C)_2] = 615 \cdot 69,$$

$$2[C-(C)_2(H)(S)]+[S-(C)_2] = 505 \cdot 29,$$

$$2[C-(C)_3(S)]+[S-(C)_2] = 396 \cdot 55.$$

(b) *The Laidler method* $E(C-S) = 70 \cdot 06.$

(c) *The Allen method* $\Gamma_{CSC} = 2 \cdot 47.$

1, 4-steric correction terms across the $-S-$ atom have been ignored yet the agreement between observed and estimated heats of atomization is excellent in most instances, in line with McCullough and Good's contention (Section 7.6.4.) that steric interactions across the $-S-$ atom are negligible.

We have now applied the Group, Laidler and Allen schemes in unified form to series of compounds for which the thermochemical data are of high quality. The gratifying measure of agreement between estimated and observed heats of atomization gives great confidence in the use of these schemes to estimate unknown heats of atomization. The only caveat to this affirmation of confidence is that the problem of estimating steric correction terms in highly branched substituted alkanes is not yet solved, so it may sometimes

TABLE 38

Estimated and observed heats of atomization of aliphatic thioethers at 298·15°K, kcal.

	ΔH_a(obs.)	ΔH_a(est.)	ΔH_a(est.) − ΔH_a(obs.)
2-Thiapropane	728·94	728·85	−0·09
2-Thiabutane	1009·32	1009·59	+0·27
2-Thiapentane	1289·83	1289·61	−0·22
3-Thiapentane	1290·14	1290·33	+0·19
3-Methyl-2-thiabutane	1291·80	1291·71	−0·09
2-Thiahexane	1569·70	1569·63	−0·07
3-Thiahexane	1570·31	1570·35	+0·04
2-Methyl-3-thiapentane	1573·35	1572·45	−0·90
2, 2-Dimethyl-3-thiabutane	1574·26	1574·66	+0·40
2-Thiaheptane	1849·55	1849·65	+0·10
3-Thiaheptane	1850·75	1850·37	−0·38
4-Thiaheptane	1850·32	1850·37	+0·05
2, 2-Dimethyl-3-thiapentane	1855·75	1855·40	−0·35
2, 4-Dimethyl-3-thiapentane	1854·18	1854·57	+0·39
5-Thianonane	2410·57	2410·41	−0·16
2, 6-Dimethyl-4-thiaheptane	2413·45	2414·47	+1·02
2, 2, 4, 4-Tetramethyl-3-thiapentane	2415·25	2420·47	+5·22
6-Thiaundecane	2969·75	2970·45	+0·70
2, 8-Dimethyl-5-thianonane	2973·75	2974·51	+0·76

be necessary to resort to inspired guesswork in applying steric corrections to estimates of ΔH_a for such compounds.

7.7. Destabilization energies

In previous Sections we have described the destabilization that results from steric interaction between hydrogen atoms attached to 1, 4-carbon atoms, but destabilization can arise from other causes, and in this Section destabilization is discussed more fully. For example, the formation of some cyclic compounds is necessarily accompanied by changes in the type of bond disposition that occurs in open-chain structures; physical methods of structure analysis reveal relatively slight changes in bond lengths but in some instances significant changes in bond angles. This effect is therefore referred to as *angular strain*. Another consequence of ring formation is that steric interaction may occur to a greater extent than in open structures, if the ring shape is such that hydrogen atoms on, say, 1, 6-carbon atoms are brought to within 2 Å of each other. Another form of destabilization arises from the mutual electrostatic repulsion† of polar groups within a molecule, open-chain or cyclic. We shall discuss these effects separately, but first we consider steric repulsion in open-chain structures arising from interactions between hydrogen atoms attached to carbon atoms further apart than 1, 4.

7.7.1. 1, 5-STERIC REPULSION ENERGIES IN ALKANES

This type of steric repulsion occurs in 2, 2, 4-trimethylpentane. (Section 7.4.4.) If the observed value of ΔH_a is subtracted from that estimated by a four-parameter scheme, a total strain energy of 3·79 kcal mol^{-1} is found. There are three 1, 4-steric repulsions in the structure and if 0·6 kcal is allowed for each (Section 7.5), 2 kcal mol^{-1} must be ascribed to 1, 5-interaction between hydrogen atoms on the t-butyl and isopropyl groups respectively. In 2, 2, 4, 4-tetramethylpentane, with two t-butyl groups, the possibility of 1, 5-interaction is increased and not surprisingly the total destabilization energy [S] is considerable; the value is shown below alongside the values for the corresponding ether and thioether.

```
    C     C              C     C              C     C
    |     |              |     |              |     |
C − C − C − C − C    C − C − O − C − C    C − C − S − C − C
    |     |              |     |              |     |
    C     C              C     C              C     C
```

[S] = 8·0 kcal mol^{-1} [S] = 10·8 kcal mol^{-1} [S] = 5·2 kcal mol^{-1}

---

† The net potential energy of a system of mutually interacting dipoles can be positive or negative depending on the vector directions of the individual dipoles, but in the instances studied so far the net effect has been positive, corresponding to destabilization.

It is seen that the smallest value of [S] is that for the thioether, probably because the relatively long $C-S$ bonds permit the interfering t-butyl groups to be further apart than in the corresponding alkane and ether. Models of the latter's molecular structure show that the normal COC angle of *ca.* 110° must be opened to almost 180° to permit this structure to exist at all. Clearly overcrowding in structures of this type results in such a large distortion of the molecular framework that *a priori* estimation of the steric repulsion energy is difficult. Values of [S] obtained by subtraction of the observed ΔH_a from the value estimated with no allowance for steric effects are listed below for some highly branched alkanes.

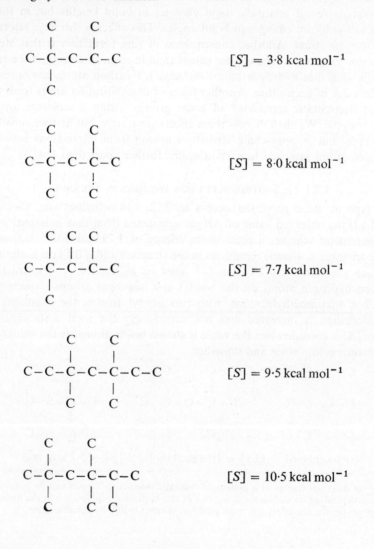

$$[S] = 3 \cdot 8 \text{ kcal mol}^{-1}$$

$$[S] = 8 \cdot 0 \text{ kcal mol}^{-1}$$

$$[S] = 7 \cdot 7 \text{ kcal mol}^{-1}$$

$$[S] = 9 \cdot 5 \text{ kcal mol}^{-1}$$

$$[S] = 10 \cdot 5 \text{ kcal mol}^{-1}$$

```
          C       C
          |       |
    C - C - C - C - C - C - C          [S] = 13·4 kcal mol⁻¹
          |       |
          C       C
          |       |
          C       C
```

$$[S] = 13\cdot4 \text{ kcal mol}^{-1}$$

7.7.2. RING STRAIN IN UNSUBSTITUTED CYCLIC COMPOUNDS

It is customary to derive the ring-strain energy of a cyclic compound from the relation

$$\text{Ring-strain energy} = [\Delta H_a \text{ (estimated)} - \Delta H_a \text{ (observed)}] \text{ (at } 298\cdot15°\text{K)}, \tag{187}$$

where ΔH_a (estimated) is derived by application of a bond energy scheme. Nelander and Sunner (66/54) have pointed out that the bond energy terms commonly used in such a calculation differ from those appropriate to a cyclic compound in respect of pressure–volume energy. They therefore suggest that calculation of ring-strain energy should be in terms of ΔU rather than ΔH, and that allowances should also be made for differences in zero-point and thermal energies between the reference compounds and ring compounds. Nelander and Sunner's suggestions are certainly important in attempts at *a priori* estimation of the strain energy of a cyclic structure. In

TABLE 39. Conventional ring-strain energies (C.R.S.E.) in cycloalkanes, kcal mol⁻¹.

	ΔH_a(obs.)	ΔH_a(est.)	C.R.S.E.
C_3H_6	812·57	840·06	27·5
C_4H_8	1093·62	1120·08	26·5
C_5H_{10}	1393·94	1400·10	6·2
C_6H_{12}	1680·10	1680·12	0·0
C_7H_{14}	1953·91	1960·14	6·2
C_8H_{16}	2230·53	2240·16	9·6
C_9H_{18}	2507·63	2520·18	12·6
$C_{10}H_{20}$	2787·88	2800·20	12·3
$C_{11}H_{22}$	3068·97	3080·22	11·3
$C_{12}H_{24}$	3356·23	3360·24	4·0
$C_{13}H_{26}$	3635·18	3640·26	5·1
$C_{14}H_{28}$	3908·53	3920·28	11·8
$C_{15}H_{30}$	4198·54	4200·30	1·8
$C_{16}H_{32}$	4478·48	4480·32	1·8
$C_{17}H_{34}$	4763·77	4760·34	−3·4

the following discussion, however, the emphasis is on the intercomparison of ring-strain energies for many structures, and for this purpose the convenience of the conventional approach is the paramount factor. We shall call the quantity given by equation (187) the *conventional ring-strain energy*; values for some cycloalkanes derived from the data of Chapter 5 are listed in Table 39. ΔH_a (estimated) was calculated using a four-parameter scheme, the Group method being particularly convenient with $[C-(C)_2(H)_2]$ = 280·02 kcal.

The large values of C.R.S.E. in cyclopropane and cyclobutane (Table 39) are presumably largely angular in origin, since considerable distortion of the CCC angles from the tetrahedral angle is inevitable in these small rings. The structures of these compounds have been described in terms of bent or "banana" bonds (49/37). It has also been suggested that the hybridization of the orbitals differs significantly from that in alkanes (63/61); if this is so, the increased s-character of the $C-H$ bonds ought to be taken into account in any attempt to calculate realistic strain energies.

For the remaining cycloalkanes strain energies are less likely to arise from distortion of valency angles; forced adoption of partially eclipsed configurations by neighbouring $-CH_2-$ groups (*torsional strain*) and in some instances the close approach of hydrogen atoms across the ring are likely to be more significant sources of strain. The net C.R.S.E. is *ca.* 6 kcal mol^{-1} in the C_5 and C_7 rings, and zero in the C_6 ring.

The conformations of the C_9, C_{10} and C_{12} cycloalkane rings in the crystalline state have been determined by X-ray diffraction (60/53, 61/61, 62/72). In the C_9 and C_{10} rings the CCC bond angles are 8° wider than the tetrahedral angle. The C_{10} ring must have six strong H $\cdots$ H repulsions due to the approach of hydrogen atoms across the ring to within 1·8 Å, and the C_9 ring has four such repulsions together with two eclipsed $-CH_2-$ groups. In the C_{12} ring the distortion of the CCC angle is only 3°, and all $-CH_2-$ groups are in the staggered configuration; there are eight H $\cdots$ H repulsions across the ring but because of the ring geometry these interactions must be weaker than those in the C_{10} compound. The relative magnitudes of the C.R.S.E.'s for the C_9–C_{12} cycloalkanes in the gas state are seen to be in line with the crystal structure data. For rings larger than C_{13} the C.R.S.E. is seen to be quite small, except for the C_{14} compound; detailed discussion is unwarranted because information on the structures is lacking and for these compounds the experimental errors attaching to ΔH_a are significant in relation to the size of the C.R.S.E.'s.

Listed in Table 40 are some C.R.S.E. values for cyclic compounds containing more than one ring; ΔH_a (estimated) was calculated using the Group method, with the parameters given in Table 49. For comparison, values of the sum of the C.R.S.E.'s for the separate rings are given in a column headed $\sum$(C.R.S.E.)$_{rings}$. As is seen, the values of C.R.S.E. are roughly approximated by those for $\sum$(C.R.S.E.)$_{rings}$; this observation provides a basis for an approximate estimate of the heat of formation of a multi-ring compound.

TABLE 40

Conventional ring-strain energies (C.R.S.E.) in some multi-ring alkanes, kcal mol^{-1}.

Carbon skeleton	ΔH_a(obs.)	ΔH_a(est.)	C.R.S.E.	$\sum$(C.R.S.E.)$_{rings}$
	944·3	1009·5	65·2	55·0
	1227·05	1291·04	64·0	55·0
	1635·3	1798·0	162·7	159·0
	1537·31	1569·58	32·3	33·7
	1821·16	1849·60	28·4	27·5
	2100·45	2129·62	29·2	33·7
	2102·99	2129·62	26·6	26·5
	2379·09	2409·64	30·6	37·1

Table 41 lists conventional ring-strain energies of three series of saturated heterocyclic compounds with (for comparison) the values for the corresponding carbocyclic compounds. Values of ΔH_a (estimated) used in calculating the C.R.S.E.'s were derived by application of the Group method, with parameters from Table 49.

TABLE 41

Conventional ring-strain energies (C.R.S.E.) in some saturated monocyclic compounds, kcal mol^{-1}.

No. of atoms in ring, n	C.R.S.E.			
	$(CH_2)_{n-1}CH_2$	$(CH_2)_{n-1}O$	$(CH_2)_{n-1}NH$	$(CH_2)_{n-1}S$
3	27·5	27·2	27·1	19·9
4	26·5	25·4	—	19·7
5	6·2	5·6	5·9	2·0
6	0·0	1·1	−0·2	−0·2

It is evident from Table 41 that the magnitudes of the C.R.S.E.'s for a given ring size differ little as between cycloalkanes, cyclic ethers and cyclic imines, though the values for the cyclic thioethers are significantly smaller. This observation is consistent with the fact that the C−S bond is longer than the C−C, C−O and C−N bonds and angular strain in the S-compounds must be less.

7.7.3. RING STRAIN IN SUBSTITUTED CYCLIC COMPOUNDS

In cyclic compounds containing alkyl substituents, destabilization can arise from the ring *per se*, from interactions of the substituents with hydrogen atoms attached to the ring, and from mutual interactions between two or more substituents. For example, in the dimethylcyclohexanes steric interactions occur between the methyl groups and ring hydrogen atoms in the axial positions. Such repulsions were considered by Skinner (62/70) and Table 42 shows the application of his arguments to the dimethylcyclohexanes. [S], the steric repulsion energy calculated by Skinner, was included in ΔH_a (estimated); the destabilization of the cyclohexane ring itself was assumed to be zero (See Table 39).

TABLE 42

Steric repulsion energies and heats of atomization in dimethylcyclohexanes, kcal.

	[S]	ΔH_a(obs.)	ΔH_a(est.)	ΔH_a(est.) − ΔH_a(obs.)
1, 1-Dimethylcyclohexane	1·24	2244·03	2244·46	+0·43
cis-1, 2-Dimethylcyclohexane	2·38	2241·93	2241·84	−0·09
trans-1, 2-Dimethylcyclohexane	0·45	2243·79	2243·77	−0·02
cis-1, 3-Dimethylcyclohexane	0·00	2244·93	2244·22	−0·71
trans-1, 3-Dimethylcyclohexane	1·16	2242·98	2243·06	+0·08
cis-1, 4-Dimethylcyclohexane	1·16	2243·00	2243·06	+0·06
trans-1, 4-Dimethylcyclohexane	0·00	2244·90	2244·22	−0·68

There is reasonable agreement between calculated and observed heats of atomization, indicating that the strain energies have been successfully estimated.

7.7.4. DESTABILIZATION DUE TO ELECTROSTATIC REPULSION

As explained in Section 7.7, electrostatic interaction between polar substituents can have either a stabilizing or a destabilizing effect, i.e. the effect can appear in either the second or the third term on the right-hand side of equation (159). In the experimental studies of this phenomenon to date, dipole–dipole repulsion has predominated over dipole–dipole attraction.

In previous Sections we have emphasised the short-range nature of steric repulsion forces ($\propto 1/r^{12}$). In contradistinction, dipole–dipole interactions operate over relatively long ranges since according to the classic equation (188), the potential energy ε_{ij} of a pair of dipoles, r_{ij} apart, varies as $1/r_{ij}{}^3$:

$$\varepsilon_{ij} = \frac{\mu_1 \mu_2}{r_{ij}{}^3} \, (3 \cos \theta_i \cos \theta_j - \cos \theta_{ij}). \tag{188}$$

The dipoles, of moments μ_1 and μ_2, are inclined at angles θ_i and θ_j to the line of centres, the angle between them being θ_{ij}. Boyd (67/45) has used equation (188) for calculations on polycyano-compounds; the results are complicated by the existence of conjugation between cyano-groups and ethylenic bonds in some of the compounds studied, but it is evident that dipole–dipole interactions can appreciably influence estimates of ΔH_a in compounds containing more than one polar group on the same atom or on neighbouring atoms. Probably equation (188) is too simple for accurate application to polar compounds other than those having complete charge separation (e.g. zwitterions); moreover a complete treatment of electrostatic effects would have to take electrostatic induction and the possible existence of multipoles into account. It seems likely that electrostatic effects could be absorbed into four-parameter bond energy schemes for groups of molecules of similar geometry and similar type of substitution, such as the perfluoroalkanes. Lacher and Skinner (68/14) have indeed proposed such a scheme, but too few experimental data exist at present to justify further discussion here.

7.8. Bond energy schemes for unsaturated compounds

Discussion of bond energies has so far been restricted to structures having sigma bonds only. In this and subsequent Sections we widen the discussion to include structures in which pi electrons participate in the bonding. For

mono-olefines the approach is an extension of the schemes already described, but for structures in which conjugation of pi electrons can occur,
new problems have to be solved.

7.8.1. BOND ENERGY SCHEMES FOR MONO-OLEFINES

The Group, Laidler and Allen schemes can be applied to mono-olefines in
an equivalent fashion, as discussed in the following paragraphs.

(a) *The Group method*

With the Benson and Buss notation (58/8), one can represent the olefinic
part of the molecule by the following seven groups, where C_d represents
a doubly-bonded carbon atom: $[C_d-(C_d)(H)_2]$, $[C_d-(C_d)(H)(C)]$,
$[C_d-(C_d)(C)_2]$, $[C-(C_d)(C)_3]$, $[C-(C_d)(C)_2(H)]$, $[C-(C_d)(C)(H)_2]$,
$[C-(C_d)(H)_3]$. In conformity with Benson and Buss's treatment we
choose to work with the six composite groups shown below; the numerical
values given (kcal) are the contributions of these groups to the heat
of atomization and were chosen to give the best fit between observed
and estimated heats of atomization (Table 43).

$$[C_d-(C_d)(H)_2] \qquad\qquad\qquad = \quad 268\cdot88$$

$$[C_d-(C_d)(C)(H)] + [C-(C_d)(H)_3] = \quad 551\cdot67$$

$$[C_d-(C_d)(C)_2] \quad +2[C-(C_d)(H)_3] = \quad 835\cdot79$$

$$[C-(C_d)(C)(H)_2] - [C-(C_d)(H)_3] = \quad -57\cdot30$$

$$[C-(C_d)(C)_2(H)] - [C-(C_d)(H)_3] = \quad -112\cdot57$$

$$[C-(C_d)(C)_3] \qquad - [C-(C_d)(H)_3] = \quad -166\cdot36$$

The first three values above are concerned with the double bond while
the remaining values are concerned with the attached alkyl groups, and
are taken equal to the corresponding terms in alkanes.

(b) *The Laidler-type scheme*

Several methods of applying a Laidler-type bond energy scheme to mono-
olefines have been proposed. There is no unambiguous way of deciding
which terms to define as independent parameters, but in keeping with the
intention of the original Laidler scheme we will suppose the C_d-H bond
energy term to depend on whether the C_d-H bond is in the group

$$=C\big<\begin{smallmatrix}H\\H\end{smallmatrix} \quad \text{or in the group} \quad =C\big<\begin{smallmatrix}H\\C\end{smallmatrix}.$$ We will symbolise these bond energy

terms as $E(C_d-H)_2$ and $E(C_d-H)_1$ respectively, the subscript indicating the number of hydrogen atoms attached to the carbon atom of the double bond. For parameter values (kcal) we select $E(C=C) = 133\cdot00$ and then derive from the equivalence relations below, together with the parameters given in Section 7.3.2: $E(C_d-H)_2 = 101\cdot19$, $E(C_d-H)_1 = 100\cdot53$, $E(C_d-C) = 90\cdot07$.

Group Parameters	*Laidler Parameters*
$[C_d-(C_d)(H)_2]$	$\equiv 2E(C_d-H)_2+\frac{1}{2}E(C=C)$
$[C_d-(C_d)(C)(H)]+[C-(C_d)(H)_3]$	$\equiv E(C_d-H)_1+E(C_d-C)$
	$\quad+\frac{1}{2}E(C=C)+3E(C-H)_p$
$[C_d-(C_d)(C)_2]+2[C-(C_d)(H)_3]$	$\equiv 2E(C_d-C)+\frac{1}{2}E(C=C)$
	$\quad+6E(C-H)_p$
$[C-(C_d)(C)(H)_2]-[C-(C_d)(H)_3]$	$\equiv \frac{1}{2}E(C-C)+2E(C-H)_s$
	$\quad-3E(C-H)_p$
$[C-(C_d)(C)_2(H)]-[C-(C_d)(H)_3]$	$\equiv E(C-C)+E(C-H)_t$
	$\quad-3E(C-H)_p$
$[C-(C_d)(C)_3]-[C-(C_d)(H)_3]$	$\equiv \frac{3}{2}E(C-C)-3E(C-H)_p$

(c) *The Allen scheme*

The Allen scheme can be applied to mono-olefines using the parameters (kcal): $B(C=C) = 140\cdot55$, $\Gamma_{CC_dC_d} = 5\cdot35$, $\Delta_{CCC_d} = -1\cdot25$, where $\Gamma_{CC_dC_d}$ is the interaction energy associated with $C\dot{-}\ddot{C}\dot{=}C$ and Δ_{CCC_d} is that

associated with the trio $\begin{matrix} C_{\diagdown} \\ | \diagdown C\equiv\!\!\!\!\cdot C \\ C_{\diagup} \end{matrix}$. These parameter values are compat-

ible with the equivalence relations given below, together with those given in Section 7.3.3.†

Group Parameters	*Allen Parameters*
$[C_d-(C_d)(H)_2]$	$\frac{1}{2}B(C=C)+2B(C-H)$
$[C_d-(C_d)(C)(H)]+[C-(C_d)(H)_3]$	$\frac{1}{2}B(C=C)+B(C-C)+4B(C-H)$
	$\quad+\Gamma_{CC_dC_d}$
$[C_d-(C_d)(C)_2]+2[C-(C_d)(H)_3]$	$\frac{1}{2}B(C=C)+2B(C-C)+6B(C-H)$
	$\quad+2\Gamma_{CC_dC_d}+\Gamma_{CCC}+\Delta_{CCC_d}$
$[C-(C_d)(C)(H)_2]-[C-(C_d)(H)_3]$	$\frac{1}{2}B(C-C)-B(C-H)+\Gamma_{CCC}$
$[C-(C_d)(C)_2(H)]-[C-(C_d)(H)_3]$	$B(C-C)-2B(C-H)+3\Gamma_{CCC}$
	$\quad+\Delta_{CCC}$
$[C-(C_d)(C)_3]-[C-(C_d)(H)_3]$	$\frac{3}{2}B(C-C)-3B(C-H)+6\Gamma_{CCC}$
	$\quad+4\Delta_{CCC}$

---

† Γ_{CCC_d}, the interaction energy associated with $C\dot{-}\overset{\frown}{C}\dot{-}C=$, is taken equal to Γ_{CCC}.

7.8.2. STERIC CORRECTION TERMS IN MONO-OLEFINES

Steric interaction in mono-olefines has been discussed by Skinner (62/70). Additional to the steric terms already encountered in alkanes, there will be repulsion terms due to alkyl groups attached *cis* to the double bond. These repulsions can be classified according to the number of alkyl groups attached to each carbon atom, e.g.

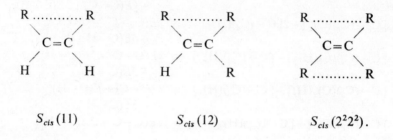

$$S_{cis}(11) \qquad\qquad S_{cis}(12) \qquad\qquad S_{cis}(2^22^2).$$

If some relief of these repulsions can occur by angle widening, such relief would be more effective in $S_{cis}(11)$ than in the more crowded situations. Skinner proposed the following parameter values, (kcal mol^{-1})]: $[S_{cis}(11)] = 1\cdot0$, $[S_{cis}(12)] = 1\cdot4$, $[S_{cis}(12^2)] = 1\cdot6$, $[S_{cis}(2^22^2)] = 2\cdot2$. When one or more of the interacting *cis* groups is a t-butyl group, steric repulsion becomes much more severe, due to the appearance of 1, 5-repulsive interactions (Cf. Section 7.7.1). From data for specific cases where this type of repulsion occurs, the following values (kcal mol^{-1}) may be deduced: $[S_{cis}(11, tBu-R)] = 6\cdot0$, $[S_{cis}(12, tBu-R] = 7\cdot2$, $[S_{cis}(11, tBu-tBu)] = 14\cdot5$.

In addition, there will be some *gauche* 1, 4-steric repulsions in structures in which hydrogen atoms attached to C_d can interact with an alkyl group. Such repulsions will have slightly lower values than in the corresponding alkanes, because the angle of the trigonal carbon atom (120°) is wider than the tetrahedral angle.

Table 43 gives the observed and estimated heats of atomization of the olefines listed in the Tables of Chapter 5. The column headed [S] records the total steric terms as given by Skinner (62/70), and these terms have been included in ΔH_a (estimated). It is seen that the general measure of agreement is good, although the deviations for some of the highly branched olefines show that Skinner's method underestimates steric interaction in these compounds.

TABLE 43. Observed and estimated heats of atomization of mono-olefines at $298 \cdot 15°K$, kcal

	ΔH_a(obs.)	[S]	ΔH_a(est.)	ΔH_a(est.) $-\Delta H_a$(obs.)
Ethylene	537·75	—	537·76	+0·01
Propylene	820·42	—	820·55	+0·13
1-Butene	1100·60	—	1100·57	−0·03
cis-2-Butene	1102·26	1·00	1102·34	+0·08
trans-2-Butene	1103·39	—	1103·34	−0·05
2-Methylpropene	1104·66	—	1104·67	+0·01
1-Pentene	1380·83	—	1380·59	−0·24
cis-2-Pentene	1382·50	1·00	1382·36	−0·14
trans-2-Pentene	1383·43	—	1383·36	−0·07
2-Methyl-1-butene	1384·05	0·23	1384·46	+0·41
3-Methyl-1-butene	1382·11	—	1382·62	+0·51
2-Methyl-2-butene	1385·62	1·40	1386·06	+0·44
1-Hexene	1660·55	—	1660·61	+0·06
cis-2-Hexene	1663·11	1·00	1662·38	−0·73
trans-2-Hexene	1663·48	—	1663·38	−0·10
cis-3-Hexene	1661·98	1·00	1662·38	+0·40
trans-3-Hexene	1663·61	—	1663·38	−0·23
2-Methyl-1-pentene	1664·79	0·23	1664·48	−0·31
3-Methyl-1-pentene	1662•43	—	1662·64	+0·21
4-Methyl-1-pentene	1662•85	—	1662·64	−0·21
2-Methyl-2-pentene	1666·58	1·40	1666·08	−0·50
3-Methyl-cis-2-pentene	1665·49	1·98	1665·50	+0·01
3-Methyl-trans-2-pentene	1665·69	1·98	1665·50	−0·19
4-Methyl-cis-2-pentene	1664·34	1·00	1664·41	+0·07
4-Methyl-trans-2-pentene	1665·30	—	1665·41	+0·11
2-Ethyl-1-butene	1663·99	0·46	1664·25	+0·26
2, 3-Dimethyl-1-butene	1665·79	1·10	1665·54	−0·25
3, 3-Dimethyl-1-butene	1665·11	0·94	1665·21	+0·10
2, 3-Dimethyl-2-butene	1667·02	4·40	1667·18	+0·16
1-Heptene	1940·51	—	1940·63	+0·12
5-Methyl-1-hexene	1941·40	0·33	1942·33	+0·93
3-Methyl-cis-3-hexene	1944·68	1·98	1945·52	+0·94
3-Methyl-trans-3-hexene	1944·06	1·98	1945·52	+1·46
2, 4-Dimethyl-1-pentene	1945·73	0·23	1946·53	+0·80
4, 4-Dimethyl-1-pentene	1945·14	0·27	1945·90	+0·76
2, 4-Dimethyl-2-pentene	1946·90	1·40	1948·13	+1·23
4, 4-Dimethyl-cis-2-pentene	1943·06	6·0	1942·94	−0·12
4,4-Dimethyl-trans-2-pentene	1946·92	1·64	1947·30	+0·38
3-Methyl-2-ethyl-1-butene	1944·71	2·56	1944·20	−0·51
2, 3, 3-Trimethyl-1-butene	1946·13	3·70	1946·57	+0·44
1-Octene	2220·21	—	2220·65	+0·44
2, 2-Dimethyl-cis-3-hexene	2222·14	6·0	2223·06	+0·82
2, 2-Dimethyl-trans-3-hexene	2226·53	1·64	2227·32	+0·73
2-Methyl-3-ethyl-1-pentene	2224·77	1·10	2225·70	+0·93
2, 4, 4-Trimethyl-2-pentene	2225·87	7·2	2225·86	+0·01
1-Decene	2780·48	—	2780·69	+0·21
1-Hexadecene	4460·69	—	4460·81	+0·12

7.8.3. RING STRAIN IN CYCLIC OLEFINES

Conventional ring-strain energies (Section 7.7.2) of unconjugated cyclic olefines can be derived from equation (187) if ΔH_a (estimated) is obtained by use of the parameters recommended in Section 7.8.1. The resulting values for some unsubstituted cycloalkenes are given in Table 44.

TABLE 44. Conventional ring strain energies (C.R.S.E.) in cycloalkenes, kcal mol^{-1}.

	ΔH_a(obs.)	ΔH_a(est.)	C.R.S.E.
C_3H_4	654·9	708·71	53·8
C_4H_6	958·75	988·73	30·0
C_5H_8	1263·07	1268·75	5·7
C_6H_{10}	1547·48	1548·77	1·3
C_7H_{12}	1823·69	1828·79	5·1
C_8H_{14}	2103·05	2108·81	5·8

Comparison of the C.R.S.E. values for cycloalkenes with those for the corresponding cycloalkanes (Table 39) shows that

$$\text{for } C_3, \qquad \text{C.R.S.E. (ene)} \gg \text{C.R.S.E. (ane),}$$
$$\text{for } C_4, \qquad \text{C.R.S.E. (ene)} > \text{C.R.S.E. (ane),}$$
$$\text{for } C_5, \qquad \text{C.R.S.E. (ene)} \approx \text{C.R.S.E. (ane),}$$
$$\text{for } C_6, \qquad \text{C.R.S.E. (ene)} > \text{C.R.S.E. (ane),}$$
$$\text{for } C_7, C_8, \qquad \text{C.R.S.E. (ene)} < \text{C.R.S.E. (ane).}$$

The greatly enhanced C.R.S.E. of the C_3-ene must be due to gross distortion of the inter-orbital angles of the two sp^2-hybridized carbon atoms. With the larger rings a more subtle interplay of factors is involved. Departures from the normal bond angles of the sp^2-carbon atoms ($120°$) may give rise to more or less angular strain than in rings containing sp^3-carbon atoms, depending on ring size; also, the differing conformations of -ane and -ene rings of a given size, and the differing numbers of hydrogen atoms in the respective structures give rise to differing torsional strains. Some alternation in the net result is apparent from Table 44. A discussion of the conformations available to cyclic compounds is outside the scope of this book; readers are referred to the monograph by Eliel et al. (65/57).

7.8.4. STABILIZATION IN CONJUGATED OLEFINES

In the estimation of ΔH_a for 1, 3-butadiene, $CH_2{=}CH{-}CH{=}CH_2$, by a bond energy scheme, the problem arises as to what value should be ascribed to the bond energy term of the central $C_d{-}C_d$ bond. One possibility would be to assume that in applying the Laidler-type scheme, for example, $E(C_d{-}C_d)$ is equal to the value for $E(C{-}C_d)$; hence with this assumption

$$\Delta H_a \text{ (estimated)} = 2E(C{=}C) + 4E(C_d{-}H)_2 + 2E(C_d{-}H)_1 + E(C_d{-}C).$$

However, application of the Allen scheme to 1, 3-butadiene would lead to

$$\Delta H_a \text{ (estimated)} = 2B(C{=}C) + 6B(C{-}H) + B(C{-}C) + 2\Gamma_{C_dC_dC_d}.$$

The parameter $\Gamma_{C_dC_dC_d}$ corresponds to the interaction $=C\overset{\frown}{-}C \doteq C$, where the three carbon atoms are trigonally hybridized; a value for this parameter cannot be obtained, so that an assumed value, say that of $\Gamma_{CC_dC_d}$, would have to be used. The numerical results of calculations on 1, 3-butadiene by the two methods would be expected to differ since the assumption made in the Laidler calculation is unrelated to that made in the Allen calculation. To illustrate this point, estimated values of ΔH_a, obtained by both methods, are compared in Table 45 with the experimental values for two conjugated 1, 3-dienes and one unconjugated 1, 4-diene.

TABLE 45. Heats of atomization of some dienes, kcal.

		Laidler Method		Allen Method	
	ΔH_a(obs.)	ΔH_a(est.)	[ΔH_a(est.)− ΔH_a(obs.)]	ΔH_a(est.)	[ΔH_a(est.)− ΔH_a(obs.)]
1, 3-Butadiene	970·09	961·89	−8·20	966·44	−3·65
trans-1, 3-Pentadiene	1253·18	1244·68	−8·50	1249·23	−3·95
1, 4-Pentadiene	1246·05	1246·48	+0·43	1246·46	+0·41

It is seen that the quantities [ΔH_a (est.)− ΔH_a (obs.)] are small for 1, 4, pentadiene but are quite large and negative for both conjugated dienes-implying the presence of a stabilization effect [equation (159)]; this stabilization is usually ascribed to the delocalisation of the π-electrons. Clearly, the magnitude of the stabilization energy calculated in this way depends crucially on the values assumed for the parameters $E(C_d-C_d)$ or $\Gamma_{C_dC_dC_d}$; such arbitrariness is unfortunate but the difficulty in ascribing unambiguous values to these parameters is insurmountable by thermochemical arguments alone.

There have been several attempts to break this impasse and one which merits particular consideration is that by Dewar and Schmeising (59/41, 60/54). They postulated a relation between a bond energy term E_i and the bond length r_i:

$$r_i = (1/b_i)\,[a_i \log(a_i + (a_i^2 - E_i^2)^{\frac{1}{2}}) - a_i \log E_i - (a_i^2 - E_i^2)^{\frac{1}{2}}]. \qquad (189)$$

Values of the parameters a_i and b_i appropriate for C−C and C−H bonds were determined from experimental data for diamond, CH_4, C_2H_6, C_2H_4 and C_2H_2. Dewar and Schmeising then derived bond energy term values for C−H and C−C bonds involving the various hybridisation states of carbon, $E(C_{sp3}-C_{sp3})$, $E(C_{sp3}-C_{sp2})$, $E(C_{sp2}-C_{sp2})$ etc.; the latter term corresponds to that designated $E(C_d-C_d)$ above. It should be noted that there is no theoretical basis for a bond-energy–bond-length relationship.

The Dewar–Schmeising scheme is therefore as empirical as the schemes described in earlier Sections and like them must be judged by its reasonableness and applicability to diverse compounds.

Numerical values of bond energies deduced from equation (189) are extraordinarily sensitive to the bond lengths taken. For example, very precise structural data on benzene have been obtained by Langseth and Stoicheff (56/48) [$r_o(C-C) = 1.397 \pm 0.001$ Å and $r_o(C-H) = 1.084 \pm 0.005$ Å], and if these bond lengths, together with their experimental errors, are substituted into equation (189), the uncertainty in ΔH_a (estimated) for benzene is no less than ± 10.8 kcal. Because of the sensitivity of their equation to errors in r_i, Dewar and Schmeising recommended use of mean values of E_i for various types of $C-H$ and $C-C$ bonds, in preference to values specific to particular molecular structures.

Dewar and Schmeising pointed out that stabilization energies as derived thermochemically (see the discussion of the data of Table 45) are composite quantities, to which π-electron delocalisation is only one contributor, and that energy effects due to hybridization changes are important. Consider the redistribution reaction

$$CH_2 = CH - CH = CH_2(g) + C_2H_6(g) = 2CH_3.CH = CH_2(g),$$

for which $\Delta H_r^\circ = +3.89 \pm 0.38$ kcal†. Let us now apply the Dewar and Schmeising approach to this reaction, labelling the sp^2-carbon atoms C*:

$$\Delta H_a \, (1,3\text{-butadiene}) = 6E(C^* - H) + 2E(C^* = C^*) + E(C^* - C^*) + R_\pi,$$

$$\Delta H_a \, (\text{ethane}) \qquad = 6E(C-H)_p + E(C-C),$$

$$\Delta H_a \, (\text{propene}) \qquad = 3E(C-H)_p + 3E(C^* - H) + E(C^* = C^*) \\ \qquad\qquad + E(C^* - C),$$

where R_π represents the π-electron delocalisation energy. Then because

$$\Delta H_r^\circ = \Delta H_a \, (1,3\text{-butadiene}) + \Delta H_a \, (\text{ethane}) - 2\Delta H_a \, (\text{propene}),$$

it follows that

$$R_\pi = 3.89 + [2E(C^* - C) - E(C-C) - E(C^* - C^*)]. \tag{190}$$

Hence derivation of R_π requires evaluation of the term in square brackets in equation (190). Dewar and Schmeising, using equation (189) and taking

† Dewar and Schmeising applied their scheme to 0°K, but as we have seen in Section 7.1.2, use of data at 298·15°K introduces no inconsistencies; in fact for this redistribution reaction, $\Delta H_r^\circ,\, 0^\circ K = +3.48$ kcal.

$r(C-C) = 1.5445$ Å, $r(C^*-C) = 1.517$ Å and $r(C^*-C^*) = 1.479$ Å, calculated that the term in brackets is sufficiently negative to make $R_\pi \approx 0$. Because of its sensitivity to error in r_t, this result must be treated with caution and indeed Bernstein (62/69) from different bond-energy–bond-length relations found $R_\pi \approx 9$ kcal mol^{-1}.

The evaluation of R_π is so important a matter as to merit discussion from other standpoints. Thus, from inorganic thermochemistry we know that the energy of a bond $A-B$ is almost invariably larger than the mean of the energies of bonds $A-A$ and $B-B$, the difference becoming greater as the difference in electronegativity of A and B increases. On this basis we would expect the term in square brackets in equation (190) to be positive, especially as a significant difference in the electronegativities of trigonal and tetrahedral carbon atoms has been claimed (62/66). We would thus expect R_π to be greater than 3.9 kcal mol^{-1}. By a totally different argument, based on the heats of isomerization of bicyclic olefines and with certain assumptions concerning hybridization states, Staley (67/14) deduced that $R_\pi > 1.9$ kcal mol^{-1}. With such a range of estimates it is obvious that the problem of evaluating R_π is not solved. The only supportable contention at the present time is that $R_\pi = 3.9 \pm x$ kcal mol^{-1}, where probably $x \not> 2$ kcal mol^{-1}. The case that $x = 0$ (i.e. that the term in square brackets in equation (190) is zero) corresponds to the assumption made in many calculations in the literature.

A more extensive discussion of stabilization energies in conjugated systems has been given by Mortimer in Chapter 3 of "Reaction Heats and Bond Strengths" (62/68).

7.8.5. STABILIZATION IN AROMATIC COMPOUNDS

The general phenomenon of π-electron delocalisation in benzene and aromatic compounds is well known. In discussions of the associated energy effect there are two fundamental problems. Firstly, if one describes a benzenoid compound as possessing stabilization energy resulting from delocalisation of π-electrons, it is necessary to decide in respect of what structure the compound is stabilized. Secondly, it is necessary to make any estimates of stabilization energies compatible with theoretical and experimental studies of the electronic structures of the molecules. Additional to the fundamental problems there is a semantic problem, since different authors have used terms such as "resonance energy", "delocalisation energy", "hybridization energy", "conjugation energy", "stabilization energy" with sometimes differing, sometimes equivalent, definitions.

It is appropriate to begin the present discussion by examining the thermochemistry of the simplest aromatic compound, benzene. From the data of

Chapter 5 and Table 24, $\Delta H_a = 1318 \cdot 19$ kcal. Now one can in principle compare the observed heat of atomization with that of a Kekulé structure

, but as with 1, 3-butadiene (Section 7.8.4) there is the problem of deciding the appropriate bond energy terms to use. From the same type of assumptions as were made in discussing 1, 3-butadiene, ΔH_a (estimated) for a Kekulé structure is calculated to be 1272·39 kcal by the Laidler scheme and 1286·07 kcal by the Allen scheme. We shall call the difference between the observed value of ΔH_a for a benzenoid compound and any estimated value for a Kekulé structure the *conventional stabilization energy*. Thus the conventional stabilization energy of benzene is 45·80 kcal mol^{-1} (Laidler) or 32·12 kcal mol^{-1} (Allen); a value as low as 22 kcal mol^{-1} is obtained from a scheme based on the assumption that $R_\pi(1, 3\text{-butadiene}) = 0$ (63/62). There is a basic defect in *all* these methods of estimating conventional stabilization energy, namely that all require values of ΔH_a (estimated) for a structure, cyclohexatriene, for which no model compound is available for experimental study. The compounding of this difficulty with that of evaluating R_π in 1, 3-butadiene (Section 7.8.4) leads inevitably to a state of disarray.

A different approach to the evaluation of the energy of π-electron delocalisation in benzene is provided by data on heats of hydrogenation:

(g) + H$_2$(g) $=$ (g), $\Delta H_r^\circ = -28 \cdot 42$ kcal

(g) + 3H$_2$(g) $=$ (g), $\Delta H_r^\circ = -49 \cdot 31$ kcal.

If benzene had a cyclohexatriene structure with localised π-electrons, one might suppose that its heat of hydrogenation would be three times that of cyclohexene. The difference, 35·95 kcal mol^{-1}, between the heat of hydrogenation so calculated and the experimental value is frequently described as the *resonance energy* of benzene. This quantity has the advantage over conventional stabilization energy that its evaluation is in terms of measured heats of reaction and is not based on the estimation of ΔH_a of a hypothetical structure.

The question of the relation between (i) resonance energy derived from heats of hydrogenation, (ii) conventional stabilization energy derived from heats of atomization, and (iii) true π-delocalisation energy is a complex one. It involves consideration of factors such as hybridization changes, ring-strain energies, energy factors involved in changing bond lengths etc. The whole problem is discussed more fully by Mortimer in Chapter 4 of "Reaction Heats and Bond Strengths" (62/68).

7.8.6. BOND ENERGY SCHEMES FOR ALKYLBENZENES

There are several possible methods for correlating the heats of atomization of benzene derivatives that avoid the vexed question of evaluating π-delocalisation energy. Alkylbenzenes can be treated by the Group method, using the following parameters (kcal):

$$[C_b - (C_b)_2(H)] \qquad\qquad\qquad = \quad 219 \cdot 70,$$

$$[C_b - (C_b)_2(C)] \quad + \; [C - (C_b)(H)_3] = \quad 502 \cdot 60,$$

$$[C - (C_b)(C)(H)_2] \; - \; [C - (C_b)(H)_3] = \quad - \; 57 \cdot 04,$$

$$[C - (C_b)(C)_2(H)] \; - \; [C - (C_b)(H)_3] = \; -113 \cdot 30,$$

$$[C - (C_b)(C)_3] \qquad - \; [C - (C_b)(H)_3] = \; -169 \cdot 30,$$

where C_b refers to a carbon atom in the benzene ring.

The Laidler method can also be employed if each alkylbenzene is regarded as a Kekulé structure and $45 \cdot 80 \, \text{kcal mol}^{-1}$ (Section 7.8.5) is added to the initial estimate of ΔH_a, to allow for the conventional stabilization energy; this method was adopted by Laidler (56/48). An alternative approach is to devise Laidler-type parameters that include the π-delocalisation energy, i.e. $\Sigma v_i \times \text{Parameter}_i = \Delta H_a$. Appropriate values (kcal) are

$$E(C_b - C_b) = 119 \cdot 17, \; E(C_b - H) = 100 \cdot 53, \; E(C_b - C) = 88 \cdot 91;$$

here $E(C_b - H)$ has been taken equal to $E(C_d - H)_1$. This approach was used by Lovering and Nor (62/12), by Mackle and O'Hare (61/39) and in effect by Cox (63/61). The Allen scheme could, if desired, be employed in a similar fashion.

It should be noted that the Laidler and Group schemes for alkyl-substituted benzenes are *not* equivalent, because the bond energy terms for the alkyl groups are taken from the alkane series in the Laidler method, but from the alkylbenzene series in the Group method.

Steric effects in di- and poly-substituted alkylbenzenes will be greatest when the alkyl groups are *ortho* to one another. Empirically we propose $0 \cdot 4$ kcal for each pair of *ortho* methyl groups and $1 \cdot 0$ kcal for each set of three buttressed *ortho* methyl groups (e.g. as in 1, 2, 3-trimethylbenzene). Observed values of ΔH_a for alkylbenzenes and values estimated with inclusion of these steric correction terms are compared in Table 46. The measure of agreement is seen to be quite good.

TABLE 46. Observed and estimated heats of atomization of alkybenzenes, kcal.

		Group method		Laidler method	
	ΔH_a(obs.)	ΔH_a(est.)	[ΔH_a(est.) − ΔH_a(obs.)]	ΔH_a(est.)	[ΔH_a(est.) − ΔH_a(obs.)]
Benzene	1318·19	1318·20	+0·01	1318·20	+0·01
Toluene	1601·11	1601·10	−0·01	1601·15	+0·05
Ethylbenzene	1881·05	1881·38	+0·33	1881·17	+0·12
o-Xylene	1883·64	1883·60	−0·04	1883·70	+0·06
m-Xylene	1884·06	1884·00	−0·06	1884·10	+0·04
p-Xylene	1883·89	1884·00	+0·11	1884·10	+0·21
n-Propylbenzene	2161·41	2161·40	−0·01	2161·19	−0·21
s-Propylbenzene	2162·34	2162·44	+0·10	2163·21	+0·87
1-Methyl-2-ethylbenzene	2162·91	2163·88	+0·97	2163·72	+0·81
1-Methyl-3-ethylbenzene	2163·73	2164·28	+0·55	2164·12	+0·39
1-Methyl-4-ethylbenzene	2164·06	2164·28	+0·22	2164·12	+0·06
1, 2, 3-Trimethylbenzene	2165·56	2165·90	+0·34	2166·05	+0·49
1, 2, 4-Trimethylbenzene	2166·61	2166·50	−0·11	2166·65	+0·04
1, 3, 5-Trimethylbenzene	2167·11	2166·90	−0·21	2167·05	−0·06
n-Butylbenzene	2441·68	2441·42	−0·26	2441·21	−0·47
Isobutylbenzene	2443·54	2443·45	−0·09	2443·23	−0·31
s-Butylbenzene	2442·55	2442·46	−0·09	2443·23	+0·68
t-Butylbenzene	2443·80	2443·76	−0·04	2446·73	+2·93
Hexamethylbenzene	3009·35	3009·60	+0·25	3009·90	+0·55

7.8.7. BOND ENERGY SCHEMES FOR SUBSTITUTED BENZENES

The heats of atomization of benzenoid compounds with other than alkyl substituents, $C_6 H_5 X$, $C_6 H_4 X_2$ etc., can be dealt with in an analogous way to the alkylbenzenes. The group method merely requires an extra term $[C_b − (C_b)_2(X)]$ and the Laidler scheme an additional bond energy term $E(C_b − X)$, derivable from data for mono-substituted benzenes. If, however, the compound in question contains two or more highly polar groups, it may be necessary to include an additional term in either scheme to allow for electrostatic interaction between the substituents. As stated in Section 7.7.4, it is common for the net electrostatic effect to be destabilizing and this is necessarily so for polyhalobenzenes, where the negative ends of the dipoles all lie outside the benzene ring. Cox, Gundry and Head (64/8) calculated the magnitude of the effect using equation (188); the results obtained are listed in Table 47. The interaction energies are seen to be quite large, especially for compounds containing *ortho* substituents. Later studies by these workers (69/1) on compounds in the series $C_6 F_5 X$ gave further evidence of the need to take account of electrostatic interactions in estimations of ΔH_a or ΔH_f°(g) for polyhalobenzenoid compounds.

TABLE 47

Calculated† dipole–dipole interaction energies for chloro-and fluoro-benzenes, kcal mol⁻¹

Compound	Interaction Energy (destabilizing)	
	for X = F	for X = Cl
1, 2-$C_6H_4X_2$	2·7	2·2
1, 3-$C_6H_4X_2$	0·7	0·6
1, 4-$C_6H_4X_2$	0·5	0·4
1, 2, 3-$C_6H_3X_3$	6·1	5·0
1, 2, 4-$C_6H_3X_3$	4·0	3·2
1, 3, 5-$C_6H_3X_3$	2·2	1·8
1, 2, 3, 4-$C_6H_2X_4$	10·1	8·2
1, 2, 3, 5-$C_6H_2X_4$	8·1	6·6
1, 2, 4, 5-$C_6H_2X_4$	7·9	6·4
C_6HX_5	15	12
C_6X_6	22	18

† The calculations are based on the assumption that the dipole is localized seven-eighths of the way along the bond, outwards from the carbon atoms of the ring. Subsequent work (69/1) showed that this assumption underestimates the destabilizing effect for X = F.

7.8.8. BOND ENERGY SCHEMES FOR CONDENSED AROMATIC HYDROCARBONS

Tatevskii, Korolov and Mendzheretskii (50/30) and McGinn (62/74) proposed to deal with aromatic hydrocarbons by a Laidler-type method, distinction being drawn between different types of $C-C$ aromatic bonds,

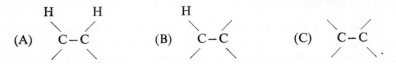

Thus for naphthalene,

$$\Delta H_a = 8E(C_b - H) + 6E(A) + 4E(B) + E(C).$$

We now propose the following parameter values for this Laidler-type scheme (kcal): $E(C_b - H) = 100\cdot53$, $E(A) = 119\cdot17$, $E(B) = 114\cdot30$, and $E(C) = 112\cdot80$. Observed and estimated heats of atomization for condensed-ring aromatic compounds are compared in Table 48. Agreement is generally good, in relation to the appreciable experimental uncertainties in $\Delta H_f^\circ(g)$ for these compounds (see Chapter 5).

Bernstein (62/69) pointed out that since the above scheme recognises three types of $C-C$ aromatic-ring bonds, one might anticipate three distinct $C-C$

bond lengths in condensed-ring aromatic compounds, whereas an almost continuous range of values from 1·36 to 1·48Å is observed. In a previous publication (61/62) Bernstein had related π-bond order to $C-C$ bond length, and later (62/69) he proposed a correlation between the bond energy term of an aromatic $C-C$ bond and the π-bond order of that and adjacent bonds. Bernstein's scheme is more complex than the simple Laidler-type scheme advocated here, yet is apparently less successful, at least when Bernstein's recommended parameters are employed (see Table 48).

TABLE 48. Observed and estimated heats of
atomization of condensed-ring aromatic hydrocarbons, kcal.

		Laidler method		Bernstein method	
	ΔH_a(obs.)	ΔH_a(est.)	$[\Delta H_a$(est.) $- \Delta H_a$(obs.)]	ΔH_a(est.)	$[\Delta H_a$(est.) $- \Delta H_a$(obs.)]
Benzene	1318·19	1318·20	0·0	1318·56	+0·4
Naphthalene	2089·75	2089·24	−0·5	2090·59	+0·8
Anthracene	2858·40	2860·28	+1·9	2859·77	+1·4
Phenanthrene	2864·10	2863·69	−0·4	2866·30	+2·2
Triphenylene	3639·5	3641·46	+2·0	3641·18	+1·7
Chrysene	3638·6	3638·12	−0·5	3640·2	+1·6
3, 4-Benzphenanthrene	3631·8	3638·12	+6·3	3639·0	+7·2
Tetracene	3631·6	3631·3	−0·3	3634·0	+2·4
Biphenyl	2528·27	2528·66	+0·4	2533·7	+5·4

7.9. Recommended values of parameters for the estimation of heats of atomization and heats of formation.

In this Section we offer selected values of parameters for the estimation of unknown heats of atomization or heats of formation, by the Group, the Laidler or the Allen method. We reiterate that by use of values given in this book identical results are obtained for a given compound whichever method is employed, except for some aromatic compounds. The choice of the method will depend on the reader's preference for thinking in terms of groups, in terms of chemical bonds of specific types, or in terms of bonds plus interactions between them. In tables 49–51 we list recommended parameters for the calculation of both ΔH_a and ΔH_f°(g); discussion in earlier Sections has been in terms of ΔH_a, but it is easy to show that a given bond energy scheme that is additive in contributions to ΔH_a must have a counterpart that is additive in contributions to ΔH_f°(g). Readers who wish to estimate values of ΔH_f°(g), and then values of ΔH_f°(l) or ΔH_f°(c), will find it simpler to carry out the calculation in terms of contributions to ΔH_f°(g), according to the same principles as have been described in earlier Sections for estimation of ΔH_a.

If the parameters from Tables 49–51 are appropriately summed they will yield a value for ΔH_a [or $\Delta H_f^\circ(g)$] that contains no allowance for steric destabilization, ring strain, or electrostatic interaction effects. These effects must be allowed for separately.

We recommend that $0.6\,\text{kcal mol}^{-1}$ should be ascribed to each 1, 4-steric repulsion in alkanes and alkyl groups, and that steric repulsion in alkenes be treated by the method described in Section 7.8.2. Steric repulsions in highly branched alkanes or alkyl groups may be estimated with the aid of parameters given in Section 7.7.1, and ring strains may be estimated from the data given in Sections 7.7.2 and 7.7.3. For estimation of steric repulsions

in molecules containing O, N— and S groups etc., consideration of

the bond lengths will assist in judging the size of the steric term (Sections 7.6.4 and 7.7.1), and for this purpose scale models are very useful.

We make no specific recommendations concerning the calculation of electrostatic interaction energies or π-electron delocalization energy. In the estimation of ΔH_a or $\Delta H_f^\circ(g)$ for a benzenoid compound the need to allow for π-electron delocalisation energy may be circumvented by the method given in Sections 7.8.6 and 7.8.7; selected parameters for use with this method are offered in respect of the Group and Laidler methods only.

TABLE 49. Recommended values of Group parameters for estimation
of ΔH_a and $\Delta H_f^o(g)$, kcal

	Parameter			Parameter	
	ΔH_a	$\Delta H_f^o(g)$		ΔH_a	$\Delta H_f^o(g)$
[C—(C)(H)$_3$]	337·32	−10·12	[O—(CO)(C)]	104·63	−45·07
[C—(C)$_2$(H)$_2$]	280·02	− 4·92	[O—(CO)$_2$]	112·02	−52·46
[C—(C)$_3$(H)]	224·75	− 1·75	[C—(CO)(H)$_3$]^{ass}	337·32	−10·12
[C—(C)$_4$]	170·96	− 0·06	[C—(CO)(C)(H)$_2$]	280·22	− 5·12
[C$_d$—(C$_d$)(H)$_2$]	268·88	+ 6·22	[C—(CO)(C)$_2$(H)]	223·74	− 0·74
[C$_d$—(C$_d$)(C)(H)]	214·35	+ 8·65	[C—(CO)(C)$_3$]	168·04	+ 2·86
[C$_d$—(C$_d$)(C)$_2$]	161·15	+ 9·75	[C$_b$—(C$_b$)$_2$(CO)]^{ass}	165·28	+ 5·62
[C—(C$_d$)(H)$_3$]^{ass}	337·32	−10·12			
[C—(C$_d$)(C)(H)$_2$]	280·02	− 4·92	[N—(C)(H)$_2$]	212·58	+ 4·62
[C—(C$_d$)(C)$_2$(H)]	224·75	− 1·75	[N—(C)$_2$(H)]	149·11	+15·99
[C—(C$_d$)(C)$_3$]	170·96	− 0·06	[N—(C)$_3$]	88·31	+24·69
[C$_t$—(C$_t$)(H)]	195·83	+27·17	[N—(N)(H)$_2$]	205·83	+11·37
[C$_t$—(C$_t$)(C)]	143·49	+27·41	[N—(N)(C)(H)]	143·85	+21·25
[C—(C$_t$)(H)$_3$]^{ass}	337·32	−10·12	[N—(N)(C)$_2$]	83·88	+29·12
[C—(C$_t$)(C)(H)$_2$]	280·02	− 4·92	[C—(N)(H)$_3$]^{ass}	337·32	−10·12
[C—(C$_t$)(C)$_2$(H)]	224·75	− 1·75	[C—(N)(C)(H)$_2$]	281·54	− 6·44
[C—(C$_t$)(C)$_3$]	170·96	− 0·06	[C—(N)(C)$_2$(H)]	227·34	− 4·34
[C$_b$—(C$_b$)$_2$(H)]	219·70	+ 3·30	[C—(N)(C)$_3$]	174·17	− 3·26
[C$_b$—(C$_b$)$_2$(C)]	165·28	+ 5·62	[N—(C$_b$)(H)$_2$]^{ass}	212·58	+ 4·62
[C—(C$_b$)(H)$_3$]^{ass}	337·32	−10·12	[N—(C$_b$)(H)]	143·28	+21·82
[C—(C$_b$)(C)(H)$_2$]	280·28	− 5·18	[C$_b$—(N)(C$_b$)$_2$]	171·21	− 0·31
[C—(C$_b$)(C)$_2$(H)]	224·02	− 1·02			
[C—(C$_b$)(C)$_3$]	168·02	+ 2·88	[C—(CN)(C)(H)$_2$]	535·84	+23·16
			[C—(CN)(C)$_2$(H)]	480·46	+26·44
[O—(C)(H)]	149·51	−37·85	[C—(CN)(C)$_3$]	426·78	+28·12
[O—(C)$_2$]	83·36	−23·80			
[O—(O)(H)]	127·91	−16·25	[C—(NO$_2$)(C)(H)$_2$]	521·72	−14·50
[O—(O)(C)]	64·69	− 5·13	[C—(NO$_2$)(C)$_2$(H)]	468·12	−13·00
[C—(O)(H)$_3$]^{ass}	337·32	−10·12	[C—(NO$_2$)(C)$_3$]	412·54	− 9·52
[C—(O)(C)(H)$_2$]	283·10	− 8·00	[C—(NO$_3$)(C)(H)$_2$]	593·38	−26·70
[C—(O)(C)$_2$(H)]	230·03	− 7·03	[C—(NO$_3$)(C)$_2$(H)]	540·08	−25·40
[C—(O)(C)$_3$]	177·56	− 6·66	[C—(NO$_3$)(C)$_3$]	483·38	−20·80
[C—(O)$_2$(H)$_2$]	290·53	−15·43			
[C—(O)$_2$(C)(H)]	238·30	−15·30	[S—(C)(H)]	113·22	+ 4·53
[C—(O)$_2$(C)$_2$]	183·21	−12·31	[S—(C)$_2$]	54·21	+11·05
[O—(C$_b$)(H)]^{ass}	149·51	−37·85	[S—(S)(C)]	58·31	+ 7·34
[O—(C$_b$)(C)]	82·02	−22·46	[C—(S)(H)$_3$]^{ass}	337·32	−10·12
[O—(C$_b$)$_2$]	78·24	−18·68	[C—(S)(C)(H)$_2$]	280·74	− 5·64
[C$_b$—(O)(C$_b$)$_2$]	172·09	− 1·19	[C—(S)(C)$_2$(H)]	225·54	− 2·54
			[C—(S)(C)$_3$]	171·17	− 0·27
[(CO)—(C)(H)]	312·17	−29·61	[S—(C$_b$)(H)]^{ass}	113·03	+ 4·72
[(CO)—(C)$_2$]	262·12	−31·66	[C$_b$—(C$_b$)$_2$(S)]	165·18	+ 5·72
[(CO)—(C$_b$)(H)]	313·48	−30·92	[(SO)—(C)$_2$]	138·82	−13·61
[(CO)—(C$_b$)(C)]	263·17	−32·71	[C—(SO)(H)$_3$]^{ass}	337·32	−10·12
[(CO)—(C$_b$)$_2$]	260·10	−29·64	[C—(SO)(C)(H)$_2$]	283·10	− 8·00
[(CO)—(CO)(H)]	307·89	−25·33			
[(CO)—(CO)(C)]	259·44	−28·98	[(SO$_2$)—(C)$_2$]	253·63	−68·86
[(CO)—(O)(H)]	312·17	−29·61	[C—(SO$_2$)(H)$_3$]^{ass}	337·32	−10·12
[(CO)—(O)(C)]	262·64	−32·18	[C—(SO$_2$)(C)(H)$_2$]	281·60	− 6·50
[(CO)—(O)(C$_b$)]	261·76	−31·30	[C—(SO$_2$)(C)$_2$(H)]	227·38	− 4·38
[O—(CO)(H)]	172·62	−60·96	[C—(SO$_2$)(C)$_3$]	174·66	− 3·76

TABLE 49—*continued*

Parameter	ΔH_a	$\Delta H_f^\circ(g)$	Parameter	ΔH_a	$\Delta H_f^\circ(g)$
[C—(F)(C)(H)₂]	346·52	−52·56	[C—(Hg)(C)(H)₂]	263·9	+19·0
[C—(F)(C)₂(H)]	289·78	−47·92	[C—(Hg)(C)₂(H)]	204·7	+25·1
[C—(F)(C)₃]	231·49	−41·73			
[C—(F)₂(C)(H)]	365·90	−105·18	[B—(C)₃]^ass	0	0
[C—(F)₂(C)₂]	306·26	−97·64	[C—(B)(H)₃]	381·2	−9·8
[C—(F)₃(C)]	388·04	−160·56	[C—(B)(C)(H)₂]	320·4	−1·1
[C—(F)(C_b)₂]	234·02	−44·26	[C—(B)(C)₂(H)]	264·8	−2·4
			[Al—(C)₃]^ass	0	0
[C—(Cl)(C)(H)₂]	320·00	−15·98	[C—(Al)(H)₃]	360·2	−7·0
[C—(Cl)(C)₂(H)]	265·28	−13·36	[C—(Al)(C)(H)₂]	307·4	−6·3
[C—(Cl)(C)₃]	212·32	−12·50	[Ga—(C)₃]^ass	0	0
[C—(Cl)₂(C)(H)]	301·51	−20·68	[C—(Ga)(H)₃]	353·9	−3·7
[C—(Cl)₂(C)₂]	248·71	−19·97	[In—(C)₃]^ass	0	0
[C—(Cl)₃(C)]	281·48	−23·82	[C—(In)(H)₃]	332·9	+13·6
[C—(Cl)(C_b)₂]	204·11	−4·29			
			[Ge—(C)₄]^ass	0	0
[C—(Br)(C)(H)₂]	306·92	−5·08	[C—(Ge)(C)(H)₂]	297·4	+0·2
[C—(Br)(C)₂(H)]	253·01	−3·26	[Sn—(C)₄]^ass	0	0
[C—(Br)(C)₃]	199·14	−1·50	[C—(Sn)(H)₃]	346·4	−1·2
[C—(Br)(C_b)₂]	188·94	+8·70	[C—(Sn)(C)(H)₂]	285·6	+7·5
			[Pb—(C)₄]	0	0
[C—(I)(C)(H)₂]	292·61	+8·02	[C—(Pb)(H)₃]	330·7	+8·2
[C—(I)(C)₂(H)]	237·66	+10·84	[C—(Pb)(C)(H)₂]	270·1	+16·7
[C—(I)(C)₃]	183·94	+12·46			
[C—(I)(C_b)₂]	173·84	+22·60	[P—(C)₃]^ass	0	0
			[C—(P)(H)₃]	359·9	−7·5
[Zn—(C)₂]^ass	0	0	[As—(C)₃]^ass	0	0
[C—(Zn)(H)₃]	336·0	+6·6	[C—(As)(H)₃]	349·3	+0·9
[C—(Zn)(C)(H)₂]	273·6	+16·9	[Sb—(C)₃]^ass	0	0
[Cd—(C)₂]^ass	0	0	[C—(Sb)(H)₃]	345·6	+2·6
[C—(Cd)(H)₃]	327·6	+12·9	[Bi—(C)₃]^ass	0	0
[C—(Cd)(C)(H)₂]	265·5	+22·9	[C—(Bi)(H)₃]	328·3	+15·4
[Hg—(C)₂]^ass	0	0			
[C—(Hg)(H)₃]	323·3	+11·2	[Se—(C)₂]^ass	0	0
			[C—(Se)(C)(H)₂]	296·5	+3·3

(i) Parameters are given for groups occurring in the more common types of organic compound and in organometallic compounds for which tolerably reliable values of $\Delta H_f^\circ(g)$ exist. The list is not, however, exhaustive, and further parameters can be derived from the data of Chapter 5.

(ii) The order in which the parameters are given corresponds approximately with the order in which compounds are listed in Chapter 5.

(iii) The following symbols are used for carbon atoms: C_d, doubly bonded; C_t, triply bonded; C_b, benzenoid.

(iv) Where only composite group parameters (e.g. $3[C—(N)(H)_3]+[N—(C)_3]$) can be unambiguously evaluated, it is nevertheless convenient to assign values to the parts; such values are indicated by the superscript "ass". Although the assigned values are arbitrary, they should not be changed, because other values depend on them.

(v) The following polyatomic combinations are treated as entities, and are shown in the group symbols enclosed in brackets: carbonyl, CO; cyano, CN; nitro, NO_2; nitrato, NO_3; sulphoxide, SO; sulphone, SO_2.

TABLE 50. Recommended values of Laidler parameters for estimation of ΔH_a and $\Delta H_f^\circ(g)$, kcal

Parameter	ΔH_a	$\Delta H_f^\circ(g)$	Parameter	ΔH_a	$\Delta H_f^\circ(g)$
$E(C-C)$	85·48	− 0·03	$E(C-NO_3)$	355·19	−20·88
$E(C-H)_p$	98·19	− 3·37	$E(C-H)_s^{NO_3}$	97·73	− 2·90
$E(C-H)_s$	97·27	− 2·44	$E(C-H)_t^{NO_3}$	99·43	− 4·60
$E(C-H)_t$	96·53	− 1·70			
$E(C=C)$	133·00	+37·90	$E(C\equiv N)$	204·00	+37·21
$E(C_d-H)_2$	101·19	− 6·36	$E(C-H)_s^{CN}$	97·26	− 2·44
$E(C_d-H)_1$	100·53	− 5·70	$E(C-H)_t^{CN}$	96·40	− 1·58
$E(C_d-C)$	90·07	− 4·62			
$E(C\equiv C)$	183·28	+73·07	$E(S-H)$	86·11	− 1·00
$E(C_t-H)$	104·19	− 9·36	$E(C-S)$	70·06	+ 5·29
$E(C_t-C)$	94·60	− 9·15	$E(C_b-S)$	73·32	+ 1·99
$E(C_b-C_b)$	119·17	+ 9·01	$E(S-S)$	62·86	+ 2·16
$E(C_b-H)$	100·53	− 5·70	$E(C-H)_p^{S}$	98·06	− 3·23
$E(C_b-C)$	88·91	− 3·46	$E(C-H)_s^{S}$	97·52	− 2·69
$E(C_b-C_b{<})$	114·30	+13·88	$E(C-H)_t^{S}$	97·11	− 2·28
$E({>}C_b-C_b{<})$	112·80	−27·35			
$E(C_6H_5-)$	1217·67	+25·56	$E(C-SO)$	112·55	− 7·24
			$E(C-H)_p^{SO}$	98·06	− 3·23
$E(O-H)$	107·83	−25·95	$E(C-H)_s^{SO}$	98·66	− 3·83
$E(C-O)$	91·02	−18·51			
$E(C_b-O)$	94·60	−22·15	$E(C-SO_2)$	169·96	−34·86
$E(C_b-O-C_b)$	184·08	−39·18	$E(C-H)_p^{SO_2}$	98·06	− 3·23
$E(O-O)$	45·58	+13·98	$E(C-H)_s^{SO_2}$	97·86	− 3·03
$E(C-H)_p^{O}$	96·00	− 1·17	$E(C-H)_t^{SO_2}$	98·17	− 3·34
$E(C-H)_s^{O}$	95·51	− 0·68			
$E(C-H)_t^{O}$	95·21	− 0·38	$E(C-F)$	114·59	−53·00
$E(C-H)_s^{OO}$	95·90	− 1·07	$E(C_b-F)$	115·24	−53·71
$E(C-H)_t^{OO}$	96·85	− 2·02	$E(C-H)_s^{F}$	94·60	+ 0·23
$E(C=O)$	158·28	−13·28	$E(C-H)_t^{F}$	89·73	+ 5·10
$E(C_{co}-H)_2$	101·19	− 6·36	$E(C-H)_s^{F_2}$	95·92	− 1·09
$E(C_{co}-H)_1$	101·97	− 7·14	$E(C-H)_t^{F_2}$	93·99	+ 0·84
$E(C_{co}-C)$	94·67	− 9·23	$E(C-H)_t^{F_3}$	102·11	− 7·28
$E(C-H)_p^{CO}$	98·19	− 3·37			
$E(C-H)_s^{CO}$	97·34	− 2·52	$E(C-Cl)$	84·13	−12·48
$E(C-H)_t^{CO}$	95·53	− 0·70	$E(C_b-Cl)$	85·33	−13·74
$E(C_b-C_{co})$	99·13	−13·68	$E(C-H)_s^{Cl}$	96·57	− 1·74
$E(C_{co}-C_{co})$	95·54	−10·14	$E(C-H)_t^{Cl}$	95·72	− 0·89
$E(COOH)$	383·34	−83·92	$E(C-H)_s^{Cl_2}$	93·82	+ 1·01
			$E(C-H)_t^{Cl_2}$	90·52	+ 4·31
$E(N-H)_2$	91·57	− 1·81	$E(C-H)_t^{Cl_3}$	82·07	+12·89
$E(N-H)_1$	90·21	− 0·47			
$E(C-N)$	74·00	+ 6·39	$E(C-Br)$	70·95	− 1·38
$E(C_b-N)$	81·87	− 1·54	$E(C_b-Br)$	70·16	− 0·75
$E(N-N)$	46·45	+28·89	$E(C-H)_s^{Br}$	96·62	− 1·79
$E(C-H)_p^{N}$	97·59	− 2·76	$E(C-H)_t^{Br}$	96·60	− 1·77
$E(C-H)_s^{N}$	97·12	− 2·29			
$E(C-H)_t^{N}$	97·32	− 2·49	$E(C-I)$	55·85	+12·37
			$E(C_b-I)$	55·06	+13·15
$E(C-NO_2)$	284·38	− 9·52	$E(C-H)_s^{I}$	97·02	− 2·19
$E(C-H)_s^{NO_2}$	97·31	− 2·48	$E(C-H)_t^{I}$	96·35	− 1·97
$E(C-H)_t^{NO_2}$	98·28	− 3·45			

Footnotes to Table 50

(i) Laidler parameters are given for organic compounds only. The values have been selected to give values of ΔH_a or $\Delta H_f^\circ(g)$ for compounds equal to those given by the Group parameters in Table 49.

(ii) The symbols C_d, C_t and C_b have the same meanings as in Table 49. Additionally, the carbon atom of a carbonyl group is distinguished by use of a subscript CO.

(iii) C–H bond energy terms are distinguished by a system of subscripts and superscripts: $E(C-H)_p$ as in a $-CH_3$ group, $E(C-H)_s$ as in a $-CH_2-$ group, $E(C-H)_t$ as in a $>CH-$ group, $E(C-H)_s^O$ as in a $-CH_2-O-$ group, $E(C-H)_s^{OO}$ as in a $-O-CH_2-O-$ group, $E(C-H)_t^{CO}$ as in a $>CH-CO-$ group, $E(C-H)_t^{F_2}$ as in a $-CHF_2$ group etc. Also, subscripts 1 and 2 are used to denote the numbers of hydrogen atoms attached to an atom, e.g. $E(N-H)_2$ refers to $-NH_2$ and $E(N-H)_1$ refers to $-NH-$.

TABLE 51. Recommended Values of Allen Parameters for Estimation of ΔH_a and $\Delta H_f^\circ(g)$, kcal

	Parameter				Parameter	
	ΔH_a	$\Delta H_f^\circ(g)$			ΔH_a	$\Delta H_f^\circ(g)$
$B(C-H)$	99·30	− 4·47		Δ_{CCN}	−1·00	+ 1·00
$B(C-C)$	78·84	+ 6·61		Δ_{CCC}^N	−1·63	+ 1·63
$B(C=C)$	140·55	+30·36		Δ_{CCN}^N	−2·30	+ 2·30
$B(C\equiv C)$	193·05	+63·30				
Γ_{CCC}	2·58	− 2·58		$B(C-NO_2)$	279·28	− 4·42
$\Gamma_{CC_dC_d}$	5·35	− 5·35		$\Gamma_{CC(NO_2)}$	4·42	− 4·42
$\Gamma_{CC_tC_t}$	7·89	− 7·89		$\Delta_{CC(NO_2)}$	−0·72	+ 0·72
Δ_{CCC}	−0·55	+ 0·55				
Δ_{CCC_d}	−1·25	+ 1·25 .		$B(C-NO_3)$	350·09	−15·67
				$\Gamma_{CC(NO_3)}$	5·27	− 5·27
$B(O-H)$	110·78	−28·90		$\Delta_{CC(NO_3)}$	−1·27	+ 1·27
$B(C-O)$	78·15	− 5·64				
$B(C=O)$	162·06	−17·04		$B(S-H)$	87·35	− 2·24
$B(O-O)$	34·26	+25·30		$B(C-S)$	65·29	+10·28
Γ_{CCO}	5·66	− 5·66		$B(S-S)$	58·27	+ 7·01
Γ_{COC}	5·90	− 5·90		Γ_{CCS}	3·30	− 3·30
Γ_{OCO}	13·09	−13·09		Γ_{CSC}	2·47	− 2·47
Γ_{COO}	8·63	− 8·63		Γ_{CSS}	3·35	− 3·35
$\Gamma_{CC_dO_d}$	11·45	−11·45		Δ_{CCS}	−1·20	+ 1·20
$\Gamma_{O_dO_d}$	34·50	−34·50				
Δ_{CCO}	−1·43	+ 1·43		$B(C-SO)$	108·83	− 3·49
Δ_{COO}	−3·67	+ 3·67		$\Gamma_{CC(SO)}$	5·74	− 5·74
Δ_{CCO_d}	−4·20	+ 4·20				
Δ_{COO_d}	−1·04	+ 1·04		$B(C-SO_2)$	166·74	−31·31
				$\Gamma_{CC(SO_2)}$	3·15	− 3·15
$B(N-H)$	93·45	− 3·68		$\Delta_{CC(SO_2)}$	−0·67	+ 0·67
$B(C-N)$	65·10	+15·30				
$B(C\equiv N)$	204·40	+36·78		$B(C-F)$	104·36	−42·84
$B(N-N)$	37·85	+37·49		Γ_{FCF}	13·64	−13·64
Γ_{CCN}	4·10	− 4·10		Γ_{CCF}	4·14	− 4·14
Γ_{CNC}	4·30	− 4·30		Δ_{FFF}	−7·42	+ 7·42
Γ_{CNN}	5·80	− 5·80		Δ_{CFF}	−3·46	+ 3·46
$\Gamma_{CC_tN_t}$	12·00	−12·00		Δ_{CCF}	−3·58	+ 3·58

TABLE 51—continued

Parameter	ΔH_a	$\Delta H_f^\circ(g)$	Parameter	ΔH_a	$\Delta H_f^\circ(g)$
$B(C-Cl)$	78·77	− 7·11	Γ_{CCAl}	−3·0	+ 3·0
Γ_{ClCCl}	−0·25	+ 0·25	$B(Ga-C)+\Gamma_{CGaC}$	56·1	+ 9·7
Γ_{CCCl}	3·20	− 3·20	$+\frac{1}{3}\Delta_{CCC}^{Ga}$		
$\Delta_{ClCClCl}$	−0·40	+ 0·40	$B(In-C)+\Gamma_{CInC}$	35·0	+27·0
Δ_{CClCl}	−0·90	+ 0·90	$+\frac{1}{3}\Delta_{CCC}^{In}$		
Δ_{CCCl}	−0·50	+ 0·50			
			$B(Ge-C)+\frac{3}{2}\Gamma_{CGeC}$	59·4	+ 5·9
$B(C-Br)$	65·14	+ 4·34	$+\Gamma_{CCGe}+\Delta_{CCC}^{Ge}$		
Γ_{CCBr}	4·30	− 4·30	$B(Sn-C)+\frac{3}{2}\Gamma_{CSnC}$	48·5	+12·3
Δ_{CCBr}	−1·45	+ 1·45	$+\Delta_{CCC}^{Sn}$		
			Γ_{CCSn}	−0·3	+ 0·3
$B(C-I)$	51·44	+16·84	$B(Pb-C)+\frac{3}{2}\Gamma_{CPbC}$	32·9	+21·6
Γ_{CCI}	3·15	− 3·15	$+\Delta_{CCC}^{Pb}$		
Δ_{CCI}	−0·80	+ 0·80	Γ_{CCPb}	−0·8	+ 0·8
$B(Zn-C)+\frac{1}{2}\Gamma_{CZnC}$	38·4	+20·0			
Γ_{CCZn}	−2·6	+ 2·6	$B(P-C)+\Gamma_{CPC}$	62·0	+ 5·9
$B(Cd-C)+\frac{1}{2}\Gamma_{CCdC}$	29·8	+26·3	$+\frac{1}{3}\Delta_{CCC}^{P}$		
Γ_{CCCd}	−2·7	+ 2·7	$B(As-C)+\Gamma_{CAsC}$	51·4	+14·4
$B(Hg-C)+\frac{1}{2}\Gamma_{CHgC}$	25·5	+24·6	$+\frac{1}{3}\Delta_{CCC}^{As}$		
Γ_{CCHg}	0·2	− 0·2	$B(Sb-C)+\Gamma_{CSbC}$	47·7	+16·0
			$+\frac{1}{3}\Delta_{CCC}^{Sb}$		
$B(B-C)+\Gamma_{CBC}$			$B(Bi-C)+\Gamma_{CBiC}$	30·5	+28·8
$+\frac{1}{3}\Delta_{CCC}^{B}$	83·3	+ 3·7	$+\frac{1}{3}\Delta_{CCC}^{Bi}$		
Γ_{CCB}	0·9	− 0·9			
Δ_{CCB}	−1·0	+ 1·0	$B(Se-C)+\frac{1}{2}\Gamma_{CSeC}$	58·5	+10·7
$B(Al-C)+\Gamma_{CAlC}$	62·3	+ 6·5	$+\Gamma_{CCSe}$		
$+\frac{1}{3}\Delta_{CCC}^{Al}$					

(i) Allen parameters are not given for aromatic compounds. For organometallic compounds composite parameters are given, because too few experimental data exist to permit evaluation of individual terms.

(ii) Γ_{XYZ} parameters are interaction terms for pairs of bonds, as defined by the three subscripts. Thus Γ_{CCO} is the energy of the interaction $-C\overset{\cdots}{-}C\overset{\cdots}{-}O-$, and $\Gamma_{CC_dO_d}$ is the energy of the interaction $-C\overset{\cdots}{-}C = O$, etc. Δ_{XYZ} are interaction terms for trios of non-bonded atoms, as defined by the three subscripts; the absence of a superscript implies that the trios are around a carbon atom, otherwise the central atom is defined by a superscript. Thus Δ_{CCC_d} is the energy of the interaction

$$\overset{-C\diagdown}{\underset{-C\diagup}{\vdots}}C\equiv C-,\text{ and }\Delta_{CCN}^{N}\text{ is the energy of the interaction }-N\overset{\diagup C-}{\underset{\diagdown C-}{-}N}\text{ etc.}$$

7.10. Some worked examples of methods of estimating $\Delta H_f^\circ(g)$.

(a) *Estimation of $\Delta H_f^\circ(g)$ for 1,4-dichlorobutane*

$\Delta H_f^\circ(g)$ for simple substituted aliphatic compounds can in principle be estimated by the methods of intra-series or inter-series increments, described in Chapter 6.

For application of the method of intra-series increments, a value of $\Delta H_f^\circ(g)$ for a homologue must be available, in this instance the value for an α, ω-dichloroalkane. From Chapter 5, $\Delta H_f^\circ(g)$, $ClCH_2.CH_2.CH_2Cl$ $= -38\cdot2\pm2\cdot0$ kcal.

$$\therefore \Delta H_f^\circ(g), ClCH_2.(CH_2)_2.CH_2Cl = -38\cdot2\pm2 -4\cdot9 \text{ (Section 6.4.1)}$$
$$= -43\cdot1\pm2.0 \text{ kcal.}$$

For application of the method of inter-series increments, a value of $\Delta H_f^\circ(g)$ for a compound of similar skeletal structure must be available; in this instance the value for either butyl chloride or butane will serve. From Chapter 5, $\Delta H_f^\circ(g)$, $n\text{-}C_4H_9Cl = -35\cdot1\pm2\cdot0$ kcal, and $\Delta H_f^\circ(g)$, $n\text{-}C_4H_{10} = -30\cdot4\pm0\cdot2$ kcal.

$$\therefore \Delta H_f^\circ(g), ClCH_2.(CH_2)_2.CH_2Cl = -35\cdot1\pm2\cdot0 -6\cdot5 \text{ (Section 6.4.2)}$$
$$= -41\cdot6\pm2\cdot0 \text{ kcal,}$$

or $\Delta H_f^\circ(g)$, $ClCH_2.(CH_2)_2.CH_2Cl$
$$= -30\cdot4\pm0\cdot2 -2\times6\cdot5$$
$$= -43\cdot4\pm0\cdot2^* \text{ kcal.}$$

The four-parameter schemes (Tables 49–51) give the estimated value $-41\cdot8$ kcal for $\Delta H_f^\circ(g)$, $ClCH_2.(CH_2)_2.CH_2Cl$.

(b) *Estimation of $\Delta H_f^\circ(g)$ for $C_2H_5.CO.CH(t\text{-}C_4H_9).CH_2.c\text{-}Pr$*

$\Delta H_f^\circ(g)$ for this compound can be estimated by any of the four-parameter schemes described earlier in this Chapter. The first step is to write the full structural formula and to enumerate the carbon atoms in any convenient sequence:

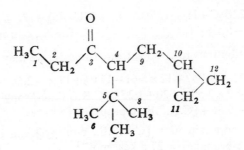

The next step is to write down the appropriate parameters in sequence, and finally to substitute numerical values.

By the Group scheme:

$$\Delta H_f^\circ(g) = [C-(C)(H)_3] + [C-(C)(CO)(H)_2] + [(CO)-(C)_2]$$
$$+ [C-(C)_2(CO)(H)] + [C-(C)_4] + 3[C-(C)(H)_3]$$
$$+ [C-(C)_2(H)_2] + [C-(C)_3(H)] + 2[C-(C)_2(H)_2]$$
$$+ [C.R.S.E., \text{ c-propyl}] + [S],$$

where $[S]$ takes account of *gauche* 1, 4-steric interactions.

Numerical values for the above Group parameters are available from Table 49 in terms of contributions to ΔH_a and contributions to $\Delta H_f^\circ(g)$; we adopt the latter in this example, and by gathering like terms we may write:

$$\Delta H_f^\circ(g) = 4 \times -10\cdot12 - 5\cdot12 - 31\cdot66 - 0\cdot74 - 0\cdot06 + 3 \times -4\cdot92$$
$$-1\cdot75 + [C.R.S.E., \text{ c-propyl}] + [S]$$
$$= -94\cdot6 + [C.R.S.E., \text{ c-propyl}] + [S] \text{ kcal.}$$

By the Laidler scheme:

$$\Delta H_f^\circ(g) = 3E(C-H)_p + E(C-C) + 2E(C-H_s^{CO}) + E(C-C_{CO}) + E(C=O)$$
$$+ E(C-C_{CO}) + E(C-H)_r^{CO} + 4E(C-C) + 9E(C-H)_p$$
$$+ 2E(C-C) + 2E(C-H)_s + E(C-H)_t + 3E(C-C)$$
$$+ 4E(C-H)_s + [C.R.S.E., \text{ c-propyl}] + [S].$$

Using contributions to $\Delta H_f^\circ(g)$ from Table 50, and gathering like terms we may write:

$$\Delta H_f^\circ(g) = 12 \times -3\cdot37 + 10 \times -0\cdot03 + 2 \times -2\cdot52 + 2 \times -9\cdot23 - 13\cdot28$$
$$-0\cdot70 + 6 \times -2\cdot44 - 1\cdot70 + [C.R.S.E., \text{ c-propyl}] + [S]$$
$$= -94\cdot6 + [C.R.S.E., \text{ c-propyl}] + [S] \text{ kcal.}$$

By the Allen scheme:

$$\Delta H_f^\circ(g) = 22B(C-H) + 12B(C-C) + B(C=O) + 17\Gamma_{CCC} + 2\Gamma_{CC_dO_d}$$
$$+ 5\Delta_{CCC} + \Delta_{CCO_d} + \Delta_{CCC_d} + [C.R.S.E., \text{ c-propyl}] + [S].$$

Using contributions to $\Delta H_f^\circ(g)$ from Table 51, we may write:

$$\Delta H_f^\circ(g) = 22 \times -4\cdot47 + 12 \times 6\cdot61 - 17\cdot04 + 17 \times -2\cdot58 + 2 \times -11\cdot45$$
$$+ 5 \times 0\cdot55 + 4\cdot20 + 1\cdot25 + [C.R.S.E., \text{ c-propyl}] + [S].$$
$$= -94\cdot6 + [C.R.S.E., \text{ c-propyl}] + [S] \text{ kcal.}$$

To a good approximation $[C.R.S.E., \text{ c-propyl}]$ can be taken equal to $[C.R.S.E., \text{ c-propane}] = 27\cdot5 \text{ kcal mol}^{-1}$.

The term $[S]$ includes contributions from two *gauche* 1, 4-interactions between hydrogens on *C9* and those on *C7* and *C8* ($0.6\,\mathrm{kcal\,mol^{-1}}$ each, Section 7.5), and one between hydrogens on *C11* (or *C12*) and that on *C4* (*ca* $0.3\,\mathrm{kcal\,mol^{-1}}$).

$$\therefore\ \Delta H_f^\circ(\mathrm{g}) = -94.6 + 27.5 + 1.5$$
$$= -65.6\ \mathrm{kcal.}$$

(c) *Estimation of $\Delta H_f^\circ(\mathrm{g})$ for 2-propenyl-4-amino-5-methylbenzenethiol*
In this example we use contributions to ΔH_a to estimate ΔH_a for the compound, and then derive $\Delta H_f^\circ(\mathrm{g})$.

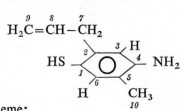

By the Laidler scheme:

$$\begin{aligned}
\Delta H_a = &\ 6E(C_b-C_b) + 2E(C_b-H) + E(C_b-S) + E(S-H) \\
&+ E(C_b-C) + 2E(C-H)_s + E(C_d-C) + E(C_d-H)_1 \\
&+ E(C=C) + 2E(C_d-H)_2 + E(C_b-N) + 2E(N-H)_2 \\
&+ E(C_b-C) + 3E(C-H)_p + [E.E.] - [S],
\end{aligned}$$

where $[E.E.]$ represents the energy resulting from electrostatic interaction and $[S]$ represents steric interaction energy. Using data from Table 50, and gathering like terms, we may write:

$$\begin{aligned}
\Delta H_a = &\ 6 \times 119.17 + 2 \times 100.53 + 73.32 + 86.11 + 2 \times 88.91 + 2 \times 97.27 \\
&+ 90.09 + 100.50 + 133.00 + 2 \times 101.19 + 81.87 + 2 \times 91.57 \\
&+ 3 \times 98.19 + [E.E.] - [S] \\
= &\ 2533.4 + [E.E.] - [S]\ \mathrm{kcal.}
\end{aligned}$$

The polar groups are sufficiently far apart that $[E.E.]$ may be neglected without serious error; interestingly, the interaction is likely to have a stabilizing effect because the interaction between the *para* SH and NH$_2$ groups is attractive. $[S]$, arising mainly from interaction between the *ortho* CH$_3$ and NH$_2$ groups may also be neglected without serious error.

$$\therefore\ \Delta H_a = 2533.4\ \mathrm{kcal.}$$
$$\therefore\ \Delta H_f^\circ(\mathrm{g}) = 10\Delta H_f^\circ,\ \mathrm{C(g)} + 13\Delta H_f^\circ,\ \mathrm{H(g)} + \Delta H_f^\circ,\ \mathrm{S(g)}$$
$$+ \Delta H_f^\circ,\ \mathrm{N(g)} - 2533.4$$
$$= +31.5\ \mathrm{kcal.}$$

Bibliography

1926

26/1.　E. W. Washburn, ed., "International Critical Tables of Numerical Data, Physics, Chemistry and Technology", McGraw-Hill, New York and London, in 8 parts, 1926-1933.

26/2.　J. H. Mathews, *J. Am. Chem. Soc.* **48**, 562 (1926).

26/3.　D. H. Andrews, G. Lynn, and J. Johnston, *J. Am. Chem. Soc.* **48**, 1274 (1926).

26/4.　P. E. Verkade, H. Hartman, and J. Coops, *Rec. Trav. Chim.* **45**, 373 (1926).

26/5.　G. W. C. Yates, *Phil. Mag.* **2**, 817 (1926).

1927

27/1.　Mme. Bérenger-Calvet, *J. Chim. Phys.* **24**, 325 (1927).

27/2.　W. H. Rodebush and C. C. Coons, *J. Am. Chem. Soc.* **49**, 1953 (1927).

27/3.　A. S. Coolidge and M. S. Coolidge, *J. Am. Chem. Soc.* **49**, 100 (1927).

27/4.　O. E. May, J. F. T. Berliner, and D. F. J. Lynch, *J. Am. Chem. Soc.* **49**, 1012 (1927).

1928

28/1.　L. E. Steiner and J. Johnston, *J. Phys. Chem.* **32**, 912 (1928).

1929

29/1.　M. S. Kharasch, *J. Res. Natl. Bur. Stand.* **2**, 359 (1929).

29/2.　K. K. Kelley, *J. Am. Chem. Soc.* **51**, 1145 (1929).

29/3.　P. Landrieu, F. Baylocq, and J. R. Johnson, *Bull. Soc. Chim. (France)*, **45**, 36 (1929).

1930

30/1.　A. Wassermann *Z. Phys. Chem.* **A151**, 113 (1930).

30/2.　R. H. Bullard and A. C. Haussmann, *J. Phys. Chem.* **34**, 743 (1930).

30/3.　R. Mecke, *Nature*, **125**, 526 (1930).

30/4.　D. P. Evans, W. C. Davies, and W. J. Jones, *J. Chem. Soc.* p.1310 (1930).

1931

31/1.   F. D. Rossini, *J. Res. Natl. Bur. Stand.* **6**, 1 (1931).
31/2.   F. D. Rossini, *J. Res. Natl. Bur. Stand.* **7**, 329 (1931).
31/3.   J. H. Mathews and P. R. Fehlandt, *J. Am. Chem. Soc.* **53**, 3213 (1931).
31/4.   K. M. Watson, *Ind. Eng. Chem.* **23**, 360 (1931).
31/5.   F. D. Rossini, *J. Res. Natl. Bur. Stand.* **6**, 37 (1931).
31/6.   L. J. P. Keffler, *J. Chim. Phys.*, **28**, 457 (1931).
31/7.   M. Beckers, *Bull. Soc. Chim. (Belg.)*, **40**, 518 (1931).
31/8.   E. F. Fiock, D. C. Ginnings, and W. B. Holton, *J. Res. Natl. Bur. Stand.* **6**, 881 (1931).
31/9.   A. Skita and W. Faust, *Ber.* **64**, 2878 (1931).
31/10.   C. E. Bills, F. G. McDonald, L. N. BeMiller, G. E. Steel, and M. Nussmeir, *J. Biol. Chem.* **93**, 775 (1931).
31/11   M. Milone and S. Allavena, *Gazz. Chim. Ital.* **61**, 75 (1931).
31/12   H.-J. Schumacher and P. Bergmann, *Z. Phys. Chem.* **B13**, 269 (1931).
31/13   E. Wiberg and W. Sütterlin, *Z. Anorg. Chem.* **202**, 1 (1931).

1932

32/1.   F. D. Rossini, *J. Res. Natl. Bur. Stand.* **8**, 119 (1932).
32/2.   K. Neumann and E. Völker, *Z. Phys. Chem.* **A161**, 33 (1932).
32/3.   P. Clausing, *Ann. Phys.* **12**, 961 (1932).
32/4.   F. D. Rossini and M. Frandsen, *J. Res. Natl. Bur. Stand.* **9**, 733 (1932).
32/5.   W. A. Roth and H. Banse, *Arch. Eisenhüttenw.* **6**, 43 (1932); C.A. **27**, 18b (1933).
32/6.   M. Milone and P. Rossignoli, *Gazz. Chim. Ital.* **62**, 644 (1932).
32/7.   A. Stern and G. Klebs, *Ann. Chem.* **500**, 91 (1932).
32/8.   E. Terres and H. Wesemann, *Angew. Chem.* **45**, 795 (1932).
32/9.   G. S. Parks and H. M. Huffman, "Free Energies of Some Organic Compounds", A. C. S. Monograph No. 60, The Chemical Catalog Co. Inc., New York, 1932.

1933

33/1.   E. W. Washburn, *J. Res. Natl. Bur. Stand.* **10**, 525 (1933).
33/2.   M. M. Popoff and P. K. Schirokich, *Z. Phys. Chem.* **A167**, 183 (1933).
33/3.   L. Ruzicka and P. Schläpfer, *Helv. Chim. Acta* **16**, 162 (1933).
33/4.   H. Banse and G. S. Parks, *J. Am. Chem. Soc.* **55**, 3223 (1933).
33/5.   W. Hückel, A. Gercke, and A. Gross, *Ber.* **66**, 563 (1933).
33/6.   F. Walker, *J. Am. Chem. Soc.* **55**, 2821 (1933).
33/7.   R. H. Newton and B. F. Dodge, *J. Am. Chem. Soc.* **55**, 4747 (1933).
33/8.   W. A. Roth and I. Meyer, *Z. Electrochem.* **39**, 35 (1933).
33/9.   P. E. Verkade and H. Hartman, *Rec. Trav. Chim.* **52**, 945 (1933).
33/10.   A. Stern and G. Klebs, *Ann. Chem.* **504**, 287 (1933).
33/11.   H. v. Wartenberg and H. Schütza, *Z. Phys. Chem.* **A164**, 386 (1933).
33/12.   A. Stern and G. Klebs, *Ann. Chem.* **505**, 295 (1933).
33/13.   P. Rothemund and H. Beyer, *Ann. Chem.* **492**, 292 (1932).

33/14. G. Becker and W. A. Roth, *Z. Phys. Chem.* **A167**, 1 (1933).

33/15. F. T. Miles and A. W. C. Menzies, *J. Phys. Chem.* **37**, 425 (1933).

33/16. E. Pohland and W. Mehl, *Z. Phys. Chem.* **A164**, 48 (1933).

33/17. N. V. Sidgwick, "The Covalent Link in Chemistry", Cornell University Press, 1933.

33/18. A. B. Burg and H. I. Schlesinger, *J. Am. Chem. Soc.* **55**, 4020 (1933).

1934

34/1.   R. S. Jessup and C. B. Green, *J. Res. Natl. Bur. Stand.* **13**, 469 (1934).

34/2.   "Premier Rapport de la Commission Permanente de Thermochimie", Union Internationale de Chimie, Paris, 1934.

34/3.   G. Becker and W. A. Roth, *Z. Phys. Chem.* **A169**, 287 (1934).

34/4.   F. D. Rossini, *J. Res. Natl. Bur. Stand.* **12**, 735 (1934).

34/5.   G. Becker and W. A. Roth, *Ber.* **67**, 627 (1934).

34/6.   F. O. Rice and J. Greenberg, *J. Am. Chem. Soc.* **56**, 2268 (1934).

34/7.   W. F. Eberz and H. J. Lucas, *J. Am. Chem. Soc.* **56**, 1230 (1934).

34/8.   L. J. P. Keffler, *J. Phys. Chem.* **38**, 717 (1934).

34/9.   H. Hirsbrunner, *Helv. Chim. Acta*, **17**, 477 (1934).

34/10. C. T. Zahn, *J. Chem. Phys.* **2**, 671 (1934).

34/11. J. H. van Vleck, *J. Chem. Phys.* **2**, 20, 297 (1934).

34/12. F. D. Rossini, *J. Res. Natl. Bur. Stand.* **13**, 21 (1934).

34/13. L. M. Dennis, R. W. Work, E. G. Rochow, and E. M. Chamot, *J. Am. Chem. Soc.* **56**, 1047 (1934).

1935

35/1.   H. M. Huffman and E. L. Ellis, *J. Am. Chem. Soc.* **57**, 41 (1935).

35/2.   G. B. Kistiakowsky, H. Romeyn, J. R. Ruhoff, H. A. Smith, and W. E. Vaughan, *J. Am. Chem. Soc.* **57**, 65 (1935).

35/3.   G. B. Kistiakowsky, J. R. Ruhoff, H. A. Smith, and W. E. Vaughan, *J. Am. Chem. Soc.* **57**, 876 (1935).

35/4.   L. Brüll, *Gazz. Chim. Ital.* **65**, 19 (1935).

35/5.   K. Fries, R. Walter, and K. Schilling, *Ann. Chem.* **516**, 248 (1935).

35/6.   J. W. Barrett and R. P. Linstead, *J. Chem. Soc.* p.436 (1935).

35/7.   F. Eisenlohr and W. Haas, *Z. Phys. Chem.* **A173**, 249 (1935).

35/8.   A. Wassermann, *J. Chem. Soc.* p.828 (1935).

35/9.   R. D. Stiehler and H. M. Huffman, *J. Am. Chem. Soc.* **57**, 1734 (1935).

35/10. A. Michael, *J. Am. Chem. Soc.* **57**, 159 (1935).

35/11. G. H. Burrows and L. A. King, *J. Am. Chem. Soc.* **57**, 1789 (1935).

35/12. G. R. Cuthbertson and G. B. Kistiakowsky, *J. Chem. Phys.* **3**, 631 (1935).

35/13. E. Schjånberg, *Z. Phys. Chem.* **A172**, 197 (1935).

35/14. W. J. Jones, D. P. Evans, T. Gulwell, and D. C. Griffiths, *J. Chem. Soc.* p. 39 (1935).

35/15. W. Hieber and E. Romberg, *Z. Anorg. Allg. Chem.* **221**, 321 (1935).

35/16. O. J. Schierholtz and M. L. Staples, *J. Am. Chem. Soc.* **57**, 2709 (1935).

35/17. E. U. Condon and G. H. Shortley, "Theory of Atomic Spectra", Cambridge University Press, 1935.

1936

36/1.   F. R. Bichowsky and F. D. Rossini, "Thermochemistry of the Chemical Substances", Reinhold, New York, 1936.
36/2.   F. D. Rossini, *Chem. Rev.* **18**, 233 (1936).
36/3.   G. B. Kistiakowsky, J. R. Ruhoff, H. A. Smith, and W. E. Vaughan, *J. Am. Chem. Soc.* **58**, 146 (1936).
36/4.   N. Bekkedahl, L. A. Wood, and M. Wojciechowski, *J. Res. Natl. Bur. Stand.* **17**, 883 (1936).
36/5.   G. B. Kistiakowsky, J. R. Ruhoff, H. A. Smith, and W. E. Vaughan, *J. Am. Chem. Soc.* **58**, 137 (1936).
36/6.   W. A. Roth and H. Pahlke, *Angew. Chem.* **49**, 618 (1936).
36/7.   H. E. Bent, G. R. Cuthbertson, M. Dorfman, and R. E. Leary, *J. Am. Chem. Soc.* **58**, 165 (1936).
36/8.   J. W. Barrett and R. P. Linstead, *J. Chem. Soc.* p.611 (1936).
36/9.   B. L. Crawford and G. S. Parks, *J. Am. Chem. Soc.* **58**, 373 (1936).
36/10. H. E. Bent and G. R. Cuthbertson, *J. Am. Chem. Soc.* **58**, 170 (1936).
36/11. J. V. Harispe, *Ann. Chim.* **6**, 249 (1936).
36/12. L. J. P. Keffler, *J. Soc. Chem. Ind.* **55**, 331 (1936).
36/13. E. Schjånberg, *Z. Phys. Chem.* **A175**, 342 (1936).
36/14. H. R. Ambler, *J. Soc. Chem. Ind.* **55**, 291 (1936).
36/15. H. M. Huffman, E. L. Ellis, and S. W. Fox, *J. Am. Chem. Soc.*, **58**, 1728 (1936).
36/16. M. Milone and G. Venturello, *Gazz. Chim. Ital.* **66**, 808 (1936).
36/17. K. Fries and F. Beyerlein, *Ann. Chem.* **527**, 71 (1937).
36/18. H. W. Thompson and J. W. Linnett, *Trans. Faraday Soc.* **32**, 681 (1936).
36/19. H. G. Trieschmann, *Z. Phys. Chem.* **B33**, 283 (1936).
36/20. H. Merten and H. Schlüter, *Ber.* **69**, 1364 (1936).
36/21. E. J. Buckler and R. G. W. Norrish, *J. Chem. Soc.* p.1567 (1936).
36/22. H. H. Voge, *J. Chem. Phys.* **4**, 581 (1936).

1937

37/1.   F. D. Rossini and J. W. Knowlton, *J. Res. Natl. Bur. Stand.* **19**, 249 (1937).
37/2.   M. A. Dolliver, T. L. Gresham, G. B. Kistiakowsky, and W. E. Vaughan, *J. Am. Chem. Soc.* **59**, 831 (1937).
37/3.   R. S. Jessup, *J. Res. Natl. Bur. Stand.* **18**, 115 (1937).
37/4.   W. Hückel, E. Kamenz, A. Gross, and W. Tappe, *Ann. Chem.* **533**, 1 (1938).
37/5.   H. Moureu and M. Dodé, *Bull. Soc. Chim. (France)*, **4**, 637 (1937).
37/6.   E. Schjånberg, *Z. Phys. Chem.* **A178**, 274 (1937).
37/7.   G. B. Bonino, R. Manzoni-Ansidei, and M. Rolla, *Ricerca Sci.* **8**, 357 (1937).
37/8.   F. Eisenlohr and A. Metzner, *Z. Phys. Chem.* **A178**, 339 (1937).
37/9.   L. J. P. Keffler, *J. Phys. Chem.* **41**, 715 (1937).
37/10. A. F. Gallaugher and H. Hibbert, *J. Am. Chem. Soc.* **59**, 2521 (1937).
37/11. M. Badoche, *Bull. Soc. Chim. (France)*, **4**, 549 (1937).
37/12. W. A. Roth and G. Becker, *Z. Phys. Chem.* **A179**, 450 (1937).
37/13. J. G. Aston, C. W. Siller, and G. H. Messerly, *J. Am. Chem. Soc.* **59**, 1743 (1937).

37/14. H. M. Huffman, S. W. Fox, and E. L. Ellis, *J. Am. Chem. Soc.* **59**, 2144 (1937).
37/15. A. Perlick, *Bull. Int. Inst. Refrig.* **18**, 1 (1937).
37/16. G. B. Kistiakowsky and C. H. Stauffer, *J. Am. Chem. Soc.* **59**, 165 (1937).
37/17. J. L. Jones and R. A. Ogg, *J. Am. Chem. Soc.* **59**, 1943 (1937).
37/18. J. E. Cline and G. B. Kistiakowsky, *J. Chem. Phys.* **5**, 990 (1937).
37/19. K. C. D. Hickman, J. C. Hecker, and N. D. Embree, *Ind. Eng. Chem.* (*Anal. Ed.*) **9**, 264 (1937).

1938

38/1. P. H. Dewey and D. R. Harper, *J. Res. Natl. Bur. Stand.* **21**, 457 (1938).
38/2. R. S. Jessup, *J. Res. Natl. Bur. Stand.* **21**, 475 (1938).
38/3. M. A. Dolliver, T. L. Gresham, G. B. Kistiakowsky, E. A. Smith, and W. E. Vaughan, *J. Am. Chem. Soc.* **60**, 440 (1938).
38/4. H. M. Huffman, *J. Am. Chem. Soc.* **60**, 1171 (1938).
38/5. W. v. Luschinsky, *Z. Phys. Chem.* **A182**, 384 (1938).
38/6. M. Enderlin, *Ann. Chim.* **10**, 5 (1938).
38/7. L. de V. Moulds and H. L. Riley, *J. Chem. Soc.* p.621 (1938).
38/8. H. Essex and M. Sandholzer, *J. Phys. Chem.* **42**, 317 (1938).
38/9. E. Schjånberg, *Z. Phys. Chem.* **A181**, 430 (1938).
38/10. E. Schjånberg, *Svensk. Kem. Tidr.* **50**, 102 (1938).
38/11. H. M. Huffman and S. W. Fox, *J. Am. Chem. Soc.* **60**, 1400 (1938).
38/12. C. J. Egan and J. D. Kemp, *J. Am. Chem. Soc.* **60**, 2097 (1938).
38/13. J. B. Conn, G. B. Kistiakowsky, and E. A. Smith, *J. Am. Chem. Soc.* **60**, 2764 (1938).
38/14. K. L. Wolf and H. Weghofer, *Z. Phys. Chem.* **B39**, 194 (1938).
38/15. J. L. Crenshaw, A. C. Cope, N. Finkelstein, and R. Rogan, *J. Am. Chem. Soc.* **60**, 2308 (1938).
38/16. R. S. Jessup, *J. Res. Natl. Bur. Stand.* **20**, 589 (1938).

1939

39/1. F. D. Rossini, *J. Res. Natl. Bur. Stand.* **22**, 407 (1939).
39/2. F. D. Rossini and W. E. Deming, *J. Wash. Acad. Sci.* **29**, 416 (1939).
39/3. N. S. Osborne, H. F. Stimson, and D. C. Ginnings, *J. Res. Natl. Bur. Stand.* **23**, 261 (1939).
39/4. A. M. Hughes, R. J. Corruccini, and E. C. Gilbert, *J. Am. Chem. Soc.* **61**, 2639 (1939).
39/5. J. B. Conn, G. B. Kistiakowsky, and E. A. Smith, *J. Am. Chem. Soc.* **61**, 1868 (1939).
39/6. J. W. Knowlton and F. D. Rossini, *J. Res. Natl. Bur. Stand.* **22**, 415 (1939).
39/7. G. E. Moore and G. S. Parks, *J. Am. Chem. Soc.* **61**, 2561 (1939).
39/8. A. L. Glasebrook and W. G. Lovell, *J. Am. Chem. Soc.* **61**, 1717 (1939).
39/9. M. Matsui and T. Abe, *Bull. Tokyo Univ. Eng.* **8**, 339 (1939): Chem. Zent. I, 804 (1941).
39/10. J. W. Richardson and G. S. Parks, *J. Am. Chem. Soc.* **61**, 3543 (1939).
39/11. G. S. Parks and G. E. Moore, *J. Chem. Phys.* **7**, 1066 (1939).
39/12. A. Skita and W. Faust, *Ber.* **72**, 1127 (1939).

39/13. E. I. Blat, M. I. Gerber, and M. B. Neumann, *Acta Physicochim. U.R.S.S.* **10**, 273 (1939).
39/14. A. Skita and R. Rössler, *Ber.* **72**, 265 (1939).
39/15. T. H. Clarke and G. Stegeman, *J. Am. Chem. Soc.* **61**, 1726 (1939).
39/16. J. G. Aston, M. L. Eidinoff, and W. S. Forster, *J. Am. Chem. Soc.* **61**, 1539 (1939).
39/17. E. Burlot, M. Thomas, and M. Badoche, *Mém. Poudres* **29**, 226 (1939).
39/18. R. A. Ruehrwein and W. F. Giauque, *J. Am. Chem. Soc.* **61**, 2940 (1939).
39/19. R. J. Corruccini and E. C. Gilbert, *J. Am. Chem. Soc.* **61**, 2925 (1939).
39/20. M. Badoche, *Bull. Soc. Chim.* (*France*), **6**, 570 (1939).
39/21. L. Riedel, *Bull. Int. Inst. Refrig.* **20**, Annex B3, 1 (1939).
39/22. K. L. Müller and H.-J. Schumacher, *Z. Phys. Chem.* **B42**, 327 (1939).
39/23. L. Pauling, "The Nature of the Chemical Bond", Cornell University Press, 1939.

1940

40/1. J. G. Aston and G. H. Messerly, *J. Am. Chem. Soc.* **62**, 1917 (1940).
40/2. J. G. Aston, R. M. Kennedy, and S. C. Schumann, *J. Am. Chem. Soc.* **62**, 2059 (1940).
40/3. G. E. Moore, M. L. Renquist, and G. S. Parks, *J. Am. Chem. Soc.* **62**, 1505 (1940).
40/4. E. C. Stathis and A. C. Egerton, *Trans. Faraday Soc.* **36**, 606 (1940).
40/5. R. Manzoni-Ansidei and T. Storto, *Atti. Acad. Italia, Rend. Classe Sci. Fis., Mat. Nat.* **1**, 465 (1940).
40/6. G. Tappi, *Gazz. Chim. Ital.* **70**, 414 (1940).
40/7. H. M. Huffman, *J. Am. Chem. Soc.* **62**, 1009 (1940).
40/8. E. J. Rosenbaum and C. R. Sandberg, *J. Am. Chem. Soc.* **62**, 1622 (1940).

1941

41/1. E. J. Prosen and F. D. Rossini, *J. Res. Natl. Bur. Stand.* **27**, 289 (1941).
41/2. "Temperature–Its Measurement and Control in Science and Industry", Reinhold, New York, 1941.
41/3. C. B. Miles and H. Hunt, *J. Phys. Chem.* **45**, 1346 (1941).
41/4. K. S. Pitzer, *J. Am. Chem. Soc.* **63**, 2413 (1941).
41/5. D. M. Yost, D. W. Osborne, and C. S. Garner, *J. Am. Chem. Soc.* **63**, 3492 (1941).
41/6. G. F. Davies and E. C. Gilbert, *J. Am. Chem. Soc.* **63**, 2730 (1941).
41/7. E. J. Prosen and F. D. Rossini, *J. Res. Natl. Bur. Stand.* **27**, 519 (1941).
41/8. E. Baur and S. Frater, *Helv. Chim. Acta*, **24**, 768 (1941).
41/9. G. F. Davies and E. C. Gilbert, *J. Am. Chem. Soc.* **63**, 1585 (1941).
41/10. M. Badoche, *Bull. Soc. Chim.* (*France*), **8**, 212 (1941).
41/11. J. W. Baker and W. T. Tweed, *J. Chem. Soc.* p.796 (1941).
41/12. M. W. Lister, *J. Am. Chem. Soc.* **63**, 143 (1941).
41/13. A. W. Laubengayer and W. F. Gilliam, *J. Am. Chem. Soc.* **63**, 477 (1941).

1942

42/1. R. B. Williams, *J. Am. Chem. Soc.* **64**, 1395 (1942).

42/2.   M. Delépine and M. Badoche, *Compt. Rend.* **214**, 777 (1942).
42/3.   R. S. Crog and H. Hunt, *J. Phys. Chem.* **46**, 1162 (1942).
42/4.   G. Jung and J. Dahmlos, *Z. Phys. Chem.* **A190**, 230 (1942).
42/5.   J. O. Halford and D. Brundage, *J. Am. Chem. Soc.* **64**, 36 (1942).
42/6.   R. S. Jessup, *J. Res. Natl. Bur. Stand.* **29**, 247 (1942).
42/7.   J. B. Conn, G. B. Kistiakowsky, R. M. Roberts, and E. A. Smith, *J. Am. Chem. Soc.* **64**, 1747 (1942).
42/8.   C. M. Anderson and E. C. Gilbert, *J. Am. Chem. Soc.* **64**, 2369 (1942).
42/9.   G. E. Williams and E. C. Gilbert, *J. Am. Chem. Soc.* **64**, 2776 (1942).
42/10. H. M. Huffman, *J. Phys. Chem.* **46**, 885 (1942).
42/11. M. Badoche, *Bull. Soc. Chim. (France)*, **9**, 86 (1942).

1943

43/1.   J. F. Walker and P. J. Carlisle, *Chem. Eng. News*, **21**, 1251 (1943).
43/2.   J. A. A. Ketelaar and S. Kruyer, *Rec. Trav. Chim.* **62**, 550 (1943).

1944

44/1.   E. J. Prosen and F. D. Rossini, *J. Res. Natl. Bur. Stand.* **33**, 439 (1944).
44/2.   E. J. Prosen, R. S. Jessup and F. D. Rossini, *J. Res. Natl. Bur. Stand.* **33**, 447 (1944).
44/3.   H. J. McDonald, *J. Phys. Chem.* **48**, 47 (1944).
44/4.   L. Smith and S. Sunner, "The Svedberg", p.352, Almqvist & Wiksells Boktryckeri AB, Uppsala, 1944.
44/5.   W. A. Roth, *Ber.* **77**, 535 (1944).
44/6.   E. J. Prosen and F. D. Rossini, *J. Res. Natl. Bur. Stand.* **33**, 255 (1944).
44/7.   J. W. Knowlton and H. M. Huffman, *J. Am. Chem. Soc.* **66**, 1492 (1944).
44/8.   R. B. Scott, W. J. Ferguson, and F. G. Brickwedde, *J. Res. Natl. Bur Stand.* **33**, 1 (1944).
44/9.   W. A. Roth and E. Rist-Schumacher, *Z. Electrochem.* **50**, 7 (1944).
44/10. T. H. Clarke and G. Stegeman, *J. Am. Chem. Soc.* **66**, 457 (1944).
44/11. J. G. Aston, M. L. Sagenkahn, G. J. Szasz, G. W. Moessen, and H. F Zuhr, *J. Am. Chem. Soc.* **66**, 1171 (1944).
44/12. A. Pongratz, S. Böhmert-Süss and K. Scholtis, *Ber.* **77**, 651 (1944).
44/13. Y. Oka, *Nippon Seirigaku Zasshi*, **9**, 365 (1944); *C.A.* **41**, 470lf (1947).
44/14. W. A. Roth and K. Isecke, *Ber.* **77**, 537 (1944).
44/15. F. R. Mayo and A. A. Dolnick, *J. Am. Chem. Soc.* **66**, 985 (1944).

1945

45/1.   E. J. Prosen and F. D. Rossini, *J. Res. Natl. Bur. Stand.* **34**, 263 (1945).
45/2.   W. Swietoslawski, "Ebulliometric Measurements", Reinhold, New York, 1945.
45/3.   H. Pines, B. Kvetinskas, L. S. Kassel, and V. N. Ipatieff, *J. Am. Chem. Soc.* **67**, 631 (1945).
45/4.   R. B. Scott, C. H. Meyers, R. D. Rands, F. G. Brickwedde, and N. Bekkedahl, *J. Res. Natl. Bur. Stand.* **35**, 39 (1945).
45/5.   L. Guttman and K.S. Pitzer, *J. Am. Chem. Soc.* **67**, 324 (1945).

45/6.   E. J. Prosen, R. Gilmont, and F. D. Rossini, *J. Res. Natl. Bur. Stand.* **34**, 65 (1945).

45/7.   R. B. Scott and F. G. Brickwedde, *J. Res. Natl. Bur. Stand.* **35**, 501 (1945).

45/8.   E. J. Prosen and F. D. Rossini, *J. Res. Natl. Bur. Stand.* **34**, 163 (1945).

45/9.   W. H. Johnson, E. J. Prosen, and F. D. Rossini, *J. Res. Natl. Bur. Stand.* **35**, 141 (1945).

45/10. N. Shlechter, D. F. Othmer, and S. Marshak, *Ind. Eng. Chem.* **37**, 900 (1945).

45/11. H. S. Davis and O. F. Wiedeman, *Ind. Eng. Chem.* **37**, 482 (1945).

45/12. A. J. Miller and H. Hunt, *J. Phys. Chem.* **49**, 20 (1945).

1946

46/1.   S. Sunner, *Svensk. Kem. Tidr.* **58**, 71 (1946).

46/2.   G. W. Thomson, *Chem. Rev.* **38**, 1 (1946).

46/3.   J. G. Aston, H. L. Fink, A. B. Bestul, E. L. Pace, and G. J. Szasz, *J. Am. Chem. Soc.* **68**, 52 (1946).

46/4.   J. Coops, D. Mulder, J. W. Dienske, and J. Smittenberg, *Rev. Trav. Chim.* **65**, 128 (1946).

46/5.   W. H. Johnson, E. J. Prosen, and F. D. Rossini, *J. Res. Natl. Bur. Stand.* **36**, 463 (1946).

46/6.   J. E. Kilpatrick and K. S. Pitzer, *J. Am. Chem. Soc.* **68**, 1066 (1946).

46/7.   E. J. Prosen, W. H. Johnson, and F. D. Rossini, *J. Res. Natl. Bur. Stand.* **36**, 455 (1946).

46/8.   G. S. Parks, T. J. West, B. F. Naylor, P. S. Fujii, and L. A. McClaine, *J. Am. Chem. Soc.* **68**, 2524 (1946).

46/9.   G. S. Parks and R. D. Rowe, *J. Chem. Phys.* **14**, 507 (1946).

46/10. M. Badoche, *Bull. Soc. Chim. (France)*, **13**, 37 (1946).

46/11. D. A. Crooks and F. M. Feetham, *J. Chem. Soc.* p.899 (1946).

46/12. A. Albert and J. B. Willis, *Nature*, **157**, 341 (1946).

46/13. T. B. Douglas, *J. Am. Chem. Soc.* **68**, 1072 (1946).

46/14. C. H. Bamford, D. L. Levi, and D. M. Newitt, *J. Chem. Soc.* p.468 (1946).

46/15. K. S. Pitzer, L. Guttman, and E. F. Westrum, *J. Am. Chem. Soc.* **68**, 2209 (1946).

46/16. J. W. Knowlton, N. C. Schieltz, and D. Macmillan, *J. Am. Chem. Soc.* **68**, 208 (1946).

46/17. E. J. Prosen, W. H. Johnson, and F. D. Rossini, *J. Res. Natl. Bur. Stand.* **37**, 51 (1946).

46/18. H. S. Booth and J. F. Suttle, *J. Am. Chem. Soc.* **68**, 2658 (1946).

1947

47/1.   F. D. Rossini, K. S. Pitzer, W. J. Taylor, J. P. Ebert, J. E. Kilpatrick, C. W. Beckett, M. G. Williams and H. G. Werner, "Selected Values of Properties of Hydrocarbons", American Petroleum Institute Research Project 44, Natl. Bur. Stand. Circular 461, U.S. Government Printing Office, Washington D.C., 1947.

47/2.   J. G. Aston, H. L. Finke, J. W. Tooke and M. R. Cines, *Anal. Chem.* **19**, 218 (1947).

47/3.   J. Coops, K. Van Nes, A. Kentie and J. W. Dienske, *Rec. Trav. Chim.* **66**, 113 *et seq.* (1947).

47/4.   D. E. Roberts, W. W. Walton, and R. S. Jessup, *J. Res. Natl. Bur. Stand.* **38**, 627 (1947).

47/5.   N. S. Osborne and D. C. Ginnings, *J. Res. Natl. Bur. Stand.* **39**, 453 (1947).

47/6.   G. Waddington, S. S. Todd, and H. M. Huffman, *J. Am. Chem. Soc.* **69**, 22 (1947).

47/7.   E. W. Balson, *Trans. Faraday Soc.* **43**, 48 (1947).

47/8.   E. W. Balson, *Trans. Faraday Soc.* **43**, 54 (1947).

47/9.   H. Wiener, *J. Chem. Phys.* **15**, 766 (1947); *J. Am. Chem. Soc.* **69**, 17, 2636 (1947).

47/10. J. R. Platt, *J. Chem. Phys.* **15**, 419 (1947).

47/11. J. G. Aston and G. J. Szasz, *J. Am. Chem. Soc.* **69**, 3108 (1947).

47/12. R. Spitzer and H. M. Huffman, *J. Am. Chem. Soc.* **69**, 211 (1947).

47/13. W. H. Johnson, E. J. Prosen, and F. D. Rossini, *J. Res. Natl. Bur. Stand.* **39**, 49 (1947).

47/14. W. H. Johnson, E. J. Prosen, and F. D. Rossini, *J. Res. Natl. Bur. Stand.* **38**, 419 (1947).

47/15. D. R. Stull, *Ind. Eng. Chem.* **39**, 517 (1947).

47/16. J. B. Willis, *Trans. Faraday Soc.* **43**, 97 (1947).

47/17. J. Taylor and C. R. L. Hall, *J. Phys. Chem.* **51**, 593 (1947).

47/18. K. J. Ivin and F. S. Dainton, *Trans. Faraday Soc.* **43**, 32 (1947).

47/19. J. A. A. Ketelaar, P. F. Van Velden and P. Zalm, *Rec. Trav. Chim.* **66**, 721 (1947).

47/20. D. J. G. Ives, R. W. Pittman, and W Wardlaw, *J. Chem. Soc.* p.1080 (1947).

47/21. J. G. M. Bremner and G. D. Thomas, *Trans. Faraday Soc.* **43**, 779 (1947).

1948

48/1.   G. W. Sears and E. R. Hopke, *J. Phys. Chem.* **52**, 1137 (1948).

48/2.   S. T. Bowden and W. J. Jones, *Phil. Mag.* **39**, 155 (1948).

48/3.   H. Wiener, *J. Phys. Chem.* **52**, 425, 1082 (1948).

48/4.   G. E. Coates and L. E. Sutton, *J. Chem. Soc.* p.1187 (1948).

48/5.   A. Leman and G. Lepoutre, *Compt. Rend.* **226**, 1976 (1948).

48/6.   H. E. Bent and R. J. Francel, *J. Am. Chem. Soc.* **70**, 634 (1948).

48/7.   J. H. Raley, F. F. Rust, and W. E. Vaughan, *J. Am. Chem. Soc.* **70**, 88 (1948).

48/8.   W. N. Hubbard, J. W. Knowlton, and H. M. Huffman, *J. Am. Chem. Soc.* **70**, 3259 (1948).

48/9.   D. E. Roberts and R. S. Jessup, *J. Res. Natl. Bur. Stand.* **40**, 281 (1948).

48/10. D. J. Salley and J. B. Gray, *J. Am. Chem. Soc.* **70**, 2650 (1948).

48/11. V. O. Glemser and V. Hausser, *Z. Naturforsch.* **B3**, 159 (1948).

48/12. T. B. Douglas, *J. Am. Chem. Soc.*, **70**, 2001 (1948).

48/13. B. Bak, *Kgl. Danske Videnskab. Selskab.* **24**, 15 (1948).

48/14. P. A. Small, K. W. Small, and P. Cowley, *Trans. Faraday Soc.* **44**, 810 (1948).

48/15. T. L. Cottrell, *J. Chem. Soc.* p.1448 (1948).

1949

49/1.   L. H. Long and R. G. W. Norrish, *Phil. Trans. Roy. Soc. London*, **A241**, 587 (1949).

49/2.   J. R. Lacher, J. J. McKinley, C. M. Snow, L. Michel, G. Nelson, and J. D. Park, *J. Am. Chem. Soc.* **71**, 1330 (1949).

49/3.   J. W. Knowlton and F. D. Rossini, *J. Res. Natl. Bur. Stand.* **43**, 113 (1949).

49/4.   F. Klages, *Chem. Ber.* **82**, 358 (1949).

49/5.   E. J. Prosen, F. W. Maron, and F. D. Rossini, *J. Res. Natl. Bur. Stand.* **42**, 269 (1949).

49/6.   M. B. Epstein, K. S. Pitzer and F. D. Rossini, *J. Res. Natl. Bur. Stand.* **42**, 379 (1949), quoting unpublished data by E. J. Prosen, F. Yenchius and F. D. Rossini.

49/7.   D. W. Scott, G. Waddington, J. C. Smith, and H. M. Huffman, *J. Am. Chem. Soc.* **71**, 2767 (1949).

49/8.   G. Waddington, J. C. Smith, D. W. Scott, and H. M. Huffman, *J. Am. Chem. Soc.* **71**, 3902 (1949).

49/9.   W. H. Johnson, E. J. Prosen, and F. D. Rossini, *J. Res. Natl. Bur. Stand.* **42**, 251 (1949).

49/10. D. W. Scott, M. E. Gross, G. D. Oliver, and H. M. Huffman, *J. Am. Chem. Soc.* **71**, 1634 (1949).

49/11. G. S. Parks and G. E. Moore, *J. Chem. Phys.* **17**, 1151 (1949).

49/12. R. S. Bradley and A. D. Shellard, *Proc. Roy. Soc. London*, **A198**, 239 (1949).

49/13. C. F. Coleman and T. de Vries, *J. Am. Chem. Soc.* **71**, 2839 (1949).

49/14. W. F. Giauque and J. Gordon, *J. Am. Chem. Soc.* **71**, 2176 (1949).

49/15. B. T. Collins, C. F. Coleman and T. de Vries, *J. Am. Chem. Soc.* **71**, 2929 (1949).

49/16. G. M. Kibler and H. Hunt, *J. Phys. Chem.* **53**, 955 (1949).

49/17. M. V. Sullivan and H. Hunt, *J. Phys. Chem.* **53**, 497 (1949).

49/18. D. E. Holcomb and C. L. Dorsey, *Ind. Eng. Chem.* **41**, 2788 (1949).

49/19. G. Waddington, J. W. Knowlton, D. W. Scott, G. D. Oliver, S. S. Todd, W. N. Hubbard, J. C. Smith, and H. M. Huffman, *J. Am. Chem. Soc.* **71**, 797 (1949).

49/20. H. Guérin, M. Marthe, M. J. Bastick and J. Adam-Gironne, *Compt. Rend.* **228**, 87 (1949).

49/21. A. S. Carson, K. Hartley and H. A. Skinner, *Proc. Roy. Soc. London*, **A195**, 500 (1949).

49/22. J. R. Lacher, J. J. McKinley, C. H. Walden, K. R. Lea, and J. D. Park, *J. Am. Chem. Soc.* **71**, 1334 (1949).

49/23. H. D. Springall and T. R. White, *Research*, **2**, 296 (1949).

49/24. O. H. Gellner and H. A. Skinner, *J. Chem. Soc.* p.1145 (1949).

49/25. R. R. Dreisbach and R. A. Martin, *Ind. Eng. Chem.* **41**, 2875 (1949).

49/26. G. W. Sears and E. R. Hopke, *J. Am. Chem. Soc.* **71**, 1632 (1949).

49/27. H. v. Wartenberg, *Z. Anorg. Chem.* **258**, 356 (1949).

49/28. A. S. Carson and H. A. Skinner, *J. Chem. Soc.* p.936 (1949).

49/29. A. S. Carson, K. Hartley, and H. A. Skinner, *Trans. Faraday Soc.* **45**, 1159 (1949).

49/30. L. F. Hatch, G. Sutherland, and W. J. Ross, *J. Org. Chem.* **14**, 1130 (1949).

49/31. M. V. Forward, S. T. Bowden, and W. J. Jones, *J. Chem. Soc.* p.S121 (1949).

49/32. J. Birks and R. S. Bradley, *Proc. Roy. Soc. London*, **A198**, 226 (1949).

49/33. L. O. Winstrom and L. Kulp, *Ind. Eng. Chem.* **41**, 2584 (1949).

49/34. H. A. Skinner, *Trans. Faraday Soc.* **45**, 20 (1949).

49/35. M. Souders, C. S. Matthews, and C. O. Hurd, *Ind. Eng. Chem.* **41**, 1048 (1949).

49/36. J. L. Franklin, *Ind. Eng. Chem.* **41**, 1070 (1949).
49/37. C. A. Coulson and W. E. Moffitt, *Phil. Mag.* **40**, 1 (1949).

1950

50/1. F. D. Rossini, D. D. Wagman, W. H. Evans, S. Levine and I. Jaffe, "Selected Values of Chemical Thermodynamic Properties", Natl. Bur. Stand. Circular 500, U.S. Government Printing Office, Washington D.C., 1950.

50/2. F. D. Rossini, "Chemical Thermodynamics", John Wiley and Sons Inc., New York, 1950.

50/3. E. F. G. Herington and R. Handley, *J. Chem. Soc.* p.199 (1950).

50/4. S. Ekegren, O. Ohrn, K. Granath and P. Kinell, *Acta Chem. Scand.* **4**, 126 (1950).

50/5. A. G. Worthing and J. Geffner, "Treatment of Experimental Data", John Wiley and Sons Inc., New York. 1950.

50/6. J. Coops and S. Kaarsemaker, *Rec. Trav. Chim.* **69**, 1364 (1950).

50/7. J. G. Aston, S. V. R. Mastrangelo and G. W. Moessen, *J. Am. Chem. Soc.* **72**, 5287 (1950).

50/8. A. F. Forziati, D. L. Camin and F. D. Rossini, *J. Res. Natl. Bur. Stand.* **45**, 406 (1950).

50/9. G. L. Humphrey and R. Spitzer, *J. Chem. Phys.* **18**, 902 (1950).

50/10. D. W. Scott, H. L. Finke, W. N. Hubbard, J. P. McCullough, M. E. Gross, K. D. Williamson, G. Waddington, and H. M. Huffman, *J. Am. Chem. Soc.* **72**, 4664 (1950).

50/11. E. J. Prosen, W. H. Johnson, and F. D. Rossini, *J. Am. Chem. Soc.* **72**, 626 (1950).

50/12. G. S. Parks and J. R. Mosley, *J. Am. Chem. Soc.* **72**, 1850 (1950).

50/13. J. Coops and G. J. Hoijtink, *Rec. Trav. Chim.* **69**, 358 (1950).

50/14. G. S. Parks, J. R. Mosley and P. V. Peterson, *J. Chem. Phys.* **18**, 152 (1950).

50/15. J. W. Breitenbach and J. Derkosch, *Monat. Chem.* **81**, 689 (1950).

50/16. C. M. Anderson, L. G. Cole, and E. C. Gilbert, *J. Am. Chem. Soc.* **72**, 1263 (1950).

50/17. G. Edwards, *Trans. Faraday Soc.* **46**, 423 (1950).

50/18. D. W. Scott, H. L. Finke, M. E. Gross, G. B. Guthrie, and H. M. Huffman, *J. Am. Chem. Soc.* **72**, 2424 (1950).

50/19. S. Sunner, *Svensk. Kem. Tidr.* **62**, 204 (1950).

50/20. J. R. Lacher, K. R. Lea, C. H. Walden, G. G. Olson, and J. D. Park, *J. Am. Chem. Soc.* **72**, 3231 (1950).

50/21. J. R. Lacher, C. H. Walden, K. R. Lea and J. D. Park, *J. Am. Chem. Soc.* **72**, 331 (1950).

50/22. H. O. Pritchard and H. A. Skinner, *J. Chem. Soc.* p.272 (1950).

50/23. H. O. Pritchard and H. A. Skinner, *J. Chem. Soc.* p.1928 (1950).

50/24. H. O. Pritchard and H. A. Skinner, *J. Chem. Soc.* p.1099 (1950).

50/25. A. S. Carson, H. O. Pritchard, and H. A. Skinner, *J. Chem. Soc.* p.656 (1950).

50/26. K. Hartley, H. O. Pritchard, and H. A. Skinner, *Trans. Faraday Soc.* **46**, 1019 (1950).

50/27. I. Nitta, S. Seki, M. Momotani, K. Suzuki, and S. Nakagawa, *Proc. Jap. Acad.* **26**, (10), 11 (1950).

50/28. I. Nitta, S. Seki, M. Momotani, and K. Sato, *J. Chem. Soc. Japan*, **71**, 378 (1950).
50/29. E. S. Naidus and M. B. Mueller, *J. Am. Chem. Soc.* **72**, 1829 (1950).
50/30. V. M. Tatevskii, V. V. Korolov, and E. A. Mendzheretskii, *Dokl. Akad. Nauk. S.S.S.R.* **74**, 743 (1950).

1951

51/1.   A. Magnus and F. Becker, *Z. Phys. Chem.* **196**, 378 (1951).
51/2.   E. J. Prosen, F. W. Maron, and F. D. Rossini, *J. Res. Natl. Bur. Stand.* **46**, 106 (1951).
51/3.   J. W. Knowlton and E. J. Prosen, *J. Res. Natl. Bur. Stand.* **46**, 489 (1951).
51/4.   D. W. H. Casey and S. Fordham, *J. Chem. Soc.* p.2513 (1951).
51/5.   L. G. Cole and E. C. Gilbert, *J. Am. Chem. Soc.* **73**, 5423 (1951).
51/6.   G. B. Kistiakowsky and A. G. Nickle, *Diss. Faraday Soc.* **10**, 175 (1951).
51/7.   D. W. Scott, J. P. McCullough, K. D. Williamson, and G. Waddington, *J. Am. Chem. Soc.* **73**, 1707 (1951).
51/8.   D. E. Roberts and R. S. Jessup, *J. Res. Natl. Bur. Stand.* **46**, 11 (1951).
51/9.   H. Hock and G. Knauel, *Chem. Ber.* **84**, 1 (1951).
51/10. G. S. Parks and L. M. Vaughan, *J. Am. Chem. Soc.* **73**, 2380 (1951).
51/11. A. Magnus, H. Hartmann and F. Becker. *Z. Phys. Chem.* **197**, 75 (1951).
51/12. J. W. Breitenbach and J. Derkosch, *Monat. Chem.* **82**, 177 (1951).
51/13. G. T. Furukawa, D. C. Ginnings, R. E. McCoskey, and R. A. Nelson, *J. Res. Natl. Bur. Stand.* **46**, 195 (1951).
51/14. W. E. Vaughan, *Diss. Faraday Soc.* **10**, 330 (1951).
51/15. G. R. Nicholson, quoted by J. W. Cook, A. R. Gibb, R. A. Raphael, and A. R. Somerville, *J. Chem. Soc.* p.503 (1951).
51/16. A. C. Egerton, W. Emte and G. J. Minkoff, *Diss. Faraday Soc.* **10**, 278 (1951).
51/17. D. J. Salley and J. B. Gray, *J. Am. Chem. Soc.* **73**, 5925 (1951).
51/18. J. G. Aston, G. J. Janz and K. E. Russell, *J. Am. Chem. Soc.* **73**, 1943 (1951).
51/19. W. S. McEwan and M. W. Rigg, *J. Am. Chem. Soc.* **73**, 4725 (1951).
51/20. D. W. Scott, H. L. Finke, J. P. McCullough, M. E. Gross, K. D. Williamson, G. Waddington, and H. M. Huffman, *J. Am. Chem. Soc.* **73**, 261 (1951).
51/21. J. H. Sullivan and N. Davidson, *J. Chem. Phys.* **19**, 143 (1951).
51/22. K. E. Howlett, *J. Chem. Soc.* p.1409 (1951).
51/23. K. Hartley, H. O. Pritchard, and H. A. Skinner, *Trans. Faraday Soc.* **47**, 254 (1951).
51/24. G. D. Oliver and J. W. Grisard, *J. Am. Chem. Soc.* **73**, 1688 (1951).
51/25. T. Charnley and H. A. Skinner, *J. Chem. Soc.* p.1921 (1951).
51/26. I. Nitta, S. Seki and K. Suzuki, *Bull. Chem. Soc. Japan*, **24**, 63 (1951).
51/27. J. G. Aston, H. L. Fink, G. J. Janz and K. E. Russell, *J. Am. Chem. Soc.* **73**, 1939 (1951).
51/28. D. Bryce-Smith and K. E. Howlett, *J. Chem. Soc.* p.1141 (1951).
51/29. A. B. Burg and C. L. Randolph, *J. Am. Chem. Soc.* **73**, 953 (1951).

1952

52/1.   G. L. Humphrey and E. G. King, *J. Am. Chem. Soc.* **74**, 2041 (1952).

52/2. J. R. Platt, *J. Phys. Chem.* **56**, 328 (1952).

52/3. S. Kaarsemaker and J. Coops, *Rec. Trav. Chim.* **71**, 261 (1952).

52/4. D. W. Scott, D. R. Douslin, M. E. Gross, G. D. Oliver, and H. M. Huffman, *J. Am. Chem. Soc.* **74**, 883 (1952).

52/5. P. Bender and J. Farber, *J. Am. Chem. Soc.* **74**, 1450 (1952).

52/6. D. S. Brackman and P. H. Plesch, *J. Chem. Soc.* p.2188 (1952).

52/7. G. B. Guthrie, D. W. Scott, W. N. Hubbard, C. Katz, J. P. McCullough, M. E. Gross, K. D. Williamson, and G. Waddington, *J. Am. Chem. Soc.* **74**, 4662 (1952).

52/8. R. W. Taft, J. B. Levy, D. Aaron, and L. P. Hammett, *J. Am. Chem. Soc.* **74**, 4735 (1952).

52/9. R. E. Rebbert and K. J. Laidler, *J. Chem. Phys.* **20**, 574 (1952).

52/10. W. N. Hubbard, C. Katz, G. B. Guthrie, and G. Waddington, *J. Am. Chem. Soc.* **74**, 4456 (1952).

52/11. J. W. C. Crawford, and S. D. Swift, *J. Chem. Soc.* p.1220 (1952).

52/12. L. Médard and M. Thomas, *Mém. Poudres*, **34**, 421 (1952).

52/13. G. S. Parks and K. E. Manchester, *J. Am. Chem. Soc.* **74**, 3435 (1952).

52/14. R. A. Nelson and R. S. Jessup, *J. Res. Natl. Bur. Stand.* **48**, 206 (1952).

52/15. G. N. Vriens and A. G. Hill, *Ind. Eng. Chem.* **44**, 2732 (1952).

52/16. J. G. Aston, E. J. Rock, and S. Isserow, *J. Am. Chem. Soc.* **74**, 2484 (1952).

52/17. J. W. Breitenbach, J. Derkosch, and F. Wessely, *Nature*, **169**, 922 (1952).

52/18. J. W. Breitenbach, J. Derkosch, and F. Wessely, *Monat. Chem.* **83**, 591 (1952).

52/19. R. M. Currie, C. O. Bennett, and D. E. Holcomb, *Ind. Eng. Chem.* **44**, 329 (1952).

52/20. R. S. Bradley, S. Cotson, and E. G. Cox, *J. Chem. Soc.* p.740 (1952).

52/21. G. Edwards, *Trans. Faraday Soc.* **48**, 513 (1952).

52/22. G. B. Guthrie, D. W. Scott, and G. Waddington, *J. Am. Chem. Soc.* **74**, 2795 (1952).

52/23. J. P. McCullough, D. W. Scott, H. L. Finke, M. E. Gross, K. D. Williamson, R. E. Pennington, G. Waddington, and H. M. Huffman, *J. Am. Chem. Soc.* **74**, 2801 (1952).

52/24. W. N. Hubbard, H. L. Finke, D. W. Scott, J. P. McCullough, C. Katz, M. E. Gross, J. F. Messerly, R. E. Pennington, and G. Waddington. *J. Am. Chem. Soc.* **74**, 6025 (1952).

52/25. D. W. Scott, H. L. Finke, W. N. Hubbard, J. P. McCullough, G. D. Oliver, M. E. Gross, C. Katz, K. D. Williamson, G. Waddington, and H. M. Huffman, *J. Am. Chem. Soc.* **74**, 4656 (1952).

52/26. D. W. Scott, H. L. Finke, J. P. McCullough, M. E. Gross, R. E. Pennington, and G. Waddington, *J. Am. Chem. Soc.* **74**, 2478 (1952).

52/27. R. J. Nichol and A. R. Ubbelohde, *J. Chem. Soc.* p.415 (1952).

52/28. C. T. Mortimer, H. O. Pritchard, and H. A. Skinner, *Trans. Faraday Soc.* **48**, 220 (1952).

52/29. J. R. Lacher, T. J. Billings, D. E. Campion, K. R. Lea, and J. D. Park, *J. Am. Chem. Soc.* **74**, 5291 (1952).

52/30. A. S. Carson, E. M. Carson, and B. Wilmshurst, *Nature*, **170**, 320 (1952).

52/31. T. F. S. Tees, *Trans. Faraday Soc.* **48**, 227 (1952).

52/32. T. N. Rezukhina and V. V. Shvyrev, *Vestnik Moskov. Univ.* **7**, 41 (1952).

52/33. F. A. Cotton and G. Wilkinson, *J. Am. Chem. Soc.* **74**, 5764 (1952).

52/34. T. A. Scott, D. Macmillan, and E. H. Melvin, *Ind. Eng. Chem.* **44**, 172 (1952).

52/35. T. Charnley, H. A. Skinner, and N. B. Smith, *J. Chem. Soc.* p.2288 (1952).
52/36. G. M. Barrow, *J. Chem. Phys.* **20**, 1739 (1952).
52/37. D. W. van Krevelen and H. A. G. Chermin, *Chem. Eng. Sci.* **1**, 66 (1952).

1953

53/1. "Selected Values of Properties of Hydrocarbons", American Petroleum Institute Research Project 44, Carnegie Institute of Technology, Pittsburgh, Pa., 1953; currently issued in loose-leaf form by A.P.I.R.P. 44, Texas A. and M., College Station, Texas.
53/2. J. T. Clarke, H. L. Johnston and W. de Sorbo, *Anal. Chem.* **25**, 1156 (1953).
53/3. J. Coops, D. Mulder, J. W. Dienske, and J. Smittenberg, *Rec. Trav. Chim.* **72**, 785 (1953).
53/4. L. Smith, L. Bjellerup, S. Krook and H. Westermark, *Acta Chem. Scand.* **7**, 65 (1953).
53/5. R. Thompson, *J. Chem. Soc.* p.1908 (1953).
53/6. D. W. Scott, H. L. Finke, W. N. Hubbard, J. P. McCullough, C. Katz, M. E. Gross, J. F. Messerly, R. E. Pennington, and G. Waddington, *J. Am. Chem. Soc.* **75**, 2795 (1953).
53/7. R. S. Bradley and A. D. Care, *J. Chem. Soc.* p.1688 (1953).
53/8. R. S. Bradley and T. G. Cleasby, *J. Chem. Soc.* p.1690 (1953).
53/9. J. A. Hall, "Fundamentals of Thermometry", The Institute of Physics, London, 1953.
53/10. J. A. Hall, "Practical Thermometry", The Institute of Physics, London, 1953.
53/11. G. W. Rathjens and W. D. Gwinn, *J. Am. Chem. Soc.* **75**, 5629 (1953).
53/12. J. Coops, G. J. Hoijtink, T. J. E. Kramer and A. C. Faber, *Rec. Trav. Chim.* **72**, 781 (1953).
53/13. J. Coops, G. J. Hoijtink, T. J. E. Kramer and A. C. Faber, *Rec. Trav. Chim.* **72**, 765 (1953).
53/14. J. Coops, C. J. Hoijtink and T. J. E. Kramer, *Rec. Trav. Chim.* **72**, 793 (1953).
53/15. L. Médard and M. Thomas, *Mém. Poudres.* **35**, 155 (1953).
53/16. A. D. Jenkins and D. W. G. Style, *J. Chem. Soc.* p.2337 (1953).
53/17. J. G. Aston, J. L. Wood, and T. P. Zolki, *J. Am. Chem. Soc.* **75**, 6202 (1953).
53/18. T. F. Fagley, E. Klein, and J. F. Albrecht, *J. Am. Chem. Soc.* **75**, 3104 (1953).
53/19. J. P. McCullough, D. W. Scott, H. L. Finke, W. N. Hubbard, M. E. Gross, C. Katz, R. E. Pennington, J. F. Messerly, and G. Waddington, *J. Am. Chem. Soc.* **75**, 1818 (1953).
53/20. J. P. McCullough, S. Sunner, H. L. Finke, W. N. Hubbard, M. E. Gross, R. E. Pennington, J. F. Messerly, W. D. Good, and G. Waddington, *J. Am. Chem. Soc.* **75**, 5075 (1953).
53/21. M. R. Lane, J. W. Linnett, and H. G. Oswin, *Proc. Roy. Soc. London,* **A216**, 361 (1953).
53/22. R. Juza and E. Hillenbrand, *Z. Anorg. Chem.* **273**, 297 (1953).
53/23. H. A. Skinner and T. F. S. Tees, *J. Chem. Soc.* p.3378 (1953).
53/24. G. L. Humphrey and C. J. O'Brien, *J. Am. Chem. Soc.* **75**, 2805 (1953).
53/25. E. R. Lippincott and M. C. Tobin, *J. Am. Chem. Soc.* **75**, 4141 (1953).
53/26. B. Stevens, *J. Chem. Soc.* p.2973 (1953).

53/27. H. Stage, *Fette Seifen*, **55**, 217 (1953).
53/28. R. S. Bradley and T. G. Cleasby, *J. Chem. Soc.* p.1681 (1953).
53/29. R. S. Bradley and S. Cotson, *J. Chem. Soc.* p.1684 (1953).
53/31. A. Aihara, *J. Chem. Soc. Japan*, **74**, 437 (1953).
53/32. A. Aihara, *J. Chem. Soc. Japan*, **74**, 634 (1953).
53/33. G. Edwards, *Trans. Faraday Soc.* **49**, 152 (1953).
53/34. S. Sunner and B. Lundin, *Acta Chem. Scand.* **7**, 1112 (1953).
53/35. H. A. Skinner and N. B. Smith, *J. Chem. Soc.* p.4025 (1953).
53/36. W. T. Grubb and R. C. Osthoff, *J. Am. Chem. Soc.* **75**, 2230 (1953).
53/37. T. Charnley, C. T. Mortimer, and H. A. Skinner, *J. Chem. Soc.* p.1181 (1953).

1954

54/1. W. N. Hubbard, C. Katz, and G. Waddington, *J. Phys. Chem.* **58**, 142 (1954).
54/2. W. N. Hubbard, J. W. Knowlton, and H. M. Huffman, *J. Phys. Chem.* **58**, 396 (1954).
54/3. L. Bjellerup and L. Smith, *Kgl. Fysiograph. Sällskap. Lund Förh.* **24**, 21 (1954).
54/4. L. Riedel, *Chem. Ing. Tech.* **26**, 679 (1954).
54/5. H. D. Springall, T. R. White, and R. C. Cass, *Trans. Faraday Soc.* **50**, 815 (1954).
54/6. J. E. Hawkins and W. T. Eriksen, *J. Am. Chem. Soc.* **76**, 2669 (1954).
54/7. J. E. Hawkins and G. T. Armstrong, *J. Am. Chem. Soc.* **76**, 3756 (1954).
54/8. G. S. Parks, K. E. Manchester, and L. M. Vaughan, *J. Chem. Phys.* **22**, 2089 (1954).
54/9. G. R. Nicholson, M. Szwarc, and J. W. Taylor, *J. Chem. Soc.* p.2767 (1954).
54/10. H. D. Springall and T. R. White, *J. Chem. Soc.* p.2764 (1954).
54/11. C. K. Hancock, G. M. Watson, and R. F. Gilby, *J. Phys. Chem.* **58**, 127 (1954).
54/12. L. Médard and M. Thomas, *Mém. Poudres*, **36**, 97 (1954).
54/13. E. Briner and P. de Chastonay, *Helv. Chim. Acta*, **37**, 626, 1904 (1954); *Compt. Rend.* **238**, 539 (1954).
54/14. J. D. Cox, A. R. Challoner, and A. R. Meetham, *J. Chem. Soc.* p.265 (1954).
54/15. T. F. Fagley and H. W. Myers, *J. Am. Chem. Soc.* **76**, 6001 (1954)
54/16. J. P. McCullough, D. W. Scott, R. E. Pennington, I. A. Hossenlopp, and G. Waddington, *J. Am. Chem. Soc.* **76**, 4791 (1954).
54/17. S. Mizushima, "Structure of Molecules and Internal Rotation", Ch. 5, Academic Press Inc., New York, 1954.
54/18. Y. Tsuzuki, S. Kata, and H. Okazaki, *Kagaku*, **24**, 523 (1954); *C. A.* **48**, 13740h (1954).
54/19. W. N. Hubbard and G. Waddington, *Rec. Trav. Chim.* **73**, 910 (1954).
54/20. J. P. McCullough, H. L. Finke, D. W. Scott, M. E. Gross, J. F. Messerly, R. E. Pennington, and G. Waddington, *J. Am. Chem. Soc.* **76**, 4796 (1954).
54/21. J. P. McCullough, H. L. Finke, W. N. Hubbard, W. D. Good, R. E. Pennington, J. F. Messerly, and G. Waddington, *J. Am. Chem. Soc.* **76**, 2661 (1954).
54/22. A. Abrams and T. W. Davis, *J. Am. Chem. Soc.* **76**, 5993 (1954).
54/23. G. Lord and A. A. Woolf, *J. Chem. Soc.* p.2546 (1954).

54/24. A. D. Mah, *J. Am. Chem. Soc.* **76**, 3363 (1954).
54/25. L. H. Long and J. F. Sackman, *Trans. Faraday Soc.* **50**, 1177 (1954).
54/26. G. Wilkinson, P. L. Pauson, and F. A. Cotton, *J. Am. Chem. Soc.* **76**, 1970 (1954).
54/27. M. Davies and J. I. Jones, *Trans. Faraday Soc.* **50**, 1042 (1954).
54/28. T. E. Jordan, "Vapour Pressure of Organic Compounds", Wiley (Interscience), New York, 1954.
54/29. H. A. Skinner, *Rec. Trav. Chim.* **73**, 991 (1954).
54/30. H. A. Skinner and N. B. Smith, *J. Chem. Soc.* p.3930 (1954).
54/31. H. A. Skinner and N. B. Smith, *J. Chem. Soc.* p.2324 (1954).

1955

55/1. D. D. Tunnicliff and H. Stone, *Anal. Chem.* **27**, 73 (1955).
55/2. A. R. Challoner, H. A. Gundry and A. R. Meetham, *Phil. Trans. Roy. Soc. London*, **A247**, 553 (1955).
55/3. G. Pilcher and L. E. Sutton, *Phil. Trans. Roy. Soc. London*, **A248**, 23 (1955).
55/4. R. S. Jessup, *J. Res. Natl. Bur. Stand.* **55**, 317 (1955).
55/5. K. J. Ivin, *Trans. Faraday Soc.* **51**, 1273 (1955).
55/6. "Temperature—Its Measurement and Control in Science and Industry", Reinhold, New York, 1955.
55/7. A. F. Trotman-Dickenson, "Gas Kinetics", Butterworths Scientific Publications, London, 1955.
55/8. R. S. Jessup, R. E. McCoskey and R. A. Nelson, *J. Am. Chem. Soc.* **77**, 244 (1955).
55/9. A. W. Searcy, and R. D. Freeman, *J. Chem. Phys.* **23**, 88 (1955).
55/10. O. P. Kharbanda, *Ind. Chem.* p.124 (1955).
55/11. S. V. R. Mastrangelo and R. W. Dornte, *J. Am. Chem. Soc.* **77**, 6200 (1955).
55/12. F. M. Fraser and E. J. Prosen, *J. Res. Natl. Bur. Stand.* **54**, 143 (1955).
55/13. D. W. Scott, H. L. Finke, J. P. McCullough, M. E. Gross, J. F. Messerly, R. E. Pennington, and G. Waddington, *J. Am. Chem. Soc.* **77**, 4993 (1955).
55/14. F. D. Rossini and K. Li, *Science*, **122**, 513 (1955).
55/15. E. Kováts, H. H. Günthard and P. A. Plattner, *Helv. Chim. Acta*, **38**, 1912 (1955).
55/16. A. Székely, *Acta Chim. Acad. Sci. Hungaricae*, **5**, 317 (1955).
55/17. F. M. Fraser and E. J. Prosen, *J. Res. Natl. Bur. Stand.* **55**, 329 (1955).
55/18. R. W. Taft and P. Riesz, *J. Am. Chem. Soc.* **77**, 902 (1955).
55/19. R. C. Cass, H. D. Springall, and T. R. White, *Chem. Ind. (London)*, p.387 (1955).
55/20. R. C. Cass, H. D. Springall, and P. G. Quincey, *J. Chem. Soc.* p.1188 (1955).
55/21. L. Médard and M. Thomas, *Mém. Poudres*, **37**, 129 (1955).
55/22. P. Tavernier and M. Lamouroux, *Mém. Poudres*, **37**, 197 (1955).
55/23. L. Médard, *J. Chim. Phys.* **52**, 467 (1955).
55/24. A. A. Strepikheev, S. M. Skuratov, O. N. Kachinskaya, R. S. Muromova, E. P. Brykina and S. M. Shtekher, *Dokl. Akad. Nauk S.S.S.R.* **102**, 105 (1955).
55/25. A. A. Strepikheev, S. M. Skuratov, S. M. Shtekher, R. S. Muromova, E. P. Brykina and O. N. Kachinskaya, *Dokl. Akad. Nauk S.S.S.R.* **102**, 543 (1955).

55/26. W. N. Hubbard, D. W. Scott, F. R. Frow, and G. Waddington, *J. Am. Chem. Soc.* **77**, 5855 (1955).

55/27. S. Sunner, *Acta Chem. Scand.* **9**, 847 (1955).

55/28. J. P. McCullough, H. L. Finke, J. F. Messerly, R. E. Pennington, I. A. Hossenlopp, and G. Waddington, *J. Am. Chem. Soc.* **77**, 6119 (1955).

55/29. S. Sunner, *Acta Chem. Scand.* **9**, 837 (1955).

55/30. H. v. Wartenberg and J. Schiefer, *Z. Anorg. Allg. Chem.* **278**, 326 (1955).

55/31. K. E. Howlett, *J. Chem. Soc.* p.1784 (1955).

55/32. E. Neale and L. T. D. Williams, *J. Chem. Soc.* p.2485 (1955).

55/33. L. H. Long and J. F. Sackman, *Research* (*Correspondence*), **8**, S23 (1955).

55/34. L. H. Long and J. F. Sackman, *Trans. Faraday Soc.* **51**, 1062 (1955).

55/35. K. W. Sykes and S. C. Townshend, *J. Chem. Soc.* p.2528 (1955).

55/36. A. Aihara, *J. Chem. Soc. Japan*, **76**, 492 (1955).

55/37. H. M. Spiers, ed., "Technical data on fuel", 5th edition, British National Committee, World Power Conference, London, 1955.

55/38. C. L. Chernick, H. A. Skinner, and C. T. Mortimer, *J. Chem. Soc.* p.3936 (1955).

1956

56/1. F. D. Rossini, ed., "Experimental Thermochemistry", Vol. I, Wiley (Interscience), New York, 1956.

56/2. W. M. Smit, *Rec. Trav. Chim.* **75**, 1309 (1956).

56/3. D. W. Scott, W. D. Good, and G. Waddington, *J. Phys. Chem.* **60**, 1080 (1956).

56/4. W. D. Good, D. W. Scott, and G. Waddington, *J. Phys. Chem.* **60**, 1090 (1956).

56/5. J. R. Lacher, E. Emery, E. Bohmfalk, and J. D. Park, *J. Phys. Chem.* **60**, 492 (1956).

56/6. T. H. Benzinger, *Proc. Natl. Acad. Sci.* **42**, 109 (1956).

56/7. G. Milazzo, *Chem. Ing. Tech.* **28**, 646 (1956).

56/8. K. J. Laidler, *Can. J. Chem.* **34**, 626 (1956).

56/9. D. L. Camin and F. D. Rossini, *J. Phys. Chem.* **60**, 1446 (1956).

56/10. H. L. Finke, D. W. Scott, M. E. Gross, J. F. Messerly, and G. Waddington, *J. Am. Chem. Soc.* **78**, 5469 (1956).

56/11. P. Tavernier and M. Lamouroux, *Mém. Poudres*, **38**, 65 (1956).

56/12. R. W. Schmid, E. Kloster-Jensen, E. Kováts, and E. Heilbronner, *Helv. Chim. Acta*, **39**, 806 (1956).

56/13. C. W. Shoppee and D. F. Williams, *J. Chem. Soc.* p.2488 (1956).

56/14. G. Pilcher and L. E. Sutton, *J. Chem. Soc.* p.2695 (1956).

56/15. A. Magnus, *Z. Phys. Chem.* (Frankfurt), **9**, 141 (1956).

56/16. W. Pritzkow and K. A. Müller, *Chem. Ber.* **89**, 2318 (1956).

56/17. J. Coops, N. Adriaanse, and K. Van Nes, *Rec. Trav. Chim.* **75**, 237 (1956).

56/18. T. Tanaka and T. Watase, *Technol. Repts. Osaka Univ.* **6**, 367 (1956).

56/19. J. A. Young, J. E. Keith, P. Stehle, W. C. Dzombak, and H. Hunt, *Ind. Eng. Chem.* **48**, 1375 (1956).

56/20. L. Médard and M. Thomas, *Mém. Poudres*, **38**, 45 (1956).

56/21. A. B. Burg and C. D. Good, *J. Inorg. Nucl. Chem.* **2**, 237 (1956).

56/22. R. E. Pennington, D. W. Scott, H. L. Finke, J. P. McCullough, J. F. Messerly, I. A. Hossenlopp, and G. Waddington, *J. Am. Chem. Soc.* **78**, 3266 (1956).

56/23. R. E. Pennington, H. L. Finke, W. N. Hubbard, J. F. Messerly, F. R. Frow, I. A. Hossenlopp, and G. Waddington, *J. Am. Chem. Soc.* **78**, 2055 (1956).

56/24. D. W. Scott, J. P. McCullough, W. N. Hubbard, J. F. Messerly, I. A. Hossenlopp, F. R. Frow, and G. Waddington, *J. Am. Chem. Soc.* **78**, 5463 (1956).

56/25. C. A. Neugebauer and J. L. Margrave, *J. Phys. Chem.* **60**, 1318 (1956).

56/26. F. W. Kirkbride, *J. App. Chem.* **6**, 11 (1956).

56/27. J. R. Lacher, L. Casali, and J. D. Park, *J. Phys. Chem.* **60**, 608 (1956).

56/28. J. R. Lacher, A. Kianpour, F. Oetting, and J. D. Park, *Trans. Faraday Soc.* **52**, 1500 (1956).

56/29. J. C. M. Li and K. S. Pitzer, *J. Am. Chem. Soc.* **78**, 1077 (1956).

56/30. J. R. Lacher, A. Kianpour, and J. D. Park, *J. Phys. Chem.* **60**, 1454 (1956).

56/31. L. Smith, *Acta Chem. Scand.* **10**, 884 (1956).

56/32. D. W. Scott, J. P. McCullough, W. D. Good, J. F. Messerly, R. E. Pennington, T. C. Kincheloe, I. A. Hossenlopp, D. R. Douslin, and G. Waddington, *J. Am. Chem. Soc.* **78**, 5457 (1956).

56/33. C. L. Chernick, H. A. Skinner, and I. Wadsö, *Trans. Faraday Soc.* **52**, 1088 (1956).

56/34. D. Brennan and A. R. Ubbelohde, *J. Chem. Soc.* p.3011 (1956).

56/35. J. C. M. Li, *J. Am. Chem. Soc.* **78**, 1081 (1956).

56/36. D. M. Fairbrother and H. A. Skinner, *Trans. Faraday Soc.* **52**, 956 (1956).

56/37. C. L. Chernick and H. A. Skinner, *J. Chem. Soc.* p.1401 (1956).

56/38. E. Neale, L. T. D. Williams and V. T. Moores, *J. Chem. Soc.* p.422 (1956).

56/39. L. H. Long and J. F. Sackman, *Trans. Faraday Soc.* **52**, 1201 (1956).

56/40. F. A. Cotton, A. K. Fischer, and G. Wilkinson, *J. Am. Chem. Soc.* **78**, 5168 (1956).

56/41. C. Spizzichino, *J. Rech. Cent. Natl. Rech. Sci., Bellevue*, **34**, 1 (1956).

56/42. K. Suzuki and T. Koide, *J. Chem. Soc. Japan*, **77**, 346 (1956).

56/43. K. Suzuki, S. Onishi, T. Koide, and S. Seki, *Bull. Chem. Soc. Japan*, **29**, 127 (1956).

56/44. W. S. Holmes and E. Tyrrall, *Chem. Ind. (London)*, p.662 (1956).

56/45. E. A. Haseley, A. B. Garrett, and H. H. Sisler, *J. Phys. Chem.* **60**, 1136 (1956).

56/46. K. S. Pitzer and E. Catalano, *J. Am. Chem. Soc.* **78**, 4844 (1956).

56/47. H. Mauras, *Bull. Soc. Chim. (France)* p.1642 (1956).

56/48. A. Langseth and B. P. Stoicheff, *Can. J. Phys.* **34**. 350 (1956).

1957

57/1. J. P. McCullough and G. Waddington, *Anal. Chim. Acta*, **17**, 80 (1957).

57/2. G. Pilcher, *Anal. Chim. Acta*, **17**, 144 (1957).

57/3. W. M. Smit and G. Kateman, *Anal. Chim. Acta*, **17**, 161 (1957).

57/4. W. M. Smit, ed., "Purity Control by Thermal Analysis", Elsevier, Amsterdam, 1957.

57/5. J. R. Lacher, A. Kianpour, and J. D. Park *J. Phys. Chem.* **61** 1124 (1957).

57/6. J. B. Pedley, H. A. Skinner, and C. L. Chernick, *Trans. Faraday Soc.* **53**, 1612 (1957).

57/7. T. L. Flitcroft, H. A. Skinner, and M. C. Whiting, *Trans. Faraday Soc.* **53**, 784 (1957).

57/8. F. S. Dainton, J. Diaper, K. J. Ivin, and D. R. Sheard, *Trans. Faraday Soc.* **53**, 1269 (1957).

57/9. R. Handley, *Anal. Chim. Acta*, **17**, 115 (1957).

57/10 A. R. Glasgow, G. S. Ross, A. T. Horton, D. Enagonio, H. D. Dixon, C. P. Saylor, G. T. Furukawa, M. L. Reilly and J. M. Henning, *Anal. Chim. Acta*, **17**, 54 (1957).

57/11. J. H. Day and C. Oestreich, *J. Org. Chem.* **22**, 214 (1957).

57/12. E. Kováts, H. H. Günthard, and P. A. Plattner, *Helv. Chim. Acta*, **40**, 2008 (1957).

57/13. J. P. McCullough, H. L. Finke, J. F. Messerly, S. S. Todd, T. C. Kincheloe, and G. Waddington, *J. Phys. Chem.* **61**, 1105 (1957).

57/14. R. E. Pennington and K. A. Kobe, *J. Am. Chem. Soc.* **79**, 300 (1957).

57/15. S. M. Skuratov, A. A. Strepikheev and M. P. Kozina, *Dokl. Akad. Nauk S.S.S.R.* **117**, 452 (1957); Eng. Ed. 711.

57/16. S. M. Skuratov, A. A. Strepikheev, S. M. Shtekher and A. V. Volokina, *Dokl. Akad. Nauk S.S.S.R.* **117**, 263 (1957); Eng. Ed. 687.

57/17. G. R. Nicholson, *J. Chem. Soc.* p.2431 (1957).

57/18. S. Sunner, *Acta Chem. Scand.* **11**, 1757 (1957).

57/19. R. Littlewood, *J. Chem. Soc.* p.2419 (1957).

57/20. E. Briner and E. Dallwigk, *Helv. Chim. Acta*, **40**, 1978 (1957).

57/21. I. Jaffe, E. J. Prosen and M. Szwarc, *J. Chem. Phys.* **27**, 416 (1957).

57/22. P. Tavernier and M. Lamouroux, *Mém. Poudres*, **39**, 335 (1957).

57/23. L. Médard and M. Thomas, *Mém. Poudres*, **39**, 195 (1957).

57/24. J. P. McCullough, D. R. Douslin, J. F. Messerly, I. A. Hossenlopp, T. C. Kincheloe, and G. Waddington, *J. Am. Chem. Soc.* **79**, 4289 (1957).

57/25. J. L. Hales and E. F. G. Herington, *Trans. Faraday Soc.* **53**, 616 (1957).

57/26. R. J. L. Andon, J. D. Cox, E. F. G. Herington, and J. F. Martin, *Trans. Faraday Soc.* **53**, 1074 (1957).

57/27. M. M. Williams, W. S. McEwan, and R. A. Henry, *J. Phys. Chem.* **61**, 261 (1957).

57/28. T. Tsuzuki and H. Hunt, *J. Phys. Chem.* **61**, 1668 (1957).

57/29. D. M. Fairbrother, H. A. Skinner, and F. W. Evans, *Trans. Faraday Soc.* **53**, 779 (1957).

57/30. J. P. McCullough, W. N. Hubbard, F. R. Frow, I. A. Hossenlopp, and G. Waddington, *J. Am. Chem. Soc.* **79**, 561 (1957).

57/31. D. W. Scott, H. L. Finke, J. P. McCullough, J. F. Messerly, R. E. Pennington, I. A. Hossenlopp, and G. Waddington, *J. Am. Chem. Soc.* **79**, 1062 (1957).

57/32. S. Sunner and I. Wadsö, *Trans. Faraday Soc.* **53**, 455 (1957).

57/33. I. Wadsö, *Acta Chem. Scand.* **11**, 1745 (1957).

57/34. S. Sunner, *Acta Chem. Scand.* **11**, 1766 (1957).

57/35. J. R. Lacher, A. Kianpour, P. Montgomery, H. Knedler, and J. D. Park, *J. Phys. Chem.* **61**, 1125 (1957).

57/36. K. D. Williamson and R. H. Harrison, *J. Chem. Phys.* **26**, 1409 (1957).

57/37. K. E. Howlett, *J. Chem. Soc.* p.2834 (1957).

57/38. S. W. Benson and J. H. Buss, *J. Phys. Chem.* **61**, 104 (1957).

57/39. A. D. Mah, *J. Phys. Chem.* **61**, 1572 (1957).
57/40. A. A. Balandin, E. I. Klabunovskii, M. P. Kozina and O. D. Ul'yanova, *Izvest. Akad. Nauk S.S.S.R. Otdel. Khim. Nauk* p.12, (1957); *C.A.* **52**, 8714i (1958).
57/41. L. H. Long and J. F. Sackman, *Trans. Faraday Soc.* **53**, 1606 (1957).
57/42. A. K. Fischer, F. A. Cotton, and G. Wilkinson, *J. Am. Chem. Soc.* **79**, 2044 (1957).
57/43. A. J. Saggiomo, *J. Org. Chem.* **22**, 1171 (1957).
57/44. P. Gray and M. W. T. Pratt, *J. Chem. Soc.* p.2163 (1957).
57/45. A. Van Kamp, *Dissertation*, Free University of Amsterdam, 1957.
57/46. F. Glaser and H. Rüland, *Chem. Ing. Tech.* **29**, 772 (1957).
57/47. L. Bjellerup, S. Sunner, and I. Wadsö, *Acta Chem. Scand.* **11**, 1761 (1957).
57/48. C. L. Chernick, J. B. Pedley, and H. A. Skinner, *J. Chem. Soc.* p.1851 (1957).

1958

58/1. O. Kubaschewski and E. L. Evans, "Metallurgical Thermochemistry", 3rd. edition, Pergamon Press, London, 1958.
58/2. W. A. Keith and H. Mackle, *Trans. Faraday Soc.* **54**, 353 (1958).
58/3. W. N. Hubbard, W. D. Good, and G. Waddington, *J. Phys. Chem.* **62**, 614 (1958).
58/4. I. Wadsö, *Acta Chem. Scand.* **12**, 630 (1958).
58/5. A. S. Carson, D. R. Stranks, and B. Wilmshurst, *Proc. Roy. Soc. London*, **A244**, 72 (1958).
58/6. T. L. Cottrell, "The Strengths of Chemical Bonds", 2nd. edition, Butterworths Scientific Publications, London, 1958.
58/7. E. A. Coulson and E. F. G. Herington, "Laboratory Distillation Practice", George Newnes Ltd., London, 1958.
58/8. S. W. Benson and J. H. Buss, *J. Chem. Phys.* **29**, 546 (1958).
58/9. G. J. Janz, "Estimation of Thermodynamic Properties of Organic Compounds", Academic Press Inc., New York, 1958.
58/10. D. P. Biddiscombe and J. F. Martin, *Trans. Faraday Soc.* **54**, 1316 (1958).
58/11. J. B. Greenshields and F. D. Rossini, *J. Phys. Chem.* **62**, 271 (1958).
58/12. T. L. Flitcroft and H. A. Skinner, *Trans. Faraday Soc.* **54**, 47 (1958).
58/13. T. P. Wilson, E. G. Caflisch, and G. F. Hurley, *J. Phys. Chem.* **62**, 1059 (1958).
58/14. W. N. Hubbard, F. R. Frow, and G. Waddington, *J. Phys. Chem.* **62**, 821 (1958).
58/15. R. C. Cass, S. E. Fletcher, C. T. Mortimer, H. D. Springall, and T. R. White, *J. Chem. Soc.* p.1406 (1958).
58/16. S. M. Skuratov and M. P. Kozina, *Dokl. Akad. Nauk S.S.S.R.* **122**, 109 (1958); Eng. Ed. 643.
58/17. S. M. Skuratov and S. M. Shtekher, *Khim. Nauka i Prom.* **3**, 688 (1958); *C.A.* **53**, 4883c (1959).
58/18. E. L. Eliel and R. G. Haber, *J. Org. Chem.* **23**, 2041 (1958).
58/19. H. A. Gundry and A. R. Meetham, *Trans. Faraday Soc.* **54**, 664 (1958).
58/20. R. C. Cass, S. E. Fletcher, C. T. Mortimer, P. G. Quincey, and H. D. Springall, *J. Chem. Soc.* p.2595 (1958).
58/21. G. C. Sinke, D. L. Hildenbrand, R. A. McDonald, W. R. Kramer and D. R. Stull, *J. Phys. Chem.* **62**, 1461 (1958).

58/22. K. Schwabe and W. Wagner, *Chem. Ber.* **91**, 686 (1958).
58/23. F. P. Chappel and F. E. Hoare, *Trans. Faraday Soc.* **54**, 367 (1958).
58/24. J. D. Cox and H. A. Gundry, *J. Chem. Soc.* p.1019 (1958).
58/25. R. C. Cass, S. E. Fletcher, C. T. Mortimer, P. G. Quincey, and H. D. Springall, *J. Chem. Soc.* p.958 (1958).
58/26. T. Tsuzuki, D. O. Harper and H. Hunt, *J. Phys. Chem.* **62**, 1594 (1958).
58/27. J. P. McCullough, H. L. Finke, D. W. Scott, R. E. Pennington, M. E. Gross, J. F. Messerly, and G. Waddington, *J. Am. Chem. Soc.* **80**, 4786 (1958).
58/28. D. W. Scott, J. P. McCullough, J. F. Messerly, R. E. Pennington, I. A. Hossenlopp, H. L. Finke, and G. Waddington, *J. Am. Chem. Soc.* **80**, 55 (1958).
58/29. W. N. Hubbard, D. R. Douslin, J. P. McCullough, D. W. Scott, S. S. Todd, J. F. Messerly, I. A. Hossenlopp, A. George, and G. Waddington, *J. Am. Chem. Soc.* **80**, 3547 (1958).
58/30. I. Wadsö, *Acta Chem. Scand.* **12**, 635 (1958).
58/31. C. A. Neugebauer and J. L. Margrave, *J. Phys. Chem.* **62**, 1043 (1958).
58/32. G. C. Sinke and D. R. Stull, *J. Phys. Chem.* **62**, 397 (1958).
58/33. W. F. Lautsch, P. Erzberger and A. Tröber, *Wiss. Z. Tech. Hochsch. Chem. Leuna-Merseburg*, **1**, 31 (1958).
58/34. L. H. Long and J. F. Sackman, *Trans. Faraday Soc.* **54**, 1797 (1958).
58/35. P. A. Fowell and C. T. Mortimer, *J. Chem. Soc.* p.3734 (1958).
58/36. E. O. Fischer and S. Schreiner, *Chem. Ber.* **91**, 2213 (1958).
58/37. W. D. Good, D. M. Fairbrother, and G. Waddington, *J. Phys. Chem.* **62**, 853 (1958).
58/38. K. W. Sykes, *J. Chem. Soc.* p.2053 (1958).
58/39. H. Hoyer and W. Peperle, *Z. Electrochem.* **62**, 61 (1958).
58/40. V. P. Klochkov, *Zh. Fiz. Khim.* **32**, 1177 (1958).
58/41. W. J. Peppel, *Ind. Eng. Chem.* **50**, 767 (1958).
58/42. F. S. Dainton and K. J. Ivin, *Quart. Rev.* **12**, 61 (1958).
58/43. G. Pilcher and H. A. Skinner, *J. Inorg. Nucl. Chem.* **7**, 8 (1958).
58/44. E. A. V. Ebsworth and H. J. Emeléus, *J. Chem. Soc.* p.2150 (1958).

1959

59/1.   J. H. Brooks and G. Pilcher, *J. Chem. Soc.* p.1535 (1959).
59/2.   W. D. Good, D. W. Scott, J. L. Lacina, and J. P. McCullough, *J. Phys. Chem.* **63**, 1139 (1959).
59/3.   S. Sunner and I. Wadsö, *Acta Chem. Scand.* **13**, 97 (1959).
59/4.   J. S. Rowlinson, "Liquids and Liquid Mixtures", Butterworths Scientific Publications, London, 1959.
59/5.   C. Kitzinger and R. Hems, *Biochem. J.* **71**, 395 (1959).
59/6.   T. Benzinger, C. Kitzinger, R. Hems, and K. Burton, *Biochem. J.* **71**, 400 (1959).
59/7.   L. Bjellerup, *Acta Chem. Scand.* **13**, 1511 (1959).
59/8.   A. Weissberger, Ed., "Physical Methods of Organic Chemistry Part I", 3rd. Edition, p.401. Interscience Publishers Inc., New York, 1959.
59/9.   W. D. Good, D. R. Douslin, D. W. Scott, A. George, J. L. Lacina, J. P. Dawson, and G. Waddington, *J. Phys. Chem.* **63**, 1133 (1959).
59/10. J. F. Cordes and H. Günzler, *Chem. Ber.* **92**, 1055 (1959).
59/11. H. A. Skinner and A. Snelson, *Trans. Faraday Soc.* **55**, 404 (1959).

59/12. J. P. McCullough, R. E. Pennington, J. C. Smith, I. A. Hossenlopp, and G. Waddington, *J. Am. Chem. Soc.* **81**, 5880 (1959).

59/13. S. M. Shtekher, S. M. Skuratov, V. K. Daukshas and R. Ya. Levina, *Dokl. Akad. Nauk S.S.S.R.* **127**, 812 (1959); Eng. Ed. 621.

59/14. F. S. Dainton, K. J. Ivin, and D. A. G. Walmsley, *Trans. Faraday Soc.* **55**, 61 (1959).

59/15. A. K. Fischer, F. A. Cotton, and G. Wilkinson, *J. Phys. Chem.* **63**, 154 (1959).

59/16. M. Colomina, M. Cambeiro, R. Pérez-Ossorio and C. Latorre, *Anal. Fisc. Quim.* **B59**, 509 (1959).

59/17. G. C. Sinke, *J. Phys. Chem.* **63**, 2063 (1959).

59/18. F. W. Evans and H. A. Skinner, *Trans. Faraday Soc.* **55**, 260 (1959).

59/19. J. E. Bennett, *M. Sc. Thesis*, University of Manchester, 1959.

59/20. S. E. Fletcher, C. T. Mortimer, and H. D. Springall, *J. Chem. Soc.* p.580 (1959).

59/21. G. Saville and H. A. Gundry, *Trans. Faraday Soc.* **55**, 2036 (1959).

59/22. F. W. Evans and H. A. Skinner, *Trans. Faraday Soc.* **55**, 255 (1959).

59/23. D. L. Hildenbrand, G. C. Sinke, R. A. McDonald, W. R. Kramer, and D. R. Stull, *J. Chem. Phys.* **31**, 650 (1959).

59/24. J. P. McCullough, D. R. Douslin, W. N. Hubbard, S. S. Todd, J. F. Messerly, I. A. Hossenlopp, F. R. Frow, J. P. Dawson, and G. Waddington, *J. Am. Chem. Soc.* **81**, 5884 (1959).

59/25. F. W. Evans, D. M. Fairbrother, and H. A. Skinner, *Trans. Faraday Soc.* **55**, 399 (1959).

59/26. M. P. Kozina and S. M. Skuratov, *Dokl. Akad. Nauk S.S.S.R.* **127**, 561 (1959); Eng. Ed. 575.

59/27. J. D. Ray and R. A. Ogg, *J. Phys. Chem.* **63**, 1522 (1959).

59/28. D. L. Hildenbrand and R. A. McDonald, *J. Phys. Chem.* **63**, 1521 (1959).

59/29. D. L. Hildenbrand, R. A. McDonald, W. R. Kramer and D. R. Stull, *J. Chem. Phys.* **30**, 930 (1959).

59/30. D. W. Scott, D. R. Douslin, J. F. Messerly, S. S. Todd, I. A. Hossenlopp, T. C. Kincheloe, and J. P. McCullough, *J. Am. Chem. Soc.* **81**, 1015 (1959).

59/31. E. R. Alton, R. D. Brown, J. C. Carter and R. C. Taylor, *J. Am. Chem. Soc.* **81**, 3550 (1959).

59/32. J. B. Pedley and H. A. Skinner, *Trans. Faraday Soc.* **55**, 544 (1959).

59/33. P. A. Fowell and C. T. Mortimer, *J. Chem. Soc.* p.2913 (1959).

59/34. F. A. Cotton, A. K. Fischer and G. Wilkinson, *J. Am. Chem. Soc.* **81**, 800 (1959).

59/35. A. J. Leadbetter and J. E. Spice, *Can. J. Chem.* **37**, 1923 (1959).

59/36. A. Aihara, *Bull. Chem. Soc. Japan*, **32**, 1242 (1959).

59/37. M. Davies, A. H. Jones, and G. H. Thomas, *Trans. Faraday Soc.* **55**, 1100 (1959).

59/38. J. Vacek and J. Staněk, *Chem. Průmsyl.* **9**, 286 (1959).

59/39. C. S. Cope and B. F. Dodge, *Am. Inst. Chem. Eng. J.* **5**, 10 (1959).

59/40. T. L. Allen, *J. Chem. Phys.* **31**, 1039 (1959).

59/41. M. J. S. Dewar and H. N. Schmeising, *Tetrahedron*, **5**, 166 (1959).

1960

60/1. A. R. Meetham and J. A. Nicholls, *Proc. Roy. Soc. London*, **A256**, 384 (1960).

60/2.   F. S. Dainton, K. J. Ivin, and D. A. G. Walmsley, *Trans. Faraday Soc.* **56**, 1784 (1960).

60/3.   L. Haraldson, C. J. Olander, S. Sunner, and E. Varde, *Acta Chem. Scand.* **14**, 1509 (1960).

60/4.   I. Wadsö, *Acta Chem. Scand.* **14**, 566 (1960).

60/5.   R. J. L. Andon, D. P. Biddiscombe, J. D. Cox, R. Handley, D. Harrop, E. F. G. Herington, and J. F. Martin, *J. Chem. Soc.* p.5246 (1960).

60/6.   "Comptes Rendus des Séances de la Onzième Conférence Générale des Poids et Mesures", Gauthier-Villars, Paris, 1960.

60/7.   R. S. Jessup, "Precise Measurement of Heat of Combustion with a Bomb Calorimeter", National Bureau of Standards Monograph No. 7, Washington D.C., 1960.

60/8.   E. Calvet, P. Chovin, H. Moureu, and H. Tachoire, *J. Chim. Phys.* **57**, 593 (1960).

60/9.   F. J. Wright, *Rec. Trav. Chim.* **79**, 784 (1960).

60/10. E. G. Lovering and K. J. Laidler, *Can. J. Chem.* **38**, 2367 (1960).

60/11. D. W. Scott, W. T. Berg, and J. P. McCullough, *J. Phys. Chem.* **64**, 906 (1960).

60/12. D. L. Camin and F. D. Rossini, *J. Chem. Eng. Data*, **5**, 368 (1960).

60/13. H. F. Bartolo and F. D. Rossini, *J. Phys. Chem.* **64**, 1685 (1960).

60/14. O. N. Kachinskaya, S. Kh. Togoeva, A. P. Meshcheryakov and S. M. Skuratov, *Dokl. Akad. Nauk S.S.S.R.* **132**, 119 (1960); Eng. Ed. 451.

60/15. C. C. Browne and F. D. Rossini, *J. Phys. Chem.* **64**, 927 (1960).

60/16. D. M. Speros and F. D. Rossini, *J. Phys. Chem.* **64**, 1723 (1960).

60/17. J. Coops, H. Van Kamp, W. A. Lambregts, B. J. Visser and H. Dekker, *Rec. Trav. Chim.* **79**, 1226 (1960).

60/18. W. G. Dauben, O. Rohr, A. Labbauf, and F. D. Rossini, *J. Phys. Chem.* **64**, 283 (1960).

60/19. M. C. Loeffler and F. D. Rossini, *J. Phys. Chem.* **64**, 1530 (1960).

60/20. J. H. S. Green, *Chem. Ind. (London)*. p.1215 (1960).

60/21. J. Tjebbes, *Acta Chem. Scand.* **14**, 180 (1960).

60/22. G. R. Nicholson, *J. Chem. Soc.* p.2377 (1960).

60/23. H. A. Skinner and A. Snelson, *Trans. Faraday Soc.* **56**, 1776 (1960).

60/24. G. R. Nicholson, *J. Chem. Soc.* p.2378 (1960).

60/25. Chung-Chêng Li, *Hua Hsüeh, Tung Pao*, p.128 (1960); *C.A.* **55**, 24167b (1961).

60/26. I. Wadsö, *Acta Chem. Scand.* **14**, 561 (1960).

60/27. V. V. Ponomarev and L. B. Migarskaya, *Zh. Fiz. Khim.* **34**, 2506 (1960); Eng. Ed. 1182.

60/28. A. A. Balandin, E. I. Klabunovskii, A. P. Oberemok-Yakubova, and I. I. Brusov, *Izvest. Akad. Nauk S.S.S.R. Otdel. Khim. Nauk* **5**, 784 (1960).

60/29. J. D. Cox, *Trans. Faraday Soc.* **56**, 959 (1960).

60/30. T. M. Donovan, C. H. Shomate and W. R. McBride, *J. Phys. Chem.* **64**, 281 (1960).

60/31. G. T. Armstrong and S. Marantz, *J. Phys. Chem.* **64**, 1776 (1960).

60/32. T. Wada, E. Kishida, Y. Tomiie, H. Suga, S. Seki, and I. Nitta, *Bull. Chem. Soc. Japan*, **33**, 1317 (1960).

60/33. R. C. Hirt, J. E. Steger, and G. L. Simard, *J. Poly. Sci.* **43**, 319 (1960).

60/34. I. Wadsö, *Acta Chem. Scand.* **14**, 903 (1960).

60/35. A. F. Vorob'ev, N. M. Privalova, L. V. Storozhenko, and S. M. Skuratov, *Dokl. Akad. Nauk S.S.S.R.* **135**, 1131 (1960); Eng. Ed. 1403.

60/36. W. D. Good, J. L. Lacina, and J. P. McCullough, *J. Am. Chem. Soc.* **82**, 5589 (1960).

60/37. A. F. Vorob'ev and S. M. Skuratov, *Zh. Neorg. Khim.* **5**, 1398 (1960); Eng. Ed. 679.

60/38. D. N. Andreevskii, *Dokl. Akad. Nauk S.S.S.R.* **135**, 312 (1960); Eng. Ed. 1255.

60/39. L. Rosenblum, *J. Org. Chem.* **25**, 1652 (1960).

60/40. Yu. M. Golutvin and T. M. Kozlovskaya, *Zh. Fiz. Khim.* **34**, 2350 (1960); Eng. Ed. 1116.

60/41. A. F. Bedford and C. T. Mortimer, *J. Chem. Soc.* p.1622 (1960).

60/42. K.-H. Birr, *Z. Anorg. Allg. Chem.* **306**, 21 (1960).

60/43. A. P. Claydon, P. A. Fowell, and C. T. Mortimer, *J. Chem. Soc.* p.3284 (1960).

60/44. A. F. Bedford and C. T. Mortimer, *J. Chem. Soc.* p.4649 (1960).

60/45. R. R. Monchamp and F. A. Cotton, *J. Chem. Soc.* p.1438 (1960).

60/46. F. A. Cotton and R. R. Monchamp, *J. Chem. Soc.* p.533 (1960).

60/47. M. Davies and G. H. Thomas, *Trans. Faraday Soc.* **56**, 185 (1960).

60/48. H. Mackle, R. G. Mayrick, and J. J. Rooney, *Trans. Faraday Soc.* **56**, 115 (1960).

60/49. S. G. Entelis, G. V. Korovina and N. M. Chirkov, *Dokl. Akad. Nauk S.S.S.R.* **134**, 856 (1960); Eng. Ed. 1115.

60/50. S. S. Penner and J. Ducarme, eds., "The chemistry of propellants", Pergamon Press Ltd., Oxford, 1960.

60/51. D. R. Lide, *J. Chem. Phys.* **33**, 1514, 1519 (1960).

60/52. J. Issoire and C. van Long, *Bull. Soc. Chim. (France)*, p.2004 (1960).

60/53. J. D. Dunitz and V. Prelog, *Angew. Chem.* **72**, 897 (1960).

60/54. M. J. S. Dewar and H. N. Schmeising, *Tetrahedron*, **11**, 96 (1960).

1961

61/1. "Landolt-Börnstein, Zahlenwerte und Funktionen", 6th edition, Band II, Teil 4, Springer-Verlag, Berlin, Göttingen, Heidelberg, 1961.

61/2. "Selected Values of Properties of Chemical Compounds", Manufacturing Chemists Association Research Project, College Station, Texas, 1961.

61/3. D. R. Stull, ed., "JANAF Thermochemical Data", Dow Chemical Co., Midland, Michigan, 1961.

61/4. P. A. Fowell and C. T. Mortimer, *J. Chem. Soc.* p.3793 (1961).

61/5. J. E. Bennett and H. A. Skinner, *J. Chem. Soc.* p.2472 (1961).

61/6. H. A. Skinner, J. E. Bennett, and J. B. Pedley, *Pure App. Chem.* **2**, 17 (1961).

61/7. A. Snelson and H. A. Skinner, *Trans. Faraday Soc.* **57**, 2125 (1961).

61/8. L. Bjellerup, *Acta Chem. Scand.* **15**, 121 (1961).

61/9. D. Ambrose and B. A. Ambrose, "Gas Chromatography", George Newnes Ltd., London, 1961.

61/10. V. V. Ponomarev and T. A. Alekseeva, *Zh. Fiz. Khim.* **35**, 1629 (1961); Eng. Ed. 800.

61/11. W. N. Hubbard, F. R. Frow, and G. Waddington, *J. Phys. Chem.* **65**, 1326 (1961).

61/12. A. Labbauf and F. D. Rossini, *J. Phys. Chem.* **65**, 476 (1961).

61/13. J. D. Rockenfeller and F. D. Rossini, *J. Phys. Chem.* **65**, 267 (1961).
61/14. M. P. Kozina, S. M. Skuratov, S. M. Shtekher, I. E. Sosnina, and M. B. Turova-Polyak, *Zh. Fiz. Khim.* **35**, 2316 (1961); Eng. Ed. 1144.
61/15. D. R. Stull, G. C. Sinke, R. A. McDonald, W. E. Hatton and D. L. Hildenbrand, *Pure App. Chem.* **2**, 315 (1961).
61/16. M. P. Kozina, M. Yu. Lukina, N. D. Zubareva, I. L. Safonova, S. M. Skuratov, and B. A. Kazanskii, *Dokl. Akad. Nauk S.S.S.R.* **138**, 843 (1961); Eng. Ed. 537.
61/17. A. Labbauf, J. B. Greenshields, and F. D. Rossini, *J. Chem. Eng. Data*, **6**, 261 (1961).
61/18. U. Krüerke, C. Hoogzand and W. Hübel, *Chem. Ber.* **94**, 2817 (1961).
61/19. C. Hoogzand and W. Hübel, *Tetrahedron Letters*, p.637 (1961).
61/20. J. H. S. Green, *Trans. Faraday Soc.* **57**, 2132 (1961).
61/21. A. E. Pope, *M. Sc. Thesis*, University of Manchester, 1961.
61/22. J. K. Nickerson, K. A. Kobe, and J. J. McKetta, *J. Phys. Chem.* **65**, 1037 (1961).
61/23. E. J. Smutny and A. Bondi, *J. Phys. Chem.* **65**, 546 (1961).
61/24. M. Colomina, C. Latorre, and R. Pérez-Ossorio, *Pure App. Chem.* **2**, 133 (1961).
61/25. M. Colomina, M. L. Boned, and C. Turrión, *Anales Fisc. Quim.* **B57**, 655 (1961).
61/26. M. Colomina, R. Pérez-Ossorio, M. L. Boned, M. Panea, and C. Turrión, *Anales Fisc. Quim.* **B57**, 665 (1961).
61/27. G. Silvestro and C. Lenshitz, *J. Phys. Chem.* **65**, 694 (1961).
61/28. K. Schwabe and W. Wagner, *Z. Electrochem.* **65**, 812 (1961).
61/29. I. B. Rabinovich, V. I. Tel'noi, P. M. Nikolaev, and G. A. Razuvaev, *Dokl. Akad. Nauk S.S.S.R.* **138**, 852 (1961); Eng. Ed. 545.
61/30. G. Geiseler and W. Thierfelder, *Z. Phys. Chem. (Frankfurt)*, **29**, 248 (1961)
61/31. W. D. Good, J. L. Lacina, and J. P. McCullough, *J. Phys. Chem.* **65**, 2229 (1961).
61/32. W. T. Berg, D. W. Scott, W. N. Hubbard, S. S. Todd, J. F. Messerly, I. A. Hossenlopp, A. G. Osborn, D. R. Douslin, and J. P. McCullough, *J. Phys. Chem.* **65**, 1425 (1961).
61/33. J. P. McCullough, H. L. Finke, W. N. Hubbard, S. S. Todd, J. F. Messerly, D. R. Douslin, and G. Waddington, *J. Phys. Chem.* **65**, 784 (1961).
61/34. J. L. Lacina, W. D. Good, and J. P. McCullough, *J. Phys. Chem.* **65**, 1026 (1961).
61/35. H. Mackle and P. A. G. O'Hare, *Trans. Faraday Soc.* **57**, 2119 (1961).
61/36. W. K. Busfield, H. Mackle, and P. A. G. O'Hare, *Trans. Faraday Soc.* **57**, 1054 (1961).
61/37. H. Mackle and P. A. G. O'Hare, *Trans. Faraday Soc.* **57**, 1873 (1961).
61/38. H. Mackle and P. A. G. O'Hare, *Trans. Faraday Soc.* **57**, 1070 (1961).
61/39. H. Mackle and P. A. G. O'Hare, *Trans. Faraday Soc.* **57**, 1521 (1961).
61/40. W. D. Good, J. L. Lacina, and J. P. McCullough, *J. Phys. Chem.* **65**, 860 (1961).
61/41. A. S. Carson, W. Carter, and J. B. Pedley, *Proc. Roy. Soc. London*, **A260**, 550 (1961).
61/42. L. Bjellerup, *Acta Chem. Scand.* **15**, 231 (1961).
61/43. P. N. Walsh and N. O. Smith, *J. Chem. Eng. Data*, **6**, 33 (1961).
61/44. J. Berkowitz, D. A. Bafus, and T. A. Brown, *J. Phys. Chem.* **65**, 1380 (1961).

61/45. L. H. Long and J. Cattanach, *J. Inorg. Nucl. Chem.* **20**, 340 (1961).
61/46. W. H. Johnson, M. V. Kilday and E. J. Prosen, *J. Res. Natl. Bur. Stand.* **65A**, 215 (1961).
61/47. W. H. Johnson, I. Jaffe and E. J. Prosen, *J. Res. Natl. Bur. Stand.* **65A**, 71 (1961).
61/48. P. A. Fowell, *Ph. D. Thesis*, University of Manchester, 1961.
61/49. A. Rose and W. R. Supina, *J. Chem. Eng. Data*, **6**, 173 (1961).
61/50. E. O. Fischer and A. Reckziegel, *Chem. Ber.* **94**, 2204 (1961).
61/51. G. Geiseler, K. Quitzsch and M. Kockert, *Z. Phys. Chem. (Leipzig)*, **218**, 367 (1961).
61/52. K.-H. Birr, *Z. Anorg. Allg. Chem.* **311**, 92 (1961).
61/53. E. Amberger, *Chem. Ber.* **94**, 1447 (1961).
61/54. E. O. Fischer, S. Schreiner and A. Reckziegel, *Chem. Ber.* **94**, 258 (1961).
61/55. M. Davies and V. E. Malpass, *J. Chem. Soc.* p.1048 (1961).
61/56. S. G. Davis, D. F. Anthrop, and A. W. Searcy, *J. Chem. Phys.* **34**, 659 (1961).
61/57. M. V. Kilday, W. H. Johnson, and E. J. Prosen, *J. Res. Natl. Bur. Stand.* **65A**, 435 (1961).
61/58. British Standard 526:1961, "Definitions of the calorific value of fuels", British Standards Institution, London, 1961.
61/59. V. M. Tatevskii, V. A. Benderskii and S. S. Yarovoi, "Rules and Methods for Calculating the Physico-chemical Properties of Paraffinic Hydrocarbons", (translation ed. B. P. Mullins), Pergamon Press, Oxford, 1961.
61/60. J. P. McCullough and W. D. Good, *J. Phys. Chem.* **65**, 1430 (1961).
61/61. E. Huber-Buser, J. D. Dunitz and K. Venkatesan, *Proc. Chem. Soc.* p.463 (1961).
61/62. H. J. Bernstein, *Trans. Faraday Soc.* **57**, 1649 (1961).
61/63. B. E. Geller, *Zh. Fiz. Khim.* **35**, 1105 (1961); Eng. Ed. 542.

1962

62/1. "Comptes Rendus de la XXI Conférence I.U.P.A.C." p.284. Butterworth & Co., London, 1962.
62/2. H. A. Skinner, ed., "Experimental Thermochemistry", Vol. II, Wiley (Interscience), New York, 1962.
62/3. J. V. Davies and S. Sunner, *Acta Chem. Scand.* **16**, 1870 (1962).
62/4. I. Wadsö, *Svensk. Kem. Tidr.* **74**, 121 (1962).
62/5. A. S. Carson, R. Cooper, and D. R. Stranks, *Trans. Faraday Soc.* **58**, 2125 (1962).
62/6. "Temperature—Its Measurement and Control in Science and Industry", Vol. III, Reinhold, New York, in three parts 1962—3.
62/7. H. A. Gundry, A. J. Head, and G. B. Lewis, *Trans. Faraday Soc.* **58**, 1309 (1962).
62/8. V. P. Kolesov, I. D. Zenkov, and S. M. Skuratov, *Zh. Fiz. Khim.* **36**, 2082 (1962); Eng. Ed. 1120.
62/9. E. Calvet and H. Tachoire, *J. Chim. Phys.* **59**, 788 (1962).
62/10. J. D. Cox and A. J. Head, *Trans. Faraday Soc.* **58**, 1839 (1962).
62/11. S. R. Gunn, *Anal. Chem.* **34**, 1292 (1962).
62/12. E. G. Lovering and O. M. Nor, *Can. J. Chem.* **40**, 199 (1962).
62/13. K. B. Wiberg, W. J. Bartley, and F. P. Lossing, *J. Am. Chem. Soc.* **84**, 3980 (1962).

62/14. P. J. C. Fierens and J. Nasielski, *Bull. Soc. Chim. (Belg.)*, 71, 187 (1962).
62/15. D. W. Scott, G. B. Guthrie, J. F. Messerly, S. S. Todd, W. T. Berg, I. A. Hossenlopp, and J. P. McCullough, *J. Phys. Chem.* 66, 911 (1962).
62/16. A. Bauder and H. H. Günthard, *Helv. Chim. Acta*, 45, 1698 (1962).
62/17. A. F. Bedford, J. G. Carey, I. T. Millar, C. T. Mortimer, and H. D. Springall, *J. Chem. Soc.* p.3895 (1962).
62/18. J. Tjebbes, *Acta Chem. Scand.* 16, 953 (1962).
62/19. G. C. Sinke and D. L. Hildenbrand, *J. Chem. Eng. Data*, 7, 74 (1962).
62/20. P. Sellers and S. Sunner, *Acta Chem. Scand.* 16, 46 (1962).
62/21. I. B. Rabinovich, V. I. Tel'noi, L. M. Terman, A. S. Kirillova and G. A. Razuvaev, *Dokl. Akad. Nauk S.S.S.R.* 143, 133 (1962); Eng. Ed. 171.
62/22. G. S. Parks and H. P. Mosher, *J. Chem. Phys.* 37, 919 (1962).
62/23. J. H. Stern and F. H. Dorer, *J. Phys. Chem.* 66, 97 (1962).
62/24. R. F. Kempa and W. H. Lee, *Talanta*, 9, 325 (1962).
62/25. I. Wadsö, *Acta Chem. Scand.* 16, 471 (1962).
62/26. W. E. Hatton, D. L. Hildenbrand, G. C. Sinke and D. R. Stull, *J. Chem. Eng. Data*, 7, 229 (1962).
62/27. A. F. Bedford, P. B. Edmondson, and C. T. Mortimer, *J. Chem. Soc.* p.2927 (1962).
62/28. J. Tjebbes, *Acta Chem. Scand.* 16, 916 (1962).
62/29. J. D. Ray and A. A. Gershon, *J. Phys. Chem.* 66, 1750 (1962).
62/30. V. V. Ponomarev, T. A. Alekseeva, and L. N. Akimova, *Zh. Fiz. Khim.* 36, 872 (1962); Eng. Ed. 457.
62/31. V. V. Ponomarev, T. A. Alekseeva, and L. N. Akimova, *Zh. Fiz. Khim.* 36, 1083 (1962); Eng. Ed. 574.
62/32. A. I. Zakharov, L. A. Aleshina, and K. Macharacek, *Chem. Průmsyl.* 12, 23 (1962).
62/33. H. Mackle and R. G. Mayrick, *Trans. Faraday Soc.* 58, 238 (1962).
62/34. D. W. Scott, D. R. Douslin, H. L. Finke, W. N. Hubbard, J. F. Messerly, I. A. Hossenlopp, and J. P. McCullough, *J. Phys. Chem.* 66, 1334 (1962).
62/35. H. Mackle and R. G. Mayrick, *Trans. Faraday Soc.* 58, 230 (1962).
62/36. D. W. Scott, W. D. Good, S. S. Todd, J. F. Messerly, W. T. Berg, I. A. Hossenlopp, J. L. Lacina, A. G. Osborn, and J. P. McCullough, *J. Chem. Phys.* 36, 406 (1962).
62/37. H. Mackle and R. T. B. McClean, *Trans. Faraday Soc.* 58, 895 (1962).
62/38. H. Mackle and R. G. Mayrick, *Trans. Faraday Soc.* 58, 33 (1962).
62/39. G. Waddington, J. C. Smith, K. D. Williamson, and D. W. Scott, *J. Phys. Chem.* 66, 1074 (1962).
62/40. M. Månsson and S. Sunner, *Acta Chem. Scand.* 16, 1863 (1962).
62/41. V. P. Kolesov, I. D. Zenkov, and S. M. Skuratov, *Zh. Fiz. Khim.* 36, 89 (1962); Eng. Ed. 45.
62/42. V. P. Kolesov, A. M. Martynov, S. M. Shtekher, and S. M. Skuratov, *Zh. Fiz. Khim.* 36, 2078 (1962); Eng. Ed. 1118.
62/43. J. R. Lacher, H. B. Gottlieb, and J. D. Park, *Trans. Faraday Soc.* 58, 2348 (1962).
62/44. S. W. Benson and A. Amano, *J. Chem. Phys.* 36, 3464 (1962).
62/45. A. M. Rozhnov and D. N. Andreevskii, *Dokl. Akad. Nauk S.S.S.R.* 147 388 (1962). Eng. Ed. 998.
62/46. D. N. Andreevskii and A. M. Rozhnov, *Neftekhimiya*, 2, 378 (1962).
62/47. S. W. Benson and A. Amano, *J. Chem. Phys.* 37, 197 (1962).

62/48. W. D. Good, J. L. Lacina, D. W. Scott, and J. P. McCullough, *J. Phys. Chem.* **66**, 1529 (1962).

62/49. D. W. Scott, J. F. Messerly, S. S. Todd, I. A. Hossenlopp, D. R. Douslin, and J. P. McCullough, *J. Chem. Phys.* **37**, 867 (1962).

62/50. Yu. A. Lebedev, E. A. Miroshnichenko and A. M. Chaikin, *Dokl. Akad. Nauk S.S.S.R.* **145**, 1288 (1962); Eng. Ed. 751.

62/51. A. M. Chaikin, *Zh. Fiz. Khim.* **36**, 130 (1962); Eng. Ed. 65.

62/52. G. L. Gal'chenko, R. M. Varushchenko, Yu. N. Bubnov, and B. M. Mikhailov, *Zh. Obshch. Khim.* **32**, 284 (1962); Eng. Ed. 278.

62/53. G. L. Gal'chenko, R. M. Varushchenko, Yu. N. Bubnov, and B. M. Mikhailov, *Zh. Obshch. Khim.* **32**, 2405 (1962); Eng. Ed. 2373.

62/54. A. D. Mah, *U.S. Bureau of Mines Rept. Invest.* 5965 (1962).

62/55. A. F. Bedford, D. M. Heinekey, I. T. Millar, and C. T. Mortimer, *J. Chem. Soc.* p.2932 (1962).

62/56. J. W. Edwards and G. L. Kington, *Trans. Faraday Soc.* **58**, 1323 (1962).

62/57. A. Ya. Kipnis, *Zh. Neorg. Khim.* **7**, 1500 (1962); Eng. Ed. 775.

62/58. M. Kraus, L. Beranek, K. Kochloeff and V. Bazant, *Chem. Prumysl.* **12**, 649 (1962).

62/59. A. P. Claydon and C. T. Mortimer, *J. Chem. Soc.* p.3212 (1962).

62/60. L. Brewer, "Electronic Structure and Alloy Chemistry of Transition Elements", A.I.M.E. Monograph Series, Wiley (Interscience), New York, 1962.

62/61. V. N. Kondratiev, "Energies of Dissociation of Chemical Bonds", Akad. Nauk S.S.S.R., Moscow, 1962.

62/62. J. F. Smith *in* "Thermodynamics of Nuclear Materials", I.A.E.A., Vienna, 1962.

62/63. D. L. Hildenbrand and W. F. Hall, *J. Phys. Chem.* **66**, 754 (1962).

62/64. M. H. Rand *in* "Thermodynamics of Nuclear Materials", I.A.E.A., Vienna, 1962.

62/65. J. Hinze and H. H. Jaffe, *J. Am. Chem. Soc.* **84, 540** (1962).

62/66. G. Pilcher and H. A. Skinner, *J. Inorg. Nucl. Chem.* **24**, 937 (1962).

62/67. J. E. Bennett and H. A. Skinner, *J. Chem. Soc.* p.2150 (1962).

62/68. C. T. Mortimer, "Reaction heats and bond strengths", Pergamon Press, Oxford, 1962.

62/69. H. J. Bernstein, *Trans. Faraday Soc.* **58**, 2285 (1962).

62/70. H. A. Skinner, *J. Chem. Soc.* p.4396 (1962).

62/71. J. D. Cox, *Tetrahedron*, **18**, 1337 (1962).

62/72. R. A. Raphael, *Proc. Chem. Soc.* p.97 (1962).

62/73. G. R. Ross and W. J. Heideger, *J. Chem. Eng. Data*, **7**, 505 (1962).

62/74. C. J. McGinn, *Tetrahedron*, **18**, 311 (1962).

62/75. H. J. Bittrich, E. Kauer, M. Kraft, G. Schoeppe, W. Soell and A. Ullrich, *J. Prakt. Chem.* **17**, 250 (1962).

1963

63/1. J. V. Davies, A. E. Pope, and H. A. Skinner, *Trans. Faraday Soc.* **59**, 2233 (1963).

63/2. G. Pilcher, H. A. Skinner, A. S. Pell, and A. E. Pope, *Trans. Faraday Soc.* **59**, 316 (1963).

63/3. A. E. Pope and H. A. Skinner. *J. Chem. Soc.* p.3704 (1963).

63/4. J. H. S. Green, *Trans. Faraday Soc.* **59**, 1559 (1963).
63/5. J. L. Hales, J. D. Cox, and E. B. Lees, *Trans. Faraday Soc.* **59**, 1544 (1963).
63/6. G. A. Miller, *J. Chem. Eng. Data*, **8**, 69 (1963).
63/7. K. W. Bentley, ed. "Technique of organic chemistry. Vol. XI. Elucidation of structures by physical and chemical means. Pt.1". John Wiley and Sons, New York, 1963.
63/8. E. F. G. Herington, "Zone melting of organic compounds," Blackwell Scientific Publications, Oxford, 1963.
63/9. G. S. Ross and L. J. Frolen, *J. Res. Natl. Bur. Stand.* **67A**, 607 (1963).
63/10. M. Månsson and S. Sunner, *Acta Chem. Scand.* **17**, 723 (1963).
63/11. H. Mackle and P. A. G. O'Hare, *Trans. Faraday Soc.* **59**, 2693 (1963).
63/12. E. Calvet and H. Prat (translated by H. A. Skinner) "Recent Progress in Microcalorimetry", Pergamon Press, Oxford, 1963.
63/13. P. Gerding, I. Leden and S. Sunner, *Acta Chem. Scand.* **17**, 2190 (1963).
63/14. S. Sunner, *Acta Chem. Scand.* **17**, 728 (1963).
63/15. E. Morawetz and S. Sunner, *Acta Chem. Scand.* **17**, 473 (1963).
63/16. S. H. Fishtine, *Ind. Eng. Chem.* **55**, (4), 20; **55**, (5), 49; **55**, (6), 47, (1963).
63/17. A. Bondi, *J. Chem. Eng. Data*, **8**, 371 (1963).
63/18. S. W. Benson and A. N. Bose, *J. Am. Chem. Soc.* **85**, 1385 (1963).
63/19. O. S. Pascual and E. Almeda, *Philippine Atomic Energy Comm. Rept. PAEC(D) CH*-634 (1963); *C.A.* **60**, 1052lg (1964).
63/20. A. F. Bedford, A. E. Beezer, C. T. Mortimer, and H. D. Springall, *J. Chem. Soc.* p.3823 (1963).
63/21. J. L. Margrave, M. A. Frisch, R. G. Bautista, R. L. Clarke, and W. S. Johnson, *J. Am. Chem. Soc.* **85**, 546 (1963).
63/22. M. A. Frisch, C. Barker, J. L. Margrave, and M. S. Newman, *J. Am. Chem. Soc.* **85**, 2356 (1963).
63/23. G. S. Parks and H. P. Mosher, *J. Poly. Sci.* **A1**, 1979 (1963).
63/24. O. S. Pascual and Z. S. Garcia, *Philippine Atomic Energy Comm. Rept. PAEC(D) CH*-635 (1963); *C.A.* **60**, 10522a (1964).
63/25. D. P. Biddiscombe, R. Handley, D. Harrop, A. J. Head, G. B. Lewis, J. F. Martin, and C. H. S. Sprake, *J. Chem. Soc.* p.5764 (1963).
63/26. N. D. Zubareva, A. P. Oberemok-Yakubova, Yu. I. Petrov, E. I. Klabunovskii and A. A. Balandin, *Izvest. Akad. Nauk S.S.S.R.*, *Ser. Khim.* p.2207 (1963); *C.A.* **60**, 8710a (1964).
63/27. H. H. Stroh and C. R. Finke, *Z. Chem.* **3**, 265 (1963).
63/28. A. P. Oberemok-Yakubova and A. A. Balandin, *Izvest. Akad. Nauk S.S.S.R.*, *Ser. Khim.* p.2210 (1963); *C.A.* **60**, 8710b (1964).
63/29. A. F. Bedford, A. E. Beezer, and C. T. Mortimer, *J. Chem. Soc.* p.2039 (1963).
63/30. D. W. Scott, W. N. Hubbard, J. F. Messerly, S. S. Todd, I. A. Hossenlopp, W. D. Good, D. R. Douslin, and J. P. McCullough, *J. Phys. Chem.* **67**, 680 (1963).
63/31. D. W. Scott, W. D. Good, G. B. Guthrie, S. S. Todd, I. A. Hossenlopp, A. G. Osborn, and J. P. McCullough, *J. Phys. Chem.* **67**, 685 (1963).
63/32. G. T. Armstrong and S. Marantz, *J. Phys. Chem.* **67**, 2888 (1963).
63/33. R. H. Boyd, *J. Chem. Phys.* **38**, 2529 (1963).
63/34. J. O. Hutchens, A. G. Cole and J. W. Stout, *J. Phys. Chem.* **67**, 1128 (1963).
63/35. V. V. Ponomarev, T. A. Alekseeva and L. N. Akimova, *Zh. Fiz. Khim.* **37**, 227 (1963); Eng. Ed. 117.

63/36. D. M. Gardner and J. C. Grigger, *J. Chem. Eng. Data*, **8**, 73 (1963).

63/37. H. Mackle and P. A. G. O'Hare, *Tetrahedron*, **19**, 961 (1963).

63/38. G. Gattow and B. Krebs, *Z. Anorg. Allg. Chem.* **322**, 113 (1963).

63/39. V. P. Kolesov, I. D. Zenkov, and S. M. Skuratov, *Zh. Fiz. Khim.* **37**, 720 (1963); Eng. Ed. 378.

63/40. P. Corbett, A. Tarr, and E. Whittle, *Trans. Faraday Soc.* **59**, 1609 (1963).

63/41. V. P. Kolesov, I. D. Zenkov, and S. M. Skuratov, *Zh. Fiz. Khim.* **37**, 224 (1963); Eng. Ed. 115.

63/42. H. Hiraoka and J. H. Hildebrand, *J. Phys. Chem.* **67**, 916 (1963).

63/43. J. Puyo, D. Balesdent, M. Niclause and M. Dzierzynski, *Compt. Rend.* **256**, 3471 (1963).

63/44. G. Ya. Kabo and D. N. Andreevskii, *Neftekhimiya*, **3**, 764 (1963).

63/45. D. W. Scott, J. F. Messerly, S. S. Todd, I. A. Hossenlopp, A. G. Osborn, and J. P. McCullough, *J. Chem. Phys.* **38**, 532 (1963).

63/46. S. J. Ashcroft, A. S. Carson, and J. B. Pedley, *Trans. Faraday Soc.* **59**, 2713 (1963).

63/47. W. D. Good, S. S. Todd, J. F. Messerly, J. L. Lacina, J. P. Dawson, D. W. Scott, and J. P. McCullough, *J. Phys. Chem.* **67**, 1306 (1963).

63/48. W. D. Good, D. R. Douslin, and J. P. McCullough, *J. Phys. Chem.* **67**, 1312 (1963).

63/49. W. F. Lautsch, A. Tröber, W. Zimmer, L. Mehner, W. Linck, H.-M. Lehmann, H. Brandenberger, H. Korner, H.-J. Metschker, K. Wagner and R. Kaden, *Z. Chem.* **3**, 415 (1963).

63/50. B. Wilmshurst, *Ph. D. Thesis*, University of Leeds, 1963, quoted by H. A. Skinner, "Advances in Organometallic Chemistry" Vol. 2, Academic Press, New York—London, 1964.

63/51. G. L. Gal chenko and R. M. Varushchenko, *Zh. Fiz. Khim.* **37**, 2513 (1963); Eng. Ed. 1355.

63/52. J. P. McCullough, J. F. Messerly, R. T. Moore, and S. S. Todd, *J. Phys. Chem.* **67**, 677 (1963).

63/53. C. T. Mortimer and P. Sellers, *J. Chem. Soc.* p.1978 (1963).

63/54. I. B. Rabinovich, V. I. Tel'noi, N. V. Karyakin and G. A. Razuvaev, *Dokl. Akad. Nauk S.S.S.R.* **149**, 324 (1963); Eng. Ed. 217.

63/55. K. Quitzsch, R. Hüttig, H.-G. Vogel, H. J. Gesemann and G. Geiseler, *Z. Phys. Chem. (Leipzig)*, **223**, 225 (1963).

63/56. S. B. Hartley, W. S. Holmes, J. K. Jacques, H. F. Mole and J. C. McCoubrey, *Quart. Rev.* **17**, 204 (1963).

63/57. O. H. Krikorian, *J. Phys. Chem.* **67**, 1586 (1963).

63/58. H. A. Skinner and G. Pilcher, *Quart. Rev.* **17**, 264 (1963).

63/59. A. L. Woodman and A. Adicoff, *J. Chem. Eng. Data*, **8**, 241 (1963).

63/60. G. R. Horton and W. W. Wendlandt, *J. Inorg. Nucl. Chem.* **25**, 241 (1963).

63/61. C. S. Foote, *Tetrahedron Letters*, p.579 (1963).

63/62. J. D. Cox, *Tetrahedron*, **19**, 1175 (1963).

1964

64/1. H. A. Skinner *in* "Advances in Organometallic Chemistry", Vol. 2, Academic Press, New York—London, 1964.

64/2. A. E. Pope and H. A. Skinner, *Trans. Faraday Soc.* **60**, 1402 (1964).

64/3.   W. D. Good, J. L. Lacina, B. L. DePrater, and J. P. McCullough, *J. Phys. Chem.* **68**, 579 (1964).

64/4.   G. Pilcher, A. S. Pell, and D. J. Coleman, *Trans. Faraday Soc.* **60**, 499 (1964).

64/5.   "Thermodynamics and Thermochemistry", Butterworths, London, 1964.

64/6.   N. K. Smith, G. Gorin, W. D. Good, and J. P. McCullough, *J. Phys. Chem.* **68**, 940 (1964).

64/7.   N. K. Smith, D. W. Scott, and J. P. McCullough, *J. Phys. Chem.* **68**, 934 (1964).

64/8.   J. D. Cox, H. A. Gundry, and A. J. Head, *Trans. Faraday Soc.* **60**, 653 (1964).

64/9.   P. Sellers, S. Sunner, and I. Wadsö, *Acta Chem. Scand.* **18**, 202 (1964).

64/10.   B. N. Oleinik and V. S. Uskov, *Zh. Fiz. Khim.* **38**, 2162 (1964); Eng. Ed. 1171.

64/11.   A. D. Buckingham, "The Laws and Applications of Thermodynamics", Pergamon Press, Oxford, 1964.

64/12.   R. C. Wilhoit and D. Shiao, *J. Chem. Eng. Data*, **9**, 595 (1964).

64/13.   D. M. Golden, K. W. Egger, and S. W. Benson, *J. Am. Chem. Soc.* **86**, 5416 (1964).

64/14.   M. P. Kozina, A. K. Mirzaeva, I. E. Sosnina, N. V. Elagina, and S. M. Skuratov, *Dokl. Akad. Nauk S.S.S.R.* **155**, 1123 (1964); Eng. Ed. 375.

64/15.   M. L. Boned, M. Colomina, R. Pérez-Ossorio, and C. Turrión, *Anal. Fisc. Quim.* **B60**, 459 (1964).

64/16.   M. A. Frisch, R. G. Bautista, J. L. Margrave, C. G. Parsons, and J. H. Wotiz, *J. Am. Chem. Soc.* **86**, 335 (1964).

64/17.   A. S. Pell, *Ph. D. Thesis*, University of Manchester, 1964.

64/18.   G. C. Sinke and F. L. Oetting, J. Phys. Chem. **68**, 1354 (1964).

64/19.   N. D. Lebedeva, *Zh. Fiz. Khim.* **38**, 2648 (1964); Eng. Ed. 1435.

64/20.   N. A. Kozlov and I. B. Rabinovich, *Tr. po Khim. Tekhnol.* p.189 (1964); *C.A.* **63**, 6387 (1964).

64/21.   J. H. Hildebrand and R. L. Scott, "The Solubility of Non-Electrolytes", 4th edition, Dover, New York, 1964.

64/22.   M. Colomina, C. Turrión, M. L. Boned, and M. Panea, *Anal. Fisc. Quim.* **B60**, 619 (1964).

64/23.   M. Colomina, R. Pérez-Ossorio, C. Turrión, M. L. Boned, and B. Pedraja, *Anal. Fisc. Quim.* **B60**, 627 (1964).

64/24.   H. A. Swain, L. S. Silbert, and J. G. Miller, *J. Am. Chem. Soc.* **86**, 2562 (1964).

64/25.   W. Cocker, T. B. H. McMurray, M. A. Frisch, T. McAllister, and H. Mackle, *Tetrahedron Letters*, **32**, 2235 (1964).

64/26.   J. L. Bills and F. A. Cotton, *J. Phys. Chem.* **68**, 806 (1964).

64/27.   L. Nelander, *Acta Chem. Scand.* **18**, 973 (1964).

64/28.   H. H. Stroh and W. Kuechenmeister, *Z. Chem.* **4**, 427 (1964).

64/29.   N. N. Goroshko, M. P. Kozina, S. M. Skuratov, N. A. Belikova, and A. F. Plate, *Vestnik Moskov. Univ. Ser. II Khim.* **19**, 3 (1964); *C.A.* **62**, 4695h (1965).

64/30.   M. Parris, P. S. Raybin, and L. C. Labowitz, *J. Chem. Eng. Data*, **9**, 221 (1964).

64/31.   T. V. Charlu and M. R. A. Rao, *Proc. Ind. Acad. Sci.* **A60**, 31 (1964).

64/32.   H. Mackle and R. T. B. McClean, *Trans. Faraday Soc.* **60**, 669 (1964).

64/33.   H. Mackle and P. A. G. O'Hare, *Trans. Faraday Soc.* **60**, 506 (1964).

64/34.   D. M. McEachern and J. E. Kilpatrick, *J. Chem. Phys.* **41**, 3127 (1964).

64/35.   J. W. Edwards and P. A. Small, *Nature*, **202**, 1329 (1964).

64/36. V. P. Kolesov, O. G. Talakin, and S. M. Skuratov, *Zh. Fiz. Khim.* **38**, 1701 (1964); Eng. Ed. 930.

64/37. S. V. Levanova and D. N. Andreevskii, *Neftekhimiya*, **4**, 447 (1964).

64/38. M. F. Zimmer, E. E. Baroody, M. Schwartz, and M. P. McAllister, *J. Chem. Eng. Data*, **9**, 527 (1964).

64/39. Yu. Kh. Shaulov, V. S. Tubyanskaya, E. V. Evstigneeva and G. O. Shmyreva, *Zh. Fiz. Khim.* **38**, 1779 (1964); Eng. Ed. 967.

64/40. A. E. Pope and H. A. Skinner, *Trans. Faraday Soc.* **60**, 1404 (1964).

64/41. J. R. Van Wazer and L. Maier, *J. Am. Chem. Soc.* **86**, 811 (1964).

64/42. W. F. Lautsch, A. Tröber, H. Korner, K. Wagner, R. Kaden and S. Blase, *Z. Chem.* **4**, 441 (1964).

64/43. C. T. Mortimer and P. Sellers, *J. Chem. Soc.* p.1965 (1964).

64/44. J. D. Kelley and F. O. Rice, *J. Phys. Chem.* **68**, 3794 (1964).

64/45. H. Mackle and R. T. B. McClean, *Trans. Faraday Soc.* **60**, 817 (1964).

64/46. R. J. Irving and I. Wadsö, *Acta Chem. Scand.* **18**, 195 (1964).

64/47. B. N. Oleinik, "Tochnaya Kalorimetriya", State Publishing House of the Committee of Standards, Measurements, and Measuring Instruments of the U.S.S.R., Moscow, 1964.

64/48. W. J. Cooper and J. F. Masi, *J. Phys. Chem.* **64**, 682 (1964).

64/49. A. Finch and P. J. Gardner, *J. Chem. Soc.* p.2985 (1964).

64/50. B. D. Sully, *Chem. Ind. (London)*, p.263 (1964).

64/51. C. S. Cope, *Am. Inst. Chem. Eng. J.* **10**, 277 (1964).

64/52. J. L. Wood and M. M. Jones, *Inorg. Chem.* **3**, 1553 (1964).

64/53. J. D. Overmars and S. M. Blinder, *J. Phys. Chem.* **68**, 1801 (1964).

1965

65/1. D. D. Wagman, W. H. Evans, I. Halow, V. B. Parker, S. M. Bailey, and R. H. Schumm, "Selected Values of Chemical Thermodynamic Properties", Natl. Bur. Stand. Tech. Notes 270-1, 270-2; U.S. Government Printing Office, Washington D.C., 1965.

65/2. J. D. Cox, H. A. Gundry, and A. J. Head, *Trans. Faraday Soc.* **61**, 1594 (1965).

65/3. A. S. Pell and G. Pilcher, *Trans. Faraday Soc.* **61**, 71 (1965).

65/4. M. Davies and B. Kybett, *Trans. Faraday Soc.* **61**, 1608 (1965).

65/5. C. A. Goy and H. O. Pritchard, *J. Phys. Chem.* **69**, 3040 (1965).

65/6. E. Buckley and E. F. G. Herington, *Trans. Faraday Soc.* **61**, 1618 (1965).

65/7. P. Anderton and P. H. Bigg, "Changing to the Metric System", Her Majesty's Stationery Office, London, 1965.

65/8. A. Weissberger, ed., "Technique of Organic Chemistry, Vol. IV, Distillation" 2nd edition, John Wiley and Sons, New York, 1965.

65/9. J. F. Counsell, J. L. Hales, and J. F. Martin, *Trans. Faraday Soc.* **61**, 1869 (1965).

65/10. S. R. Gunn, *J. Phys. Chem.* **69**, 2902 (1965).

65/11. D. M. Golden, R. Walsh, and S. W. Benson, *J. Am. Chem. Soc.* **87**, 4053 (1965).

65/12. D. R. Douslin and A. G. Osborn, *J. Sci. Inst.* **42**, 369 (1965).

65/13. N. H. Chen, *J. Chem. Eng. Data*, **10**, 207 (1965).

65/14. H. S. Hull, A. F. Reid and A. G. Turnbull, *Austral. J. Chem.* **18**, 249 (1965).

65/15. K. W. Egger and S. W. Benson, *J. Am. Chem. Soc.* **87**, 3311 (1965).

65/16. J. F. Messerly, S. S. Todd, and H. L. Finke, *J. Phys. Chem.* **69**, 4304 (1965).

65/17. H. L. Finke, J. F. Messerly, and S. S. Todd, *J. Phys. Chem.* **69**, 2094 (1965).

65/18. R. H. Boyd, R. L. Christensen, and R. Pua, *J. Am. Chem. Soc.* **87**, 3554 (1965).

65/19. H. A. Karnes, B. D. Kybett, M. H. Wilson, J. L. Margrave, and M. S. Newman, *J. Am. Chem. Soc.* **87**, 5554 (1965).

65/20. A. E. Beezer, C. T. Mortimer, H. D. Springall, F. Sondheimer, and R. Wolovsky, *J. Chem. Soc.* p.216 (1965).

65/21. J. Chao and F. D. Rossini, *J. Chem. Eng. Data*, **10**, 374 (1965).

65/22. M. Colomina, A. S. Pell, H. A. Skinner, and D. J. Coleman, *Trans. Faraday Soc.* **61**, 2641 (1965).

65/23. G. Baker, J. H. Littlefair, R. Shaw and J. C. J. Thynne, *J. Chem. Soc.* p.6970 (1965).

65/24. B. D. Kybett, G. K. Johnson, C. K. Barker, and J. L. Margrave, *J. Phys. Chem.* **69**, 3603 (1965).

65/25. L. A. McDougall and J. E. Kilpatrick, *J. Chem. Phys.* **42**, 2311 (1965).

65/26. N. Adriaanse, H. Dekker and J. Coops, *Rec. Trav. Chim.* **84**, 393 (1965).

65/27. L. A. McDougall and J. E. Kilpatrick, *J. Chem. Phys.* **42**, 2307 (1965).

65/28. A. A. Yaprintseva and A. V. Finkel'shtein, *Materialy Konf. po Itogam Nauchn. Issled. Rabot za* 1964 *god, Sibirsk. Tekhnol. Inst. Krasnoyarsk, U.S.S.R.* p.33 (1965); *C.A.* **65**, 5345e (1966).

65/29. M. P. Kozina, D. N. Shigorin, A. P. Skoldinov, and S. M. Skuratov. *Dokl. Akad. Nauk S.S.S.R., Fiz. Khim.*, **160**, 1114 (1965); Eng. Ed. 135.

65/30. R. C. Wilhoit and I. Lei, *J. Chem. Eng. Data*, **10**, 166 (1965).

65/31. L. S. Silbert, B. F. Daubert, and L. S. Mason, *J. Phys. Chem.* **69**, 2887 (1965).

65/32. L. B. Clark, G. G. Peschel, and I. Tinoco, *J. Phys. Chem.* **69**, 3615 (1965).

65/33. I. Wadsö, *Acta Chem. Scand.* **19**, 1079 (1965).

65/34. E. Z. Zhuravlev, S. Ya. Serukhina and I. I. Konstantinov, *Zh. Prikl. Khim.* **38**, 2855 (1965); *C.A.* **64**, 16717b (1966).

65/35. L. A. Kalashnikova, E. G. Rozantsev and A. M. Chaikin, *Izv. Akad. Nauk S.S.S.R.* p.2525 (1965).

65/36. H. L. Finke, I. A. Hossenlopp, and W. T. Berg, *J. Phys. Chem.* **69**, 3030 (1965).

65/37. E. Bechtold, *Ber. Phys. Chem.* **69**, 326 (1965).

65/38. S. W. Benson, *J. Chem. Phys.* **43**, 2044 (1965).

65/39. P. Fowell, J. R. Lacher, and J. D. Park, *Trans. Faraday Soc.* **61**, 1324 (1965).

65/40. V. P. Kolesov, A. M. Martynov, and S. M. Skuratov, *Zh. Fiz. Khim.* **39**. 435 (1965); Eng. Ed. 223.

65/41. S. J. Ashcroft, A. S. Carson, W. Carter, and P. G. Laye, *Trans. Faraday Soc.* **61**, 225 (1965).

65/42. J. V. Davies, J. R. Lacher, and J. D. Park, *Trans. Faraday Soc.* **61**, 2413 (1965).

65/43. G. Ya. Kabo and D. N. Andreevskii, *Neftekhimiya*, **5**, 132 (1965).

65/44. J. F. Counsell, J. H. S. Green, J. L. Hales, and J. F. Martin, *Trans. Faraday Soc.* **61**, 212 (1965).

65/45. V. P. Kolesov, I. D. Zenkov, and S. M. Skuratov, *Zh. Fiz. Khim.* **39**, 2474 (1965); Eng. Ed. 1320.

65/46. G. L. Gal'chenko, M. M. Ammar, S. M. Skuratov, Yu. N. Bubnov and B. M. Mikhailov, *Vestnik Mosk. Univ. Ser. II Khim.* **20**, 3 (1965); *C.A.* **63**, 7703e (1965).

65/47. G. L. Gal'chenko, M. M. Ammar, S. M. Skuratov, Yu. N. Bubnov and B. M. Mikhailov, *Vestnik Mosk. Univ. Ser. II Khim.* **20**, 10 (1965); *C.A.* **63**, 14145h (1965).

65/48. Yu. Kh. Shaulov, G. O. Shmyreva and V. S. Tubyanskaya, *Zh. Fiz. Khim.* **39**, 105 (1965); Eng. Ed. 51.

65/49. V. I. Tel'noi and I. B. Rabinovich, *Zh. Fiz. Khim.* **39**, 2134 (1965); Eng. Ed. 1239.

65/50. G. A. Nash, H. A. Skinner and W. F. Stack, *Trans. Faraday Soc.* **61**, 640 (1965).

65/51. E. V. Evstigneeva and G. O. Shmyreva, *Zh. Fiz. Khim.* **39**, 1000 (1965); Eng. Ed. 529.

65/52. M. Frankowsky and J. G. Aston, *J. Phys. Chem.* **69**, 3126 (1965).

65/53. H. J. Svec, and D. D. Clyde, *J. Chem. Eng. Data.* **10**, 151 (1965).

65/54. V. P. Glushko, ed., "Termicheskie Konstanty Veshchestv", Akademiya Nauk S.S.S.R., VINITI, Moscow: Part I (1965): Part II (1966): Part III (1968).

65/55. J. C. Greenbank and B. B. Argent, *Trans. Faraday Soc.* **61**, 655 (1965).

65/56. S. W. Benson, *J. Chem. Educ.* **42**, 502 (1965).

65/57. E. L. Eliel, N. L. Allinger, S. J. Angyal and G. A. Morrison, "Conformational Analysis", J. Wiley & Sons Inc., New York, 1965.

1966

66/1. P. Hawtin, J. B. Lewis, N. Moul, and R. H. Phillips, *Phil. Trans. Roy. Soc. London*, **A261**, 67 (1966).

66/2. W. D. Good and B. L. DePrater, *J. Phys. Chem.* **70**, 3606 (1966).

66/3. D. J. Coleman and H. A. Skinner, *Trans. Faraday Soc.* **62**, 1721 (1966).

66/4. W. D. Good and M. Månsson, *J. Phys. Chem.* **70**, 97 (1966).

66/5. A. G. Osborn and D. R. Douslin, *J. Chem. Eng. Data*, **11**, 502 (1966).

66/6. G. W. C. Kaye and T. H. Laby, "Tables of Physical and Chemical Constants", 13th. edition, Longmans Green and Co., London, 1966.

66/7. J. A. Kerr, *Chem. Rev.* **66**, 465 (1966).

66/8. A. P. Gray, *Instrument News, Perkin-Elmer Corporation*, **16**, (3), 9 (1966).

66/9. E. C. W. Clarke and D. N. Glew, *Trans. Faraday Soc.* **62**, 539 (1966).

66/10. P. Hawtin, *Nature*, **210**, 411 (1966).

66/11. J-H. Hu, H-K. Yen, and Y-L. Geng, *Acta Chimica Sinica*, **32**, 242 (1966).

66/12. R. C. Reid and T. K. Sherwood, "Properties of gases and liquids: their estimation and correlation", McGraw Hill, New York, 1966.

66/13. I. Wadsö, *Acta Chem. Scand.* **20**, 536 (1966).

66/14. I. Wadsö, *Acta Chem. Scand.* **20**, 544 (1966).

66/15. K. W. Egger and S. W. Benson, *J. Am. Chem. Soc.* **88**, 236 (1966).

66/16. A. E. Beezer, W. Lüttke, A. de Meijere, and C. T. Mortimer, *J. Chem. Soc.* (*B*) p.648 (1966).

66/17. B. D. Kybett, S. Carroll, P. Natalis, D. W. Bonnell, J. L. Margrave, and J. L. Franklin, *J. Am. Chem. Soc.* **88**, 626 (1966).

66/18. D. J. Coleman and G. Pilcher, *Trans. Faraday Soc.* **62**, 821 (1966).

66/19. K. W. Sadowska, *Prezmysl. Chem.* **45**, 66 (1966): *C.A.* **64**, 14072f (1966).

66/20. G. Geiseler, K. Quitzsch, H.-J. Rauh, H. Schaffernicht, and H. J. Walther, *Ber. Phys. Chem.* **70**, 551 (1966).

66/21. R. H. Boyd, *Tetrahedron*, **22**, 119 (1966).

66/22. B. Börjesson, Y. Nakase and S. Sunner, *Acta Chem. Scand.* **20**, 803 (1966).

66/23. M. F. Zimmer, R. A. Robb, E. E. Baroody, and G. A. Carpenter, *J. Chem. Eng. Data*, **11**, 577 (1966).

66/24. P. K. Shirokikh, V. M. Bystrov, V. V. Ponomarev, and V. A. Solnster, *Zh. Fiz. Khim.* **40**, 1650 (1966); Eng. Ed. 894.

66/25. N. D. Lebedeva, *Zh. Fiz. Khim.* **40**, 2725 (1966); Eng. Ed. 1465.

66/26. D. J. Coleman and H. A. Skinner, *Trans. Faraday Soc.* **62**, 2057 (1966).

66/27. Yu. A. Lebedev, E. G. Rozantsev, L. A. Kalashnikova, V. P. Lebedev, M. B. Ne'man and A. Ya. Apin, *Dokl. Akad. Nauk S.S.S.R.* **168**, 104 (1966); Eng. Ed. 460.

66/28. S. M. Skuratov and A. K. Bonetskaya, *Vysokomolek. Soyedin.* **8**, 1591 (1966).

66/29. D. L. Ornellas, J. H. Carpenter, and S. R. Gunn, *Rev. Sci. Inst.* **37**, 907 (1966).

66/30. G. P. Adams, A. S. Carson, and P. G. Laye, *Trans. Faraday Soc.* **62**, 1447 (1966).

66/31. M. Månsson and S. Sunner, *Acta Chem. Scand.* **20**, 845 (1966).

66/32. G. C. Sinke, *J. Phys. Chem.* **70**, 1326 (1966).

66/33. A. S. Rogers, D. M. Golden, and S. W. Benson, *J. Am. Chem. Soc.* **88**, 3194 (1966).

66/34. R. Walsh and S. W. Benson, *J. Phys. Chem.* **70**, 3751 (1966).

66/35. V. P. Kolesov and I. D. Zenkov, *Russ. J. Phys. Chem.* **40**, 743 (1966).

66/36. P. B. Howard and H. A. Skinner, *J. Chem. Soc.* (*A*) p.1536 (1966).

66/37. M. F. Zimmer, E. E. Baroody, M. G. Graff, G. A. Carpenter, and R. A. Robb, *J. Chem. Eng. Data*, **11**, 579 (1966).

66/38. G. L. Gal'chenko, N. S. Zaugol'nikova, S. M. Skuratov, L. S. Vasil'ev, Yu. N. Bubnov and B. M. Mikhailov, *Dokl. Akad. Nauk S.S.S.R.* **169**, 587 (1966); Eng. Ed. 715.

66/39. G. L. Gal'chenko, N. S. Zaugol'nikova, S. M. Skuratov, L. S. Vasil'ev, A. Ya. Bezmanov, and B. M. Mikhailov, *Dokl. Akad. Nauk S.S.S.R.* **166**, 12 (1966).

66/40. A. E. Beezer and C. T. Mortimer, *J. Chem. Soc.* (*A*) p.514 (1966).

66/41. M. A. Ring, H. E. O'Neal, A. H. Kadhim, and F. Jappe, *J. Organomet. Chem.* **5**, 124 (1966).

66/42. P. Gross, C. Hayman, and J. T. Bingham, *Trans. Faraday Soc.* **62**, 2388 (1966).

66/43. V. I. Tel'noi and I. B. Rabinovich, *Zh. Fiz. Khim.* **40**, 1556 (1966); Eng. Ed. 842.

66/44. A. D. Starostin, A. V. Nikolaev and Yu. A. Afanas'ev, *Izvest. Akad. Nauk S.S.S.R., Ser. Khim.* p.1303 (1966); Eng. Ed. 1255

66/45. G. Gattow and M. Dräger, *Z. Anorg. Allg. Chem.* **343**, 232 (1966).

66/46. Yu. E. Bronshtein, V. Yu. Gankin, D. P. Krinkin and D. M. Rudkovskii, *Zh. Fiz. Khim.* **40**, 1475 (1966); Eng. Ed. 802.

66/47. G. Geiseler and R. Jannasch, *Z. Phys. Chem.* (*Leipzig*), **233**, 42 (1966).

66/48. S. Gladstone and H. Y. Chang, *J. Chem. Eng. Data*, **11**, 238 (1966).

66/49. C. E. Vanderzee and A. S. Quist, *Inorg. Chem.* **5**, 1238 (1966).

66/50. A. Finch and P. J. Gardner, *Trans. Faraday Soc.* **62**, 3314 (1966).

66/51. A. Finch, P. J. Gardner, and E. J. Pearn, *Trans. Faraday Soc.* **62**, 1072 (1966).

66/52. R. F. Gould, ed., "Advanced propellant chemistry", American Chemical Society Publications, Washington D.C., 1966.

66/53. G. R. Somayajulu and B. J. Zwolinski, *Trans. Faraday Soc.* **62**, 2327 (1966).

66/54. B. Nelander and S. Sunner, *J. Chem. Phys.* **44**, 2476 (1966).

1967

67/1.   E. Buckley and J. D. Cox, *Trans. Faraday Soc.* **63**, 895 (1967).
67/2.   G. Geiseler and H. Schaffernicht, *Z. Phys. Chem. (Frankfurt)*, **52**, 329 (1967).
67/3.   E. S. Domalski and G. T. Armstrong, *J. Res. Natl. Bur. Stand.* **71A**, 105 (1967).
67/4.   M. Zief and W. R. Wilcox, "Fractional Solidification, Vol. I", Marcel Dekker Inc., New York, 1967.
67/5.   H. A. Gundry, D. Harrop and A. J. Head, "Thermodynamik-Symposium", paper II 4, Heidelberg, 1967.
67/6.   S. N. Hajiev and M. J. Agarunov, "Thermodynamik-Symposium", paper IV 13, Heidelberg, 1967.
67/7.   E. Morawetz, "Thermodynamik-Symposium", paper IV 6, Heidelberg, 1967.
67/8.   G. Pilcher and J. D. M. Chadwick, *Trans. Faraday Soc.* **63**, 2357 (1967).
67/9.   H. Lehmann and E. Ruschitzky, *Chem. Tech. (Berlin)*, **19**, 226 (1967).
67/10. R. H. Boyd, C. Shieh, S. Chang, and D. McNally, "Thermodynamik-Symposium", paper II 7, Heidelberg, 1967.
67/11. E. F. Westrum and S. Wong, "Thermodynamik-Symposium", paper II 10, Heidelberg, 1967.
67/12. P. Sellers, "Thermodynamik-Symposium", paper II 2, Heidelberg, 1967.
67/13. K. Pihlaja and J. Heckkila, *Acta Chem. Scand.* **21**, 2390 (1967).
67/14. S. W. Staley, *J. Am. Chem. Soc.* **89**, 1532 (1967).
67/15. G. P. Adams, D. H. Fine, P. Gray, and P. G. Laye, *J. Chem. Soc. (B)* p.720 (1967).
67/16. N. K. Smith and W. D. Good, *J. Chem. Eng. Data*, **12**, 572 (1967).
67/17. D. W. Scott, W. T. Berg, I. A. Hossenlopp, W. N. Hubbard, J. F. Messerly, S. S. Todd, D. R. Douslin, J. P. McCullough, and G. Waddington, *J. Phys. Chem.* **71**, 2263 (1967).
67/18. E. A. Miroshnichenko, Yu. A. Lebedev, S. A. Shevelev, V. I. Gulevskaya, A. A. Fainzil'berg and A. Ya. Apin, *Zh. Fiz. Khim.* **41**, 1477 (1967); Eng. Ed. 783.
67/19. D. W. Scott and G. A. Crowder, *J. Chem. Phys.* **46**, 1054 (1967).
67/20. J. L. Wood, R. J. Lagow, and J. L. Margrave, *J. Chem. Eng. Data*, **12**, 255 (1967).
67/21. A. Lord, C. A. Goy, and H. O. Pritchard, *J. Phys. Chem.* **71**, 2705 (1967).
67/22. J. W. Coomber and E. Whittle, *Trans. Faraday Soc.* **63**, 2656 (1967).
67/23. J. W. Coomber and E. Whittle, *Trans. Faraday Soc.* **63**, 608 (1967).
67/24. C. A. Goy, A. Lord, and H. O. Pritchard, *J. Phys. Chem.* **71**, 1086 (1967).
67/25. J. R. Lacher, A. Amador, and J. D. Park, *Trans. Faraday Soc.* **63**, 1608 (1967).
67/26. J. W. Coomber and E. Whittle, *Trans. Faraday Soc.* **63**, 1394 (1967).
67/27. V. P. Kolesov, O. G. Talakin and S. M. Skuratov, *Vestnik Moscov. Univ. (Khim.)*, **5**, 60 (1967).
67/28. V. P. Kolesov, A. M. Martynov and S. M. Skuratov, *Zh. Fiz. Khim.* **41**, 913 (1967); Eng. Ed. 482.
67/29. V. P. Kolesov, E. M. Tomareva, S. M. Skuratov and S. P. Alekhin, *Zh. Fiz. Khim.* **41**, 1528 (1967); Eng. Ed. 817.

67/30. L. C. Walker and H. Prophet, *Trans. Faraday Soc.* **63**, 879 (1967).
67/31. G. C. Sinke, C. J. Thompson, R. E. Jostad, L. C. Walker, A. C. Swanson, and D. R. Stull, *J. Chem. Phys.* **47**, 1852 (1967).
67/32. H. S. Hull, A. F. Reid and A. G. Turnbull, *Inorg. Chem.* **6**, 805 (1967).
67/33. G. L. Gal'chenko and N. S. Zaugol'nikova, *Zh. Fiz. Khim.* **41**, 1018 (1967); Eng. Ed. 538.
67/34. N. K. Smith and W. D. Good, *J. Chem. Eng. Data*, **12**, 570 (1967).
67/35. S. Pawlenko, *Chem. Ber.* **100**, 3591 (1967).
67/36. H. S. Hull and A. G. Turnbull, *Inorg. Chem.* **6**, 2020 (1967).
67/37. J. O. Hill and R. J. Irving, *J. Chem. Soc.* (*A*) p.1413 (1967).
67/38. A. G. Turnbull, *Austral. J. Chem.* **20**, 2757 (1967).
67/39. N. Wakayama and H. Inokuchi, *Bull. Chem. Soc. Japan*, **40**, 2267 (1967).
67/40. E. A. Boucher, *J. Chem. Educ.* **44**, A935 (1967).
67/41. J. L. Settle, E. Greenberg, and W. N. Hubbard, *Rev. Sci. Inst.* **38**, 1805 (1967).
67/42. H. Peters and E. Tappe, *Mber. Deut. Akad. Wiss. Berlin*, **9**, 828 (1967).
67/43. H. Peters and E. Tappe, *Mber. Deut. Akad. Wiss. Berlin*, **9**, 901 (1967).
67/44. A. Bondi, *Chem. Rev.* **67**, 565 (1967).
67/45. R. H. Boyd, K. R. Guha, and R. Wuthrich, *J. Phys. Chem.* **71**, 2187 (1967).
67/46. G. P. Adams, A. S. Carson, and P. G. Laye, *J. Chem. Soc.* (*A*) p.1832 (1967).
67/47. G. Geiseler, H. Schaffernicht, and H. J. Walther, *Ber. Phys. Chem.* **71**, 196 (1967).
67/48. J. L. Hales, E. B. Lees, and D. J. Ruxton, *Trans. Faraday Soc.* **63**, 1876 (1967).
67/49. A. Finch, P. J. Gardner, E. J. Pearn, and G. B. Watts, *Trans. Faraday Soc.* **63**, 1880 (1967).
67/50. A. Finch, P. J. Gardner, and G. B. Watts, *Trans. Faraday Soc.* **63**, 1603 (1967).
67/51. A. Finch, P. J. Gardner, and G. B. Watts, *Chem. Comm.* p.1054 (1967).
67/52. J. C. Baldwin, M. F. Lappert, J. B. Pedley, and J. A. Treverton, *J. Chem. Soc.* (*A*) p.1980 (1967).
67/53. A. J. Gait, "Heavy organic chemicals", Pergamon Press, Oxford, 1967.
67/54. K. W. Schneider, *Chem. Ind.* (*London*), p.1764 (1967).
67/55. J. S. S. Brame and J. G. King, "Fuels–solid, liquid and gaseous", 6th. edition, Edward Arnold (Publishers) Ltd., London, 1967.
67/56. M. Sasiadek, *Bull. Soc. Chim.* (*France*), p.4437 (1967).
67/57. Z. A. Musharov, I. G. Kaufman and S. G. Entelis, *Zh. Fiz. Khim.* **41**, 858 (1967); Eng. Ed. 445.

1968

68/1. E. Greenberg and W. N. Hubbard, *J. Phys. Chem.* **72**, 222 (1968).
68/2. S. N. Hajiev and M. J. Agarunov, *J. Organomet. Chem.* **11**, 415 (1968).
68/3. D. P. Baccanari, J. A. Novinski, Y.-C. Pan, M. M. Yevitz, and H. A. Swain, *Trans. Faraday Soc.* **64**, 1201 (1968).
68/4. D. D. Wagman, W. H. Evans, V. B. Parker, I. Halow, S. M. Bailey, and R. H. Schumm, "Selected Values of Chemical Thermodynamic Properties" Natl. Bur. Stand. Tech. Note 270-3, U.S. Government Printing Office, Washington D.C., 1968.

68/5.   W. J. Evans, E. J. McCourtney and W. B. Carney, *Anal. Chem.* **40**, 262 (1968).
68/6.   I. Wadsö, *Acta Chem. Scand.* **22**, 2438 (1968).
68/7.   R. A. Robb and M. F. Zimmer, *J. Chem. Eng. Data*, **13**, 200 (1968).
68/8.   D. Ambrose, *J. Sci. Inst. (Ser.* 2), **1**, 41 (1968).
68/9.   K. B. Wiberg and R. A. Fenoglio, *J. Am. Chem. Soc.* **90**, 3395 (1968).
68/10. S. Marantz and G. T. Armstrong, *J. Chem. Eng. Data*, **13**, 118, 455 (1968).
68/11. S. W. Benson, "Thermochemical Kinetics", J. Wiley and Sons Inc., New York, 1968.
68/12. W. D. Clark and S. J. W. Price, *Can. J. Chem.* **46**, 1633 (1968).
68/13. J. O. Hill and R. J. Irving, *J. Chem. Soc. (A)* p.1052 (1968).
68/14. J. R. Lacher and H. A. Skinner, *J. Chem. Soc. (A)* p.1034 (1968).
68/15. E. Morawetz, *Acta Chem. Scand.* **22**, 1509 (1968).
68/16. K. F. Zmbov, O. M. Uy, and J. L. Margrave, *J. Am. Chem. Soc.* **90**, 5090 (1968).
68/17. J. O. Hill and R. J. Irving, *J. Chem. Soc. (A)* p.3116 (1968).
68/18. V. P. Kolesov, S. N. Shtekher, A. M. Martynov and S. M. Skuratov, *Zhur. Fiz. Khim.* **42**, 1847 (1968); Eng. Ed. 975.
68/19. E. F. G. Herington and I. J. Lawrenson, *Nature*, **219**, 928 (1968).
68/20. K. L. Churney and G. T. Armstrong, *J. Res. Natl. Bur. Stand.* **72A**, 453 (1968).
68/21. V. P. Kolesov, O. G. Talakin and S. M. Skuratov, *Zhur. Fiz. Khim.* **42**, 2307 (1968); Eng. Ed. 1218.
68/22. J. E. Mark, *J. Phys. Chem.* **72**, 2941 (1968).

1969

69/1.   J. D. Cox, H. A. Gundry, D. Harrop and A. J. Head, *J. Chem. Thermodyn.* **1**, 77 (1969).
69/2.   "The International Practical Temperature Scale of 1968", Her Majesty's Stationery Office, London, 1969.
69/3.   C. Mosselman and H. Dekker, *Rec. Trav. Chim.* **88**, 161 (1969).
69/4.   H. A. Gundry, D. Harrop, A. J. Head and G. B. Lewis, *J. Chem. Thermodyn.* **1**, 321 (1969).

Miscellaneous unpublished work, and private communications.

M1.   J. Jaffe
M2.   H. Mackle, W. V. Steele and D. V. McNally.
M3.   L. C. Walker.
M4.   D. Stull.
M5.   C. R. Patrick.
M6.   P. Sellers.
M7.   A. J. Head.
M8.   H. L. Finke.
M9.   G. Pilcher and R. A. Fletcher.
M10. W. D. Good.

Index